Essays

Chapter 1	Pheromones: Chemical Communications in the Insect World
Chapter 2	Water-Soluble and Fat-Soluble Vitamins Redox Reactions in Biochemistry
Chapter 3	Steroids Freons, Free Radicals, and the Ozone Layer
Chapter 4	Allylic Oxidation in Metabolic Reactions
Chapter 5	Substituent Effects on Metabolism of Benzene Derivatives
Chapter 6	Chirality and Your Senses Metabolism May Vary By Species and Within Species
Chapter 7	Halogen Compounds in Ocean Organisms Biological Substitution Reactions with Sulfur Compounds
Chapter 8	Toxicity of Alcohols Industrial Synthesis of Alcohols Phenols are Germicides
Chapter 9	Polyethers in the Laboratory and in Nature Ring Opening of Biological Epoxides
Chapter 10	Oral Contraceptives Addition Reactions and Vision
Chapter 11	Metabolism of Lactose Human Blood Groups
Chapter 12	Industrial Synthesis of Acetic Acid Thioesters are Nature's Active Acyl Compounds Polyesters Esters and Anhydrides of Phosphoric Acid
Chapter 13	Prostaglandins Sleep Disorders and Brain Lipids
Chapter 14	Amphetamines Polyamides
Chapter 15	Essential Amino Acids and Dietary Protein Cholesterol and Lipoproteins
Chapter 16	Chemistry and Medicine
Chapter 17	Polyurethanes in Treatment of Cancer
Chapter 18	Magnetic Resonance Imaging

Common Functional Groups

Class	*Functional group structure*	*IUPAC prefix or suffix*	*Example*	*Name*
Alkane	C—H and C—C single bonds	-ane	$CH_3—CH_3$	ethane
Alkene	$>C=C<$	-ene	$CH_2=CHCH_3$	propene
Alkyne	$—C\equiv C—$	-yne	$CH_3C\equiv CH$	propyne
Arene	C_6H_6 (benzene ring with H on each carbon)	none	$C_6H_5CH_3$ (benzene ring with CH_3)	toluene
Alcohol	$—C—\ddot{O}H$	-ol	$CH_3CH_2—OH$	ethanol
Ether	$—C—\ddot{O}—C—$	alkoxy-	$CH_3—O—CH_2CH_3$	methoxyethane
Halide	$—C—\ddot{X}:$	halo-	$CH_3—Br$	bromomethane
Amine	$—C—\ddot{N}—$	-amine	$CH_3CH_2—NH_2$	ethylamine
Nitrile	$—C\equiv N:$	-nitrile	$CH_3C\equiv N$	ethanenitrile
Sulfide	$—C—\ddot{S}—C—$	sulfide	$CH_3—S—CH_3$	dimethyl sulfide
Thiol	$—C—\ddot{S}H$	-thiol	$CH_3CH_2—SH$	ethanethiol
Aldehyde	$—C—C(=\ddot{O}:)—H$	-al	$CH_3—C(=O)—H$	ethanal
Ketone	$—C—C(=\ddot{O}:)—C—$	-one	$CH_3—C(=O)—CH_3$	propanone
Carboxylic acid	$—C—C(=\ddot{O}:)—\ddot{O}—H$	-oic acid	$CH_3—C(=O)—OH$	ethanoic acid
Ester	$—C—C(=\ddot{O}:)—\ddot{O}—C—$	-oate	$CH_3—C(=O)—OCH_3$	methyl ethanoate
Acid chloride	$—C—C(=\ddot{O}:)—\ddot{Cl}:$	-oyl chloride	$CH_3—C(=O)—Cl$	ethanoyl chloride
Amide	$—C—C(=\ddot{O}:)—\ddot{N}—$	-amide	$CH_3—C(=O)—NH_2$	ethanamide

Organic Chemistry

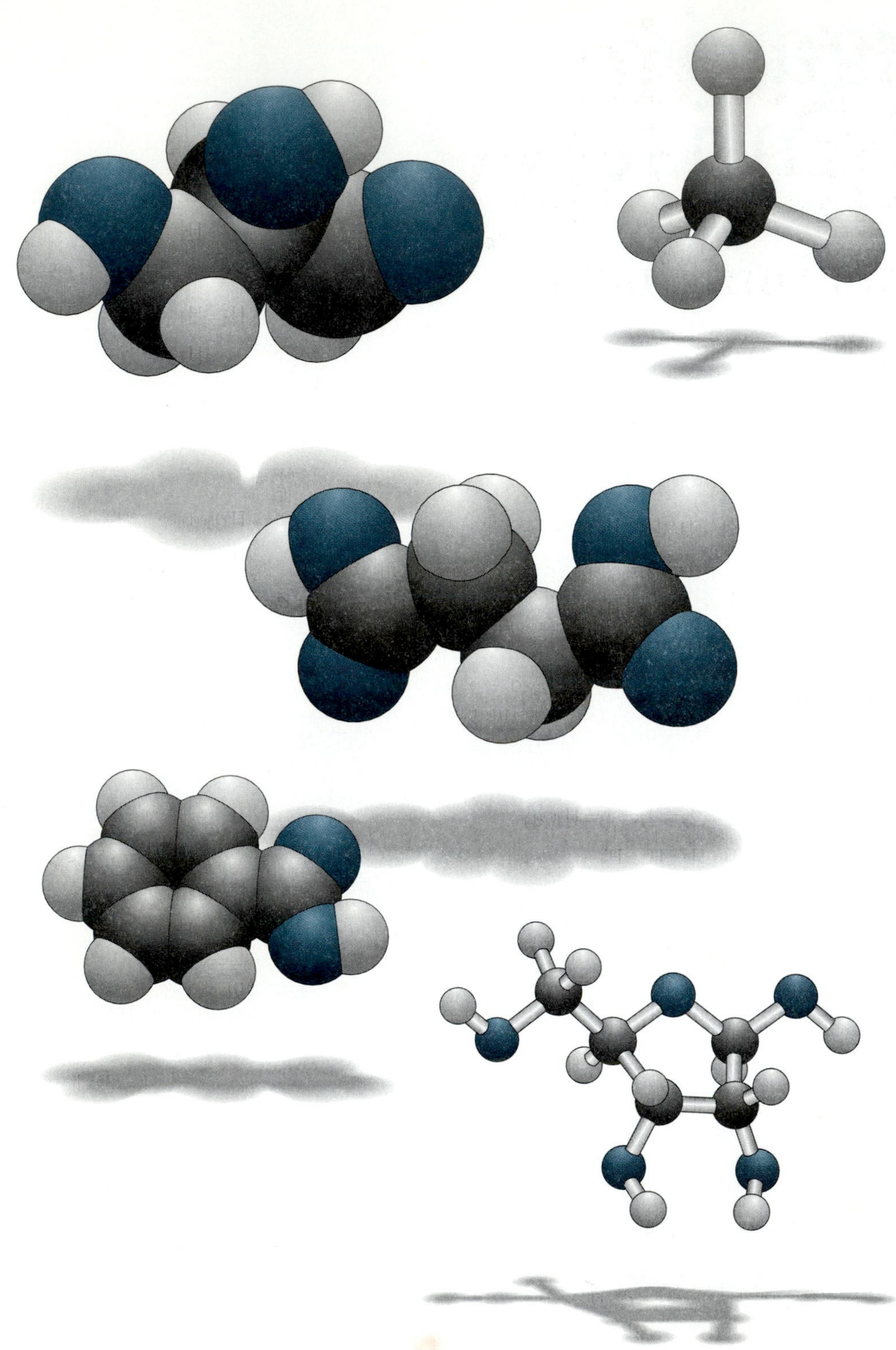

Organic Chemistry

A Brief Introduction

Second Edition

Robert J. Ouellette

The Ohio State University

PRENTICE HALL

Upper Saddle River, New Jersey 07458

Library of Congress Cataloging-in-Publication Data

Ouellette, Robert J.
Organic chemistry : a brief introduction / Robert J. Ouellette. —
2nd ed.
p. cm.
Includes bibliographical references and index.
ISBN 0-13-841933-7
1. Chemistry, Organic. I. Title.
QD253.088 1998
547—dc21 97-25850
CIP

Acquisitions Editor: John Challice
Editor in Chief: Paul F. Corey
Editorial Director: Tim Bozik
Assistant Vice President of Production and Manufacturing: David W. Riccardi
Executive Managing Editor: Kathleen Schiaparelli
Assistant Managing Editor: Shari Toron
Production Editor: Rose Kernan
Creative Director: Paula Maylahn
Art Director: Joseph Sengotta
Interior Design: Amy Rosen
Art Manager: Gus Vibal
Manufacturing Manager: Trudy Pisciotti
Manufacturing Buyer: Ben Smith
Photo Research: Rona Tucillo
Illustrations: Rolando Corujo
Copy Editor: Brian Baker
Editorial Assistant: Betsy Williams
Proofreader: Edith Baker
Text Composition/Prepress: Progressive Information Technologies
Cover Design: Joseph Sengotta
Cover Photos: Enantomiers: Caraway and Spearmint Leaves/Paul Silverman, Fundamental Photos; Red Blood Cells, Dr. Gopal Murti/Science Photo Library, Custom Medical Stock Photo; Gypsy Moth Larva/Runk/Schoenberg, Grant Heilman Photography, Inc.

Printed in the United States of America
10 9 8 7 6 5 4 3

ISBN 0-13-841933-7

Prentice-Hall International (UK) Limited, *London*
Prentice-Hall of Australia Pty. Limited, *Sydney*
Prentice-Hall Canada Inc., *Toronto*
Prentice-Hall Hispanoamericana, S. A., *Mexico*
Prentice-Hall of India Private Limited, *New Delhi*
Prentice-Hall of Japan, Inc., *Tokyo*
Simon & Schuster Asia Pte. Ltd., *Singapore*
Editora Prentice-Hall do Brasil, Ltda., *Rio de Janeiro*

CONTENTS

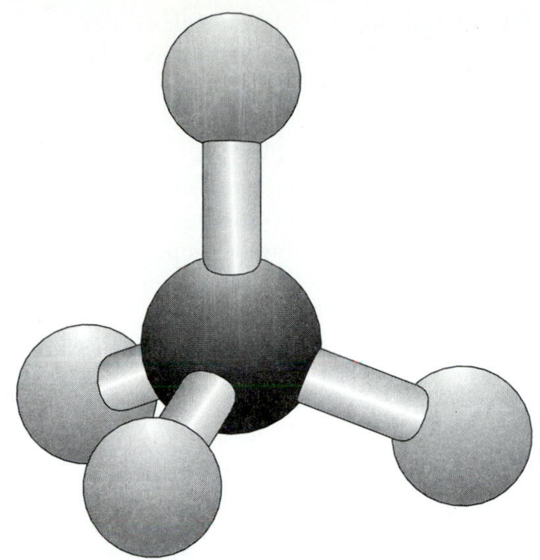

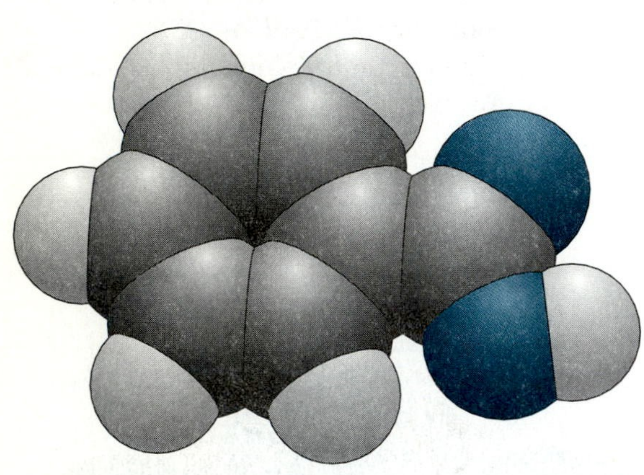

5 Aromatic Compounds

6 Stereochemistry

7 Nucleophilic Substitution and Elimination Reactions

8 Alcohols and Phenols

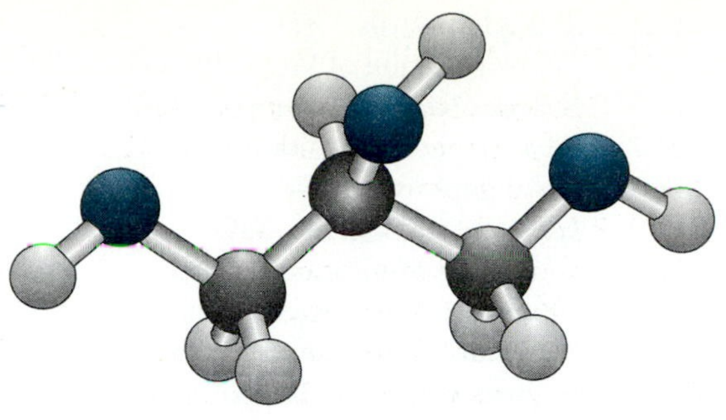

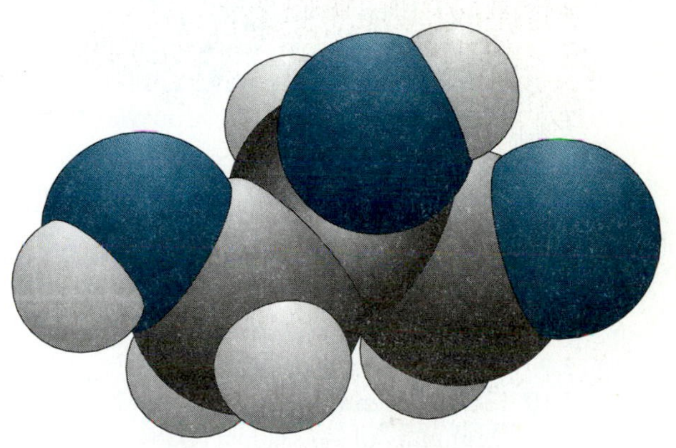

Essays

12 Carboxylic Acids and Esters

Essays

13 Lipids

Essays

14 Amines and Amides

Essays

15 Amino Acids, Peptides, and Proteins

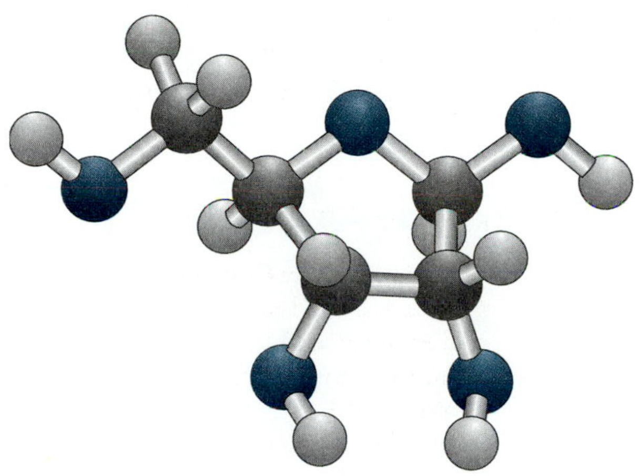

16 Nucleic Acids

17 Synthetic Polymers

18 Spectroscopy

PREFACE

I prepared this second edition of Organic Chemistry: A Brief Introduction with the aim of retaining and refining what worked in the first edition as well as including additional features that will enhance the presentation of organic chemistry to students who are not going to be chemistry majors. Accordingly, the final product is the result of considering the opinions of two groups of faculty reviewers. The first group consisted of faculty who had used the first edition in their classes. The second group of reviewers had not used the first edition but had taught one-term organic chemistry courses. As a result of the feedback from both groups, this text now features several enhancements. Foremost among these is the increased use of color to highlight chemical reactions. This text now has the modern appearance associated with the texts available for the full-year organic chemistry course. While not every equation uses color (because students need to learn to interpret chemical equations without such highlights), many reactions in the text are highlighted. The decision was made in conjunction with the editor not to use four colors in spite of its cosmetic appeal. The process required for full color is expensive, and the cost is inevitably passed on to students.

The artwork in the book has also been dramatically improved. More molecules are now represented in figures as three-dimensional structures, and virtually all of the artwork has been redone with the goal of enhancing its clarity as well as increasing the students' understanding of the geometric changes associated with the mechanistic representation of the reaction. Reactions such as addition to alkenes and carbonyl compounds now show the reacting centers of the reactants, intermediates, and products with appropriate geometry at the sp^3- and sp^2-hybridized carbon atoms.

Comments I received from faculty reviewers were most helpful in the preparation of this edition. Many of the changes may not be easily identified without a page-by-page comparison of both editions. They are minor refinements that greatly enhance the clarity and simplicity of the presentation of organic chemistry at a level appropriate for this audience. However, one general comment by one of the reviewers was intriguing. It was suggested that material using molecular models should be developed to enhance student understanding of structure and reactivity. The feature titled "Explorations with Molecular Models" is the result of that suggestion. These exercises are located at the end of each chapter.

Retained Features

Two major features of the first edition are retained. These are the two introductory chapters that review concepts of general chemistry and the early inclusion of the bioorganic chapters. Although I and the majority of the reviewers feel that these features are important, both are presented so that each can be tailored to fit the needs of a particular course.

Introductory Chapters

The length of time assigned to the coverage of Chapters 1 and 2 can be modified based on the backgrounds of the students and the strength of their general chemistry course. These chapters present the principles of general chemistry required for a brief introduction to organic chemistry. Each of the principles is presented using organic molecules as examples. The review provided in the first two chapters gives students a chance to revisit basic topics, both building up their confidence and ensuring that they are up to speed before moving on.

Bioorganic Chapters

I have moved the bioorganic chapters forward from their traditional location, preferring instead to present them immediately after discussing the chemistry of the relevant functional groups (for example, Carbohydrates follows immediately after Aldehydes and Ketones). However, these chapters are modular and may be presented at the end of the course or omitted entirely. In those cases, there is no difficulty in omitting these chapters and proceeding to the next chapter, because subsequent organic chapters are not based on any chemistry in the bioorganic chapters.

Exercises

The arrangement of Exercises by concepts at the end of each chapter is also retained in the second edition. This feature allows for flexibility when assigning problems. The 777 Exercises in the text and the additional 355 Examples in the Study Guide provide adequate practice for students to determine their understanding of the material in each chapter. A significant decrease in the number of text pages has resulted from the careful recasting of Exercises to better utilize space. As a result, the added features of this edition were acheived without an increase in the total number of pages. I have maintained the paired Exercise approach of the first edition. Each even-numbered Exercise explores the same concept and is of equal difficulty as the odd-numbered Exercise that precedes it.

Essays

Essays on bioorganic topics are still a feature of this text. Some new essays on industrial applications have been added. The 33 essays illustrate interesting material that relates chemistry to the real world as well as to areas of interest to students.

New Features

Molecular Models

The incorporation of an additional 175 ball-and-stick and space-filling models significantly alters the appearance of this edition and reinforces the importance of using molecular models—a feature often neglected in other texts after the initial chapters. These models are distributed within the text as well as in figures. In addition, they appear in a new feature titled Explorations with Molecular Models. This unique feature provides a hands-on way to aid the development of visual skill pertaining to structure and its relationship to physical and chemical properties of organic molecules. These explorations, located before the Exercises at the end of the chapter, are programmed to be done in sequential order.

Summary of Reactions

A new Summary of Reactions section is now located at the end of each chapter. Each type of reaction is referenced by the section number where the material is presented within the chapter. I discourage students from depending primarily on these summaries as the organic chemistry equivalent of Cliff Notes. An understanding of organic chemistry, which is a vertically integrated study, requires that students develop an understanding of the underlying principles of the science.

New Chapters

I gave considerable priority to the revision of Chapter 7. The material dealing with the structure, properties, and nomenclature of haloalkanes as well as the preparation and reactions of Grignard reagents is now found in other chapters. As a result, Chapter 7 now deals exclusively with nucleophilic substitution and elimination reactions.

Addition and condensation polymers are still discussed in several chapters. However, a chapter on Synthetic Polymers is new to this edition. This chapter provides more details about the relationship of structure to the properties of polymers as well as how polymers are formed. The effects of chain length, chain branching, cross-linking, and stereochemistry are considered in conjunction with the properties of specific polymers.

Spectroscopy

Based on reviewers' comments, the location of the spectroscopy chapter is unchanged. A majority feel that the core chapters detailing chemical reactivity clearly constitute a formidable body of information to cover in one term. However, some faculty prefer to cover spectroscopy along with functional groups. Accordingly, one exercise at the end of each chapter on functional groups refers students to those Exercises in Chapter 18 that may be done if the course instructor is integrating spectroscopy throughout the course. In addition, the chapter now includes a brief introduction to ^{13}C NMR spectroscopy.

Answers to Exercises

As in the first edition, full, worked-out solutions to all Exercises are found in the Solutions Manual. However, brief answers to the odd-numbered Exercises are now located in an Appendix of this text. This accommodation reflects a compromise based on the differing philosophies of instructors on the availability of answers. The answers to half of the Exercises allow students to easily access the information required to check their answers. For full solutions, I recommend that students consult the Solutions Manual, but only after making an honest effort to solve the problem without sneaking a peek.

Examples and Problems

Many of the solved Examples located within the chapters have been revised, and new Examples have been added. Now accompanying each Example is an unsolved Problem that is based on the same concept as the Example. As faculty know, success in organic chemistry is based on practice, and these additional problems provide practice within an established framework using concepts presented in close proximity to text material. Answers to the Problems are available in an Appendix.

The Study Guide and Solutions Manual

The Study Guide and Solutions Manual was carefully revised over a period of one year as the manuscript for the second edition itself was developed. Both authors read the complete manuscript several times, and we rechecked all answers each time. The period over which the material was examined and the multiple revisions reflect our desire to ensure accuracy to the greatest extent possible.

As in the first edition, the Study Guide is meant to complement the text, and it follows closely the explanation and Examples given in the text. In this way the Study Guide provides students with a reinforcement of what to learn. If they do not quite understand the material in the text, then the Study Guide provides additional discussion of the subject matter as well as more solved Examples. There are 355 Examples in the Study Guide that contain material similar to the 108 Examples in the text. Thus, students have many ways to test their comprehension of the material presented in the text and to reinforce their knowledge by repetition of the techniques of solving chemistry problems.

The Solutions Manual portion of this book gives full, step-by-step solutions to all Exercises at the end of the chapters in Organic Chemistry: A Brief Introduction.

Acknowledgments

I am pleased to thank the many reviewers who evaluated the first edition as well as those who examined the revised manuscript for this edition. Their helpful comments were given special consideration in developing this second edition. They include

Bob Allen
Arkansas Tech University

Dana Chatellier
University of Delaware

Jeff Elbert
South Dakota State University

Morris Fishman
New York University

Chip Frasier
Virginia Polytechnic University

Joe Gandler
California State University, Fresno

Don Hopwood
Grand Rapids Community College

Robert Huffman
Northern Arizona University

Norman Hunter
University of Manitoba

William Jenks
Iowa State University

Elizabeth Larson
Grand Canyon University

William Meena
Rock Valley College

Mark Meier
University of Kentucky

David Nelson
University of Wyoming, Laramie

Ron Starkey
University of Wisconsin, Green Bay

Jason Stenzel
University of Idaho

I thank the people of Prentice Hall for their efforts in moving this product along. They include John Challice, the acquisitions editor; Betsy Williams, the editorial assistant; Rose Kernan, the production editor; and the artist Rolando Corujo.

I am grateful for the expert computer advice from my colleague Robert Tatz. He was always willing to fix both hardware and software problems, thus facilitating the preparation of the manuscript for this text. In addition he assisted in the conversion and electronic transmission of molecular models to Prentice Hall.

Finally, I gratefully acknowledge the dedicated work of my colleague Mary Bailey. She patiently read several versions of the manuscript and made numerous suggestions for changes that improved the clarity and consistency of the text. Her comments on how students would most likely react to both textual as well as schematic representations of chemistry were immensely helpful. She kindly assumed the responsibility of reviewing all copy-edited manuscript and figures when the production schedule was advanced and coincided with my long planned trip to New Zealand. Finally she read both the first and second sets of page proofs. Her concern for the quality of the text is greatly appreciated. Students who use this text will benefit from her numerous contributions at each stage of the project.

ROBERT J. OUELLETTE
Columbus, Ohio

PHOTO CREDITS

Chapter 1	Apollo II Earth View. Photo courtesy of NASA Headquarters. Selenium. Photo by Robert J. Ouellette. Shaving cream. Photo by Robert J. Ouellette. Shampoo. Photo by Robert J. Ouellette.
Chapter 2	Vitamins C + E. Photo by Robert J. Ouellette. Lighter Fluid. Photo by Robert J. Ouellette. Vinegar. Photo by Robert J. Ouellette. Oregon coast. Photo by Peter Menzel, Stock Boston. Kuwait oil well fires: Greater al Burgan Field. Photo by Peter Menzel, Stock Boston. Cortaid. Photo by Robert J. Ouellette.
Chapter 3	Afternoon sun picks up details of an oil refinery near Concord, California. Photo by Spencer Grant, Stock Boston. Propane for drying grain. Photo by Thomas Hovland, Grant Heilman Photography, Inc.
Chapter 4	Crisco. Photo by Robert J. Ouellette. Glad bags. Photo by Robert J. Ouellette. Great Smokies in Tennessee. Photo by Hugh Rogers, Monkmeyer Press. Acetylene torch cutting steel. Photo by Charles D. Winters, Photo Researchers, Inc.
Chapter 5	Pain Relievers. Photo by Robert J. Ouellette. Mothballs: Napthalene (C10H8). Photo by Fundamental Photographs.
Chapter 6	Chambered nautilus; nautilus pomilius. Photo by Bruce Coleman, Inc.
Chapter 7	CD-Roms. Photo by The Image Bank.
Chapter 8	Listerine. Photo by Robert J. Ouellette. Woman adds anti-freeze to car to protect engine.Photo by Bob Daemmrich, Stock Boston. Grapes. Photo by Christi Carter, Grant Heilman Photography, Inc.
Chapter 9	Operating room anesthetist with child. Photo by Photo Researchers, Inc.
Chapter 10	A southern California teenager titivates in her room. Photo by Stock Boston. Long-eared owl. Photo by Monkmeyer Press.
Chapter 11	Lactaid. Photo by Robert J. Ouellette. Rotini pasta. Photo by Robert J. Ouellette. Cows. Photo by Grant Heilman Photography, Inc.
Chapter 12	Desenex. Photo by Robert J. Ouellette. Bayer aspirin. Photo by Robert J. Ouellette. Windsurfer. Photo by Christopher Brown, Stock Boston. Casco Bay Marathon, Maine. Photo by Dean Abramson, Stock Boston. Baked goods store, Buena Park, CA. Photo by Lawrence Migdale, Stock Boston.
Chapter 13	Tide. Photo by Robert J. Ouellette. Spray oils. Photo by Robert J. Ouellette. Bees on honeycomb partially filled with honey. Photo by Mark Antman, The Image Works. King penguins. Photo by Ira Kirschenbaum, Stock Boston. Mongolia. Bactrian camel. Steppe grasslands in background. Photo by George Holton, Photo Researchers, Inc.
Chapter 14	Scope. Photo by Robert J. Ouellette. Benadryl. Photo by Robert J. Ouellette. Federal Agents ATF, FBI, etc., Branch Davidian standoff. Waco, TX, David Koresh raid. Photo by Stock Boston.
Chapter 15	L-Lysine. Photo by Robert J. Ouellette. Brown/grizzly bear (Ursus arctos). Photo by Stock Boston. Goat. Photo by Stock Boston. 10K run. Photo by Stock Boston.
Chapter 16	Asian elephants in zoo. Photo by Stock Boston. Biotechnology DNA testing. Photo by Photo Researchers, Inc. A 25 year old bodybuilder poses for the judges during a competition. Photo by Mitch Wojnarowicz, The Image Works.
Chapter 17	Hefty bags. Photo by Robert J. Ouellette. Gatorade. Photo by Robert J. Ouellette. Parachute preparing to land. Photo by A. De Menil, Photo Researchers, Inc
Chapter 18	Medical-diagnostic testing technician studies MRI scan of human brain, N.Y.U. Med. Center. Photo by Photo Researchers, Inc.

CHAPTER 1

STRUCTURE OF ORGANIC COMPOUNDS

Overview

The earth, as seen from the moon, is characterized by the chemistry of organic compounds.

1.1 Inorganic and Organic Compounds

In the late eighteenth century, substances were divided into two classes called inorganic and organic compounds. Inorganic compounds were derived from mineral sources, whereas organic compounds were obtained only from plants or animals. Organic compounds were more difficult to work with in the laboratory, and decomposed more easily, than inorganic compounds. The differences between inorganic and organic compounds were attributed to a "vital force" associated with organic compounds. This unusual attribute was thought to exist only in living matter. It was believed that without the vital force, organic compounds could not be synthesized in the laboratory. However, by the mid-nineteenth century, chemists had learned not only how to work with organic compounds in the laboratory but how to synthesize them as well.

Organic compounds always contain carbon and a limited number of other elements, such as hydrogen, oxygen, and nitrogen. Compounds containing sulfur, phosphorus, and halogens are known but are less prevalent. Most organic compounds contain many more atoms per structural unit than inorganic compounds and have more complex structures. Common examples of organic compounds include the sugar sucrose ($C_{12}H_{22}O_{11}$), vitamin B_2 ($C_{17}H_{20}N_4O_6$), cholesterol ($C_{27}H_{46}O$), and the fat glycerol tripalmitate ($C_{51}H_{98}O_6$). Some organic molecules are gigantic. DNA, which stores genetic information, has molecular weights that range from 3 million in *Escherichia coli (E. coli)* to 2 billion for mammals.

Based on the physical characteristics of compounds, such as solubility, melting point, and boiling point, chemists have proposed that the atoms of the elements are bonded in compounds in two principal ways—ionic bonds and covalent bonds. Both types of bonds result from a change in the electronic structure of atoms as they associate with each other. Thus, the number and type of bonds formed and the resultant shape of the molecule depend on the electron configuration of the atoms. Therefore, we will review some of the electronic features of atoms and the periodic properties of the elements before describing the structures of organic compounds.

1.2 Atomic Structure

Each atom has a central, small, dense nucleus that contains protons and neutrons; electrons are located outside the nucleus. Protons have a +1 charge; electrons have a −1 charge. The number of protons, which determines the identity of an atom, is given as its **atomic number.** Since atoms have an equal number of protons and electrons and are electrically neutral, the atomic number also indicates the number of electrons in the atom. The number of electrons in the hydrogen, carbon, nitrogen, and oxygen atoms are one, six, seven, and eight, respectively.

The periodic table of the elements is arranged by atomic number. The elements are arrayed in horizontal rows called **periods** and vertical columns called **groups.** In this text we will emphasize hydrogen in the first period and the elements carbon, nitrogen, and oxygen in the second period. The electronic structure of these atoms is the basis for their chemical reactivity.

Atomic Orbitals

Electrons about the nucleus of an atom are found in **atomic orbitals.** Each orbital can contain a maximum of two electrons. The orbitals, designated by the letters *s*, *p*, *d*, and *f*, differ in energy, shape, and orientation. We need to consider only the *s* and *p* orbitals for elements such as carbon, oxygen, and nitrogen.

Orbitals are grouped in shells of increasing energy designated by the integers $n = 1, 2, 3, 4, \cdots n$. These integers are called **principal quantum numbers.** With few exceptions, we need consider only the orbitals of the first three shells for the common elements found in organic compounds.

Each shell contains a unique number and type of orbitals. The first shell contains only one orbital—the *s* orbital. It is designated 1*s*. The second shell contains two types of orbitals—one *s* orbital and three *p* orbitals.

An *s* orbital is a spherical region of space centered around the nucleus (Figure 1.1). The electrons in a 2*s* orbital are higher in energy than those in a 1*s* orbital. The 2*s* orbital is larger than the 1*s* orbital, and its electrons on average are farther from the nucleus. The three *p* orbitals in a shell are shaped like "dumb bells." However, they have different orientations with respect to the nucleus (Figure 1.1). The orbitals are often designated p_x, p_y, and p_z to emphasize that they are mutually perpendicular to one another. Although the orientations of the *p* orbitals are different, the electrons in each *p* orbital have equal energies.

Orbitals of the same type within a shell are often considered as a group called a **subshell.** There is only one orbital in an *s* subshell. An *s* subshell can contain only two electrons, but a *p* subshell can contain a total of six electrons within its p_x, p_y, and p_z orbitals.

Electrons are located in subshells of successively higher energies so that the total energy of all electrons is as low as possible. The order of increasing energy of subshells is $1s < 2s < 2p < 3s < 3p$ for elements of low atomic number. If there is more than one orbital in a subshell, one electron occupies each with parallel spins until all are half full. A single electron within an orbital is unpaired; two electrons with opposite spins within an orbital are paired and constitute an electron pair. The number and location of electrons for the first 18 elements are given in Table 1.1. The location of electrons in atomic orbitals is the **electron configuration** of an atom.

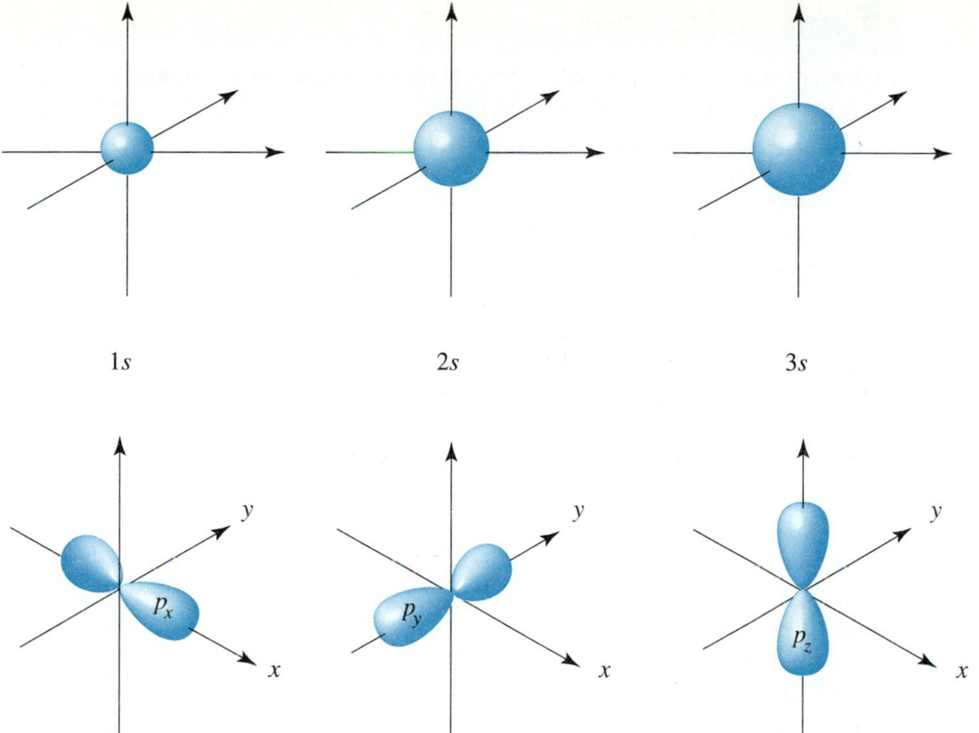

FIGURE 1.1
The *s* and *p* Orbitals
Electrons are pictured within a volume called an orbital. A "cloud" of negative charge surrounds the nucleus, which is located at the origin of the intersecting axes. The *s* orbital is pictured as a sphere. The three orbitals of the *p* subshell are arranged perpendicular to one another. Each orbital may contain two electrons.

Valence Shell Electrons

Electrons in filled, lower energy shells of atoms have no role in determining the structure of molecules, nor do they participate in chemical reactions. Only the higher energy electrons located in the outermost shell, the **valence shell,** participate in chemical reactions. Electrons in the valence shell are **valence electrons.** For example, the single electron of the hydrogen atom is a valence electron. The number of valence electrons for the common atoms contained in organic molecules is given by their group number in the periodic table. Thus carbon, nitrogen, and oxygen atoms have four, five, and six valence electrons, respectively. With this information we can understand how these elements combine to form the structure of organic compounds.

1.3 Atomic Properties

The physical and chemical properties of an element may be estimated from its position in the periodic table. Two properties that help us explain the properties of organic compounds are atomic radius and electronegativity.

The overall shape of an isolated atom is spherical, and the volume of the atom depends on the number of electrons and the energies of the electrons in

Table 1.1 Electron Configuration

Atomic Number	Element	Number of Electrons in Subshells				
		1*s*	2*s*	2*p*	3*s*	3*p*
1	H	1				
2	He	2				
3	Li	2	1			
4	Be	2	2			
5	B	2	2	1		
6	C	2	2	2		
7	N	2	2	3		
8	O	2	2	4		
9	F	2	2	5		
10	Ne	2	2	6		
11	Na	2	2	6	1	
12	Mg	2	2	6	2	
13	Al	2	2	6	2	1
14	Si	2	2	6	2	2
15	P	2	2	6	2	3
16	S	2	2	6	2	4
17	Cl	2	2	6	2	5
18	Ar	2	2	6	2	6

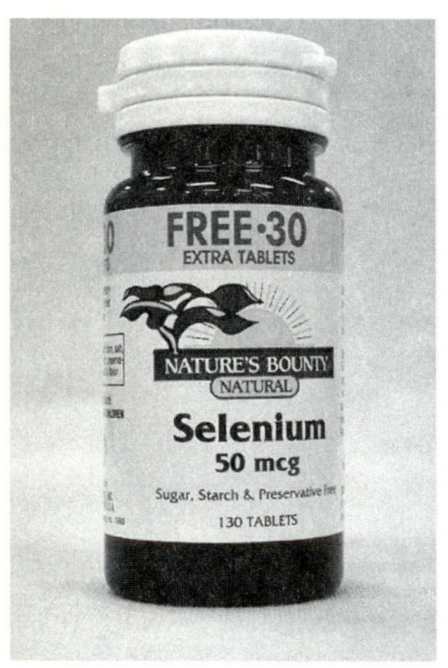

Selenium, a member of Group VI, has some properties related to sulfur, an important element found in compounds such as amino acids.

occupied orbitals. The sizes of some atoms expressed as the **atomic radius,** in picometers, are given in Figure 1.2. The atomic radius for an atom does not vary significantly from one compound to another.

Atomic radii increase from top to bottom in a group of the periodic table. Each successive member of a group has one additional energy level containing electrons located at larger distances from the nucleus. Thus, the atomic radius of sulfur is greater than that of oxygen, and the radii of the halogens increase in the order F < Cl < Br.

The atomic radius decreases from left to right across a period. Although electrons are located in the same energy level within the *s* and *p* orbitals of the elements, the nuclear charge increases from left to right within a period. As a consequence, the nucleus draws the electrons inward and the radius decreases. The radii of the common elements in organic compounds are in the order C > N > O.

Electronegativity

Electronegativity is a measure of the attraction of an atom for bonding electrons in molecules compared to that of other atoms. The electronegativity values devised by Linus Pauling, an American chemist, are dimensionless quantities that range from slightly less than one for the alkali metals to a maximum of four for fluorine. Large electronegativity values indicate a stronger attraction for electrons than small electronegativity values.

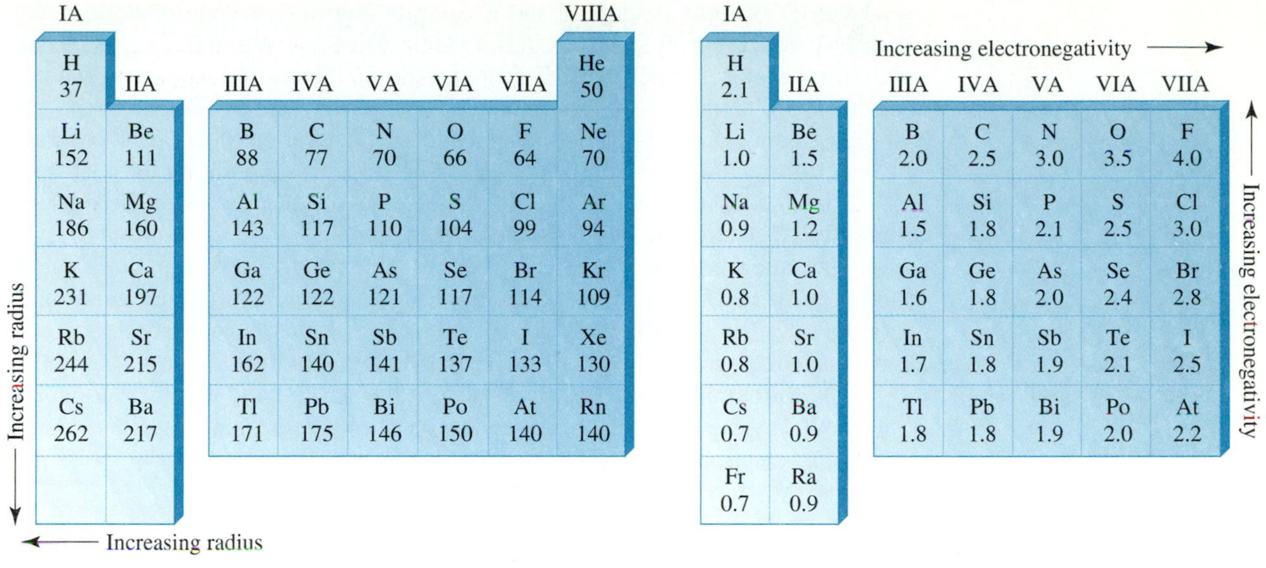

Atomic radii (pm)

IA	IIA	IIIA	IVA	VA	VIA	VIIA	VIIIA
H 37							He 50
Li 152	Be 111	B 88	C 77	N 70	O 66	F 64	Ne 70
Na 186	Mg 160	Al 143	Si 117	P 110	S 104	Cl 99	Ar 94
K 231	Ca 197	Ga 122	Ge 122	As 121	Se 117	Br 114	Kr 109
Rb 244	Sr 215	In 162	Sn 140	Sb 141	Te 137	I 133	Xe 130
Cs 262	Ba 217	Tl 171	Pb 175	Bi 146	Po 150	At 140	Rn 140

Electronegativities

IA	IIA	IIIA	IVA	VA	VIA	VIIA
H 2.1						
Li 1.0	Be 1.5	B 2.0	C 2.5	N 3.0	O 3.5	F 4.0
Na 0.9	Mg 1.2	Al 1.5	Si 1.8	P 2.1	S 2.5	Cl 3.0
K 0.8	Ca 1.0	Ga 1.6	Ge 1.8	As 2.0	Se 2.4	Br 2.8
Rb 0.8	Sr 1.0	In 1.7	Sn 1.8	Sb 1.9	Te 2.1	I 2.5
Cs 0.7	Ba 0.9	Tl 1.8	Pb 1.8	Bi 1.9	Po 2.0	At 2.2
Fr 0.7	Ra 0.9					

FIGURE 1.2
Atomic Radii and Electronegativity Values of the Elements

Electronegativities increase from left to right across the periodic table (Figure 1.2). Elements on the left of the periodic table have low electronegativities and are often called **electropositive** elements. The order of electronegativities F > O > N > C is an important property that we will use to explain the chemical properties of organic compounds.

Electronegativities decrease from top to bottom within a group of elements. The order of decreasing electronegativities F > Cl > Br > I is another sequence that we will use to interpret the chemical and physical properties of organic compounds.

1.4 Types of Bonds

In 1916, the American chemist G. N. Lewis proposed that second period elements tend to react to obtain an electron configuration of eight electrons so that they electronically resemble the inert gases. This hypothesis is summarized in the **Lewis octet rule:** Second period atoms tend to combine and form bonds by transferring or sharing electrons until each atom is surrounded by eight electrons in its highest energy shell. Note that hydrogen requires only two electrons to complete its valence shell.

Ionic Bonds

Ionic bonds are formed between two or more atoms by the transfer of one or more electrons between atoms. Electron transfer produces negative ions called **anions** and positive ions called **cations.** These ions attract each other.

Let us examine the ionic bond in sodium chloride. A sodium atom, which has 11 protons and 11 electrons, has a single valence electron in its 3*s* subshell. A chlorine atom, which has 17 protons and 17 electrons, has seven valence electrons in its third shell, represented as $3s^23p^5$. In forming an ionic bond, the sodium atom, which is electropositive, loses its valence electron to chlorine. The resulting sodium ion has the same electron configuration as neon ($1s^22s^22p^6$) and has a +1 charge, because there are 11 protons in the nucleus, but only 10 electrons about the nucleus of the ion.

The chlorine atom, which has a high electronegativity, gains an electron and is converted into a chloride ion that has the same electron configuration as argon ($1s^22s^22p^63s^23p^6$). The chloride ion has a −1 charge because there are 17 protons in the nucleus, but there are 18 electrons about the nucleus of the ion.

The formation of sodium chloride from the sodium and chlorine atoms can be shown by Lewis structures. **Lewis structures** represent only the valence electrons; electron pairs are shown as pairs of dots.

$$\mathrm{Na}\cdot + :\ddot{\underset{..}{\mathrm{Cl}}}\cdot \longrightarrow \mathrm{Na}^+ + :\ddot{\underset{..}{\mathrm{Cl}}}:^-$$

Note that by convention, the complete octet is shown for anions formed from electronegative elements. However, the filled outer shell of cations that results from loss of electrons by electropositive elements is not shown.

Metals are electropositive and tend to lose electrons, whereas nonmetals are electronegative and tend to gain electrons. A metal atom loses one or more electrons to form a cation with an octet. The same number of electrons are accepted by the appropriate number of atoms of a nonmetal to form an octet in the anion, producing an ionic compound. In general, ionic compounds result from combinations of metallic elements, located on the left side of the periodic table, with nonmetals, located on the upper right side of the periodic table.

Covalent Bonds

A **covalent bond** consists of the mutual sharing of one or more pairs of electrons between two atoms. These electrons are simultaneously attracted by the two atomic nuclei. A covalent bond forms when the difference between the electronegativities of two atoms is too small for an electron transfer to occur to form ions. Shared electrons located in the space between the two nuclei are called **bonding electrons.** The bonded pair is the "glue" that holds the atoms together in molecular units.

The hydrogen molecule is the simplest substance having a covalent bond. It forms from two hydrogen atoms, each with one electron in a 1*s* orbital. Both hydrogen atoms share the two electrons in the covalent bond, and each acquires a helium-like electron configuration.

$$\mathrm{H}\cdot + \cdot\mathrm{H} \longrightarrow \mathrm{H{-}H}$$

A similar bond forms in F_2. The two fluorine atoms in the fluorine molecule are joined by a shared pair of electrons. Each fluorine atom has seven valence electrons in the second energy level and requires one more electron to form a neon-like electron configuration. Each fluorine atom contributes one electron to the bonding pair shared by the two atoms. The remaining six valence electrons of each fluorine atom are not involved in bonding and are concentrated around their respective atoms. These valence electrons, customarily shown as pairs of electrons, are variously called **nonbonding electrons, lone pair electrons, or unshared electron pairs.**

:F̤· + ·F̤: ⟶ :F̤—F̤: ← nonbonding electrons

The covalent bond is drawn as a dash in a **Lewis structure** to distinguish the bonding pair from the lone pair electrons. Lewis structures show the nonbonding electrons as pairs of dots located about the atomic symbols for the atoms. The Lewis structures of four simple organic compounds—methane, methylamine, methanol, and chloromethane—are drawn here to show both bonding and nonbonding electrons. In these compounds carbon, nitrogen, oxygen, and chlorine atoms have four, three, two, and one bonds, respectively.

H_3C—H	H_3C—N̈—H (N bearing H)	H_3C—Ö: (O bearing H)	H_3C—C̈l:
methane	methylamine	methanol	chloromethane

The hydrogen atom and the halogen atoms form only one covalent bond to other atoms in most stable neutral compounds. However, the carbon, oxygen, and nitrogen atoms can simultaneously bond to more than one atom. The number of such bonds is the **valence** of the atom. The valences of carbon, nitrogen, and oxygen are four, three, and two, respectively.

Multiple Covalent Bonds

In some molecules more than one pair of electrons is shared between pairs of atoms. If four electrons (two pairs) or six electrons (three pairs) are shared, the bonds are called **double** and **triple** bonds, respectively. A carbon atom can form single, double, or triple bonds with other carbon atoms as well as with atoms of some other elements. Single, double, and triple covalent bonds link two carbon atoms in ethane, ethylene, and acetylene, respectively. Each carbon atom in these compounds shares one, two, and three electrons, respectively, with the other. The remaining valence electrons of the carbon atoms are contained in the single bonds with hydrogen atoms.

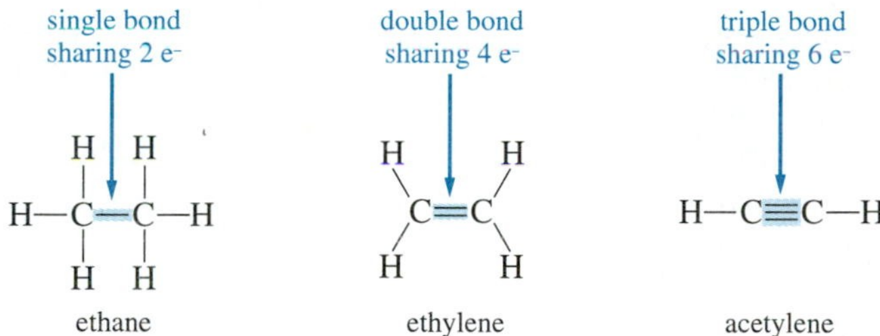

triple bond sharing 6 e⁻

H—C≡C—H

acetylene

Polar Covalent Bonds

A **polar covalent bond** exists when atoms with different electronegativities share electrons in a covalent bond. Consider the hydrogen chloride molecule (HCl). Each atom in HCl requires one more electron to form an inert gas electron configuration. Chlorine has a higher electronegativity than hydrogen, but the chlorine atom's attraction for electrons is not sufficient to remove an electron from hydrogen. Consequently, the bonding electrons in hydrogen chloride are shared unequally in a polar covalent bond. The molecule is represented by the conventional Lewis structure, even though the shared electron pair is associated to a larger extent with chlorine than with hydrogen. The unequal sharing of the bonding pair results in a partial negative charge on the chlorine atom and a partial positive charge on the hydrogen atom. The symbol δ (Greek lowercase delta) denotes these fractional charges.

$$\overset{\delta+}{\text{H}}-\overset{\delta-}{\ddot{\underset{..}{\text{Cl}}}}:$$

The hydrogen chloride molecule has a **dipole** (two poles), which consists of a pair of opposite charges separated from each other. The dipole is shown by an arrow with a cross at one end. The cross is near the end of the molecule that is partially positive, and the arrowhead is near the partially negative end of the molecule.

$$\overset{+\!\!\longrightarrow}{\text{H}-\ddot{\underset{..}{\text{Cl}}}:}$$

Single or multiple bonds between carbon atoms are nonpolar. Hydrogen and carbon have similar electronegativity values, so the C—H bond is not normally considered a polar covalent bond. Thus ethane, ethylene, and acetylene have nonpolar covalent bonds, and the compounds are nonpolar.

Bonds between carbon and other elements such as oxygen and nitrogen are polar. The polarity of a bond depends on the electronegativities of the bonded atoms. Large differences between the electronegativities of the bonded atoms increase the polarity of bonds. The direction of the polarity of common bonds found in organic molecules is easily predicted. The common nonmetals are more electronegative than carbon. Therefore, when a carbon atom is bonded to common nonmetal atoms, it has a partial positive charge.

$$\overset{+\!\!\longrightarrow}{\text{C}-\text{N}} \quad \overset{+\!\!\longrightarrow}{\text{C}-\text{O}} \quad \overset{+\!\!\longrightarrow}{\text{C}-\text{F}}$$

Hydrogen is also less electronegative than the common nonmetals. Therefore, when a hydrogen atom is bonded to common nonmetals, the resulting polar bond has a partial positive charge on the hydrogen atom.

$$\overset{+\!\!\longrightarrow}{\text{H}-\text{N}} \quad \overset{+\!\!\longrightarrow}{\text{H}-\text{O}} \quad \overset{+\!\!\longrightarrow}{\text{H}-\text{F}}$$

The magnitude of the polarity of a bond is the **bond moment.** This value is reported in Debye units (D). The bond moments of several bond types are given in Table 1.2. The bond moment of a specific bond is relatively constant from compound to compound.

Table 1.2 Bond Moments (Debye)

Structural Unit*	Bond Moment (D)
H—C	0.4
H—N	1.3
H—O	1.5
H—F	1.7
H—S	0.7
H—Cl	1.1
H—Br	0.8
H—I	0.4
C—C	0.0
C—N	0.2
C—O	0.7
C—F	1.4
C—Cl	1.5
C—Br	1.4
C—I	1.2
C=O	2.3
C≡N	3.5

* The more electronegative part of the unit is to the right of the bond.

When carbon forms multiple bonds to atoms other than carbon, these bonds are polar. Both the carbon-oxygen double bond in formaldehyde and the carbon-nitrogen triple bond in acetonitrile are polar.

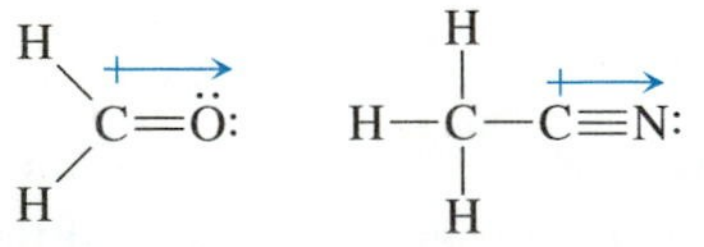

$$\text{H}_2\overset{+\!\!\longrightarrow}{\text{C}=\ddot{\text{O}}}: \qquad \text{H}_3\text{C}-\overset{+\!\!\longrightarrow}{\text{C}\equiv\text{N}}:$$

formaldehyde acetonitrile

1.5 Formal Charge

Although most organic molecules are represented by Lewis structures containing the "normal" number of bonds, some organic ions and even some molecules contain less than or more than the customary number of bonds. First let's review the structures of some "inorganic" ions. The valence of the oxygen atom is two—it normally forms two bonds. However, there are three bonds in the hydronium ion and one in the hydroxide ion.

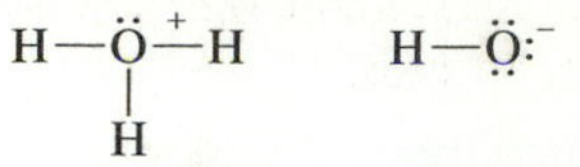

H—Ö—H (with + on O, H below O) hydronium ion

H—Ö:⁻ hydroxide ion

How do we predict the charge of the ions? Second, what atoms bear the charge? There is a useful formalism for answering both of these question. Each atom is assigned a formal charge by a bookkeeping method that involves counting electrons. The method is also used for neutral molecules that have unusual numbers of bonds. In such cases, centers of both positive and negative charge are located at specific atoms.

The **formal charge** of an atom is equal to the number of its valence electrons as a free atom minus the number of electrons that it "owns" in the Lewis structure.

$$\text{formal charge} = \left[\begin{array}{l}\text{number of valence}\\ \text{electrons in free atom}\end{array}\right] - \left[\begin{array}{l}\text{number of valence}\\ \text{electrons in bonded atom}\end{array}\right]$$

The question of ownership is decided by two simple rules. Unshared electrons belong exclusively to the parent atom. One-half of the bonded electrons between a pair of atoms is assigned to each atom. Thus, the total number of electrons "owned" by an atom in the Lewis structure equals the number of nonbonding electrons plus half the number of bonding electrons. Therefore, we write

$$\begin{array}{l}\text{formal}\\ \text{charge}\end{array} = \left[\begin{array}{l}\text{number of}\\ \text{valence electrons}\end{array}\right] - \left[\begin{array}{l}\text{number of}\\ \text{nonbonding electrons}\end{array}\right] - \frac{1}{2}\left[\begin{array}{l}\text{number of}\\ \text{bonding electrons}\end{array}\right]$$

The formal charge of each atom is zero in most organic molecules. However, the formal charge may also be negative or positive. The sum of the formal charges of each atom in a molecule equals zero; the sum of the formal charges of each atom in an ion equals the charge of the ion. Let's consider the molecule methyl isocyanide and calculate the formal charges of the carbon and nitrogen atoms bonded in a triple bond.

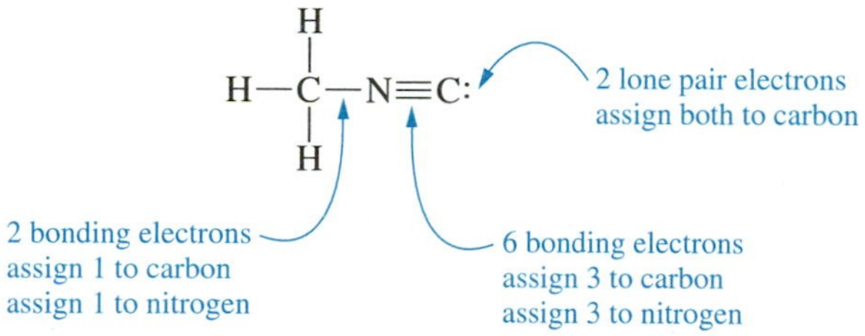

$H_3C—N{\equiv}C:$

2 lone pair electrons assign both to carbon

2 bonding electrons assign 1 to carbon assign 1 to nitrogen

6 bonding electrons assign 3 to carbon assign 3 to nitrogen

The formal charge of each atom may be calculated by substitution into the formula.

$$\text{formal charge of carbon} = 4 - 2 - \tfrac{1}{2}(6) = -1$$
$$\text{formal charge of nitrogen} = 5 - 0 - \tfrac{1}{2}(8) = +1$$

The formal charges of the carbon and nitrogen atoms in this compound are not zero. However, note that the sum of the formal charges of the atoms equals the net charge of the species, which in this case is zero.

There are often important chemical consequences when a neutral molecule contains centers whose formal charges are not zero. It is important that you be able to recognize these situations so you can understand the chemical reactivity of such molecules.

1.6 Resonance Structures

In the Lewis structures for the molecules shown to this point, the electrons have been pictured as either between two nuclei or about a specific atom. These electrons are **localized.** The electronic structures of molecules are written to be consistent with their physical properties. However, the electronic structures of some molecules cannot be represented adequately by a single Lewis structure. For example, the Lewis structure of the acetate ion has one double bond and one single bond to oxygen atoms. Note that the formal charge of the single-bonded oxygen atom is −1 whereas that of the double-bonded oxygen atom is zero.

```
     H      Ö:
     |     //
 H—C—C
     |     \
     H      :Ö:⁻
```

However, single and double bonds are known to have different bond lengths—a double bond between two atoms is shorter than a single bond. The Lewis structure shown implies that there is one "long" C—O bond and a "short" C═O bond in the acetate ion. But both carbon-oxygen bond lengths in the acetate ion have been shown experimentally to be equal. Moreover, both oxygen atoms bear equal amounts of negative charge. Therefore, the preceding Lewis structure with single and double bonds does not accurately describe the acetate ion. Under these circumstances, the concept of **resonance** is used. We say that a molecule is **resonance stabilized** if two or more Lewis structures can be written that have identical arrangements of atoms, but different arrangements of electrons. The real structure of the acetate ion can be represented better as a **hybrid** of two Lewis structures, neither of which is completely correct.

```
     H      Ö:                 H      :Ö:⁻
     |     //                  |     /
 H—C—C          ←→     H—C—C
     |     \                   |     \\
     H      :Ö:⁻               H      Ö:
```

A double-headed arrow between two Lewis structures indicates that the actual structure is similar in part to the two simple structures, but lies somewhere between them. The individual Lewis structures are called contributing structures or **resonance structures.**

Curved arrows can be used to keep track of the electrons when writing resonance structures. The tail of the arrow is located near the bonding or nonbonding pair of electrons to be "moved" or "pushed," and the arrowhead shows the final destination of the electron pair.

```
     H      Ö:                                        H      :Ö:⁻
     |     //                                         |     /
 H—C—C        "pushing" electrons gives       H—C—C
     |     \                                          |     \\
     H      :Ö:⁻                                      H      Ö:
   structure 1                                    structure 2
```

In resonance structure 1, the nonbonding pair of electrons on the bottom oxygen atom is moved to form a double bond with the carbon atom. A bonding pair of electrons of the carbon-oxygen double bond is also moved to form a nonbonding pair of electrons on the top oxygen atom. The result is resonance structure 2. This procedure of "pushing" electrons from one position to another is only a bookkeeping formalism. Electrons do not really move this way! The actual ion has **delocalized** electrons distributed over three atoms—a phenomenon that cannot be shown by a single Lewis structure.

Electrons can be delocalized over many atoms. For example, benzene, C_6H_6, consists of six equivalent carbon atoms contained in a ring in which all carbon-carbon bonds are identical. Each carbon atom is bonded to a hydrogen atom. A single Lewis structure containing alternating single and double bonds can be written to satisfy the Lewis octet requirements.

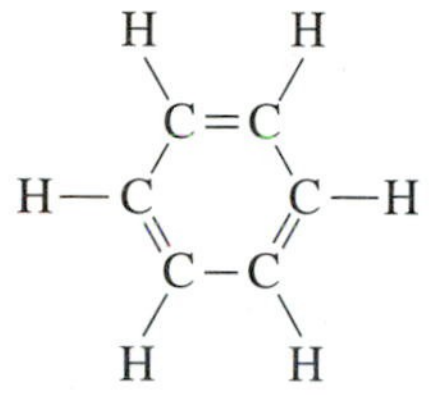

```
      H     H
       \   /
        C=C
       /   \
   H—C       C—H
       \\   //
        C—C
       /   \
      H     H
```

However, single and double bonds have different bond lengths. In benzene, all carbon-carbon bonds have been shown to be the same length. Like the acetate ion, benzene is represented by two contributing resonance structures separated by a double-headed arrow. The positions of the alternating single and double bonds are interchanged in the two resonance structures.

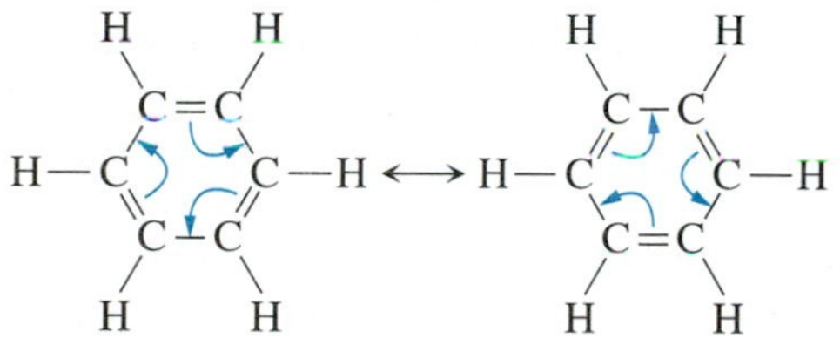

```
      H     H                 H     H
       \   /                   \   /
        C=C                     C—C
       /   \                   //   \\
   H—C       C—H  ⟷  H—C       C—H
       \\   //                 \   /
        C—C                     C=C
       /   \                   /   \
      H     H                 H     H
```

The electrons in benzene are delocalized over the six carbon atoms in the ring, resulting in a unique structure. There are no carbon-carbon single or double bonds in benzene; its bonds are of an intermediate type that cannot be represented with a single structure.

EXAMPLE 1-1

Consider the structure of nitromethane, a compound used to increase the power in specialized race car engines. A nitrogen-oxygen single bond length is 136 pm; a nitrogen-oxygen double bond length is 114 pm. The nitrogen-oxygen bonds in nitromethane are equal and are 122 pm. Explain.

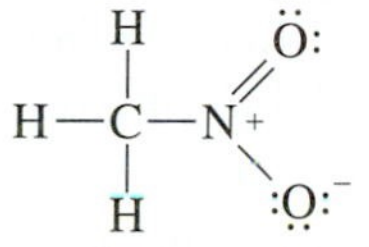

```
      H       Ö:
      |      //
   H—C—N+
      |      \
      H      :Ö:⁻
```

Solution

The actual nitrogen-oxygen bonds are neither single nor double bonds. Two resonance forms can be written to represent nitromethane. They result from "moving" a nonbonding pair of electrons from the single-bonded oxygen atom to form a double bond with the nitrogen atom. One of the bonding pairs of electrons from the nitrogen-oxygen double bond is moved to the other oxygen atom. The structures differ only in the location of the single and double bonds.

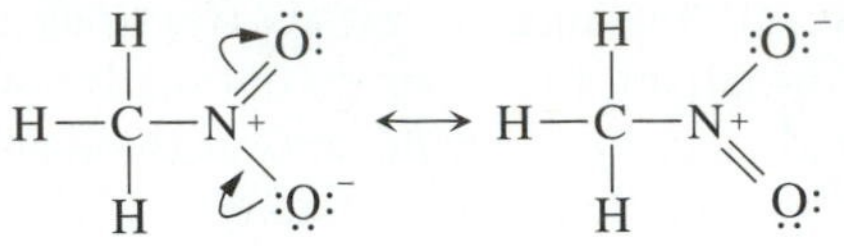

The two structures are contributing resonance forms. The actual molecule has delocalized electrons distributed over the nitrogen atom and the two oxygen atoms.

Problem 1.1

Nitrites (NO_2^-) are added as antioxidants in some processed meats. Both nitrogen-oxygen bonds are of equal lengths. Write resonance structures for the nitrite ion.

1.7 Predicting the Shapes of Simple Molecules

Up to this point, we have considered the distribution of bonding electrons and nonbonding electrons within molecules without regard to their location in three-dimensional space. But molecules have characteristic shapes that reflect the spatial arrangement of electrons in bonds. For example, the shapes of carbon dioxide, formaldehyde, and methane are linear, trigonal planar, and tetrahedral, respectively. (Note that wedge-shaped bonds are used to show the location of atoms above the plane of the page and dashed lines to indicate the location of atoms behind the plane of the page.)

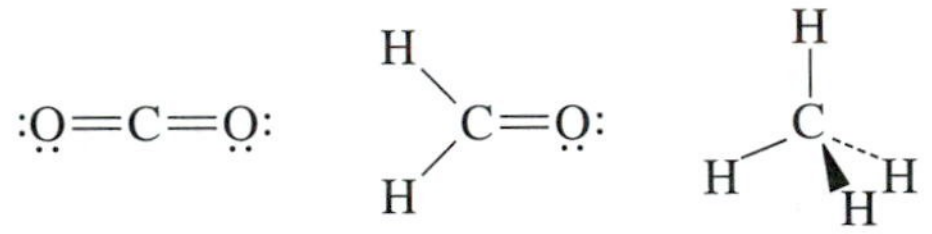

carbon dioxide formaldehyde methane

We can "predict" the geometry of these simple molecules and approximate the bond angles using **valence-shell electron-pair repulsion** (abbreviated VSEPR) theory. This theory is based on the idea that bonding and nonbonding electron pairs about a central atom repel each other. VSEPR theory predicts that electron pairs in molecules should be arranged as far apart as possible. Thus, two electron pairs should be arranged at 180° to each other; three pairs should be at 120° in a common plane; four electron pairs should have a tetrahedral arrangement with angles of 109.5°.

All of the valence electrons about the central carbon atom in carbon dioxide, formaldehyde, and methane are in bonds. Each type of bond may be regarded as a region that contains electrons that should be arranged as far apart as possible. Carbon dioxide has two double bonds; the double bonds are separated by the maximum distance, and the resulting angle between the bonds is 180°. Formaldehyde has a double bond and two single bonds to the central carbon atom; these bonds correspond to three regions containing electrons and are separated by the maximum distance in a trigonal planar arrangement with bond angles of 120°. Methane has four bonding electron pairs, and they are best located in a tetrahedral arrangement. Each H—C—H bond angle is predicted to be 109.5°, in agreement with the experimental value.

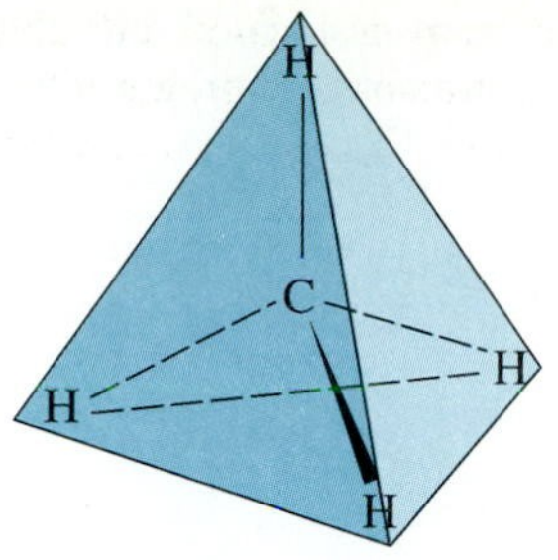

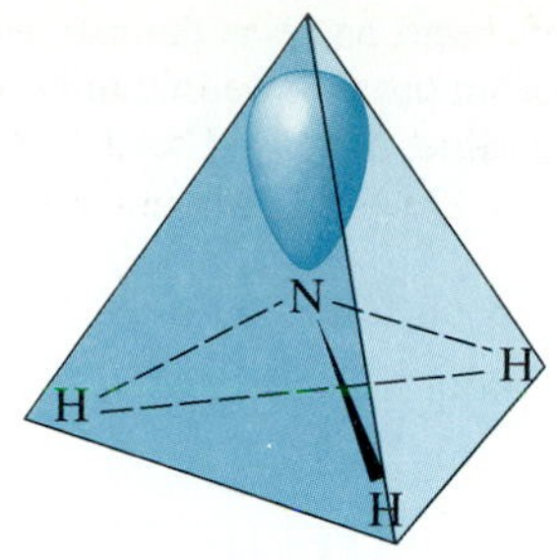

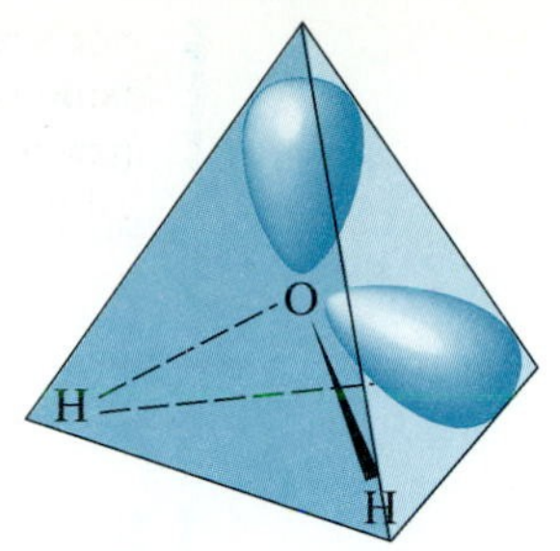

FIGURE 1.3
VSEPR Models to Predict Geometry of Molecules
All electron pairs in methane, ammonia, and water are directed to the corners of a tetrahedron. However, the ammonia molecule is described as trigonal pyramidal; the water molecule is angular.

Now let us consider molecules that have both bonding and nonbonding pairs of electrons in the valence shell of the central atom. Water and ammonia have experimentally determined shapes described as angular and trigonal pyramidal, respectively. Both have four electron pairs about the central atom, as does methane. They both have central atoms bonded to hydrogen atoms, but there are also unshared electron pairs.

H—Ö:
 \
  H
angular molecule

N̈ (bonded to H, H, H)
trigonal pyramidal molecule

VSEPR theory describes the distribution of electron pairs, including the nonbonding pairs. However, molecular structure is defined by the positions of the nuclei. Although the four pairs of electrons in both water and ammonia are tetrahedrally arranged, water and ammonia are angular and pyramidal molecules, respectively (Figure 1.3).

The arrangements of bonds to the oxygen atom and the nitrogen atom in organic molecules are similar to those in water and ammonia, respectively. The groups bonded to the oxygen atom of an alcohol or an ether (Section 1.9) are arranged to form angular molecules. The groups bonded to the nitrogen atom of an amine (Section 1.9) are arranged to form a pyramid.

EXAMPLE 1-2

The electronic structure of allyl isothiocyanate, a flavor ingredient in horseradish, is shown here. What are the C—N=C and N=C=S bond angles?

$$CH_2{=}CH{-}CH_2{-}\ddot{N}{=}C{=}\ddot{S}:$$

Solution

The C—N=C bond angle is determined by the electrons associated with the nitrogen atom. This atom has a single bond, a double bond, and a nonbonding pair of electrons. These three regions containing electrons have a trigonal planar arrangement. Only two of the electron-containing regions are bonding, but the C—N=C bond angle must still be 120°.

$$\overset{..}{N}{=}C{=}\ddot{S}:$$
C (bonded to N)

The N=C=S bond angle is determined by the electrons associated with the carbon atom. This atom has two double bonds. These electron-containing regions are arranged in a straight line; the N=C=S bond angle is 180°.

Problem 1.2

Using one of the resonance forms for the nitrite ion (NO_2^-), determine the shape of this ion.

1.8 Orbitals and Molecular Shapes

Because electrons form bonds between atoms, the shapes of molecules depend on the location of the electrons in the orbitals of the various atoms. Two electrons in a covalent bond are shared in a region of space common to the bonding atoms. This region of space is pictured as an overlap or merging of two atomic orbitals. For example, the covalent bond in H_2 results from the overlap of two *s* orbitals to give a sigma (σ) bond (Figure 1.4). This bond is symmetrical around an axis joining the two nuclei. Viewed along the interatomic axis, the σ bond looks like an *s* orbital. All single bonds are also σ bonds regardless of the component orbitals used in forming the bond.

The simple picture of bonding described for H_2 has to be modified somewhat for carbon-containing compounds. Carbon has the electronic configuration $1s^2 2s^2 2p^2$, which suggests that only the two electrons in the $2p$ orbitals would be available to form two covalent bonds. If this were so, the molecular formula for a compound of carbon and hydrogen would be CH_2 and the carbon atom would not have a Lewis octet.

```
:C—H
 |
 H
```

However, there are four equivalent C—H bonds in methane, CH_4. All carbon compounds presented in this chapter have a Lewis octet about the carbon atoms, and each carbon atom has four bonds. The difference between these structural facts and predictions based on the atomic orbitals of carbon is explained using the concept of **hybrid orbitals,** which result from the "mixing" of two or more orbitals in the bonded atoms. This mixing process, called **orbital hybridization,** was proposed by Linus Pauling to account for the formation of bonds by using orbitals having the geometry appropriate for the actual molecule. As a result of hybridization, two or more hybrid orbitals can be formed from the appropriate number of atomic orbitals. The number of hybrid orbitals created equals the number of atomic orbitals used in hybridization.

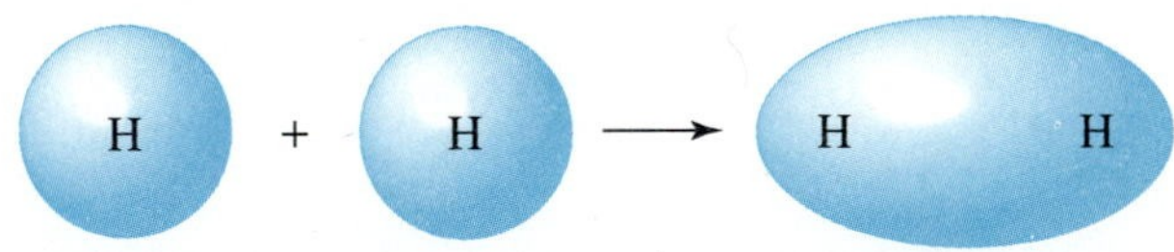

FIGURE 1.4
The Sigma Bond of the Hydrogen Molecule
The region occupied by the electron pair is symmetrical about both hydrogen nuclei. Although the two electrons may be located anywhere within the volume shown, it is most probable that they are between the two nuclei.

sp^3 Hybridization of Carbon

Pauling suggested that the tetrahedral geometry of methane results from hybridization of the $2s$ and the three $2p$ orbitals of carbon, which combine to form four equivalent hybrid orbitals. Each hybrid orbital contains one electron. These orbitals extend toward the corners of a tetrahedron so there is maximum separation of the electrons. Each hybrid atomic orbital then overlaps with a hydrogen $1s$ orbital to form a σ bond.

The formation of the hybrid orbitals is illustrated in Figure 1.5. The four new orbitals are called sp^3 hybrid orbitals because they result from the combination of one s and three p orbitals. Each sp^3 orbital has the same shape, and the electrons in each orbital have the same energy. The orbitals differ only in their position in space.

sp^2 Hybridization of Carbon

Now let us consider the bonding electrons in the double bond of ethylene, in which each carbon atom is bonded to three atoms. All six nuclei lie in a plane, and all the bond angles are close to 120°. Each carbon atom in ethylene is pictured with three **sp^2 hybrid orbitals** and one remaining $2p$ orbital. The three sp^2 hybrid orbitals result from "mixing" a single $2s$ orbital and two $2p$ orbitals. Each sp^2 orbital has the same shape, and the electrons in each orbital have the same energy. The orbitals differ only in their position in space. They are separated by 120° and are directed to the corners of a triangle to have maximum separation of the electrons. The four valence electrons are distributed as indicated in Figure 1.6. The three sp^2 hybridized orbitals are used to make σ bonds. Two of the sp^2 orbitals, containing one electron each, form σ bonds with hydrogen. The third sp^2 orbital, which also contains one electron, forms a σ bond with the other carbon atom in ethylene.

The second bond of the double bond in ethylene results from a lateral (side-by-side) overlap of the p orbitals of each carbon atom. Each p orbital is perpendicular to the plane containing the sp^2 orbitals. The $2p$ orbital of each atom provides one electron to the electron pair for the second bond. A bond

$2p_x$ $2p_y$ $2p_z$

$2s$

Isolated C atom

sp^3 sp^3 sp^3 sp^3

Hybridized C atom in CH_4

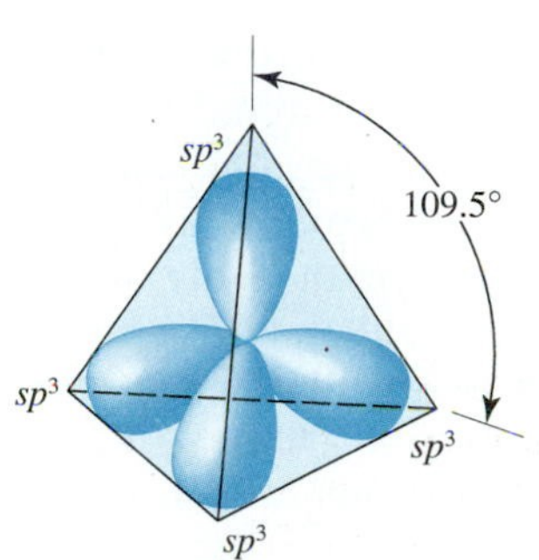

FIGURE 1.5
Hybridized Atomic Orbitals in Methane

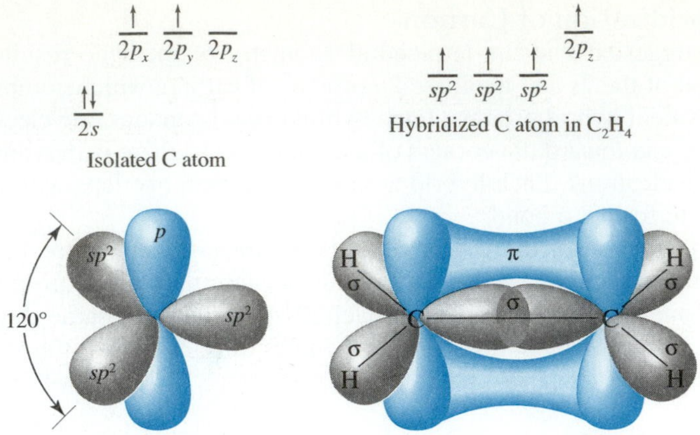

The three sp^2 hybrid orbitals lie in a plane with a 120° angle between them. The remaining p orbital is perpendicular to the plane of the sp^2 orbitals.

The σ bond between the carbon atoms is formed by overlap of sp^2 orbitals. The σ bonds to hydrogen are formed by overlap of the sp^2 orbitals of carbon with the s orbital of individual hydrogen atoms. The π bond is formed by sideways overlap of parallel p orbitals.

FIGURE 1.6
Hybridization and the Double Bond of Ethylene

formed by sideways overlap of p orbitals is a π **(pi)** bond. Viewed along the carbon-carbon internuclear axis, a π bond resembles a p orbital. Note that the electrons in the π bond are not concentrated along an axis between the two atoms but are shared in regions of space both above and below the plane defined by the sp^2 orbitals. It is, nevertheless, only one bond.

sp Hybridization of Carbon

Now let us consider the triple bond of acetylene, in which each carbon atom is bonded to two atoms. All four nuclei are arranged in a line, and all the bond angles are 180°. In acetylene, we mix a 2s orbital with a 2p orbital to give two ***sp* hybrid orbitals** of equal energy. The remaining two 2p orbitals are unchanged (Figure 1.7). The sp orbitals have the same shape, and the electrons in each orbital have the same energy. The orbitals differ only in their position in space; they are at 180° angles to each other—again to provide for maximum separation of the electrons.

Each carbon atom in acetylene has four valence electrons. The two sp hybrid orbitals of each carbon atom each contain one electron, and the two 2p orbitals of each carbon atom each contain one electron. The carbon atoms in acetylene are linked by one σ bond and two π bonds to give a triple bond. One sp orbital and its electron form a bond with hydrogen; the other sp orbital forms a σ bond with the second carbon atom. The second and third bonds between carbon atoms result from sideways overlap of 2p orbitals. One set of 2p orbitals overlaps in front and back of the molecule to form one π bond. The second set of 2p orbitals overlaps above and below the molecule to form the second π bond.

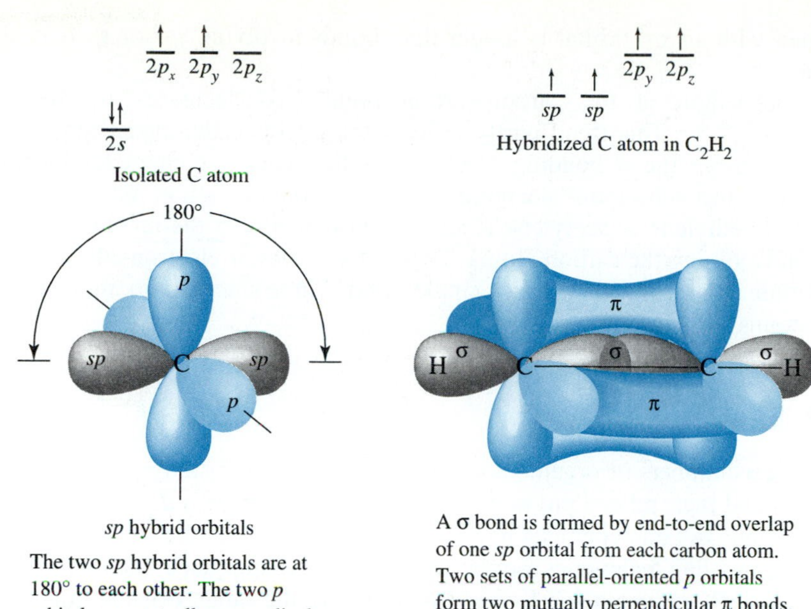

sp hybrid orbitals

The two *sp* hybrid orbitals are at 180° to each other. The two *p* orbitals are mutually perpendicular to each other and to the axis of the *sp* orbitals.

A σ bond is formed by end-to-end overlap of one *sp* orbital from each carbon atom. Two sets of parallel-oriented *p* orbitals form two mutually perpendicular π bonds. The remaining *sp* orbital of each carbon atom forms a σ bond to a hydrogen atom.

FIGURE 1.7
Hybridization of the Triple Bond of Acetylene

Effect of Hybridization on Bond Length

The hybridization of carbon in methane, ethylene, and acetylene affects the C—H and C—C bond lengths. Note in Table 1.3 that the length of the C—H bond decreases in the order $sp^3 > sp^2 > sp$. This order reflects the lower energy of the 2*s* orbital compared to the energy of the 2*p* orbital and the fact that, on average, the 2*s* orbital is closer to the nucleus than the 2*p* orbital. The average distance of hybrid orbitals from the nucleus depends on the percent contribution of the *s* and *p* orbitals. The contribution of the *s* orbital is 25% in an sp^3 hybrid orbital, because one *s* and three *p* orbitals are replaced by the four hybrid orbitals. Similarly, the contribution of the *s* orbital is 33% and 50% for the sp^2 and *sp* hybrid orbitals, respectively. Because an sp^3 hybrid orbital has a smaller *s* character than an sp^2 or *sp* hybrid orbital, the electrons in an sp^3 orbital are in general farther from the nucleus. As a consequence, a bond

Table 1.3 Effect of Hybridization on Bond Lengths

	$H_3C—CH_3$	$H_2C=CH_2$	H—C≡C—H
C—H bond	110 pm	108 pm	106 pm
C—C bond	154 pm	133 pm	120 pm

formed with an sp^3 orbital is longer than bonds involving sp^2 or sp hybrid orbitals.

The length of the carbon-carbon bond also decreases in the order $sp^3 > sp^2 > sp$. This trend partly reflects the effect of the closer approach to the nucleus of the σ bonding electrons as the percent s character increases. However, the substantial decrease in the carbon to carbon bond length of ethane > ethylene > acetylene is also a consequence of the increased number of bonds joining the carbon atoms. Two shared pairs of electrons draw the carbon atoms closer together than a single bond. Three shared pairs move the carbon atoms still closer.

1.9 Functional Groups

The sheer numbers of organic compounds can make the study of organic compounds and their related physical and chemical properties a daunting objective. Fortunately, organic chemists have found ways to handle the immense number of facts by using techniques similar to those used in inorganic chemistry. Just as the elements are organized by groups in a periodic table, organic compounds are organized into families of compounds. Families of organic compounds are organized by functional groups. Atoms or groups of bonded atoms responsible for similar physical and chemical properties in a family of compounds are **functional groups.** Thus, the study of 10 million compounds is organized into more manageable groups of compounds whose reactivity is predictable. A summary of the more common functional groups is give in Table 1.4.

Some functional groups are a part of the carbon skeleton. These include the carbon-carbon double bond in compounds called alkenes, such as ethylene, and the carbon-carbon triple bond in compounds called alkynes, such as acetylene. Although benzene, a member of a class called arenes, is represented as a series of alternating carbon-carbon single and double bonds, it reacts differently from ethylene (Chapter 5).

```
H       H                                    H     H
 \     /                                      \   /
  C=C                                          C=C
 /     \          H—C≡C—H                 H—C     C—H
H       H                                      \\  //
                                               C—C
                                              /   \
                                             H     H
```

ethylene (an alkene) — acetylene (an alkyne) — benzene (an arene)

Functional groups can contain a variety of elements. The most common elements in functional groups are oxygen and nitrogen, although sulfur or the halogens may also be present. **Alcohols** and **ethers** are two classes of compounds that contain carbon-oxygen single bonds. The —OH unit in alcohols is the **hydroxyl group.**

```
    H                 H       H
    |                 |       |
H—C—Ö—H           H—C—Ö—C—H
    |                 |       |
    H                 H       H
```

methanol (an alcohol) — dimethyl ether (an ether)

Table 1.4 Classes and Functional Groups of Organic Compounds

Class	Functional Group	Example of Expanded Structural Formula
Alkene	>C=C<	H, H C=C H, H
Alkyne	—C≡C—	H—C≡C—H
Alcohol	—O—H	H H H—C—C—O—H H H
Ether	| | —C—O—C— | |	H H H—C—O—C—H H H
Aldehyde	O ‖ —C—H	H H O H—C—C—C—H H H
Ketone	O ‖ —C—	H O H H—C—C—C—H H H
Carboxylic acid	O ‖ —C—O—H	H O H—C—C—O—H H
Ester	O ‖ —C—O—C—	H O H H—C—C—O—C—H H H
Amine	H —N—H	H H H—C—N—H H
Amide	O H —C—N—H	H O H H—C—C—N—H H
Halide	—X (X = F, Cl, Br, I)	H H H—C—C—Br H H
Mercaptan or thiol	—S—H	H H H—C—C—S—H H H

Aldehydes and ketones contain double bonds to oxygen. The unit C═O is called the **carbonyl group.** The carbon atom of the carbonyl group is called the **carbonyl carbon atom,** and the oxygen atom is called the **carbonyl oxygen atom.** Note that an **aldehyde** has at least one hydrogen atom bonded to the carbonyl carbon atom. In **ketones,** the carbonyl carbon atom is bonded to two other carbon atoms.

```
    H  Ö:              H  Ö: H
    |  ‖               |  ‖  |
 H—C—C—H            H—C—C—C—H
    |                  |     |
    H                  H     H
```

acetaldehyde
(an aldehyde)

acetone
(a ketone)

Carboxylic acids and esters contain both single and double bonds from a carbon atom to oxygen atoms. In a **carboxylic acid,** the carbonyl group is bonded to a hydroxyl group and either a hydrogen or a carbon atom. In an **ester,** the carbonyl group is bonded to an "O—R" group, where "R" contains one or more carbon atoms, and to either a hydrogen or a carbon atom.

```
    H  Ö:                 H  Ö:     H
    |  ‖                  |  ‖      |
 H—C—C—Ö—H            H—C—C—Ö—C—H
    |                     |         |
    H                     H         H
```

acetic acid
(a carboxylic acid)

methyl acetate
(an ester)

A nitrogen atom can form single, double, or triple bonds to a carbon atom. Compounds with one or more carbon-nitrogen single bonds are **amines.** The remaining bonds to nitrogen can be to hydrogen or carbon atoms. Compounds with carbon-nitrogen double and triple bonds are **imines** and **nitriles,** respectively.

```
    H                 H  N̈—H          H
    |                 |  ‖            |
 H—C—N̈—H          H—C—C—H        H—C—C≡N:
    |  |              |               |
    H  H              H               H
```

methylamine
(an amine)

ethylimine
(an imine)

acetonitrile
(a nitrile)

Amides are functional groups in which a carbonyl carbon atom is linked by a single bond to a nitrogen atom and either a hydrogen or a carbon atom. The remaining two bonds to the nitrogen atom may be to either hydrogen or carbon atoms.

```
    H  Ö:
    |  ‖
 H—C—C—N̈—H
    |     |
    H     H
```

acetamide
(an amide)

Sulfur forms single bonds to carbon in two classes of compounds. **Thiols** (also called mercaptans) and **thioethers** (also called sulfides) structurally resemble alcohols and ethers, which contain oxygen, another element in the same group of the periodic table as sulfur.

```
    H                 H     H
    |                 |     |
 H—C—S̈—H          H—C—S̈—C—H
    |                 |     |
    H                 H     H
```

methanethiol
(a thiol)

dimethyl sulfide
(a thioether)

The halogens form single bonds to carbon. Chlorine and bromine are the more common halogens in organic compounds. Compounds with halogens bonded to a carbon atom containing only single bonds to carbon or hydrogen are haloalkanes (alkyl halides). Compounds with halogens bonded to a carbonyl carbon atom are acyl halides.

$$H_3C-\ddot{\underset{\cdot\cdot}{Br}}: \qquad H_3C-\overset{\ddot{O}:}{\overset{\|}{C}}-\ddot{\underset{\cdot\cdot}{Cl}}:$$

bromomethane (a haloalkane) acetyl chloride (an acyl halide)

1.10 Structural Formulas

Propane and butane are two propellants used in products such as shaving cream.

The **molecular formula** of a compound indicates its atomic composition. For example, the molecular formula of butane is C_4H_{10}. However, to understand the chemistry of organic compounds, it is necessary to represent the structure of a molecule by a **structural formula** that shows the arrangement of atoms and bonds.

To save time and space, chemists draw abbreviated or condensed versions of structural formulas. **Condensed structural formulas** show only specific bonds; other bonds are left out but implied. The degree of condensation depends on which bonds are shown and which are implied. For example, because hydrogen forms only a single bond to carbon, the C—H bond need not be shown in the condensed structure of a molecule such as butane.

$$CH_3-CH_2-CH_2-CH_3$$

butane

One carbon-carbon bond is shown between a terminal carbon atom and an internal carbon atom. The terminal carbon atoms are understood to have single bonds to three hydrogen atoms. Each carbon atom in the interior of the molecule has the two carbon-carbon bonds shown; the two carbon-hydrogen bonds are implied but not shown. Note that by convention the symbol for the hydrogen atom is written to the right of the symbol for the carbon atom.

In a further condensation of a structural formula, the C—C bonds are left out.

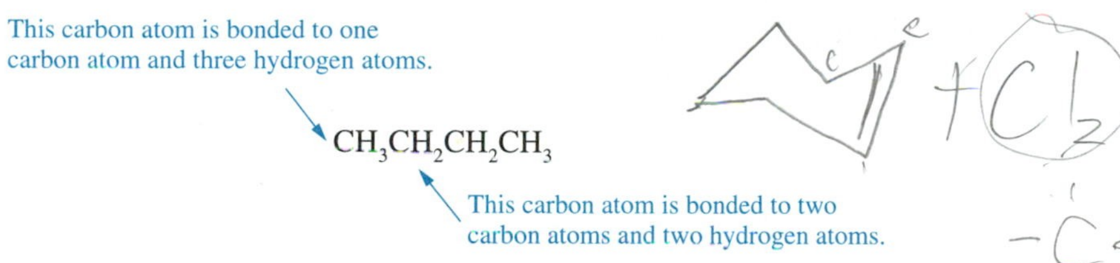

In this representation, the carbon atom on the left is understood to be bonded to the three hydrogen atoms and the carbon atom to the right. The second carbon atom from the left is bonded to the two hydrogen atoms to the right. That carbon atom is also bonded to a carbon atom to its immediate right and a carbon atom to its left.

Large structures may have repeated structural subunits that are represented by grouping the subunits within parentheses. The number of times the unit is repeated is given by a subscript after the closing parenthesis. For example, butane is represented by

$$CH_3(CH_2)_2CH_3$$

Pheromones: Chemical Communications in the Insect World

The scope of organic chemistry is rapidly changing and contributes to many fields. For example, we cannot understand modern biology without a foundation in organic chemistry and indirectly without an understanding of functional groups. Organic chemistry underlies all life forms. As an example, we will consider the structure and functional groups of some pheromones. Pheromones (Gk. *pherein,* "to transfer," + *hormon,* "to excite") are compounds (occasionally mixtures of compounds) that insects and other animals use to communicate. The major identified species that use pheromones are insects, but even higher animals including mammals also emit pheromones.

Pheromones are used to mark trails, warn of dangers, cause aggregation of species, defend against danger, and attract members of the opposite sex. The whip scorpion ejects a defensive spray that it uses to ward off predators. Some species of ants warn other ants of danger by an alarm pheromone. Bark beetles responsible for Dutch Elm disease emit an aggregation pheromone that results in the gathering of a large number of beetles on the trees. This species carries and transmits a fungus that kills the tree. The sex attractants, usually emitted by the female of the species, attract members of the opposite sex. They are signals that the female is ready to mate. They also aid the male in locating the female, often from great distances.

All moths that have been studied have sex attractants that are species specific. The compounds usually are derived from long chains of carbon atoms. However, the functional groups in the pheromones vary considerably. Two examples are the sex attractants of the gypsy moth and the grape berry moth. Their structures are shown below. The oxygen atom in the three-membered ring of the sex attractant of the gypsy moth is part of an ether functional group. The oxygen atoms in the sex attractant of the grape berry moth are part of an ester functional group. Note that this compound also contains a carbon-carbon double bond.

When the structures of sex attractants were determined, some scientists predicted that it might be possible to bait traps with the compound and, by removal of one sex, break the reproductive cycle. Unfortunately, this "ideal" way to control insects and eliminate the use of pesticides has not proved effective for most species. However, there is some evidence that pheromone traps cause some confusion and may inhibit mating. Also, pheromone traps serve as an early warning of possible infestation by that species. The ultimate goal of replacing pesticides with pheromones may yet be possible. ■

$$CH_3CH_2CH_2CH_2CH_2CH_2CH_2CH_2CH_2CH_2CH_2{-}\underset{\diagdown\,O\,\diagup}{CH{-}CH}{-}CH_2CH_2CH_2CH_2CH(CH_3)_2$$

sex attractant of the gypsy moth

$$CH_3CH_2{-}CH{=}CH{-}CH_2CH_2CH_2CH_2CH_2CH_2CH_2CH_2{-}O{-}\overset{\displaystyle O}{\overset{\|}{C}}{-}CH_3$$

sex attractant of the grape berry moth

The $-CH_2-$ unit is a **methylene** group. It occurs twice in butane. Because the methylene units are linked in a repeating chain, they are placed within the parentheses.

Two or more identical groups of atoms bonded to a common central atom may also be represented within parentheses with an appropriate subscript in a condensed formula. The parentheses may be placed to the right or left in the condensed structure depending on the way in which the molecule is drawn.

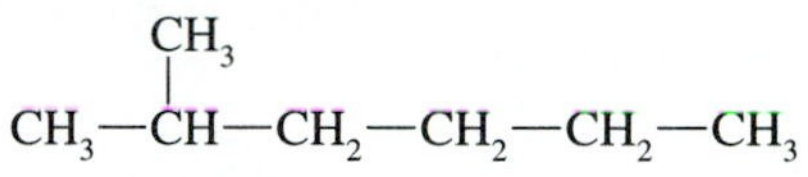

$$CH_3-\underset{}{\overset{CH_3}{\overset{|}{C}H}}-CH_2-CH_2-CH_2-CH_3 \quad \text{is} \quad (CH_3)_2CHCH_2CH_2CH_2CH_3$$

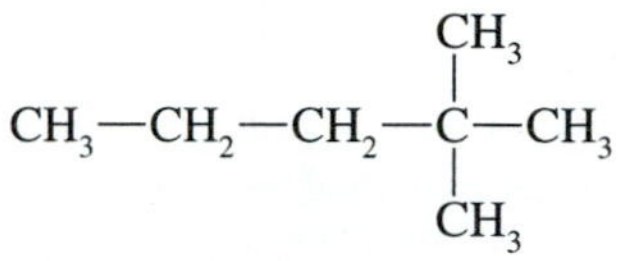

$$CH_3-CH_2-CH_2-\underset{CH_3}{\underset{|}{\overset{CH_3}{\overset{|}{C}}}}-CH_3 \quad \text{is} \quad CH_3CH_2CH_2C(CH_3)_3$$

Bond-Line Structures

Condensed structural formulas are convenient but still require considerable time to draw compared to yet another shorthand method using **bond-line structures.** The bond-line structure method also results in a less cluttered drawing. However, you have to mentally add many more features to understand the structure. The rules for drawing bond-line structures are as follows:

1. Carbon atoms are not shown unless needed for special emphasis or clarity.
2. All atoms other than carbon and hydrogen are explicitly shown.
3. A carbon atom is assumed to be located at the end of each line segment or at the intersection of two or more lines, which are used to depict bonds.
4. Multiple bonds are shown with multiple lines.
5. The proper number of hydrogen atoms to provide four bonds to each carbon atom is implied. Hydrogen atoms on other atoms such as oxygen and nitrogen are explicitly indicated.

To draw a bond-line structure, it is best to arrange the carbon atoms in a zigzag manner and then mentally remove the carbon and hydrogen atoms.

$CH_3-CH_2-CHBr-CH_3$ is CH_3 CH_2 $\overset{Br}{\overset{|}{C}H}$ CH_3 is Br

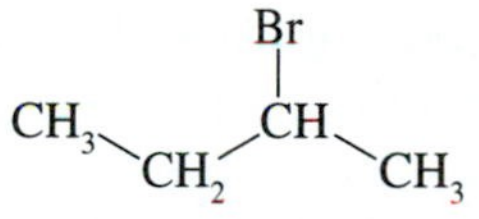

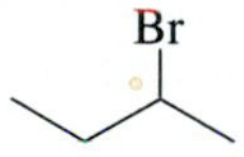

Bond-line formulas are also used to show cyclic structures. Rings of carbon atoms are shown by polygons such as an equilateral triangle, square, pentagon, and hexagon.

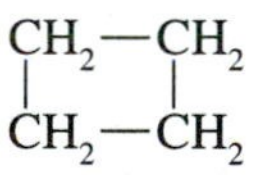

$$\begin{matrix} CH_2-CH_2 \\ | \quad\quad | \\ CH_2-CH_2 \end{matrix} \quad \text{is} \quad \square$$

There are differences in the representation of multiple bonds. Atoms such as oxygen must be shown in carbonyl groups, but a double-bonded carbon atom is not shown.

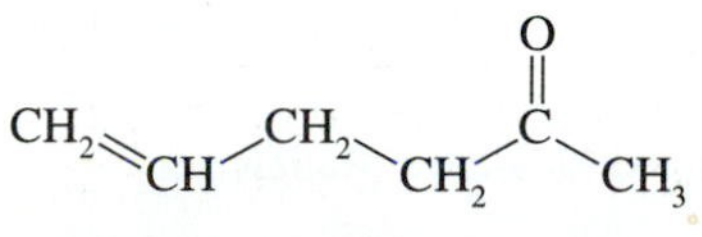

CH_2 CH CH_2 CH_2 $\overset{O}{\overset{||}{C}}$ CH_3 is O

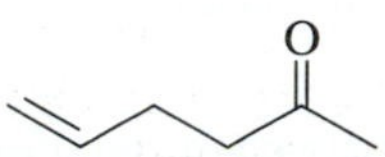

It is important to remember the normal number of bonds formed by each common atom in an organic compound. Carbon, nitrogen, and oxygen form four, three, and two bonds, respectively.

There is a carbon atom and two hydrogen atoms at this point.

There is a carbon atom and one hydrogen atom at this point.

There is a carbon atom and one hydrogen atom at each of these two points.

—OH

There is a carbon atom and two hydrogen atoms at this point.

Three-Dimensional Structures and Models

Because structure is so important to understanding chemical reactions, chemists construct models of molecules that can be viewed from a variety of angles. You may find it useful to purchase a molecular model kit to help you understand the structures of organic molecules.

Ball-and-stick models and space-filling models are two types of molecular models; each has certain advantages and disadvantages. Ball-and-stick models show the molecular framework and bond angles: the balls represent the nuclei of the atoms, and the sticks represent the bonds (Figure 1.8). The actual volume occupied by the molecule is not shown realistically. Space-filling models show the entire volume occupied by the electrons surrounding each atom, but as a consequence the carbon skeleton and its bond angles are obscured.

On paper, the three-dimensional shape of molecules is shown by a wedge and a dashed line (Figure 1.8). The wedge is viewed as a bond extending out of the plane of the page toward the reader. The dashed lines represent a bond directed behind the plane of the page. The other line is a bond in the plane of the page. Three-dimensional representations of molecules using wedges and dashed lines are **perspective structural formulas.**

Recognizing Structural Features

The structural features that allow chemists to predict the physical and chemical properties of naturally occurring molecules are often only a small part of a larger structure. These large structures are written in condensed forms that are meaningful because certain conventions are used. Regardless of the size and complexity of a molecule, you should glance over the molecule, ignore the many lines indicating the carbon-carbon bonds, and learn to focus on the important parts. Are there multiple bonds? Are there atoms, such as oxygen and nitrogen, that are part of functional groups? How are these atoms bonded, and

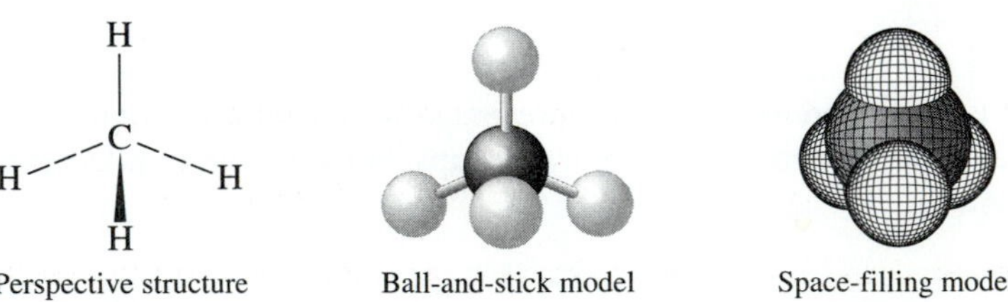

FIGURE 1.8
Perspective Structural Formulas and Molecular Models

what other atoms are nearby? For example, if a C═O group is present, it may be part of an aldehyde, ketone, acid, ester, or amide. The distinction between these functional groups can be decided by looking at the atoms bonded to the carbonyl carbon atom.

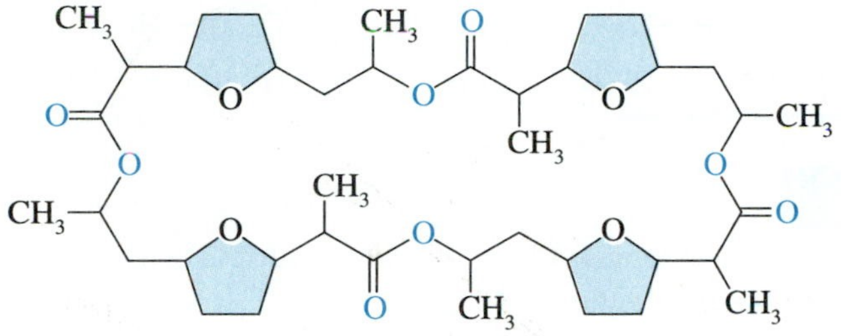

nonactin

Consider the structure for nonactin, an ionophorous antibiotic. It binds potassium ions through the many oxygen atoms in the large ring of atoms. It transports potassium ions across bacterial cell membranes, and the cells die. What are the oxygen-containing functional groups in this complex structure? Concentrate on one oxygen atom at a time. Some oxygen atoms are part of a C═O group, a carbonyl group; in fact, there are four such locations. Now look at the atoms bonded to the carbonyl carbon atom of the C═O groups. One bond is to carbon and the other to oxygen. Both carboxylic acids and esters have such features. The oxygen atom of carboxylic acids is in an —OH group, whereas the oxygen atom of esters is bonded to another carbon atom. Convince yourself that four ester groups are in this molecule.

Now concentrate on the second type of oxygen-containing functional group in the molecule. There are oxygen atoms contained as part of a five-membered ring—four times, in fact. Each oxygen atom is bonded by single bonds to two carbon atoms. The functional groups are ethers.

EXAMPLE 1-3

A species of cockroach secretes the following substance as a signal for other cockroaches to congregate. Write three condensed structural formulas for the substance.

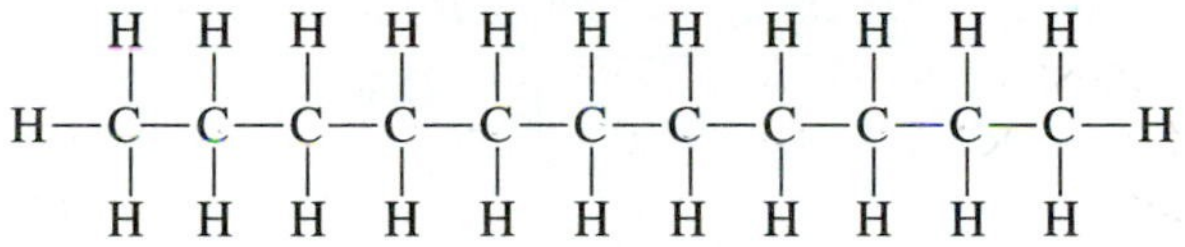

$$\begin{array}{ccccccccccccc} & H & H & H & H & H & H & H & H & H & H & H & \\ & | & | & | & | & | & | & | & | & | & | & | & \\ H- & C- & C- & C- & C- & C- & C- & C- & C- & C- & C- & C & -H \\ & | & | & | & | & | & | & | & | & | & | & | & \\ & H & H & H & H & H & H & H & H & H & H & H & \end{array}$$

Solution

With the C—H bonds understood, we write

$$CH_3{-}CH_2{-}CH_2{-}CH_2{-}CH_2{-}CH_2{-}CH_2{-}CH_2{-}CH_2{-}CH_2{-}CH_3$$

With both the C—H and C—C bonds understood, we write

$$CH_3CH_2CH_2CH_2CH_2CH_2CH_2CH_2CH_2CH_2CH_3$$

In the most condensed version, the nine methylene units are represented within parentheses.

$$CH_3(CH_2)_9CH_3$$

Problem 1.3

Hexamethylenediamine, a compound used to produce nylon, has the following structural formula. Write three condensed structural formulas for the substance.

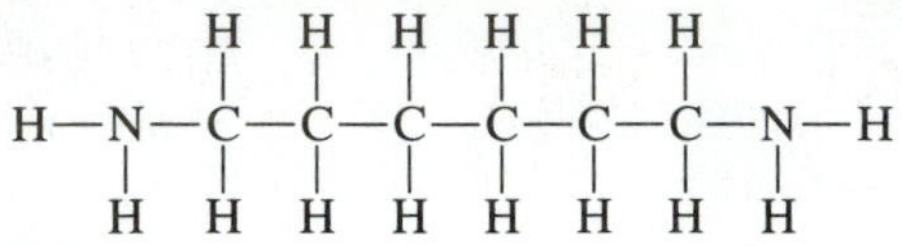

```
        H   H   H   H   H   H
        |   |   |   |   |   |
H—N—C—C—C—C—C—C—N—H
    |   |   |   |   |   |   |   |
    H   H   H   H   H   H   H   H
```

EXAMPLE 1-4

What is the molecular formula of carvone, which is found in oil of caraway?

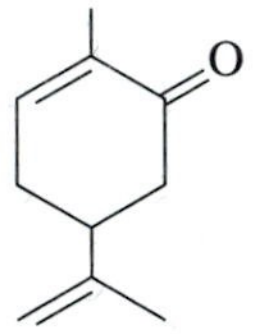

Solution

There are 10 carbon atoms in the structure, located at the ends or intersections of line segments. An oxygen atom is located at the end of a segment representing the double bond of a carbonyl group. Hydrogen atoms are counted by determining the number of bonds from each carbon atom to other atoms. Note that three carbon atoms have no hydrogen atoms. The molecular formula is $C_{10}H_{14}O$.

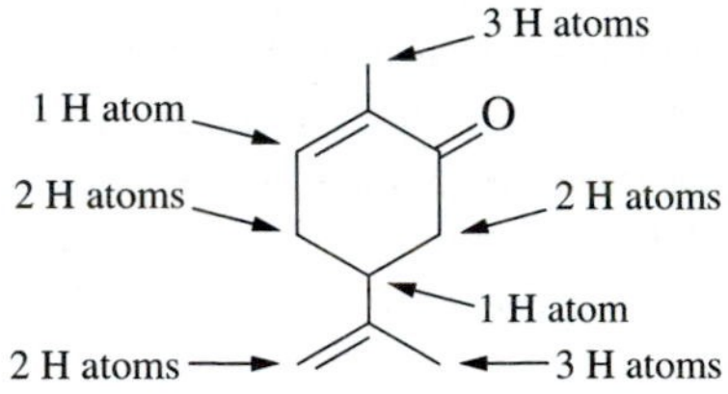

3 H atoms
1 H atom
2 H atoms
2 H atoms
1 H atom
2 H atoms
3 H atoms

Problem 1.4

What is the molecular formula of indoleacetic acid, a plant growth hormone that promotes shoot growth?

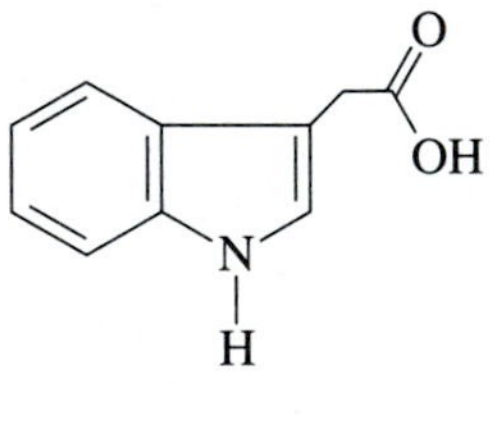

1.11 Isomers

Compounds that have the same molecular formula but different structures are **isomers.** Structure refers to the linkage of the atoms. As we examine the structure of organic compounds in increasing detail, you will learn how subtle structural differences in isomers affect the physical and chemical properties of compounds.

There are several types of isomers. Isomers that differ in their carbon skeleton are **skeletal isomers.** Consider the structural differences in the two isomers of C_4H_{10}, butane and isobutane. Butane has an uninterrupted chain of four carbon atoms (Figure 1.9), but isobutane has only three carbon atoms connected in

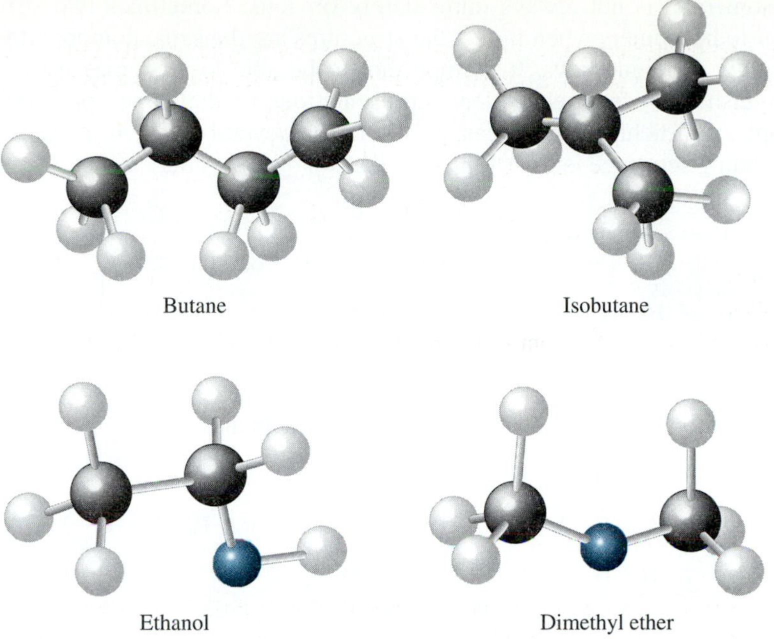

FIGURE 1.9
Structures of Isomers

sequence and a fourth carbon atom appended to the chain. The boiling points (bp) of butane and isobutane are −1 °C and −12 °C, respectively; the chemical properties of the two compounds are similar but different.

Isomers that have different functional groups are **functional group isomers.** The molecular formula for both ethyl alcohol and dimethyl ether is C_2H_6O (Figure 1.9). Although the compositions of the two compounds are identical, their functional groups differ. The atomic sequence is C—C—O in ethyl alcohol, and the oxygen atom is present as an alcohol. The C—O—C sequence in the isomer corresponds to an ether.

$$CH_3—CH_2—OH \qquad CH_3—O—CH_3$$

ethyl alcohol (bp 78.5°C) dimethyl ether (bp −24°C)

The physical properties of these two functional group isomers, as exemplified by their boiling points, are very different. These substances also have different chemical properties because their functional groups differ.

Positional isomers are compounds that have the same functional groups in different positions on the carbon skeleton. For example, the isomeric alcohols 1-propanol and 2-propanol differ in the location of the hydroxyl group. The chemical properties of these two compounds are similar because they both contain the same type of functional group.

—OH group is on the end of the chain.

$$CH_3—CH_2—CH_2—OH$$

1-propanol
(bp 97.4°C)

—OH group is in the middle of the chain.

$$CH_3—\underset{}{\overset{OH}{\overset{|}{C}}}H—CH_3$$

2-propanol
(bp 82.4°C)

Isomerism is not always immediately obvious. Sometimes two structures appear to be isomers when in fact the structures are the same compound written in slightly different ways. It is important to be able to recognize isomers and distinguish them from equivalent representations of the same compound. For example, 1,2-dichloroethane can be written in several ways. In each formula, the bonding sequence is Cl—C—C—Cl.

$$\begin{array}{ccccccc} & & H & & H & & \\ & & | & & | & & \\ Cl & — & C & — & C & — & Cl \\ & & | & & | & & \\ & & H & & H & & \end{array} \quad \text{or} \quad \begin{array}{ccccccc} & & H & & Cl & & \\ & & | & & | & & \\ Cl & — & C & — & C & — & H \\ & & | & & | & & \\ & & H & & H & & \end{array} \quad \text{or} \quad \begin{array}{ccccccc} & & Cl & & H & & \\ & & | & & | & & \\ H & — & C & — & C & — & H \\ & & | & & | & & \\ & & H & & Cl & & \end{array}$$

1,2-dichloroethane (CH_2ClCH_2Cl)
(bp 83.5°C)

The isomer of 1,2-dichloroethane is 1,1-dichloroethane.

$$\begin{array}{ccccccc} & & Cl & & H & & \\ & & | & & | & & \\ Cl & — & C & — & C & — & H \\ & & | & & | & & \\ & & H & & H & & \end{array} \quad \text{or} \quad \begin{array}{ccccccc} & & H & & H & & \\ & & | & & | & & \\ Cl & — & C & — & C & — & H \\ & & | & & | & & \\ & & Cl & & H & & \end{array} \quad \text{or} \quad \begin{array}{ccccccc} & & Cl & & H & & \\ & & | & & | & & \\ H & — & C & — & C & — & H \\ & & | & & | & & \\ & & Cl & & H & & \end{array}$$

1,1-dichloroethane ($CHCl_2CH_3$)
(bp 57.3°C)

In 1,1-dichloroethane, the two chlorine atoms are bonded to the same carbon atom. In 1,2-dichlorethane, the two chlorine atoms are bonded to different carbon atoms. The different condensed structural formulas, $CHCl_2CH_3$ and CH_2ClCH_2Cl, also convey information about the different structures, which are shown as space-filling models in Figure 1.10.

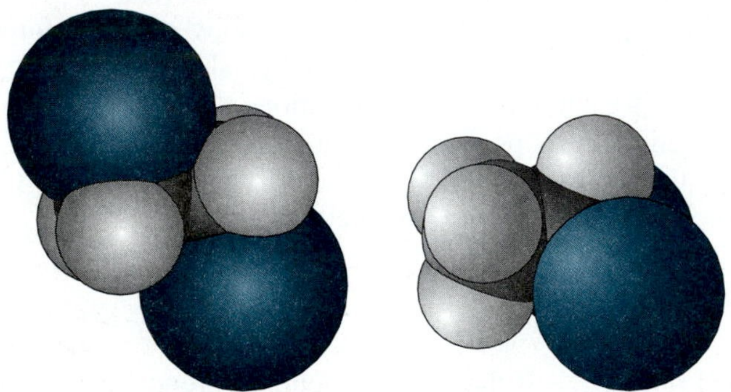

FIGURE 1.10
Molecular Models of $C_2H_4Cl_2$ Isomers

EXAMPLE 1-5

Consider the following structural formulas for two compounds used as general anesthetics. Do they represent isomers? How do they differ?

$$\begin{array}{ccccccccc} & & F & & F & & & & F & & \\ & & | & & | & & & & | & & \\ Cl & — & C & — & C & — & O & — & C & — & F \\ & & | & & | & & & & | & & \\ & & H & & F & & & & H & & \end{array} \qquad \begin{array}{ccccccccc} & & F & & Cl & & & & H & & \\ & & | & & | & & & & | & & \\ F & — & C & — & C & — & O & — & C & — & F \\ & & | & & | & & & & | & & \\ & & F & & H & & & & F & & \end{array}$$

Solution

The atomic compositions given in these structural formulas are identical; the molecular formula is $C_3H_2F_5ClO$. Therefore, the compounds are isomers. The carbon skeletons are identical and the compounds are both ethers.

Both isomers have a CHF_2 unit on the right side of the ether oxygen atom in spite of the different ways in which the fluorine and hydrogen are written—this is not the basis for isomerism. The two-carbon unit on the left of the oxygen atom has the halogen atoms distributed in two different ways. That is, they are positional isomers. The structure on the left has two fluorine atoms bonded to the carbon atom appended to the oxygen atom. The carbon atom on the left has a fluorine and a chlorine atom bonded to it. The structure on the right has one chlorine atom bonded to the carbon atom appended to the oxygen atom. The carbon atom on the left has three fluorine atoms bonded to it.

Problem 1.5

Compare the following structures of two intermediates involved in the metabolism of glucose. Do they represent isomers? How do they differ?

$$HO{-}CH_2{-}CH(OH){-}C(=O){-}H \qquad HO{-}CH_2{-}C(=O){-}CH_2{-}OH$$

1.12 Nomenclature

Nomenclature refers to a systematic method of naming materials. In chemistry, the nomenclature of compounds is exceedingly important. The phenomenon of isomerism easily illustrates this point. The common names *butane* and *isobutane* of the two isomeric C_4H_{10} compounds are easy to learn. However, there are 75 isomers of $C_{10}H_{22}$ and 62,491,178,805,831 isomers of $C_{40}H_{82}$. Without a system of naming compounds, organic chemistry would be difficult, if not impossible, to comprehend.

The list of ingredients on many commercial products contains common and systematic chemical names.

At a meeting in Geneva, Switzerland, in 1892 chemists devised a systematic nomenclature for all compounds, including organic compounds. Compounds are now named by rules developed by the International Union of Pure and Applied Chemistry (IUPAC). The rules result in a clear and definitive name for each compound. A universal system for naming organic compounds was needed because different names had often been given to the same compound. For example, CH_3CH_2OH had been called not only alcohol but also spirits, grain alcohol, ethyl alcohol, methyl carbinol, and ethanol. Furthermore, a variety of names developed in each language.

A chemical name consists of three parts: prefix, parent, and suffix. The parent indicates how many carbon atoms are in the main carbon skeleton. The suffix identifies most of the functional groups present in the molecule. Examples of suffixes are *-ol* for alcohols, *-al* for aldehydes, and *-one* for ketones. The prefix specifies the location of the functional group designated in the suffix as well as some other types of substituents on the main parent chain.

prefix — parent — suffix

Once the rules are applied, there is only one name for each structure, and one structure for each name. For example, a compound that is partly responsible for the odor of a skunk is 3-methyl-1-butanethiol.

$$\begin{array}{c} \quad CH_3 \\ \quad | \\ CH_3-CH-CH_2-CH_2-SH \end{array}$$

Butane is the parent name of the four-carbon unit that is written horizontally. The prefix 3-methyl refers to the $-CH_3$ written above the chain of carbon atoms. The prefix 1- and the suffix *-thiol* refer to the position and identity of the $-SH$ group. This method of assigning numbers to the carbon chain and other features of the IUPAC system will be discussed further in subsequent chapters.

In spite of the IUPAC system, many common names are so well established that both common and IUPAC names must be recognized. The IUPAC name for CH_3CH_2OH is ethanol, but the common name ethyl alcohol is still used. In addition, many complex compounds, particularly ones of biological origin, retain common names partly due to the unwieldy nature of their IUPAC names.

Explorations with Molecular Models

1. Construct two models of CH_4. Place them on the desk. Orient the models as shown so that one of the hydrogen atoms of the tripod is pointing to the right. One of the remaining hydrogen atoms will be directed to the left and the front of the desk, whereas the other will be directed to the left and the back of the desk.

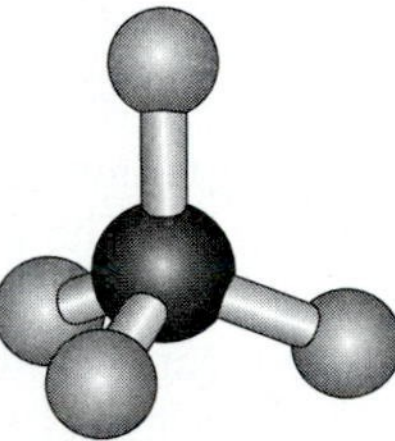

Place a piece of masking tape on the "right" hydrogen atom on one model and a piece of masking tape on the "left and front" hydrogen atom on the other model. Keeping the tripod in contact with the desk, rotate one of the models so that the taped atom of model one is in the same relative position as the taped atom in model two. What does this experiment show about the equivalence of the hydrogen atoms forming the tripod?

2. Remove the masking tape from one of the models prepared in Exploration Exercise 1 and place it on the hydrogen atom pointing up. Now pick up this model and reposition it next to the other model so that both taped hydrogen atoms are oriented identically on the table top. What does this experiment show about the equivalence of the hydrogen atoms in methane?

3. Construct the following model of methanol. Replace one of the hydrogen atoms bonded to the carbon atom with a carbon atom bonded to three hydrogen atoms. What is the molecular formula of the compound? To what class of compounds does the structure belong?

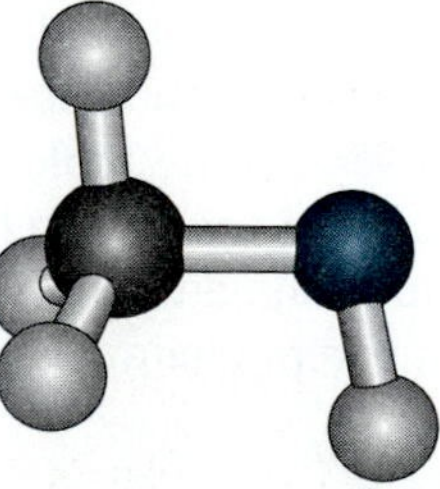

4. Construct the model of methanol again. Replace the hydrogen atom bonded to the oxygen atom with a carbon atom bonded to three hydrogen atoms. What is the molecular formula of the compound? To what class of compounds does the structure belong? What relationship exists between this structure and the structure constructed in Exploration Exercise 3?

Exercises

Atomic Properties

1.1 How many valence shell electrons are in each of the following elements?

(a) N (b) F (c) C (d) O (e) Cl (f) Br (g) S (h) P

1.2 Which of the following atoms has the higher electronegativity? Which has the larger atomic radius?

(a) Cl or Br (b) O or S (c) C or N (d) N or O (e) C or O

Lewis Structures of Covalent Compounds

1.3 Write a Lewis structure for each of the following compounds

(a) NH_2OH (b) CH_3CH_3 (c) CH_3OH (d) CH_3NH_2 (e) CH_3Br (f) CH_3SH

1.4 Write a Lewis structure for each of the following compounds

(a) HCN (b) HNNH (c) CH_2NH (d) CH_3NO (e) CH_2NOH (f) CH_2NNH_2

1.5 Place any required unshared pairs of electrons that are missing from the following formulas

(a) $CH_3-\overset{O}{\overset{\|}{C}}-OH$ (b) $CH_3-\overset{O}{\overset{\|}{C}}-O-CH_3$ (c) $CH_3-\overset{O}{\overset{\|}{C}}-NH-CH_3$ (d) $CH_3-S-CH=CH_2$

1.6 Place any required unshared pairs of electrons that are missing from the following formulas

(a) $CH_3-\overset{O}{\overset{\|}{C}}-Cl$ (b) $CH_3-O-CH=CH_2$ (c) $CH_3-\overset{O}{\overset{\|}{C}}-SH$ (d) $NH_2-\overset{O}{\overset{\|}{C}}-O-CH_3$

1.7 Using the number of valence electrons in the constituent atoms and the given arrangement of atoms in the compound, write the Lewis structure for each of the following molecules. Add multiple bonds where necessary and show the lone pair electrons

(a) $Cl-\overset{O}{\overset{|}{C}}-Cl$ (b) $H-\underset{\underset{H}{|}}{N}-\overset{O}{\overset{|}{C}}-\underset{\underset{H}{|}}{N}-H$ (c) $H-\overset{H}{\overset{|}{\underset{\underset{H}{|}}{C}}}-\overset{O}{\overset{|}{C}}-S-H$

1.8 Using the number of valence electrons in the constituent atoms and the given arrangement of atoms in the compound, write the Lewis structure for each of the following molecules. Add multiple bonds where necessary and show the lone pair electrons

(a) $H-\overset{H}{\overset{|}{\underset{\underset{H}{|}}{C}}}-S-S-\overset{H}{\overset{|}{\underset{\underset{H}{|}}{C}}}-H$ (b) $Cl-\overset{H}{\overset{|}{\underset{\underset{H}{|}}{C}}}-\overset{O}{\overset{|}{C}}-O-H$ (c) $H-\overset{H}{\overset{|}{\underset{\underset{H}{|}}{C}}}-\overset{O}{\overset{|}{C}}-\underset{\underset{Cl}{|}}{N}-H$

1.9 Two compounds used as dry cleaning agents have the molecular formulas C_2Cl_4 and C_2HCl_3. Write the Lewis structure for each compound.

1.10 Acrylonitrile, a compound used to produce polymeric fibers for rugs, is represented by the condensed formula CH_2CHCN. Write the Lewis structure for the compound.

Formal Charge

1.11 Assign the formal charges for the atoms other than hydrogen in each of the following species

(a) $H-\underset{..}{\ddot{O}}-C\equiv N:$ (b) $H-\underset{..}{\ddot{O}}-N\equiv C:$ (c) $CH_3-\ddot{N}=N=\ddot{N}:$

1.12 Each of the following species is isoelectronic (that is, it has the same number of electrons bonding the same number of atoms). Determine which atoms have a formal charge. Calculate the net charge for each species

(a) $:C\equiv O:$ (b) $:N\equiv O:$ (c) $:C\equiv N:$ (d) $:C\equiv C:$ (e) $:N\equiv N:$

1.13 Acetylcholine, a compound involved in the transfer of nerve impulses, has the following structure. What is the formal charge of the nitrogen atom?

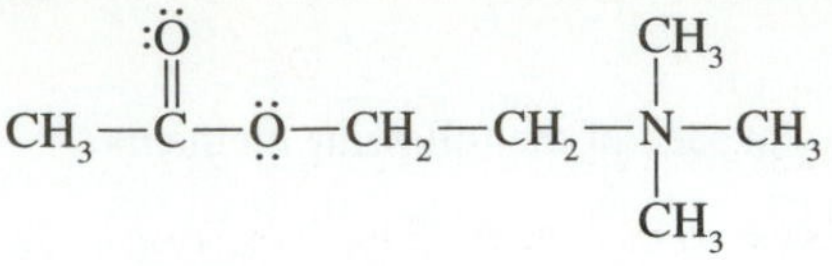

$$CH_3-\overset{\overset{:\ddot{O}}{\|}}{C}-\ddot{\underset{\cdot\cdot}{O}}-CH_2-CH_2-\overset{\overset{CH_3}{|}}{\underset{\underset{CH_3}{|}}{N}}-CH_3$$

1.14 Sarin, a nerve gas, has the following structure. What is the formal charge of the phosphorus atom?

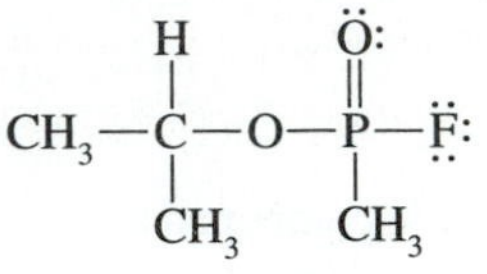

$$CH_3-\overset{\overset{H}{|}}{\underset{\underset{CH_3}{|}}{C}}-O-\overset{\overset{\ddot{O}:}{\|}}{\underset{\underset{CH_3}{|}}{P}}-\ddot{\underset{\cdot\cdot}{F}}:$$

Resonance

1.15 The small amounts of cyanide ion contained in the seeds of some fruits are eliminated from the body as SCN^-. Draw two possible resonance forms for the ion. Which atom has the formal negative charge in each form?

1.16 Are the following pairs contributing resonance forms of a single species or not? Explain. Indicate the formal charges where appropriate

(a) $:\ddot{\underset{\cdot\cdot}{N}}-N\equiv N:$ $\quad :\underset{\cdot\cdot}{N}=N=\underset{\cdot\cdot}{N}:$ (b) $H-C\equiv N-\ddot{\underset{\cdot\cdot}{O}}:$ $\quad H-\ddot{C}=N=\underset{\cdot\cdot}{O}:$

1.17 Write the resonance structure that results when electrons are moved in the direction indicated by the curved arrows for the following amide. Calculate any formal charges that may result

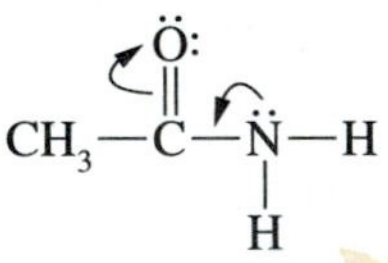

$$CH_3-\overset{\overset{\ddot{O}:}{\|}}{C}-\ddot{N}-H \quad (\text{with } \underset{|}{N}-H)$$

1.18 Write the resonance structure that results when electrons are moved in the direction indicated by the curved arrows for ozone. Calculate the formal charges of the oxygen atoms in each structure

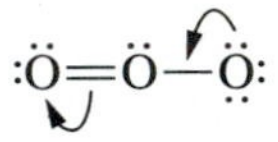

$$:\ddot{O}=\ddot{O}-\ddot{\underset{\cdot\cdot}{O}}:$$

Molecular Shapes

1.19 Based on VSEPR theory, what is the expected value of the indicated bond angle in each of the following compounds?

(a) C—C—N in $CH_3-C\equiv N$ (b) C—O—C in CH_3-O-CH_3

(c) C—N—C in $CH_3-NH-CH_3$ (d) C—C—C in $CH_2=C=CH_2$

1.20 Based on VSEPR theory, what is the expected value of the indicated bond angle in each of the following ions?

(a) C—O—H in $CH_3-OH_2^+$ (b) C—N—H in $CH_3-NH_3^+$

(c) O—C—O in $CH_3CO_2^-$ (d) C—O—C in $(CH_3)_2OH^+$

1.21 What is the expected C—N=N bond angle in Prontosil, an antibiotic?

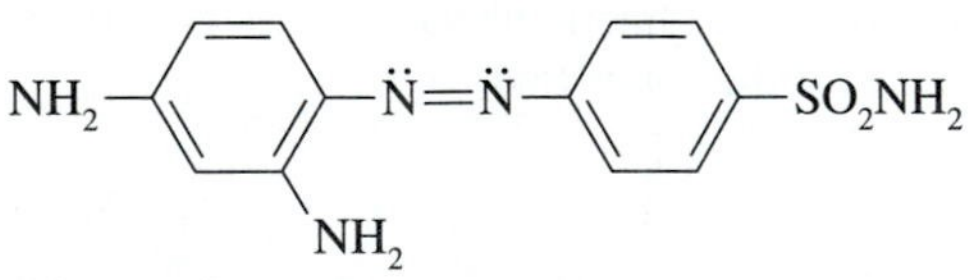

$$NH_2-C_6H_3(NH_2)-\ddot{N}=\ddot{N}-C_6H_4-SO_2NH_2$$

1.22 What is the expected S—C—S bond angle in dibenzthiozole disulfide, a catalyst used in the vulcanization of rubber?

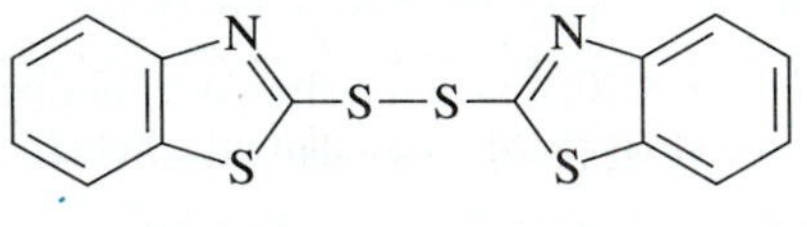

$$C_6H_4(N{=}C\text{—}S)-S-S-(C\text{—}S{=}N)C_6H_4$$

Hybridization

1.23 What is the hybridization of each carbon atom in each of the following compounds?

(a) $CH_3-\overset{\overset{\large O}{\|}}{C}-H$ (b) $CH_3-O-CH{=}CH_2$ (c) $CH_3-\overset{\overset{\large O}{\|}}{C}-SH$

1.24 What is the hybridization of each carbon atom in each of the following compounds?

(a) $CH_3-\overset{\overset{\large O}{\|}}{C}-NH-CH_3$ (b) $CH_3-S-CH{=}CH_2$ (c) $CH_3-\overset{\overset{\large NH}{\|}}{C}-CH_3$

1.25 What is the hybridization of each of the carbon atoms bonded to two oxygen atoms in aspirin?

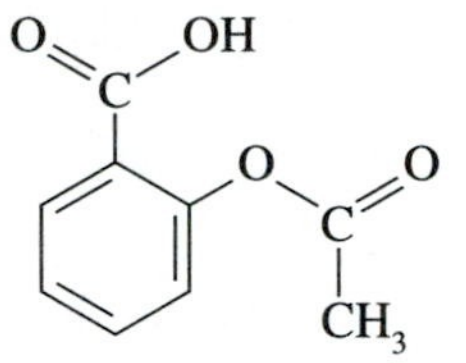

1.26 What is the hybridization of the carbon atom bonded to the nitrogen atom and of the carbon atom bonded to two oxygen atoms in L-dopa, a drug that is used in the treatment of Parkinson's disease?

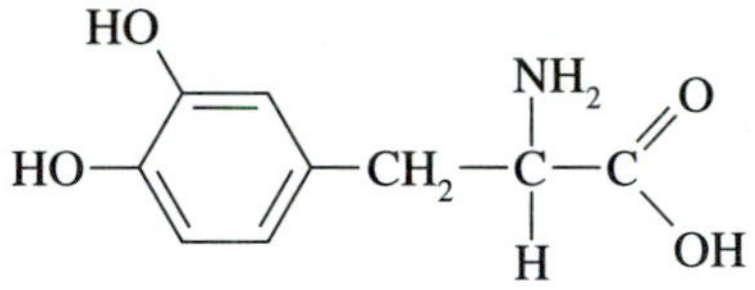

Molecular Formulas

1.27 Write the molecular formula for each of the following

(a) $CH_3-CH_2-CH_2-CH_2-CH_3$ (b) $CH_3-CH_2-CH_2-CH_3$

(c) $CH_2{=}CH-CH_2-CH_3$ (d) $CH_3-CH_2-C{\equiv}C-H$

1.28 Write the molecular formula for each of the following

(a) $CH_3CH_2CH_2CH_2CH_2CH_3$ (b) $CH_3CH{=}CHCH_3$ (c) $CH_3CH_2C{\equiv}CCH_3$ (d) $CH_3C{\equiv}CCH_2CH{=}CHCH_3$

1.29 Write the molecular formula for each of the following

(a) $CH_3CH_2CHCl_2$ (b) $CH_3CCl_2CH_3$ (c) $BrCH_2CH_2Br$

1.30 Write the molecular formula for each of the following

(a) $CH_3CH_2CH_2OH$ (b) CH_3CH_2SH (c) $CH_3CH_2CH_2NH_2$

Condensed Structural Formulas

1.31 Write condensed structural formulas in which only the bonds to hydrogen are not shown

(a) 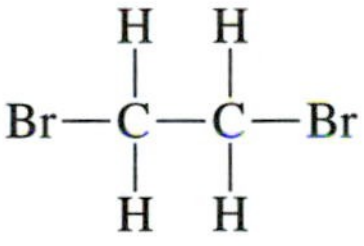 $\begin{array}{c} \quad H \;\; H \\ \quad | \;\;\; | \\ Br-C-C-Br \\ \quad | \;\;\; | \\ \quad H \;\; H \end{array}$ (b) 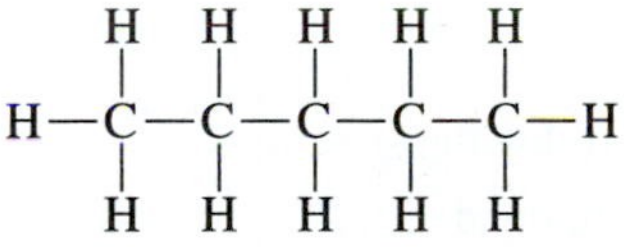 $\begin{array}{c} \; H \;\; H \;\; H \;\; H \;\; H \\ \; | \;\;\; | \;\;\; | \;\;\; | \;\;\; | \\ H-C-C-C-C-C-H \\ \; | \;\;\; | \;\;\; | \;\;\; | \;\;\; | \\ \; H \;\; H \;\; H \;\; H \;\; H \end{array}$ (c) $\begin{array}{c} H \;\; H \;\; H \quad\;\; \\ | \;\;\; | \;\;\; | \quad\;\; \\ H-C-C-C-S-H \\ | \;\;\; | \;\;\; | \quad\;\; \\ H \;\; H \;\; H \quad\;\; \end{array}$

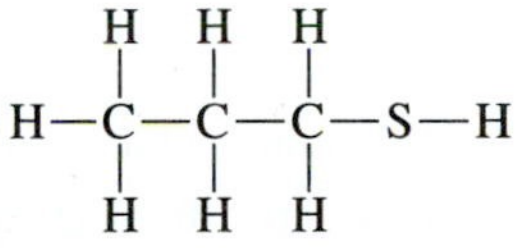

1.32 Write condensed structural formulas in which only the bonds to hydrogen are not shown

(a) 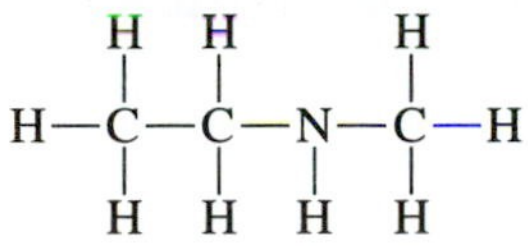 $\begin{array}{c} H \;\; H \quad\;\; H \\ | \;\;\; | \quad\;\;\; | \\ H-C-C-N-C-H \\ | \;\;\; | \;\;\; | \;\;\; | \\ H \;\; H \;\; H \;\; H \end{array}$ (b) 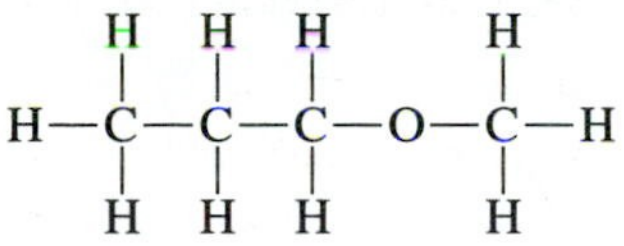 $\begin{array}{c} H \;\; H \;\; H \quad\;\; H \\ | \;\;\; | \;\;\; | \quad\;\;\; | \\ H-C-C-C-O-C-H \\ | \;\;\; | \;\;\; | \quad\;\;\; | \\ H \;\; H \;\; H \quad\;\; H \end{array}$ (c) $\begin{array}{c} H \;\; H \;\; H \;\; Cl \\ | \;\;\; | \;\;\; | \;\;\; | \\ H-C-C-C-C-Cl \\ | \;\;\; | \;\;\; | \;\;\; | \\ H \;\; H \;\; H \;\; Cl \end{array}$

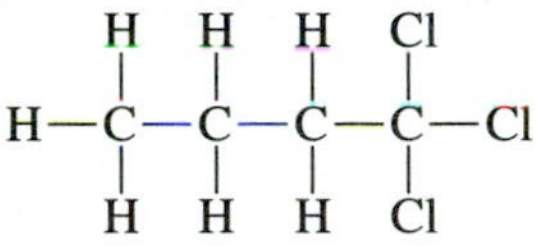

1.33 Write a condensed structural formula in which no bonds are shown for each substance in 1.31.

1.34 Write a condensed structural formula in which no bonds are shown for each substance in 1.32.

1.35 Write a complete structural formula showing all bonds for each of the following condensed formulas

(a) $CH_3CH_2CH_2CH_3$ (b) $CH_3CH_2CH_2Cl$ (c) $CH_3CHClCH_2CH_3$
(d) $CH_3CH_2CHBrCH_3$ (e) $CH_3CH_2CHBr_2$ (f) $CH_3CBr_2CH_2CH_2CH_3$

1.36 Write a complete structural formula showing all bonds for each of the following condensed formulas

(a) $CH_3CH_2CH_3$ (b) $CH_3CH_2CHCl_2$ (c) $CH_3CH_2CH_2CH_2SH$
(d) $CH_3CH_2C{\equiv}CCH_3$ (e) $CH_3CH_2OCH_2CH_2CH_3$ (f) $CH_3CH_2CH_2C{\equiv}CH$

Bond-Line Structures

1.37 What is the molecular formula for each of the following bond-line representations?

(a) (b) (c)

1.38 What is the molecular formula for each of the following bond-line representations?

(a) (b) (c)

1.39 What is the molecular formula for each of the following bond-line representations?

(a) a scent marker of the red fox (b) a compound responsible for the odor of the iris

1.40 What is the molecular formula for each of the following bond-line representations?

(a) a compound found in clover and grasses (b) an oil found in citrus fruits

Functional Groups

1.41 Identify the functional groups contained in each of the following structures.

(a) caprolactam, a compound used to produce a type of nylon (b) civetone, a compound in the scent gland of the civet cat

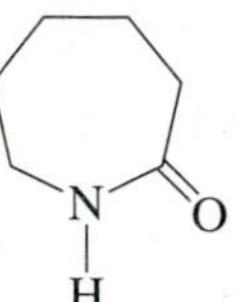

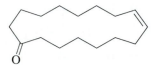

1.42 Identify the oxygen-containing functional groups in each of the following compounds.

(a) isoimpinellin, a carcinogen found in diseased celery (b) rotenone, an insecticide

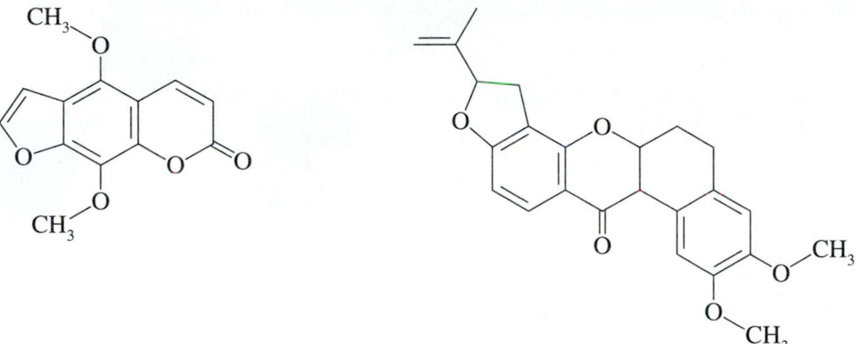

Isomerism

1.43 Indicate whether the following pairs of structures are isomers or different representations of the same compound

(a) $\mathrm{H{-}C(Br)(H){-}C(H)(H){-}Br}$ and $\mathrm{Br{-}C(H)(H){-}C(H)(H){-}Br}$

(b) $\mathrm{CH_3{-}CH_2(CH_2{-}Cl)}$ and $\mathrm{CH_3{-}CH_2{-}CH_2{-}Cl}$

(c) $\mathrm{CH_3{-}CH(CH_3){-}Cl}$ and $\mathrm{CH_3{-}CH_2{-}CH_2{-}Cl}$

1.44 Indicate whether the following pairs of structures are isomers or different representations of the same compound

(a) $\mathrm{H{-}C(H)(H){-}C(Cl)(H){-}Br}$ and $\mathrm{Cl{-}C(H)(H){-}C(H)(H){-}Br}$

(b) $\mathrm{CH_3{-}CH_2(CH_2{-}Cl)}$ and $\mathrm{CH_3{-}CH(Cl){-}CH_3}$

(c) $\mathrm{CH_3{-}CH(CH_3){-}CH_2{-}Cl}$ and $\mathrm{CH_3{-}CH(CH_2{-}Cl){-}CH_3}$

1.45 There are two isomers for each of the following molecular formulas. Draw their structural formulas

(a) $C_2H_4Br_2$ (b) C_2H_6O (c) C_2H_4BrCl (d) C_3H_7Cl (e) C_2H_7N

1.46 There are three isomers for each of the following molecular formulas. Draw their structural formulas

(a) $C_2H_3Br_2Cl$ (b) C_3H_8O (c) C_3H_8S

CHAPTER 2

PROPERTIES OF ORGANIC COMPOUNDS

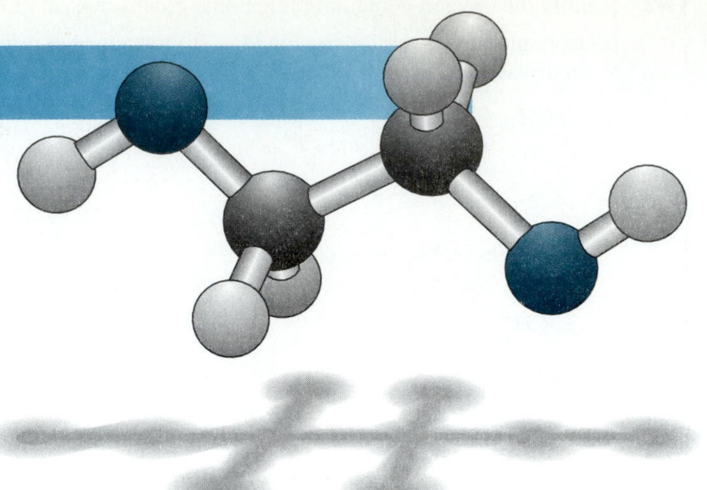

Overview

2.1 Structure and Physical Properties

Each of the millions of organic compounds has unique physical and chemical properties. Thus, we might expect that understanding the relationships between the structure of compounds and their physical properties, such as melting point, boiling point, and solubility, would be a difficult task. Yet we can make reasonable guesses about the physical properties of a compound based on its structure, because organic compounds belong to a small number of classes of substances characterized by their functional groups. These structural units within a molecule are largely responsible for its properties. These properties reflect the attractive **intermolecular** (between molecules) forces attributable to the functional groups. Intermolecular forces are of three types: **dipole-dipole forces, London forces,** and **hydrogen-bonding forces.**

Dipole-Dipole Forces

The bonding electrons in polar covalent bonds are not shared equally, and a bond moment results. However, a molecule may be polar or nonpolar depending on its geometry. For example, tetrachloromethane (carbon tetrachloride, CCl_4) has polar C—Cl bonds, but the tetrahedral arrangement of the four bonds about the central carbon atom causes the individual bond moments to cancel. In contrast, dichloromethane (methylene chloride, CH_2Cl_2) is a polar molecule with a net polarity away from the partially positive carbon atom toward the partially negative chlorine atoms.

The bond moments cancel and there is no net polarity.

The bond moments do not cancel and a net polarity results.

Polar molecules have a negative "end" and a positive "end" and tend to associate. The positive end of one molecule attracts the negative end of another molecule. The physical properties of polar molecules reflect this association. An increased association between molecules decreases their

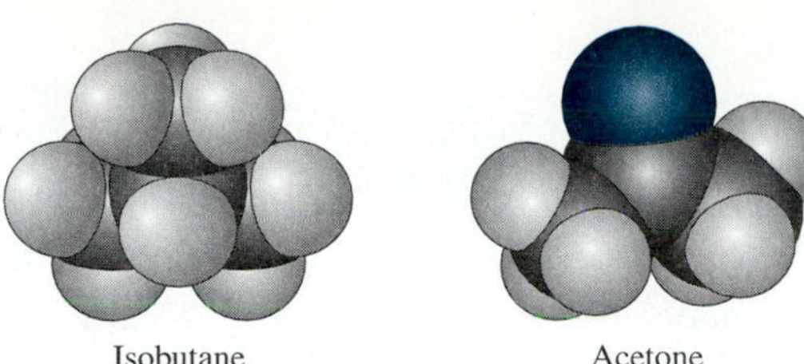

FIGURE 2.1
Molecular Models of Isobutane and Acetone

vapor pressure, which in turn results in a higher boiling point, because more energy is required to vaporize the molecules.

The molecular weights and molecular shapes of acetone and isobutane are similar (Figure 2.1), but acetone boils at a higher temperature than isobutane. Acetone contains a polar carbonyl group, whereas isobutane is a nonpolar molecule. The higher boiling point of acetone results from the dipole-dipole interaction as a result of the polar carbonyl group.

$(CH_3)_3CH$
isobutane (nonpolar)
(bp −11.7°C)

$CH_3C^{\delta+}(=O^{\delta-})CH_3$
acetone (polar)
(bp 56.2°C)

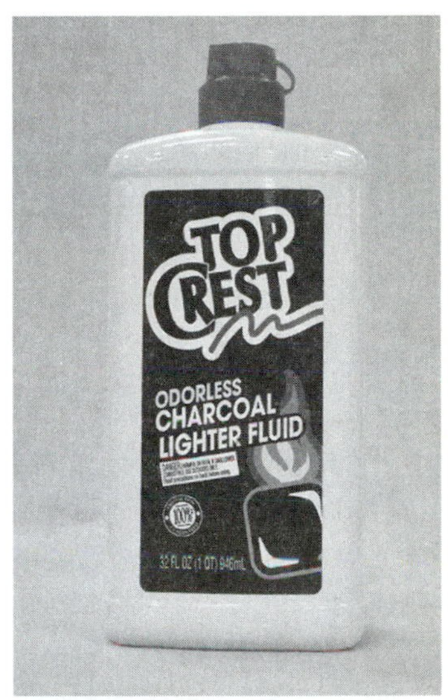

The components of this lighter fluid are selected to be moderately volatile and not to form explosive gaseous mixtures with oxygen.

London Forces

In a nonpolar molecule, the electrons, on average, are distributed uniformly about the molecule. However, the electrons at some instant may be distributed closer to one atom in a molecule or toward one side of a molecule. At that instant, a **temporary dipole** is present (Figure 2.2). A temporary dipole exerts an influence on nearby molecules; it polarizes neighboring molecules and results in an **induced dipole.** The resultant attractive forces between a temporary dipole and an induced dipole are called **London forces.** The ease with which an electron cloud is distorted by nearby charges or dipoles is called **polarizability.**

The attractive forces between the temporary dipoles in otherwise nonpolar molecules are small and have a short lifetime at any given site in the sample. However, the cumulative effect of these attractive forces holds a collection of molecules together in the condensed state. The strength of London forces depends on the number of electrons in a molecule and on the types of atoms containing those electrons. Electrons that are far from atomic nuclei are more easily distorted or polarizable than electrons that are closer to atomic nuclei. For example, the polarizability of the halogens increases in the order $F < Cl < Br < I$. London forces also depend on the size and shape of a molecule.

The boiling point of bromoethane is higher than the boiling point of chloroethane. Because a C—Cl bond is more polar than a C—Br bond, we might have expected the more polar chloroethane to have a higher boiling point than bromoethane. But polarity isn't everything. The molecular weights of the two compounds are substantially different and the electrons of the bromine atom are more polarizable than the electrons of the chlorine atom. Thus, the or-

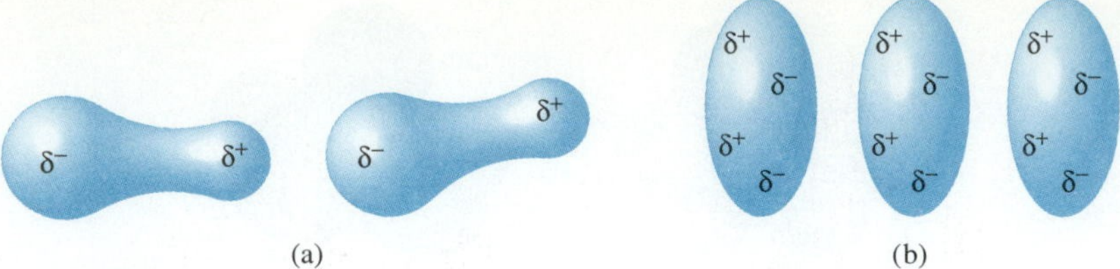

FIGURE 2.2
London Forces in Nonpolar Molecules

In a molecule represented by (a), the electrons are distorted toward one end of the molecule. In a molecule represented by (b), the electrons are distorted toward one side of the molecule.

der of boiling points reflects the polarizability of the molecules and the larger London attractive forces of bromoethane.

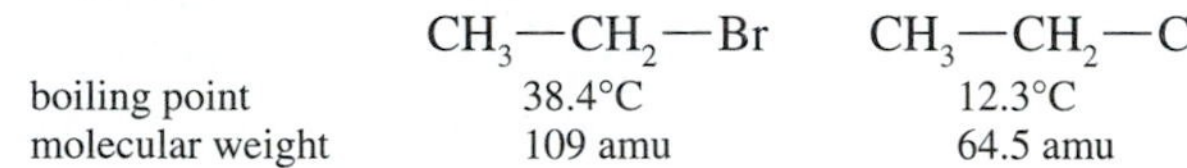

CH3—CH2—Br
CH3—CH2—Cl
boiling point
38.4°C
12.3°C
molecular weight
109 amu
64.5 amu

The properties of water are a result of strong attractive hydrogen-bonding forces.

Even when the types of atoms in molecules are the same, London forces differ when the molecular weights are different. For example, the boiling points of pentane and hexane are 36°C and 69°C, respectively. These two nonpolar molecules contain the same types of atoms, but the numbers of atoms differ. Hexane is a larger molecule whose chain has more surface area to interact with neighboring molecules. As a result, the London forces are stronger in hexane than in pentane. This increased attraction between molecules decreases the vapor pressure of hexane, and its boiling point is higher than the boiling point of pentane.

London forces also depend on molecular shape. For example, the boiling point of 2,2-dimethylpropane is lower than that of pentane. 2,2-Dimethylpropane is more spherical, and it therefore has less surface area than the more ellipsoidal-shaped pentane molecule (Figure 2.3). As a consequence, there is less effective contact between 2,2-dimethylpropane molecules, and the London forces are weaker and less effective.

Hydrogen-Bonding Forces

Compounds containing hydrogen bonded to oxygen or nitrogen, such as H_2O and NH_3, interact by very strong intermolecular forces. This interaction is called a **hydrogen bond.**

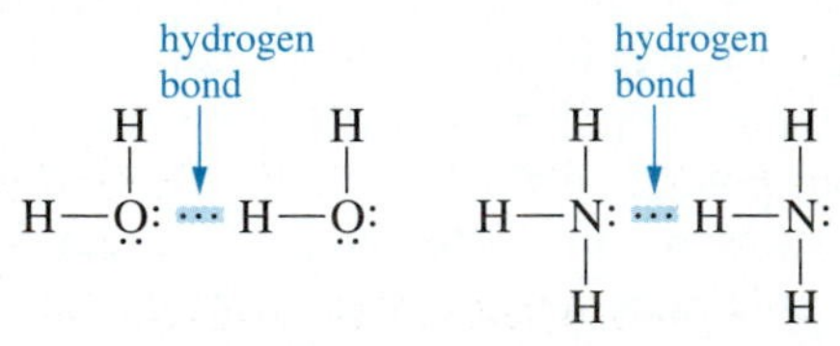

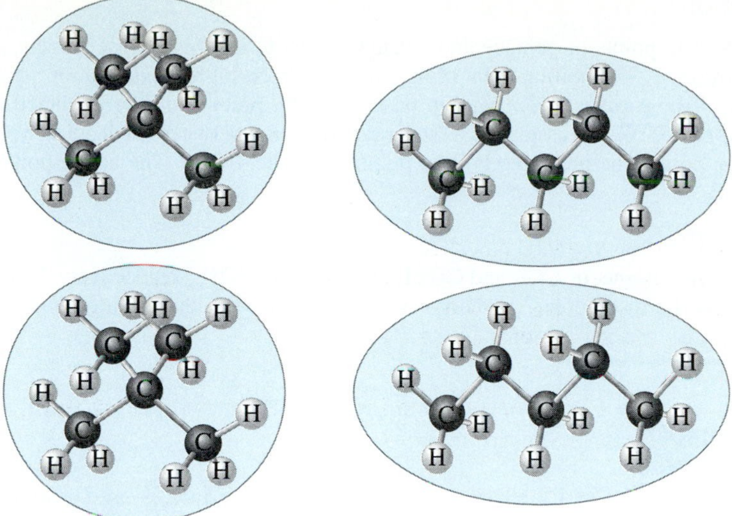

FIGURE 2.3
Shapes of Molecules and Intermolecular Forces
The shape of 2,2-dimethylpropane is nearly spherical. Neighboring molecules cannot approach one another closely, and the points of contact are limited. Pentane is an extended molecule with an ellipsoid shape, so neighboring molecules can approach each other side by side. As a consequence, the intermolecular attractive forces are stronger.

The hydrogen atom in a polar covalent bond to an electronegative element has a partial positive charge. As a result, there is an attraction between the hydrogen atom and the lone pair electrons of another molecule.

The O—H or N—H groups in organic compounds can form hydrogen bonds. For example, the physical properties of alcohols and amines are strongly affected by hydrogen bonds. The boiling point of ethanol, an alcohol, is substantially higher than the boiling point of dimethyl ether, which has the same molecular weight.

	$CH_3—CH_2—OH$	$CH_3—O—CH_3$
boiling point	78.5°C	–24°C
	ethanol	dimethyl ether

Because the numbers of atoms are the same and the shapes of the molecules are similar, the boiling point difference cannot be due to differences in London forces. Both molecules have polar bonds, and the dipole-dipole forces should be similar. The higher boiling point of the ethanol results from the hydrogen bonding between hydroxyl groups of neighboring molecules, which are much stronger interactions than London forces (Figure 2.4).

EXAMPLE 2-1

Based on the boiling points of pentane and hexane, predict the boiling point of heptane.

$$CH_3CH_2CH_2CH_2CH_2CH_2CH_3$$

Solution

The boiling points of pentane and hexane are 36°C and 69°C, respectively, a difference of 33°C. The boiling point of heptane should be higher than that of hexane. The effect of the extra $—CH_2—$ group on the boiling point could be predicted to be an additional 33°C, by assuming a linear relationship between molecular weight and boiling point. The predicted boiling point would be 102°C. The actual boiling point is 98°C.

Problem 2.1

The boiling points of CCl_4 and $CHCl_3$ are 77°C and 62°C, respectively. Which compound is the more polar? Is the polarity consistent with the boiling points? Why or why not?

EXAMPLE 2-2

The boiling point of 1,2-ethanediol (ethylene glycol), which is used as antifreeze, is 190°C. Why is the boiling point higher than that of 1-propanol (97°C)?

$$HO—CH_2—CH_2—OH \qquad CH_3—CH_2—CH_2—OH$$

1,2-ethanediol 1-propanol

Solution

Both molecules have similar molecular weights and should have comparable London forces. However, 1,2-ethanediol has two hydroxyl groups per molecule, compared to only one per molecule in 1-propanol. As a consequence, liquid 1,2-ethanediol can form twice as many hydrogen bonds. The increased number of hydrogen bonds decreases the vapor pressure of 1,2-ethanediol, which leads to a higher boiling point.

Problem 2.2

Explain why the boiling points of ethanethiol and dimethyl sulfide are very similar.

$$CH_3—CH_2—SH \qquad CH_3—S—CH_3$$

ethanethiol (bp 35°C) dimethyl sulfide (bp 37°C)

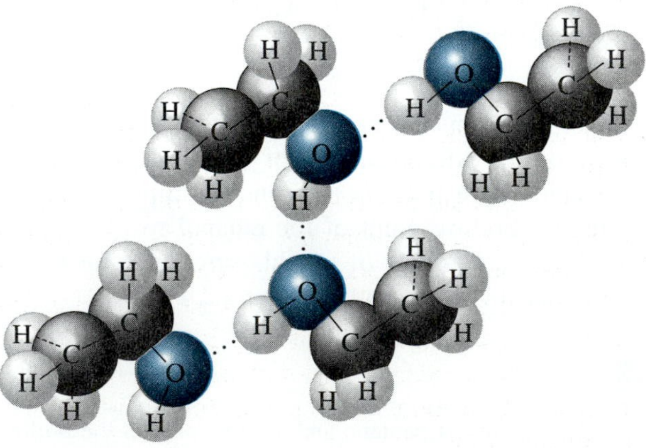

FIGURE 2.4
Hydrogen Bonding in Ethanol

Solubility

A maxim of the chemistry laboratory is that "like dissolves like." This generalization is reasonable because molecules of solute that are similar to molecules of solvent should interact by similar intermolecular attractive forces. Carbon tetrachloride, CCl_4, a nonpolar substance, does not dissolve ionic compounds such as sodium chloride. However, this nonpolar compound is a good solvent for nonpolar compounds such as fats and waxes. Water, which is quite polar, is a good solvent for ionic compounds and substances that can produce ions in water. Water dissolves a limited number of low molecular weight organic compounds if they are sufficiently polar or can form hydrogen bonds with water.

Liquids that dissolve in each other in all proportions are said to be **miscible.** Liquids that do not dissolve in each other are **immiscible.** Immiscible liquids form separate layers in a container. For example, ethyl alcohol is miscible with water, but carbon tetrachloride is immiscible with water. The miscibility of ethyl alcohol with water is explained by its structure.

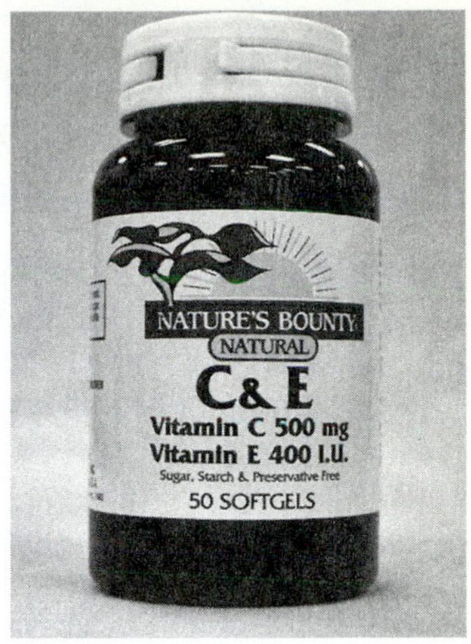

This vitamin supplement contains both water-soluble and fat-soluble vitamins.

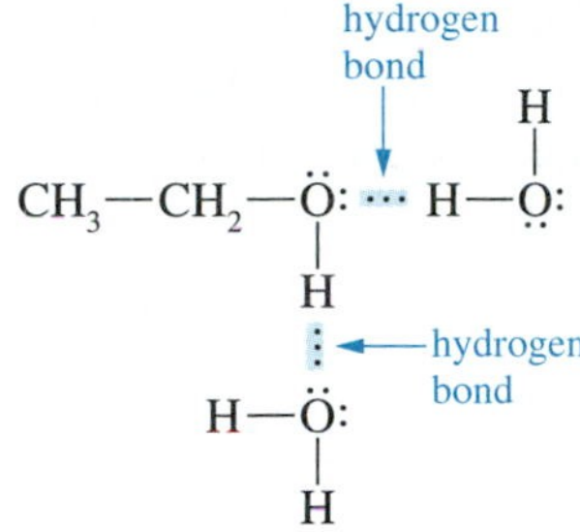

Ethyl alcohol, like water, has an —OH group. Ethyl alcohol is polar, as is water. The nonbonding electron pairs on the oxygen atom in ethyl alcohol and the hydroxyl hydrogen atom form hydrogen bonds with water and thus make it soluble.

2.2 Chemical Reactions

The number of known and potential reactions among the dozens of functional groups in the millions of organic compounds is astronomically large. However, we can understand these myriad reactions by learning the fundamental concepts that underlie all organic chemical reactions. In other words, we can discern patterns of chemical behavior that unify many facts into a few classes of chemical reactions.

We will review acid-base and oxidation-reduction reactions in Sections 2.3 and 2.4 and illustrate how these concepts apply to organic chemical reactions. Several other classes of organic reactions will be briefly illustrated in Section 2.5.

All chemical reactions are reversible to some degree, and some reactions result in an equilibrium mixture containing substantial amounts of reactants as well as products. There are important reasons to determine the conditions that control chemical equilibria. For example, industrial processes must convert as much reactant to product as possible. Not only is the inefficient conversion of chemicals costly, but the unwanted material must be removed to purify the product. Impure materials cannot be tolerated for many products, especially those for human consumption.

The study of the kinetics of chemical reactions is also of concern in industry, where it is important to understand how to form a desired chemical product

Water-Soluble and Fat-Soluble Vitamins

The different solubilities of vitamins, characterized as water soluble and fat soluble, illustrate the maxim that "like dissolves like." Water-soluble vitamins have large numbers of functional groups that can hydrogen bond with water. Water-insoluble vitamins are essentially nonpolar structures that can "dissolve in" the nonpolar fatty tissue of the body.

Water-soluble vitamins are not stored in the body and should be taken in as part of one's daily diet. Unneeded water-soluble vitamins are excreted. Fat-soluble vitamins are stored by the body. If excessive quantities are consumed in vitamin supplements, illness can result. The condition is known as hypervitaminosis.

The structures of several water-soluble vitamins are shown in the accompanying diagrams. Note that the relatively small vitamin C molecule has a high proportion of —OH groups that can form hydrogen bonds to water. In contrast, vitamin A is not "like" water. It contains an —OH group, but that single functional group is insufficient to allow the relatively large nonpolar portion of the molecule to be accommodated within water. Vitamin B_6 and riboflavin contain not only —OH groups but also nitrogen-containing functional groups that can also hydrogen bond to water. Vitamins E and D_3 are nonpolar compounds that are not soluble in water. ■

Water-soluble Vitamins

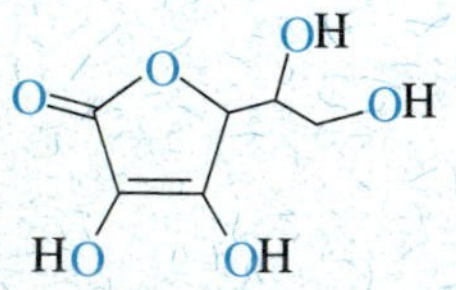

vitamin C

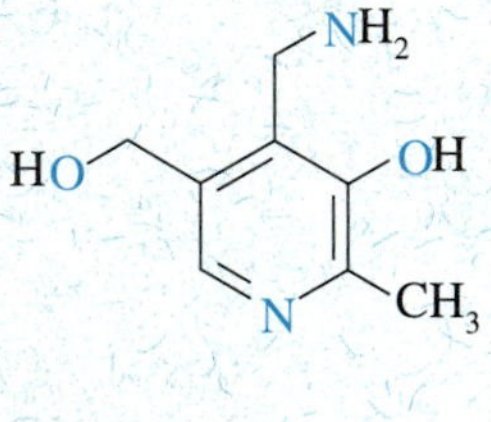

vitamin B_6

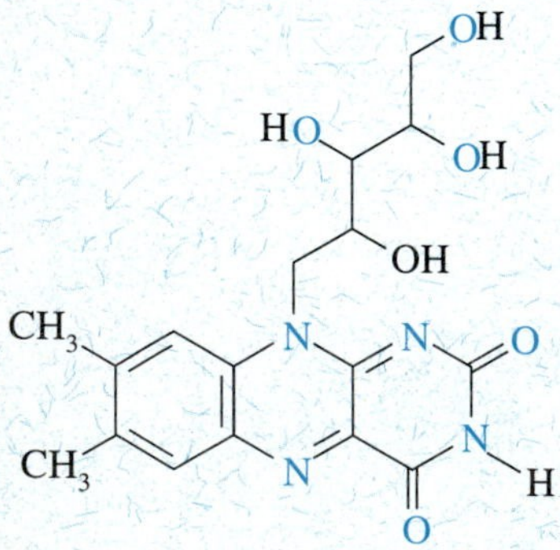

riboflavin

Fat-soluble Vitamins

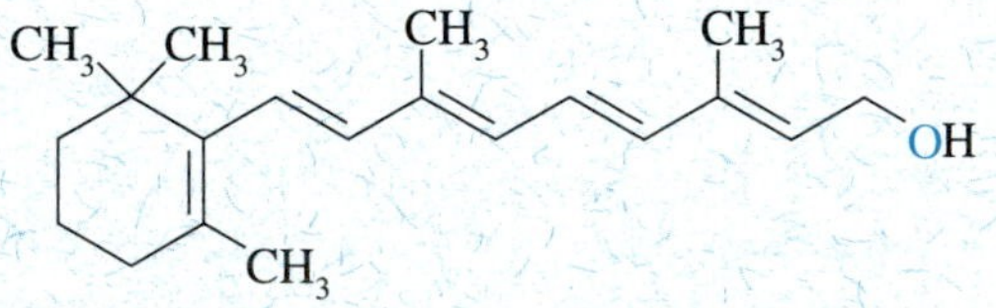

vitamin A

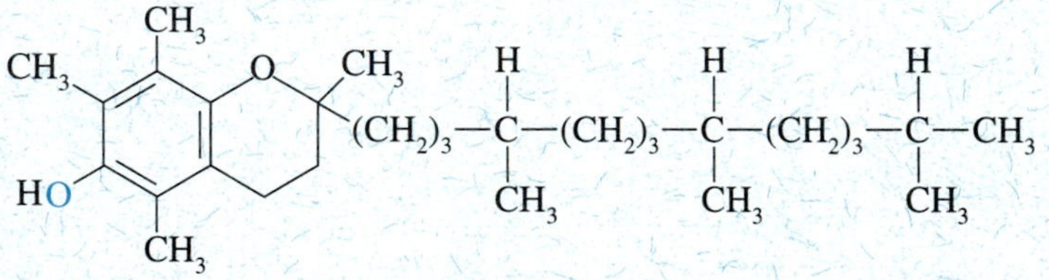

vitamin E

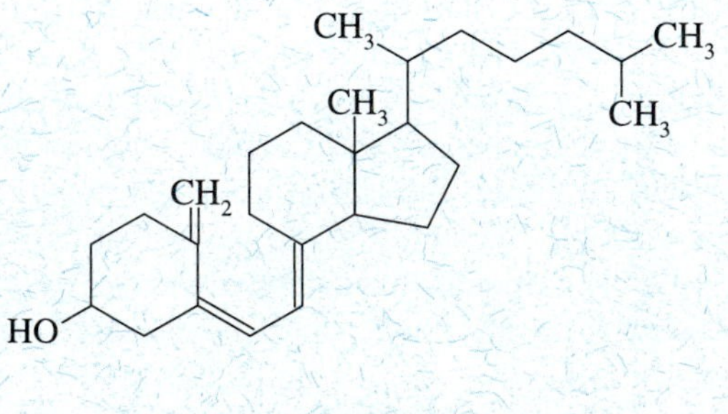

vitamin D_3

rapidly and in preference to other products. A reaction may be too slow to be economically practical, or a reaction may be so fast that it is dangerous and difficult to control. In biological systems, virtually all reactions produce only a single product, and the reactions are very fast.

The study of the kinetics of a reaction helps us determine the **mechanism** of the process. A mechanism details the order of bond cleavage and bond formation that occurs during the reaction. This information establishes general guidelines, which allow chemists to extrapolate a few observations on selected reactions to many other reactions.

2.3 Acid-Base Reactions

According to the Brønsted-Lowry theory, an **acid** is a substance that can donate a proton (H^+); a **base** is a substance that can accept a proton. For example, when gaseous hydrogen chloride dissolves in water, virtually all of the HCl molecules transfer a proton to water, and a solution of hydronium ions and chloride ions results.

$$H_2\ddot{O}: + H{-}\ddot{Cl}: \longrightarrow H_3O^+ + :\ddot{Cl}:^-$$

This reaction is illustrated by curved arrows. Electrons are pictured as flowing from the start of the arrow toward the arrowhead. The nonbonding pair of electrons of the oxygen atom forms a bond to the hydrogen atom, and the bonding pair of electrons in HCl is fully transferred to the chlorine atom.

The hydroxide ion, which exists as an ion in compounds such as NaOH, KOH, and $Ca(OH)_2$, is a base because it has nonbonding pairs of electrons that can accept a proton from an acid such as a hydronium ion. Ammonia is also a base because it has a nonbonding pair of electrons on its nitrogen atom, which can form a bond to a hydrogen atom of H_3O^+. A curved arrow shows the movement of the pair of electrons from the nitrogen atom toward a hydrogen atom.

$$H_3N: + H{-}\overset{+}{O}H_2 \longrightarrow NH_4^+ + :\ddot{O}H_2$$

Vinegar is a 5% aqueous solution of acetic acid.

Organic acids and bases behave similarly. Carboxylic acids contain a carboxyl group, which can donate a proton to a base such as water. The curved arrow formalism used to depict the movement of electrons shows how the nonbonding electron pair of the water molecule forms a bond with the hydrogen atom of the carboxyl group.

$$CH_3{-}\overset{O}{\overset{\|}{C}}{-}\ddot{O}{-}H + H{-}\ddot{O}{-}H \rightleftharpoons CH_3{-}\overset{O}{\overset{\|}{C}}{-}\ddot{O}:^- + H{-}\overset{+}{O}H{-}H$$

acetic acid (an acid)	water (a base)	acetate ion (a conjugate base)	hydronium ion (a conjugate acid)

Amines are a class of organic bases whose acid-base chemistry is like that of ammonia. Thus, methylamine behaves as a base because the nonbonding electron pair of the nitrogen atom can accept a proton from an acid such as the hydronium ion.

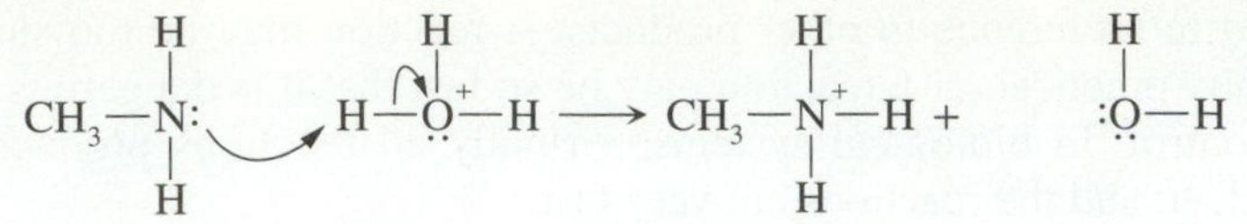

$$CH_3{-}\ddot{N}H_2 + H_3O^+ \longrightarrow CH_3{-}NH_3^+ + H_2\ddot{O}$$

methylamine (a base) | hydronium ion (an acid) | methylammonium ion (a conjugate acid) | water (a conjugate base)

When an acid transfers a proton to a base, another base and acid are produced. The acid loses a proton and becomes a **conjugate base.** For example, the conjugate base of acetic acid is acetate ion. When a base accepts a proton, the substance formed is a **conjugate acid.** Thus, the conjugate acid of methylamine is the methylammonium ion.

Lewis Acids and Bases

Some chemical reactions that do not occur with proton transfer are also regarded as acid-base reactions. These reactions can be explained in terms of Lewis acids and Lewis bases, which are defined based on electron pairs. A **Lewis acid** is a substance that accepts an electron pair; a **Lewis base** is a substance that donates an electron pair. Thus HCl, which is an acid in the Brønsted-Lowry sense, is also a Lewis acid because it contains a proton that can "accept" an electron pair. Similarly, ammonia is a Lewis base as well as a Brønsted-Lowry base because it can donate an electron pair. However, the Lewis classification of acids and bases is more extensive because it is not restricted to protons.

Boron trifluoride (BF_3) and aluminum trichloride ($AlCl_3$) are two common Lewis acids encountered in organic chemical reactions. Each has only six electrons in its valence shell, and each can thus accept an electron pair from a Lewis base.

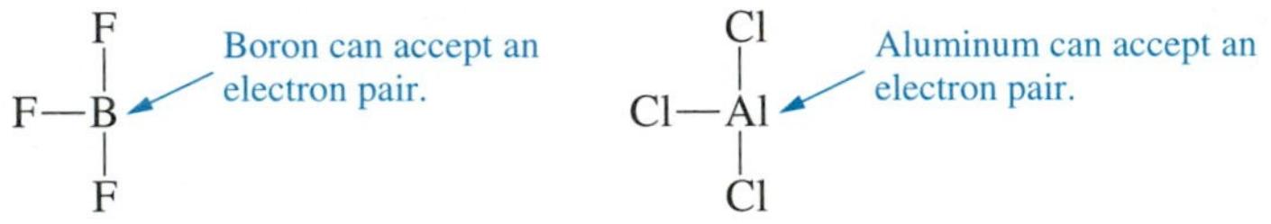

BF_3 — Boron can accept an electron pair.

boron trifluoride

$AlCl_3$ — Aluminum can accept an electron pair.

aluminum trichloride

Other Lewis acids include transition metal compounds, such as $FeBr_3$, that react by accepting a pair of electrons. For example, $FeBr_3$ reacts with molecular bromine to accept a bromide ion via a pair of electrons. In this reaction, $FeBr_3$ behaves as a Lewis acid and bromine behaves as a Lewis base.

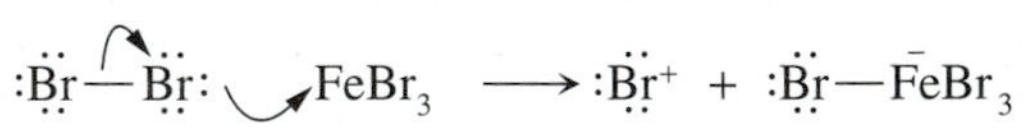

$$:\ddot{Br}{-}\ddot{Br}: + FeBr_3 \longrightarrow :\ddot{Br}^+ + :\ddot{Br}{-}\overset{-}{Fe}Br_3$$

Lewis base | Lewis acid

Many organic compounds that contain oxygen and nitrogen atoms can act as Lewis bases because these atoms have nonbonding electrons that can react with Lewis acids. For example, ethers react with boron trifluoride to give a product with a bond between boron and oxygen.

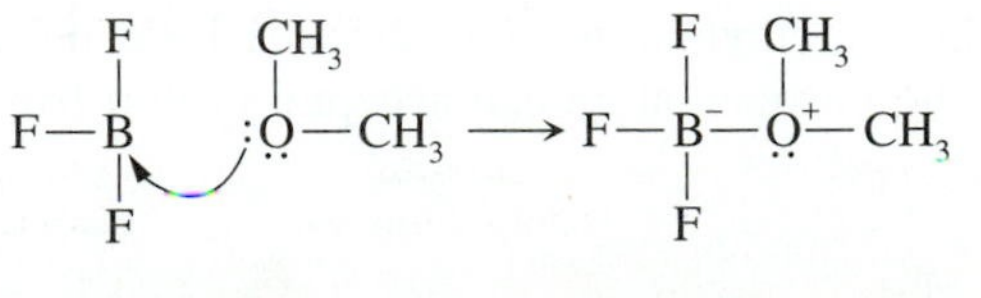

$$BF_3 + :\ddot{O}(CH_3)_2 \longrightarrow F_3\overset{-}{B}{-}\overset{+}{O}(CH_3)_2$$

EXAMPLE 2-3

Assume that ethanol behaves as a Brønsted-Lowry acid in a reaction. What is the conjugate base of ethanol?

$$CH_3-CH_2-\ddot{\underset{..}{O}}-H$$

Solution

Loss of a proton (H^+) from an electrically neutral acid results in a conjugate base with a negative charge. The electron pair of the O—H bond remains with the oxygen atom.

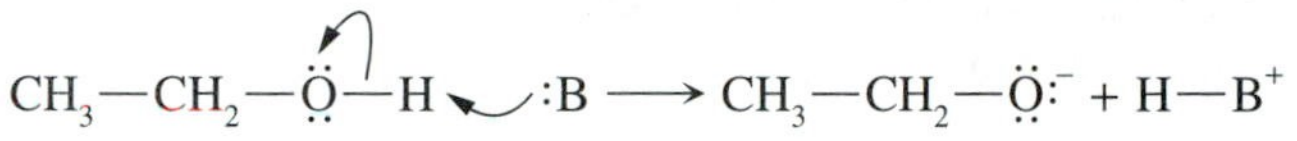

$$CH_3-CH_2-\ddot{\underset{..}{O}}-H \; :B \longrightarrow CH_3-CH_2-\ddot{\underset{..}{O}}:^- + H-B^+$$

Problem 2.3

Consider the reaction of Br^+ with ethylene to give a charged intermediate called a carbocation. Classify the reactants according to Lewis acid-base nomenclature.

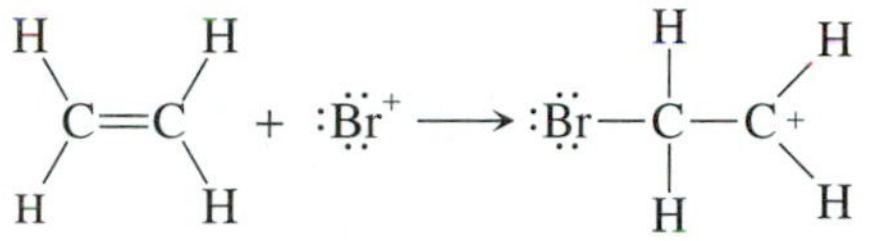

$$H_2C=CH_2 + :\ddot{\underset{..}{Br}}^+ \longrightarrow :\ddot{\underset{..}{Br}}-CH_2-CH_2^+$$

a carbocation

2.4 Oxidation-Reduction Reactions

Oxidation is the loss of electrons by a substance or an increase in its oxidation number. **Reduction** is the gain of electrons by a substance or a decrease in its oxidation number. From a slightly different point of view, it follows that when a substance is reduced, it gains the electrons from a substance that becomes oxidized.

The relationship between oxidation and reduction is emphasized further in the terms oxidizing agent and reducing agent. In an oxidation-reduction reaction, the substance that is reduced is the **oxidizing agent** because, by gaining electrons, it causes oxidation of another substance. The substance that is oxidized is called the **reducing agent** because, by losing its electrons, it causes the reduction of another substance.

In organic chemistry, oxidation numbers are not as easily assigned as in inorganic chemistry. However, we can decide on the change in the oxidation state of a compound in many chemical reactions by accounting for the number of hydrogen atoms or oxygen atoms gained or lost. The oxidation state of a molecule increases (oxidation) if its hydrogen content decreases or its oxygen content increases. Conversely, the oxidation state of a molecule decreases (reduction) if its hydrogen content increases or its oxygen content decreases. For example, the reaction of methanol (CH_3OH) to produce methanal (formaldehyde, CH_2O) is an oxidation because methanol loses two hydrogen atoms. Further reaction of methanal to produce methanoic acid (formic acid, HCO_2H) occurs with an increase in the oxygen content and is also an oxidation process.

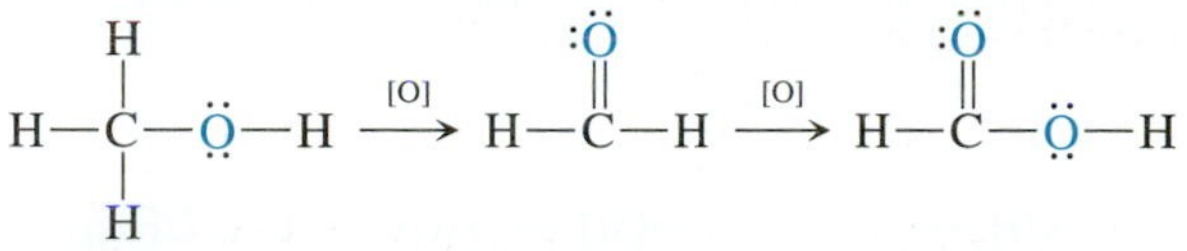

$$H_3C-\ddot{\underset{..}{O}}-H \xrightarrow{[O]} H-\underset{}{C}(=\ddot{O}:)-H \xrightarrow{[O]} H-C(=\ddot{O}:)-\ddot{\underset{..}{O}}-H$$

methanol — methanal (formaldehyde) — methanoic acid (formic acid)

The symbol [O] represents an unspecified oxidizing agent. Note also that the equations are not balanced. The focus in organic chemistry is on the conversion of one organic compound to another. Oxidizing agents such as potassium dichromate, which might be used in an oxidation reaction, are seldom balanced in writing an equation but instead are written above the reaction arrow. The following conversion of an alcohol into an acid is an oxidation, and the substance above the arrow is the oxidizing agent.

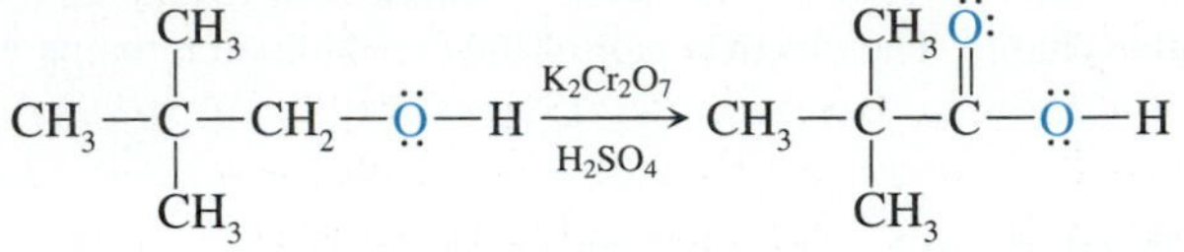

$$CH_3-C(CH_3)_2-CH_2-\ddot{O}-H \xrightarrow[H_2SO_4]{K_2Cr_2O_7} CH_3-C(CH_3)_2-C(=\ddot{O}:)-\ddot{O}-H$$

The conversion of a carbon-carbon triple bond into a double bond and finally into a single bond involves reduction because the hydrogen content increases in each step. The symbol [H] represents an unspecified reducing agent, such as hydrogen gas in the presence of a platinum catalyst.

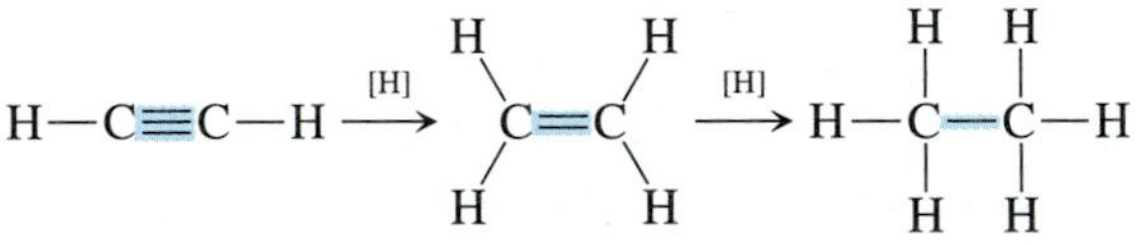

$$H-C\equiv C-H \xrightarrow{[H]} H_2C=CH_2 \xrightarrow{[H]} H_3C-CH_3$$

ethyne (acetylene) — ethene (ethylene) — ethane

The simultaneous increase or decrease of two hydrogen atoms and one oxygen atom in a reactant is neither reduction nor oxidation. Thus, the conversion of ethene to ethanol is not an oxidation-reduction reaction.

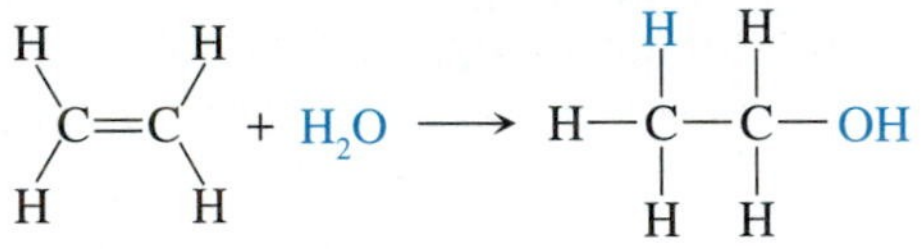

$$H_2C=CH_2 + H_2O \longrightarrow H-CH_2-CH_2-OH$$

ethene — ethanol

EXAMPLE 2-4

Ethylene oxide is used to sterilize medical equipment that is temperature sensitive and cannot be heated in an autoclave. It is produced from ethylene by the following process. Classify the type of reaction. Is an oxidizing or reducing agent required?

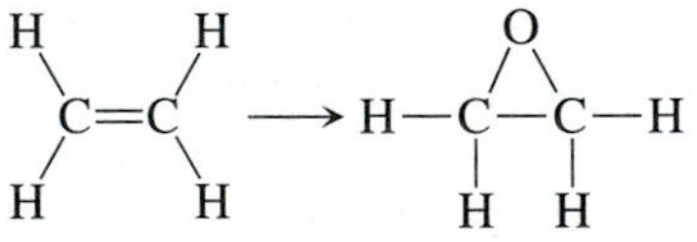

$$H_2C=CH_2 \longrightarrow H-CH(-O-)CH-H \text{ (ethylene oxide, three-membered ring)}$$

Solution

The oxygen content of the ethylene is increased, and thus the reaction involves oxidation. An oxidizing agent is required for the reaction.

Problem 2.4

Consider the following reaction and determine if an oxidation or reduction reaction occurs.

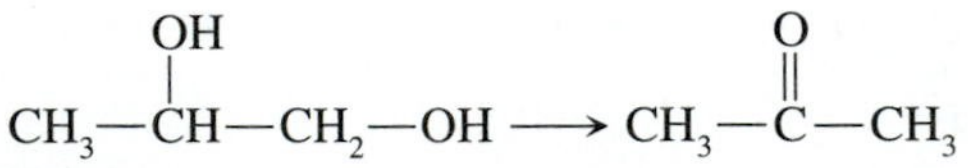

$$CH_3-CH(OH)-CH_2-OH \longrightarrow CH_3-C(=O)-CH_3$$

Redox Reactions in Biochemistry

Metabolic reactions provide energy in multiple-step reactions in which metabolites are oxidized. Biosynthetic reactions, which build the necessary compounds to maintain organisms, are usually reduction reactions. Thus, organisms require both oxidizing and reducing agents. Furthermore, because there are so many diverse reactions that occur in oxidative degradation and reductive biosynthesis, these oxidizing and reducing agents must have a wide range of reactivity.

Nature has designed chemical reactions that are catalyzed by enzymes—most frequently a specific enzyme for each type of reaction. However, these redox reactions also require a coenzyme that serves as the oxidizing or reducing agent. Nature is efficient in this regard because there are a limited number of these coenzymes. Two of these compounds, nicotinamide adenine dinucleotide and flavin adenine dinucleotide, are represented by the shorthand NAD^+ and FAD, respectively. Each of these coenzymes oxidizes biological molecules by removing two hydrogen atoms from covalent bonds. These reactions, using hydrogen atoms for simplicity, are represented as

$$\underset{\text{oxidized form}}{NAD^+} + 2H\cdot \longrightarrow \underset{\text{reduced form}}{NADH} + H^+$$

$$\underset{\text{oxidized form}}{FAD} + 2H\cdot \longrightarrow \underset{\text{reduced form}}{FADH_2}$$

The reduced form of each coenzyme can serve as a reducing agent and be oxidized to regenerate the oxidized form.

NAD^+ and FAD oxidize different classes of compounds. NAD^+ removes hydrogen atoms from a C—H and an O—H bond, whereas FAD removes hydrogen atoms from two C—H bonds. For example, NAD^+ oxidizes malic acid to oxaloacetic acid in one of the steps of the citric acid cycle, which is ultimately responsible for the oxidation of metabolites to carbon dioxide and water.

$$NAD^+ + \underset{\text{malic acid}}{HO_2C{-}CH_2{-}\overset{\overset{\large OH}{|}}{\underset{\underset{\large H}{|}}{C}}{-}CO_2H} \longrightarrow NADH + H^+ + \underset{\text{oxaloacetic acid}}{HO_2C{-}CH_2{-}\overset{\overset{\large O}{\|}}{C}{-}CO_2H}$$

FAD oxidizes fatty acids such as stearic acid, which is derived from fats—which are our long-term sources of energy.

$$FAD + \underset{\text{stearic acid}}{CH_3(CH_2)_{14}{-}\overset{\overset{\large H}{|}}{C}H{-}\overset{\overset{\large H}{|}}{C}H{-}CO_2H} \longrightarrow FADH_2 + CH_3(CH_2)_{14}{-}CH{=}CH{-}CO_2H$$

Foreign compounds (xenobiotics) and many common drugs are eliminated from the body by oxidative reactions. Water-soluble substances are easily excreted, but most organic compounds are nonpolar and are **lipid** soluble (that is, they dissolve in the fatty components of cells). If **lipophilic** (lipid loving) xenobiotics or drugs were not eliminated, they would accumulate and an organism would eventually become a living (and soon dead) "toxic dump."

Organisms ordinarily transform lipophilic substances into more polar water-soluble products that can be excreted. The liver is the most important organ for the oxidation of xenobiotics and drugs. The oxidation of compounds, represented as R—H, in the liver requires molecular oxygen and the coenzyme nicotinamide adenosine dinucleotide phosphate, NADPH. One of the oxygen atoms is incorporated in the reactant and the other oxygen atom in water.

$$R{-}H + NADPH + O_2 + H^+ \longrightarrow R{-}OH + NADP^+ + H_2O$$

The enzyme responsible for catalyzing the reaction is cytochrome *P*-450. It contains iron surrounded by heme, a complex nitrogen-containing compound, and the amino acid lysine, which is part of a protein structure. Substrates are accommodated at a site within the protein that allows them to be oxidized by the coenzyme.

The oxidation of the oral hypoglycemic drug tolbutamide (Orenase) to an alcohol is one example of the metabolism of drugs by the liver. ■

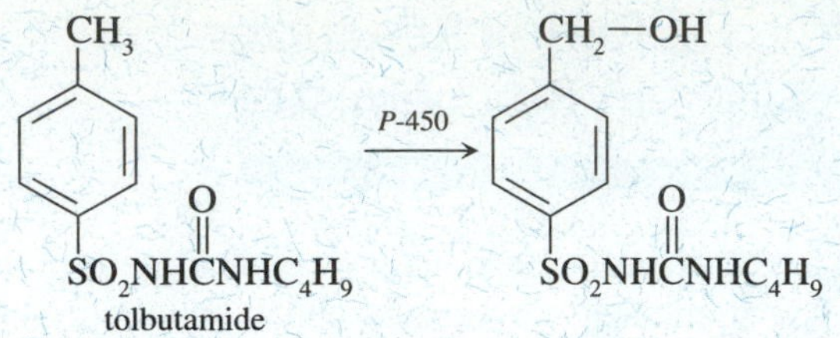

CH_3–C₆H₄–$SO_2NHC(=O)NHC_4H_9$ (tolbutamide) $\xrightarrow{P\text{-}450}$ CH_2—OH–C₆H₄–$SO_2NHC(=O)NHC_4H_9$

tolbutamide

2.5 Classification of Organic Reactions

In the preceding two sections we have reviewed two classes of reactions that you learned about in your first course in chemistry. Now we will look at several more common examples of organic reactions. These reactions will be discussed in greater detail in subsequent chapters.

Addition reactions occur when two reactants combine to give a single product. An example of an addition reaction is the reaction of ethylene with HBr to form bromoethane. The hydrogen and bromine atoms are added to adjacent atoms, a common characteristic of addition reactions.

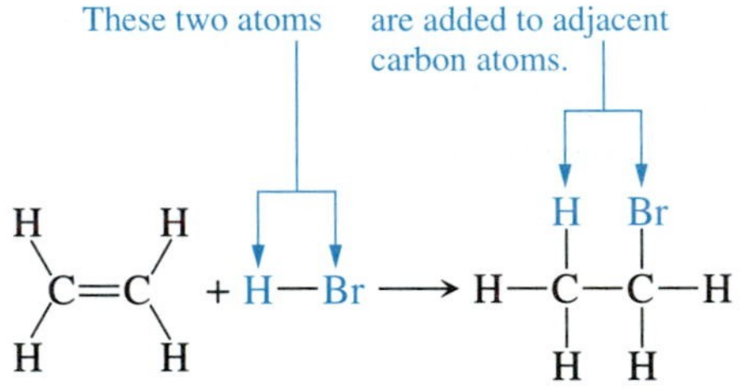

These two atoms are added to adjacent carbon atoms.

$$H_2C{=}CH_2 + H{-}Br \longrightarrow H{-}CH_2{-}CH_2{-}Br$$

bromoethane

Elimination reactions involve the splitting apart of a single compound into two compounds. Most elimination reactions form a product with a double bond containing the majority of the atoms in the reactant, and a second smaller molecule such as H_2O or HCl. The atoms eliminated to form the smaller molecule are usually located on adjacent carbon atoms in the reactant. For example, 2-propanol reacts with concentrated sulfuric acid to produce propene; water is eliminated in this reaction.

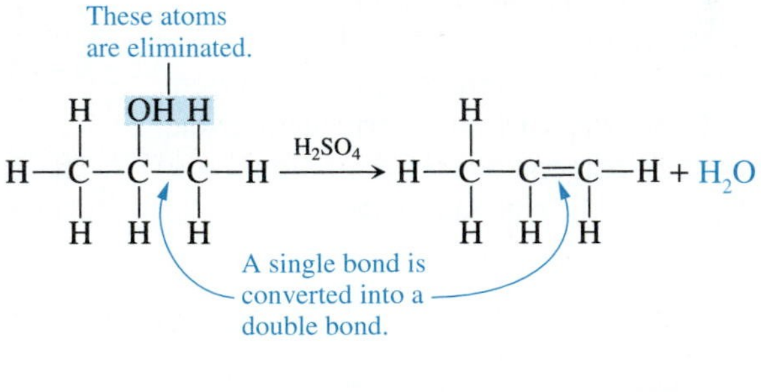

These atoms are eliminated.

$$CH_3{-}CH(OH){-}CH_3 \xrightarrow{H_2SO_4} CH_3{-}CH{=}CH_2 + H_2O$$

A single bond is converted into a double bond.

Substitution reactions involve replacement of one atom or group of atoms by a second atom or group of atoms. A generalized reaction in which Y substitutes for X is as follows.

$$A{-}X + Y \longrightarrow A{-}Y + X$$

An example of a substitution reaction is the conversion of bromomethane into methanol.

$$\underset{\text{bromomethane}}{CH_3{-}Br} + OH^- \longrightarrow \underset{\text{methanol}}{CH_3{-}OH} + Br^-$$

In **hydrolysis reactions** (Greek *hydro,* "water" + *lysis,* "splitting") water splits a large reactant molecule into two smaller product molecules. The generalized reaction is

$$A{-}B + H_2O \longrightarrow A{-}H + HO{-}B$$

One product molecule is bonded to a hydrogen atom derived from water. The other product is bonded to an —OH group derived from water. The hydrolysis of an amide to produce a carboxylic acid and an amine is an example of this process.

This N—C bond is cleaved. This N—H bond is formed. The —OH group is bonded to the carbon atom.

$$\underset{\text{an amide}}{CH_3{-}\overset{H}{\overset{|}{N}}{-}\overset{O}{\overset{\|}{C}}{-}CH_3} + H_2O \longrightarrow \underset{\text{an amine}}{CH_3{-}\overset{H}{\overset{|}{N}}{-}H} + \underset{\text{a carboxylic acid}}{HO{-}\overset{O}{\overset{\|}{C}}{-}CH_3}$$

In **condensation reactions,** two reactants combine to form one larger product with the simultaneous formation of a second, smaller product such as water. When the second product is water, the reaction is the reverse of a hydrolysis reaction. The general reaction is

$$A{-}H + HO{-}B \longrightarrow A{-}B + H_2O$$

The formation of an ester from an alcohol and a carboxylic acid is an example of this process.

This C—O bond is formed.

$$\underset{\text{a carboxylic acid}}{CH_3{-}\overset{O}{\overset{\|}{C}}{-}OH} + \underset{\text{an alcohol}}{H{-}O{-}CH_3} \longrightarrow \underset{\text{an ester}}{CH_3{-}\overset{O}{\overset{\|}{C}}{-}O{-}CH_3} + H_2O$$

Rearrangement reactions result from the reorganization of bonds within a single reactant to give an isomeric product. This type of reaction will not be encountered as frequently in this text as the other reactions described in this section. One example is a rearrangement in which the location of a double bond changes to give an isomer. Note that the bromine atom is also relocated in the product.

$$CH_3{-}\overset{Br}{\overset{|}{C}H}{-}CH{=}CH_2 \longrightarrow CH_3{-}CH{=}CH{-}CH_2{-}Br$$

EXAMPLE 2-5

Classify the following reaction.

$$CH_3-\underset{|}{\overset{Br}{CH}}-CH_2-Br + Zn \longrightarrow CH_3-CH=CH_2 + ZnBr_2$$

Solution

An elimination of the two bromine atoms located on adjacent carbon atoms occurs. The organic product contains the majority of the atoms of the reactant, and the byproduct is $ZnBr_2$.

Problem 2.5

Classify the following reaction.

$$(H)_2C=C(OH)(H) \longrightarrow H-C(H)(H)-C(=O)H$$

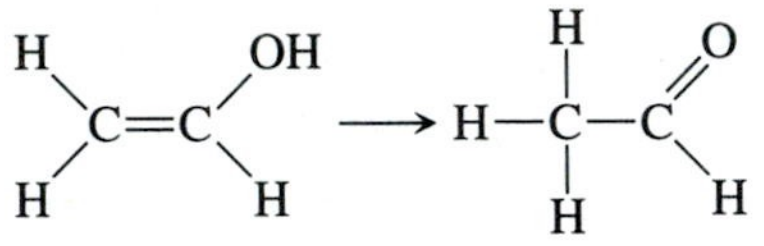

2.6 Chemical Equilibrium

Chemical reactions do not proceed in only one direction. As a reaction occurs, product molecules can revert to reactant molecules. Thus, two opposing reactions occur. When the rate of product formation is equal to the rate of reactant formation, an equilibrium is established.

Consider the general equation for a reaction at equilibrium, where A and B are reactants, X and Y are products, and *m*, *n*, *p*, and *q* are coefficients.

$$m\,A + n\,B \rightleftharpoons p\,X + q\,Y$$

For this reaction, the following ratio is a constant at a specific temperature. The brackets indicate concentration in mole/liter.

$$\frac{[X]^p[Y]^q}{[A]^m[B]^n} = K$$

Consider the equilibrium constant expression for the addition reaction of gaseous ethylene with gaseous hydrogen bromide.

$$CH_2=CH_2 + HBr \rightleftharpoons CH_3CH_2Br$$

$$\frac{[CH_3CH_2Br]}{[CH_2=CH_2][HBr]} = K = 10^8$$

Because the equilibrium constant is very large, the reaction "goes to completion." That is, for all practical purposes, no reactant remains at equilibrium.

Now let's consider the condensation reaction of acetic acid and ethyl alcohol to produce ethyl acetate and the related equilibrium constant expression.

$$CH_3-\overset{O}{\overset{\|}{C}}-OH + HO-CH_2CH_3 \rightleftharpoons CH_3-\overset{O}{\overset{\|}{C}}-O-CH_2CH_3 + H_2O$$

$$\frac{[CH_3CO_2CH_2CH_3][H_2O]}{[CH_3CO_2H][CH_3CH_2OH]} = K = 4.0$$

In this reaction, significant concentrations of reactants are present at equilibrium. Thus, the product yield is less than 100% based on the balanced equation.

The position of a chemical equilibrium and the value of the equilibrium constant are not affected by catalysts. A **catalyst** (see Section 2.10) increases the rates of the forward and reverse reactions equally, and K does not change. In the reaction of ethanol and acetic acid, the reaction is acid catalyzed. Thus, the equilibrium is established in a shorter time period at the same temperature in the presence of an acid such as HCl.

Le Châtelier's Principle

Most reactions discussed in this text have large equilibrium constants. For those that do not, reaction conditions are selected to "force" the reaction in the direction favoring product based on Le Châtelier's principle. **Le Châtelier's principle** states that a change in the conditions of a chemical equilibrium causes a shift in the concentration of reactants and products to result in a new equilibrium system. If additional reactant is added to a chemical system at equilibrium, the concentrations of both reactants and products change to establish a new equilibrium system, but the equilibrium constant is unchanged. After adding reactant, the total concentration of reactant is initially increased, but then decreases to establish a new equilibrium. As a result, the concentration of the products increases. In short, the change imposed on the system by adding reactants is offset when added reactants are converted to product. If a product is removed from a chemical system at equilibrium, the forward reaction occurs to give more product. Regardless of the condition imposed on the system at equilibrium, the concentrations change to maintain the same value of the equilibrium constant. Consider the equilibrium in the formation of ethyl acetate.

$$CH_3{-}\overset{\overset{\displaystyle O}{\|}}{C}{-}OH + HO{-}CH_2CH_3 \rightleftharpoons CH_3{-}\overset{\overset{\displaystyle O}{\|}}{C}{-}O{-}CH_2CH_3 + H_2O$$

Adding ethyl alcohol "pushes" the reaction to the right.

Removing water "pulls" the reaction to the right.

If water is removed from the system by some means, the equilibrium is disturbed and the equilibrium position of the reaction would shift to the right to produce more water and ethyl acetate. If the amount of alcohol is increased, a larger amount of the carboxylic acid will be converted into product.

2.7 Equilibria in Acid-Base Reactions

Water is the reference solvent commonly used to compare the strengths of acids or bases. The strengths of acids are measured by their tendencies to transfer protons to water.

$$HA + H_2O \rightleftharpoons H_3O^+ + A^-$$

A quantitative measure of the acidity of an acid with the general formula HA is given by the equilibrium constant for ionization, which is obtained from the equation for ionization.

$$K = \frac{[H_3O^+][A^-]}{[HA][H_2O]}$$

The concentration of water, about 55 M, is so large compared to that of the other components of the equilibrium system that its value changes very little when the acid HA is added. Therefore, the concentration of water is included in the acid ionization constant K_a.

$$K_a = K[H_2O] = \frac{[H_3O^+][A^-]}{[HA]}$$

Acids with $K_a > 10$ are strong acids. Most organic acids have $K_a < 10^{-4}$ and are weak acids. Acid dissociation constants are conveniently expressed as pK_a values. The pK_a expressed as the logarithm of base 10 is

$$pK_a = -\log K_a$$

Note that the pK_a values listed in Table 2.1 increase as the K_a decreases.

Weak acids do not completely transfer their protons to water, and few ions are produced. An example of a weak acid is acetic acid, which ionizes in water to give acetate ions and hydronium ions.

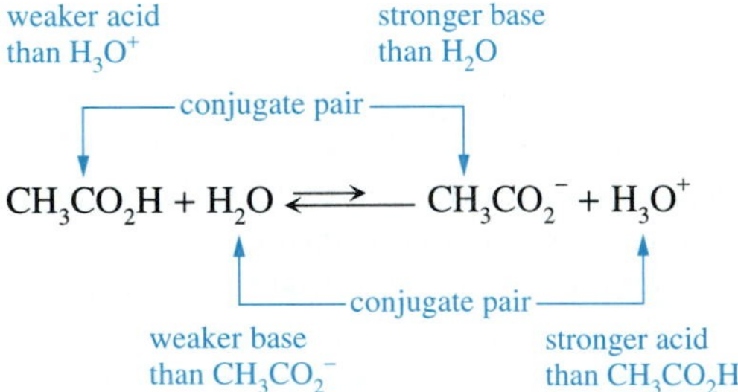

The equilibria between acids and bases and their conjugate bases and acids can be viewed as a "contest" for protons. The equilibrium position favors the side containing the weaker acid and weaker base. Acetic acid is a weaker acid than H_3O^+, and $CH_3CO_2^-$ is a stronger base than H_2O. Note that a strong acid, with its great tendency to lose protons, is paired with a weak conjugate base that has

TABLE 2.1 K_a and pK_a Values

Acid	K_a	pK_a
HBr	10^9	–9
HCl	10^7	–7
H_2SO_4	10^5	–5
HNO_3	10^1	–1
HF	6×10^{-4}	3.2
CH_3CO_2H	2×10^{-5}	4.7
$(CF_3)_3COH$	2×10^{-5}	4.7
CH_3CH_2SH	3×10^{-11}	10.6
CF_3CH_2OH	4×10^{-13}	12.4
CH_3OH	3×10^{-16}	15.5
$(CH_3)_3COH$	1×10^{-18}	18
CCl_3H	10^{-25}	25
$HC{\equiv}CH$	10^{-25}	25
NH_3	10^{-36}	36
$CH_2{=}CH_2$	10^{-44}	44
CH_4	10^{-49}	49

a low affinity for protons. Thus, as the tendency of an acid to lose a proton increases, the tendency of its conjugate base to accept a proton decreases.

There is a close relationship between the acidity of acids and the basicity of bases. When an acid dissociates, a base is formed that can react in the reverse direction by accepting a proton. Thus, we can discuss the acidity of the acid HA or the basicity of the base A^-. As in the case of acids, the basicity of bases is both qualitatively and quantitatively compared to the properties of water. A base, A^-, removes a proton from water to form hydroxide ion and the conjugate acid HA. The base dissociation constant, K_b, for the reaction is

$$H_2O + A^- \rightleftharpoons HA + OH^-$$

$$K_b = \frac{[HA][OH^-]}{[A^-]}$$

The K_b values of bases are conveniently expressed as pK_b values. The pK_b is defined as

$$pK_b = -\log K_b$$

pK_b values increase with decreasing basicity. The pK_b values of some organic bases are listed in Table 2.2.

A strong base has a large K_b (small pK_b) and completely removes the proton of an acid. The most common strong base is hydroxide ion, which will remove and accept protons from even weak acids such as acetic acid.

stronger acid than H_2O — conjugate pair — weaker base than OH^-

$$CH_3CO_2H + OH^- \rightleftharpoons CH_3CO_2^- + H_2O$$

stronger base than $CH_3CO_2^-$ — conjugate pair — weaker acid than CH_3CO_2H

Weak bases do not have a large attraction for the protons of an acid. Only a small fraction of the molecules of a weak base in a sample will accept protons

TABLE 2.2 K_b and pK_b Values

Base	K_b	pK_b
$C_6H_5NH_2$ (aniline)	4×10^{-10}	9.4
$CH_3CO_2^-$	5×10^{-10}	9.3
CN^-	1.6×10^{-5}	4.8
NH_3	1.7×10^{-5}	4.8
CH_3NH_2	4.3×10^{-4}	3.4
CH_3O^-	3.3×10^{1}	−1.5

at equilibrium. For example, methylamine is a weak base. When it dissolves in water, a low concentration of methylammonium ions forms.

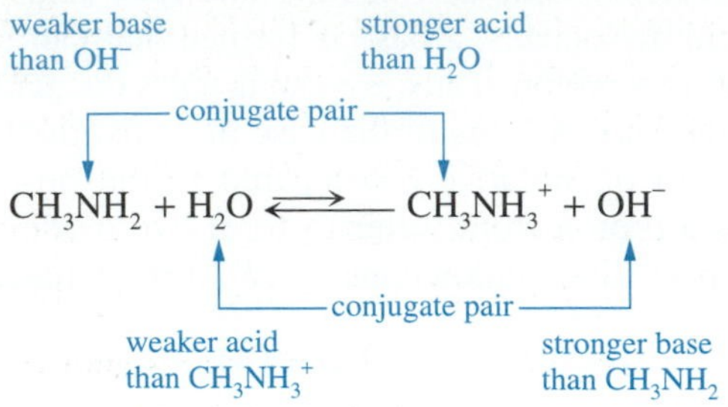

weaker base than OH^- — stronger acid than H_2O (conjugate pair)

$$CH_3NH_2 + H_2O \rightleftharpoons CH_3NH_3^+ + OH^-$$

weaker acid than $CH_3NH_3^+$ — stronger base than CH_3NH_2 (conjugate pair)

2.8 Structure and Acidity

Removing a proton from an electrically neutral acid in a solvent requires breaking a bond to hydrogen and generates a negative charge on the resulting conjugate base. Thus, K_a values depend on both the strength of the H—A bond and the stability of A^- in the solvent.

The acidity of simple inorganic acids is related to the position in the periodic table of the atom bonded to hydrogen. The acidities of acids, HA, increase as we move down a column of the periodic table. For example, the acidities of the halogen acids increase in the order HF < HCl < HBr < HI. Similarly, for the same reasons, H_2O is also a weaker acid than H_2S. In a given row of the periodic table, acidity increases from left to right. The order of increasing acidity is $CH_4 < NH_3 < H_2O < HF$. This trend reflects the stability of the negative charge on the electronegative element of the conjugate base. That is, the order of increasing strength of conjugate bases is $F^- < OH^- < NH_2^- < CH_3^-$.

Many organic compounds are structurally related to inorganic acids and bases. As a consequence, we can predict the acid-base properties by making an appropriate comparison. For example, methanesulfonic acid has an O—H bond that is structurally similar to the O—H bond in sulfuric acid. Because sulfuric acid is a strong acid, it is reasonable to expect methanesulfonic acid to be a strong acid.

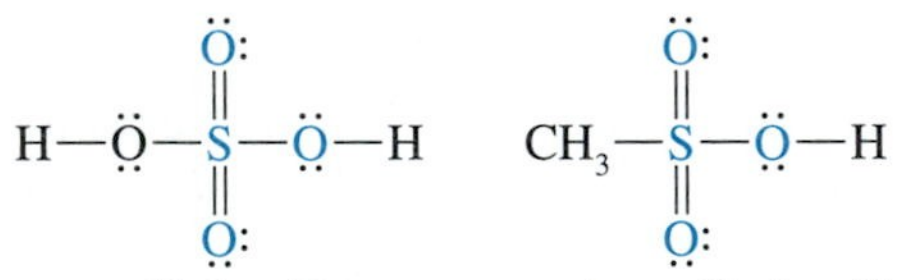

H—O—S(=O)(=O)—O—H — sulfuric acid

CH_3—S(=O)(=O)—O—H — methanesulfonic acid

Ethylamine, $CH_3CH_2NH_2$, is structurally related to ammonia, which is a weak base. Thus, ethylamine and other amines are weak bases. As we examine various functional groups in detail, we will find that the acid-base properties do vary somewhat with structure.

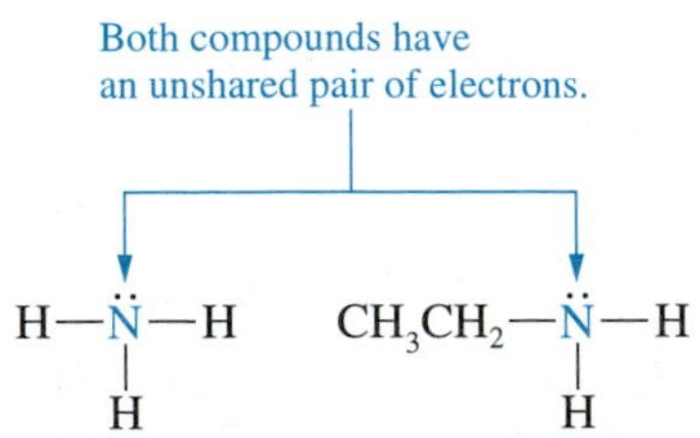

Both compounds have an unshared pair of electrons.

H—N(H)—H — ammonia

CH_3CH_2—N(H)—H — ethylamine

A reaction in which relatively unstable reactants are converted to more stable products has a large equilibrium constant. Thus, stabilizing the negative charge in the conjugate base formed from an acid increases K_a. When an anion produced by ionization of an acid has resonance stabilization, acid strength increases by a substantial amount. For example, both methanol and acetic acid ionize to form conjugate bases that have a negative charge on oxygen. However, acetic acid is about 10^{10} times more acidic than methanol.

$$CH_3OH + H_2O \rightleftharpoons CH_3O^- + H_3O^+ \qquad K_a = 10^{-16}$$
$$CH_3CO_2H + H_2O \rightleftharpoons CH_3CO_2^- + H_3O^+ \qquad K_a = 1.8 \times 10^{-5}$$

The greater acidity of acetic acid is the result of resonance stabilization of the negative charge in the conjugate base, acetate ion. In the methoxide ion (CH_3O^-) the negative charge is concentrated on a single oxygen atom.

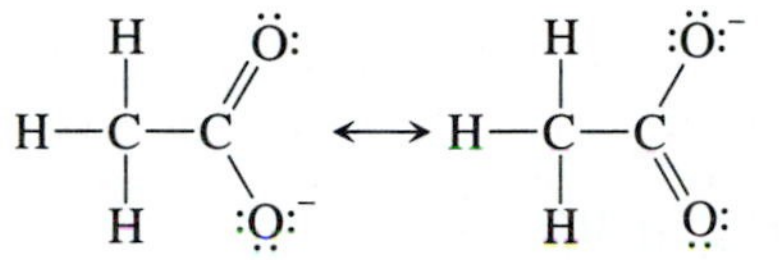

```
    H    Ö:              H   :Ö:⁻
    |   //               |   /
H—C—C        ⟷     H—C—C
    |   \                |   \\
    H   :Ö:⁻             H    Ö:
```

Acidity also reflects the ability of an atom to polarize neighboring bonds by an **inductive effect.** For example, chloroacetic acid is a stronger acid than acetic acid.

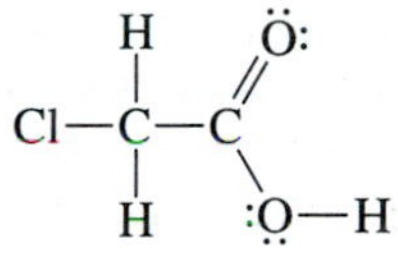

```
     H    Ö:              H    Ö:
     |   //               |   //
Cl—C—C               H—C—C
     |   \                |   \
     H   :Ö—H             H   :Ö—H
   pKa = 2.9            pKa = 4.7
```

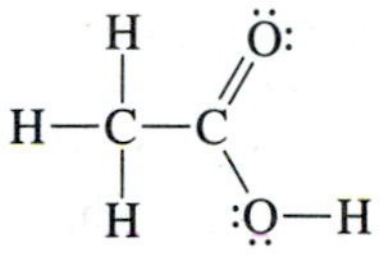

The electrons in the C—Cl bond are "pulled" toward the more electronegative chlorine atom and away from the carbon skeleton. As a result, the electrons of the oxygen atom are drawn away from the O—H bond, and the proton can therefore ionize more easily. This shifting of electrons through a network of sigma bonds is called **inductive withdrawal** of electrons.

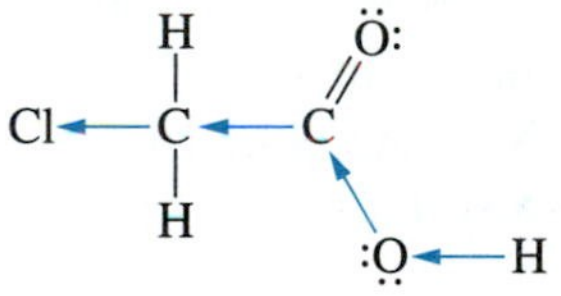

```
       H       Ö:
       |      //
Cl ←— C ←— C
       |      ↖
       H        :Ö ←— H
```

Any structural feature that withdraws electron density from the bond between hydrogen and another atom causes an increase in its acidity. Inductive effects as well as resonance effects play major roles in chemical reactions. We will employ these two concepts many times throughout this text.

EXAMPLE 2-6

The pK_a values of ethanol and 2,2,2-trifluoroethanol are 15.9 and 12.4, respectively. What is responsible for this difference?

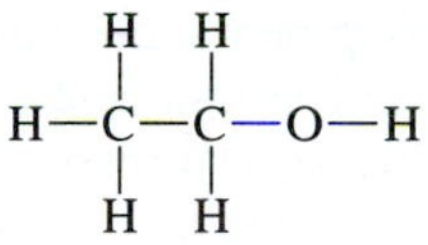

```
   H  H               F  H
   |  |               |  |
H—C—C—O—H        F—C—C—O—H
   |  |               |  |
   H  H               F  H
  ethanol      2,2,2-trifluoroethanol
```

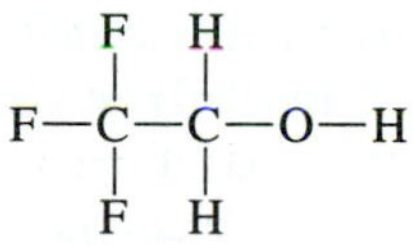

Solution

2,2,2-Trifluoroethanol is the stronger acid. The carbon atom with three fluorine atoms has a partial positive charge as a consequence of inductive electron withdrawal by the three polar covalent C—F bonds. This carbon atom, in turn, inductively attracts electrons from the other carbon atom and, indirectly, from the oxygen atom. Thus, the oxygen-hydrogen bond is more strongly polarized, and the compound is more acidic.

Problem 2.6

The pK_a of the C—H bond of nitromethane is 10.2, whereas the pK_a of methane is approximately 49. Explain why nitromethane is so much more acidic.

$$H_3C-N^{+}(=\ddot{O}:)(-:\ddot{O}:^{-})$$

2.9 Reaction Mechanisms

The description of the individual steps of a reaction, showing the order in which bonds are broken in the reactant and formed in the product, is called the **reaction mechanism.** Some reactions occur in a single step, and bonds form and break simultaneously. Such processes are **concerted reactions.** The reaction mechanism thus resembles that of an ordinary chemical equation.

$$A_{\text{(reactant)}} \longrightarrow B_{\text{(product)}}$$

Many reactions occur in a series of steps. For example, the conversion of reactant A into product B may involve two steps, in which an intermediate M is formed and then reacts.

$$A_{\text{(reactant)}} \xrightarrow{\text{step 1}} M_{\text{(intermediate)}}$$

$$M_{\text{(intermediate)}} \xrightarrow{\text{step 2}} B_{\text{(product)}}$$

The rate of the slowest step in a sequence of reactions is called the **rate-determining step** because the overall rate of conversion of reactant into product can occur no faster than this slowest step.

Types of Bond Cleavage and Formation

When a bond is broken so that one electron remains with each of the two fragments, the process is **homolytic** cleavage. Consider R—Y, where R usually represents the major portion of an organic molecule and Y is an atom or group of atoms that may or may not contain carbon atoms. Homolysis of the bond produces a **carbon radical.**

$$R-Y \longrightarrow \underset{\text{radical}}{R\cdot} + Y\cdot \quad \text{homolytic cleavage}$$

When a bond is broken so that one fragment gains both bonding electrons, the process is called **heterolytic** cleavage. The fragment produced by heterolysis that gains electrons has a negative charge. The second fragment is electron deficient and has a positive charge. If the bond to carbon breaks so that its electrons remain with the carbon atom, a negatively charged **carbanion** results. The carbanion has an octet of electrons around the carbon atom. If the bond breaks so that its electrons are lost by the carbon atom, a positively charged **carbocation** results. The carbocation has a sextet of electrons around the carbon atom and is an electron-deficient species.

$$\underset{}{R{-}Y} \longrightarrow \underset{\text{carbanion}}{R{:}^-} + Y^+ \qquad \text{heterolytic cleavage}$$

$$R{-}Y \longrightarrow \underset{\text{carbocation}}{R^+} + {:}Y^- \qquad \text{heterolytic cleavage}$$

The mode of heterolytic cleavage of a C—Y bond depends on the electronegativity of Y. If Y is a more electronegative element than carbon, such as a halogen atom, the carbon atom bears a partial positive charge, and the carbon bond tends to break heterolytically to form a carbocation. Conversely, if Y is a less electronegative element, such as a metal, the bond has the opposite polarity and tends to break heterolytically to form a carbanion.

$$\overset{\delta+}{R}{-}\overset{\delta-}{Br} \qquad \overset{\delta-}{R}{-}\overset{\delta+}{Li}$$

There are two ways to form two-electron covalent bonds from fragments. These processes are the reverse of the two cleavage reactions. Bond formation from fragments that each contain one electron is a **homogenic** process. Formation of a two-electron bond from oppositely charged fragments is a **heterogenic** process.

$$X\cdot + Y\cdot \longrightarrow X{-}Y \qquad \text{homogenic bond formation}$$

$$X^+ + {:}Y^- \longrightarrow X{-}Y \qquad \text{heterogenic bond formation}$$

Heterogenic reactions are more common than homogenic reactions in organic chemistry. In organic reactions, a carbocation behaves as an **electrophile** (electron-loving species). It seeks a negatively charged center to neutralize its positive charge and to obtain a stable octet of electrons. On the other hand, a carbanion has an electron pair that causes it to react as a **nucleophile** (nucleus-loving species). It seeks a positively charged center to neutralize its negative charge.

Many organic reactions can be depicted by the following equation, in which E^+ and Nu^- represent an electrophile and nucleophile, respectively.

$$\underset{\text{electrophile}}{E^+} + \underset{\text{nucleophile}}{{:}Nu^-} \longrightarrow E{-}Nu$$

The curved arrow notation shows the movement of a pair of electrons from the nucleophile to the electrophile. This notation is exactly like that used to show the reaction between a Lewis base and a Lewis acid.

Free Radical Substitution Reactions

Methane reacts with chlorine gas at elevated temperatures or in the presence of ultraviolet light as an energy source. In this reaction, a chlorine atom replaces a hydrogen atom.

$$CH_3{-}H + Cl_2 \longrightarrow CH_3{-}Cl + H{-}Cl$$

The mechanism of this reaction involves homolytic bond cleavage and homogenic bond formation. In the first step, a chlorine molecule absorbs either heat or light energy and the Cl—Cl bond is broken to give two chlorine atoms.

They are electron-deficient radicals and are highly reactive. This step starts the reaction and is called the **initiation step.**

Step 1 $\quad :\ddot{Cl}-\ddot{Cl}: \longrightarrow :\ddot{Cl}\cdot + \cdot\ddot{Cl}:$

Two steps, collectively known as **propagation steps,** that involve radicals as reactants and products, then occur.

Step 2 $\quad CH_3-H + \cdot\ddot{Cl}: \longrightarrow CH_3\cdot + H-\ddot{Cl}:$

Step 3 $\quad CH_3\cdot + :\ddot{Cl}-\ddot{Cl}: \longrightarrow CH_3-\ddot{Cl}: + :\ddot{Cl}\cdot$

In step 2, a C—H bond is broken and an H—Cl bond is produced; in step 3, a Cl—Cl bond is broken and a C—Cl bond is formed. Furthermore, in each step a radical reacts and a radical is produced. One radical generates another in this **chain propagation** sequence. The process continues as long as radicals and a supply of both reactants are present.

Nucleophilic Substitution Reactions

Reactions in which a nucleophile "attacks" a carbon atom and replaces another group are very common. The "leaving group" displaced from the carbon center is symbolized by L. A leaving group is invariably an electronegative atom or a group that can exist as a stable anion.

$$\underset{\text{nucleophile}}{Nu:^-} + R-L \longrightarrow R-Nu + \underset{\text{leaving group}}{:L^-}$$

Note that the nucleophile has an unshared pair of electrons that bonds to the carbon residue. Thus, bond formation is a heterogenic process. The leaving group departs with an electron pair, and cleavage of the bond between the leaving group and carbon is a heterolytic process.

An example of this type of nucleophilic substitution process is the reaction of chloromethane with hydroxide ion.

$$\underset{\text{nucleophile}}{H\ddot{O}:^-} + \overset{\delta+}{C}H_3-\overset{\delta-}{\ddot{Cl}:} \longrightarrow CH_3-\ddot{O}H + \underset{\text{leaving group}}{:\ddot{Cl}:^-}$$

In this reaction, the nucleophile approaches the carbon atom, which is made somewhat positive by the electronegative chlorine atom. The nucleophile has a nonbonding pair of electrons that begins to bond to the carbon atom. As the nucleophile approaches the carbon, the bond between carbon and the chloride ion, a leaving group, weakens. The entire process is **concerted**—that is, both bond breaking and bond formation occur simultaneously. The mechanism of nucleophilic substitution depends on many factors to be discussed in Chapter 7.

2.10 Reaction Rates

In a reaction, the reactant molecules collide with each other, some bonds rupture, and others form. The factors that affect the rate of a reaction are (1) the nature of the reactants, (2) the concentration of the reactants, (3) temperature, and (4) the presence of substances called catalysts.

The nature of the reactants is the most important feature controlling a chemical reaction. For example, in the addition reaction of ethylene (C_2H_4)

with HBr, a bond must be broken between the hydrogen and bromine atoms. Bonds must be formed between a carbon and hydrogen atom and between a carbon and bromine atom.

This bond is broken.

These bonds are formed.

$$H_2C{=}CH_2 + H{-}Br \longrightarrow H{-}CH_2{-}CH_2{-}Br$$

A double bond is converted into a single bond.

Although you may not be able to predict the order of reactivity, you should expect that the addition reaction of HCl to ethylene would occur at a rate different from that of the analogous reaction with HBr. The bonding that occurs now involves carbon and chlorine atoms, and the bond broken is between hydrogen and chlorine atoms. The energy requirements associated with the reorganization of these atoms and bonds must be different.

As the concentration of reactants is increased, the reaction velocity increases because reactant molecules are more likely to collide. The rates of chemical reactions increase with a rise in temperature because the reactant molecules collide more frequently and with greater energy. As a rule of thumb, the rate approximately doubles for a 10° rise in temperature.

A **catalyst** is a substance that increases a reaction rate. A catalyst is said to catalyze the reaction, and its effect is known as catalysis. Catalysts are usually required only in small amounts. The catalyst is present in the same amount before and after the reaction takes place. Although a catalyst increases the rate of a reaction, it does not change the equilibrium constant for the reaction.

The fires set in the Gulf war in Kuwait resulted from rapid reactions between petroleum and oxygen.

Reaction Rate Theory

Collisions between molecules that cause a chemical reaction are called **effective collisions,** and the minimum energy required for an effective collision is the **activation energy.** The activation energy for a given reaction depends on the type of bonds broken and formed in the reaction. During a reaction, the arrangement of the atoms changes as bonds are distorted and eventually broken while new bonds form. During this process, some repulsion occurs when reactant atoms move close together. This repulsion results from the proximity of the electrons surrounding each atom. During a reaction, each specific arrangement of atoms has an associated energy. All such arrangements of atoms have an energy higher than the initial energy of the reactants. The atomic arrangement lying along the lowest energy pathway connecting reactants and products whose structure has the maximum energy is the **transition state.**

The transition state in the nucleophilic substitution of hydroxide ion with chloromethane has both the hydroxide and chloride ions bonded to some degree to the carbon atom.

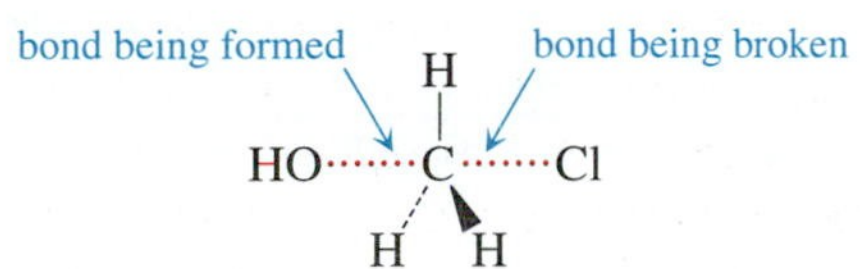

The carbon-chlorine bond breaks on one side of the transition state structure, while the carbon-oxygen bond forms on the other side. Transition states exist for periods of time as short as femtoseconds (10^{-15} s) during a reaction. However, the structure of the transition state for a reaction can be inferred from various kinds of experimental data.

Reaction Coordinate Diagrams

Reaction coordinate diagrams are used to represent the progress of a reaction. The vertical axis gives the total energy of the reacting system; the horizontal axis qualitatively represents the progress of a reaction from reactants (left) to the products (right). Figure 2.5 depicts an exothermic reaction ($\Delta H° < 0$). The difference between the energy of the reactants and the transition state is the activation energy ($E_a > 0$). In the transition state for the nucleophilic substitution of chloromethane by hydroxide ion, both hydroxide and chloride are partially bonded to the carbon atom.

A large activation energy results in a slow reaction, because only a small fraction of the molecules collide with sufficient energy to reach the transition state. At this point, energy is released as the reaction proceeds to form products. The energy released is equal to the activation energy originally added plus an amount equal to that characteristic for the exothermic reaction.

The kinetic energy of molecules increases with increasing temperature. As the kinetic energy increases, the chances increase for molecular collisions equal in energy to the activation energy, so the rate of reaction increases.

Some reactions occur in two or more steps, as in the case of the two-step addition reaction of HBr with ethylene.

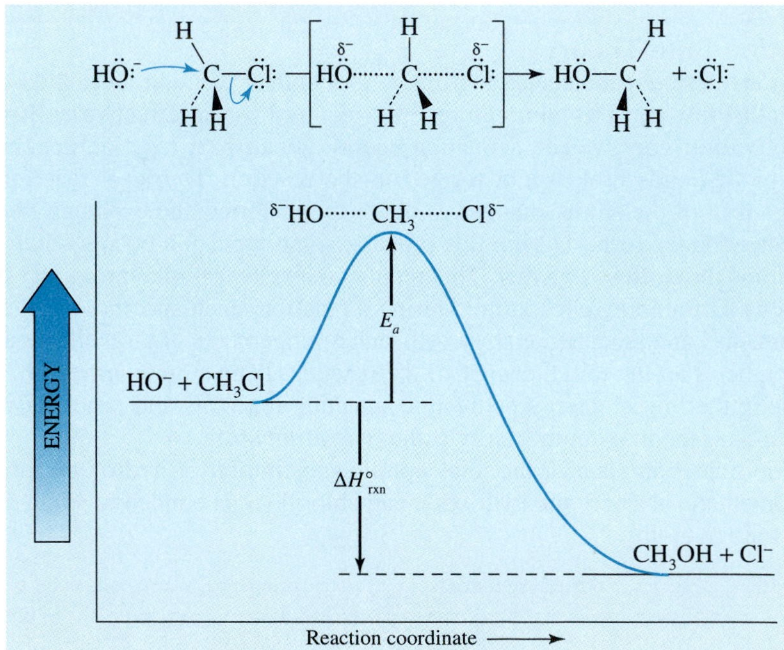

FIGURE 2.5
Reaction Coordinate Diagram for a Substitution Reaction

$$H_2C{=}CH_2 + H^+ \longrightarrow H_2\overset{+}{C}{-}CH_3$$

nucleophile electrophile carbocation

$$:\ddot{\underset{..}{Br}}:^- + H_2\overset{+}{C}{-}CH_3 \longrightarrow :\ddot{\underset{..}{Br}}{-}CH_2{-}CH_3$$

nucleophile electrophile bromoethane

In the first step, a proton acts an electrophile. It forms a bond to carbon using the π electrons of the double bond. As a consequence, an intermediate carbocation is formed. It then reacts in a second step with the bromide ion, which is a nucleophile. Each step is shown in the reaction coordinate diagram in Figure 2.6.

In the first transition state, a hydrogen ion begins to bond to carbon as π electrons are removed from the double bond. The energy then decreases until an intermediate carbocation is formed. In the second step, which has its own activation energy, the carbocation starts to bond to the nucleophilic bromide ion. Note that the energy of the carbocation is lower than the energy of the two transition states. Finally, as the carbon-bromine bond becomes fully formed, the reaction coordinate diagram shows that the energy of the product is lower than the energy of the reactants.

The Function of Catalysts

A catalyst provides a path for the progress of the reaction that is different from the path of the uncatalyzed reaction. The path starts at the same reactants and

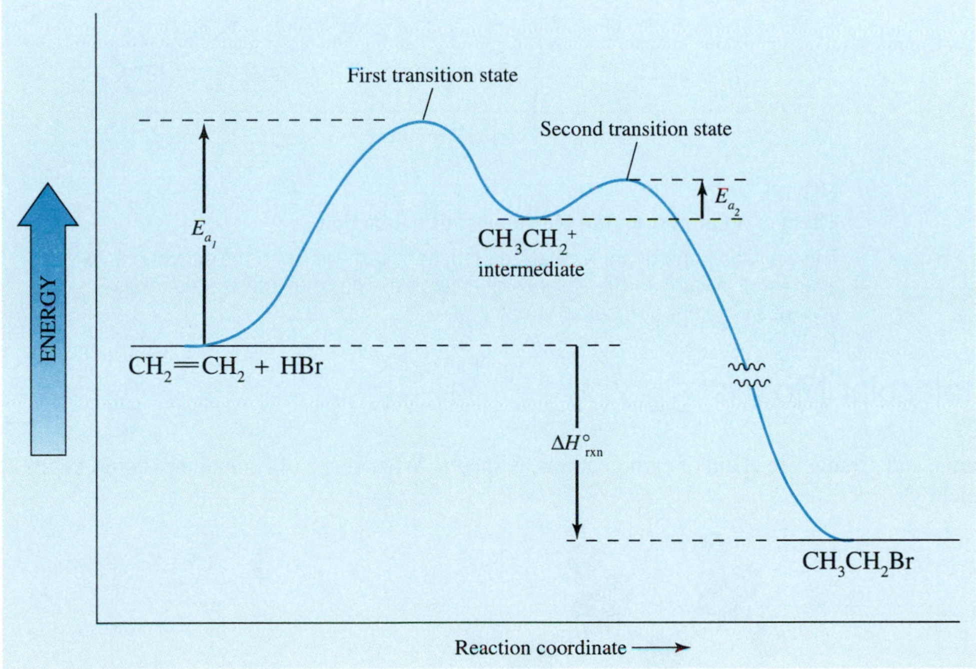

FIGURE 2.6
Reaction Mechanism for an Addition Reaction

concludes at the same products. However, the path for the catalyzed reaction has a different, lower activation energy (Figure 2.7).

To illustrate the effect of a catalyst on the path of a reaction, consider the hypothetical concerted reaction of A and B.

$$A + B \longrightarrow X$$

The activation energy required for this reaction is available in a small fraction of high-energy molecular collisions. However, in the presence of a catalyst, represented by C, the following reactions may occur:

Step 1: $A + C \longrightarrow A{-}C$

Step 2: $A{-}C + B \longrightarrow A{-}B + C$

The catalyst may combine with A in a reaction with a lower activation energy. Similarly, the reaction of A—C with B may have a low activation energy. If the activation energy of each step is low, a larger fraction of molecules will be able to react faster via this catalyzed pathway than could react without the catalyst at the same temperature.

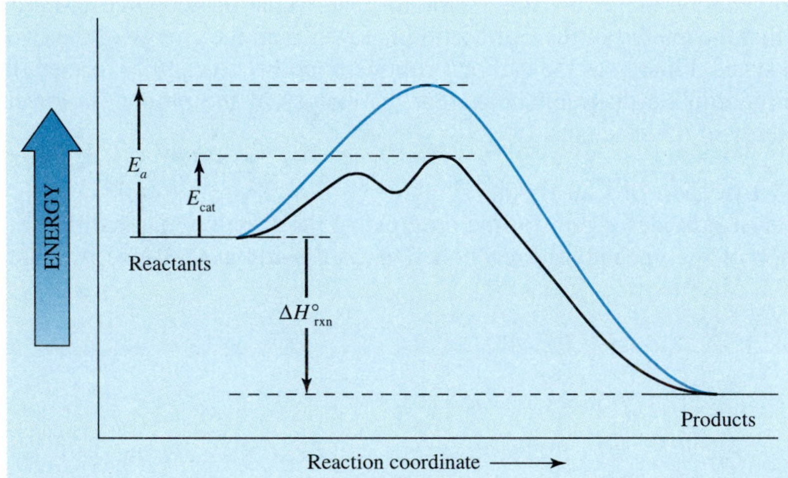

FIGURE 2.7
Effect of a Catalyst on the Mechanism of a Reaction
The activation energy for a catalyzed reaction is smaller than the activation energy for reaction in the absence of a catalyst. The catalyzed reaction may involve a different number of steps.

Explorations with Molecular Models

1. Construct a model for pentane and arrange the chain of carbon atoms as shown. What shape of "container" could be used to store the structure most efficiently?

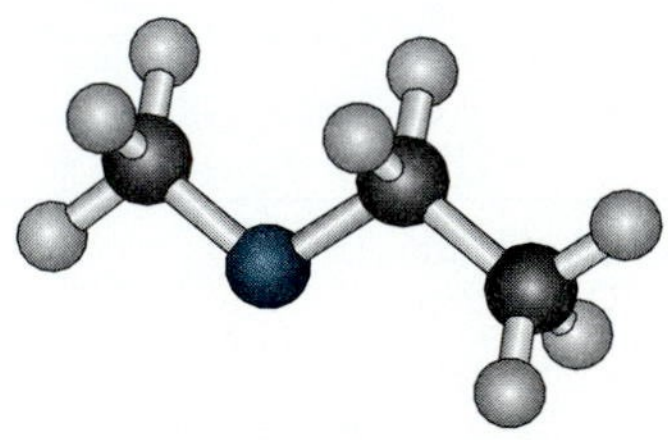

2. Construct a model for 2,2-dimethylpropane and arrange the chain of carbon atoms as shown. What shape of "container" could be used to store the structure most efficiently?

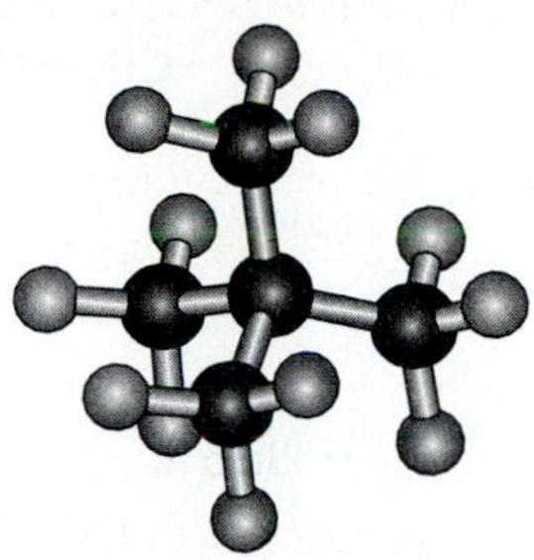

3. Some molecules can form intramolecular (within the molecule) hydrogen bonds. Construct a model for 1,3-propanediol based on the following structure. Rotate atoms about carbon-carbon bonds and arrange the structure so that an intramolecular hydrogen bond could be formed. If a structure contained a larger number of carbon atoms with an —OH group at the end of each chain, would the likelihood of forming an intramolecular hydrogen bond increase or decrease?

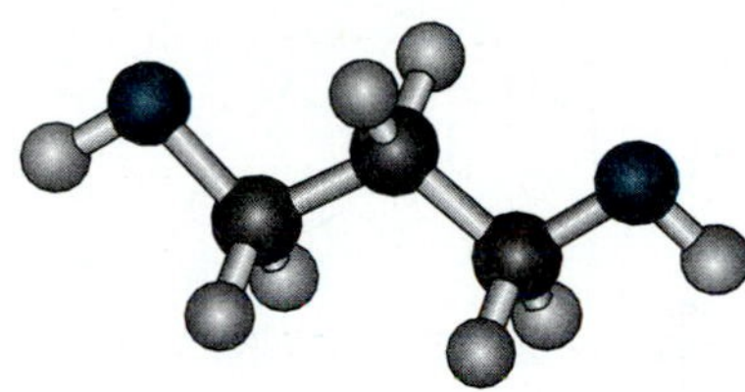

Exercises

Physical Properties

2.1 Suggest a reason for the difference in boiling points between the following pairs of isomeric compounds. (Several structural features may be responsible.)

(a)

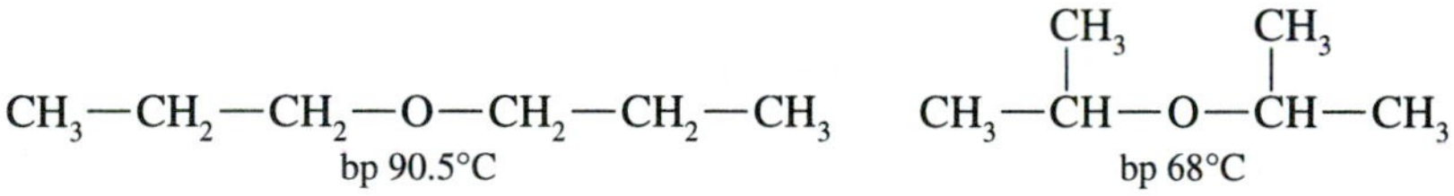

$CH_3—CH_2—CH_2—O—CH_2—CH_2—CH_3$
bp 90.5°C

$CH_3—CH(CH_3)—O—CH(CH_3)—CH_3$
bp 68°C

(b)

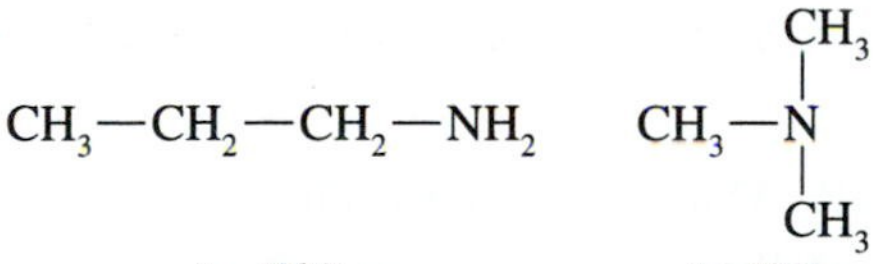

$CH_3—CH_2—CH_2—NH_2$
bp 49°C

$CH_3—N(CH_3)—CH_3$
bp 3°C

(c)

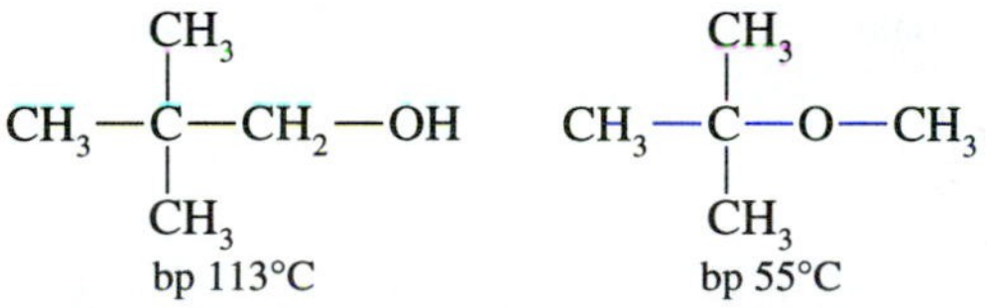

$CH_3—C(CH_3)_2—CH_2—OH$
bp 113°C

$CH_3—C(CH_3)_2—O—CH_3$
bp 55°C

2.2 The boiling points of the following pairs of isomeric compounds do not differ very much. Explain why.

(a) $CH_3—CH_2—CH_2—S—CH_3$ bp 95.5°C $CH_3—CH_2—S—CH_2—CH_3$ bp 92.1°C

(b) $CH_3—CH_2—CH_2—CH(Cl)—CH_3$ bp 96.9°C $CH_3—CH_2—CH(Cl)—CH_2—CH_3$ bp 97.8°C

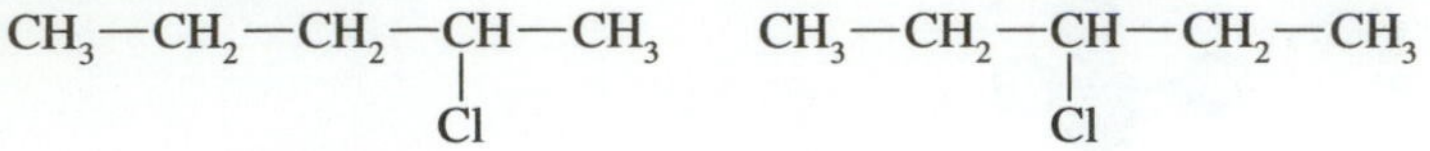

(c) $CH_3—CH_2—N(H)—CH_2—CH_3$ bp 56°C $CH_3—CH_2—CH_2—N(H)—CH_3$ bp 61°C

2.3 The ethene molecule is planar, and all bond angles are close to 120°. There are three isomeric dichloroethenes. Two isomers have dipole moments and the third does not. Which of the following three is nonpolar? Explain why.

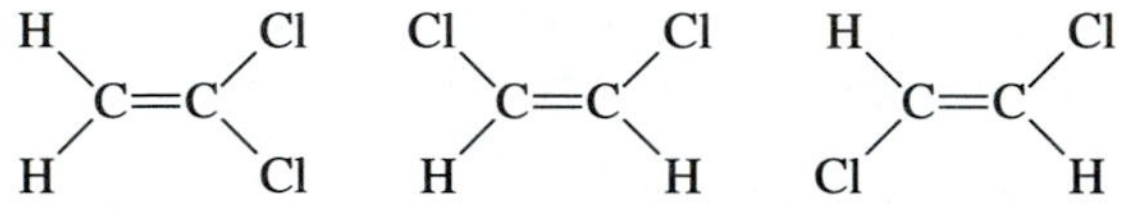

$H_2C{=}CCl_2$ (1,1-: H, H on one carbon; Cl, Cl on the other) $ClHC{=}CHCl$ (Cl upper left, H lower left; Cl upper right, H lower right) $HClC{=}CHCl$ (H upper left, Cl lower left; Cl upper right, H lower right)

2.4 One of the following compounds has a dipole moment and the other does not. Select the polar compound and explain why the other compound has no dipole moment.

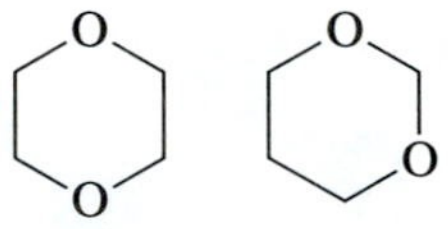

[1,4-dioxane ring]   [1,3-dioxane ring]

2.5 Propylene glycol is miscible with water, but the solubility of 1-butanol is only 7.9 g/100 mL of water. Explain why.

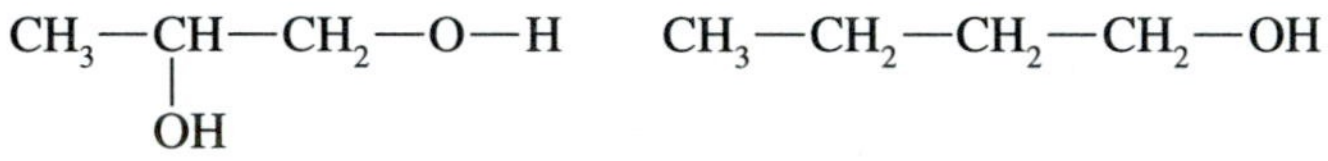

$CH_3—CH(OH)—CH_2—O—H$ propylene glycol $CH_3—CH_2—CH_2—CH_2—OH$ 1-butanol

2.6 Butanoic acid is miscible with water, but ethyl ethanoate is not. Explain why.

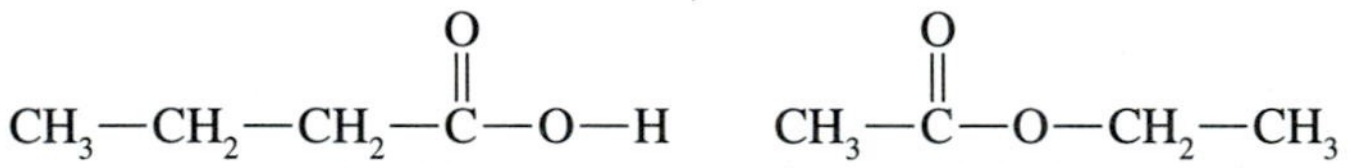

$CH_3—CH_2—CH_2—C({=}O)—O—H$ butanoic acid $CH_3—C({=}O)—O—CH_2—CH_3$ ethyl ethanoate

Acids and Bases

2.7 Write the structure of the conjugate acid of each of the following species.

(a) $CH_3—S—CH_3$ (b) $CH_3—O—CH_3$ (c) $CH_3—NH_2$ (d) $CH_3—OH$

2.8 Write the structure of the conjugate base of each of the following species.

(a) $CH_3—SH$ (b) $CH_3—NH_2$ (c) CH_3SO_3H (d) $HC{\equiv}CH$

2.9 Write the structure of the two conjugate acids of hydroxylamine, $NH_2—OH$. Which is the more acidic?

2.10 Write the structure of the two conjugate bases of hydroxylamine, $NH_2—OH$. Which is the more basic?

2.11 Identify the Lewis acid and Lewis base in each of the following reactions.

(a) $CH_3—CH_2—Cl + AlCl_3 \longrightarrow CH_3—CH_2^+ + AlCl_4^-$

(b) $CH_3—CH_2—SH + CH_3—O^- \longrightarrow CH_3—CH_2—S^- + CH_3—OH$

(c) $CH_3—CH_2—OH + NH_2^- \longrightarrow CH_3—CH_2—O^- + NH_3$

2.12 Identify the Lewis acid and Lewis base in each of the following reactions.

(a) $(CH_3)_2O + HBr \longrightarrow (CH_3)_2OH^+ + Br^-$
(b) $CH_3CH_2^+ + H_2O \longrightarrow CH_3CH_2OH_2^+$
(c) $CH_3CH{=}CH_2 + HBr \longrightarrow (CH_3)_2CH^+ + Br^-$

pK_a and Acid Strength

2.13 The approximate pK_a values of CH_4 and CH_3OH are 49 and 16, respectively. Which is the stronger acid? Will the equilibrium position of the following reaction lie to the left or right?

$$CH_4 + CH_3O^- \rightleftharpoons CH_3^- + CH_3OH$$

2.14 The approximate pK_a values of NH_3 and CH_3OH are 36 and 16, respectively. Which is the stronger acid? Will the equilibrium position of the following reaction lie to the left or right?

$$CH_3OH + NH_2^- \rightleftharpoons NH_3 + CH_3O^-$$

2.15 The pK_a of acetic acid ($CH_3{-}CO_2H$) is 4.8. Using the concept of inductive effects, explain why the carboxylic acid group of amoxicillin (pK_a = 2.4), a synthetic penicillin, is more acidic than acetic acid.

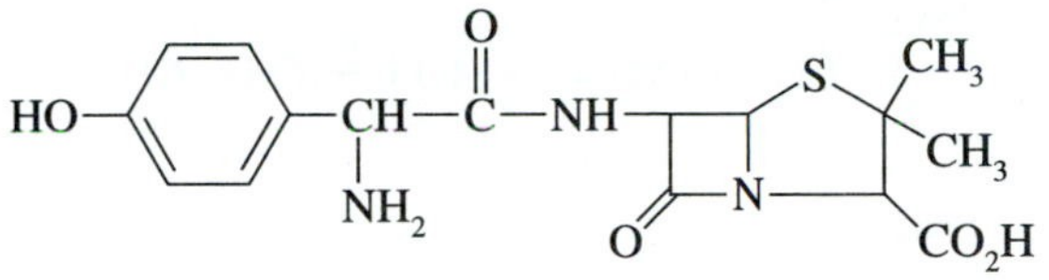

[Structure of amoxicillin: $HO{-}C_6H_4{-}CH(NH_2){-}C(=O){-}NH{-}$ β-lactam/thiazolidine ring system bearing S, N, CH_3, CH_3, CO_2H, and ring C=O]

2.16 The pK_a of the OH group of phenobarbital is 7.5, whereas the pK_a of CH_3OH is 16. Using the concept of resonance, explain why phenobarbital is significantly more acidic.

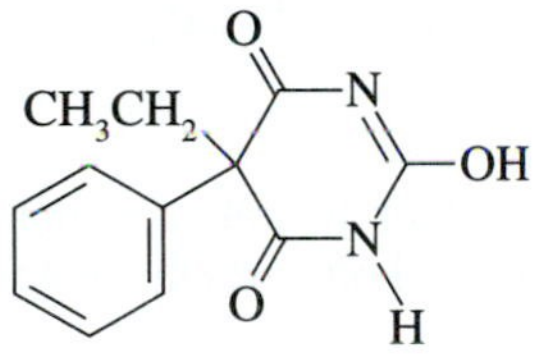

[Structure of phenobarbital (enol form): ring carbon bearing CH_3CH_2 and phenyl, two C=O groups, N=C(OH), and N–H]

Oxidation-Reduction Reactions

2.17 Determine whether each of the following transformations given by unbalanced equations involves oxidation, reduction, or neither.

(a) $CH_3{-}C{\equiv}N \longrightarrow CH_3{-}CH_2{-}NH_2$
(b) $2\ CH_3{-}SH \longrightarrow CH_3{-}S{-}S{-}CH_3$
(c) $CH_3{-}S{-}CH_3 \longrightarrow CH_3{-}S(=O){-}CH_3$

2.18 None of the following reactions involves oxidation or reduction, although they may appear to be redox reactions. Explain why.

(a) $CH_3{-}CH{=}CH_2 \longrightarrow CH_3{-}CH(OH){-}CH_3$
(b) $CH_3{-}C{\equiv}CH \longrightarrow CH_3{-}C(=O){-}CH_3$
(c) $CH_3{-}C(=NH){-}CH_3 \longrightarrow CH_3{-}C(=O){-}CH_3$

2.19 Consider each of the following reactions for the metabolism of drugs. What type of reaction occurs?

(a) tolmetin, an anti-inflammatory drug

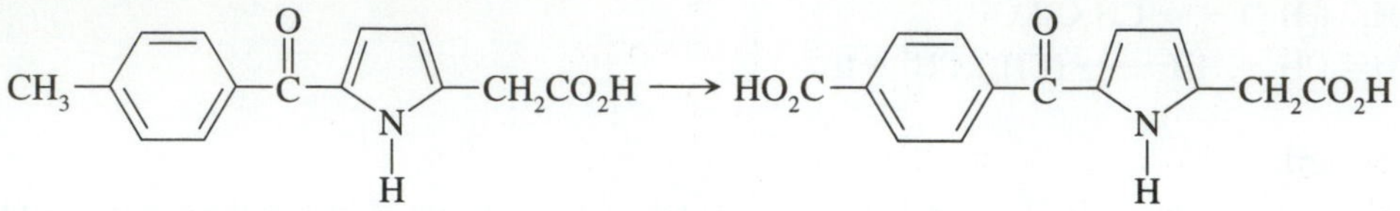

CH_3—(benzene ring)—C(=O)—(pyrrole ring, N—H)—CH_2CO_2H ⟶ HO_2C—(benzene ring)—C(=O)—(pyrrole ring, N—H)—CH_2CO_2H

(b) dantrolene, a muscle relaxant

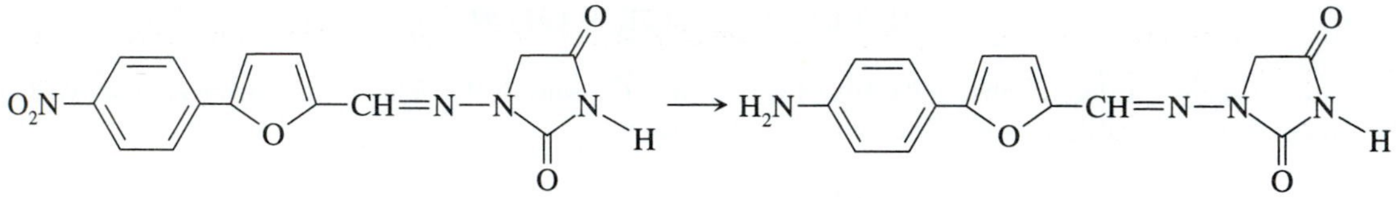

O_2N—(benzene ring)—(furan ring, O)—CH=N—N(imidazolidinedione ring, O, O, N—H) ⟶ H_2N—(benzene ring)—(furan ring, O)—CH=N—N(imidazolidinedione ring, O, O, N—H)

2.20 Consider each of the following reactions for the metabolism of drugs. What type of reaction occurs?

(a) ibuprofen, an analgesic

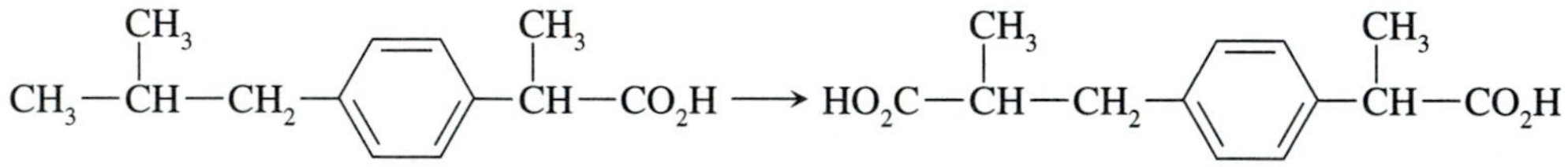

$CH_3—CH(CH_3)—CH_2$—(benzene ring)—$CH(CH_3)—CO_2H$ ⟶ $HO_2C—CH(CH_3)—CH_2$—(benzene ring)—$CH(CH_3)—CO_2H$

(b) Disulfiram, an drug used in treating alcoholism

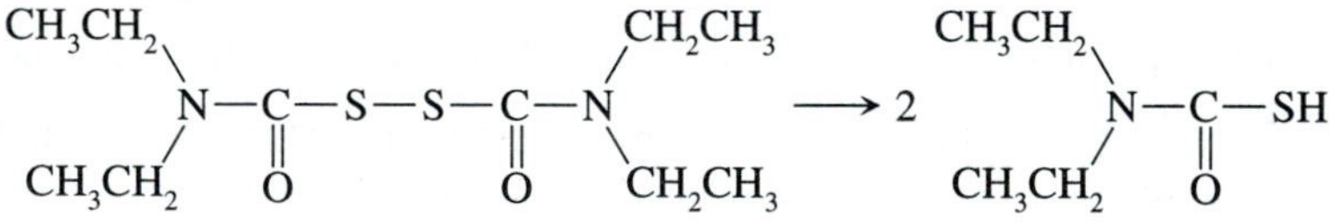

$(CH_3CH_2)_2N—C(=O)—S—S—C(=O)—N(CH_2CH_3)_2 \longrightarrow 2\ (CH_3CH_2)_2N—C(=O)—SH$

Types of Organic Reactions

2.21 Classify the type of reaction represented by each of the following unbalanced equations. Identify any additional reagents required for the reaction or any additional products that are formed.

(a) $CH_3—C(=O)—CH_3 \longrightarrow CH_3—C(OH)(OCH_3)—CH_3$

(b) OH ⟶

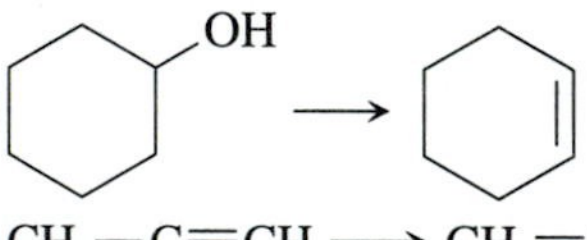

(c) $CH_3—C{\equiv}CH \longrightarrow CH_2{=}C{=}CH_2$

2.22 Classify the type of reaction represented by each of the following unbalanced equations. Identify any additional reagents required for the reaction or any additional products formed.

(a) $2\ CH_3—CH_2—OH \longrightarrow CH_3—CH_2—O—CH_2—CH_3$

(b) $CH_3—C(=O)—S—CH_3 \longrightarrow CH_3—C(=O)—O—H + CH_3SH$

(c) (cyclohexene ring) ⟶ (cyclohexane ring)—OCH_3

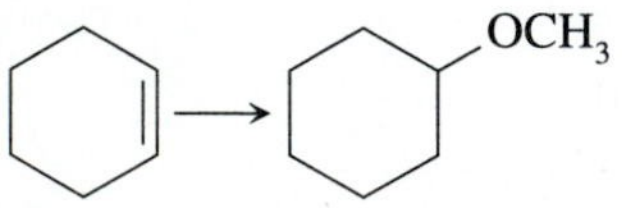

2.23 The metabolism of fatty acids (long-chain carboxylic acids) involves several steps. Indicate the type of reaction involved in each step. (The R represents a chain of carbon atoms. The CoA represents coenzyme A.)

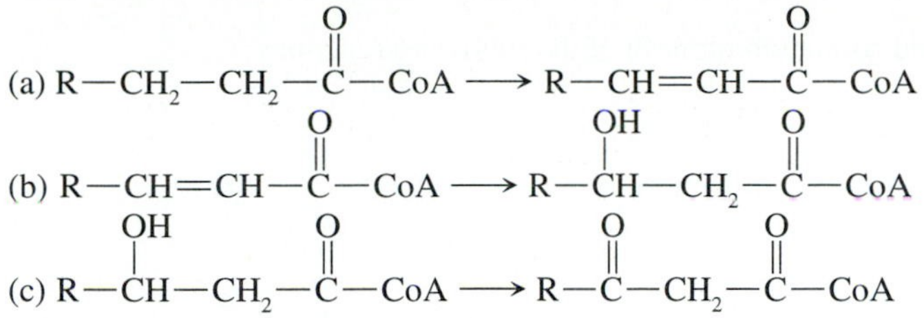

(a) $R-CH_2-CH_2-C(=O)-CoA \longrightarrow R-CH=CH-C(=O)-CoA$

(b) $R-CH=CH-C(=O)-CoA \longrightarrow R-CH(OH)-CH_2-C(=O)-CoA$

(c) $R-CH(OH)-CH_2-C(=O)-CoA \longrightarrow R-C(=O)-CH_2-C(=O)-CoA$

2.24 A series of 10 steps involved in glycolysis (metabolism of glucose) includes the following three steps. Indicate the type of reaction involved in each step.

(a) $(CH_2OH)(CH_2OPO_3^{2-})C=O \longrightarrow H-C(CH_2OH)(CH_2OPO_3^{2-})-OH$

(b) $H-C(CO_2^-)(CH_2OPO_3^{2-})-OH \longrightarrow H-C(CO_2^-)(CH_2OH)-OPO_3^{2-}$

(c) $H-C(CO_2^-)(CH_2OH)-OPO_3^{2-} \longrightarrow (CO_2^-)C(=CH_2)-OPO_3^{2-}$

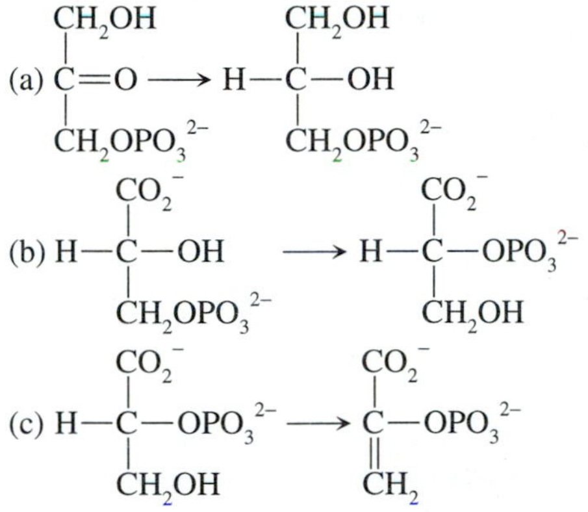

2.25 Chloroform is metabolized via an intermediate to phosgene, a compound that causes liver damage. What type of reactions are involved in the formation and decomposition of the intermediate?

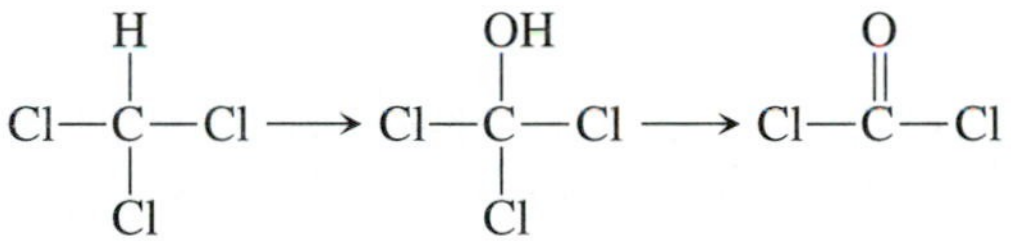

$Cl-CH(Cl)-Cl \longrightarrow Cl-C(OH)(Cl)-Cl \longrightarrow Cl-C(=O)-Cl$

chloroform phosgene

2.26 The sedative-hypnotic chloral hydrate is metabolized as follows. What type of reaction occurs in each step?

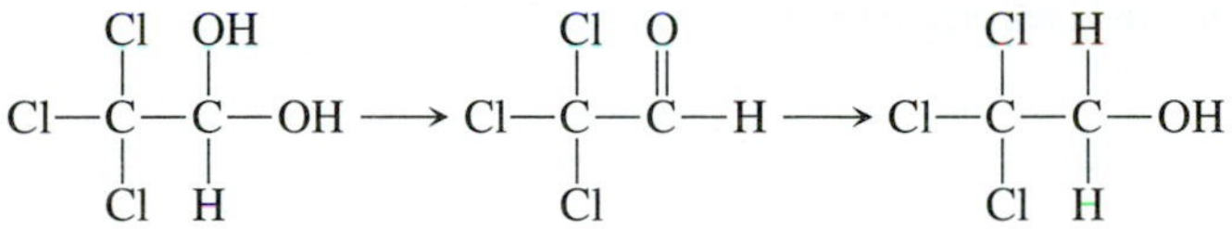

$Cl-C(Cl)_2-CH(OH)-OH \longrightarrow Cl-C(Cl)_2-C(=O)-H \longrightarrow Cl-C(Cl)_2-CH_2-OH$

Equilibria and Rates of Reactions

2.27 A reaction has $K = 1 \times 10^{-5}$. Are the products more or less stable than the reactants?

2.28 Could a reaction have $K = 1$? What relationship would exist between the energies of the reactants and products as shown in the reaction progress diagram?

2.29 Consider the following information about two reactions. Which reaction will occur at the faster rate at a common temperature?

reaction	$\Delta H°$	E_a
$A \longrightarrow X$	–30 kcal/mole	+25 kcal/mole
$B \longrightarrow Y$	–25 kcal/mole	+30 kcal/mole

2.30 Consider the information given in Exercise 2.29. Which reaction is more exothermic?

Reaction Rates and Mechanisms

2.31 Identify the processes of bond cleavage and bond formation for each of the following reactions.

(a) $H{-}\overset{\displaystyle H}{\underset{\displaystyle H}{\overset{|}{\underset{|}{C}}}}{-}H + \cdot\ddot{\underset{..}{Br}}: \longrightarrow H{-}\overset{\displaystyle H}{\underset{\displaystyle H}{\overset{|}{\underset{|}{C}}}}\cdot + H{-}\ddot{\underset{..}{Br}}:$

(b) $H{-}\overset{\displaystyle H}{\underset{\displaystyle H}{\overset{|}{\underset{|}{C}}}}\cdot + :\ddot{\underset{..}{Br}}{-}\ddot{\underset{..}{Br}}: \longrightarrow H{-}\overset{\displaystyle H}{\underset{\displaystyle H}{\overset{|}{\underset{|}{C}}}}{-}\ddot{\underset{..}{Br}}: + \cdot\ddot{\underset{..}{Br}}:$

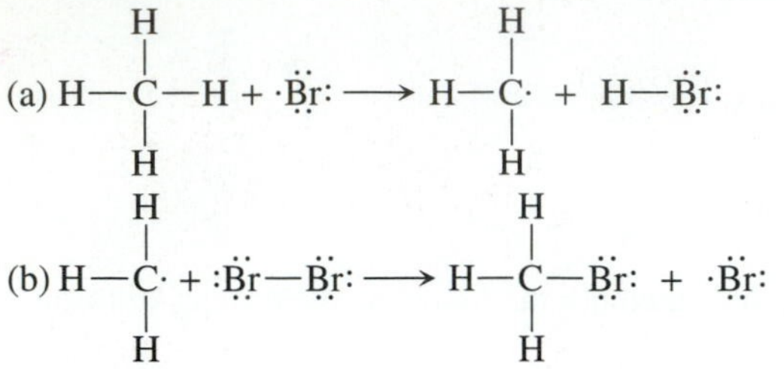

2.32 Identify the processes of bond cleavage and bond formation for each of the following reactions.

(a) $CH_3{-}\overset{\displaystyle CH_3}{\underset{\displaystyle CH_3}{\overset{|}{\underset{|}{C^+}}}} + OH^- \longrightarrow CH_3{-}\overset{\displaystyle CH_3}{\underset{\displaystyle CH_3}{\overset{|}{\underset{|}{C}}}}{-}OH$

(b) $CH_3{-}\overset{\displaystyle CH_3}{\underset{\displaystyle CH_3}{\overset{|}{\underset{|}{C}}}}{-}Cl \longrightarrow CH_3{-}\overset{\displaystyle CH_3}{\underset{\displaystyle CH_3}{\overset{|}{\underset{|}{C^+}}}} + Cl^-$

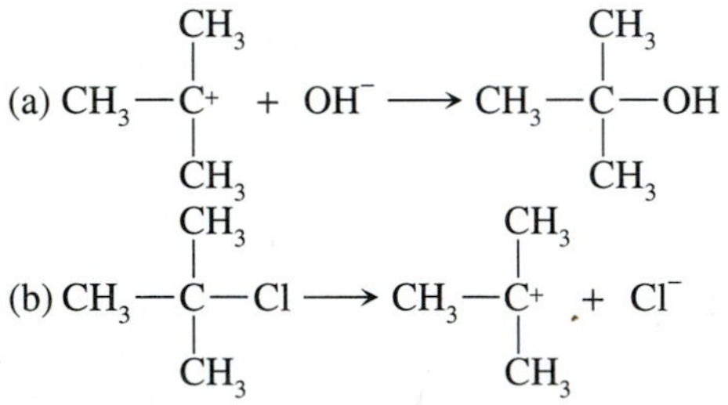

2.33 Benzoyl peroxide is used in creams to control acne. It is an irritant that causes proliferation of epithelial cells. It undergoes a homolytic cleavage of the oxygen-oxygen bond. Write the structure of the product, indicating all of the electrons present on the oxygen atom. What type of reactions might the product initiate?

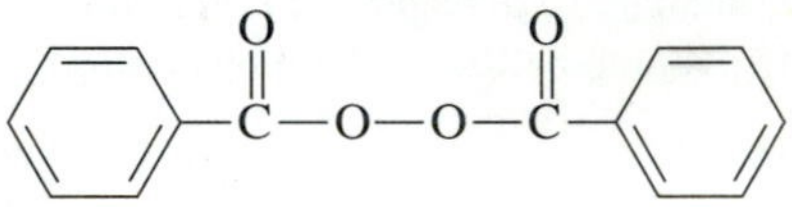

$C_6H_5{-}\overset{\displaystyle O}{\overset{\|}{C}}{-}O{-}O{-}\overset{\displaystyle O}{\overset{\|}{C}}{-}C_6H_5$

2.34 The oxygen-chlorine bond of hypochlorites, such as $CH_3{-}O{-}Cl$, can cleave heterolytically. Based on the electronegativity values of chlorine and oxygen, predict the charges on the cleavage products.

2.35 Hydrogen peroxide (H—O—O—H) reacts with a proton to give a conjugate acid that undergoes heterolytic oxygen-oxygen bond cleavage to yield water. What is the second product?

2.36 Chloromethane ($CH_3{-}Cl$) reacts with the Lewis acid $AlCl_3$ to give $AlCl_4^-$ and a carbon intermediate. What is the intermediate? What type of bond cleavage occurred to form it?

CHAPTER 3

ALKANES AND CYCLOALKANES

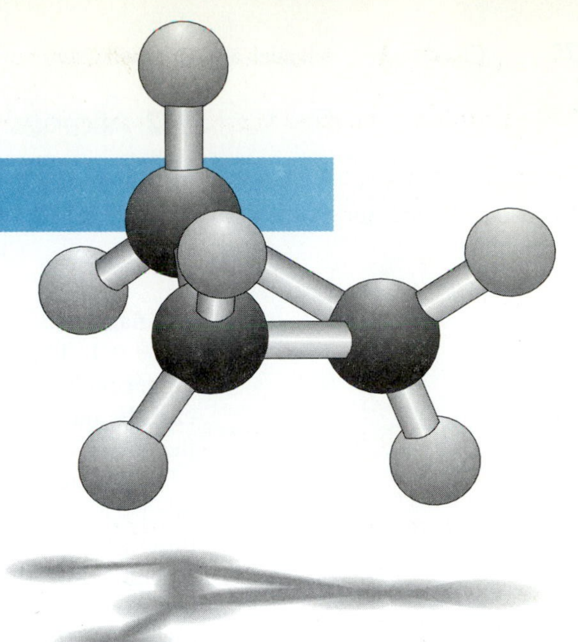

Overview

3.1 Classes of Hydrocarbons

Hydrocarbons are compounds that contain only hydrogen and carbon. They occur as mixtures in natural gas, petroleum, and coal, which are collectively known as fossil fuels. Hydrocarbons fall into two broad classes based on the types of bonds between the carbon atoms. A hydrocarbon that has only carbon-carbon single bonds is **saturated.** Hydrocarbons that contain carbon-carbon multiple bonds are **unsaturated.** Alkanes and cycloalkanes are two types of saturated hydrocarbons. **Alkanes** have carbon atoms bonded in chains; **cycloalkanes** have carbon atoms bonded to form a ring.

$CH_3-CH_2-CH_2-CH_3$

butane
(an alkane)

$$\begin{array}{c} CH_2-CH_2 \\ | \quad\quad | \\ CH_2-CH_2 \end{array}$$

cyclobutane
(a cycloalkane)

Compounds that have a chain of carbon atoms, some of which are attached to functional groups, are called **acyclic** compounds, meaning "not cyclic". Compounds that contain rings of carbon atoms, and that may also contain functional groups, are **carbocyclic** compounds, commonly called cyclic compounds. Some cyclic compounds contain at least one atom in the ring that is not a carbon atom; those atoms are called **heteroatoms.** Cyclic compounds containing one or more heteroatoms are called **heterocyclic** compounds.

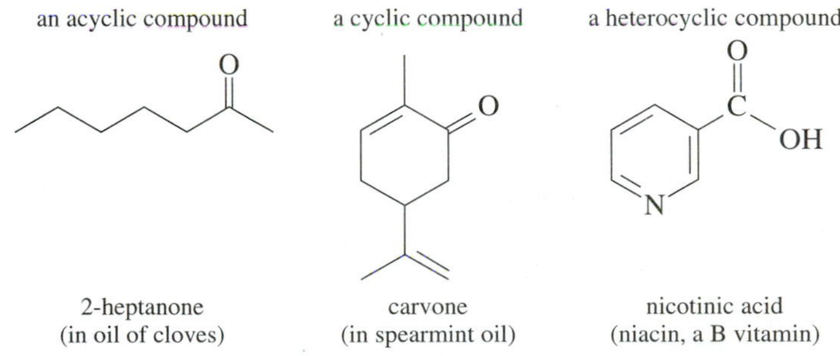

3.2 Alkanes

Saturated hydrocarbons with a continuous chain of carbon atoms are **normal alkanes.** Their structures are drawn with the carbon chain in a horizontal line.

$$CH_3{-}CH_2{-}CH_2{-}CH_2{-}CH_2{-}CH_2{-}CH_2{-}CH_3$$

octane (a normal alkane)

The names and condensed structural formulas of 20 normal alkanes are given in Table 3.1. The first four compounds have common names. The names of the higher molecular weight compounds are derived from Greek numbers that indicate the number of carbon atoms. Each name has the suffix *-ane,* which identifies the compound as an alkane.

Saturated hydrocarbons that have carbon atoms bonded to more than two other carbon atoms are called **branched alkanes.** The carbon atom bonded to three or four other carbon atoms is the branching point. The carbon atom attached to the chain of carbon atoms at the branching point is part of an **alkyl group.** The branched alkane isobutane has three carbon atoms in the main chain and one branch, a $-CH_3$ group.

$$CH_3{-}CH_2{-}CH_2{-}CH_3 \qquad CH_3{-}\underset{\displaystyle CH_3}{\underset{|}{CH}}{-}CH_3$$

butane isobutane

Both normal and branched alkanes have the general molecular formula C_nH_{2n+2}. For example, the molecular formula of hexane is C_6H_{14}.

$$\begin{array}{c} C_nH_{2n} \\ \begin{array}{ccccccccccccc} & & H & & H & & H & & H & & H & & H \\ & & | & & | & & | & & | & & | & & | \\ H & - & C & - & C & - & C & - & C & - & C & - & C & - H \\ & & | & & | & & | & & | & & | & & | \\ & & H & & H & & H & & H & & H & & H \end{array} \end{array}$$

$$H_1 + C_6H_{12} + H_1 = C_6H_{14}$$

Each carbon atom in this normal alkane, where $n = 6$, has at least two hydrogen atoms bonded to it, which accounts for the $2n$ in the general formula. Each of the two terminal carbon atoms has another hydrogen atom bonded to it, which accounts for the $+2$ in the subscript on hydrogen in the general formula.

Table 3.1 Names of Normal Alkanes

Number of Carbon Atoms	Names	Molecular Formula	Number of Carbon Atoms	Names	Molecular Formula
1	methane	CH_4	11	undecane	$C_{11}H_{24}$
2	ethane	C_2H_6	12	dodecane	$C_{12}H_{26}$
3	propane	C_3H_8	13	tridecane	$C_{13}H_{28}$
4	butane	C_4H_{10}	14	tetradecane	$C_{14}H_{30}$
5	pentane	C_5H_{12}	15	pentadecane	$C_{15}H_{32}$
6	hexane	C_6H_{14}	16	hexadecane	$C_{16}H_{34}$
7	heptane	C_7H_{16}	17	heptadecane	$C_{17}H_{36}$
8	octane	C_8H_{18}	18	octadecane	$C_{18}H_{38}$
9	nonane	C_9H_{20}	19	nonadecane	$C_{19}H_{40}$
10	decane	$C_{10}H_{22}$	20	eicosane	$C_{20}H_{42}$

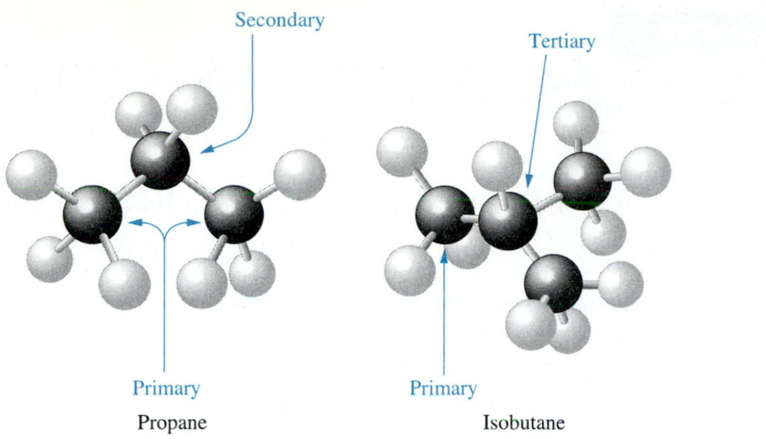

FIGURE 3.1
Classification of Carbon Atoms

Propane is a storage fuel used in some rural areas.

The molecular formulas of the alkanes in the series differ from one another in the number of $—CH_2—$ units. A series of compounds whose members differ from adjacent members by a repeating unit is called a **homologous series.**

Classification of Carbon Atoms

Hydrocarbon structures are classified according to the number of carbon atoms directly bonded to a specific carbon atom. This classification is used in later chapters to describe the reactivity of functional groups attached at the various carbon atoms in a structure.

A carbon atom bonded to only one other carbon atom is called a **primary carbon atom.** A primary carbon atom is designated by the symbol 1°. The carbon atom at each end of a carbon chain is primary. For example, propane has two primary carbon atoms. In contrast, the middle carbon atom in propane is not primary because it is bonded to two other carbon atoms (Figure 3.1).

A carbon atom that is bonded to two other carbon atoms is a **secondary carbon atom,** designated by the symbol 2°. For example, the middle carbon atom of propane is secondary. A **tertiary carbon atom** is bonded to three other carbon atoms and is designated by 3°. For example, when we examine the structure of isobutane, we see that one of the four carbon atoms is tertiary; the other three are primary (Figure 3.1). A **quaternary carbon atom** (4°) is bonded to four other carbon atoms.

EXAMPLE 3-1

One of the components of the wax of a cabbage leaf is a normal alkane containing 29 carbon atoms. What is the molecular formula of the compound?

Solution

The value of n is 29. There must be $(2 \times 29) + 2$ hydrogen atoms. The molecular formula is $C_{29}H_{60}$.

Problem 3.1

Hectane is a normal alkane with 100 carbon atoms. What is the molecular formula of hectane?

EXAMPLE 3-2

The following compound is a sex attractant released by the female tiger moth. Classify the carbon atoms in this compound as primary, secondary, or tertiary.

$$\begin{array}{l}\quad CH_3 \\ \quad | \\ CH_3CHCH_2CH_2CH_2CH_2CH_2CH_2CH_2CH_2CH_2CH_2CH_2CH_2CH_2CH_2CH_3\end{array}$$

Solution

Each of the two terminal carbon atoms and the branching $—CH_3$ group are primary carbon atoms, because each is bonded to only one other carbon atom. The second carbon atom from the left is bonded to two atoms in the chain as well as to the branching $—CH_3$ group, so it is tertiary. All 14 remaining carbon atoms are bonded to two carbon atoms, so they are secondary.

Problem 3.2

Pentaerythritol tetranitrate is used to reduce the frequency and severity of angina attacks. Classify the carbon atoms in this compound.

$$\begin{array}{c} CH_2—ONO_2 \\ | \\ O_2NO—CH_2—C—CH_2—ONO_2 \\ | \\ CH_2—ONO_2 \end{array}$$

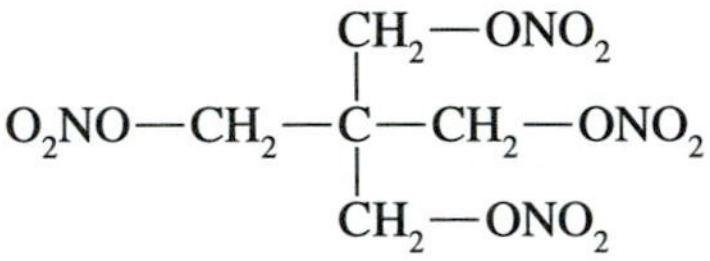

3.3 Nomenclature of Alkanes

Alkanes are named by the rules set forth by the International Union of Pure and Applied Chemistry (IUPAC). When these rules are followed, a unique name describes each compound. The IUPAC name is constructed of three parts: prefix, parent, and suffix.

prefix — parent — suffix

The **parent** is the longest continuous carbon chain in a molecule. A parent alkane has the ending *-ane*. Other suffixes identify functional groups, such as *-ol* for a hydroxyl group. Some functional groups, such as the halogens, are identified in the prefix. For example, the prefixes *chloro* and *bromo* identify chlorine and bromine, respectively.

The prefix also indicates the identity and location of any branching alkyl groups and some functional groups on the parent chain. An alkane that has "lost" one hydrogen atom is called an **alkyl** group. Alkyl groups are named by replacing the *-ane* ending of an alkane with *-yl*. The parent name of CH_4 is methane. Thus, $CH_3—$ is a methyl group. The parent name of C_2H_6 is ethane, so $CH_3CH_2—$ is an ethyl group.

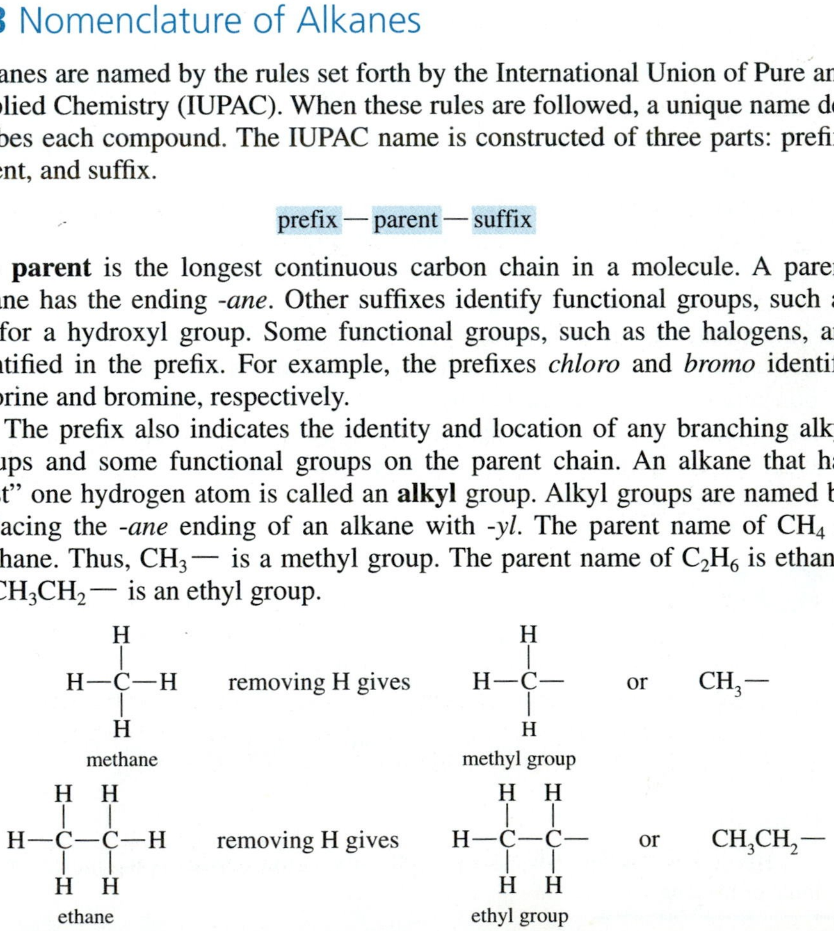

$$\underset{\text{methane}}{\begin{array}{c} H \\ | \\ H—C—H \\ | \\ H \end{array}} \quad \text{removing H gives} \quad \underset{\text{methyl group}}{\begin{array}{c} H \\ | \\ H—C— \\ | \\ H \end{array}} \quad \text{or} \quad CH_3—$$

$$\underset{\text{ethane}}{\begin{array}{c} H\;\;H \\ |\;\;\;| \\ H—C—C—H \\ |\;\;\;| \\ H\;\;H \end{array}} \quad \text{removing H gives} \quad \underset{\text{ethyl group}}{\begin{array}{c} H\;\;H \\ |\;\;\;| \\ H—C—C— \\ |\;\;\;| \\ H\;\;H \end{array}} \quad \text{or} \quad CH_3CH_2—$$

The general shorthand representation of an alkyl group is R—, which stands for the "rest" or "remainder" of the molecule.

The names of alkanes specify the length of the carbon chain and the location and identity of alkyl groups attached to it. The IUPAC rules for naming alkanes are as follows.

1. The longest continuous chain of carbon atoms is the parent. Alkane names are listed in Table 3.1. The longest chain may not always be immediately apparent.

$$\begin{array}{l} \qquad\qquad\quad CH_2-CH_3 \\ \qquad\qquad\quad | \\ CH_3-CH_2-CH-CH_3 \end{array}$$

There are five carbon atoms in the longest carbon chain — not four.

If two possible chains have the same number of carbon atoms, the parent is the one with the larger number of branch points.

$$\begin{array}{l} \qquad\qquad\quad CH_2-CH_3 \\ \qquad\qquad\quad | \\ CH_3-CH-CH-CH_2-CH_2-CH_3 \\ \qquad\quad | \\ \qquad\quad CH_3 \end{array}$$

The compound should be considered a six-carbon parent chain with two branches, a methyl group and an ethyl group.

$$\begin{array}{l} \qquad\qquad\quad CH_2-CH_3 \\ \qquad\qquad\quad | \\ CH_3-CH-CH-CH_2-CH_2-CH_3 \\ \qquad\quad | \\ \qquad\quad CH_3 \end{array}$$

A six-carbon parent chain with only a single three-carbon alkyl group is not a correct choice.

2. Number the carbon atoms in the longest continuous chain starting from the end of the chain nearer the first branch.

$$\begin{array}{l} \qquad\qquad\qquad\qquad CH_2-CH_3 \\ 1 \qquad\quad 2 \qquad\qquad\qquad | \\ CH_3-CH-CH_2-CH-CH_2-CH_3 \\ \qquad\quad | \qquad 3 \qquad\quad 4 \qquad 5 \qquad 6 \\ \qquad\quad CH_3 \end{array}$$

This substituted hexane has a methyl group at C-2 and an ethyl group at C-4, not an ethyl group at C-3 and a methyl group at C-5.

If the first branch occurs at an equal distance from each end of the chain, number from the end that is nearer the second branch.

$$\begin{array}{l} \qquad\qquad\qquad\qquad CH_3-CH_2 \qquad\qquad CH_3 \\ 8 \qquad\quad 7 \qquad\quad 6 \qquad\qquad\qquad | \qquad\qquad\qquad | \\ CH_3-CH-CH_2-CH_2-CH-CH_2-CH-CH_3 \\ \qquad\quad | \qquad\qquad\qquad 5 \qquad\quad 4 \qquad\quad 3 \qquad\quad 2 \qquad 1 \\ \qquad\quad CH_3 \end{array}$$

The ethyl group is closest to the right side of the molecule.

3. Each branch or substituent has a number that indicates its location on the parent chain. When two substituents are located on the same carbon atom, each must be assigned the same number.

$$\begin{array}{l} \qquad\qquad\qquad\qquad CH_3-CH_2 \qquad\quad CH_3 \\ 8 \qquad\quad 7 \qquad\quad 6 \qquad\qquad\qquad | \qquad\qquad\quad | \\ CH_3-CH_2-CH-CH_2-C-CH_2-CH-CH_3 \\ \qquad\qquad\qquad | \qquad 5 \qquad 4 | \qquad 3 \qquad\quad 2 \qquad 1 \\ \qquad\qquad\quad CH_3 \qquad\quad CH_3 \end{array}$$

This octane has methyl groups on the C-2, C-4, and C-6 atoms and an ethyl group on the C-4 atom.

4. The number for the position of each alkyl group is placed immediately before the name of the group and is joined to the name by a hyphen. Alkyl groups as well as halogen atoms are listed in alphabetical order.

$$\begin{array}{l} \qquad\qquad\qquad\qquad CH_2-CH_3 \\ 1 \qquad\quad 2 \qquad\qquad\qquad | \\ CH_3-CH-CH_2-CH-CH_2-CH_3 \\ \qquad\quad | \qquad 3 \qquad\quad 4 \qquad 5 \qquad 6 \\ \qquad\quad CH_3 \end{array}$$

This is 4-ethyl-2-methylhexane, not 2-methyl-4-ethylhexane.

Two or more groups of the same type are indicated by the prefixes *di-*, *tri-*, *tetra-*, and so forth. The numbers that indicate the locations of the branches are separated by commas.

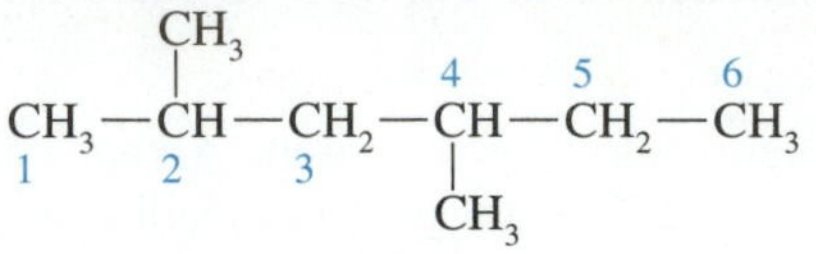

$$\underset{1}{CH_3}-\underset{2}{\overset{\overset{\displaystyle CH_3}{|}}{CH}}-\underset{3}{CH_2}-\overset{4}{\underset{\underset{\displaystyle CH_3}{|}}{CH}}-\overset{5}{CH_2}-\overset{6}{CH_3}$$

This is 2,4-dimethylhexane.

5. The prefixes *di-*, *tri-*, *tetra-*, and so forth do not alter the alphabetical ordering of the alkyl groups.

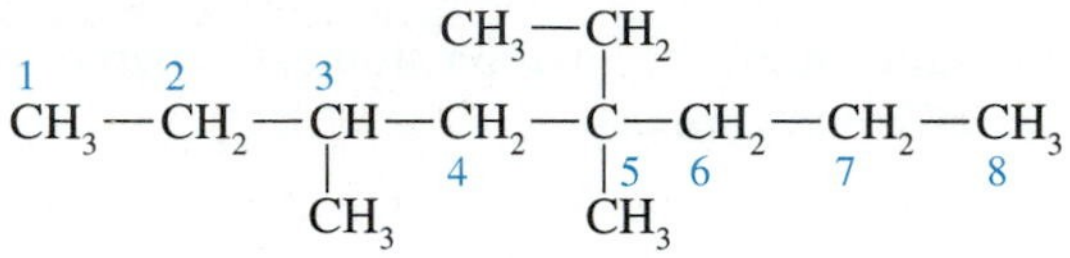

$$\overset{1}{CH_3}-\overset{2}{CH_2}-\overset{3}{\underset{\underset{\displaystyle CH_3}{|}}{CH}}-\underset{4}{CH_2}-\overset{\overset{\displaystyle CH_3-CH_2}{|}}{\underset{\underset{\displaystyle CH_3}{|5}}{C}}-\underset{6}{CH_2}-\underset{7}{CH_2}-\underset{8}{CH_3}$$

This is 5-ethyl-3,5-dimethyloctane, not 3,5-dimethyl-5-ethyloctane.

Names of Alkyl Groups

There is only one alkyl group each derived from methane and ethane. However, for a longer chain of carbon atoms, there are usually several isomeric alkyl groups depending on which carbon atom "loses" a hydrogen atom. Many of these alkyl groups are known by their common names. For example, propane has two primary carbon atoms and a secondary carbon atom. If a primary carbon atom loses a hydrogen atom, a primary alkyl group, *propyl,* is produced. Propyl and other primary alkyl groups derived from normal alkanes are **normal alkyl groups.** If the 2° carbon atom of propane loses a hydrogen atom, a secondary alkyl group known as the *isopropyl* group is formed.

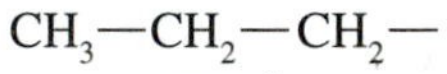

$$\underset{\text{propyl}}{CH_3-CH_2-CH_2-} \qquad \underset{\text{isopropyl}}{CH_3-\overset{\overset{\displaystyle CH_3}{|}}{CH}-}$$

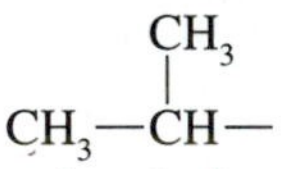

Next, let's look at the alkyl groups that can be derived from the two isomers of butane, C_4H_{10}. These alkyl groups all have the formula C_4H_9. Two alkyl groups are derived from butane and two from isobutane. If a primary carbon atom of butane loses a hydrogen atom, a *butyl* group results; if a secondary carbon atom of butane loses a hydrogen atom, a secondary alkyl group called the *sec-butyl* group forms.

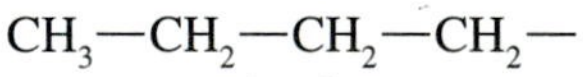

$$\underset{\text{butyl}}{CH_3-CH_2-CH_2-CH_2-} \qquad \underset{sec\text{-butyl}}{CH_3-CH_2-\overset{\overset{\displaystyle CH_3}{|}}{CH}-}$$

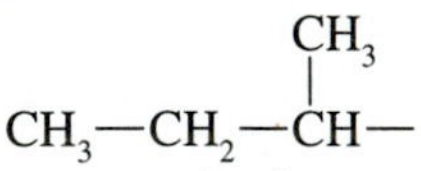

Removal of a hydrogen atom from a primary carbon atom of isobutane gives a primary alkyl group called the *isobutyl* group. Removal of a hydrogen atom from the tertiary carbon atom of isobutane gives a tertiary alkyl group called the *tert-butyl* (*t*-butyl) group. Thus, there are four isomeric C_4H_9— alkyl groups.

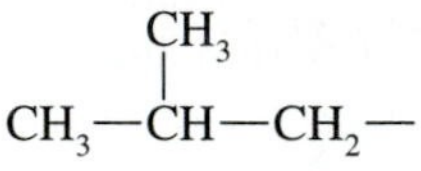

$$\underset{\text{isobutyl}}{CH_3-\overset{\overset{\displaystyle CH_3}{|}}{CH}-CH_2-} \qquad \underset{tert\text{-butyl}}{CH_3-\overset{\overset{\displaystyle CH_3}{|}}{\underset{\underset{\displaystyle CH_3}{|}}{C}}-}$$

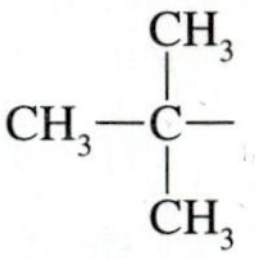

Alkyl groups are named by an IUPAC procedure similar to that used to name alkanes. These alkyl groups are named using the longest continuous chain beginning at the branch point. Thus, the IUPAC name for an isopropyl group is 1-methylethyl, and the IUPAC name for an isobutyl group is 2-methylpropyl. The point of attachment of the alkyl group is numbered carbon 1.

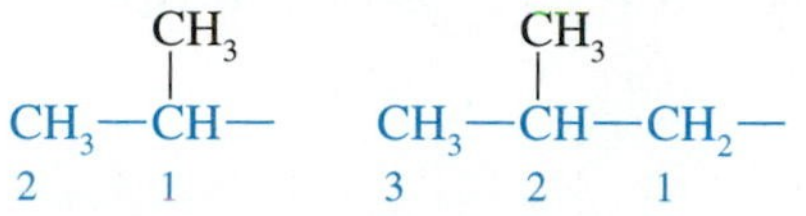

$$\underset{2}{CH_3}-\underset{1}{\overset{\overset{CH_3}{|}}{CH}}- \qquad \underset{3}{CH_3}-\underset{2}{\overset{\overset{CH_3}{|}}{CH}}-\underset{1}{CH_2}-$$

1-methylethyl 2-methylpropyl

Complex alkyl groups are enclosed within parentheses when used to name hydrocarbons. Thus 4-isopropylheptane is also 4-(1-methylethyl)heptane. The methyl within parentheses shows that it modifies ethyl, not heptane. The nonsystematic names for the alkyl groups containing three and four carbon atoms are commonly used, and the IUPAC rules allow for their continued use.

EXAMPLE 3-3

Name the following compound, which is produced by the alga *Spirogyra*.

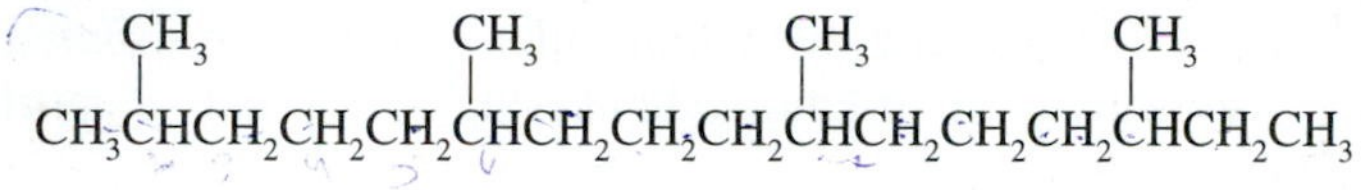

$$CH_3\overset{\overset{CH_3}{|}}{C}HCH_2CH_2CH_2\overset{\overset{CH_3}{|}}{C}HCH_2CH_2CH_2\overset{\overset{CH_3}{|}}{C}HCH_2CH_2CH_2\overset{\overset{CH_3}{|}}{C}HCH_2CH_3$$

Solution

The longest continuous chain has 16 carbon atoms and is named as a substituted hexadecane. The chain is numbered from left to right to locate the four methyl groups at positions 2, 6, 10, and 14. The compound is 2,6,10,14-tetramethylhexadecane.

Problem 3.3

Name the following compound.

$$CH_3-CH_2-\overset{\overset{CH_3}{|}}{\underset{\underset{CH_3}{|}}{C}}-CH_2-\underset{\underset{CH_2-CH_3}{|}}{CH}-CH_3$$

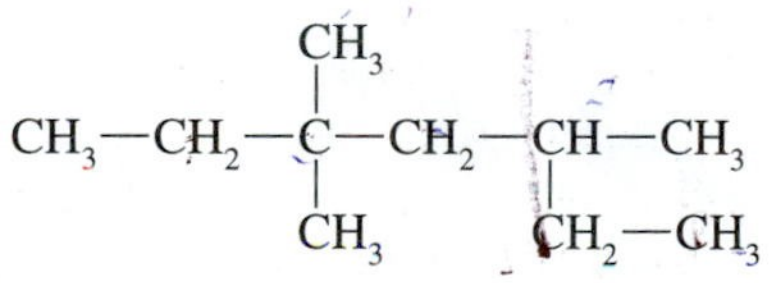

EXAMPLE 3-4

Identify the alkyl group on the left of the benzene ring in ibuprofen, an analgesic present in Nuprin®, Advil®, and Motrin®.

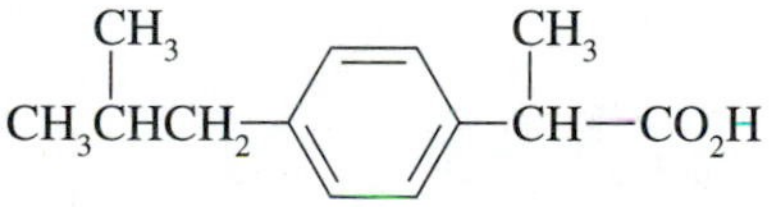

$$CH_3\overset{\overset{CH_3}{|}}{C}HCH_2-C_6H_4-\overset{\overset{CH_3}{|}}{CH}-CO_2H$$

Solution

There are four carbon atoms in the alkyl group, which is derived from isobutane—not butane. The benzene ring is bonded to the terminal carbon atom—not the internal carbon atom. This group is the isobutyl group.

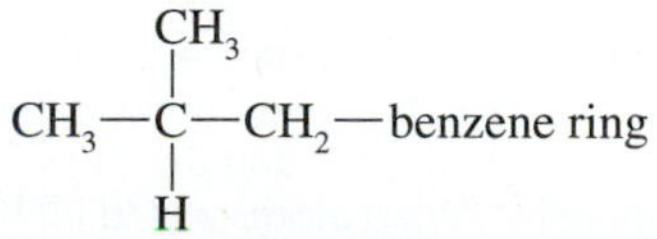

$$CH_3-\overset{\overset{CH_3}{|}}{\underset{\underset{H}{|}}{C}}-CH_2-\text{benzene ring}$$

Problem 3.4

The food preservative BHT has the following structure. Identify the alkyl groups bonded to the benzene ring.

OH

$(CH_3)_3C$ $C(CH_3)_3$

CH_3

3.4 Conformations of Alkanes

Sometimes we can construct two molecular models that appear to be different but are not isomers. Consider the models of ethane shown in Figure 3.2. In both examples, the two carbon atoms are bonded to each other, and each carbon atom is bonded to three hydrogen atoms. The two representations differ in the positions of the hydrogen atoms of one carbon atom relative to those of the other carbon atom. Which form represents ethane? The answer is that both do, to some extent. Ethane and other molecules can exist in different orientations or **conformations** by rotation about single σ bonds requiring small amounts of energy.

Rotation around the carbon-carbon σ bond occurs constantly in alkanes such as ethane and alters the spatial positions of the hydrogen atoms. However, this rotation does not alter the connectivity of the carbon-carbon or carbon-hydrogen bonds. The motion is like the twisting and turning of your body while

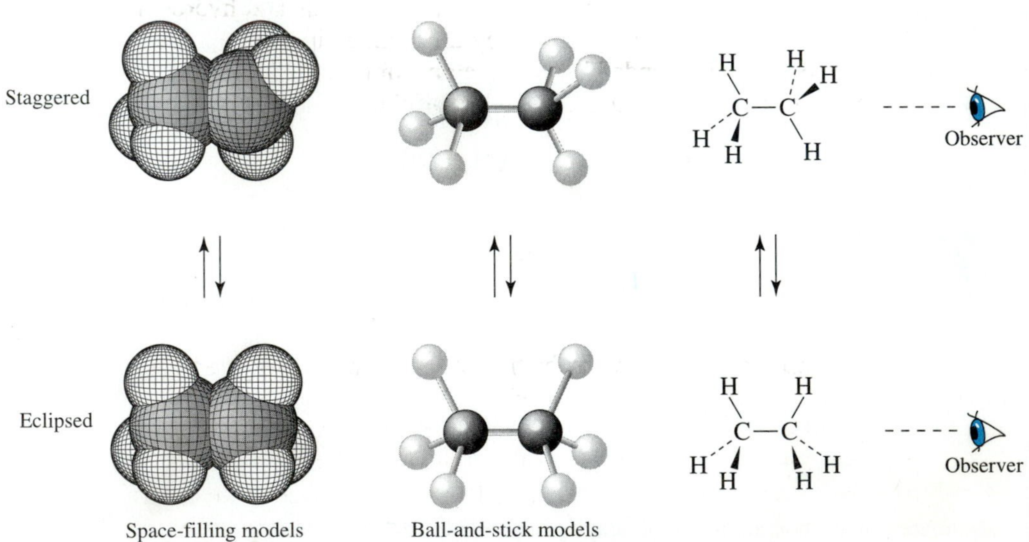

FIGURE 3.2
Conformation of Ethane

Rotation of the methyl group on the right by 60° converts a staggered conformation into an eclipsed conformation. Viewing the carbon-carbon bond end-on in the eclipsed conformation, the observer would see only the carbon atom and the three hydrogen atoms on the right. The left carbon atom and its three hydrogen atoms would be hidden.

dancing. You may look different, but the parts of your body are still connected to the normal places. Only the orientation of your limbs is changing.

Ethane can exist in many conformations. The conformation in which the hydrogen atoms and the bonding electrons are the farthest away from one another has the lowest energy. This conformation is said to be **staggered.** The conformation in which the hydrogen atoms are closest to each other has the highest energy. This conformation is said to be **eclipsed.** In the eclipsed conformation each C—H bond on one carbon atom eclipses a C—H bond on another carbon atom, as the moon sometimes eclipses the sun. Any other intermediate conformation is a **skew** conformation.

Newman Projection Formulas

Conformations of alkanes are often depicted by a representation called a **Newman projection** formula. This structure concentrates on the two carbon atoms about which rotation may occur. The two atoms are viewed end-on. The front atom—that nearest the viewer—is represented by a point with three bonds. The back atom is represented by a circle with three bonds that reach only to the perimeter of the circle. Although there is a bond between the two carbon atoms, it is hidden because it is located along the viewing axis.

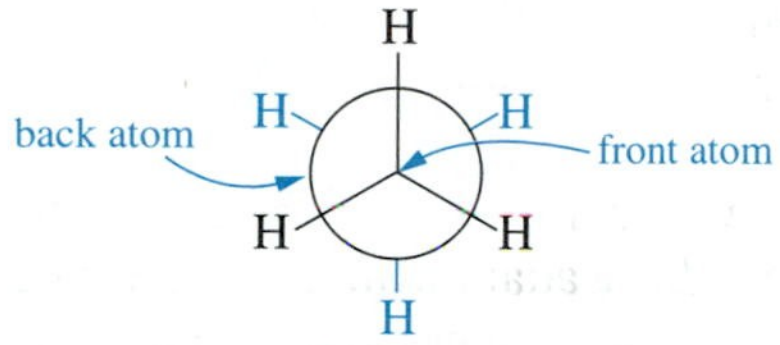

Newman projection of staggered ethane conformation

A Newman projection of the eclipsed conformation of ethane shows only the three C—H bonds of the front carbon atom. The rear bonds and hydrogen atoms are not visible because they are hidden by the front eclipsing bonds and hydrogen atoms. However, the bonded hydrogen atoms of the back carbon atom can be shown by viewing the conformation slightly off the bond axis.

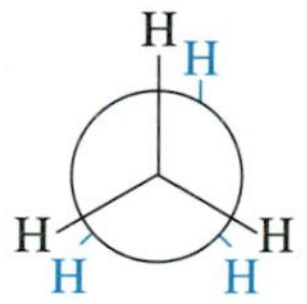

Newman projection of eclipsed ethane conformation

Why is the eclipsed conformation of higher energy than the staggered conformation? We recall that the VSEPR model predicts that electron pairs and bonded atoms should be separated by the maximum distance. Consider the positions of the bonding pairs of electrons of the C—H bonds on the front and back atoms in the Newman projection formula. The angle between these bonds is called the **torsional** or **dihedral angle.** In the eclipsed conformation the torsional angle is 0° whereas in the staggered conformation the torsional angle is 60°. The eclipsed conformation has a higher energy because of **torsional strain,** which is due to the small repulsion between the bonding electrons in the C—H bonds. Each hydrogen-hydrogen eclipsing interaction amounts to 1 kcal/mol. The total difference in energy between the eclipsed and staggered conformations (3 kcal/mol) is small, and there is rapid interconversion among these conformations. Thus, we say that rotation about the C—C bond is virtually free or unrestricted.

Different conformations of ethane are not different compounds but different forms of a single molecule. Ethane is a mixture of conformations, but they cannot be separated from each other. However, at room temperature the staggered conformation is the major form and any given molecule spends most of its time in a staggered conformation.

Conformations of Butane

All acyclic alkanes exist as mixtures of conformations that result from rotation about every single bond in their structures. In each case, the staggered conformations are more stable than eclipsed conformations. For butane and higher molecular weight alkanes there is more than one staggered conformation (Figure 3.3).

For butane the most stable staggered conformation has a zigzag arrangement of the carbon-carbon bonds.

H H H H
H H
H H H H

anti conformation of butane

A Newman projection viewed along the C(2)—C(3) bond shows that the two methyl groups in this staggered conformation are the maximum distance apart in an **anti conformation.** A second staggered conformation called the **gauche conformation** is also possible.

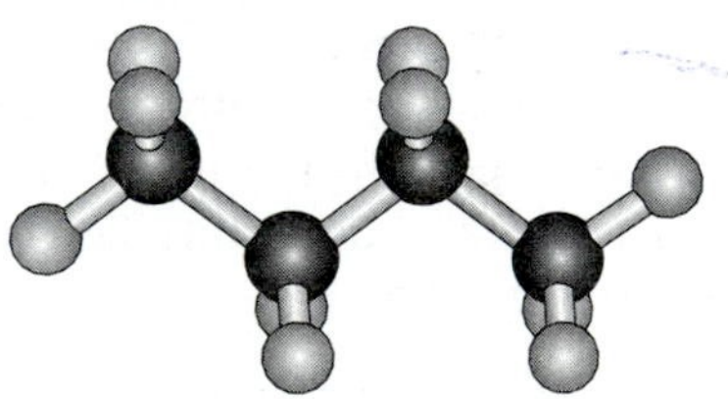

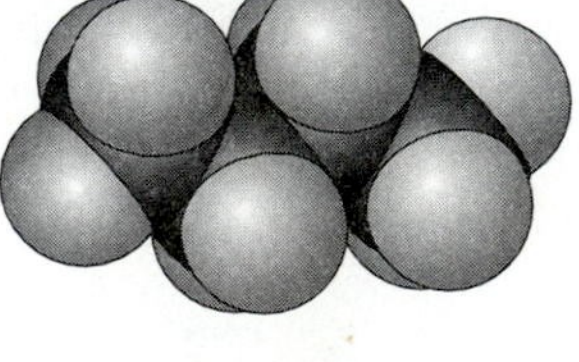

Anti conformation

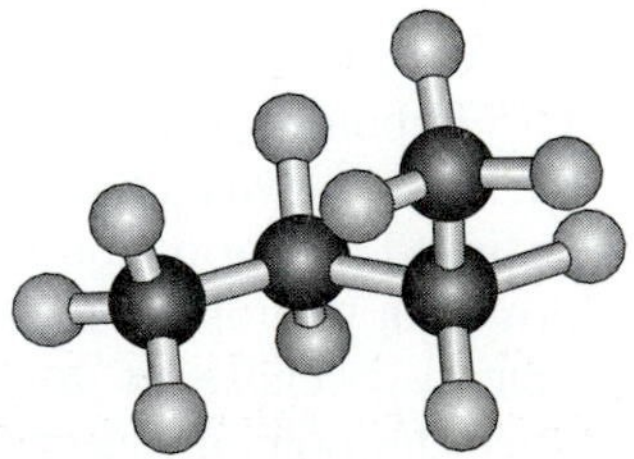

Gauche conformation

FIGURE 3.3
Staggered Conformations of Butane

anti conformation

gauche conformation

There is no torsional strain in either of these two conformations because the torsional angles between the bonding pairs is 60° in both cases. However, the methyl groups are closer to each other in the gauche than in the anti conformation. The interference and repulsion of the electron clouds of the atoms of the two methyl groups with each other is called **steric strain.** The anti conformation of butane is more stable than the gauche conformation by 0.9 kcal/mol. The anti and gauche conformations interconvert rapidly, but the ratio of anti to gauche conformation is about 2:1. The same types of interactions occur in higher alkanes.

3.5 Cycloalkanes

Cycloalkanes contain only carbon-carbon single bonds and are saturated hydrocarbons. These compounds containing a single ring have the general formula C_nH_{2n}, two fewer hydrogen atoms than alkanes, because another carbon-carbon bond is needed to form the ring. Cycloalkanes are drawn as simple polygons in which the sides represent the carbon-carbon bonds. It is understood that each corner of the polygon is a carbon atom attached to two hydrogen atoms.

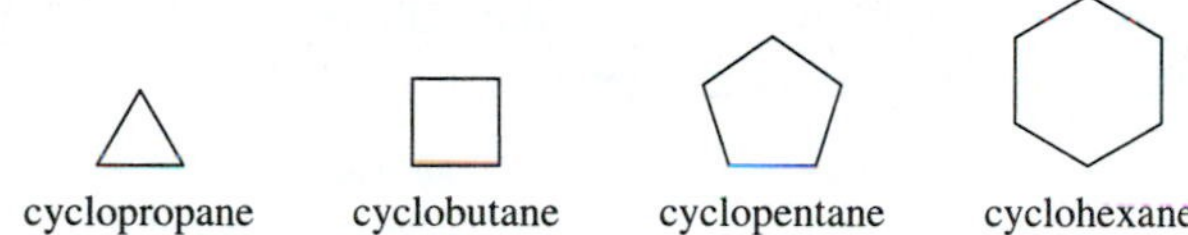

cyclopropane cyclobutane cyclopentane cyclohexane

Multiple rings in a molecule can share one or more common atoms. **Spirocyclic compounds** share one carbon atom between two rings. These compounds are relatively rare in nature. **Fused ring compounds** share two common atoms and the bond between them. These compounds are prevalent in steroids, which contain four fused rings. **Bridged ring compounds** share two nonadjacent carbon atoms, which are called the **bridgehead** carbon atoms. These compounds are less prevalent in nature than fused ring compounds.

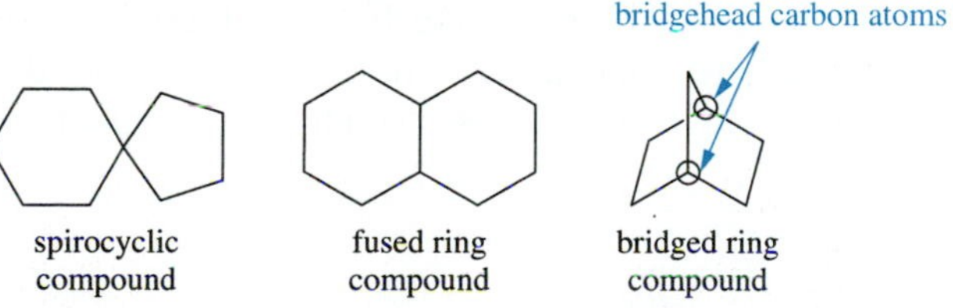

spirocyclic compound fused ring compound bridged ring compound

Geometric Isomerism

Isomers exist with different carbon skeletons, different functional groups, and different functional group locations. These isomers have different sequential arrangements of atoms. Now let us consider a different type of isomerism. Compounds that have the same sequential arrangement of atoms but different spatial arrangements are **geometric isomers.**

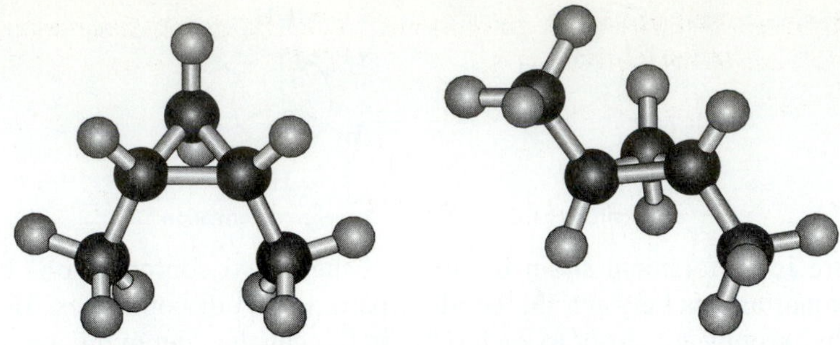

FIGURE 3.4
Geometric Isomers of 1,2-Dimethylcyclopropane

Consider cyclopropane, whose three carbon atoms define a plane. Any group attached to the ring may be held "above" or "below" the plane of the ring. If we attach two methyl groups on adjacent carbon atoms on the same side of the plane of the ring, the substance is called a **cis** isomer; it is *cis*-1,2-dimethylcyclopropane. If the two methyl groups are attached on the opposite sides of the plane of the ring, the compound is the **trans** isomer. Thus, 1,2-dimethylcyclopropane can exist as both cis and trans isomers; these isomers are geometric isomers (Figure 3.4). Note that cis and trans compounds are not two conformations of the same molecule but are isomeric substances that have different physical properties. It is impossible to convert one isomer into the other without breaking a bond. In the following structures the cyclopropane ring is viewed as perpendicular to the plane of the page, and the $—CH_2—$ is pointed away from the viewer.

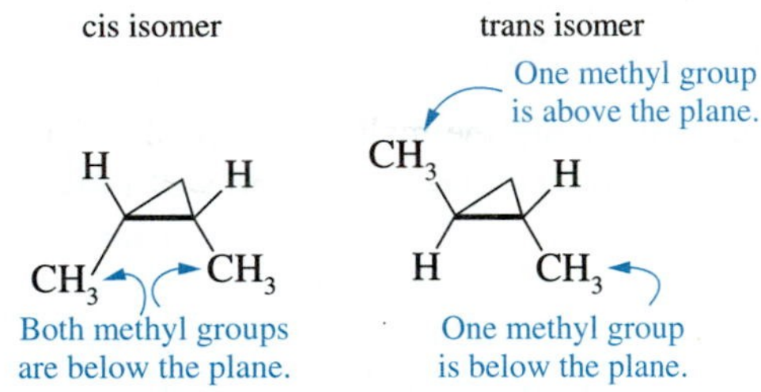

Nomenclature of Cycloalkanes

Cycloalkanes are named according to the IUPAC system by using the prefix *cyclo-*. When only one position contains a functional group or alkyl group, only one compound is possible, as in the cases of ethylcyclopentane and isopropylcyclobutane.

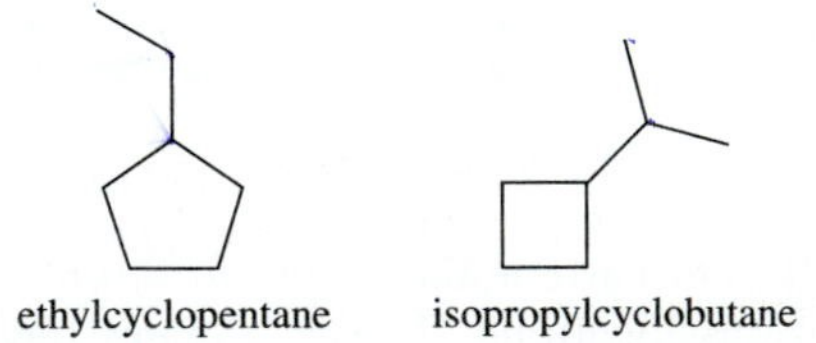

When more than one group is attached to the ring, the ring atoms are numbered. One substituent is at position 1, and the ring is numbered in a clockwise or counterclockwise direction to give the lower number to the position with the next substituent attached to the ring, as in 1,1,4-trichlorocyclodecane and *trans*-1-bromo-3-ethylcyclohexane.

1,1,4-trichlorocyclodecane

trans-1-bromo-3-ethylcyclohexane

EXAMPLE 3-5

Adamantane has a carbon skeleton that is also found as part of the structure of diamond. Amantadine, which contains an amino group (—NH_2) bonded to the adamantane structure, is useful in the prevention of infection by influenza A viruses. What are the molecular formulas of adamantane and amantadine?

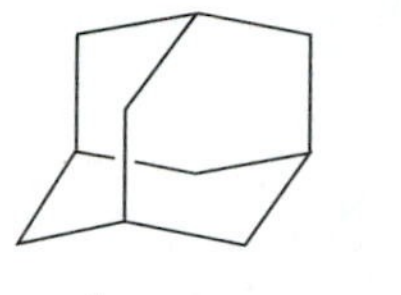
adamantane

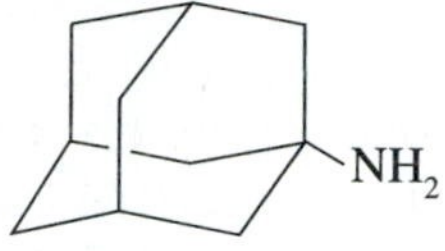

amantadine

Solution

Adamantane has 10 carbon atoms. Four of these carbon atoms are tertiary; they have bonds to three other carbon atoms and one bond to a hydrogen atom. The remaining six carbon atoms are secondary; they have bonds to two other carbon atoms and two bonds to hydrogen atoms. The total number of hydrogen atoms is $4(1) + 6(2) = 16$. The molecular formula of adamantane is $C_{10}H_{16}$.

Amantadine has an amino group (—NH_2) in place of a hydrogen atom at one of the tertiary carbon atoms. Thus, the molecular formula of amantadine differs from adamantane by one nitrogen atom and one hydrogen atom. The molecular formula is $C_{10}H_{17}N$.

Problem 3.5

Determine the molecular formula of menthol based on the following bond-line structure:

EXAMPLE 3-6

Disparlure, the sex attractant pheromone of the female gypsy moth, has the following general structure. Are geometric isomers possible for this structure?

$$(CH_3)_2CHCH_2CH_2CH_2CH_2CH{-}CHCH_2CH_2CH_2CH_2CH_2CH_2CH_2CH_2CH_2CH_2CH_3 \quad (\text{CH—CH bridged by O})$$

Solution

The three-membered heterocyclic ring contains two carbon atoms and one oxygen atom. Each carbon atom of the ring has a hydrogen atom and a large alkyl group bonded to it. These alkyl groups could be located cis or trans with respect to the plane of the ring. The cis isomer is the biologically active compound.

Problem 3.6

Brevicomin, the sex attractant of a species of pine beetle, has the following structure. Write the structure of a geometric isomer of brevicomin.

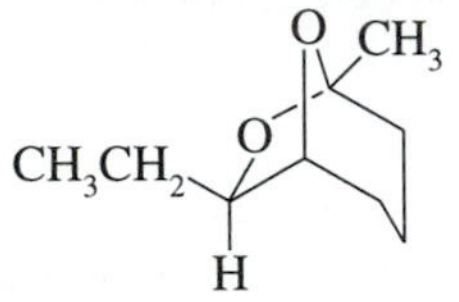

O CH_3
CH_3CH_2 O
H

EXAMPLE 3-7

What is the name of the following compound?

Cl Cl

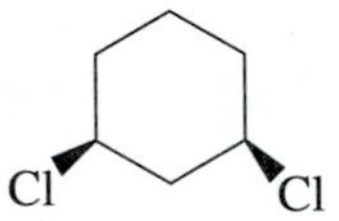

Solution

The ring must be numbered starting from one carbon atom with a chlorine atom and counting toward the other carbon atom with a chlorine atom in the direction that will give the lower position number. Starting with the carbon atom at the "4 o'clock" position and numbering clockwise gives the number 3 to the atom at the "8 o'clock" position.

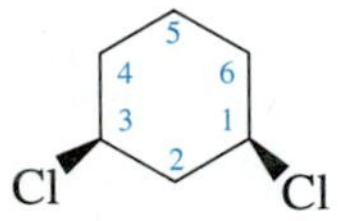

5
4 6
3 1
2
Cl Cl

The two chlorine atoms are on the same side of the plane of the ring, and the correct name is *cis*-1,3-dichlorocyclohexane.

Problem 3.7

What is the name of each of the following compounds?

(a) 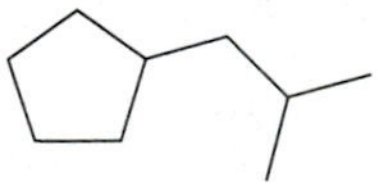 (b) 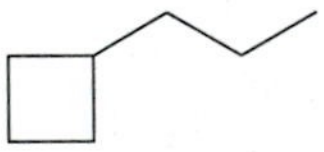 (c) 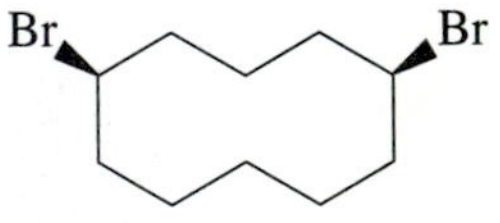 Br Br

3.6 Conformations of Cycloalkanes

Cyclopropane has only three carbon atoms, so it is a planar molecule (Figure 3.5). The hydrogen atoms of cyclopropane lie above and below the plane of the ring of carbon atoms, and they eclipse hydrogen atoms on adjacent carbon atoms. All of the C—C—C bond angles are only 60° because the carbon atoms form an equilateral triangle. Cyclobutane and cyclopentane are not planar and exist in slightly "puckered" conformations, which reduce some of the eclipsing of hydrogen atoms on adjacent carbon atoms. The conformations of these compounds will not be considered further. For the purposes of this text, the two molecules will be depicted as planar rings of carbon atoms.

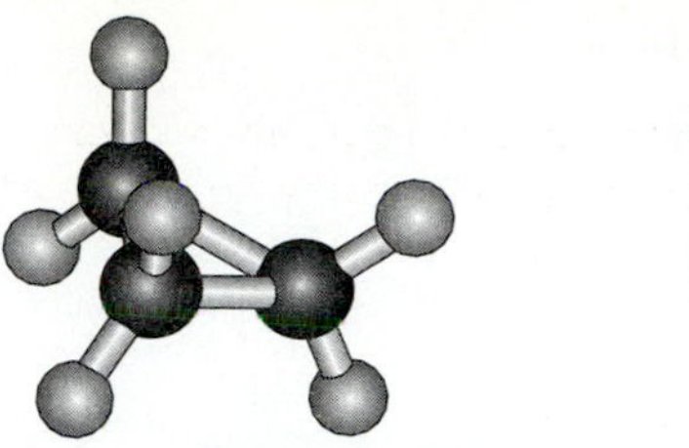
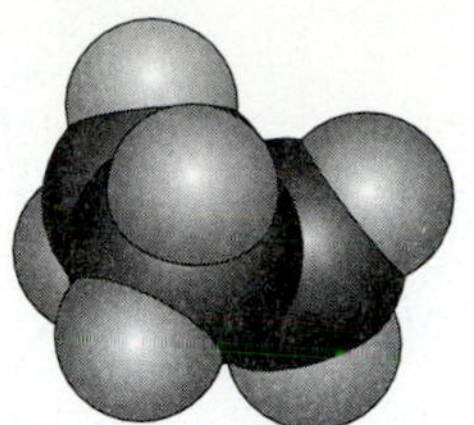

FIGURE 3.5
Molecular Models of Cyclopropane

The six-membered ring of cyclohexane is not planar; it exists in a puckered conformation in which all C—H bonds on neighboring carbon atoms are staggered. Figure 3.6 shows a bond-line representation of the **chair conformation** of cyclohexane. Note that the hydrogen atoms in this conformation fall into two sets. Six of the hydrogen atoms are **axial;** they point up or down with respect to the average plane of the ring of carbon atoms. Three of the axial hydrogen atoms point up and the other three point down. The up-down relationship alternates from one carbon atom to the next. The remaining six hydrogen atoms—

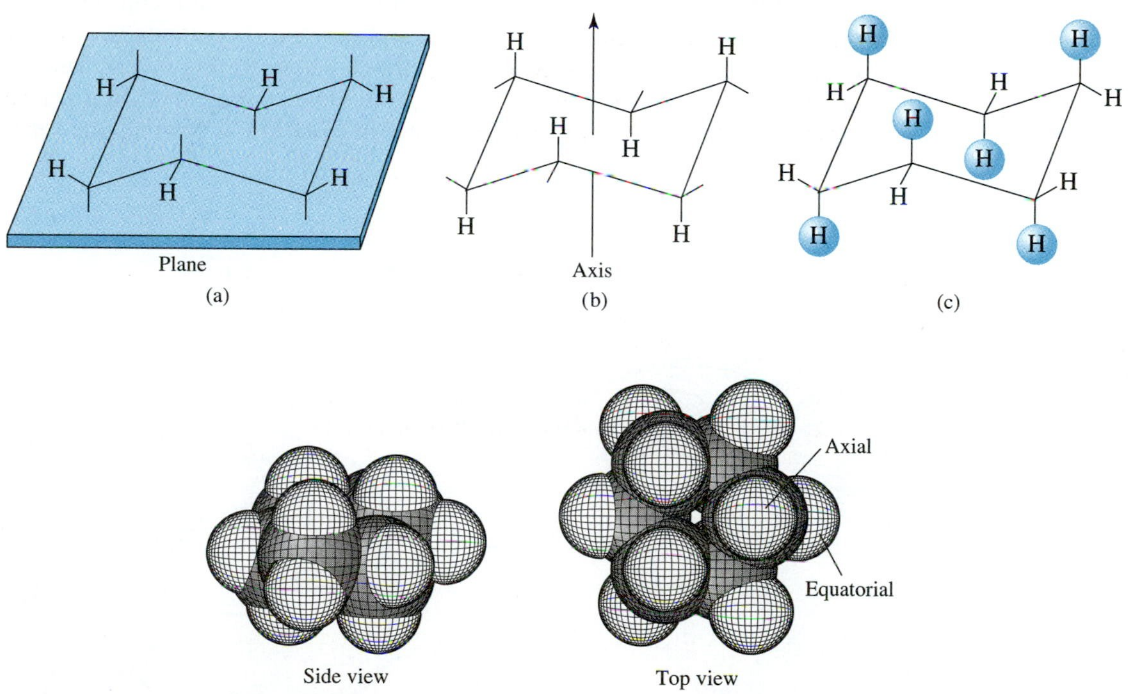

FIGURE 3.6
Axial and Equatorial Hydrogen Atoms in Cyclohexane

The equatorial C—H bonds shown in (a) are located in a band around the "equator" of the ring. Each carbon atom has one equatorial hydrogen atom. The six axial C—H bonds are parallel to the axis shown perpendicular to the plane in (b). Three hydrogen atoms are pointed up from the average plane of the ring; three hydrogen atoms are pointed down. The axial hydrogen atoms are located in an alternating up-down relationship. All hydrogen atoms are shown in (c). The axial hydrogen atoms are circled.

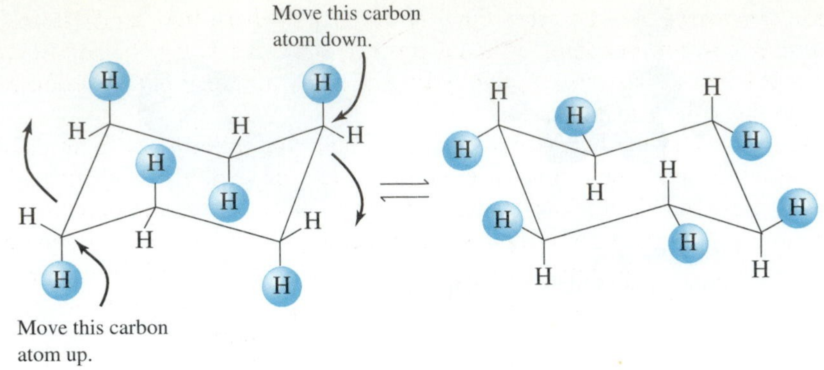

Move this carbon atom down.

Move this carbon atom up.

FIGURE 3.7
Conformational Mobility of the Cyclohexane Ring

The interconversion of chair conformations of cyclohexane is called a ring flip. When the atoms are moved in the indicated direction, one conformation is converted into another conformation. An axial position in one chair conformation becomes an equatorial position in the other conformation and vice versa. The interconversion is rapid.

called **equatorial** atoms—lie approximately in the average plane of the ring. Each carbon atom has one equatorial and one axial C—H bond.

Cyclohexane is conformationally mobile. Different chair conformations can interconvert and, as a result, the axial and equatorial positions become interchanged. This process, known as a **ring flip,** is shown in Figure 3.7. You can see this process more clearly by practicing with molecular models. To flip a cyclohexane ring, hold the four "middle" atoms in place while pushing one "end"

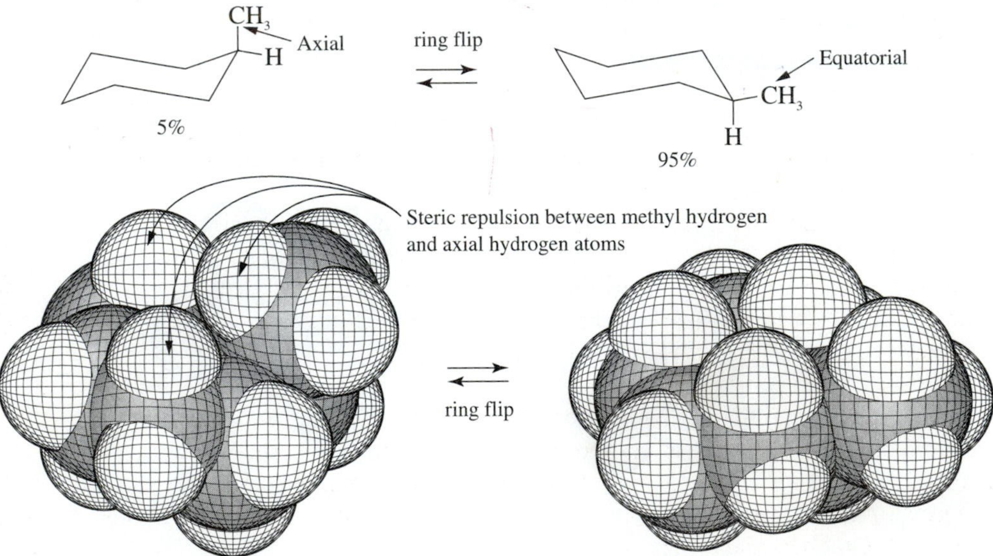

FIGURE 3.8
Conformations of Methylcyclohexane

Methylcyclohexane interconverts rapidly between two conformations of unequal energy. At any time there is 95% of the equatorial and 5% of the axial conformation. The axial conformation has unfavorable steric interactions between the methyl group hydrogen atoms and the axial hydrogen atoms on the C-3 and C-5 carbon atoms.

carbon downward and the other upward. At each of these two "end" atoms, an equatorial position becomes an axial position and vice versa. In addition, although it is not as easy to see, the hydrogen atoms on every other carbon atom undergo the same transformation.

Now consider methylcyclohexane in a chair conformation with an equatorial methyl group. When the ring flips, the equatorial methyl group moves into an axial position (Figure 3.8). These two structures are different conformations, not isomers. The chair-chair interconversion occurs very rapidly. An equatorial methyl group is more stable than an axial methyl group, and 95% of the mixture at equilibrium has an equatorial methyl group. The conformation with an axial methyl group is less stable because there is steric strain between the methyl group and the axial hydrogen atoms at the C-3 and C-5 atoms. All substituted cyclohexanes behave similarly. Thus, the equatorial conformation is always more stable than the axial one, although the relative amounts of the two conformations vary depending on the substituent.

3.7 Physical Properties of Saturated Hydrocarbons

Alkanes have densities between 0.6 and 0.8 g/cm^3, so they are less dense than water. Thus gasoline, which is largely a mixture of alkanes, is less dense than water and will float on water. Pure alkanes are colorless, tasteless, and nearly odorless. However, gasoline does have an odor and some color because dyes are added to gasoline by refiners to indicate its source and composition. Gasoline also contains aromatic compounds (Chapter 5), which have characteristic odors.

Alkanes contain only carbon-carbon and carbon-hydrogen bonds. Because carbon and hydrogen have similar electronegativity values, the C—H bonds are essentially nonpolar. Thus, alkanes are nonpolar, and they interact only by weak London forces. These forces govern the physical properties of alkanes, such as solubility and boiling point.

Alkanes are not soluble in water, a polar substance. The two substances do not meet the usual criterion of solubility: "Like dissolves like." Water molecules are too strongly attracted to each other by hydrogen bonds to allow nonpolar alkanes to slip in between them and dissolve.

The boiling points of the normal alkanes increase with increasing molecular weight (Table 3.2). As the molecular weight increases, London forces increase because more atoms are present to increase the surface area of the molecules. Simply put, there are more points of contact between neighboring molecules, and the London forces are stronger.

Normal alkanes have efficient contact between chains, and the molecules can move close together. Branching in alkanes increases the distance between molecules, and the chains of carbon atoms are less able to come close to one another. A branched alkane is more compact and has a smaller surface area than a normal alkane. The order of boiling points of the isomeric C_5H_{12} compounds illustrates this phenomenon.

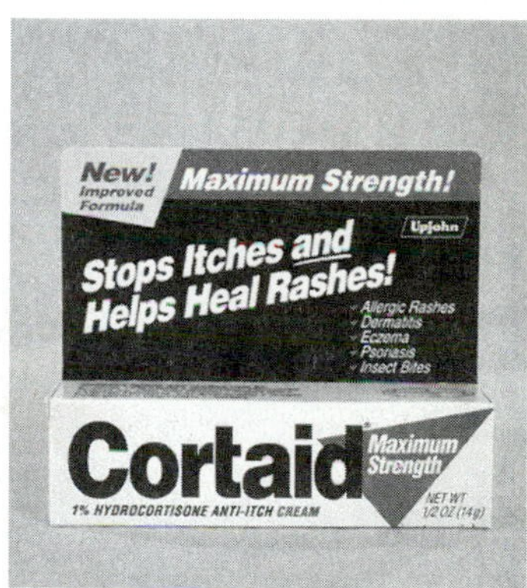

This product, used to heal rashes, contains hydrocortisone.

$CH_3—CH_2—CH_2—CH_2—CH_3$
pentane (bp 36 °C)

$CH_3—CH(CH_3)—CH_2—CH_3$
isopentane (bp 28 °C)

$CH_3—C(CH_3)_2—CH_3$
neopentane (bp 10 °C)

Steroids

Steroids are tetracyclic compounds containing three six-membered rings and a five-membered ring. Each ring is designated by a letter, and the carbon atoms are numbered by the standard system shown. These compounds contain a variety of functional groups such as hydroxyl, carbonyl, and carbon-carbon double bonds.

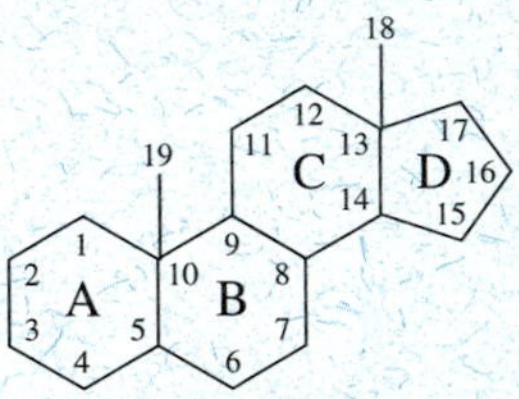

steroid ring system

The several sites of fusion of two rings may be either cis or trans, but trans fusion is more common. As a result of the trans fusion of rings, the cyclohexane rings cannot undergo a ring flipping process. As a consequence, functional groups bonded to the rings have well-defined positions. The biological activity of steroid hormones depends on the functional groups, their locations on the ring, and whether the groups are axial or equatorial. For example, the hydroxyl group at the 3-position in androsterone (a male sex hormone) is located in the axial position. The isomeric compound with an equatorial hydroxyl group, epiandrosterone, has significantly less physiological activity. The activity is related to the spatial arrangement of the hydroxyl group in ring A and the carbonyl group in ring D, which determines whether or not the steroid can bond to specific receptors.

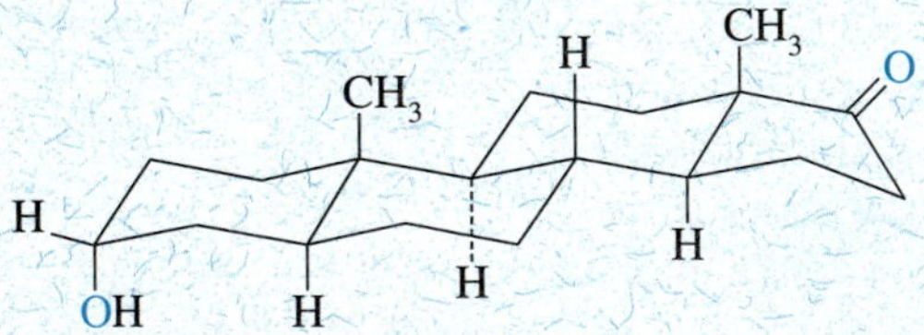

androsterone

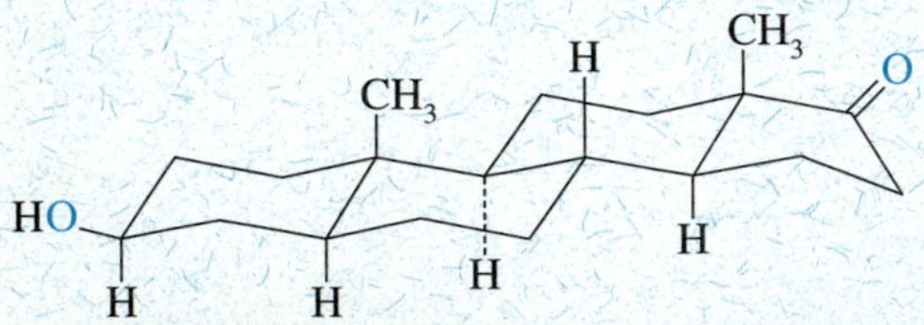

epiandrosterone

Cholesterol is one of the steroids best known to the public. Although cholesterol in the diet is thought to affect the cholesterol level in blood in a detrimental way, it is synthesized by most animals and is required in various cell membranes, including those of the brain.

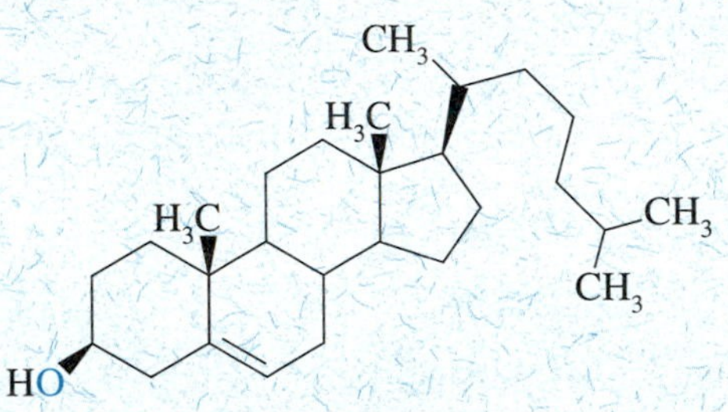

cholesterol

Cholesterol plays a vital biological role in the formation of other steroid hormones; it is converted to progesterone (a female sex hormone) by several reactions, including the shortening of the chain attached at the C-17 position (ring D).

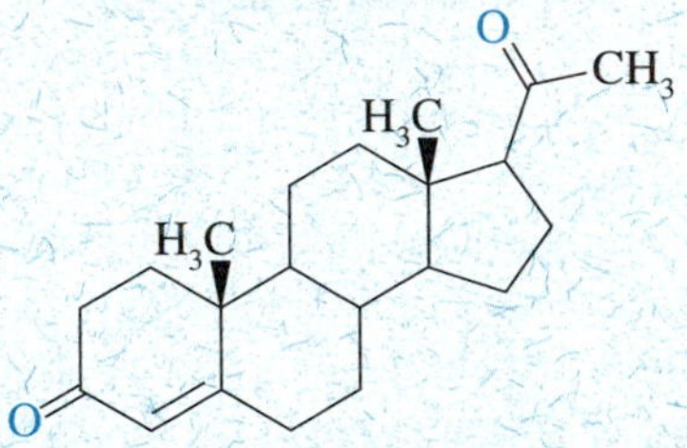

progesterone

The centrality of progesterone in biological reactions is illustrated by its conversion into cortisone as well as testosterone (a male sex hormone). ■

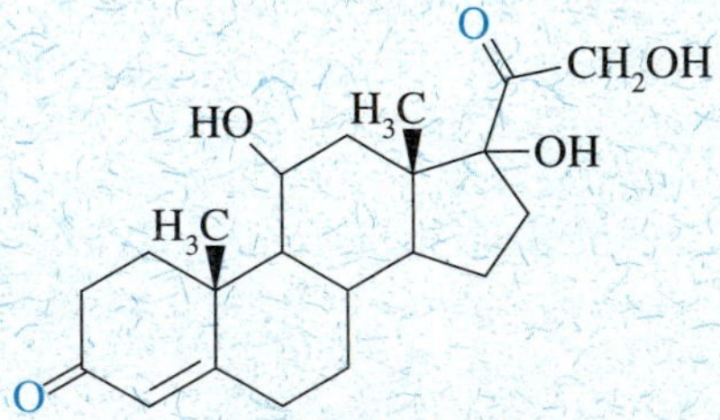

cortisone

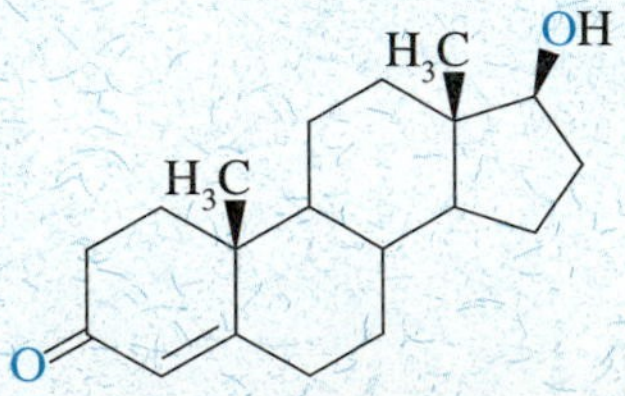

testosterone

Table 3.2 Boiling Points of Alkanes

Name	Boiling Point (°C)	Name	Boiling Point(°C)
methane	−162	hexane	69
ethane	−89	3-methylpentane	63
propane	−42	2-methylpentane	60
butane	−1	2,3-dimethylbutane	58
2-methylpropane	−12	2,2-dimethylbutane	50
pentane	36	heptane	98
2-methylbutane	28	octane	126
2,2-dimethylpropane	10	nonane	151
		decane	174

Table 3.3 Physical Properties of Cycloalkanes

Alkane	Boiling Point (°C)	Density (g/mL)
cyclopropane	−32.7	
cyclobutane	12.0	
cyclopentane	49.3	0.7547
cyclohexane	80.7	0.7786
cycloheptane	118.5	0.8098
cyclooctane	148.5	0.8349

The physical properties of a series of cycloalkanes of increasing molecular weight are similar to those of a series of alkanes. The densities increase, as do the boiling points (Table 3.3). The boiling points of the cycloalkanes are higher than those of the alkanes containing the same number of carbon atoms.

3.8 Oxidation of Alkanes and Cycloalkanes

The carbon-carbon and carbon-hydrogen bonds of alkanes and cycloalkanes are not very reactive. Alkanes are also called **paraffins** (from the Latin *parum affinis,* "little activity"). The carbon-carbon bonds are σ bonds, and the bonding electrons are tightly held between the carbon atoms. The carbon-hydrogen bonds are located about the carbon skeleton and are more susceptible to reaction but usually do so only under extreme conditions. One such process is oxidation.

Methane, the major component of natural gas, yields 212 kcal/mol when burned. Although the reaction is spontaneous, a small spark or flame is required to provide the activation energy for the reaction. Thus, natural gas can accumulate from a gas leak and not explode. But the gas-oxygen mixture is very dangerous.

$$CH_4 + 2\,O_2 \longrightarrow CO_2 + 2\,H_2O \quad \Delta H^\circ = -212\text{ kcal}$$

A petroleum refinery separates petroleum into its constituent hydrocarbons.

As in the case of natural gas leaks, vapors of any alkane can form explosive mixtures with air. Gasoline can safely be stored in closed containers such as gas tanks and pipelines. However, a spark or flame in the vicinity can cause the hydrocarbon vapors to explode. For this reason, you should not smoke while filling your automobile gas tank, lawn mower, or outboard motor with gasoline.

Octane Numbers

In an automobile engine the fuel and air are drawn into the cylinder on its downward stroke, and the piston compresses the mixture on the upward stroke. Ideally, the mixture ignites at the top of the stroke. The resulting explosion drives the piston downward. Normal alkanes are not suitable as fuel in an automobile engine because they tend to ignite prematurely and uncontrollably. Their use results in a knocking or pinging sound, which indicates that a force is resisting the upward motion of the piston. Branched hydrocarbons burn more smoothly and are the more efficient fuels.

The burning efficiency of gasoline is rated by an octane number scale (Table 3.4). An octane number of 100 is assigned to 2,2,4-trimethylpentane (also known as isooctane), which is an excellent fuel. Heptane, a poor fuel, has an octane number of zero. Gasoline with the same burning characteristics as a 90% mixture of 2,2,4-trimethylpentane and 10% heptane is rated at 90 octane. Compounds that burn more efficiently than 2,2,4-trimethylpentane have octane numbers greater than 100. Compounds that burn less efficiently than heptane have negative octane numbers. Octane numbers decrease with increasing molecular weight. In isomeric compounds, increased branching increases the octane number.

Table 3.4 Octane Numbers of Alkanes

Formula	Compound	Octane Number
C_4H_{10}	butane	94
C_5H_{12}	pentane	62
	2-methylbutane	94
C_6H_{14}	hexane	25
	2-methylpentane	73
	2,2-dimethylbutane	92
C_7H_{16}	heptane	0
	2-methylhexane	42
	2,3-dimethylpentane	90
C_8H_{18}	octane	−19
	2-methylheptane	22
	2,3-dimethylhexane	71
	2,2,4-trimethylpentane	100

3.9 Halogenation of Saturated Hydrocarbons

Alkanes react with halogens at high temperature or in the presence of light to give a substitution product that has a halogen atom in place of a hydrogen atom. For example, methane reacts with chlorine when heated to a high temperature or when exposed to ultraviolet light.

$$CH_4 + Cl_2 \longrightarrow \underset{\substack{\text{chloromethane}\\\text{(methyl chloride)}}}{CH_3Cl} + HCl$$

The reaction is difficult to control because the product also has carbon-hydrogen bonds that can continue to react with additional chlorine to produce several substitution products.

$$CH_3Cl + Cl_2 \longrightarrow \underset{\substack{\text{dichloromethane}\\\text{(methylene chloride)}}}{CH_2Cl_2} + HCl$$

$$CH_2Cl_2 + Cl_2 \longrightarrow \underset{\substack{\text{trichloromethane}\\\text{(chloroform)}}}{CHCl_3} + HCl$$

$$CHCl_3 + Cl_2 \longrightarrow \underset{\substack{\text{tetrachloromethane}\\\text{(carbon tetrachloride)}}}{CCl_4} + HCl$$

Chlorination of higher molecular weight alkanes yields many monosubstituted products. The chlorine atom is so reactive that it is not selective in its substitution for hydrogen atoms. Thus, reaction with 2-methylpropane yields 63% 1-chloro-2-methylpropane and 37% 2-chloro-2-methylpropane. A large number of polysubstituted products is also possible. Bromination of alkanes is a very much more selective reaction than the chlorination reaction. The reactivity of C—H bonds decreases in the order $3° > 2° > 1°$. Thus reaction of 2-methylpropane yields 99% 2-bromo-2-methylpropane.

$$CH_3-\underset{\displaystyle H}{\overset{\displaystyle CH_3}{\overset{|}{\underset{|}{C}}}}-CH_3 + Br_2 \longrightarrow \underset{\text{2-bromo-2-methylpropane}}{CH_3-\underset{\displaystyle Br}{\overset{\displaystyle CH_3}{\overset{|}{\underset{|}{C}}}}-CH_3} + HBr$$

Halogenated hydrocarbons are used for many industrial purposes. Unfortunately, many of these compounds can cause liver damage and cancer. In the past, chloroform was used as an anesthetic, and carbon tetrachloride was used as a dry-cleaning solvent. They are no longer used for these purposes.

Mechanism of Chlorination of Alkanes

The mechanism of chlorination involves homolytic bond cleavage and homogenic bond formation. In the first step the chlorine molecule absorbs either heat energy or light energy and the bond breaks to give two chlorine atoms that are radicals. These species are electron deficient and highly reactive. The chlorination reaction can occur only after this initial homolytic bond cleavage, which has a large activation energy. The step that starts the reaction is the **initiation step.**

$$\text{Step 1} \quad :\ddot{\underset{..}{Cl}}-\ddot{\underset{..}{Cl}}: \longrightarrow :\ddot{\underset{..}{Cl}}\cdot + \cdot\ddot{\underset{..}{Cl}}:$$

Two steps collectively known as **propagation steps,** which involve radicals as reactants and products, then occur.

$$\text{Step 2} \quad CH_3-H + \cdot\ddot{\underset{..}{Cl}}: \longrightarrow CH_3\cdot + H-\ddot{\underset{..}{Cl}}:$$

$$\text{Step 3} \quad CH_3\cdot + :\ddot{\underset{..}{Cl}}-\ddot{\underset{..}{Cl}}: \longrightarrow CH_3-\ddot{\underset{..}{Cl}}: + :\ddot{\underset{..}{Cl}}\cdot$$

In step 2, a C—H bond is broken and an H—Cl bond is produced; in step 3, a Cl—Cl bond is broken and a C—Cl bond is formed. Furthermore, in each step a radical reacts and a radical is produced. One radical generates another in this **chain propagation** sequence. Therefore, only a small number of chlorine atoms are required to initiate the process.

Many chains are started by the chain-initiating step, which produces many chlorine atoms. As a result, there are many methyl radicals as well in the reaction mixture. Sometimes, steps occur in which two radicals combine. These **chain-terminating steps** are so named because each destroys two radicals in a combination reaction and no radicals are formed. Thus, subsequent chain reactions derived from those radicals cannot occur. Three possible chain-terminating steps are

$$:\ddot{\underset{..}{Cl}}\cdot + \cdot\ddot{\underset{..}{Cl}}: \longrightarrow :\ddot{\underset{..}{Cl}}-\ddot{\underset{..}{Cl}}:$$

$$CH_3\cdot + CH_3\cdot \longrightarrow CH_3-CH_3$$

$$CH_3\cdot + \cdot\ddot{\underset{..}{Cl}}: \longrightarrow CH_3-\ddot{\underset{..}{Cl}}:$$

Physical Properties of Haloalkanes

The atomic radii of the halogens increase going from top to bottom in the periodic table. This trend is reflected in the bond lengths of the carbon-halogen bond (Table 3.5).

As a result of the greater electronegativity of the halogens, the carbon atom of the carbon-halogen bond bears a partial positive charge and the halogen atom has a partial negative charge.

$$\overset{\delta^+}{C}-\overset{\delta^-}{X} \qquad \text{where X = F, Cl, Br, I}$$

Table 3.5 Properties of Haloalkanes

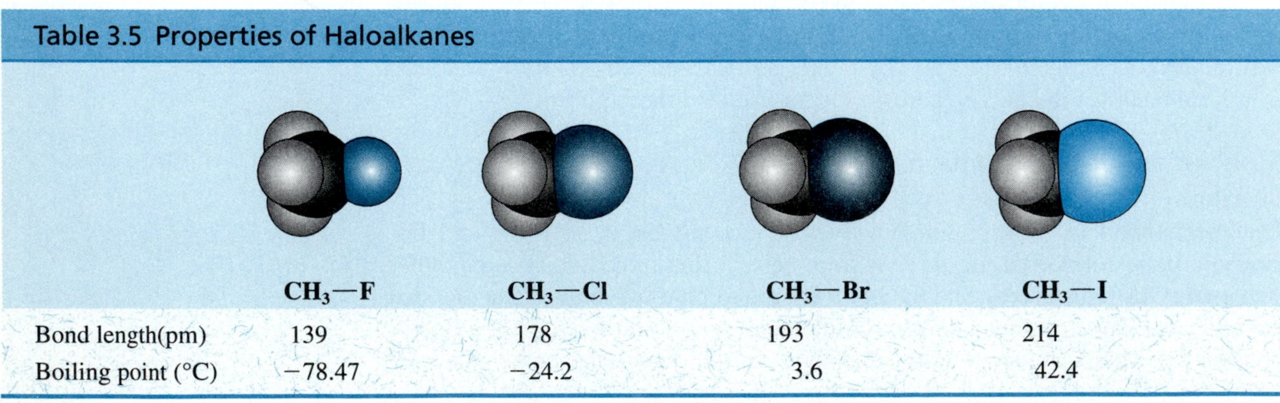

	CH_3-F	CH_3-Cl	CH_3-Br	CH_3-I
Bond length(pm)	139	178	193	214
Boiling point (°C)	−78.47	−24.2	3.6	42.4

The polarizability of the halogen atoms—the ease with which electron density changes because of interactions with other atoms or molecules—increases as we move down the periodic table: $F < Cl < Br < I$. Because highly polarizable atoms interact more strongly by London forces than less polarizable atoms, the intermolecular forces for haloalkanes increase in the order $RF < RCl < RBr < RI$. The effect of intermolecular forces is reflected in the boiling points of haloalkanes, which increase in the same order as the polarizability of their halogen components (Table 3.5).

EXAMPLE 3-8

How many mono-, di-, and trichlorinated compounds result from the chlorination of ethane?

Solution

The two carbon atoms in ethane are equivalent, and only one monochlorinated compound (CH_3CH_2Cl) can result. The two carbon atoms in this product are not equivalent. Substitution by a second chlorine atom results in two isomers.

CH_3CHCl_2	$ClCH_2CH_2Cl$
1,1-dichloroethane	1,2-dichloroethane

In subsequent reactions of these products, three chlorine atoms may be located on a single carbon atom. Another product has two chlorine atoms on the same carbon atom and one on the other.

CH_3CCl_3	$ClCH_2CHCl_2$
1,1,1-trichloroethane	1,1,2-trichloroethane

Problem 3.8

How many different mono- and dichlorinated compounds can result from chlorination of cyclobutane?

3.10 Nomenclature of Haloalkanes

Low molecular weight haloalkanes are often named using the name of the alkyl group followed by the name of the halide.

CH_3-CH_2-Br	$CH_3-CH(CH_3)-F$	$CH_3-C(CH_3)_2-Cl$
ethyl bromide	isopropyl fluoride	*tert*-butyl chloride

Haloalkanes are named in the IUPAC system by an extension of the rules outlined in Section 3.3 for alkanes. Halogen atoms are identified and located using appropriate prefixes. The IUPAC rules are as follows:

1. A parent chain having no branching alkyl groups is numbered so that the carbon atom bearing the halogen atom has the lowest number. If the parent chain has branching alkyl groups, number the chain from the end nearer the first substituent, regardless of whether it is an alkyl group or a halogen atom.

Freons, Free Radicals, and the Ozone Layer

Many polyhalogenated hydrocarbons have been synthesized by chemical industry for a variety of commercial uses. Based on the individual physical properties, polyhalogenated alkanes have been designed to serve as solvents, dry-cleaning agents, anesthetics, and refrigerants. The extreme stability of these compounds also makes them commercially appealing. For example, halogenated alkanes are less flammable than alkanes. Extensively halogenated compounds such as carbon tetrachloride will not burn at all. At one time carbon tetrachloride was used in fire extinguishers to provide an inert atmosphere to prevent oxygen from reaching the flames. Today, CF_3Br is one of the gaseous compounds known as halons that are used in environmental fire suppression systems where the use of water is not advised, such as in large computer facilities. They are released in much the same way that water is released in automatic sprinkler systems.

Reduced combustibility makes haloalkanes useful for many purposes. For example, hydrocarbons have been used as refrigerants and as aerosol propellants, but the danger of combustion is a serious drawback. Thus, halogenated alkanes were developed to avoid the danger of explosions.

Fluoroalkanes, manufactured under the trade name **Freons,** were found to have a wide range of useful physical properties. These compounds also contain chlorine and are thus also known as chlorofluorocarbons or **CFCs.** In general, Freons are nonflammable, odorless, noncorrosive, and nontoxic. Some Freons have been designed for use as refrigerants in refrigerators and air conditioners. Because the Freons are gases, they have been used as propellants in spray cans. Freons are also used as blowing agents to produce rigid foams, which are used in insulation for ice chests, and flexible foams, which are used in pillows and cushions. The wetting properties of some Freons make them suitable as cleaning fluids for printed circuit boards in computers.

The use of these gaseous compounds was so extensive that it is estimated that more than a billion pounds of CFCs were released into the atmosphere in the 1970s. CFC-11 (trichlorofluoromethane) and CFC-12 (dichlorodifluoromethane) were two of the most widely used Freons. Currently over one hundred million cars on the road in the United States use CFC-12 in air conditioners.

Cl
|
Cl—C—F
|
Cl

trichlorofluoromethane

Cl
|
F—C—F
|
Cl

dichlorodifluoromethane

The Freons are not biodegraded in the lower atmosphere and were once thought to present no environmental hazard. Unfortunately, although inert on Earth, these Freons eventually reach the stratosphere, where they absorb ultraviolet radiation from the sun and decompose to produce radicals resulting from cleavage of the carbon-chlorine bond.

$$CF_2Cl_2 \xrightarrow[\text{light}]{\text{UV}} :\ddot{\underset{..}{Cl}}\cdot + \cdot CF_2Cl$$

This process is partially responsible for the destruction of the ozone layer in the stratosphere a region from 8 to 30 miles above the Earth's surface. Ozone in the stratosphere protects us from solar ultraviolet radiation, which splits ozone molecules into molecular oxygen and atomic oxygen. These products then recombine and release heat energy.

$$:\ddot{O}{=}\ddot{O}{-}\ddot{\underset{..}{O}}: \xrightarrow{\text{UV light}} :\ddot{O}{=}\ddot{O}: + \cdot\ddot{\underset{..}{O}}\cdot$$

$$:\ddot{O}{=}\ddot{O}: + \cdot\ddot{\underset{..}{O}}\cdot \longrightarrow :\ddot{O}{=}\ddot{O}{-}\ddot{\underset{..}{O}}: + \text{heat energy}$$

As a result of the combination of the two reactions, the Earth is protected from extensive doses of ultraviolet radiation that can be harmful to life and could increase the incidence of skin cancer in humans. Although not yet well documented, the increased ultraviolet radiation could adversely affect plant life as well as aquatic life at the surface of the world's oceans.

The chlorine radical from CFC-12 (CF_2Cl_2) reacts with ozone in the stratosphere, producing ClO, which reacts with atomic oxygen.

$$:\ddot{O}{=}\ddot{O}{-}\ddot{\underset{..}{O}}: + \cdot\ddot{\underset{..}{Cl}}: \longrightarrow :\ddot{O}{=}\ddot{O}: + \cdot\ddot{\underset{..}{O}}{-}\ddot{\underset{..}{Cl}}:$$

$$\cdot\ddot{\underset{.}{O}}\cdot + \cdot\ddot{\underset{..}{O}}{-}\ddot{\underset{..}{Cl}}: \longrightarrow :\ddot{O}{=}\ddot{O}: + \cdot\ddot{\underset{..}{Cl}}:$$

Note that a chlorine radical reacts in the first equation and is a product in the second equation. Thus, the chlorine radical is a catalyst for the destruction of ozone. The net result of these two steps is the destruction of an ozone molecule for each cycle initiated by the chlorine atom.

$$:\ddot{O}{=}\ddot{O}{-}\ddot{\underset{..}{O}}: + \cdot\ddot{\underset{.}{O}}\cdot \longrightarrow 2\,:\ddot{O}{=}\ddot{O}:$$

These reactions effectively remove ozone and atomic oxygen from the atmosphere, and the protection provided by the ozone layer is diminished. The destruction of the ozone layer is most pronounced in the Antarctic; a similar effect has recently been detected in the Arctic region.

There is clear evidence of an increase in skin cancer in both Australia and New Zealand. Both countries have programs in place to encourage their citizens to avoid excessive exposure to the sun. A measure of the "burn time" is as much a part of the standard weather report as the predicted temperature and weather patterns.

Continued destruction of the ozone layer now appears to be happening at mid-latitudes as well, where more of the Earth's inhabitants live. Ultraviolet radiation can disrupt the structure of DNA, causing genetic damage. As the amount of radiation reaching the Earth's surface increases, the incidence of skin cancer may also increase in these areas.

Concern for the environmental impact of the use of Freons eventually led to the "Montreal Protocol" developed at a meeting in Montreal in 1987. This protocol called for a reduction of the use of Freons and a search for substitute compounds. In addition, the protocol set 1996 as the date for the complete phaseout of CFC-11 and CFC-12. Unfortunately, that goal has not yet been reached, because some countries are not signatories to the protocol and a black market in these compounds provides a lucrative business for some countries. Nevertheless, there has been a pronounced reduction in the release of these Freons into the atmosphere, although many years will be required until the effect on the ozone layer is reversed. It should be noted that even if CFC production and use is completely stopped, natural release of halogen compounds still occurs. Marine organisms and some terrestrial sources release several million tons of chloromethane yearly. Bromomethane is also released in amounts estimated as several hundred thousand tons by kelp, algae, and even volcanoes. (For information about the occurrence of halogenated compounds in marine organisms, see Section 7.1.)

It has been found that Freons containing one or more hydrogen atoms are reasonably stable and can be used as refrigerants. However, they do react at sufficiently rapid rates in the lower atmosphere to destroy them and thus prevent their diffusion into the stratosphere. As is often the case when substitutes must be found, these Freons are more difficult to synthesize and as a result are more expensive. CFC-22 ($CHCl_2F$) has replaced CFC-12 in refrigerators and air conditioners. HCFC-123 (CF_3CHCl_2) is now used as a substitute for CFC-11 as a blowing agent in the production of foams.

$$\begin{array}{c} Cl \\ | \\ H{-}C{-}F \\ | \\ Cl \end{array}$$

Freon 22

$$\begin{array}{c} F \quad Cl \\ | \quad\;\; | \\ F{-}C{-}C{-}H \\ | \quad\;\; | \\ F \quad Cl \end{array}$$

HCFC-123

Recently produced refrigeration units as well as air conditioners used the newly developed Freons. However, older models cannot simply be refilled with new Freons, because they were made to operate based on the physical properties of the original Freons. The cost of retro-fitting an older car's air conditioner may be several hundred dollars. As a result, the total cost for all US cars could be over 10 billion dollars.

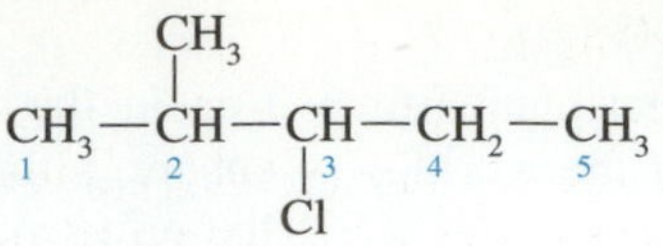

$$CH_3-CH(CH_3)-CH(Cl)-CH_2-CH_3$$

3-chloro-2-methylpentane

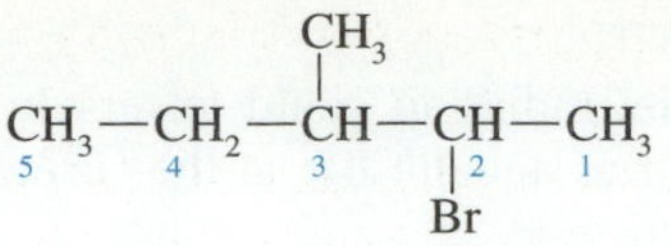

$$CH_3-CH_2-CH(CH_3)-CH(Br)-CH_3$$

2-bromo-3-methylpentane

2. Two or more halogen atoms of the same type are indicated with the prefixes *di-*, *tri-*, etc. Each halogen atom is given a number that corresponds to its position in the parent chain.

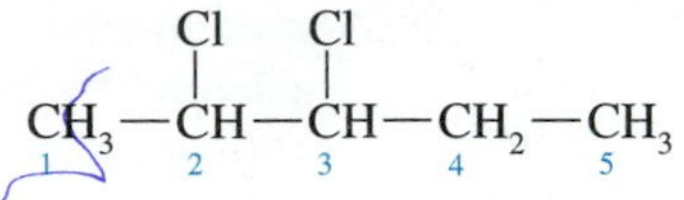

$$CH_3-CH(Cl)-CH(Cl)-CH_2-CH_3$$

2,3-dichloropentane

$$CH_3-CH_2-CH(Br)-C(Br)_2-CH_2-CH_3$$

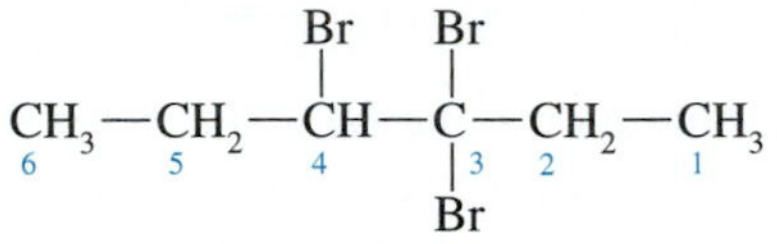

3,3,4-tribromohexane

3. Compounds containing different halogen atoms are numbered according to their positions on the chain and the atoms are listed in alphabetical order. If the chain can be numbered from either end based on the location of the substituents, the C-1 atom is selected to be at the end nearer the substituent that has alphabetical precedence, whether it is an alkyl group or a halogen atom.

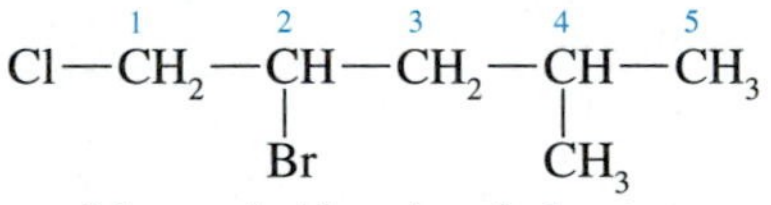

$$Cl-CH_2-CH(Br)-CH_2-CH(CH_3)-CH_3$$

2-bromo-1-chloro-4-methylpentane

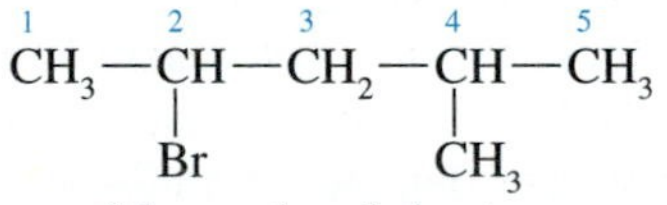

$$CH_3-CH(Br)-CH_2-CH(CH_3)-CH_3$$

2-bromo-4-methylpentane

4. Halocycloalkanes are numbered from the carbon atom bearing the halogen atom unless a higher priority group, such as a double bond, takes precedence (Section 4.4). Carbon atoms in the ring are numbered to give the lower number to the substituent.

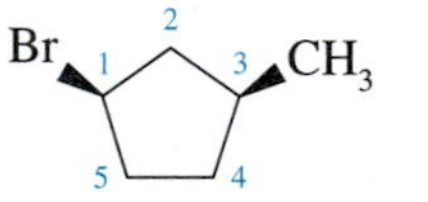

cis-1-bromo-3-methylcyclopentane

Summary of Reactions

1. Free Radical Chlorination (Section 3.9)

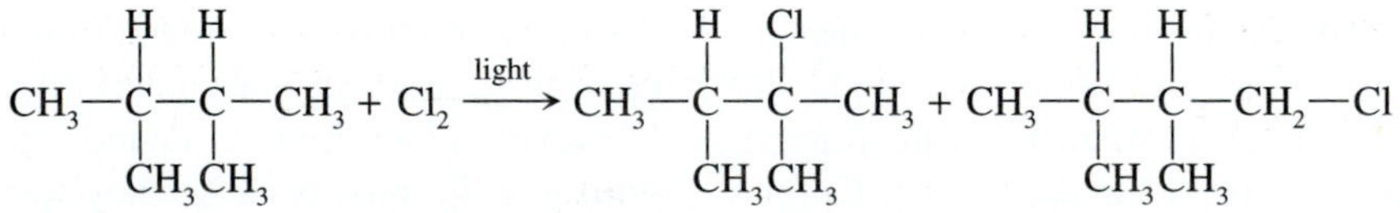

$$CH_3-CH(CH_3)-CH(CH_3)-CH_3 + Cl_2 \xrightarrow{\text{light}} CH_3-CH(CH_3)-C(Cl)(CH_3)-CH_3 + CH_3-CH(CH_3)-CH(CH_3)-CH_2-Cl$$

2. Free Radical Bromination (Section 3.9)

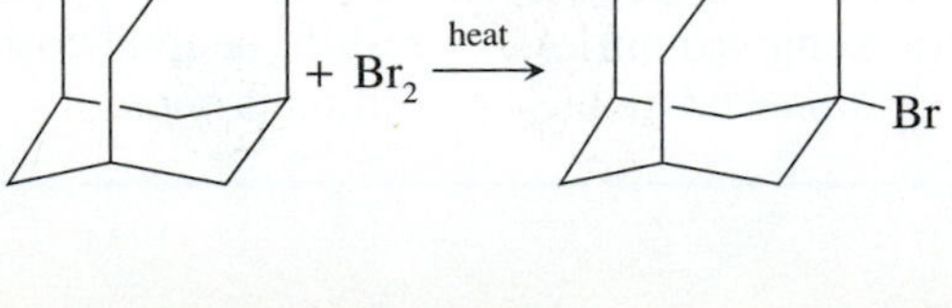

$$\text{adamantane} + Br_2 \xrightarrow{\text{heat}} \text{1-bromoadamantane}$$

Explorations with Molecular Models

1. Construct the following model for propane. Classify each carbon atom starting from left to right. Replace the hydrogen atom that is pointed toward you on the center carbon atom with a methyl group. How does this substitution affect the degree of substitution of the central carbon atom? Are the three methyl groups of the compound equivalent or nonequivalent? Draw a standard two-dimensional representation of the molecule using 90° and 180° angles.

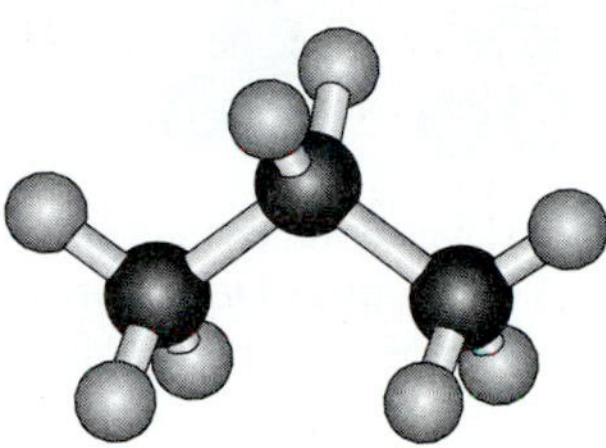

2. Construct a second model for propane as in Exploration Exercise 1. Replace the hydrogen atom that is pointed away from you on the center carbon atom with a methyl group. Answer the same questions as in Exercise 1. What relationship exists between this structure and the one derived in Exercise 1?

3. Start with the model for propane constructed for Exploration Exercise 1. Replace a hydrogen atom on the carbon atom on the left with a methyl group. Classify each carbon atom starting from left to right. Draw a standard two-dimensional representation of the molecule using 90° and 180° angles. How many sets of nonequivalent carbon atoms are represented?

4. Start with the model for propane constructed for Exploration Exercise 1. Replace a hydrogen atom on the carbon atom on the right with a methyl group. Classify each carbon atom starting from left to right. Draw a standard two-dimensional representation of the molecule using 90° and 180° angles. How many sets of nonequivalent carbon atoms are represented? What relationship exists between this structure and the one derived in Exercise 3?

5. Construct the following model of pentane, which has a zigzag arrangement of carbon atoms. Classify each carbon atom starting from left to right. Draw a standard two-dimensional representation of the molecule using 90° and 180° angles. How many sets of nonequivalent carbon atoms are represented?

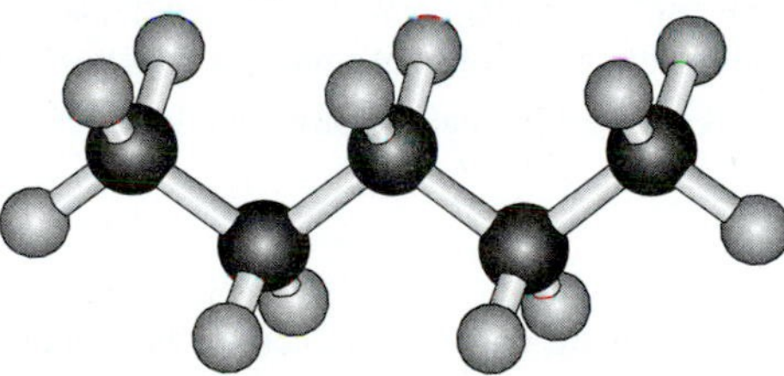

6. Using the model of pentane constructed in Exploration Exercise 5, rotate about as many carbon-carbon bonds as necessary to move the terminal carbon atoms as close to one another as possible. Remove one hydrogen atom from each terminal carbon atom and bond the carbon atoms to one another to give cyclopentane. How does its molecular formula differ from that of pentane?

7. Using the model of cyclopentane generated in Exploration Exercise 6, determine the number of sets of equivalent carbon atoms and the number of sets of equivalent hydrogen atoms.

8. Substitute a hydrogen atom of the cyclopentane model generated in Exploration Exercise 6 with a chlorine atom. Determine the number of sets of equivalent carbon atoms. Determine the number of sets of equivalent hydrogen atoms.

9. Construct the following model of 1,2-dichloroethane. What term is used to describe this conformation? Rotate one of the chlorine atoms about the carbon-carbon bond by 120°. What term is used to describe this conformation? Which of these two conformations is the more stable?

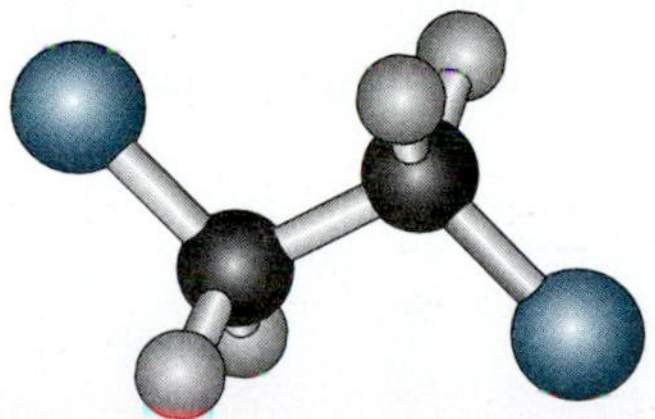

10. Prepare the following model of 2-methylbutane. Arrange the model to sight along the C-2 to C-3 bond. Draw the Newman projection formula.

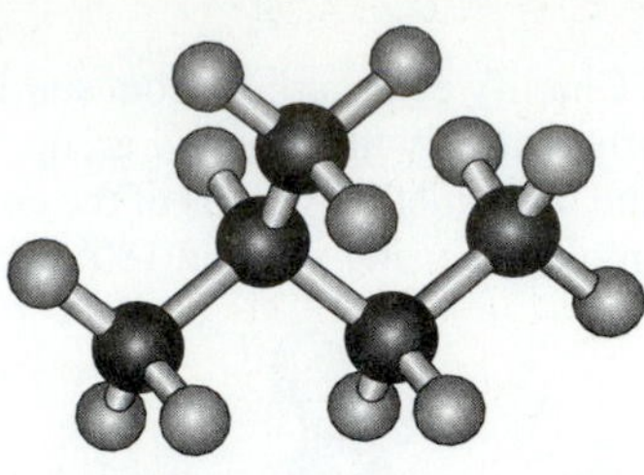

11. Construct the following model of chlorocyclohexane. What term is used to describe the location of the chlorine atom? Push the C-1 atom downward while pushing the C-4 atom upward to obtain a second chair conformation. What term is used to describe the new location of the chlorine atom? Which of these two conformations is the more stable?

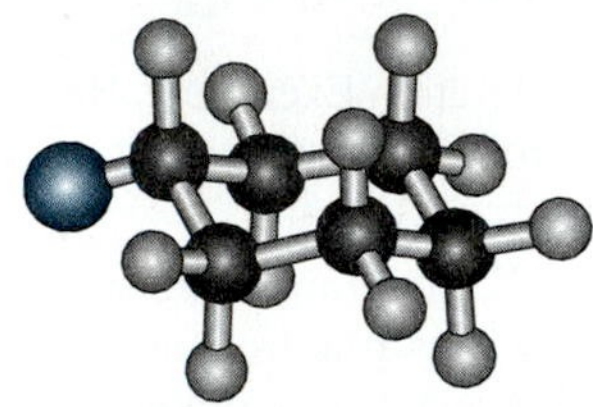

Exercises

Molecular Formulas

3.1 Beeswax contains approximately 10% hentriacontane, which is a normal alkane with 31 carbon atoms. What is the molecular formula for hentriacontane? Write a completely condensed formula for hentriacontane.

3.2 Currently, the largest known normal alkane has 390 carbon atoms. What is the molecular formula for this alkane? Write a completely condensed formula for it.

Structural Formulas

3.3 Redraw each of the following so that the longest continuous chain is written horizontally.

(a) $\begin{matrix} CH_3- & CH_2 & \\ & | & \\ & CH_2 & -CH_3 \end{matrix}$ (b) $\begin{matrix} CH_2 & -CH_2- & CH & -CH_2-CH_3 \\ | & & | & \\ CH_3 & & CH_2 & -CH_3 \end{matrix}$ (c) $\begin{matrix} CH_3- & CH & -CH_2-CH_3 \\ & | & \\ CH_3- & CH_2 & \end{matrix}$

3.4 Redraw each of the following so that the longest continuous chain is written horizontally.

(a) $\begin{matrix} CH_3- & CH & - & CH_2 \\ & | & & | \\ & CH_3 & & CH_3 \end{matrix}$ (b) $\begin{matrix} CH_3- & CH & -CH_2- & CH_2 \\ & | & & | \\ CH_3- & CH_2 & & CH_3 \end{matrix}$ (c) $\begin{matrix} CH_3- & CH & -CH_2-CH_3 \\ & | & \\ CH_3- & CH & -CH_2-CH_3 \end{matrix}$

Alkyl Groups

3.5 What is the common name for each of the following alkyl groups?

(a) $CH_3-CH_2-CH_2-$ (b) $\begin{matrix} CH_3-CH_2- & CH & - \\ & | & \\ & CH_3 & \end{matrix}$ (c) $\begin{matrix} CH_3- & CH & -CH_2- \\ & | & \\ & CH_3 & \end{matrix}$

3.6 What is the common name for each of the following alkyl groups?

(a) $\begin{matrix} CH_3- & CH & - \\ & | & \\ & CH_3 & \end{matrix}$ (b) $CH_3-CH_2-CH_2-CH_2-$ (c) $\begin{matrix} & CH_3 & \\ & | & \\ CH_3- & C & -CH_3 \\ & | & \end{matrix}$

3.7 What is the IUPAC name for each of the following alkyl groups?

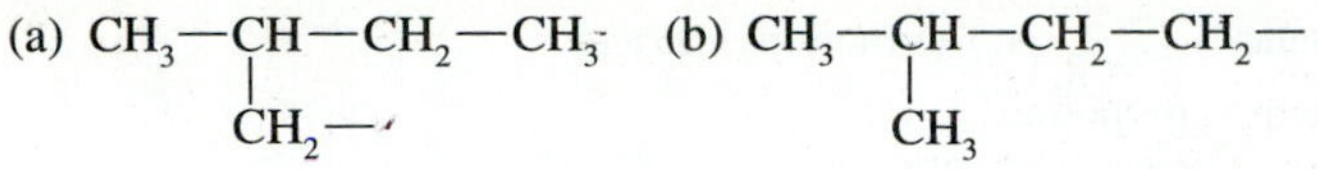

3.8 What is the IUPAC name for each of the following alkyl groups?

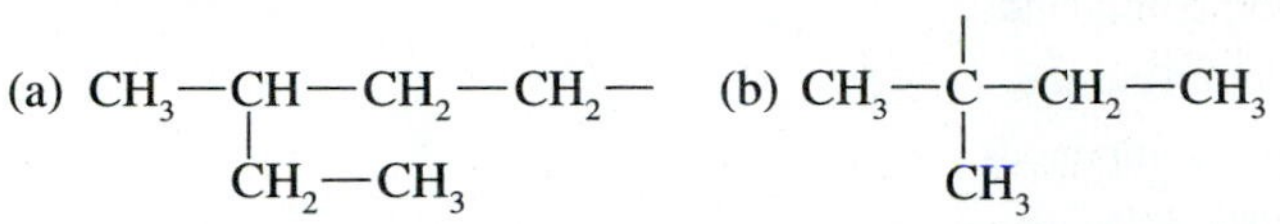

3.9 The spermicide octoxynol-9 is used in diverse contraceptive products. Name the alkyl group to the left of the benzene ring.

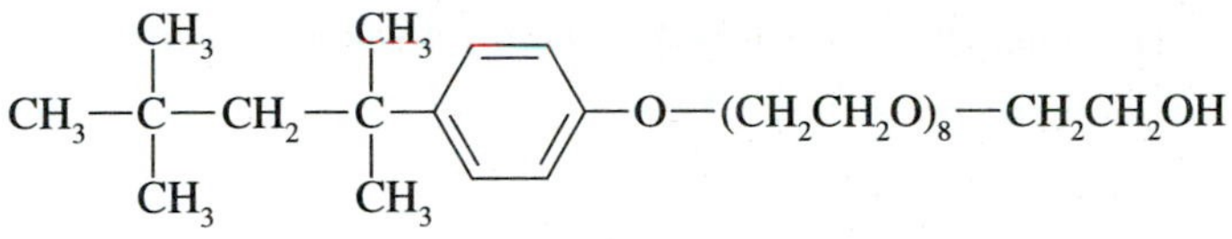

3.10 Vitamin E actually represents a series of closely related compounds called tocopherols. Name the complex alkyl group present in α-tocopherol.

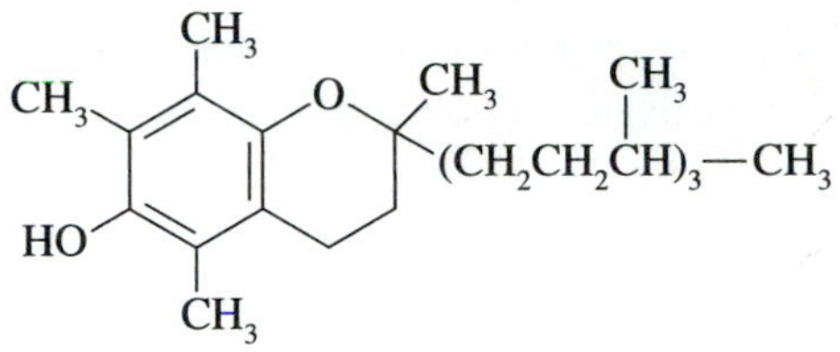

Nomenclature of Alkanes

3.11 Give the IUPAC name for each of the following compounds.

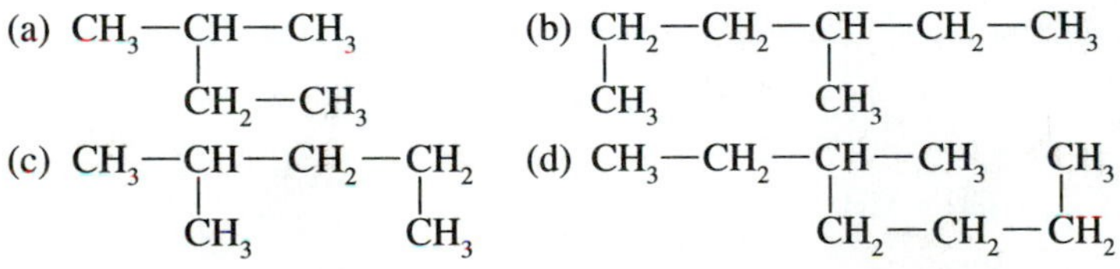

3.12 Give the IUPAC name for each of the following compounds.

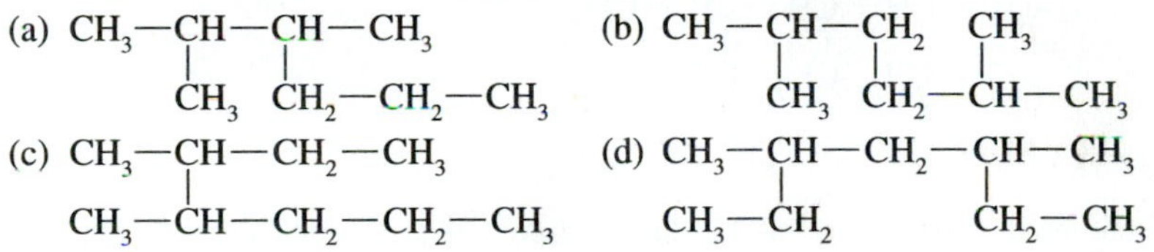

3.13 Give the IUPAC name for the following compound.

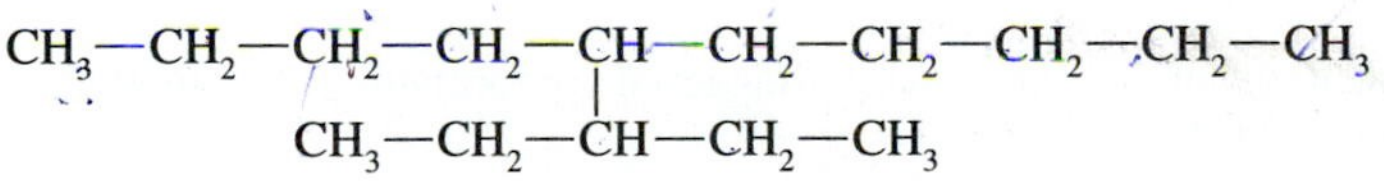

3.14 Give the IUPAC name for the following compound.

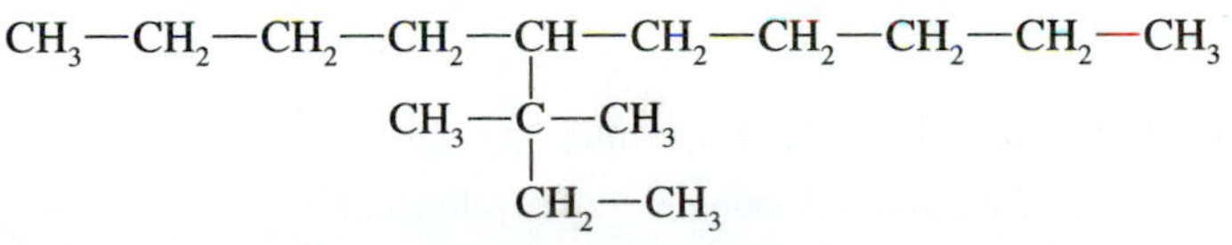

3.15 Write the structural formula for each of the following compounds.

(a) 3,4-dimethylhexane (b) 2,2,3-trimethylpentane (c) 2,3,4,5-tetramethylhexane

3.16 Write the structural formula for each of the following compounds.

(a) 3-ethylhexane (b) 2,2,4-trimethylhexane (c) 2,2,3,3-tetramethylpentane

3.17 Write the structural formula for each of the following compounds.

(a) 4-(1-methylethyl)heptane (b) 4-(1,1-dimethylethyl)octane

3.18 Write the structural formula for each of the following compounds.

(a) 4-(2-methylpropyl)octane (b) 5-(2,2-dimethylpropyl)nonane

Isomers

3.19 There are nine isomeric C_7H_{16} compounds. Which isomers have a single methyl group as a branch?

3.20 There are nine isomeric C_7H_{16} compounds. Which isomers have two methyl groups as branches and are named as dimethyl substituted pentanes?

Classification of Carbon Atoms

3.21 Draw the structure of a compound with molecular formula C_5H_{12} that has one quaternary and four primary carbon atoms.

3.22 Draw the structure of a compound with molecular formula C_6H_{14} that has two tertiary and four primary carbon atoms.

3.23 Determine the number of primary, secondary, tertiary, and quaternary carbon atoms in each of the following compounds.

(a) $$\begin{array}{ccccc} & & CH_3 & & \\ & & | & & \\ CH_3 & - & C & - & CH_3 \\ & & | & & \\ & & CH_3 & & \end{array}$$

(b) $$\begin{array}{ccccccc} CH_3 & - & CH & - & CH_2 & - & CH_3 \\ & & | & & & & \\ & & CH_3 & & & & \end{array}$$

(c) $$\begin{array}{ccccccccc} CH_3 & - & CH & - & CH_2 & - & CH_2 & - & CH_3 \\ & & | & & & & & & \\ & & CH_3 & & & & & & \end{array}$$

(d) $$\begin{array}{ccccccc} CH_3 & - & CH & - & CH & - & CH_3 \\ & & | & & | & & \\ & & CH_3 & & CH_3 & & \end{array}$$

3.24 Determine the number of primary, secondary, tertiary, and quaternary carbon atoms in each of the following compounds:

(a) $$\begin{array}{ccccccccc} & & CH_3 & & & & CH_3 & & \\ & & | & & & & | & & \\ CH_3 & - & C & - & CH_2 & - & C & - & CH_3 \\ & & | & & & & | & & \\ & & CH_3 & & & & CH_3 & & \end{array}$$

(b) $$\begin{array}{ccccccccc} CH_3 & - & CH & - & CH_2 & - & CH & - & CH_3 \\ & & | & & & & | & & \\ & & CH_3 & & & & CH_3 & & \end{array}$$

(c) $$\begin{array}{ccccccccc} CH_3 & - & CH_2 & - & CH & - & CH_2 & - & CH_3 \\ & & & & | & & & & \\ & & & & CH_3 & & & & \end{array}$$

(d) $$\begin{array}{ccccccccc} CH_3 & - & CH & - & CH & - & CH & - & CH_3 \\ & & | & & | & & | & & \\ & & CH_3 & & CH_3 & & CH_3 & & \end{array}$$

Conformations of Alkanes

3.25 Draw the Newman projection of the staggered conformation of 2,2-dimethylpropane about the C(1)—C(2) bond.

3.26 Draw the Newman projections of the two possible staggered conformations of 2,3-dimethylbutane about the C(2)—C(3) bond.

3.27 Draw the Newman projections of the two possible staggered conformations of 2-methylbutane about the C(2)—C(3) bond. Which is the more stable?

3.28 Draw the Newman projections of the two possible staggered conformations of 2,2-dimethylpentane about the C(3)—C(4) bond. Which is the more stable?

Cycloalkanes

3.29 Write fully condensed planar formulas for each of the following compounds.

(a) chlorocyclopropane (b) 1,1-dimethylcyclobutane (c) cyclooctane

3.30 Write fully condensed planar formulas for each of the following compounds.

(a) bromocyclopentane (b) 1,1-dichlorocyclopropane (c) cyclopentane

3.31 Name each of the following compounds.

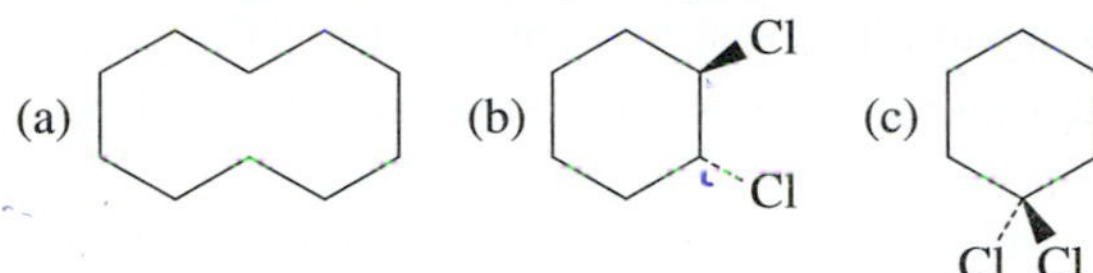

3.32 Name each of the following compounds.

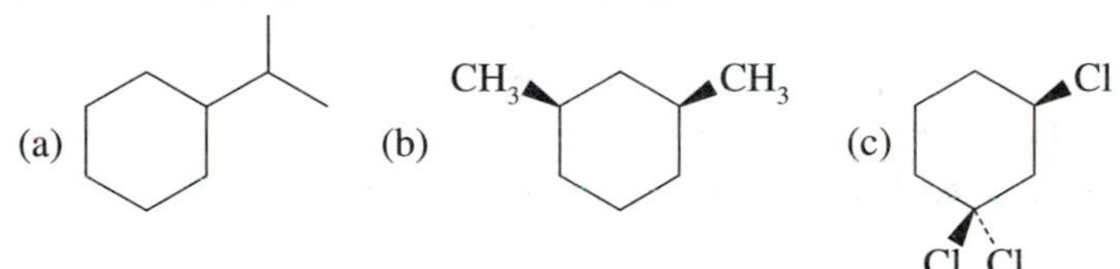

3.33 What is the molecular formula of each of the following compounds?

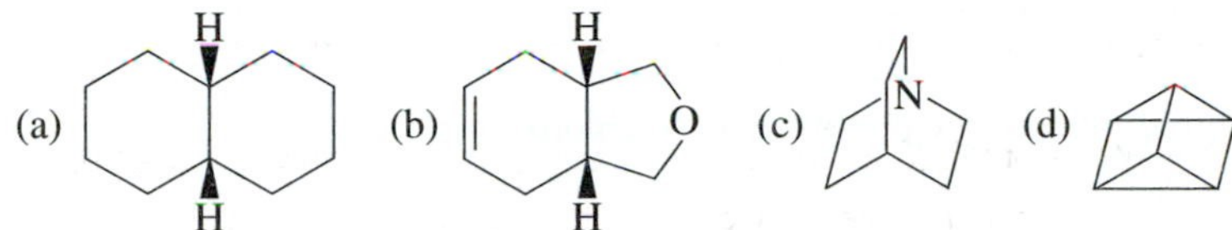

3.34 What is the molecular formula of each of the following compounds?

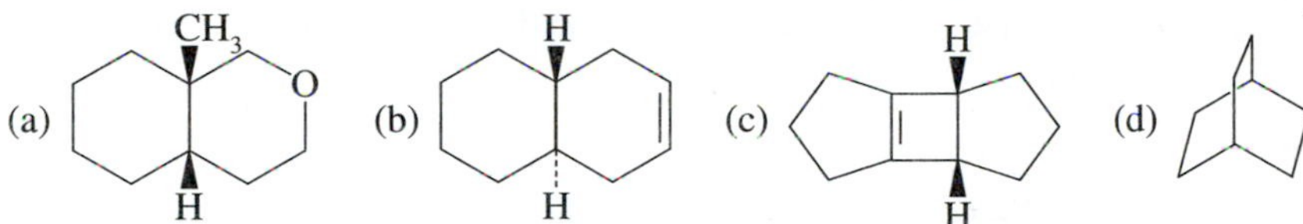

Conformations of Cyclohexanes

3.35 Draw the two chair conformations of fluorocyclohexane. Would you expect the energy difference between these two conformations to be greater or less than the energy difference between the two conformations of methylcyclohexane? Why?

3.36 Draw the two chair conformations of *tert*-butylcyclohexane. Would you expect the energy difference between these two conformations to be greater or less than the energy difference between the two conformations of methylcyclohexane? Why?

3.37 Draw the most stable conformation of each of the following compounds.

(a) *trans*-1,4-dimethylcyclohexane (b) *cis*-1,3-dimethylcyclohexane

3.38 Draw the most stable conformation of each of the following compounds.

(a) 1,1,4-trimethylcyclohexane (b) 1,1,3-trimethylcyclohexane

Properties of Hydrocarbons

3.39 Cyclopropane is an anesthetic, but it cannot be used in operations in which electrocauterization of tissue is done. Why?

3.40 Which compound should have the higher octane number, cyclohexane or methylcyclopentane?

3.41 Which of the isomeric C_8H_{18} compounds has the highest boiling point? Which has the lowest boiling point?

3.42 The boiling point of methylcyclopentane is lower than the boiling point of cyclohexane. Suggest a reason why.

Halogenation of Hydrocarbons

3.43 How many products can result from the substitution of a chlorine atom for one hydrogen atom in each of the following compounds?

(a) propane (b) butane (c) methylpropane (d) cyclohexane

3.44 How many products can result from the substitution of a chlorine atom for one hydrogen atom in each of the following compounds?

(a) 2-methylbutane (b) 2,2-dimethylbutane (c) 2,3-dimethylbutane (d) cyclopentane

3.45 Write the structure of the radicals that can result from abstraction of a hydrogen atom from pentane.

3.46 Write the structure of the hydrocarbon product that can result from a termination step in the free radical chlorination of ethane.

3.47 Halothane, an anesthetic, has the formula C_2HF_3ClBr. Draw structural formulas for the four possible isomers of this molecular formula.

3.48 A saturated refrigerant has the molecular formula C_4F_8. Draw structural formulas for two possible isomers of this compound.

Properties of Haloalkanes

3.49 Which compound is more polar, methylene chloride (CH_2Cl_2) or carbon tetrachloride (CCl_4)?

3.50 Tribromomethane is more polar than tetrabromomethane, but their boiling points are 150 and 189°C, respectively. Explain why the more polar compound has the lower boiling point.

3.51 The densities of chloroiodomethane and dibromomethane are 2.42 and 2.49 g/mL, respectively. Why are these values similar?

3.52 The density of 1,2-dichloroethane is 1.26 g/mL. Predict the density of 1,1-dichloroethane.

Spectroscopy

3.53 If you have read Chapter 18 and the material has been assigned by your professor, then Exercises 18.11, 18.12, 18.13, 18.14, 18.15, 18.16, 18.21, 18.22, 18.24, 18.30, and 18.31, which are based on the structure of hydrocarbons and haloalkanes, may be done.

CHAPTER 4

ALKENES AND ALKYNES

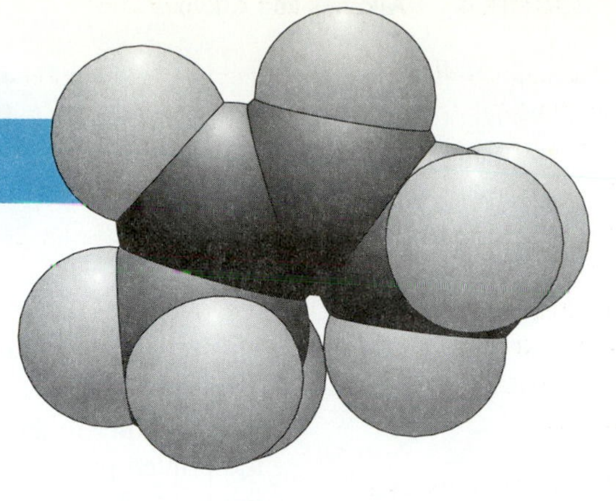

Overview

4.1 Unsaturated Hydrocarbons

Organic compounds with carbon-carbon multiple bonds contain fewer hydrogen atoms than structurally related alkanes or cycloalkanes. For this reason, these compounds are said to be **unsaturated.** In this chapter we will focus on two classes of unsaturated compounds: alkenes and alkynes. **Alkenes** contain a carbon-carbon double bond; **alkynes** contain a carbon-carbon triple bond. **Aromatic** hydrocarbons, compounds that contain a benzene ring or structural units that resemble a benzene ring, will be discussed in the next chapter.

The π bond of a carbon-carbon double bond and the two π bonds of a carbon-carbon triple bond are the sites of specific reactions. We recall that π bonds involve a side-by-side overlap of $2p$ orbitals on adjacent carbon atoms. The hybridization of carbon and the model of σ and π bonds in ethylene (an alkene) and acetylene (an alkyne) were introduced in Section 1.8.

In this chapter the chemistry of some alkadienes, informally called dienes, will also be examined. We will focus on **conjugated dienes,** compounds in which two double bond units are joined by one single bond. Conjugated dienes undergo reactions that differ from those of individual double bonds. Natural rubber is a polymer of isoprene, a conjugated diene. Some synthetic rubbers called neoprenes are produced from chloroprene and are used in products such as industrial hoses.

$$CH_2{=}\overset{\displaystyle CH_3}{\overset{|}{C}}{-}CH{=}CH_2 \qquad\qquad CH_2{=}\overset{\displaystyle Cl}{\overset{|}{C}}{-}CH{=}CH_2$$

isoprene chloroprene

Alkenes

The simplest alkene, C_2H_4, is commonly called ethylene. Its IUPAC name is ethene. The IUPAC names of alkenes use the suffix *-ene*. In the structure of ethene shown in Figure 4.1, all six atoms, two carbon atoms and four hydrogen atoms, are located in the same plane. The plane may be

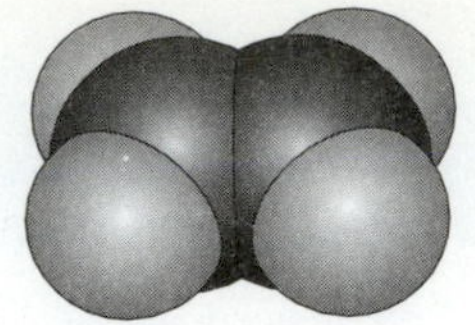

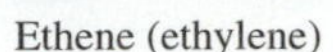

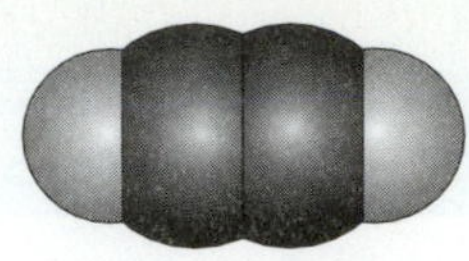

Ethene (ethylene) Ethyne (acetylene)

FIGURE 4.1
Molecular Models of Unsaturated Hydrocarbons

written either in the page or perpendicular to it. If the plane is perpendicular to the printed page, the carbon-hydrogen bonds project in front of and in back of the page. As before, wedge-shaped lines represent bonds in front of the page and dashed lines those behind the page.

representation in the plane of the page

representation perpendicular to the plane of the page

A double bond decreases the number of hydrogen atoms in a molecule by two compared to the number in alkanes, so the general formula for an alkene is C_nH_{2n}. Each additional double bond reduces the number of hydrogen atoms by another two.

Alkynes

Acetylene gas burns with enough heat to melt metals.

The simplest alkyne, C_2H_2, is commonly called acetylene. Unfortunately, the common name ends in *-ene,* which seems to suggest that the compound contains a double bond. Such confusion is one reason why IUPAC names are so important for clear communication in chemistry. The IUPAC name for C_2H_2 is ethyne.

The structure of ethyne is shown in Figure 4.1. All four atoms lie in a straight line. Each H—C≡C bond angle is 180°. In other alkynes also, the two triple-bonded carbon atoms and the two atoms directly attached to them all lie in a straight line.

The triple bond in an alkyne decreases the number of hydrogen atoms in the molecule by four compared to alkanes. As a result, the general molecular formula for alkynes is C_nH_{2n-2}.

Classification of Alkenes and Alkynes

It is often useful to describe alkenes and alkynes based on the number of alkyl groups attached to the double or triple bond unit. In the case of double bonds, we speak of the degree of substitution at the site of the double bond. A **monosubstituted** alkene has a single alkyl group attached to one sp^2 hybridized carbon atom of the double bond. An alkene whose double bond is at the end of a chain of carbon atoms is also sometimes called a terminal alkene. Alkenes with two, three, and four alkyl groups bonded to the double bond unit are **disubstituted, trisubstituted,** and **tetrasubstituted,** respectively.

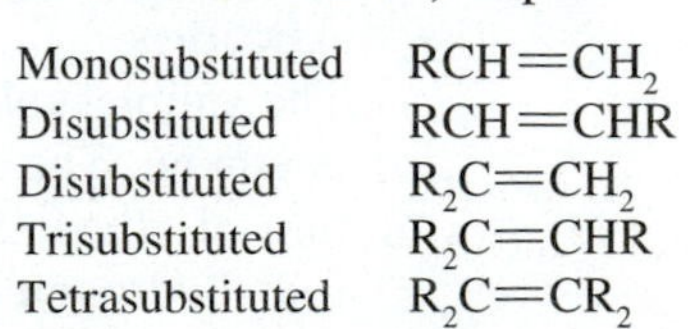

Monosubstituted	$RCH{=}CH_2$
Disubstituted	$RCH{=}CHR$
Disubstituted	$R_2C{=}CH_2$
Trisubstituted	$R_2C{=}CHR$
Tetrasubstituted	$R_2C{=}CR_2$

Table 4.1 Boiling Points of Alkanes, Alkenes, and Alkynes

Alkanes	Boiling Point (°C)	Alkenes	Boiling Point (°C)	Alkynes	Boiling Point (°C)
pentane	36	1-pentene	30	1-pentyne	40
hexane	69	1-hexene	63	1-hexyne	71
heptane	98	1-heptene	94	1-heptyne	100
octane	126	1-octene	121	1-octyne	125

In general, alkyl groups increase the stability of a double bond in an alkene. Thus, a disubstituted alkene is more stable than a monosubstituted alkene. As a consequence, if a chemical reaction can lead to either of these two products, the disubstituted alkene predominates.

The classes of alkynes are more limited. Only one alkyl group can be bonded to each of the two carbon atoms of the triple bond. If one alkyl group is bonded to one *sp* hybridized carbon atom of the triple bond, the compound is a **monosubstituted** alkyne (R—C≡C—H). It is also called a **terminal** alkyne, because the triple bond is on the end of the carbon chain. When alkyl groups are bonded to each carbon atom of the triple bond, the compound is **disubstituted** or an **internal** alkyne (R—C≡C—R).

Physical Properties of Alkenes and Alkynes

The physical properties of the homologous series of alkenes (C_nH_{2n}) and alkynes (C_nH_{2n-2}) are similar to those of the homologous series of alkanes (C_nH_{2n+2}). The compounds in both classes of unsaturated hydrocarbons are nonpolar. The members of each series containing fewer than five carbon atoms are gases at room temperature. As in the case of alkanes, the boiling points of the alkenes and alkynes increase with an increase in the number of carbon atoms in the molecule because the London forces increase (Table 4.1).

EXAMPLE 4-1

Caryophyllene, which is responsible for the odor of oil of cloves, contains 15 carbon atoms. The compound has two rings and two double bonds. What is the molecular formula for caryophyllene?

Solution

Each ring decreases the number of hydrogen atoms by two compared to an alkane. A compound with two rings would have the formula C_nH_{2n-2}. (Remember that for n carbon atoms in an alkane, there are $2n + 2$ hydrogen atoms.) For each double bond, another two hydrogen atoms must be subtracted, to give C_nH_{2n-6} for two double bonds as well as the two rings. For 15 carbon atoms, the molecular formula must be $C_{15}H_{24}$.

Problem 4.1

β-Carotene, found in carrots, has 40 carbon atoms and contains two rings and 11 double bonds. What is its molecular formula?

EXAMPLE 4-2

The urine of the red fox contains a scent marker that is an unsaturated thioether. Classify the double bond of the scent marker.

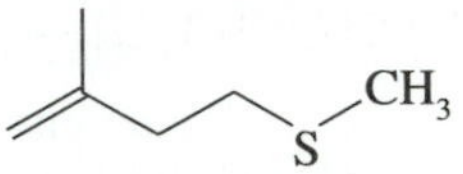

CH_3 S

Solution

The left carbon atom of the double bond is a CH_2 unit. The right carbon atom of the double bond has two bond lines emanating from it. One of these is attached to a methyl group; the other to a methylene unit that is part of the main chain. Therefore, two alkyl groups are bonded to the atoms of the double bond, and the compound is a disubstituted alkene of the general class $CH_2{=}CR_2$.

Problem 4.2

Tremorine is used to treat Parkinson's disease. Classify this alkyne.

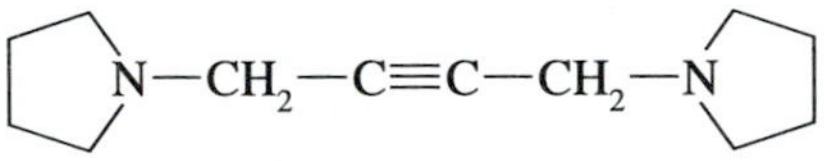

$$N{-}CH_2{-}C{\equiv}C{-}CH_2{-}N$$

Naturally Occurring Unsaturated Compounds

Alkenes and their chemical cousins, the cycloalkenes, are very common in nature. In fact, one is important to the common housefly *(Musca domestica)*. Muscalure, an unbranched alkene containing 23 carbon atoms, is released by the female to attract males. The compound has been synthesized in the laboratory and can be used to lure male flies to traps.

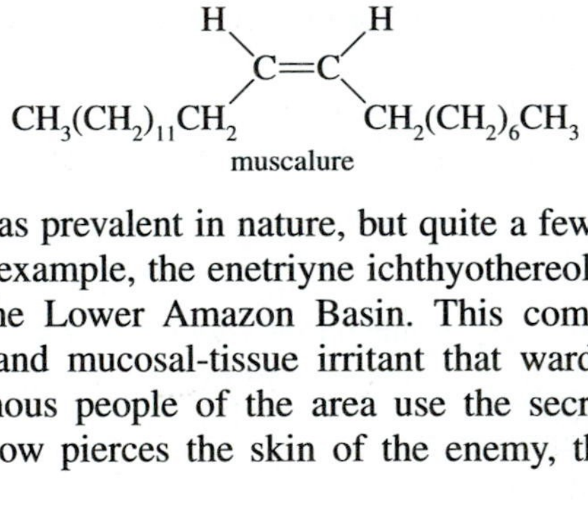

$$\underset{CH_3(CH_2)_{11}CH_2}{\overset{H}{}}\!\!\!\!\!\!\!\!\!\!\!\;C{=}C\;\underset{CH_2(CH_2)_6CH_3}{\overset{H}{}}$$

muscalure

Alkynes are not as prevalent in nature, but quite a few of them are physiologically active. For example, the enetriyne ichthyothereol is secreted from the skin of a frog in the Lower Amazon Basin. This compound is apparently a defensive venom and mucosal-tissue irritant that wards off mammals and reptiles. The indigenous people of the area use the secretion to coat arrowheads. When the arrow pierces the skin of the enemy, the compound causes convulsions.

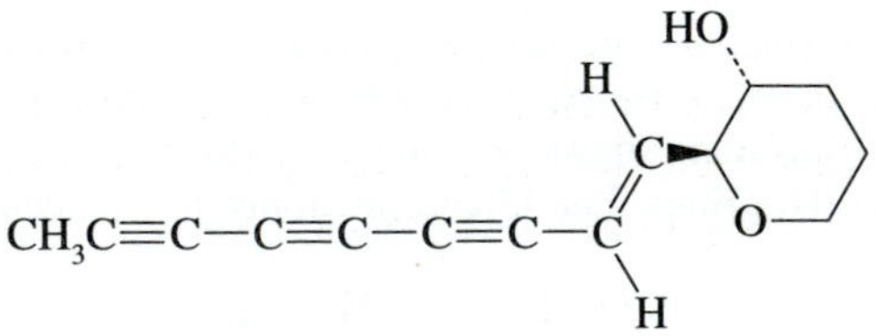

HO H

$$CH_3C{\equiv}C{-}C{\equiv}C{-}C{\equiv}C{-}C(H){=}C(H){-}$$ O

ichthyothereol

You may have encountered the term *polyunsaturated fat* (or *oil*) in various advertisements. A molecule is **polyunsaturated** if it contains several multiple bonds. For example, polyunsaturated oils contain several double bonds; these oils will be discussed in Chapter 13.

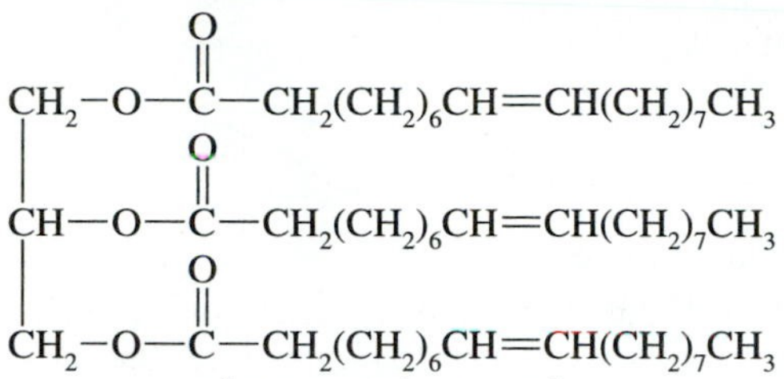

$$
\begin{array}{l}
CH_2-O-\overset{\overset{\displaystyle O}{\|}}{C}-CH_2(CH_2)_6CH{=}CH(CH_2)_7CH_3 \\
| \\
CH-O-\overset{\overset{\displaystyle O}{\|}}{C}-CH_2(CH_2)_6CH{=}CH(CH_2)_7CH_3 \\
| \\
CH_2-O-\overset{\overset{\displaystyle O}{\|}}{C}-CH_2(CH_2)_6CH{=}CH(CH_2)_7CH_3
\end{array}
$$

a polyunsaturated compound

Polyunsaturated compounds are common in nature. For example, vitamin A contains five double bonds.

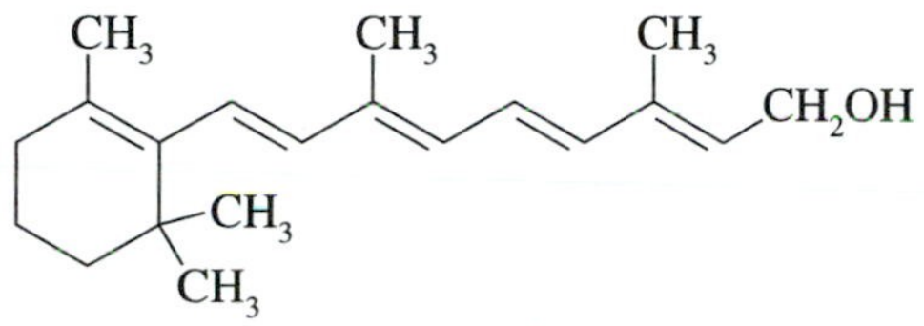

CH_3 CH_3 CH_3 CH_2OH CH_3 CH_3

vitamin A

4.2 Geometric Isomerism

We know that there is free rotation about carbon-carbon single bonds—about 3 kcal/mol is sufficient to rotate about the σ bond of ethane. Free rotation cannot occur around the carbon-carbon double bond of an alkene because of its electronic structure. As a result, the groups bonded to the carbon atoms of the double bond can exist in different spatial or geometric arrangements. These isomers have the same connectivity of atoms, but differ from each other in the geometry about the double bond. Hence, these compounds are called **geometric isomers** or *cis-trans* isomers.

Consider a general alkene whose formula is CXY=CXY. When we draw a more detailed structural formula, we find that two representations are possible.

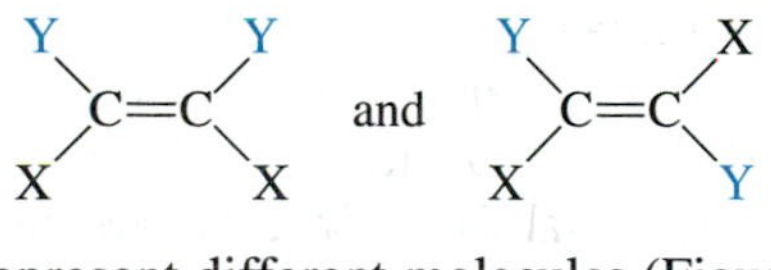

$$
\underset{X}{\overset{Y}{\diagdown}}C{=}C\underset{X}{\overset{Y}{\diagup}} \quad \text{and} \quad \underset{X}{\overset{Y}{\diagdown}}C{=}C\underset{Y}{\overset{X}{\diagup}}
$$

These two structures represent different molecules (Figure 4.2). In the structure on the left, two Y groups are on the same "side" of the molecule; this is the cis isomer. The Y's are on opposite "sides" of the molecule in the structure on the right, which is called the trans isomer.

Cis and trans isomers are possible only if an alkene has two different atoms or groups of atoms attached to each double-bonded carbon atom. Each unsaturated carbon atom in 1,2-dichloroethene has a chlorine atom and a hydrogen atom attached to it. These groups are different.

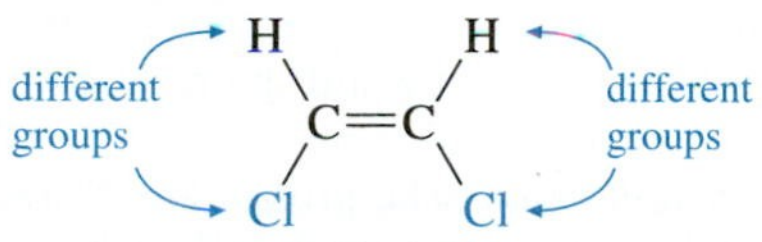

$$
\underset{Cl}{\overset{H}{\diagdown}}C{=}C\underset{Cl}{\overset{H}{\diagup}}
$$

different groups — different groups

1,2-dichloroethene

FIGURE 4.2
Geometric Isomerism of Alkenes

(a) The two X groups are on the same side of the plane that is placed perpendicular to the plane containing the molecule. This is the cis isomer. (b) The two X groups are on the opposite sides of the plane that is placed perpendicular to the plane containing the molecule. This is the trans isomer.

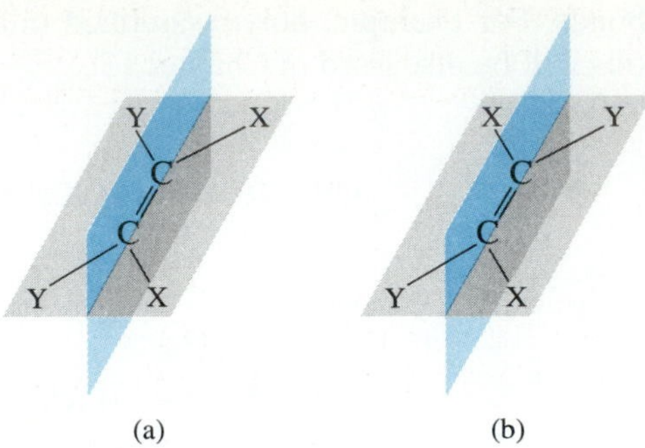

If one of the unsaturated carbon atoms is attached to two identical groups, cis-trans isomerism is not possible. Neither chloroethene nor 1,1-dichloroethene can exist as a cis-trans pair of geometric isomers. There is only one chloroethene and one 1,1-dichloroethene.

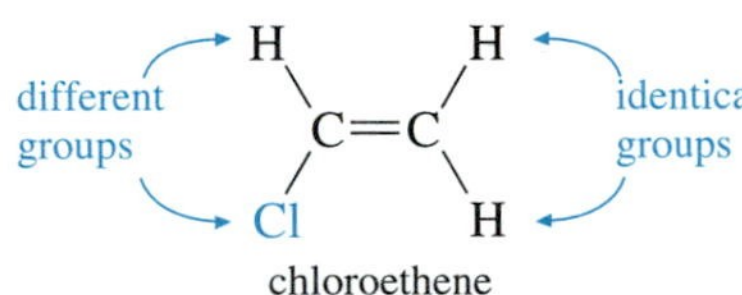

chloroethene

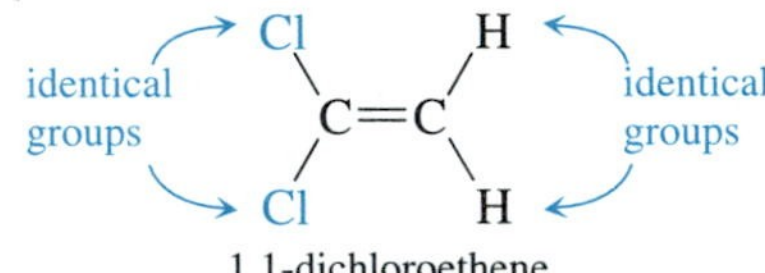

1,1-dichloroethene

cis-1,2-Dichloroethene

trans-1,2-Dichloroethene

FIGURE 4.3
Geometric Isomers of 1,2-Dichloroethene

Geometric isomers, like all isomers, have different physical properties. For example, the boiling points of *cis*- and *trans*-1,2-dichloroethene (Figure 4.3) are 60 and 47°C, respectively. The two C—Cl bond moments of the trans isomer cancel each other and the compound has no dipole moment. The cis compound has a dipole moment because the two C—Cl bond moments reinforce each other. As a consequence, the cis isomer is polar and has the higher boiling point.

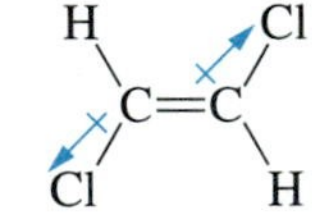

trans-1,2-dichloroethene

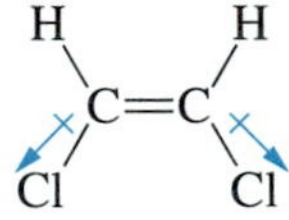

cis-1,2-dichloroethene

EXAMPLE 4-3

Is cis-trans isomerism possible about either of the double bonds of geraniol, a natural oil?

$$\begin{array}{c} \phantom{CH_3-{}}CH_3 \phantom{{}=CH-CH_2-CH_2-{}} CH_3 \phantom{{}=CH-CH_2OH} \\ CH_3-\overset{|}{C}=CH-CH_2-CH_2-\overset{|}{C}=CH-CH_2OH \end{array}$$

Solution

First, consider the double bond near the left end of the molecule. The leftmost carbon atom of that double bond has two $—CH_3$ units bonded to it. On this basis alone geometrical isomerism is not possible, regardless of the groups bonded to the right atom of the double bond.

Next, consider the double bond toward the right end of the molecule. The leftmost carbon atom of that double bond is bonded to a $—CH_3$ group and a $—CH_2—$ unit that is part of the parent chain. The groups are different. The rightmost carbon atom of that double bond is bonded to two different groups: a hydrogen atom and a —CH_2— unit. Thus, geometrical isomerism is possible. The naturally occurring isomer of geraniol has a trans arrangement in the parent chain.

OH

Problem 4.3

Is cis-trans isomerism possible about either of the double bonds of bombykol, a pheromone secreted by the female silkworm moth?

$$CH_3(CH_2)_2{-}CH{=}CH{-}CH{=}CH{-}(CH_2)_9OH$$

4.3 The E,Z Designation of Geometric Isomers

In the previous section, the terms cis and trans were applied to denote the relationship of two substituents in 1,2-dichloroethene—a disubstituted alkene. This type of nomenclature can easily be used for any disubstituted alkenes, as in the case of *cis*- and *trans*-2-butene (Figure 4.4).

CH_3		H		H		H
	C=C				C=C	
H		CH_3		CH_3		CH_3
	trans-2-butene				*cis*-2-butene	

However, the cis and trans notation fails to describe isomeric trisubstituted and tetrasubstituted alkenes because there is no longer a simple reference giving the relationship of groups one to another. For example, even the following relatively simple compounds cannot be designated as cis or trans isomers.

Br		H		Cl		H
	C=C				C=C	
Cl		F		Br		F

We can distinguish these isomers and all other tri- and tetrasubstituted alkenes by the **E,Z system** of nomenclature. The E,Z system uses **sequence**

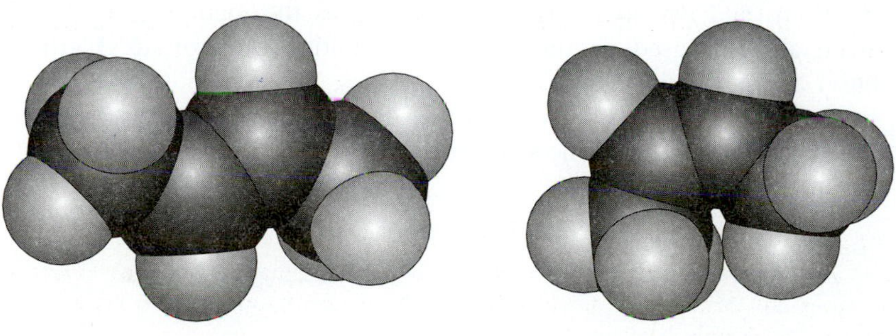

trans-2-Butene *cis*-2-Butene

FIGURE 4.4
Geometric Isomers of 2-Butene

rules to assign priorities to the groups bonded to the atoms of the double bond of any alkene. The two groups bonded to one carbon atom are ranked by their respective priorities and designated low and high priority. The same consideration is given to the two groups bonded to the other carbon atom. If the higher priority groups on each carbon atom are on the same side of the double bond, the alkene is the Z isomer (German *zusammen,* together). If the higher priority groups on each carbon atom are on opposite sides of the double bond, the alkene is the E isomer (German *entgegen,* opposite).

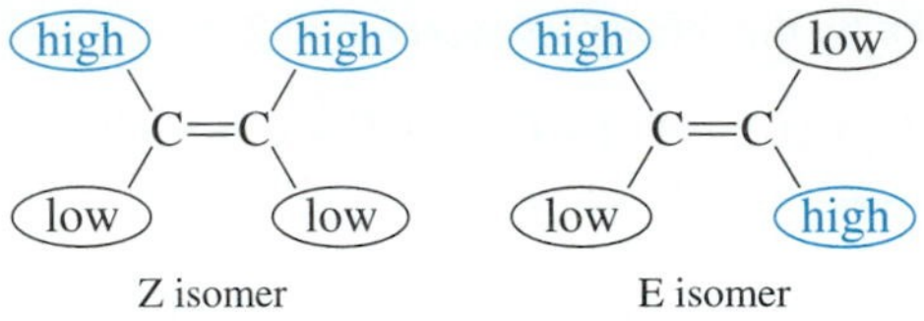

The first rule of assigning priorities to groups is based on the atomic number of the atom directly attached to the carbon atom of the double bond. A high-atomic-number atom receives a higher priority than a low-atomic-number atom. Thus for some common atoms typically attached to a double bond, the priority order is Br > Cl > F > O > N > C > H. Applying these priorities to the following alkene, which contains several halogen atoms, allows us to make the E,Z assignment.

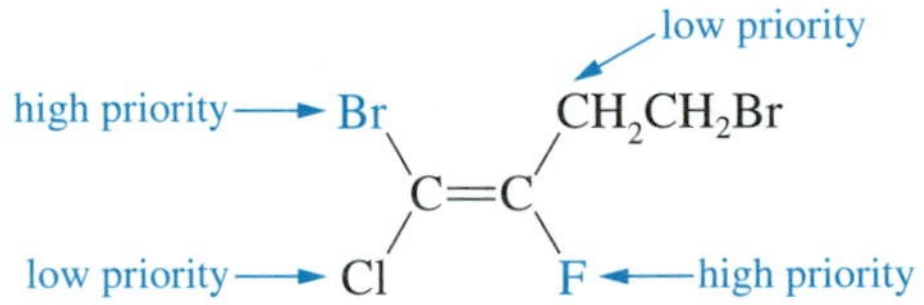

The higher priority of bromine compared to chlorine follows directly from their atomic numbers. The reason for the lower priority of the —CH_2CH_2Br group compared to fluorine may not be as evident. However, the first atom bonded to the double-bonded carbon atom is considered in assigning priorities. In this case we have to compare carbon to fluorine. Because fluorine has a higher atomic number than carbon, it has the higher priority.

If the first atoms have the same priority, then the second, third, and farther atoms are considered until the point of first difference is found. Thus, a methyl and an ethyl group are equivalent by the first rule of the directly bonded atom. However, the ethyl group has a higher priority because the second atom is a carbon atom (and two hydrogen atoms), whereas the methyl group has only hydrogen atoms. Based on this consideration, the order of alkyl groups is *tert*-butyl > isopropyl > ethyl > methyl.

In assigning priorities, a double bond is counted as two single bonds for both of the atoms involved. The same principle is used for a triple bond.

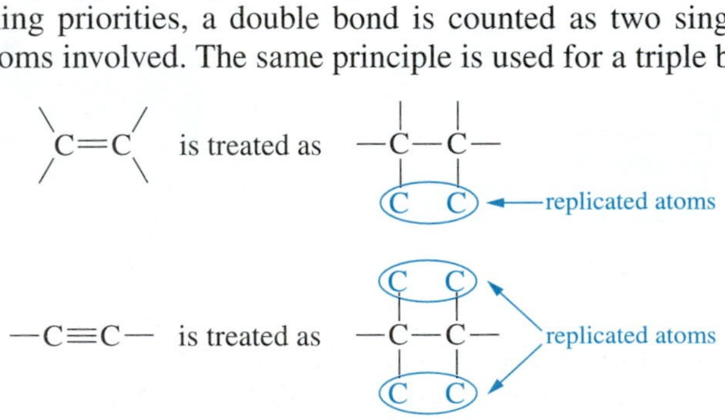

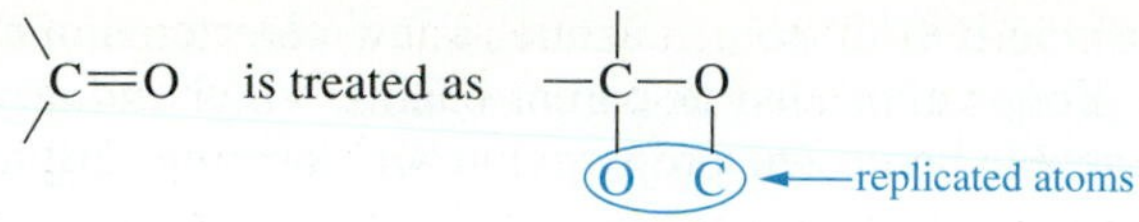

>C=O is treated as —C(—O)(O C) ← replicated atoms

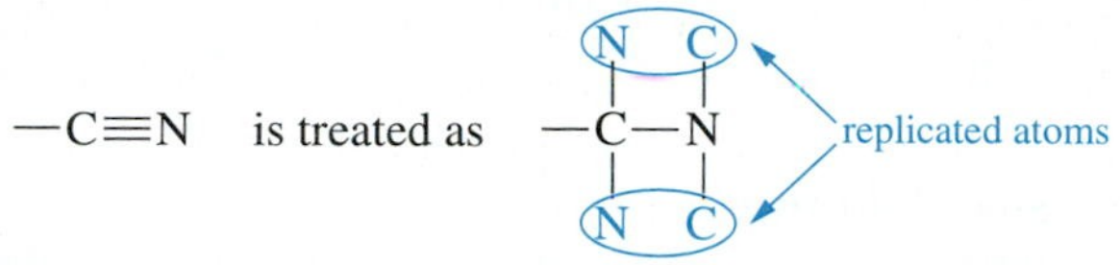

—C≡N is treated as —C(N C)—N(N C) ← replicated atoms

EXAMPLE 4-4

Tiglic acid, found in some natural oils, has the following structure. Designate the compound as an E or Z isomer.

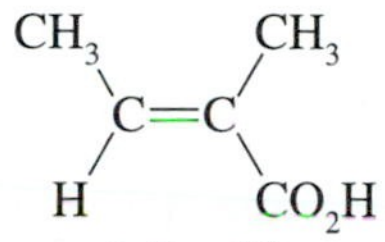

CH_3(H)C=C(CH_3)(CO_2H)

tiglic acid

Solution

The double-bonded carbon atom on the left is bonded to a hydrogen atom and a methyl group, whose priorities are low and high, respectively. The right double-bonded carbon atom is bonded to a methyl group and a carboxylic acid (CO_2H) group. The oxygen atoms of the carboxylic acid group give it a higher priority than the methyl group, which has only hydrogen atoms bonded to the carbon atom. Because the higher priority groups are on opposite sides of the double bond, the compound is E.

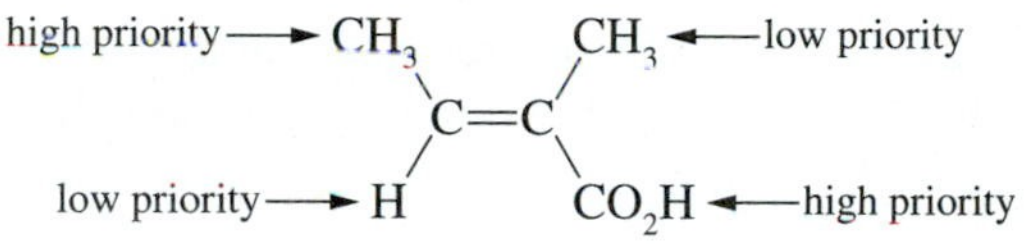

high priority → CH_3 ; CH_3 ← low priority

low priority → H ; CO_2H ← high priority

Note that if we consider only the positions of the methyl groups, the compound looks like a cis isomer. Thus, one could erroneously classify the compound as the Z isomer. This example illustrates the value of the unambiguous E,Z system.

Problem 4.4

Assign the E or Z configuration to tamoxifen, a drug used in the treatment of breast cancer.

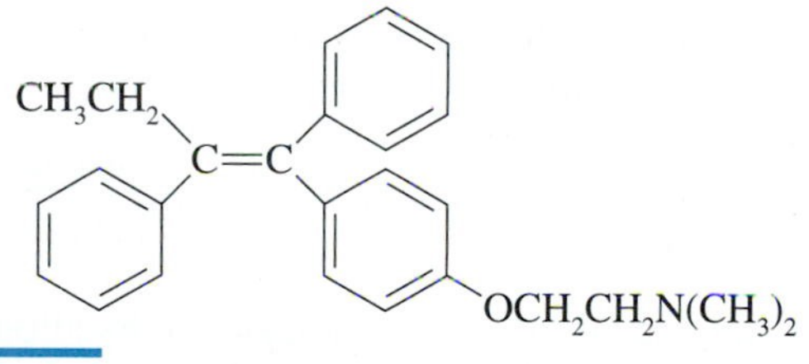

CH_3CH_2 ; C=C ; $OCH_2CH_2N(CH_3)_2$

4.4 Nomenclature of Alkenes and Alkynes

The IUPAC rules for naming alkenes and alkynes are similar to those for alkanes, but the position of the double or triple bond in the chain and the geometric arrangement of substituents about the double bond must be indicated.

Along with the IUPAC nomenclature, a few older common names are still used to name groups appended to parent chains. Vinyl and allyl are examples that contain a double bond. The propargyl group contains a triple bond.

$CH_2{=}CH{-}$ (vinyl) $CH_2{=}CHCH_2{-}$ (allyl) $HC{\equiv}C{-}CH_2{-}$ (propargyl)

IUPAC Names of Alkenes

1. The longest continuous chain containing the double bond is the parent.

$$\overset{1}{CH_3}{-}\overset{2}{CH_2}(\overset{3}{C}H_3\text{-substituted } C3)\overset{3}{C}{=}\overset{4}{C}(CH_3){-}\overset{5}{CH_2}{-}\overset{6}{CH}(CH_3){-}\overset{7}{CH_2}{-}\overset{8}{CH_3}$$

There are 8 carbon atoms in this chain.

2. The longest chain is given the same stem name as an alkane, but *-ene* replaces *-ane*. The parent name of the structure shown in rule 1 is octene.
3. The carbon atoms are numbered consecutively from the end nearer the double bond. The number of the first carbon atom of the double bond is used as a prefix to the parent name and is separated from it by a hyphen.

$$\overset{1}{CH_3}{-}\overset{2}{CH_2}{-}\overset{3}{C}(CH_3){=}\overset{4}{C}(CH_3){-}\overset{5}{CH_2}{-}\overset{6}{CH_2}{-}\overset{7}{CH_3}$$

This is a substituted 3-heptene, not a substituted 4-heptene.

4. Alkyl groups and other substituents are named and their positions on the chain are identified according to the numbering established by rule 3. Names and numbers are prefixed to the parent name.

$$\overset{5}{CH_3}{-}\overset{4}{CH_2}{-}\overset{3}{C}(CH_3){=}\overset{2}{C}(CH_3){-}\overset{1}{CH_3}$$

This is 2,3-dimethyl-2-pentene, not 3,4-dimethyl-3-pentene.

5. If the compound can exist as an E or Z isomer, the appropriate prefix followed by a hyphen is placed within parentheses in front of the name.

$$\overset{1}{CH_3}{-}\overset{2}{CH_2}{-}\overset{3}{C}(CH_3){=}\overset{4}{C}(CH_3){-}\overset{5}{CH_2}{-}\overset{6}{CH_3}$$

This is (*E*)-3,4-dimethyl-3-hexene.

6. If there is more than one double bond, indicate the location of each double bond by a number. A prefix to *-ene* indicates the number of double bonds.

$$\overset{1}{CH_2}{=}\overset{2}{CH}{-}\overset{3}{CH}{=}\overset{4}{CH}{-}\overset{5}{CH}{=}\overset{6}{CH}{-}\overset{7}{CH_3}$$

1,3,5-heptatriene

7. Name cycloalkenes by numbering the ring to give the double-bonded carbon atoms the numbers 1 and 2. Choose the direction of numbering so that the first substituent on the ring receives the lower number. The position of the

double bond is not given because it is known to be between the C-1 and C-2 atoms.

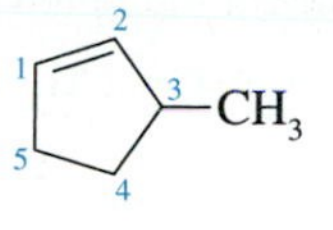

3-methylcyclopentene 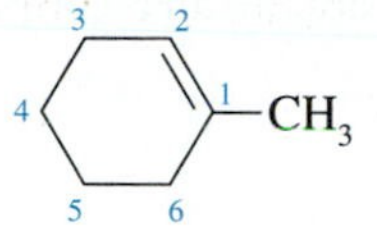 1-methylcyclohexene

IUPAC Names of Alkynes

The IUPAC rules for naming alkynes are similar to those for naming alkenes, using *-yne* to indicate the triple bond instead of *-ene* for the double bond, as illustrated by the following examples.

$$\overset{6}{CH_3}-\overset{5}{CH_2}-\overset{4}{CH_2}-\overset{3}{C}\equiv\overset{2}{C}-\overset{1}{CH_3}$$

2-hexyne

$$\overset{6}{CH_3}-\overset{5}{CH_2}-\overset{4}{C}\equiv\overset{3}{C}-\overset{2}{CH}(CH_3)-\overset{1}{CH_3}$$

2-methyl-3-hexyne

Compounds with multiple triple bonds are diynes, triynes, and so on. Compounds with both double and triple bonds are enynes, not ynenes. Numbering of compounds with both double and triple bonds starts from the end nearest the first multiple bond regardless of type. When a choice is possible, double bonds are assigned lower numbers than triple bonds.

$$\overset{6}{CH_3}-\overset{5}{C}\equiv\overset{4}{C}-\overset{3}{CH_2}-\overset{2}{C}\equiv\overset{1}{C}-H$$

1,4-hexadiyne

$$\overset{1}{CH_2}=\overset{2}{CH}-\overset{3}{CH_2}-\overset{4}{C}\equiv\overset{5}{C}-H$$

1-penten-4-yne

EXAMPLE 4-5

Name the following compound.

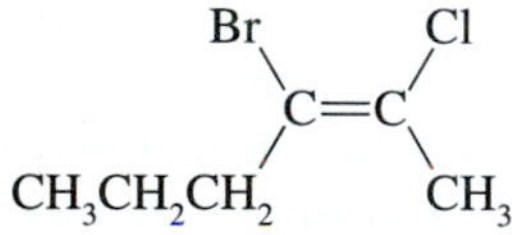

$$(Br)(CH_3CH_2CH_2)C=C(Cl)(CH_3)$$

Solution

There are six carbon atoms in the longest chain. It is numbered from right to left so that the double bond is at the carbon atom in position 2. The parent is then 2-hexene. The chlorine and bromine atoms are at the 2- and 3-positions, respectively.

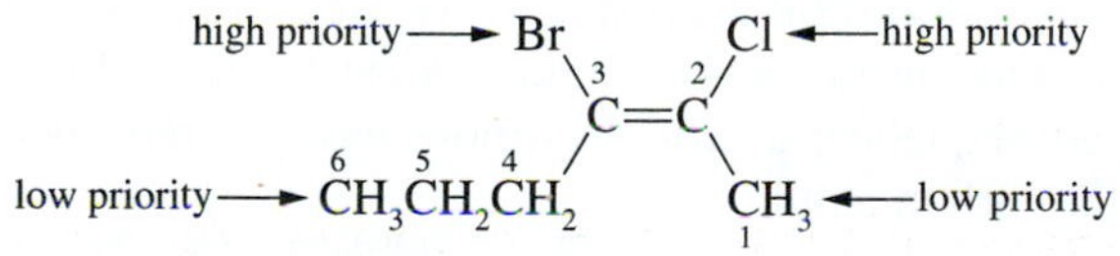

high priority → Br (C-3), Cl (C-2) ← high priority

low priority → $\overset{6}{CH_3}\overset{5}{CH_2}\overset{4}{CH_2}$, $\underset{1}{CH_3}$ ← low priority

The two different groups of atoms on each unsaturated carbon atom make geometric isomers possible. Because both bromine and chlorine have higher priorities than carbon, this is (*Z*)-3-bromo-2-chloro-2-hexene.

Problem 4.5

Draw the structures of the following isomeric compounds.

(a) 5-methyl-1,3-cyclohexadiene (b) 3-methyl-1,4-cyclohexadiene

EXAMPLE 4-6

Why is 2-bromo-4-hexyne an incorrect name for the following compound?

$$CH_3{-}\underset{\displaystyle Br}{\underset{|}{CH}}{-}CH_2{-}C{\equiv}C{-}CH_3$$

Solution

The chain must be numbered from the end nearer the triple bond. Thus the chain is numbered from right to left. The correct name is 5-bromo-2-hexyne.

$$\overset{6}{C}H_3{-}\underset{\displaystyle Br}{\underset{|}{\overset{5}{C}H}}{-}\overset{4}{C}H_2{-}\overset{3}{C}{\equiv}\overset{2}{C}{-}\overset{1}{C}H_3$$

Problem 4.6

1,3,11-Tridecatriene-5,7,9-triyne is a compound found in safflowers that is a chemical defense against nematode infestations. Write the structure of the compound.

4.5 Acidity of Alkenes and Alkynes

Although hydrocarbons are extremely weak acids, a very strong base can remove a proton from an alkane, alkene, or alkyne to produce a **carbanion,** an anion with a negative charge on the carbon atom. Hydrocarbons are weaker acids than many acids that you encountered in your first chemistry course. Those acids had the hydrogen atom bonded to an electronegative atom. Because carbon is less electronegative, carbanions are not as stable as the conjugate bases of inorganic acids.

$CH_3{-}CH_3$	$CH_2{=}CH_2$	$HC{\equiv}CH$
$K_a = 10^{-49}$	$K_a = 10^{-44}$	$K_a = 10^{-25}$

The acidity of hydrocarbons is related to the hybridization of the carbon atom. The K_a increases for carbon atoms in the order $sp^3 < sp^2 < sp$. The order of acidities parallels the contribution of the lower energy of the 2*s* orbital to the hybrid orbitals in the σ bond. The energy of a 2*s* orbital is lower than that of a 2*p* orbital, and on average a 2*s* orbital is closer to the nucleus than a 2*p* orbital. The average distance of hybrid orbitals from the nucleus depends on the percent contribution of the *s* and *p* orbitals. For an sp^3 hybrid orbital, the contribution of the *s* orbital is 25%, because one *s* and three *p* orbitals contribute to the four hybrid orbitals. Similarly, the contribution of the *s* orbital is 33% and 50% for the sp^2 and *sp* hybrid orbitals, respectively. Because an *sp* hybrid orbital has more *s* character than an sp^3 or sp^2 orbital, its electrons are located closer to the nucleus. As a consequence, a proton is more easily removed and the electron pair is left on the carbon atom.

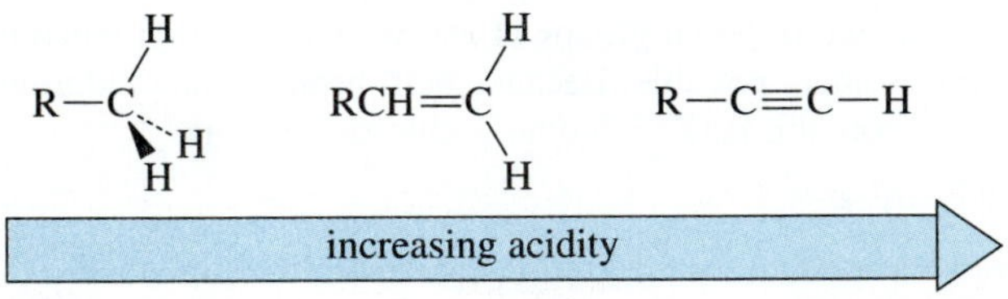

	$R{-}CH_3$ (R—C with three H)	$RCH{=}CH_2$	$R{-}C{\equiv}C{-}H$
	increasing acidity →		
hybridization	sp^3	sp^2	sp
% *s* character	25	33	50

For all practical purposes, only acetylene and terminal alkynes are sufficiently strong acids that their conjugate bases can be produced using conventional bases. However, even hydroxide ion is not a strong enough base to remove a proton from an alkyne. In fact, the conjugate base of an alkyne is rapidly and quantitatively converted to the alkyne whenever it reacts with compounds containing hydroxyl groups (such as water, alcohols, and carboxylic acids).

$$R{-}C{\equiv}C{:}^- + H_2O \rightleftarrows R{-}C{\equiv}C{-}H + OH^-$$

Recall the periodic trends of acidity discussed in Chapter 2. An N—H bond is a weaker acid than an O—H bond. Therefore NH_2^-, the conjugate base of ammonia—a very weak acid—is a stronger base than OH^-, the conjugate base of water—a weak acid. The K_a of ammonia is 10^{-36}. The K_a of a terminal alkyne is about 10^{-25}. Thus, amide ion quantitatively removes a proton from any terminal alkyne.

$$R{-}C{\equiv}C{-}H + NH_2^- \rightleftarrows R{-}C{\equiv}C{:}^- + NH_3$$

4.6 Hydrogenation of Alkenes and Alkynes

We expect alkenes and alkynes to have chemical similarities because both have π bonds. In general, that expectation is correct. Because alkynes have two π bonds, they often react with twice the amount of reagent that reacts with alkenes, which have only one π bond. Both classes of compounds are unsaturated and react with hydrogen gas to give more saturated compounds.

Reduction of Alkenes

The reaction of hydrogen gas with an alkene (or cycloalkene) yields a saturated compound. The process is a reduction, but the reaction is also called **hydrogenation.** The hydrogenation of 1-octene yields octane. Hydrogenation requires a catalyst. The catalyst is usually finely divided platinum on finely divided carbon, but nickel and palladium can also be used. The hydrogenation process is heterogeneous; that is, it occurs on the surface of the solid catalyst. The catalyst is symbolized Pt/C, where Pt is the metal and C is carbon.

This double bond is changed into a single bond.

$$\underset{\text{1-octene}}{CH_3(CH_2)_5CH{=}CH_2} + H_2 \xrightarrow{Pt/C} \underset{\text{octane}}{CH_3(CH_2)_5CH_2{-}CH_3}$$

Although functional groups such as ketones or esters also have multiple bonds, they are not normally reduced under the mild conditions used to hydrogenate a carbon-carbon double bond.

$$CH_3{-}\overset{\overset{O}{\parallel}}{C}{-}CH_2{-}CH_2{-}CH{=}CH_2 + H_2 \xrightarrow{Pt/C} CH_3{-}\overset{\overset{O}{\parallel}}{C}{-}CH_2{-}CH_2{-}CH_2{-}CH_3$$

Hydrogenation is used commercially to convert liquid vegetable oils into semisolid fats (Chapter 13). Fats and oils are structurally related esters and differ in the degree of saturation. The difference between Crisco® oil and its com-

The difference in the physical properties of these two products is the result of the degree of saturation.

panion solid Crisco is just the degree of saturation of the carboxylic acids in the esters: The solid is saturated.

$$\begin{array}{l} CH_2-O-\overset{O}{\overset{\|}{C}}(CH_2)_7CH{=}CH(CH_2)_7CH_3 \\ | \\ CH-O-\overset{O}{\overset{\|}{C}}(CH_2)_7CH{=}CH(CH_2)_7CH_3 \\ | \\ CH_2-O-\overset{O}{\overset{\|}{C}}(CH_2)_7CH{=}CH(CH_2)_7CH_3 \end{array} \xrightarrow[Ni]{3\ H_2} \begin{array}{l} CH_2-O-\overset{O}{\overset{\|}{C}}(CH_2)_7CH_2{-}CH_2(CH_2)_7CH_3 \\ | \\ CH-O-\overset{O}{\overset{\|}{C}}(CH_2)_7CH_2{-}CH_2(CH_2)_7CH_3 \\ | \\ CH_2-O-\overset{O}{\overset{\|}{C}}(CH_2)_7CH_2{-}CH_2(CH_2)_7CH_3 \end{array}$$

a liquid oil → a solid fat

Hydrogenation of an alkene occurs by the addition of hydrogen atoms to the same face of the double bond. This fact is established by the hydrogenation of 1,2-dimethylcyclopentene to produce *cis*-1,2-dimethylcyclopentane.

$$\text{1,2-dimethylcyclopentene} + H_2 \xrightarrow{Pd/C} \textit{cis}\text{-1,2-dimethylcyclopentane}$$

The finely divided metal catalyst adsorbs hydrogen gas on the surface, and the hydrogen-hydrogen bond is broken. When the alkene approaches the surface, the hydrogen atoms must add to the same face of the double bond.

Reduction of Alkynes

Alkynes can be completely reduced to alkanes by reaction with two molar equivalents of hydrogen gas in the presence of catalysts such as platinum or palladium.

This triple bond is changed into a single bond.

$$\underset{\text{1-nonyne}}{CH_3(CH_2)_6C{\equiv}CH} + 2\ H_2 \xrightarrow{Pd/C} \underset{\text{nonane}}{CH_3(CH_2)_6CH_2{-}CH_3}$$

The reaction can be stopped after adding one molar equivalent of hydrogen gas to form an alkene if the palladium is specially prepared. In the Lindlar catalyst, palladium is coated on calcium carbonate that contains a small amount of lead acetate. Hydrogenation of an alkyne using the Lindlar catalyst gives cis alkenes. This form of palladium does not reduce the alkene produced.

$$\underset{\text{5-decyne}}{CH_3(CH_2)_3C{\equiv}C(CH_2)_3CH_3} \xrightarrow[\text{Lindlar catalyst}]{H_2} \underset{\textit{cis}\text{-5-decene}}{(CH_3(CH_2)_3)(H)C{=}C(H)((CH_2)_3CH_3)}$$

In contrast, reduction of an alkyne with lithium metal as the reducing agent in liquid ammonia as the solvent gives the trans isomer. Water is added to the reaction mixture in a second step to consume excess reagent.

$$\underset{\text{5-decyne}}{CH_3(CH_2)_3C{\equiv}C(CH_2)_3CH_3} \xrightarrow[2.\ H_2O]{1.\ Li/NH_3} \underset{\textit{trans}\text{-5-decene}}{(CH_3(CH_2)_3)(H)C{=}C(H)((CH_2)_3CH_3)}$$

EXAMPLE 4-7

How many moles of hydrogen gas will react with cembrene, which is contained in pine oil? What is the molecular formula of the product?

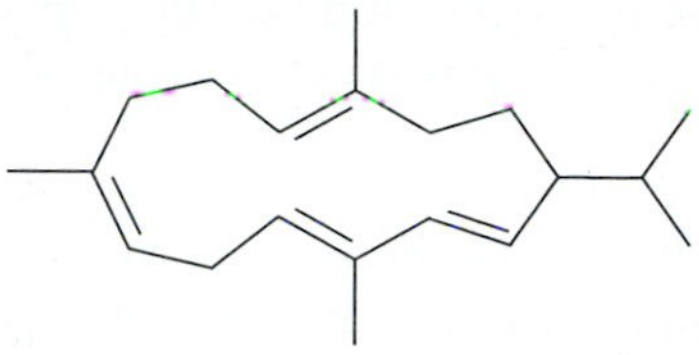

cembrene

Solution

There are four double bonds in the compound. One molar equivalent of the compound will react with four molar equivalents of hydrogen gas. The product will be a cycloalkane. There are 14 carbon atoms in the ring, three methyl groups, and an isopropyl group for a total of 20 carbon atoms. Because the general molecular formula for a cycloalkane is C_nH_{2n}, the molecular formula of the product is $C_{20}H_{40}$.

Problem 4.7

Write the structure obtained by complete hydrogenation of ipsdienol, a pheromone of the Norwegian spruce beetle.

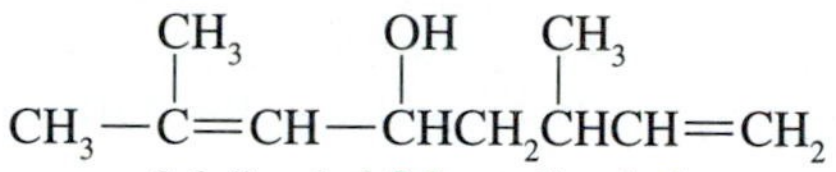

$$CH_3-\overset{\overset{CH_3}{|}}{C}=CH-\overset{\overset{OH}{|}}{C}HCH_2\overset{\overset{CH_3}{|}}{C}HCH=CH_2$$

2,6-dimethyl-2,7-octadien-4-ol
(ipsdienol)

EXAMPLE 4-8

The IUPAC name of muscalure, the sex attractant of the housefly, is (*Z*)-9-tricosene. How can this compound be synthesized in the laboratory from an alkyne? Name the alkyne.

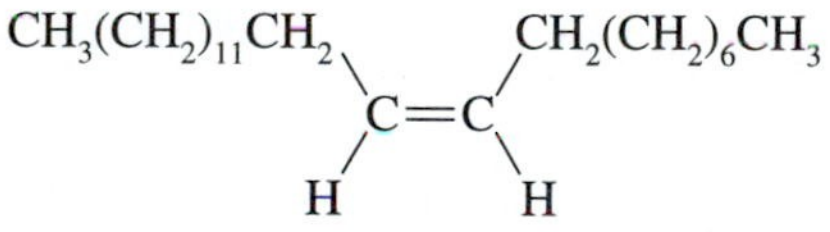

$CH_3(CH_2)_{11}CH_2$ and $CH_2(CH_2)_6CH_3$ on C=C, each carbon bearing H (cis)

Solution

Alkenes with the Z configuration can be prepared by hydrogenation of an alkyne using the Lindlar catalyst. The required alkyne must have the triple bond at the C-9 position and is named 9-tricosyne.

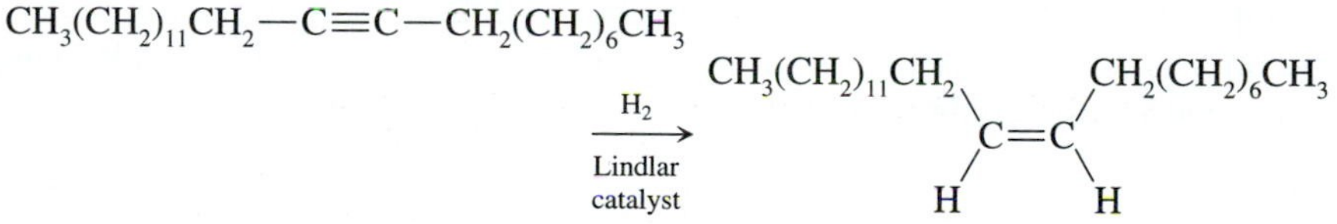

$$CH_3(CH_2)_{11}CH_2-C\equiv C-CH_2(CH_2)_6CH_3 \xrightarrow[\text{Lindlar catalyst}]{H_2} \text{(Z)-}CH_3(CH_2)_{11}CH_2CH=CHCH_2(CH_2)_6CH_3$$

Problem 4.8

(*E*)-11-Tetradecen-1-ol is an intermediate required to synthesize the sex attractant of the spruce budworm moth. How can this compound be prepared from an alkyne? Suggest a name for the alkyne. (The *-ol* ending refers to the hydroxyl group.)

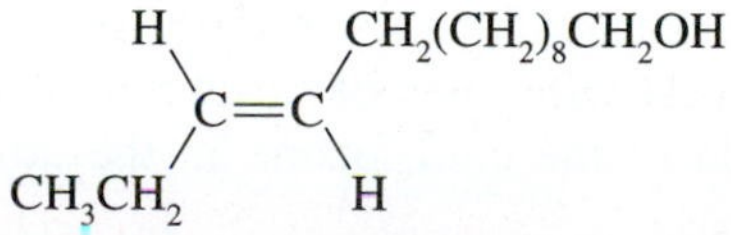

CH_3CH_2 and $CH_2(CH_2)_8CH_2OH$ on C=C, each carbon bearing H (trans)

4.7 Oxidation of Alkenes and Alkynes

Oxidizing agents react with the π electrons of the double bond of alkenes and the triple bond of alkynes more easily than the σ electrons of carbon-carbon single bonds and carbon-hydrogen bonds. Thus, in contrast to alkanes, both alkenes and alkynes can easily be oxidized without destroying the carbon chain.

Hydroxylation of Alkenes

Reaction of an alkene with potassium permanganate ($KMnO_4$) in basic solution yields a product that contains a hydroxyl group on each carbon atom of the original double-bond unit. Thus the alkene is oxidized. Permanganate is reduced to MnO_2, a brown solid.

$$\underset{\text{1-octene}}{CH_3(CH_2)_5CH{=}CH_2} \xrightarrow[NaOH/H_2O]{KMnO_4} \underset{\text{1,2-octanediol}}{CH_3(CH_2)_5\underset{OH}{\underset{|}{CH}}{-}\underset{OH}{\underset{|}{CH_2}}}$$

Potassium permanganate is purple in aqueous solution. Manganese dioxide (MnO_2), the product of the reaction, is a brown solid that precipitates from solution. The color change in oxidation with potassium permanganate allows the reaction to be used to test visually for the presence of a double bond. Alkanes and cycloalkanes are not oxidized by $KMnO_4$, so the purple color remains. Alkenes are oxidized by $KMnO_4$, so the purple color fades as a brown precipitate appears.

Ozonolysis of Alkenes and Alkynes

Alkenes and alkynes react rapidly with ozone, O_3. The reaction is carried out in an inert solvent such as dichloromethane. The oxidation occurs in two steps. In the first, an unstable intermediate called an ozonide forms. The intermediate is not isolated but is treated in solution by reacting it with zinc and acetic acid. The net result of the two reactions is the cleavage of the carbon-carbon double bond to produce two carbonyl compounds. The overall process is called **ozonolysis.**

$$>C{=}C< \xrightarrow[2.\ Zn/H_3O^+]{1.\ O_3} \underbrace{>C{=}O + O{=}C<}_{\text{carbonyl compounds}}$$

Ozonolysis can be used to determine the position of a double bond in an alkene. If one of the double-bonded carbon atoms has two hydrogen atoms bonded to it, its ozonolysis product is methanal (formaldehyde). If an alkyl group and a hydrogen atom are bonded to the double-bonded carbon atom, the product is an aldehyde; if two alkyl groups are bonded to the double-bonded carbon atom, the product is a ketone.

$$\underset{\text{methanal}}{(H)(H)C{=}O} \qquad \underset{\text{an aldehyde}}{(H)(R)C{=}O} \qquad \underset{\text{a ketone}}{(R)(R)C{=}O}$$

Ozone is used in water treatment plants in Los Angeles and some other cities in the United States as well as some cities in Europe as an alternative to chlorine. The ozone reacts with sites of unsaturation in bacteria and destroys them. On the other side of the coin, ozone in the atmosphere of some cities reacts with double bonds in some materials such as rubber and shortens their useful lifetime.

Although less commonly used, the ozonolysis of alkynes also results in cleavage products. Carboxylic acids are obtained from internal alkynes. A terminal alkyne forms one molar equivalent of CO_2.

$$R{-}C{\equiv}C{-}R' \xrightarrow[\text{2. Zn/H}_3\text{O}^+]{\text{1. O}_3} R{-}CO_2H + R'{-}CO_2H$$

$$R{-}C{\equiv}C{-}H \xrightarrow[\text{2. Zn/H}_3\text{O}^+]{\text{1. O}_3} R{-}CO_2H + CO_2$$

4.8 Addition Reactions of Alkenes and Alkynes

As first described in Section 2.5, an **addition reaction** occurs when two reactants combine to form a single product. No atoms are "left over." Examples of addition reactions of ethene (C_2H_4) with some common reagents are

$$CH_2{=}CH_2 + Br{-}Br \longrightarrow Br{-}CH_2{-}CH_2{-}Br$$
$$CH_2{=}CH_2 + H{-}Cl \longrightarrow H{-}CH_2{-}CH_2{-}Cl$$
$$CH_2{=}CH_2 + H{-}OH \longrightarrow H{-}CH_2{-}CH_2{-}OH$$

Reagents that add to alkenes are classified as symmetrical or unsymmetrical. **Symmetrical reagents** consist of two identical groups, as in bromine. **Unsymmetrical reagents** consist of different groups, as in HCl and H_2O.

Addition of Halogens

The reaction of ethene with Br_2 to form 1,2-dibromoethane is an addition reaction. The atoms that add to the double bond are located on adjacent carbon atoms, a common characteristic of addition reactions of alkenes.

These two atoms are added to adjacent carbon atoms.

$$H_2C{=}CH_2 + Br{-}Br \longrightarrow H{-}\underset{H}{\overset{Br}{C}}{-}\underset{H}{\overset{Br}{C}}{-}H$$

1,2-dibromoethane

Bromine is reddish; it reacts with alkenes to give a colorless product. Hence, the disappearance of the reddish color of bromine can be used to determine if a compound is unsaturated. Drops of Br_2 dissolved in CCl_4 are added to a compound. If the bromine color disappears, the compound is unsaturated.

$$\underset{\text{colorless}}{CH_3(CH_2)_5CH{=}CH_2} + \underset{\text{red}}{Br_2} \xrightarrow{CCl_4} \underset{\text{colorless}}{CH_3(CH_2)_5\underset{Br}{CH}{-}\underset{Br}{CH_2}}$$

Chlorine also adds to a carbon-carbon double bond, but iodine does not give a good yield of addition product. Fluorine is too reactive to control, and several competing reactions also occur.

Alkynes react with chlorine or bromine to produce tetrahaloalkanes, which contain two halogen atoms on each of the original carbon atoms of the triple bond. Thus, two molar equivalents of the halogen are consumed in the reaction.

$$\underset{\text{1-hexyne}}{CH_3(CH_2)_3-C\equiv C-H} + 2\,Cl_2 \xrightarrow{CCl_4} \underset{\text{1,1,2,2-tetrachlorohexane}}{CH_3(CH_2)_3-CCl_2-CCl_2-H}$$

If only one molar equivalent of the halogen is used, the reaction product has the halogen atoms on the opposite sides of the double bond.

$$\underset{\text{1-hexyne}}{CH_3(CH_2)_3-C\equiv C-H} + Cl_2 \xrightarrow{CCl_4} \underset{(E)\text{-1,2-dichloro-1-hexene}}{(CH_3(CH_2)_3)(Cl)C=C(Cl)(H)}$$

Addition of Hydrogen Halides

The addition of a symmetrical reagent to an alkene yields only one possible product. It makes no difference which bromine atom bonds to which carbon atom. However, the situation is quite different for an unsymmetrical reagent, such as a hydrogen halide, HX. (Usually only HCl or HBr is used.) With a symmetrical alkene, such as ethylene, only one product is possible because the two carbon atoms are identical.

$$CH_2=CH_2 + H-Br \longrightarrow H-CH_2-CH_2-Br + Br-CH_2-CH_2-H$$

equivalent atoms

identical structures

Although two products could conceivably result from the addition of HBr to an unsymmetrical alkene, only one is actually formed. Addition of HBr to propene could yield either 1-bromopropane or 2-bromopropane, but only the latter is formed. The X written through one reaction arrow indicates that the reaction does not occur. Thus, the addition of an unsymmetrical reagent to an alkene is a highly selective process; that is, only one product is formed. (A similar selectivity is observed for the reaction of alkynes with hydrogen halides.)

$$\underset{\text{propene}}{CH_3-CH=CH_2} + H-Br \longrightarrow \underset{\text{2-bromopropane}}{CH_3-CH(Br)-CH_2-H}$$

$$\underset{\text{propene}}{CH_3-CH=CH_2} + H-Br \not\longrightarrow \underset{\text{1-bromopropane}}{CH_3-CH(H)-CH_2-Br}$$

Markovnikov's Rule

The Russian chemist Vladimir Markovnikov observed that unsymmetrical reagents add to unsymmetrical double bonds in a specific way. **Markovnikov's rule** states that a molecule of the general formula HX adds to a double bond so that the hydrogen atom bonds to the unsaturated carbon atom that has the greater number of directly bonded hydrogen atoms. This is the least substituted carbon atom.

No hydrogen atoms on this carbon atom — Two hydrogen atoms on this carbon atom

$$(CH_3)_2C{=}CH_2 + H{-}Cl \longrightarrow CH_3{-}C(Cl)(CH_3){-}CH_2{-}H$$

2-methyl-1-propene — 2-chloro-2-methylpropane

EXAMPLE 4-9

Predict what product will be formed when HBr is added to 2-methyl-2-butene.

Solution

One of the unsaturated carbon atoms has one attached hydrogen atom, the other no attached hydrogen atoms. The hydrogen atom of HBr bonds to the carbon atom having the greater number of directly bonded hydrogen atoms. The bromine atom bonds to the other carbon atom of the original double bond. The predicted product is 2-bromo-2-methylbutane.

No hydrogen atoms on this carbon atom — One hydrogen atom on this carbon atom

$$(CH_3)_2C{=}CH{-}CH_3 + H{-}Br \longrightarrow CH_3{-}C(Br)(CH_3){-}CH(H){-}CH_3$$

2-methyl-2-butene — 2-bromo-2-methylbutane

Problem 4.9

Predict the product of the addition of HCl to 1-methylcyclohexene.

4.9 Mechanism of Addition Reactions

An understanding of the specificity of the addition reaction as originally discovered by Markovnikov is provided by the mechanism that has now been well established. The mechanism is based on the fact that the π electrons in an alkene serve as a nucleophile (nucleus lover) and accept an electrophile (electron-loving species). Consider the reaction of propene with HBr, which for simplicity may be considered as the H^+ and Br^- ions when writing the reaction mechanism. The first step is written by using a curved arrow to show the movement of two electrons in the π bond to form a σ bond to the hydrogen atom. As a result, a positively charged species called an isopropyl carbocation forms.

$$CH_3(H)C{=}CH_2 + H^+ \longrightarrow CH_3(H)C^+{-}CH_3$$

nucleophile — electrophile — isopropyl carbocation

The movement of the π electrons in this step is like the movement of a swinging gate. One end of the gate stays attached, and the other end "swings" free to bond to the electrophile. The electron pair stays as part of a bond on one of the two carbon atoms.

In the second step of the addition reaction, the isopropyl carbocation acts as an electrophile and accepts an electron pair from the bromide ion, which acts as a nucleophile.

$$:\ddot{\underset{..}{Br}}:^{-} + (CH_3)(H)\overset{+}{C}{-}CH_3 \longrightarrow CH_3{-}\underset{|}{\overset{:\ddot{Br}:}{CH}}{-}CH_2{-}H$$

nucleophile electrophile 2-bromopropane

In the first step of this reaction, the hydrogen atom is attached to one of the two possible carbon atoms of the original π bond; this placement accounts for the product predicted by Markovnikov's rule. Why is this particular carbocation, the isopropyl carbocation, formed? If the hydrogen atom had bonded to the interior carbon atom, the electrons would have had to "swing" in the other direction, and a propyl carbocation would have formed.

$$H^{+} + (H)(CH_3)C{=}CH_2 \longrightarrow (H)(H)(CH_3)C{-}\overset{+}{C}H_2$$

electrophile nucleophile propyl carbocation

Both the isopropyl carbocation and the propyl carbocation are unstable intermediates. However, the isopropyl carbocation is more stable and is formed preferentially. Note that as a result of the addition of the proton to a carbon-carbon double bond, the structure of the final product is determined. The second step merely combines the nucleophilic halide ion with the carbocation at the site of the charge generated in the first step.

Why is the isopropyl carbocation formed in preference to the propyl carbocation? Alkyl groups attached to a positively charged carbon atom help stabilize the charge because the electrons in the carbon-carbon bonds are polarized toward the positive center. The isopropyl carbocation has the charge on a secondary carbon atom, one that has two alkyl groups attached. The propyl carbocation has the charge on a primary carbon atom, one that has only one alkyl group attached.

By the same reasoning, it follows that a tertiary carbocation is more stable than a secondary carbocation, because it has three alkyl groups attached to the positive carbon atom. The order of stability of carbocations is

$$\underset{\text{methyl}}{H{-}\overset{H}{\underset{H}{C^{+}}}} < \underset{\text{primary}}{R{-}\overset{H}{\underset{H}{C^{+}}}} < \underset{\text{secondary}}{R{-}\overset{H}{\underset{R}{C^{+}}}} < \underset{\text{tertiary}}{R{-}\overset{R}{\underset{R}{C^{+}}}}$$

increasing stability

This order of stability "explains" Markovnikov's rule. Addition of the electrophile always occurs to give the most stable carbocation, which controls the product formed.

4.10 Hydration of Alkenes and Alkynes

Water was one of the unsymmetrical reagents listed in Section 4.8 that can add to a π bond. In fact, H_2O fits the general class of HX compounds (H—OH). Thus, water adds to the double bond of an alkene in accordance with Markovnikov's rule to give 2-propanol, but none of the isomeric 1-propanol.

$$\underset{\text{propene}}{CH_3{-}CH{=}CH_2} + H{-}OH \longrightarrow \underset{\text{2-propanol}}{CH_3{-}\overset{OH}{\overset{|}{C}}H{-}CH_2{-}H}$$

$$\underset{\text{propene}}{CH_3{-}CH{=}CH_2} + H{-}OH \not\longrightarrow \underset{\text{1-propanol}}{CH_3{-}\overset{H}{\overset{|}{C}}H{-}CH_2{-}OH}$$

The reverse of the hydration reaction is dehydration, which is an example of an elimination reaction as discussed in Section 2.5. Alcohols can be dehydrated (lose water) to produce alkenes.

$$H{-}\underset{|}{\overset{|}{C}}{-}\underset{|}{\overset{|}{C}}{-}OH \xrightarrow{H_3O^+} \;\rangle C{=}C\langle\; + H_2O$$

The direction of the reaction—hydration or dehydration—is controlled by conditions predictable by Le Châtelier's principle. Conversion of an alkene to an alcohol requires an excess of water. Dehydration occurs if the water concentration is very low, as in concentrated sulfuric acid, which is about 98% H_2SO_4.

Hydration of Alkynes Produces Carbonyl Compounds

Water adds to one of the π bonds of a triple bond in aqueous sulfuric acid in the presence of mercuric sulfate catalyst. However, the alcohol that forms has its —OH group bonded to the double-bonded carbon atom of an alkene. This type of compound is called an **enol,** a name that includes both the *-ene* suffix of a double bond and the alcohol suffix *-ol*.

$$R{-}C{\equiv}C{-}H + H{-}OH \xrightarrow[HgSO_4]{H_2SO_4} \left[\underset{\text{an enol}}{R{-}\overset{OH}{\overset{|}{C}}{=}CH_2}\right]$$

Enols are unstable compounds and are rapidly converted to carbonyl compounds in a rearrangement reaction. We will discuss this reaction further in Chapter 10.

$$\left[\underset{\text{an enol}}{R{-}\overset{OH}{\overset{|}{C}}{=}CH_2}\right] \longrightarrow \underset{\text{a ketone}}{R{-}\overset{O}{\overset{\|}{C}}{-}CH_3}$$

The final product of hydration of an alkyne is a ketone. The more substituted carbon atom of the alkyne is converted into a carbonyl carbon atom.

more substituted carbon atom — less substituted carbon atom

$$C_5H_9{-}C{\equiv}C{-}H + H{-}OH \xrightarrow[HgSO_4]{H_2SO_4} C_5H_9{-}\overset{O}{\overset{\|}{C}}{-}CH_3$$

EXAMPLE 4-10

What product(s) will result from the hydration of 2-decyne?

Solution

Note that the initial hydration of an alkyne places the hydroxyl group on the more substituted carbon atom. This group is then converted into a carbonyl group of a ketone. In 2-decyne both the C-2 and C-3 atoms are substituted to the same degree.

C-3 is bonded to one alkyl group. C-2 is bonded to one alkyl group.

$$CH_3(CH_2)_6{-}C{\equiv}C{-}CH_3$$

2-decyne

As a consequence, hydration can occur either of two ways, and two isomeric products, 3-decanone and 2-decanone, are produced.

$$CH_3(CH_2)_6{-}\overset{\overset{\large O}{\|}}{C}{-}CH_2{-}CH_3 \qquad CH_3(CH_2)_6{-}CH_2{-}\overset{\overset{\large O}{\|}}{C}{-}CH_3$$

3-decanone 2-decanone

Problem 4.10

What product results from the hydration of 1-methylcyclopentene?

4.11 Polymerization of Alkenes

Several types of plastic bags for home use consist of polyethylene.

A **polymer** is a high molecular weight molecule created by the repetitive reaction of many hundreds to thousands of low molecular weight molecules called **monomers.** The process in which monomers join to produce polymers is called **polymerization.** Naturally occurring polymers—**biopolymers**—include carbohydrates, proteins, and nucleic acids. Cellulose, a polymer of glucose, is the major structural material of plants. Lignin, another polymer, occurs within the spaces between the long fibers of cellulose. Proteins, polymers of amino acids, are major constituents of living matter. Some serve a structural role; others are catalysts for virtually all metabolic reactions. Nucleic acids, which are polymers of nucleotides, are carriers of genetic information. Each compound class will be studied later in this text.

Synthetic polymers made in the laboratory are also familiar because we are literally surrounded with products containing these substances. Examples include some of the clothes we wear, containers for foods such as milk, and the computers that we use. Football players depend on polymers in helmets for protection; police use bullet-proof vests made of a polymer; our cars have many polymers in the interior as well as in the bumpers. The structures of polymers and the reactions that join together monomers to provide these diverse products are discussed in Chapter 17.

There are two classes of polymers: addition polymers and condensation polymers. These names indicate the type of reaction used to join the monomers. Only addition polymerization will be illustrated in this section. Condensation polymerization to give polyesters and polyamides is described in Chapters 11 and 13, respectively. Polymers are discussed in detail in Chapter 17.

Alkenes can be polymerized by a multiple addition reaction. For example, when $CH_2{=}CHCl$ (vinyl chloride) is polymerized, the product is polyvinyl chloride, commonly known as PVC.

$$\mathrm{H_2C{=}CHCl} \longrightarrow \mathrm{-CH_2CHCl-CH_2CHCl-CH_2CHCl-CH_2CHCl-}$$

An exact formula for a polymer cannot be written, because the size of the molecule depends on how it forms. There is no single "polyvinyl chloride" molecule; it is really a mixture of compounds with a range of molecular weights. For this reason, polymers are represented by placing the repeating unit derived from the monomer within a set of parentheses. For polyvinyl chloride, the unit is (—CH_2CHCl—). To show that a large number of units are present, the subscript n is used. For the polymer of vinyl chloride we write

$$\left[\mathrm{-CH_2-CHCl-}\right]_n$$

The properties of polymers depend on the monomer used and the molecular weight of the product. A list of some useful addition polymers is given in Table 4.2.

Table 4.2 Uses and Structures of Polymers

Monomer	Polymer	Uses
propylene $CH_2{=}CHCH_3$	polypropylene $-CH_2CH(CH_3)CH_2CH(CH_3)-$	carpet fibers, heart valves, bottles
vinyl chloride $CH_2{=}CHCl$	polyvinyl chloride (PVC) $-CH_2CH(Cl)CH_2CH(Cl)CH_2CH(Cl)-$	floor covering, garden hoses
dichloroethylene $CH_2{=}CCl_2$	polydichloroethylene $-CH_2CCl_2CH_2CCl_2CH_2CCl_2-$	plastic food wrap
tetrafluoroethylene $CF_2{=}CF_2$	polytetrafluoroethylene $-CF_2CF_2CF_2CF_2CF_2CF_2-$	Teflon, bearings
acrylonitrile $CH_2{=}CHCN$	polyacrylonitrile $-CH_2CH(CN)CH_2CH(CN)CH_2CH(CN)-$	Orlon, Acrilan
styrene $CH_2{=}CHC_6H_5$	polystyrene $-CH_2CH(C_6H_5)-CH_2CH(C_6H_5)-CH_2CH(C_6H_5)-$	toys, styrofoam
methyl methacrylate $CH_2{=}C(CH_3)-C(=O)OCH_3$	polymethyl methacrylate $-CH_2-C(CH_3)(COOCH_3)-CH_2-C(CH_3)(COOCH_3)-CH_2-C(CH_3)(COOCH_3)-$	Plexiglas, Lucite

4.12 Preparation of Alkenes and Alkynes

Alkenes are prepared from either alcohols or haloalkanes (alkyl halides) by elimination reactions. We recall from Section 2.5 that a single compound splits into two products in an **elimination reaction.** One product usually contains most of the atoms in the reactant, and the remaining atoms are found in a second smaller molecule. The atoms eliminated to form the smaller molecule are usually located on adjacent carbon atoms in the reactant. The mechanism of elimination reactions will be discussed in Chapter 7.

Dehydration of Alcohols

The dehydration of 2-propanol produces propene. The reaction requires concentrated acids, such as sulfuric acid, H_2SO_4, or phosphoric acid, H_3PO_4. The reaction is pulled to completion because the water formed in the reaction is solvated with the concentrated acid.

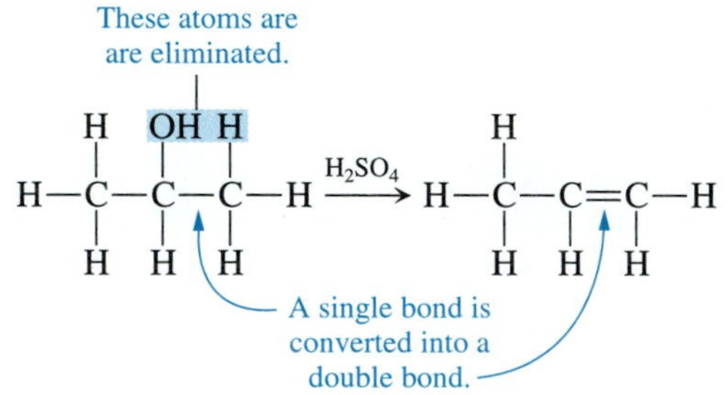

These atoms are are eliminated.

```
    H  OH  H                   H
    |   |  |      H2SO4        |
 H—C—C—C—H   ———————→   H—C—C=C—H
    |   |  |                   |  |  |
    H   H  H                   H  H  H
```

A single bond is converted into a double bond.

The elimination reaction requires breaking the carbon-oxygen bond and a carbon-hydrogen bond on an adjacent carbon atom. For alcohols such as 2-butanol, two different carbon atoms are adjacent to the OH-bearing carbon atom. Each could potentially release a hydrogen atom to form water.

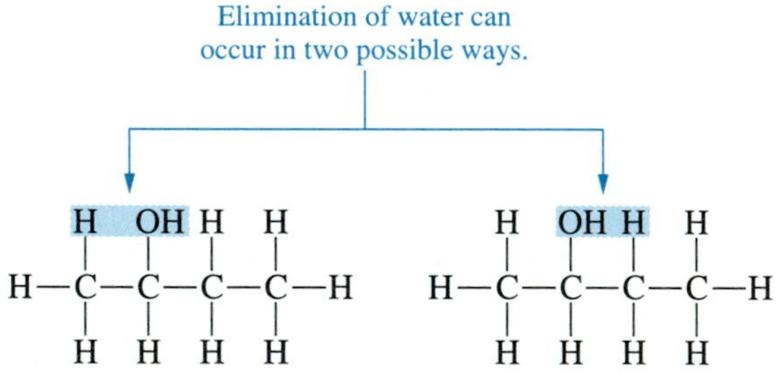

Elimination of water can occur in two possible ways.

```
    H  OH  H  H               H  OH  H  H
    |   |  |  |               |   |  |  |
 H—C—C—C—C—H          H—C—C—C—C—H
    |   |  |  |               |   |  |  |
    H   H  H  H               H   H  H  H
```

Thus, dehydration produces a mixture of products. The isomer that contains the greater number of alkyl groups attached to the double bond (the more substituted alkene) predominates in the mixture.

Elimination reactions that give the more substituted double bond are said to obey **Zaitsev's rule.** This generalization was discovered by Alexander Zaitsev, a nineteenth-century Russian chemist. We recall from Section 4.1 that increasing the number of alkyl groups bonded to unsaturated carbon atoms of an alkene increases its stability. Thus, Zaitsev simply observed that the major product of an elimination reaction is the more stable isomer. Zaitsev's rule also applies to mixtures of geometric isomers. For example, 3-pentanol yields a mixture of *cis*- and *trans*-2-pentene. The trans isomer is the major product.

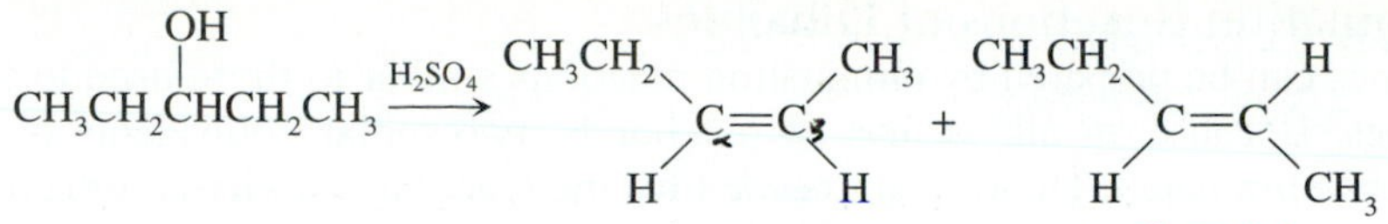

$$CH_3CH_2\underset{\displaystyle |}{\overset{\displaystyle OH}{C}}HCH_2CH_3 \xrightarrow{H_2SO_4} \underset{\text{cis-2-pentene (25\%)}}{(CH_3CH_2)(H)C{=}C(CH_3)(H)} + \underset{\text{trans-2-pentene (75\%)}}{(CH_3CH_2)(H)C{=}C(H)(CH_3)}$$

In the trans isomer the alkyl groups are well separated, whereas in the cis isomer the alkyl groups are near each other. The proximity of the alkyl groups causes steric hindrance; that is, the groups repel each other. The trans isomer is more stable, so it predominates (Figure 4.5).

Dehydrohalogenation of Alkyl Halides

The elimination of the elements H and X as in HCl or HBr from adjacent carbon atoms in an alkyl halide is called **dehydrohalogenation.** The product of the reaction is an alkene.

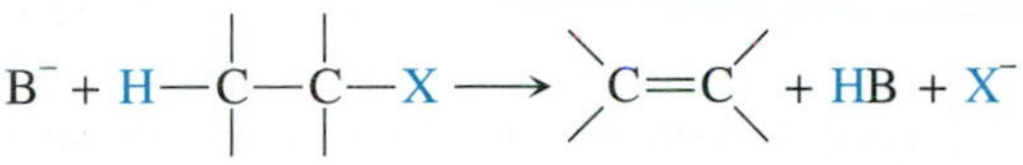

$$B^- + H{-}\overset{|}{\underset{|}{C}}{-}\overset{|}{\underset{|}{C}}{-}X \longrightarrow \rangle C{=}C\langle + HB + X^-$$

A base, represented by B^-, is required for the reaction. Although hydroxide ion is sufficiently basic for this reaction, it is usual to use an alkoxide, the conjugate base of an alcohol, as the base. Sodium ethoxide is commonly used in combination with ethanol as the solvent. The mechanism of this reaction will be presented in Chapter 7.

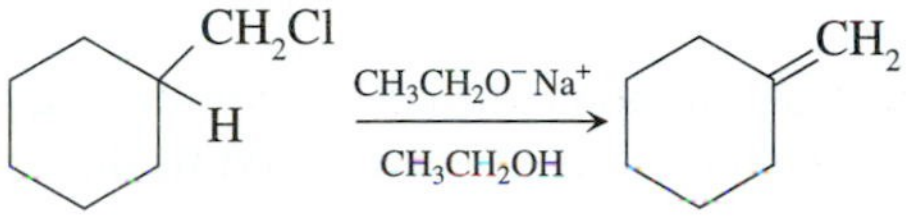

$$\text{cyclohexyl}(CH_2Cl)(H) \xrightarrow[CH_3CH_2OH]{CH_3CH_2O^-\,Na^+} \text{cyclohexylidene}{=}CH_2$$

As in the case of dehydration, the more highly substituted alkene predominates when two or more products are possible. When geometric isomers are possible, trans isomers are favored over cis isomers.

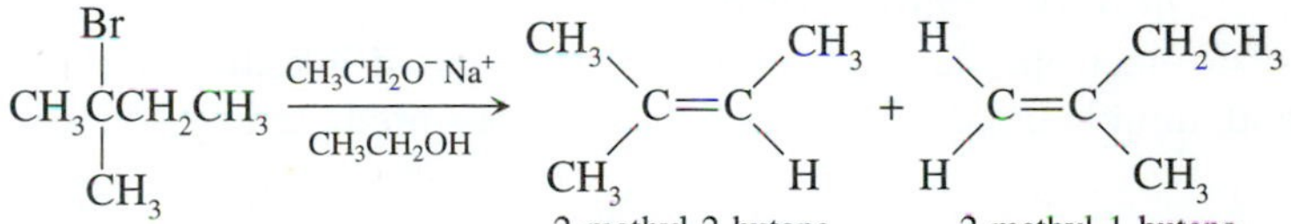

$$CH_3\underset{\displaystyle CH_3}{\overset{\displaystyle Br}{C}}CH_2CH_3 \xrightarrow[CH_3CH_2OH]{CH_3CH_2O^-\,Na^+} \underset{\text{2-methyl-2-butene (70\%)}}{(CH_3)(CH_3)C{=}C(CH_3)(H)} + \underset{\text{2-methyl-1-butene (30\%)}}{(H)(H)C{=}C(CH_2CH_3)(CH_3)}$$

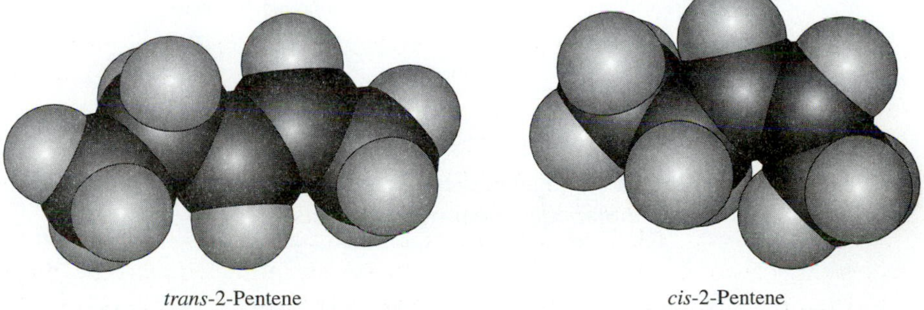

trans-2-Pentene *cis*-2-Pentene

FIGURE 4.5
Geometric Isomers of 2-Pentene

Elimination Reactions of Dihalides

Alkynes can be prepared by elimination reactions similar to those used to form alkenes. Because an alkyne has two π bonds, two molar equivalents of HX must be eliminated. The reactant needed for the reaction is a **vicinal** dihalide, a compound with halogen atoms on adjacent carbon atoms. A stronger base than an alkoxide ion is required. The most commonly used base is sodium amide ($NaNH_2$) in liquid ammonia as the solvent.

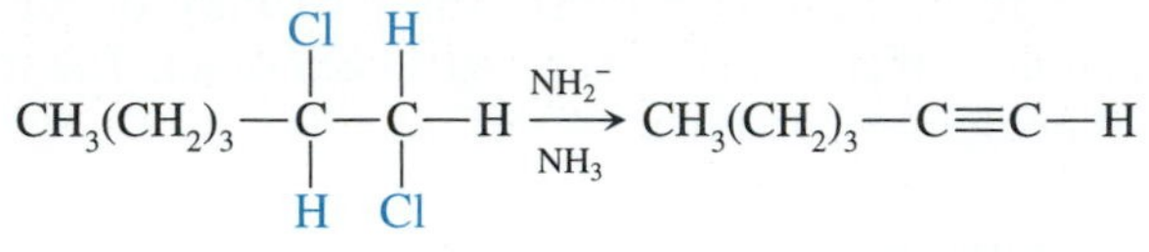

$$CH_3(CH_2)_3-\underset{H}{\overset{Cl}{C}}-\underset{Cl}{\overset{H}{C}}-H \xrightarrow[NH_3]{NH_2^-} CH_3(CH_2)_3-C\equiv C-H$$

1,2-dichlorohexane — 1-hexyne

4.13 Alkadienes

Compounds with two double bonds are **alkadienes,** commonly called dienes. When one single bond is located between the two double-bonded units, the compounds are chemically different from simple alkenes. These compounds, which are said to be **conjugated,** are the subject of this section. When more than one single bond is located between the two double-bonded units, the compounds are not chemically different from alkenes. The double bonds are said to be **isolated** or **nonconjugated.**

$$CH_3-CH=CH-CH=CH_2 \qquad CH_2=CH-CH_2-CH=CH_2$$

1,3-pentadiene (a conjugated diene) — 1,4-pentadiene (a nonconjugated diene)

Double bonds that share a common atom are said to be **cumulated.** These compounds are relatively rare and will not be discussed further.

$$CH_2=C=CH-CH_2-CH_3$$

1,2-pentadiene
(a cumulated diene)

Electrophilic Conjugate Addition

Addition of electrophilic reagents to nonconjugated alkadienes can occur at one or both double bonds. The products are those predicted by Markovnikov's rule.

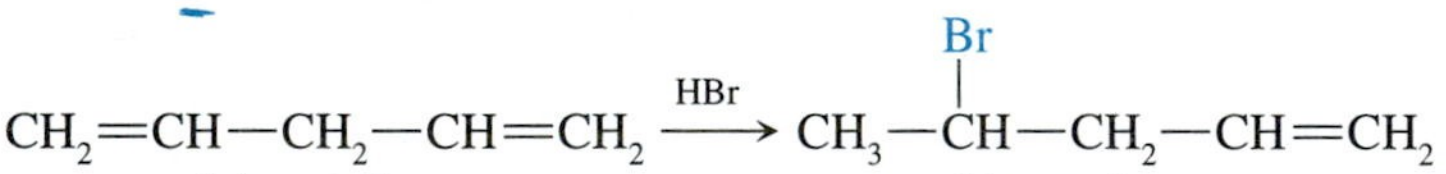

$$CH_2=CH-CH_2-CH=CH_2 \xrightarrow{HBr} CH_3-\overset{Br}{CH}-CH_2-CH=CH_2$$

1,4-pentadiene — 4-bromo-1-pentene

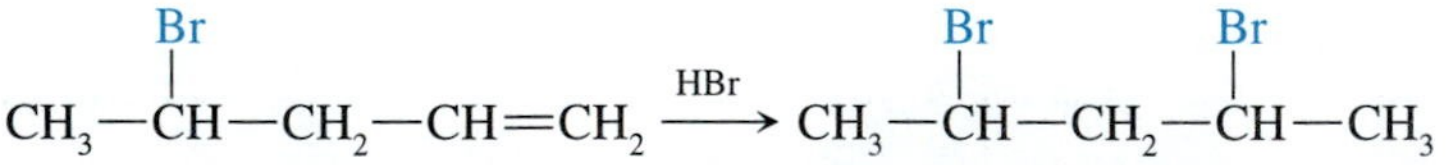

$$CH_3-\overset{Br}{CH}-CH_2-CH=CH_2 \xrightarrow{HBr} CH_3-\overset{Br}{CH}-CH_2-\overset{Br}{CH}-CH_3$$

2,4-dibromopentane

Addition of HBr to a conjugated diene is strikingly different. Two products are obtained when one molar equivalent of HBr reacts.

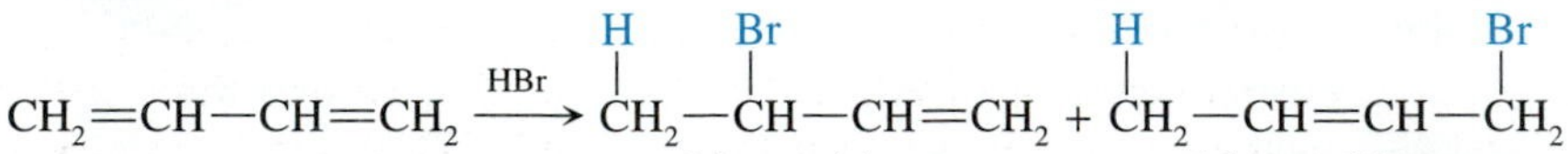

$$CH_2=CH-CH=CH_2 \xrightarrow{HBr} \overset{H}{CH_2}-\overset{Br}{CH}-CH=CH_2 + \overset{H}{CH_2}-CH=CH-\overset{Br}{CH_2}$$

3-bromo-1-butene (1,2-addition, 70%) — 1-bromo-2-butene (1,4-addition, 30%)

Allylic Oxidation in Metabolic Reactions

There are many diverse metabolic oxidation reactions catalyzed by cytochrome *P*-450, found in the liver. One of the mechanisms of this detoxifying agent involves oxidation at allylic sites to produce allylic alcohols.

$$>C{=}C(CH_3)- \xrightarrow{P\text{-}450} >C{=}C(CH_2{-}OH)-$$

an allylic alcohol

Although the nature of the intermediates is not always well known, the ready oxidation of allylic C—H bonds by cytochrome *P*-450 must involve intermediates with some resonance stabilization.

The liver ordinarily transforms organic compounds into oxidized products that are more water soluble and that can be excreted. However, one of the concerns in the design of drugs is the reactivity of metabolites produced by these metabolic oxidation reactions. Obviously, it is intolerable to have metabolites cause damage to cells and disrupt other life processes. In other words, the cure should not be worse than the disease.

The metabolism of hexobarbital (a sedative hypnotic) occurs by allylic oxidation of a methylene group in the cyclohexene ring. Subsequent reactions of this metabolite occur to allow ready elimination of the drug.

allylic position

The antiarrhythmic drug quinidine is metabolized at the indicated allylic position to give an allylic alcohol. In this case, the metabolite interestingly shows antiarrhythmic activity as well.

allylic position

Another example of a reaction producing a metabolite having biological activity is the allylic oxidation of marijuana. Marijuana contains Δ^1-tetrahydrocannabinol (Δ^1-THC), which has allylic centers at C-3, C-6, and C-7.

Δ^1-THC

Allylic oxidation does not occur at C-3 because of steric hindrance caused by the methyl groups and the aromatic ring. Of the other possible sites, the C-7 product predominates over the C-6 product. Interestingly, the C-7 allylic alcohol formed has been shown to be even more psychoactive than Δ^1-THC. ■

The 3-bromo-1-butene is the product of direct addition to a double bond, as predicted by Markovnikov's rule. The 1-bromo-2-butene is an unusual product that results from the addition of HBr to the C-1 and C-4 atoms. Note that the double bond in the product is between the C-2 and C-3 atoms. This product results from a **1,4-addition reaction.**

Before considering the origin of the 1,4-addition product derived from a conjugated diene, let's examine the structure of the **allylic carbocation.** This cation can be represented by two contributing resonance structures.

$$\left[CH_2{=}CH{-}\overset{+}{C}H_2 \longleftrightarrow \overset{+}{C}H_2{-}CH{=}CH_2\right]$$

Thus, the positive charge is distributed between both terminal carbon atoms. Either terminal carbon atom could react with a nucleophile, but the product of the reaction would be the same.

Now let's consider the electrophilic addition of a proton to a conjugated diene to give a carbocation.

$$H^+ + CH_2{=}CH{-}CH{=}CH_2 \longrightarrow \left[\underset{(1)}{\overset{H}{CH_2}{-}\overset{+}{C}H{-}CH{=}CH_2} \longleftrightarrow \underset{(2)}{\overset{H}{CH_2}{-}CH{=}CH{-}\overset{+}{C}H_2}\right]$$

The carbocation intermediate is an allylic-type ion that can be represented by two contributing resonance structures. In the next step in the addition reaction, the nucleophilic bromide ion can bond to either of the two carbon atoms bearing the positive charge. Attachment at the secondary carbon atom (see resonance structure 1) gives the 1,2-addition product. However, if the bromide ion bonds to the primary carbon atom (see resonance structure 2), the 1,4-addition product results.

The mist in the Smoky Mountains is due to the terpenes released from trees.

4.14 Terpenes

Terpenes, which are abundant in the oils of plants and flowers, have distinctive odors and flavors. They are responsible for the odors of pine trees and for the colors of carrots and tomatoes. Terpenes consist of two or more isoprene units joined together, usually head to tail. These compounds may have different degrees of unsaturation and a variety of functional groups. The structures may be acyclic or cyclic. Nevertheless, the isoprene units are usually easy to identify.

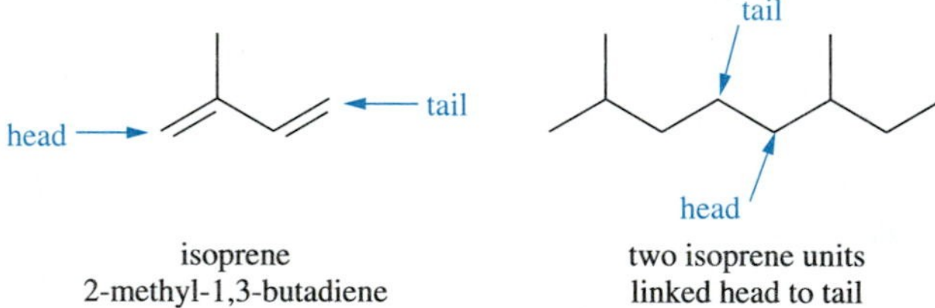

Farnesol is an acyclic terpene that has three isoprene units joined head to tail. Carvone has two isoprene units but contains a ring. Dashed lines indicate where the isoprene units are joined.

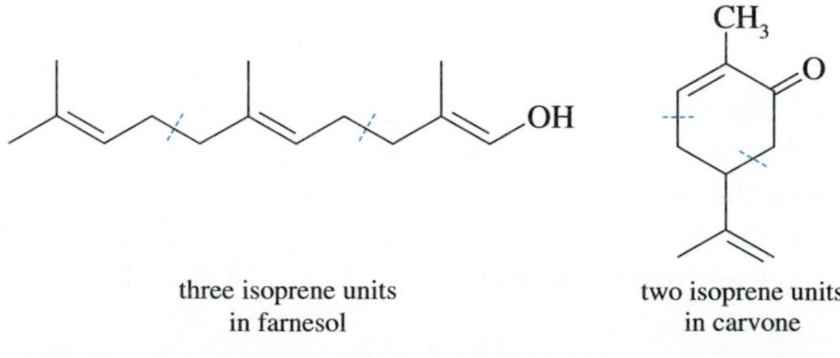

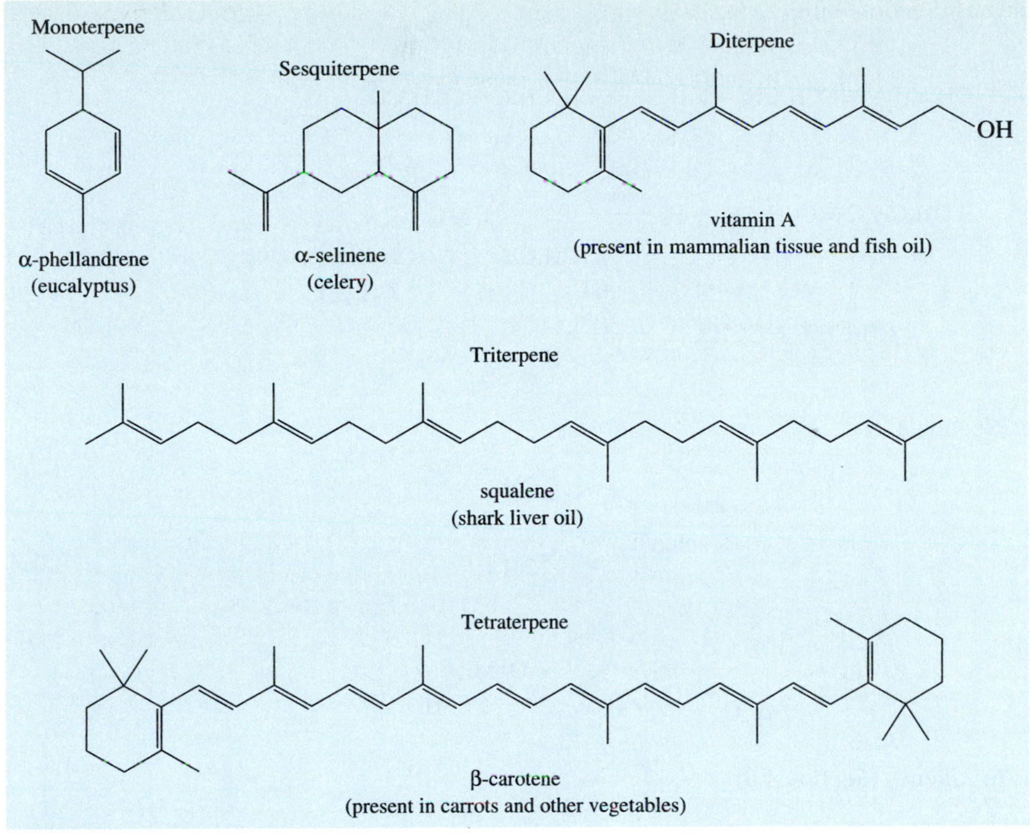

FIGURE 4.6
Classification of Terpenes

Terpenes are classified by the number of isoprene units they contain. The **monoterpenes,** the simplest terpene class, contain two isoprene units, and **sesquiterpenes** have three isoprene units. Examples of these structures are shown in Figure 4.6 using bond-line structures. **Diterpenes, triterpenes,** and **tetraterpenes** contain 4, 6, and 8 isoprene units, respectively.

Summary of Reactions

1. Catalytic Hydrogenation of Alkenes (Section 4.6)

CH_3 CH_3 $+ H_2 \xrightarrow{Pd/C}$ H H CH_3 CH_3

$+ 2\,H_2 \xrightarrow{Pt/C}$

O $+ H_2 \xrightarrow[1\ atm]{Pd/C}$ O

2. Hydrogenation of Alkynes (Section 4.6)

$$CH_3CH_2C{\equiv}CCH_2CH_3 + 2\,H_2 \xrightarrow{Pd/C} CH_3CH_2CH(H)\text{—}CH(H)CH_2CH_3$$

$$CH_3CH_2C{\equiv}CCH_2CH_3 + H_2 \xrightarrow{\text{Lindlar catalyst}} \textit{cis-}(CH_3CH_2)HC{=}CH(CH_2CH_3)$$

$$CH_3CH_2C{\equiv}CCH_2CH_3 \xrightarrow[2.\ H_2O]{1.\ Li/NH_3} \textit{trans-}(CH_3CH_2)HC{=}CH(CH_2CH_3)$$

3. Oxidation of Alkenes (Section 4.7)

Cyclohexene $\xrightarrow[NaOH/H_2O]{KMnO_4}$ cis-1,2-cyclohexanediol (H, OH; OH, H)

Methylenenorbornane ($=CH_2$) $\xrightarrow[2.\ Zn/H^+]{1.\ O_3}$ norbornanone ($=O$) $+ O{=}CH_2$ (H, H)

4. Addition of Halogens to Alkenes (Section 4.8)

$$CH_3(CH_2)_4CH{=}CH_2 + Cl_2 \longrightarrow CH_3(CH_2)_4CH(Cl)\text{—}CH_2(Cl)$$

5. Addition of Hydrogen Halides to Alkenes (Section 4.8)

Methylenecyclohexane ($=CH_2$) $+ HBr \longrightarrow$ 1-bromo-1-methylcyclohexane (CH_3, Br)

6. Addition of Water (Hydration) to Alkenes (Section 4.10)

$$(CH_3CH_2)(CH_3)C{=}CH_2 + H_2O \xrightarrow{H_2SO_4} CH_3CH_2\text{—}C(HO)(CH_3)\text{—}C(H)(H)\text{—}H$$

7. Addition of Halogens to Alkynes (Section 4.8)

$$CH_3CH_2C{\equiv}CCH_2CH_3 \xrightarrow{Br_2} (Br)(CH_3CH_2)C{=}C(CH_2CH_3)(Br) \xrightarrow{Br_2} CH_3CH_2\text{—}CBr_2\text{—}CBr_2\text{—}CH_2CH_3$$

8. Addition of Hydrogen Halides to Alkynes (Section 4.8)

$$CH_3CH_2C{\equiv}CCH_2CH_3 + 2\,HBr \longrightarrow CH_3CH_2C(Br)(Br)\text{—}C(H)(H)CH_2CH_3$$

9. **Addition of Water (Hydration) to Alkynes (Section 4.10)**

$$C_6H_{11}\text{—}C\equiv C\text{—}H \xrightarrow[HgSO_4]{H_2O/H_2SO_4} \left[C_6H_{11}\text{—}\overset{OH}{\overset{|}{C}}=\overset{H}{\overset{|}{C}}\text{—}H \right] \longrightarrow C_6H_{11}\text{—}\overset{O}{\overset{\|}{C}}\text{—}\underset{H}{\underset{|}{\overset{H}{\overset{|}{C}}}}\text{—}H$$

10. **Synthesis of Alkenes by Dehydration of Alcohols (Section 4.12)**

$$CH_3CH_2\underset{CH_2CH_3}{\underset{|}{\overset{OH}{\overset{|}{C}}}}CH_2CH_3 \xrightarrow{H_2SO_4} (CH_3CH_2)_2C=C(H)CH_3$$

11. **Synthesis of Alkenes by Dehydrohalogenation of Haloalkanes (Section 4.12)**

$$\text{bromocyclodecane (}C_{10}H_{19}Br\text{)} \xrightarrow[CH_3CH_2OH]{CH_3CH_2O^-} \text{cyclodecene}$$

12. **Synthesis of Alkynes by Dehydrohalogenation of Dihalides (Section 4.12)**

$$CH_3CH_2\underset{H}{\underset{|}{\overset{Br}{\overset{|}{C}}}}\text{—}\underset{H}{\underset{|}{\overset{Br}{\overset{|}{C}}}}CH_2CH_3 \xrightarrow{NH_2^-/NH_3} CH_3CH_2C\equiv CCH_2CH_3$$

13. **Conjugate Addition Reaction of Dienes (Section 4.13)**

$$CH_2=CHCH=CH_2 \xrightarrow{HBr} CH_3\underset{Br}{\underset{|}{C}}HCH=CH_2 + CH_3CH=CHCH_2Br$$

Explorations with Molecular Models

1. Prepare a ball-and-stick model of ethene (ethylene) corresponding to the following space-filling model. How many sets of equivalent carbon atoms are there? How many sets of equivalent hydrogen atoms are there? Substitute the hydrogen atom on the right carbon atom that is pointed toward the front with a chlorine atom. Now how many sets of equivalent carbon atoms are there? How many sets of equivalent hydrogen atoms are there? How many isomers can be generated by replacement of any of the hydrogen atoms with a second chlorine atom?

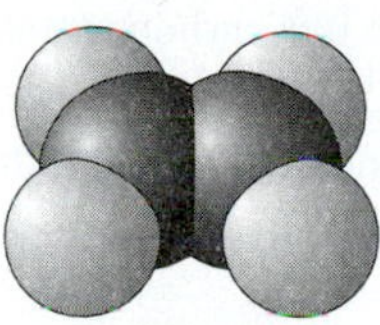

2. Prepare a ball-and-stick model of propene (propylene), which is shown in the following drawing as a space-filling model. How many sets of equivalent carbon atoms are there? How many sets of equivalent hydrogen atoms are there? Substitute the hydrogen atom on the right carbon atom of the double-bonded unit with a methyl group. Now how many sets of equivalent carbon atoms are there? How many sets of equivalent hydrogen atoms are there? Draw a standard two-dimensional representation of the molecule using 90°, 120°, and 180° angles. Name the compound.

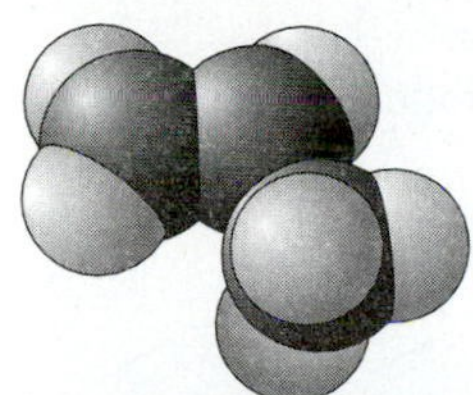

3. Prepare a ball-and-stick model of propene (propylene) as given in Exploration Exercise 2. Substitute the hydrogen atom on the left carbon atom that is pointed toward the front with a methyl group. How many equivalent carbon atoms are there? How many sets of equivalent hydrogen atoms are there? Draw a standard two-dimensional representation of the molecule using 90°, 120°, and 180° angles. Name the compound.

4. Prepare a ball-and-stick model of propene (propylene) as given in Exploration Exercise 2. Substitute the hydrogen atom on the left carbon atom that is pointed toward the back with a methyl group. How many sets of equivalent carbon atoms are there? How many sets of equivalent hydrogen atoms are there? Draw a standard two-dimensional representation of the molecule using 90°, 120°, and 180° angles. Name the compound. Considering the fact that nonbonded atoms in close proximity may sterically repel one another, is this molecule more stable or less stable than the molecule represented by the structure prepared in Exercise 3?

5. Prepare a model of ethylene and remove a pair of cis hydrogen atoms. Connect the two carbon atoms by a chain of methylene ($—CH_2—$) units. How many atoms are in the smallest ring that can be formed without straining the model? Draw a standard two-dimensional structure of the molecule using a bond-line representation.

6. Prepare a model of ethylene and remove a pair of trans hydrogen atoms. Connect the two carbon atoms by a chain of methylene units. How many atoms are in the smallest ring that can be formed without straining the model? (*Hint:* It is necessary to pass the chain over the top of the plane of the structure to reach the other side.) Name this compound.

7. Prepare a ball-and-stick model of propyne corresponding to the following space-filling model. How many sets of equivalent carbon atoms are there? How many sets of equivalent hydrogen atoms are there? Determine the number of isomeric compounds that may be formed by substituting a carbon atom containing three hydrogen atoms for one hydrogen atom in each of the possible sets of equivalent hydrogen atoms. Draw a standard two-dimensional representation of each molecule. Name each compound.

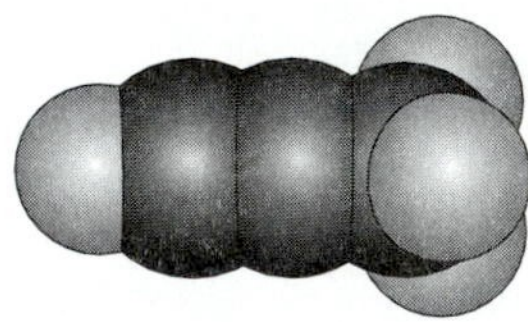

8. Prepare a model of ethyne (acetylene) and remove the hydrogen atoms. Connect the two carbon atoms by a chain of methylene units. How many atoms are in the smallest ring that can be formed without straining the model? Name this compound.

Exercises

Molecular Formulas

4.1 What is the molecular formula for the compounds with each of following structural features?

(a) six carbon atoms and one double bond (b) five carbon atoms and two double bonds
(c) seven carbon atoms, a ring, and one double bond

4.2 What is the molecular formula for the compounds with each of the following structural features?

(a) four carbon atoms and two triple bonds (b) four carbon atoms, a double bond, and a triple bond
(c) ten carbon atoms and two rings

4.3 Write the molecular formula for each of the following compounds.

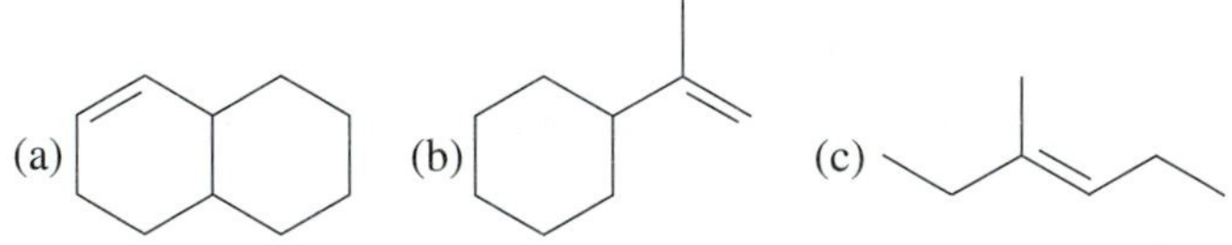

4.4 Write the molecular formula for each of the following compounds.

(a) (b) (c)

Classification of Alkenes and Alkynes

4.5 Classify the double bond in each alkene in Exercise 4.3 by its substitution pattern.

4.6 Classify the double bond in each alkene in Exercise 4.4 by its substitution pattern.

4.7 Indicate the degree of substitution of the double bond in each of the following compounds.

(a) cholesterol, a steroid required for growth in almost all life forms

(b) tamoxifen, a drug used in treatment of breast cancer

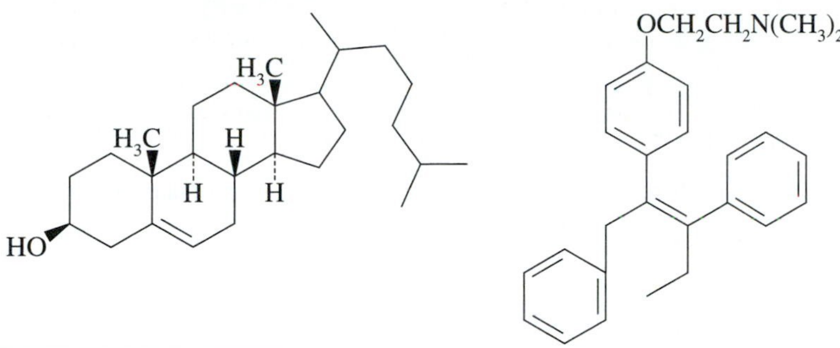

4.8 Indicate the degree of substitution of the triple bond in each of the following compounds.

(a) MDL, a drug used in breast cancer therapy

(b) RU-486, a drug used to induce abortion

Geometric Isomers

4.9 Which of the following molecules can exist as cis and trans isomers?

(a) $CH_3CH{=}CHBr$ (b) $CH_2{=}CHCH_2Br$ (c) $CH_3CH{=}CHCH_2Cl$

4.10 Which of the following molecules can exist as cis and trans isomers?

(a) $CH_3CH{=}CBr_2$ (b) $CH_2{=}CHCHBr_2$ (c) $CH_3CH{=}CHCHCl_2$

4.11 Which of the following molecules can exist as cis and trans isomers?

(a) 1-hexene (b) 3-heptene (c) 4-methyl-2-pentene

4.12 Which of the following molecules can exist as cis and trans isomers?

(a) 3-methyl-1-hexene (b) 3-ethyl-3-heptene (c) 2-methyl-2-pentene

E,Z System of Nomenclature

4.13 Select the group with the highest priority in each of the following sets.

(a) $-CH(CH_3)_2$, $-CHClCH_3$, $-CH_2CH_2Br$ (b) $-CH_2CH{=}CH_2$, $-CH_2CH(CH_3)_2$, $-CH_2C{\equiv}CH$

(c) $-OCH_3$, $-N(CH_3)_2$, $-C(CH_3)_3$

4.14 Select the group with the highest priority in each of the following sets.

(a) $-\overset{\overset{O}{\|}}{C}-CH_3$ $-\overset{\overset{O}{\|}}{C}-OH$ $-\overset{\overset{O}{\|}}{C}-F$ (b) $-\overset{\overset{O}{\|}}{C}-NH_2$ $-\overset{\overset{O}{\|}}{C}-OCH_3$ $-\overset{\overset{O}{\|}}{C}-N(CH_3)_2$

(c) $-\overset{\overset{O}{\|}}{C}-SCH_3$ $-\overset{\overset{O}{\|}}{C}-OCH_2CH_3$ $-\overset{\overset{O}{\|}}{C}-Cl$

4.15 Assign the E or Z configuration to each of the following antihistamines.

(a) pyrrobutamine (b) triprolidine

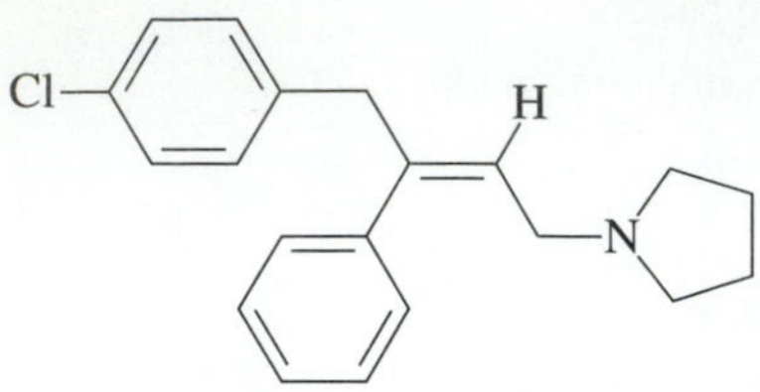

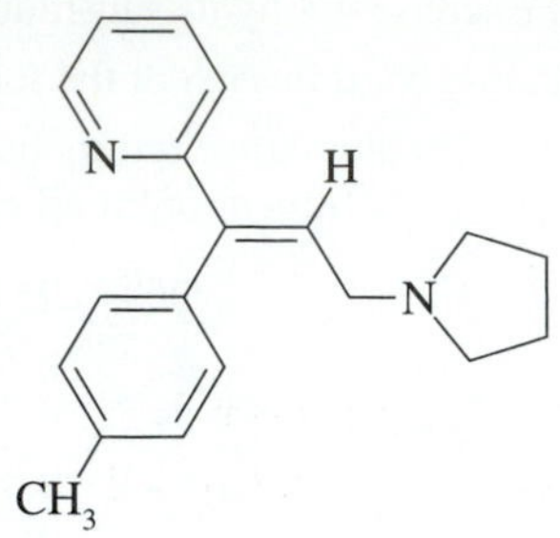

Cl H N H N N CH_3

4.16 Assign the E or Z configuration to each of the following hormone antagonists used to control cancer.

(a) clomiphene (b) nitromifene

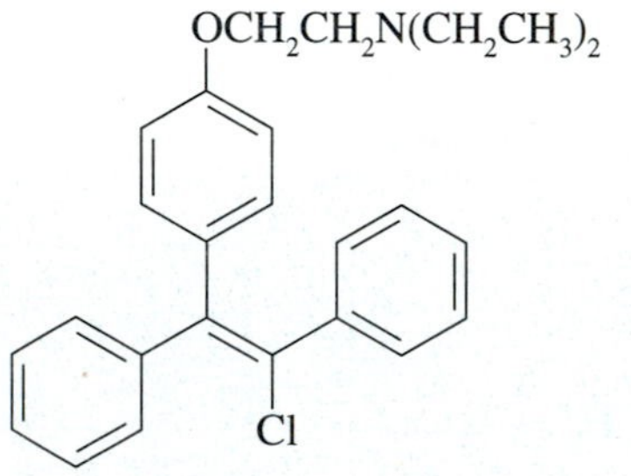

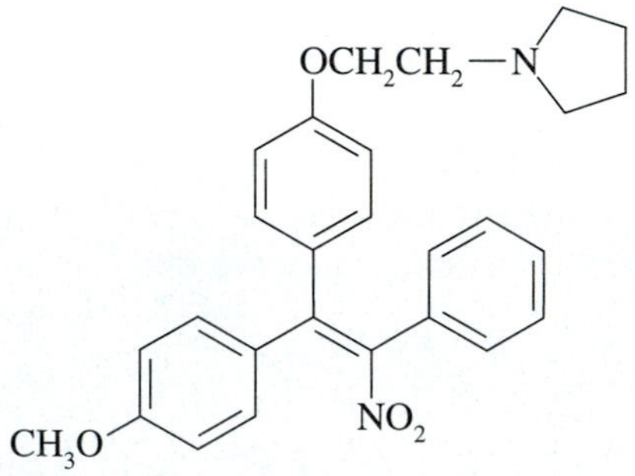

$OCH_2CH_2N(CH_2CH_3)_2$ Cl

$OCH_2CH_2—N$ CH_3O NO_2

4.17 Draw the structural formula for each of the following pheromones with the indicated configuration.

(a) sex pheromone of Mediterranean fruit fly, E isomer

$CH_3CH_2CH{=}CH(CH_2)_4CH_2OH$

(b) defense pheromone of termite, E isomer

$CH_3(CH_2)_{12}CH{=}CHNO_2$

4.18 Assign the configuration at all double bonds where geometrical isomerism is possible in each of the following pheromones.

(a) European vine moth (b) pink bollworm moth

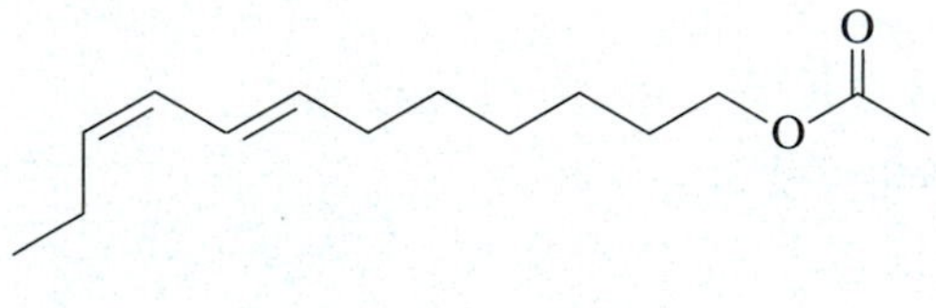

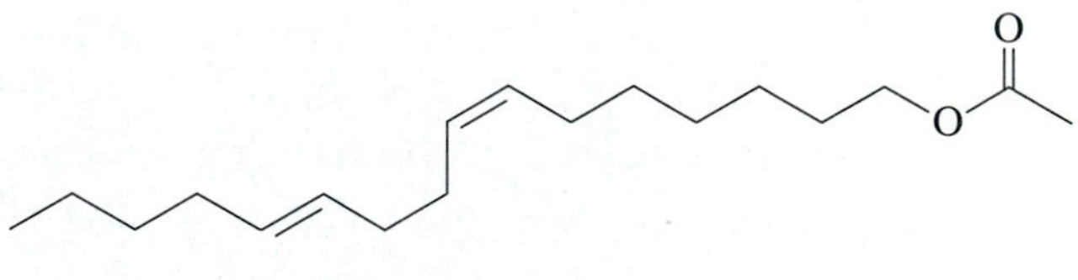

O O O O

Nomenclature of Alkenes

4.19 Name each of the following compounds.

(a) CH_3, CH_3 / C=C / CH_2CH_3, H

(b) CH_3, CH_3CH_2 / C=C / CH_2CH_3, CH_3

(c) CH_3, CH_3CH_2 / C=C / CH_3, H

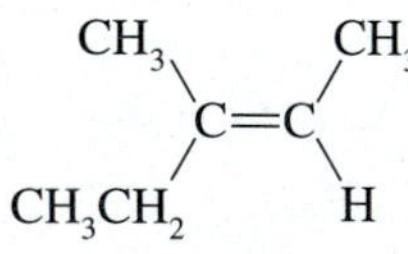

4.20 Name each of the following compounds.

(a) Br, CH_3 / C=C / CH_2CH_3, Br

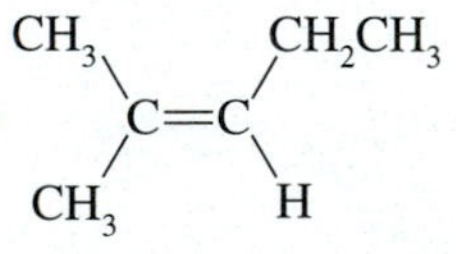

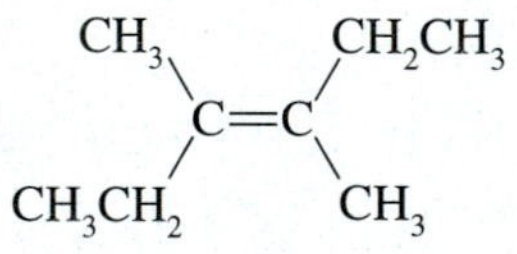

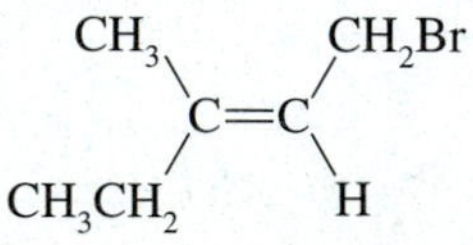

(b) CH_3, CH_3CH_2 / C=C / CH_2Br, H

(c) Cl, CH_3CH_2 / C=C / CH_3, Br

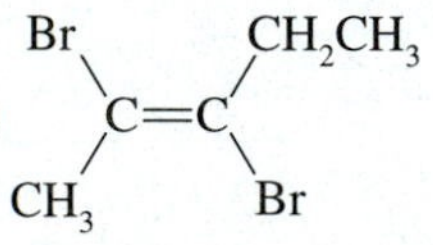

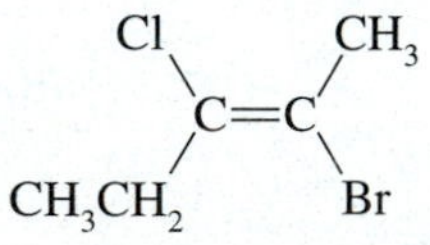

4.21 Name each of the following compounds.

(a) 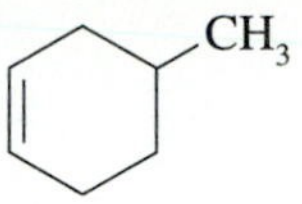 (b) 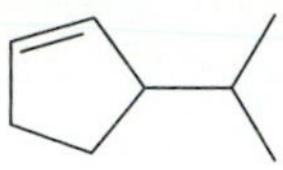 (c)

CH_3 $CH{=}CH_2$

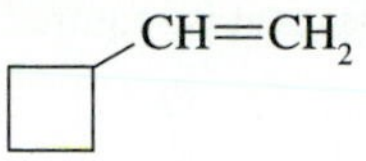

4.22 Name each of the following compounds.

(a) 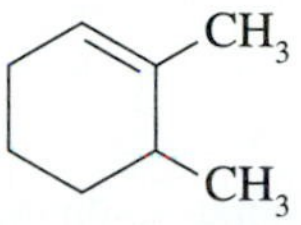 (b) 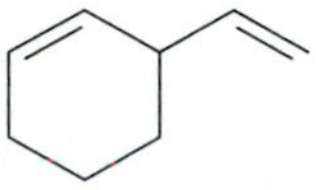 (c)

CH_3 CH_3

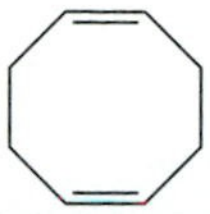

4.23 Draw a structural formula for each of the following compounds.

(a) 2-methyl-2-pentene (b) *cis*-2-methyl-3-hexene (c) *trans*-5-methyl-2-hexene

4.24 Draw a structural formula for each of the following compounds.

(a) *trans*-1-chloropropene (b) *cis*-2,3-dichloro-2-butene (c) 2,4-dimethyl-2-hexene

4.25 Draw a structural formula for each of the following compounds.

(a) 1-methylcyclopentene (b) 1,2-dibromocyclohexene (c) 4,4-dimethylcyclohexene

4.26 Draw a structural formula for each of the following compounds.

(a) 3-methylcyclohexene (b) 1,3-dibromocyclopentene (c) 3,3-dichlorocyclopentene

Nomenclature of Alkynes

4.27 Name each of the following compounds.

(a) $CH_3CH_2CH_2C{\equiv}CH$ (b) $(CH_3)_3C{-}C{\equiv}C{-}CH_2CH_3$ (c) 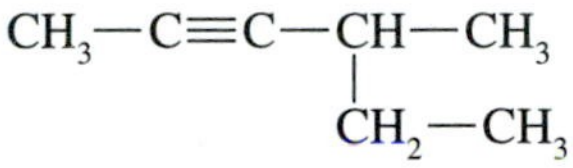 $CH_3{-}C{\equiv}C{-}CH(CH_2{-}CH_3){-}CH_3$

4.28 Name each of the following compounds.

(a) $CH_3CHBrCHBrC{\equiv}CCH_3$ (b) $Cl(CH_2)_2C{\equiv}C(CH_2)_3CH_3$ (c) 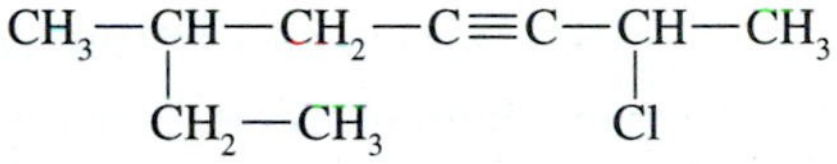 $CH_3{-}CH(CH_2{-}CH_3){-}CH_2{-}C{\equiv}C{-}CH(Cl){-}CH_3$

4.29 Write the structural formula for each of the following compounds.

(a) 2-hexyne (b) 3-methyl-1-pentyne (c) 5-ethyl-3-octyne

4.30 Write the structural formula for each of the following compounds.

(a) 3-heptyne (b) 4-methyl-1-pentyne (c) 5-methyl-3-heptyne

Hydrogenation of Alkenes and Alkynes

4.31 How many moles of hydrogen gas will react with each of the following compounds?

(a) $CH_3{-}CH{=}CH{-}C{\equiv}CH$ (b) $HC{\equiv}C{-}C{\equiv}CH$ (c) $CH_2{=}CH{-}C{\equiv}C{-}CH{=}CH_2$

4.32 How many moles of hydrogen gas will react with each of the following compounds?

(a) ichthyothereol, a convulsant

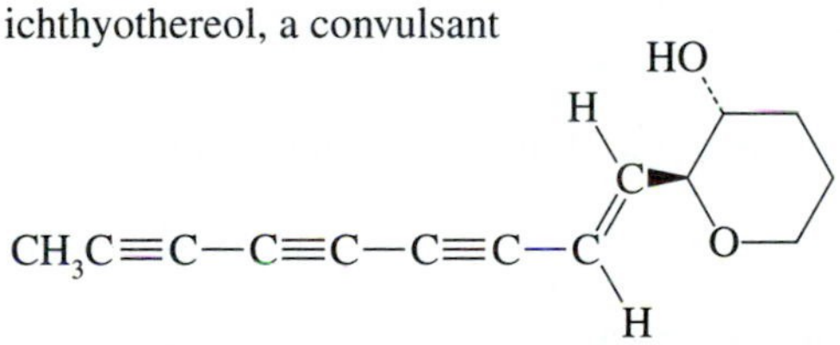

HO H O H

$CH_3C{\equiv}C{-}C{\equiv}C{-}C{\equiv}C{-}CH{=}CH{-}$

(b) mycomycin, an antibiotic

$HC{\equiv}C{-}C{\equiv}C{-}CH{=}C{=}CH{-}CH{=}CH{-}CH{=}CH{-}CH_2CO_2H$

4.33 How could the following unsaturated carboxylic acid be prepared from a structurally related alkyne? What reagent is required?

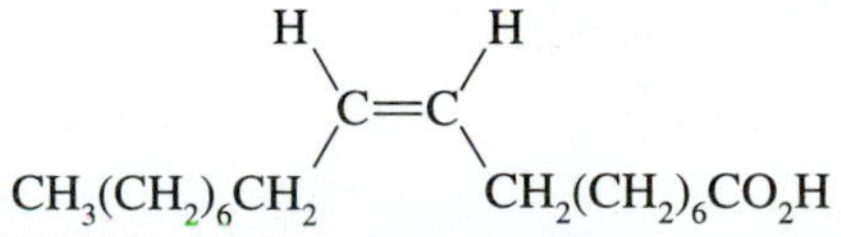

H H

$C{=}C$

$CH_3(CH_2)_6CH_2$ $CH_2(CH_2)_6CO_2H$

4.34 How could the following compound, which is a constituent of the sex pheromone of the male oriental fruit moth, be prepared from a structurally related alkyne? What reagent is required?

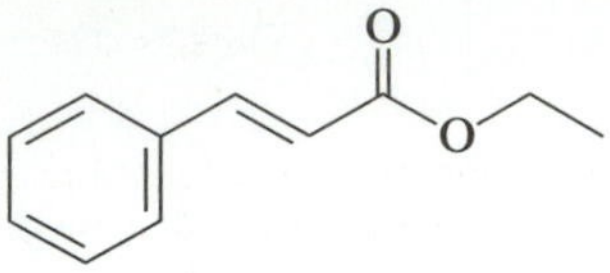

Oxidation of Alkenes

4.35 Describe the observation that is made when *cis*-2-pentene reacts with potassium permanganate. How could this reagent be used to distinguish between *cis*-2-pentene and cyclopentane?

4.36 Write the products of the reaction of vinylcyclohexane as well as allylcyclopentane with potassium permanganate.

4.37 Write the product(s) of the ozonolysis of each of the following compounds.

(a) 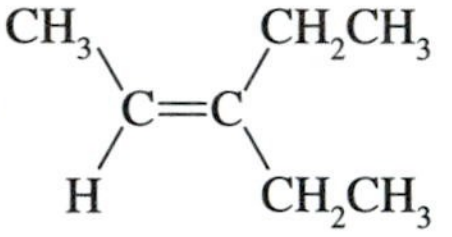 $(CH_3)(H)C{=}C(CH_2CH_3)(CH_2CH_3)$ (b) 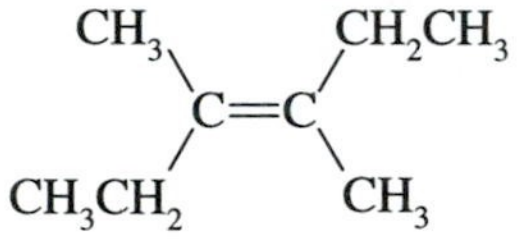 $(CH_3)(CH_3CH_2)C{=}C(CH_2CH_3)(CH_3)$ (c) $(H)(CH_3CH_2)C{=}C(CH_3)(H)$

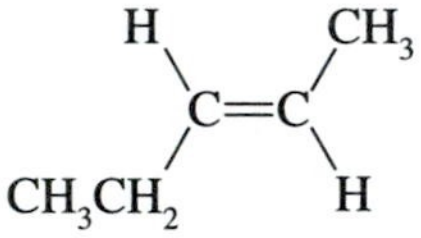

4.38 Write the product(s) of the ozonolysis of each of the following compounds.

(a) 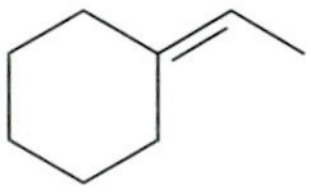 (b) 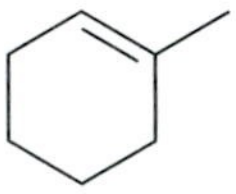 (c)

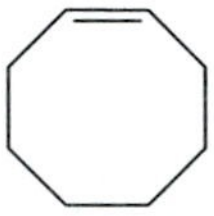

Addition Reactions

4.39 Which of the compounds in Exercise 4.37 would give a single product when reacted with HBr? Why?

4.40 Write the product of the reaction of HBr with each of the compounds in Exercise 4.38.

4.41 The addition of HBr to an alkyne results in the trans addition of hydrogen and bromine. Write the product(s) of the reaction of one mole of HBr with each of the following alkynes.

(a) $CH_3CH_2C{\equiv}CH$ (b) $CH_3CH_2C{\equiv}CCH_3$ (c) $CH_3CH_2C{\equiv}CCH_2CH_3$

4.42 Write the product(s) of the reaction of two moles of HBr with each of the compounds in Exercise 4.41.

4.43 Write the product of hydration of each of the compounds in Exercise 4.38.

4.44 Hydration of one of the following two compounds yields a single ketone product. The other compound yields a mixture of ketones. Which one yields only one ketone product? Why?

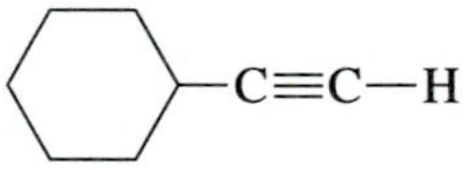

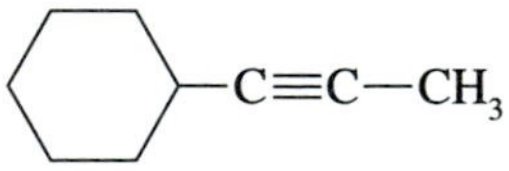

$C_6H_{11}{-}C{\equiv}C{-}H$ $C_6H_{11}{-}C{\equiv}C{-}CH_3$

Preparation of Alkenes and Alkynes

4.45 How many alkenes would be formed by dehydrohalogenation of each of the following alkyl bromides? Which compound should be the major isomer?

(a) $CH_3CH_2CH(Br)CH_2CH_3$ (b) $(CH_3)_3CCHBrCH_3$ (c) $CH_3(CH_2)_3CHBrCH_3$

4.46 Write the structure of a bromo compound that gives exclusively each of the following alkenes by dehydrohalogenation.

(a) 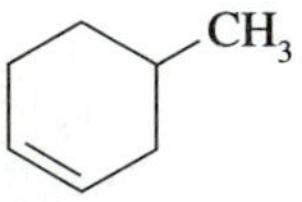 CH_3 (b) 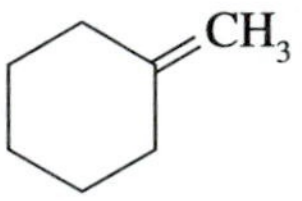 CH_3 (c) 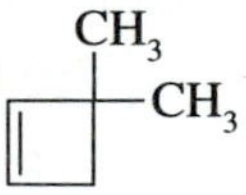 CH_3, CH_3

4.47 Write the structure of a compound that would yield the following alkyne upon dehydrohalogenation.

$$\text{C}_6\text{H}_{11}\text{—C}\equiv\text{C—H}$$

4.48 Alkynes can be prepared by elimination reactions of geminal dihalides—compounds with two halogen atoms bonded to the same carbon atom. Would the following reaction provide a good yield of the indicated product? Explain.

$$CH_3CH_2CH_2CBr_2CH_3 \xrightarrow[NH_3]{NaNH_2} CH_3CH_2C{\equiv}CCH_3$$

Polyunsaturated Compounds

4.49 Which of the following compounds has conjugated double bonds?

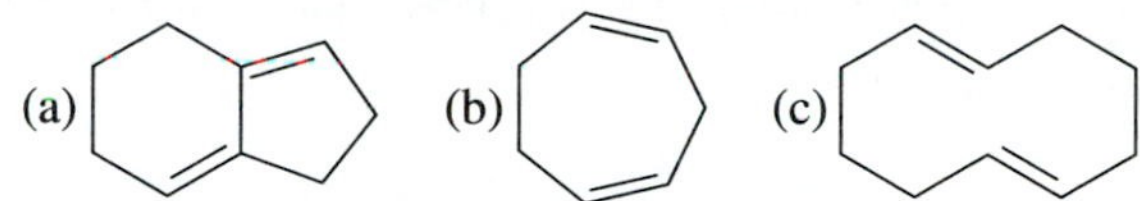

4.50 Which of the following compounds has conjugated double bonds?

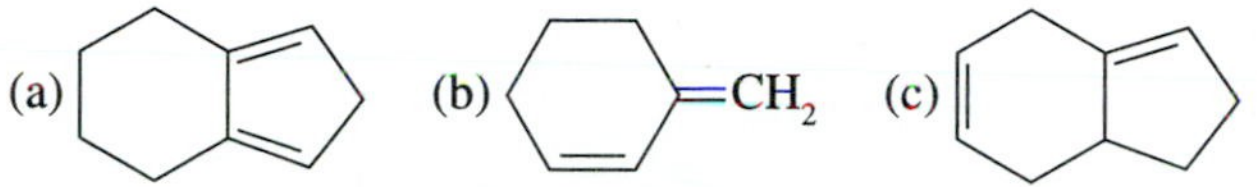

Terpenes

4.51 Classify each of the following terpenes and divide it into isoprene units.

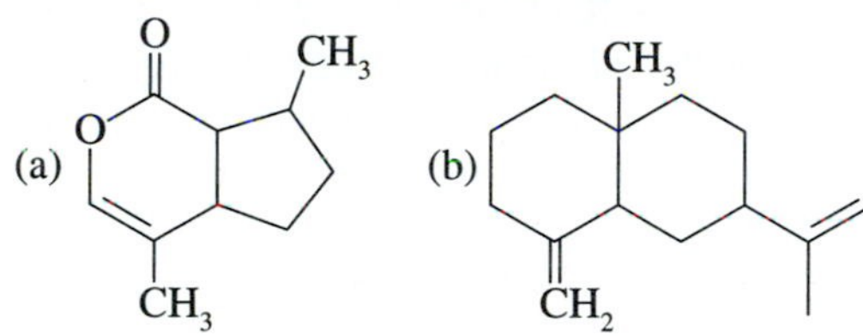

4.52 Classify each of the following terpenes and divide it into isoprene units.

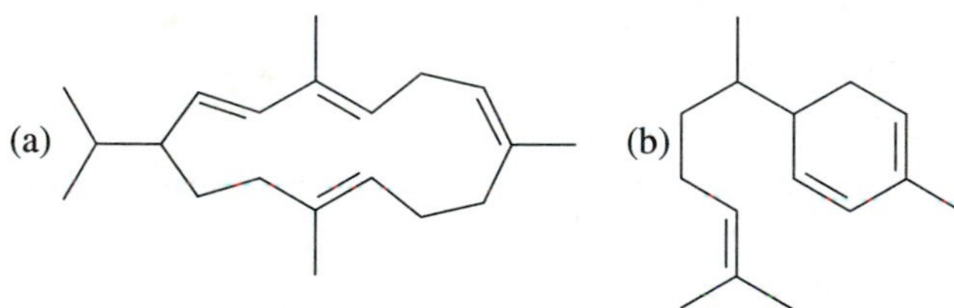

Polymers

4.53 Draw a representation of the polymer produced from $CH_2{=}CHCO_2CH_3$.

4.54 Draw a representation of the polymer produced from $CH_2{=}CCl_2$.

Spectroscopy

4.55 If you have read Chapter 18 and the material has been assigned by your professor, then Exercises 18.2, 18.3, 18.6, 18.10, 18.27, 18.28, and 18.33, which are based on the structure of alkenes and alkynes, may be done.

CHAPTER 5

Aromatic Compounds

Overview

5.1 Substituted Benzene Compounds

The term *aromatic* means "fragrant." For this reason many fragrant substances, known from earliest time, were classified as aromatic compounds. It turns out that many of these compounds contain a benzene ring (Section 1.6) that is bonded to one or more substituents. Oil of wintergreen and vanilla are two well-known fragrant compounds containing a benzene ring.

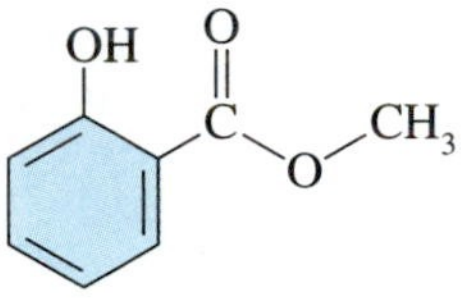

methyl salicylate (oil of wintergreen)

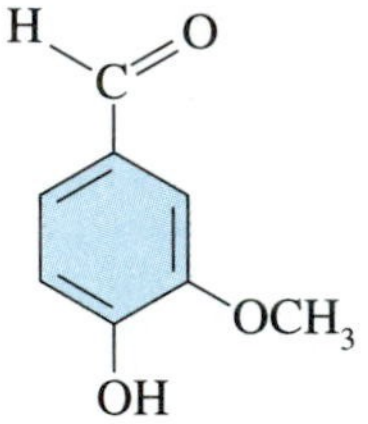

vanillin (vanilla)

Today, the classification of aromatic compounds is no longer based on odor because many compounds containing a benzene ring are not in fact fragrant. Some aromatic compounds, including the analgesics aspirin and ibuprofen, are solids that have little or no odor.

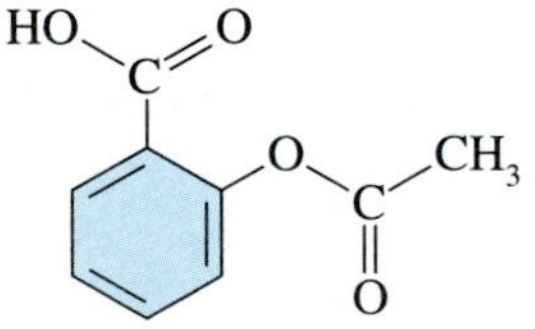

aspirin

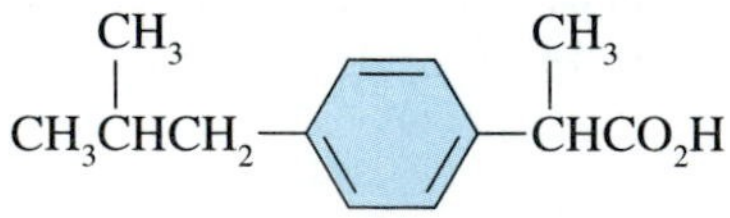

ibuprofen

Thus, the common feature of aromatic compounds is not their odor but the benzene ring. This six-carbon unit is usually not affected by reactants that alter the rest of the structure. In other words, it is the lack of reactivity of the benzene ring that characterizes these compounds.

5.2 Aromaticity

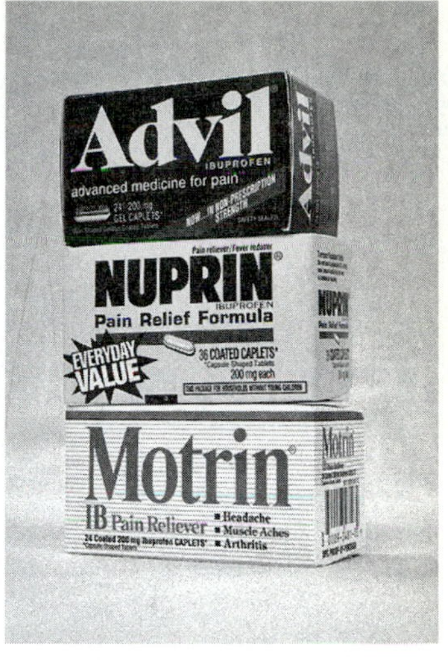

Several different brands of analgesics contain the same ingredient, ibuprofen.

Benzene, C_6H_6, is highly unsaturated—it has six fewer hydrogen atoms than cyclohexane, C_6H_{12}, its cyclic saturated counterpart. Although benzene is represented by a hexagon containing three double bonds, it does not undergo addition reactions with reagents such as bromine, HBr, or H_2O as alkenes do. Thus, the lack of reactivity of benzene contradicts what we know about unsaturated compounds. That is to say, it does not behave like the "triene" depicted by its Lewis structure. In addition, benzene typically undergoes substitution reactions, a reaction not typical of alkenes. Benzene reacts with bromine, in the presence of iron(III) bromide as a catalyst, to give a single monosubstituted product, C_6H_5Br.

$$C_6H_5\text{—H} + Br_2 \xrightarrow{FeBr_3} C_6H_5\text{—Br} + \text{H—Br}$$

This result indicates that all six hydrogen atoms of benzene are chemically equivalent. It also turns out that there are three possible isomeric dibromobenzenes, $C_6H_4Br_2$. These facts are accommodated by the electronic structure of benzene, as it is now understood.

Kekulé's Concept of Benzene

In 1865, a German chemist, August Kekulé, suggested that benzene is a ring of six carbon atoms linked by alternating single and double bonds. He proposed that benzene actually exists as two structures differing only in the arrangement of the single and double bonds, which oscillate around the ring.

$$\text{Kekulé structure I (C}_6\text{H}_6\text{)} \quad \text{or} \quad \text{Kekulé structure II (C}_6\text{H}_6\text{)}$$

Kekulé proposed that the rapid oscillation of single and double bonds somehow made benzene resist addition reactions. This concept "explained" why only one bromobenzene (C_6H_5Br) forms in the substitution reaction of benzene with Br_2 in the presence of an iron(III) catalyst. He reasoned that the rapid oscillation of single and double bonds around the ring makes all six carbon atoms, and therefore all six hydrogen atoms, equivalent. However, no evidence has been found that benzene exists as a mixture of equilibrating structures. Quite to the contrary, a single structure is observed.

Resonance Theory and Benzene

The two Kekulé structures for benzene differ only in the arrangement of electron pairs, a feature that we associate with resonance structures. The modern picture of benzene is that of a resonance hybrid that is represented by two contributing resonance structures. We indicate the relationship between the contributing structures by a single double-headed arrow.

Benzene is a planar molecule in which all carbon-carbon bonds are the same; the bond angles of the ring are all 120°. Thus, the σ bonds in benzene are made with sp^2-hybridized carbon atoms. Each carbon atom shares one electron in each of its three σ bonds: two σ bonds are to adjacent carbon atoms; the third σ bond is to a hydrogen atom. The fourth electron is in a $2p$ orbital perpendicular to the plane of the benzene ring (Figure 5.1). A set of six $2p$ orbitals (one from each carbon atom) overlap to share their six electrons in a π system that extends over the entire ring of carbon atoms. The electrons are located both above and below the plane of the ring. The sharing of electrons over many atoms is called **delocalization.** This delocalization of electrons accounts for the unique chemical stability of benzene.

The structure of benzene is usually represented in chemical equations as one of the two possible Kekulé structures. Each corner of the hexagon represents a carbon atom with one attached hydrogen atom, which is often not shown.

The Hückel Rule

Most aromatic compounds contain a benzene ring or a related structure (Section 5.3). What is responsible for the characteristic stability of benzene and its unique reactivity? Three general criteria must be met if a molecule is to be aromatic. First, the molecule must be cyclic and planar. Second, the ring must contain only sp^2-hybridized atoms containing $2p$ orbitals that can form a delocalized system of π

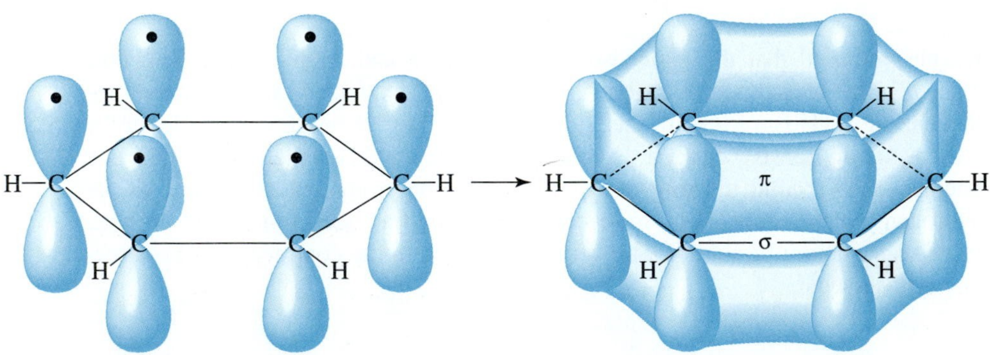

FIGURE 5.1
Bonding in the Benzene Ring

The lines between the carbon atoms represent σ bonds of the benzene ring. In addition, each carbon atom has one $2p$ orbital that contributes one electron to the π system. Overlap of the six $2p$ orbitals that are maintained mutually parallel results in a delocalized system that distributes the electrons over the entire carbon framework.

electrons. (There can be no interruption by sp^3-hybridized atoms.) Third, the number of π electrons must be equal to $4n + 2$, where n is an integer. (Note that n is not the number of carbon atoms in the ring.) The "$4n + 2$ rule" was proposed by E. Hückel and is known as the **Hückel rule.** The theoretical basis for this rule is beyond the scope of this text. However, based on the Hückel rule, cyclic π systems with 6 ($n = 1$), 10 ($n = 2$), and 14 ($n = 3$) electrons are aromatic. Benzene meets the criteria for aromaticity for $n = 1$. Other examples will be presented in the following section for various 6, 10, and 14 π electron systems. We will also see that aromatic compounds can contain atoms other than carbon (Section 5.3).

Some cyclic polyenes with alternating single and double bonds are not aromatic. These compounds do not obey the Hückel rule. Two examples are cyclobutadiene (four π electrons) and cyclooctatetraene (eight π electrons). Both compounds are reactive toward addition reactions and have none of the characteristics of benzene.

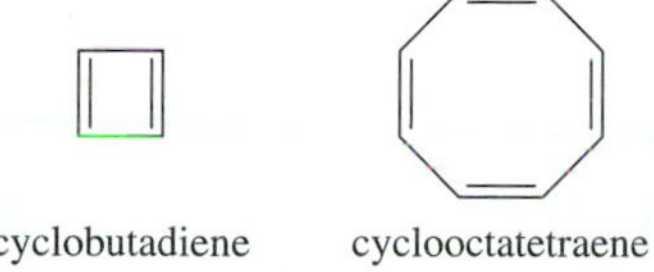

cyclobutadiene cyclooctatetraene

5.3 Polycyclic and Heterocyclic Aromatic Compounds

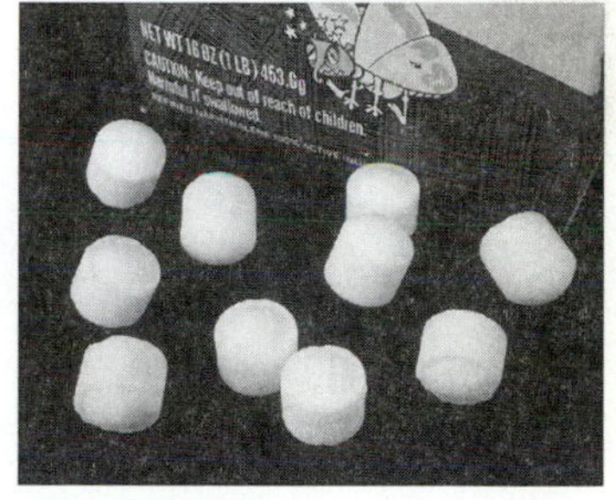

One type of moth balls contains naphthalene.

Some aromatic compounds contain two or more rings that are said to be "fused"; that is, two carbon atoms are common to two rings. Compounds of this type are called **polycyclic aromatic hydrocarbons** (Figure 5.2). These molecules are planar; that is, all atoms in the rings and those directly attached to the rings are in a plane. Several examples of polycyclic aromatic compounds are shown in the following drawings. Note that all the carbon atoms have a bond to a hydrogen atom, except those at the points of fusion.

no bonded hydrogen atoms

naphthalene anthracene phenanthrene

Cyclic compounds that have one or more atoms other than carbon within the ring are said to be **heterocyclic compounds;** those that have $4n + 2$ π electrons are **heterocyclic aromatic compounds.** The hetero atoms most commonly encountered in naturally occurring compounds are nitrogen and oxygen; sulfur-containing compounds also exist. The structures of a few heterocyclic aromatic compounds are as follows:

pyridine pyrrole furan thiophene

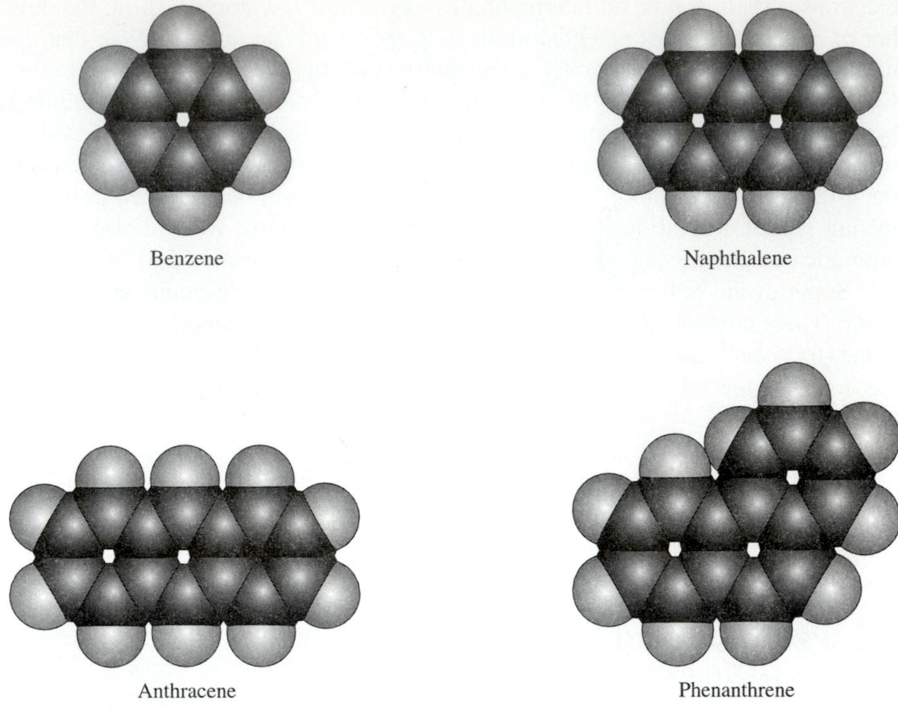

FIGURE 5.2
Models of Aromatic Hydrocarbons

Pyridine closely resembles benzene: it is planar, each of its ring atoms, including the nitrogen atom, is sp^2-hybridized, and each ring atom contributes one electron in a $2p$ orbital to an aromatic system of six π electrons. The sp^2-hybridized nitrogen atom has five valence electrons: one is contributed to the aromatic sextet. The remaining four valence electrons of nitrogen are distributed in the three sp^2 orbitals. Two valence electrons form σ bonds to two carbon atoms, and two valence electrons are present as a nonbonding pair in an sp^2-hybridized orbital. The nonbonding pair projects out from the plane of the ring in a manner similar to the carbon-hydrogen bonding electrons of benzene (Figure 5.3). This nonbonding electron pair allows pyridine to act as a base.

The nitrogen atom in pyrrole is also sp^2-hybridized. However, the electrons are distributed differently than in pyridine. The nitrogen atom contributes one electron to each of the three sp^2 orbitals; two of them form σ bonds to carbon atoms; the third orbital forms a bond with the hydrogen atom (Figure 5.3). The nitrogen atom's remaining two valence electrons are located in a $2p$ orbital. These two electrons added to the four electrons in the $2p$ orbitals of the four carbon atoms provide a six-electron π system—again an aromatic ring like benzene.

Furan and thiophene are heterocyclic aromatic compounds whose heteroatoms are oxygen and sulfur, respectively. The oxygen atom of furan and the sulfur atom of thiophene are both sp^2-hybridized. Oxygen has six valence electrons. Two are in a $2p$ orbital that can interact with the four π electrons of the four carbon atoms in

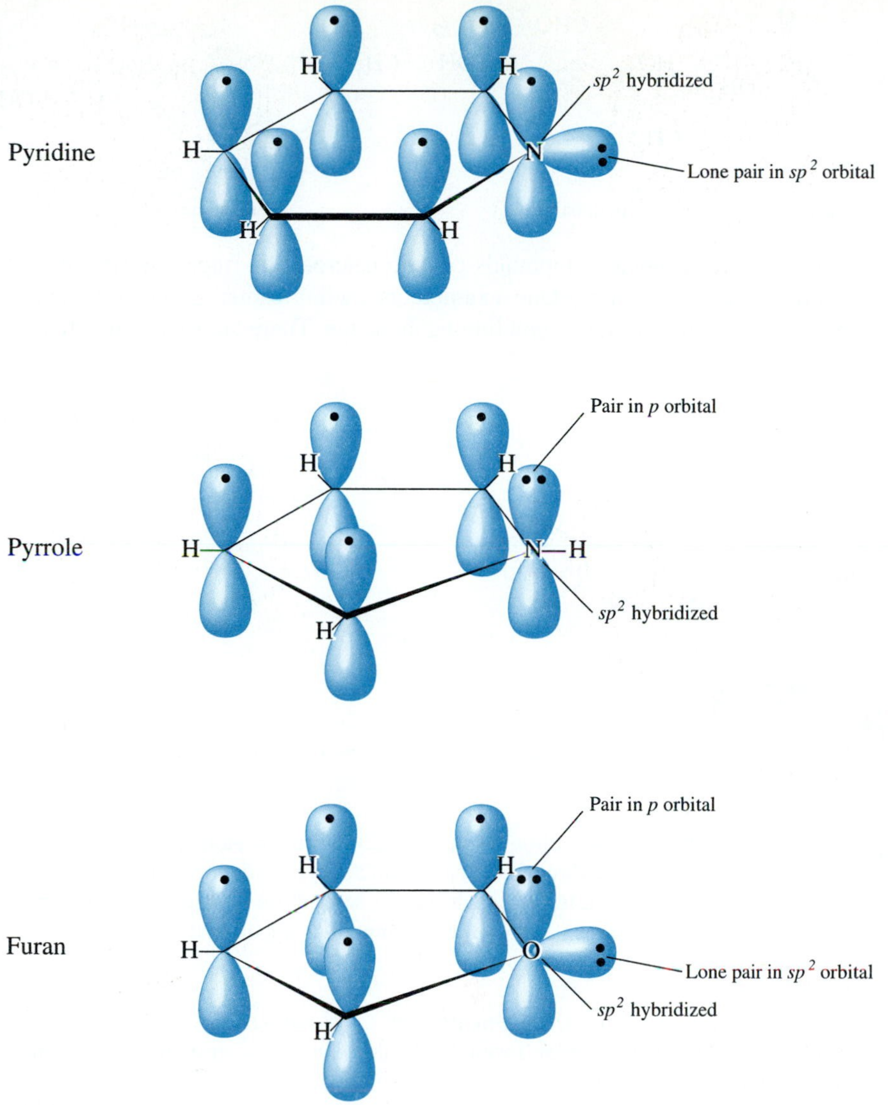

FIGURE 5.3
Bonding in Heterocyclic Aromatic Hydrocarbons

the ring to provide a six-electron π system. The remaining four valence electrons of oxygen are distributed in three sp^2 orbitals. Two of the orbitals have one electron each, and these form σ bonds to two carbon atoms. The remaining sp^2 orbital contains two electrons, and, as in pyridine, this unshared pair of electrons has the same relationship to the ring as the carbon-hydrogen bonding electrons of benzene (Figure 5.3).

Many classes of naturally occurring biologically important compounds—such as vitamins B_1, B_3, and B_6—have one or more aromatic heterocyclic rings of five or six atoms. Vitamin B_1 has two hetero atoms in each ring.

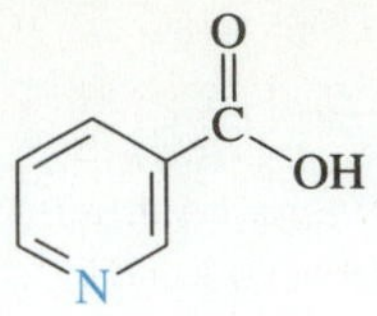

niacin (B_3) 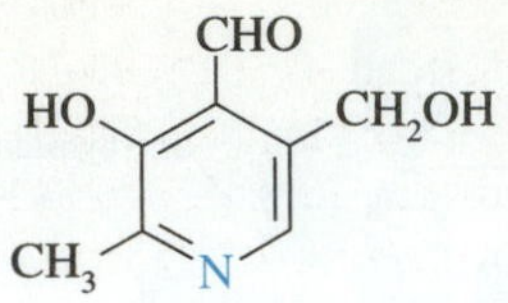 pyridoxine (B_6) 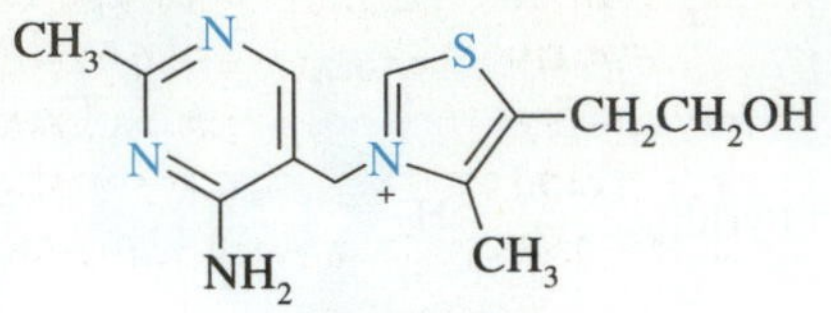 thiamine (B_1)

Many pharmaceutical compounds contain heterocyclic rings that contribute to their effectiveness as drugs. One example is methotrexate, a chemotherapeutic agent, which contains a ring resembling naphthalene. There are four nitrogen atoms in the rings.

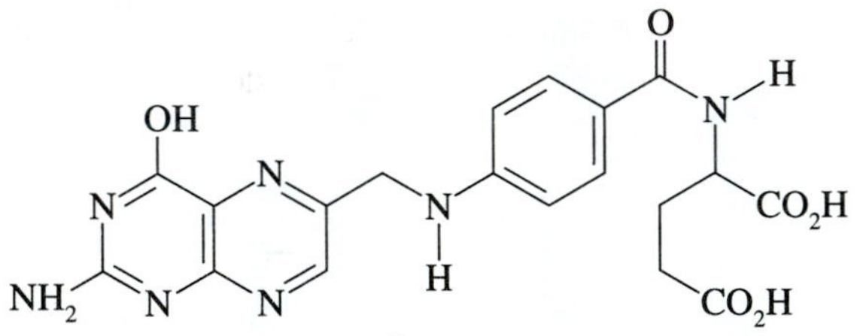

methotrexate

EXAMPLE 5-1

What are the structural and electronic similarities of anthracene and phenanthrene?

Solution

Both compounds have 14 carbon atoms and 10 hydrogen atoms; they are isomers. Both molecules consist of three fused rings that are depicted with a series of alternating single and double bonds, and each has 14 π electrons—a number that fits the Hückel rule for $n = 3$. Thus, both of these compounds are aromatic.

Problem 5.1

1,2-Benzanthracene is a carcinogen present as a combustion product in tobacco smoke. What is its molecular formula? Is the compound aromatic based on the Hückel rule?

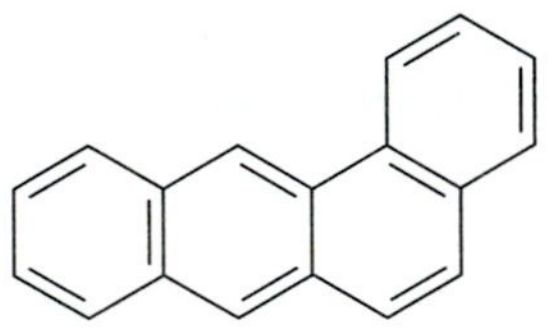

EXAMPLE 5-2

Histamine is released in the body in persons with allergic hypersensitivities, such as hay fever. Describe the distribution of the electrons in orbitals on the two nitrogen atoms in the heterocyclic ring.

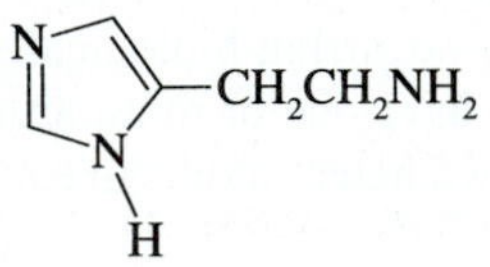

Solution

The ring nitrogen atom bonded to a hydrogen atom resembles the nitrogen atom of pyrrole. Three σ bonds are shown. Each has one valence electron contributed from the nitrogen atom. Thus, the remaining two valence electrons of nitrogen are located in a $2p$ orbital. These two electrons, along with the four electrons of the two π bonds shown in the structure, account for six electrons of an aromatic system.

The other nitrogen atom in the ring has one single and one double bond as shown. This nitrogen atom resembles the nitrogen atom of pyridine. Two of its electrons participate in the two σ bonds to carbon atoms; one electron is contributed to the π bond with a carbon atom. The remaining two electrons are in an sp^2 hybrid orbital projecting out from the plane of the ring.

Problem 5.2

Adenine is present in nucleic acids (DNA and RNA). Which nitrogen atoms in the rings have nonbonding electrons in sp^2 hybrid orbitals? How many electrons does each nitrogen atom contribute to the aromatic π system?

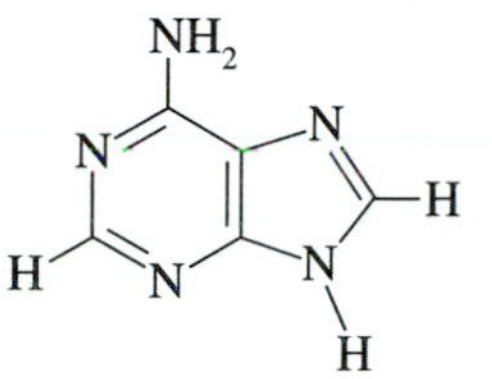

NH_2 N N H H N N H

5.4 Nomenclature of Benzene Compounds

Many compounds with a benzene ring have well-established, nonsystematic names. Such names are often based on sources of the compounds and have been used for so long that they have become accepted by IUPAC.

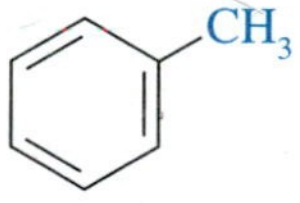

CH_3 toluene

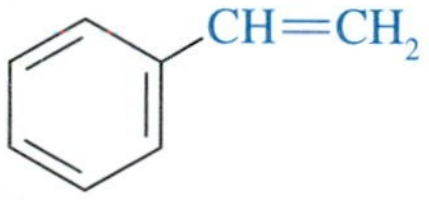

$CH{=}CH_2$ styrene

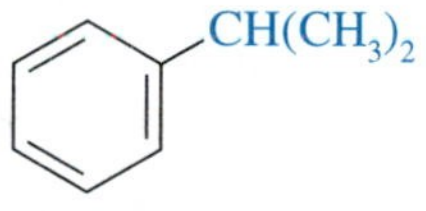

$CH(CH_3)_2$ cumene

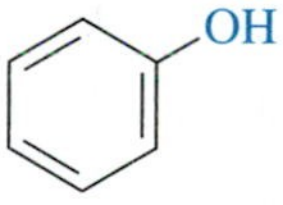

OH phenol

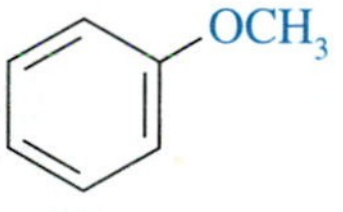

OCH_3 anisole

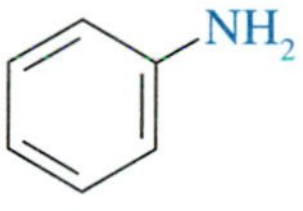

NH_2 aniline

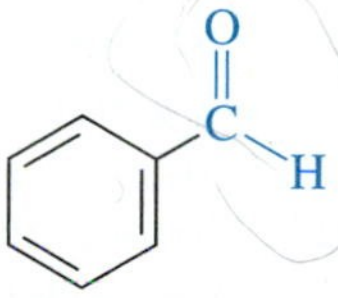

O C H benzaldehyde

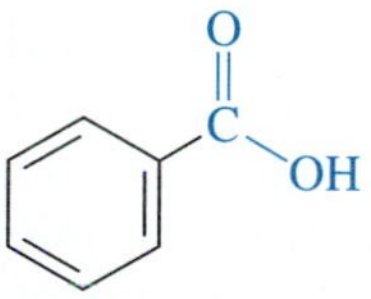

O C OH benzoic acid

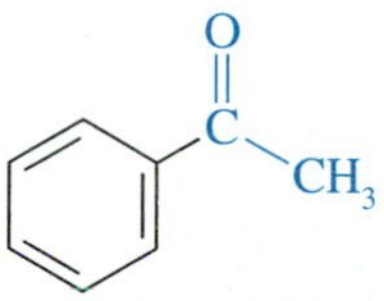

O C CH_3 acetophenone

The IUPAC system of naming substituted aromatic hydrocarbons uses the names of the substituents as prefixes to benzene. Examples include

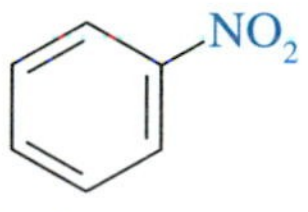

NO_2 nitrobenzene

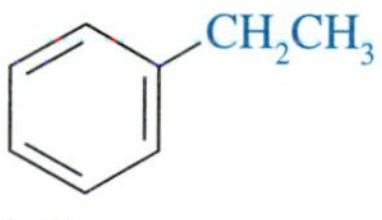

CH_2CH_3 ethylbenzene

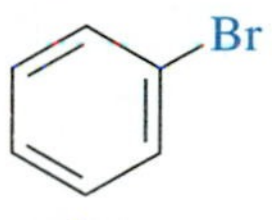

Br bromobenzene

The two substituents of disubstituted benzene compounds can be arranged in a benzene ring in three different ways to give three different isomers. These isomers are designated *ortho, meta,* and *para.* As prefixes these terms are abbreviated as *o*-, *m*-, and *p*-, respectively.

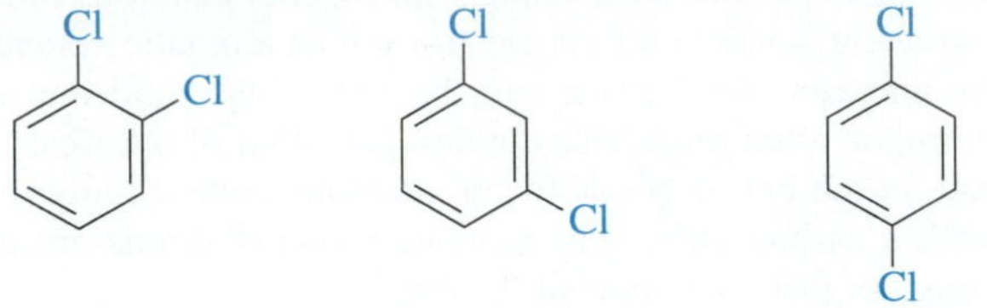

o-dichlorobenzene
1,2-dichlorobenzene

m-dichlorobenzene
1,3-dichlorobenzene

p-dichlorobenzene
1,4-dichlorobenzene

The ortho isomer has two groups on adjacent carbon atoms—that is, in a 1,2 relationship. In the meta and para isomers, the two groups are in a 1,3 and a 1,4 relationship, respectively. The IUPAC names of disubstituted aromatic compounds are obtained by numbering the benzene ring to give the lowest possible numbers to the carbon atoms bearing the substituents. When three or more substituents are present, the ring carbon atoms must be numbered.

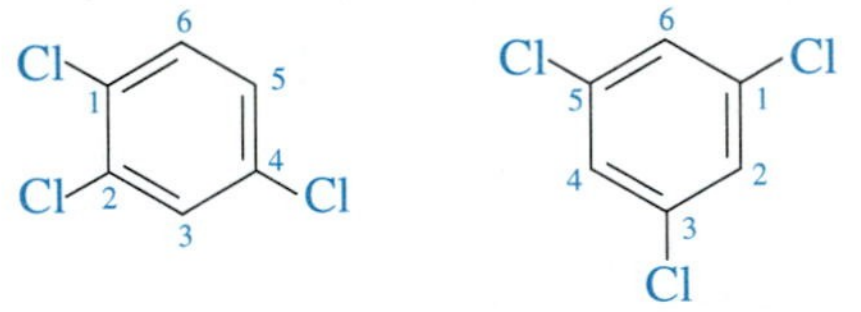

1,2,4-trichlorobenzene

1,3,5-trichlorobenzene

Many derivatives of benzene are named with the common name of a monosubstituted aromatic compound as the parent. The position of the substituent of the "parent" is automatically designated 1, but the number is not used in the name. The remaining substituents are prefixed in alphabetical order to the parent name along with numbers indicating their locations.

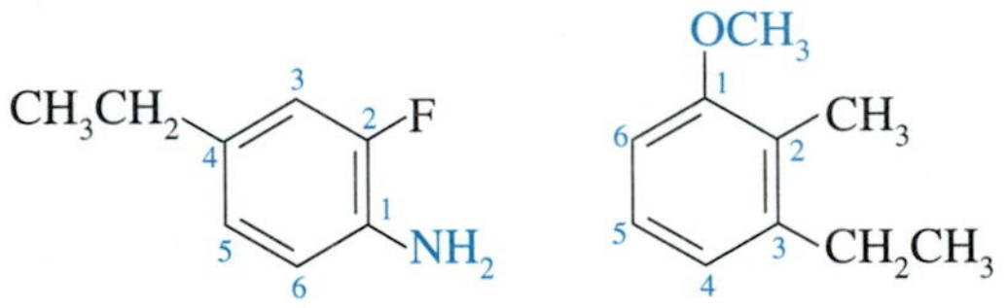

4-ethyl-2-fluoroaniline

3-ethyl-2-methylanisole

Aromatic hydrocarbons belong to a general class called **arenes.** An aromatic ring residue attached to a larger parent structure is an **aryl group;** it is symbolized as Ar (not to be confused with argon), just as the symbol R is used for an alkyl group. Two groups whose names unfortunately do not make much sense are the phenyl (fen′-nil) and benzyl groups. We might reasonably expect that the aryl group derived from benzene (C_6H_5—) would be named benzyl. Alas, it is a phenyl group. A benzyl group, which is derived from toluene, has the formula $C_6H_5CH_2$—.

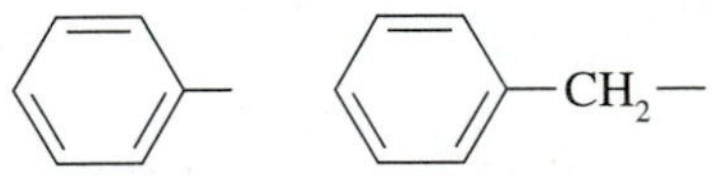

phenyl group

benzyl group

EXAMPLE 5-3

Indicate how a name could be derived for the following trisubstituted compound, known as BHA, which is used as an antioxidant in some food products.

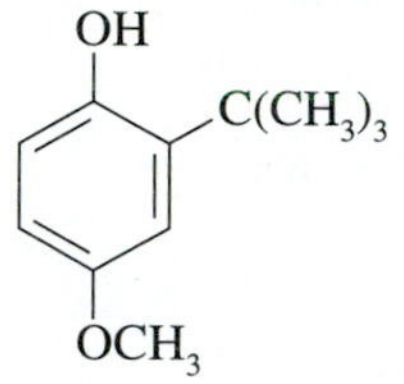

OH

$C(CH_3)_3$

OCH_3

Solution

Either the OH group or the OCH_3 group could provide the parent name, which would be phenol or anisole, respectively. Let's assume that the compound is a substituted anisole. When we assign the number 1 to the carbon atom bearing the OCH_3 group at the "six o'clock" position, we must number the ring in a counterclockwise direction. A *tert*-butyl group is located at the 3 position and a hydroxyl group at the 4 position, so the compound is 3-*tert*-butyl-4-hydroxyanisole.

Problem 5.3

Name the following compound, which is used to make a local anesthetic.

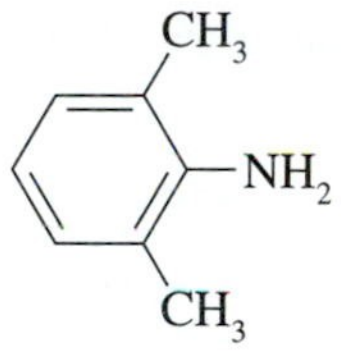

CH_3

NH_2

CH_3

EXAMPLE 5-4

What is the name of the following compound?

$CH_3CH_2CH_2$ H

C=C

CH_2CH_3

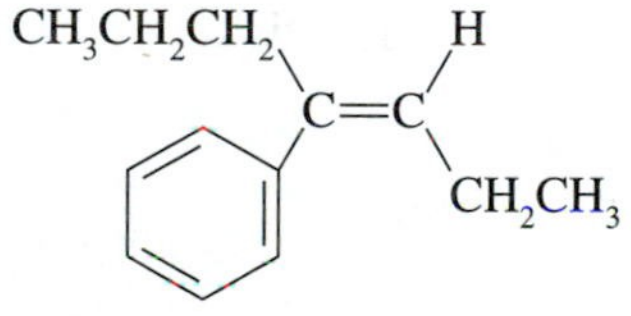

Solution

The compound is an alkene with an aromatic ring as a substituent. First, we determine that the chain has seven carbon atoms. Next, we number the chain from right to left so that the double bond is assigned to the C-3 atom. The phenyl group is then located on the C-4 atom. The phenyl group has a higher priority than the propyl group. At the C-3 atom, the ethyl group has higher priority than the hydrogen atom. The compound is the Z isomer because the higher priority phenyl group is on the same side of the double bond as the ethyl group. The complete name is (*Z*)-4-phenyl-3-heptene.

Problem 5.4

Name the following compound as a substituted pyridine.

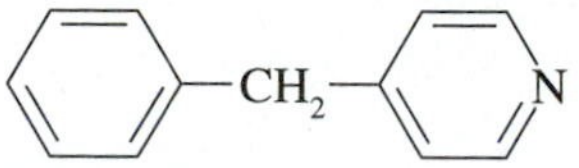

CH_2 N

5.5 Electrophilic Aromatic Substitution

Aromatic rings react with electrophiles, prepared using certain catalysts, to give a substitution product. In these reactions an electrophile (E^+) substitutes for H^+. The general process is as follows:

$$C_6H_5\text{—H} + \overset{\delta+}{E}\text{—}\overset{\delta-}{Nu} \longrightarrow C_6H_5\text{—E} + H\text{—}Nu$$

Mechanism of Electrophilic Aromatic Substitution

The first step of electrophilic aromatic substitution is similar to that discussed for the addition of electrophiles to alkenes. The electrophile accepts an electron pair from the aromatic ring.

$$C_6H_5\text{—H} + \overset{\delta+}{E}\text{—}\overset{\delta-}{Nu} \longrightarrow [C_6H_6E]^+ + :Nu^-$$

The carbocation is resonance stabilized. However, the carbocation is not aromatic—it has only four π electrons and it has an sp^3-hybridized carbon atom. Therefore, it is less stable than the original aromatic ring, which had a full array of six π electrons.

In the second step of the reaction, the proton bonded to the same carbon atom as the newly arrived electrophile is lost, and the aromatic π system is restored. The proton that leaves in this step is extracted by a nucleophile acting as a base.

$$[C_6H_6E]^+ + :Nu^- \longrightarrow C_6H_5\text{—E} + H\text{—}Nu$$

Typical Electrophilic Substitution Reactions

In the preceding discussion we used a generic electrophile, E^+. In this section we will consider some specific examples of electrophiles that react with aromatic rings. Our first examples are bromination and chlorination. Bromination requires both Br_2 and a Lewis acid catalyst, $FeBr_3$. Chlorination proceeds in the same manner as bromination but using the Lewis acid catalyst, $FeCl_3$. In bromination the catalyst forms an intermediate that has a polarized bromine-bromine bond.

$$:\ddot{B}r\text{—}\ddot{B}r: + FeBr_3 \rightleftharpoons :\ddot{B}r\text{—}\overset{\delta+}{\ddot{B}r}\text{—}\overset{\delta-}{FeBr_3}$$

Although this intermediate is the electrophilic species, it is convenient to use just the Br^+ formed by heterolytic cleavage to write the mechanism of the substitution reaction.

$$:\ddot{Br}-\overset{\delta^+}{\ddot{Br}}-\overset{\delta^-}{FeBr_3} \rightleftharpoons :\ddot{Br}^+ + :\ddot{Br}-\overset{-}{FeBr_3}$$

$$C_6H_6 + {}^+\ddot{Br}: \longrightarrow C_6H_6Br^+$$

$$C_6H_6Br^+ + :\ddot{Br}-\overset{-}{FeBr_3} \longrightarrow C_6H_5Br + H-Br + FeBr_3$$

In **nitration,** a nitro group, $—NO_2$, is introduced onto an aromatic ring using nitric acid, HNO_3, with sulfuric acid as the catalyst. The electrophile is the nitronium ion, $NO_2{}^+$, which is produced by the reaction of nitric acid with sulfuric acid.

$$HO-NO_2 + H-OSO_3H \rightleftharpoons NO_2{}^+ + HSO_4{}^- + H_2O$$

The net reaction for nitration of benzene is the result of electrophilic attack by the nitronium ion followed by loss of a proton.

$$C_6H_5-H + HNO_3 \xrightarrow{H_2SO_4} C_6H_5-NO_2 + H_2O$$

nitrobenzene

A sulfonic acid group, $—SO_3H$, can also be introduced onto an aromatic ring by electrophilic aromatic substitution. The process, called **sulfonation,** requires a mixture of SO_3 and sulfuric acid, called fuming sulfuric acid, to form ${}^+SO_3H$.

$$SO_3 + H_2SO_4 \rightleftharpoons {}^+SO_3H + HSO_4{}^-$$

The net reaction for sulfonation of benzene is the result of electrophilic attack by ${}^+SO_3H$ followed by loss of a proton.

$$C_6H_5-H + SO_3 \xrightarrow{H_2SO_4} C_6H_5-SO_3H$$

benzenesulfonic acid

An alkyl group can be substituted for a hydrogen atom of an aromatic ring by a reaction called the **Friedel-Crafts alkylation.** This reaction requires an alkyl halide with the corresponding aluminum trihalide as the catalyst. The catalyst produces an alkyl carbocation, which is the electrophile.

$$(CH_3)_2CHCl + AlCl_3 \longrightarrow (CH_3)_2CH^+ + AlCl_4{}^-$$

The reaction is commonly carried out only with alkyl bromides or alkyl chlorides. The net reaction for alkylation of benzene by 2-chloropropane is

$$C_6H_5-H + (CH_3)_2CH-Cl \xrightarrow{AlCl_3} C_6H_5-CH(CH_3)_2 + HCl$$

isopropylbenzene

The Friedel-Crafts alkylation does not occur on aromatic rings that have one of the groups $—NO_2$, $—SO_3H$, $—C{\equiv}N$, or any carbonyl-containing group (aldehydes, ketones, carboxylic acids, and esters) bonded directly to the aromatic ring. All of these substituents make the benzene ring less reactive, a subject to be discussed in the next section.

The Friedel-Crafts alkylation is also limited by the possible structural rearrangement of the alkyl carbocation generated from the alkyl halide. For example, the reaction with 1-chloropropane in the presence of $AlCl_3$ yields a small amount of propylbenzene but a larger amount of the isomer, isopropylbenzene.

$$C_6H_5{-}H + CH_3CH_2CH_2{-}Cl \xrightarrow{AlCl_3} C_6H_5{-}CH(CH_3)_2 + C_6H_5{-}CH_2CH_2CH_3$$

isopropylbenzene (major product) propylbenzene (minor product)

Isomerization of carbocations occurs in the Friedel-Crafts reaction. Some isomerizations occur by a **hydride (H:$^-$) shift,** which converts a less stable carbocation into a more stable one. We recall from Section 4.9 that the order of carbocation stability is tertiary > secondary > primary. Thus, if the Friedel-Crafts reaction is carried out with 1-chloropropane, a relatively unstable propyl carbocation forms initially. But this intermediate rearranges by shifting a hydrogen atom along with the bonding electron pair, H:$^-$, from the C-2 to the C-1 atom.

$$CH_3{-}CH(H){-}CH_2^+ \longrightarrow CH_3{-}\overset{+}{C}H{-}CH_2(H)$$

An acyl group can replace hydrogen in an aromatic ring by a reaction called the **Friedel-Crafts acylation.** The reaction requires an acyl halide and the corresponding aluminum trihalide. It is commonly performed only with acyl chlorides. The reaction yields a ketone.

$$C_6H_6 + CH_3CH_2\overset{O}{\overset{\|}{C}}{-}Cl \xrightarrow{AlCl_3} C_6H_5{-}\overset{O}{\overset{\|}{C}}{-}CH_2CH_3$$

propanoyl chloride propiophenone

The electrophile is the acylium ion that results from an acid-base reaction between the acyl chloride and aluminum trichloride.

$$CH_3CH_2\overset{:\ddot{O}}{\overset{\|}{C}}{-}\ddot{\underset{..}{Cl}}: \; AlCl_3 \longrightarrow CH_3CH_2\overset{:\ddot{O}}{\overset{\|}{C}}{-}\overset{+}{\underset{..}{\ddot{Cl}}}{-}\overset{-}{Al}Cl_3$$

(an acyl chloride)

$$CH_3CH_2\overset{:\ddot{O}}{\overset{\|}{C}}{-}\overset{+}{\underset{..}{\ddot{Cl}}}{-}\overset{-}{Al}Cl_3 \longrightarrow CH_3CH_2C{\equiv}\overset{+}{O}: + AlCl_4^-$$

(an acylium ion)

Like the Friedel-Crafts alkylation, the Friedel-Crafts acylation reaction does not occur on aromatic rings that have one of the groups $—NO_2$, $—SO_3H$,

—C≡N, or any carbonyl-containing group (aldehydes, ketones, carboxylic acids, and esters) bonded directly to the aromatic ring. However, acylium ions produced in the Friedel-Crafts reaction do not rearrange. The acyl group in the product can be reduced by using zinc/mercury amalgam and HCl (a reaction called a Clemmensen reduction) to produce an alkylbenzene. By this means, the rearrangement of primary alkyl groups that occurs in the Friedel-Crafts alkylation reaction is circumvented. For example, propylbenzene can be synthesized by reduction of propiophenone, the product of the acylation of benzene with propanoyl chloride.

propiophenone —Zn(Hg) / HCl→ propylbenzene

5.6 Structural Effects in Electrophilic Aromatic Substitution

To this point, we have discussed only electrophilic substitution reactions of benzene itself. Now let's examine the effect that a substituent already bonded to the aromatic ring has on the introduction of a second substituent. The substituent could affect the rate of the reaction compared to benzene. In addition, any of three possible products—the ortho, meta, and para isomers—could be produced.

Effects of Ring Substituents on Reaction Rate

The relative rates of nitration of benzene and several substituted benzenes are given in the following list. The difference in rate between substituting a nitro group onto phenol and substituting a nitro group onto nitrobenzene is a phenomenal factor of 10^{10}. (For comparison, this rate difference is like the difference between the speed of light and the speed of walking!)

	OH	CH_3	H	Cl	NO_2
	phenol	toluene	benzene	chlorobenzene	nitrobenzene
relative rate of nitration	1×10^3	25	1	3×10^{-2}	1×10^{-7}

The hydroxyl group and the methyl group, which make the aromatic ring more reactive compared to benzene, are called **activating groups.** The chloro and nitro groups make the aromatic ring less reactive and are called **deactivating groups.** Table 5.1 lists some common substituents and divides them into activating and deactivating groups toward electrophilic aromatic substitution.

Orientation Effect of Substituents

Now let's consider the distribution of products formed in the nitration of toluene. The ortho and para isomers predominate, and very little of the meta isomer forms. The methyl group is said to be an **ortho,para director.** That is, the methyl group directs or orients the incoming substituent into positions ortho

Table 5.1 Effect of Ring Substituents on Aromatic Substitution

Strongly activating
—NH_2, —NHR, —NR_2
—OH, —OCH_3

Weakly activating
—CH_3, —CH_2CH_3, —R

Weakly deactivating
—F, —Cl, —Br

Strongly deactivating
—C(=O)—R, —C(=O)—OH
—C≡N, —NO_2,
—CF_3, —CCl_3

and para to itself. It turns out that all activating groups are ortho,para directors. Halogens, which are weakly deactivating, are also ortho,para directors.

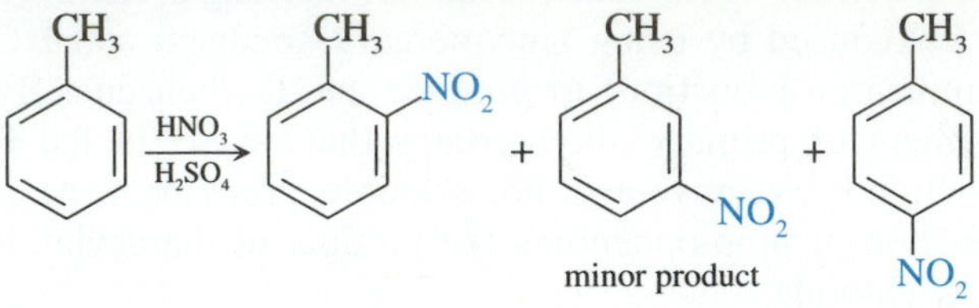

CH_3 (toluene) $\xrightarrow[H_2SO_4]{HNO_3}$ ortho-NO_2 + meta-NO_2 (minor product) + para-NO_2

A second class of ring substituents directs incoming substituents into the meta position. These groups, known as **meta directors,** include nitro, trifluoromethyl, cyano, sulfonic acid, and all carbonyl-containing groups. For a nitration reaction, the trifluoromethyl group orients the incoming nitro group into a position meta position to itself. Very small amounts of the ortho and para isomers are formed. All deactivating groups (except halogens) are meta directing groups.

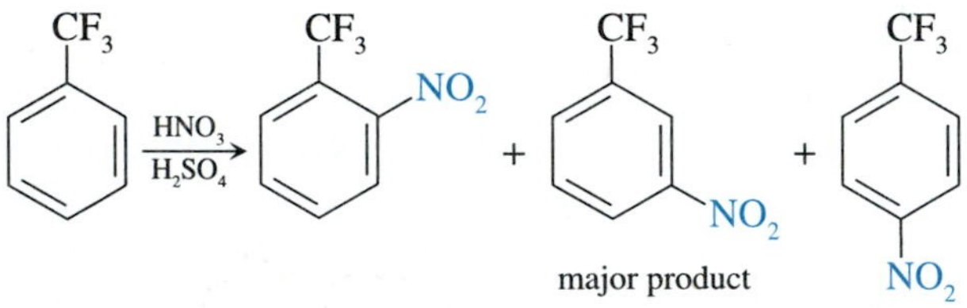

CF_3 (benzotrifluoride) $\xrightarrow[H_2SO_4]{HNO_3}$ ortho-NO_2 + meta-NO_2 (major product) + para-NO_2

EXAMPLE 5-5

Predict whether the following compounds react faster or slower than benzene in a reaction with bromine and $FeBr_3$.

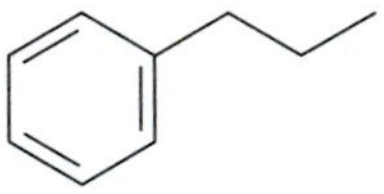

propylbenzene 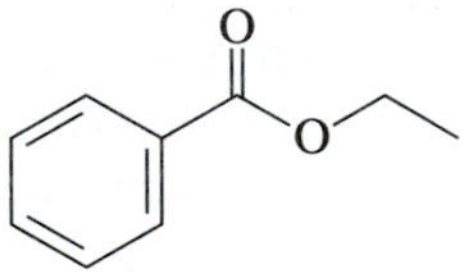 ethyl benzoate

Solution

Propylbenzene contains an alkyl substituent. As in the case of a methyl group, the propyl group is slightly activating. Thus, propylbenzene is more reactive than benzene. Ethyl benzoate has a carbonyl carbon atom bonded directly to the aromatic ring. As a result, its rate of bromination will be significantly slower than that of benzene.

Problem 5.5

Predict whether the following compounds react faster or slower than benzene in a reaction with bromine and $FeBr_3$.

(a) 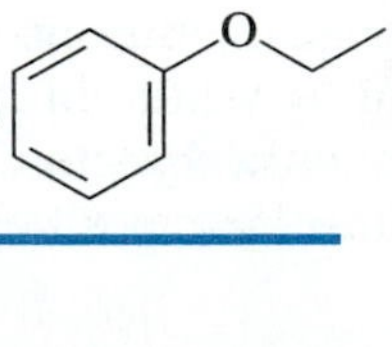 (b) 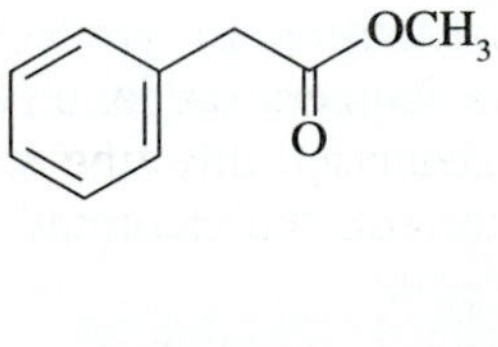 OCH_3 O (c) 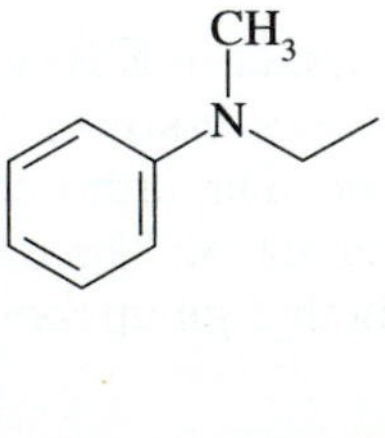 CH_3 N

EXAMPLE 5-6

Predict the structure of the product(s) formed in the bromination of each of the following compounds.

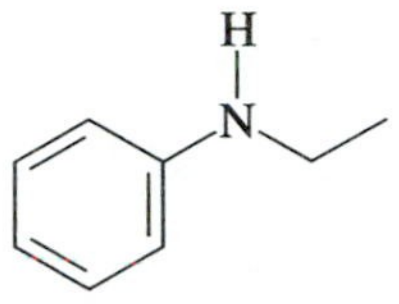

N-ethylaniline 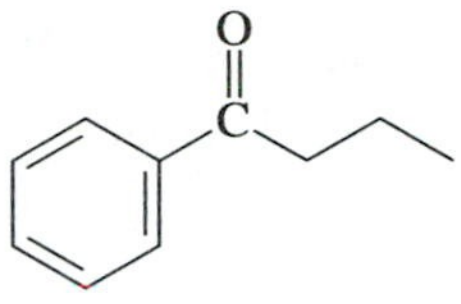 butyrophenone

Solution

N-Ethylaniline has a nitrogen atom with lone pair electrons. The compound resembles aniline, and its substituent should direct the bromine to the ortho or para positions. Two isomeric compounds should result. Butyrophenone has a carbonyl group bonded to the benzene ring. Therefore, the substituent should direct the bromine to the meta position.

Problem 5.6

Predict the structure of the product(s) formed in the bromination of each of the following compounds.

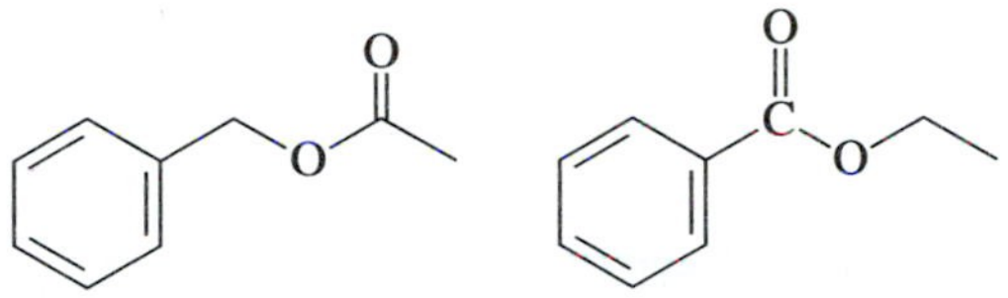

5.7 Interpretation of Rate Effects

The influence of a substituent on both the rate and distribution of products in electrophilic aromatic substitution reactions can be understood with one model based on the ability of the substituents to either donate or withdraw electrons from the aromatic ring. Let us consider the effect of two types of groups on the electron density of the benzene ring.

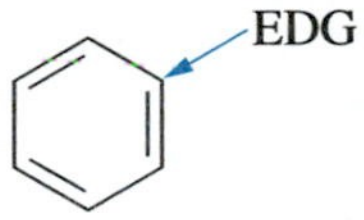

EDG EWG

If the group is electron donating (EDG), the ring gains electron density and is more reactive.

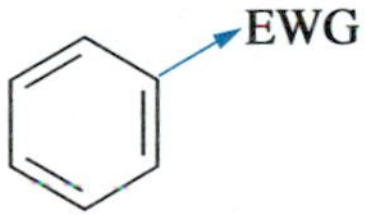

If the group is electron withdrawing (EWG), the ring loses electron density and is less reactive.

All activating groups listed in Table 5.1 are electron-donating groups; deactivating groups are electron-withdrawing groups. Substituents can donate or withdraw electron density by inductive or resonance effects or a combination of these effects.

Inductive Effects of Substituents

Inductive effects are perhaps more easily visualized than resonance effects because they are related to the concept of electronegativity. As discussed in Chapter 4, alkyl groups are electron donating relative to hydrogen and tend to stabilize double bonds; they also stabilize carbocations. Hence, alkyl groups transfer electron density through the σ bonds to sp^2-hybridized carbon atoms. Thus, they also donate electron density to the benzene ring by an inductive effect. The trifluoromethyl group, whose fluorine atoms pull electron density away from the carbon atom to which they are bonded, is inductively electron withdrawing.

CH_3 CF_3

methyl group activates the aromatic ring

trifluoromethyl group deactivates the aromatic ring

The halogens directly bonded to an aromatic ring withdraw electron density and deactivate the compound toward electrophilic aromatic substitution. Any functional group that has a formal positive charge on the atom bonded to the aromatic ring, such as the nitro group, withdraws electron density from the ring. Groups such as the carbonyl group or the nitrile group have a partial positive charge on the atom bonded to the aromatic ring, so they also withdraw electron density from the ring. All such groups deactivate the aromatic ring.

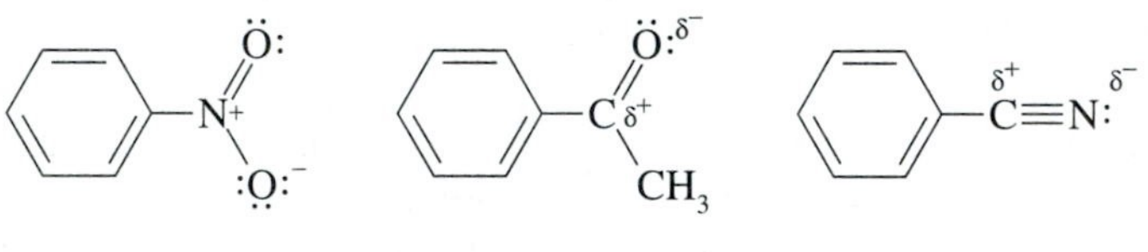

Resonance Effects of Substituents

Next, let's consider how electron density is shifted into or out of a benzene ring by resonance effects. We do this by "moving" electrons and drawing alternate resonance forms. Consider the nitro group or any carbonyl-containing groups. They have sp^2-hybridized atoms that are conjugated with the ring. The oxygen atoms of the nitro group are more electronegative than nitrogen. Hence, we can write resonance forms by "shifting" an electron pair in a nitrogen-oxygen double bond onto the oxygen atom followed by "shifting" an electron pair out of the ring to make a carbon-nitrogen double bond, leaving a positive charge on the aromatic ring. Because a positive charge develops in the ring, the ring is less reactive toward electrophiles. A similar effect explains why the acyl group also makes the ring less reactive toward electrophiles.

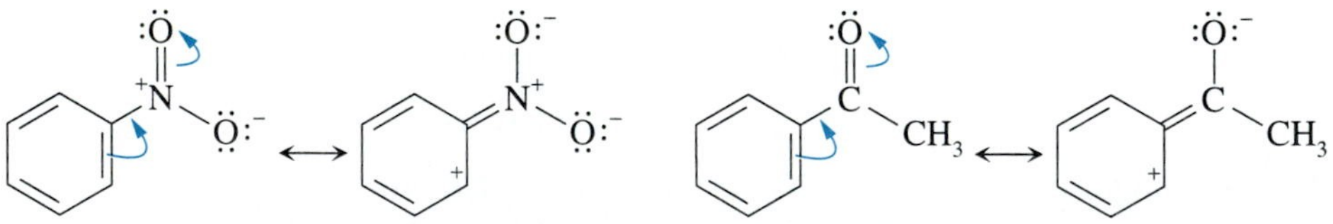

Now let's consider ring substituents that have atoms with lone pair electrons that can be donated to the ring by resonance. As a consequence, the ring develops a negative charge and is more reactive toward electrophiles. Groups that have an unshared electron pair on the atom attached to the ring include the hydroxyl (—OH); alkoxy, such as methoxy (CH_3O—); and amino (—NH_2) or any substituted amino group (—NHR, —NR_2). All of these groups donate electrons to the aromatic ring by resonance.

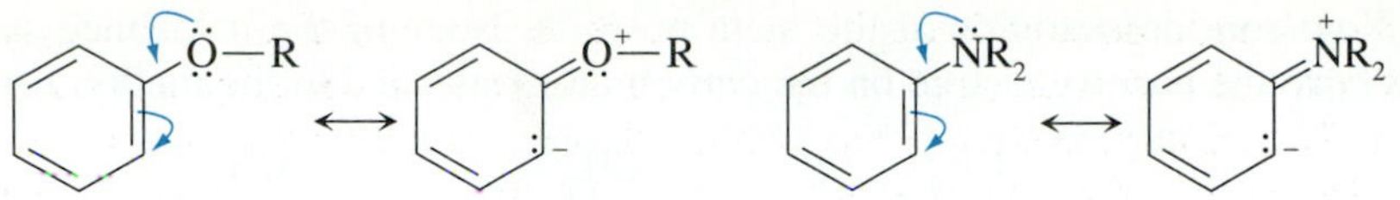

The groups are also electronegative. Therefore, they also withdraw electron density from the ring by an inductive effect. Thus, these substituents inductively take electron density from the ring while giving electron density back by resonance. A group that donates electrons by resonance, such as an amino or hydroxyl group, interacts with the ring through its 2*p* orbital, which very effectively overlaps with the 2*p* orbital of a ring carbon atom. Thus, donation of electrons by resonance is very effective and is more important than inductive electron withdrawal. As a result, these groups activate the aromatic ring. This situation, however, does not hold for chlorine or bromine. These electronegative atoms pull electron density out of the aromatic ring by an inductive effect. However, neither the 3*p* orbital of chlorine nor the 4*p* orbital of bromine overlaps effectively with the 2*p* orbital of carbon, so electron donation by resonance is less effective. As a result, the halogens have the net effect of deactivating the aromatic ring.

5.8 Interpretation of Directing Effects

We noted earlier that, with the exception of the halogens, ortho,para directors activate the ring toward electrophilic substitution because they supply electrons to the ring. But why are the ortho and para positions especially susceptible to attack? To answer this question, let's examine the resonance forms of the intermediate carbocations resulting from reaction at the ortho and para positions and compare them with those resulting when an electrophile reacts at the meta position.

First, consider the resonance forms for the intermediate formed by nitration of toluene at the ortho and para positions.

most stable resonance form

most stable resonance form

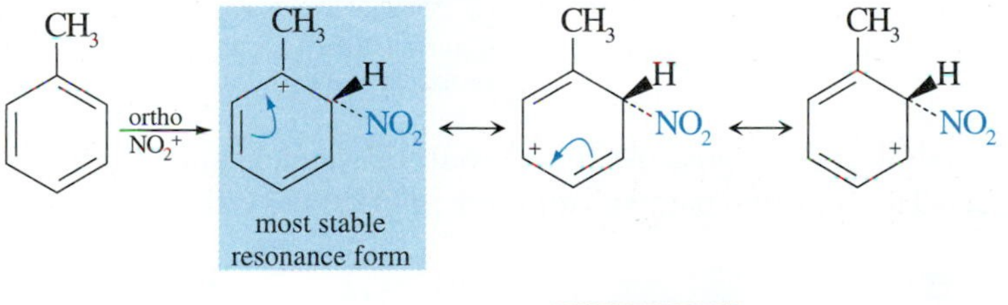

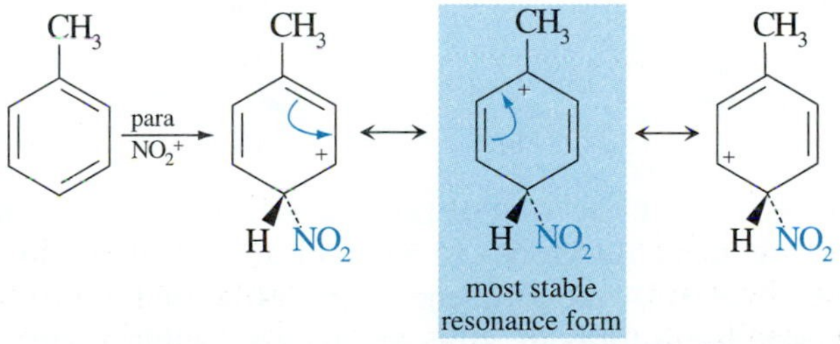

Attack at either the ortho or the para position results in one resonance structure with a positive charge on the ring carbon atom bonded to the methyl group. This tertiary carbocation makes a major contribution to the stability of the resonance hybrid.

Now consider nitration at the meta position. None of the resonance structures have the positive charge on the carbon atom attached to the methyl group.

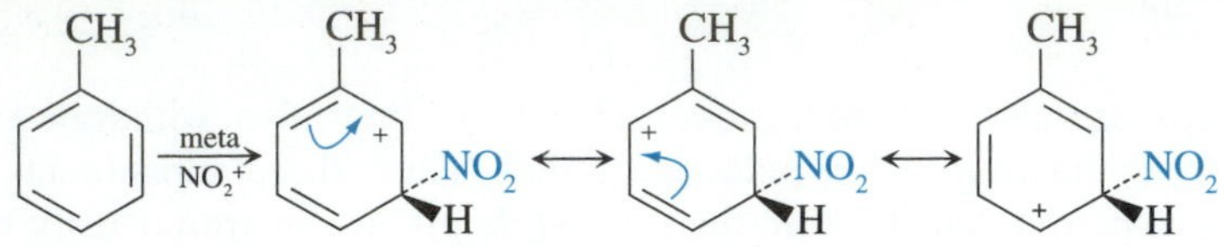

All resonance forms are secondary carbocations. As a result, this intermediate is less stable than the intermediates formed from attack at either the ortho or para position. As a consequence, reaction at the ortho or para position is favored over a reaction at the meta position.

Next, let's consider the ortho,para-directing effect of a group that can donate an unshared pair of electrons by resonance. An attack either ortho or para to the hydroxyl group by an electrophile, such as the bromonium ion, leads to an intermediate that is resonance stabilized by the lone pair electrons donated by the oxygen atom. The most stable resonance form has a positive charge on the oxygen atom, but all atoms have an octet of electrons.

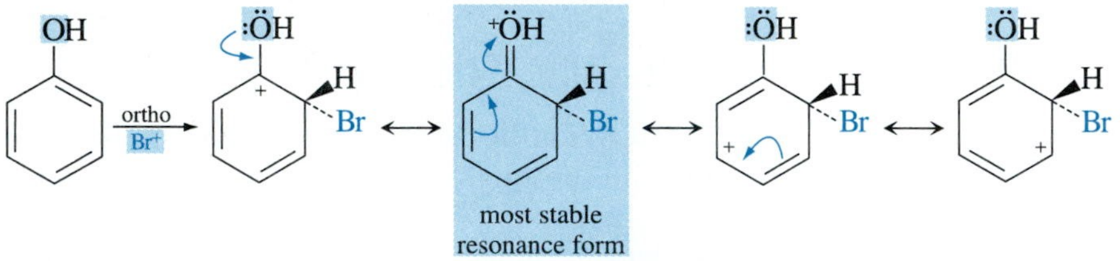

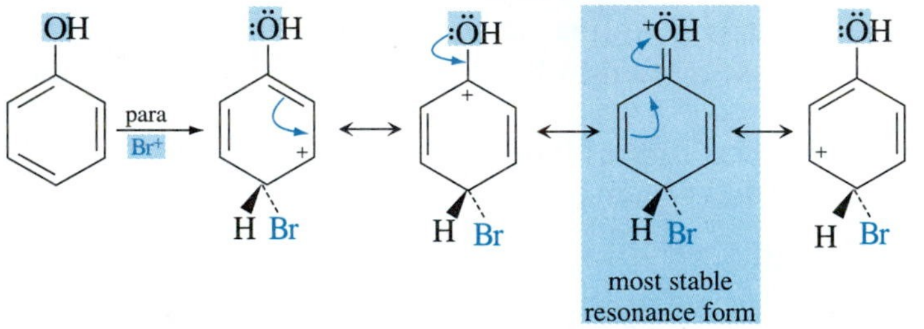

most stable resonance form

most stable resonance form

No such stabilization is possible for a group that attacks meta to the hydroxyl substituent. Hence, ortho,para substitution is preferred over meta substitution.

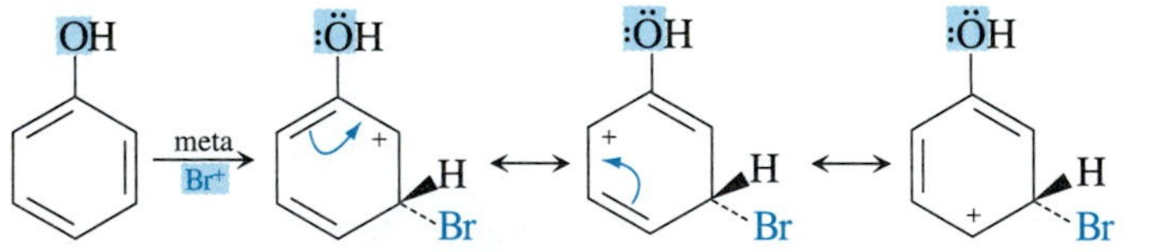

Substituents that strongly deactivate the ring toward electrophilic aromatic substitution are meta directors. To explain this relationship, let's consider the possible nitration of nitrobenzene at the ortho and para positions. In one of the resonance forms for both ortho- and para-substituted intermediates, a positive charge is located at a carbon atom containing the original nitro group. The nitrogen atom of the nitro group also has a formal positive charge, and such resonance forms containing positive charges on adjacent atoms are not favorable. As a result, these intermediates, formed by attack at either the ortho or para position, are less stable than those formed in the attack of benzene itself.

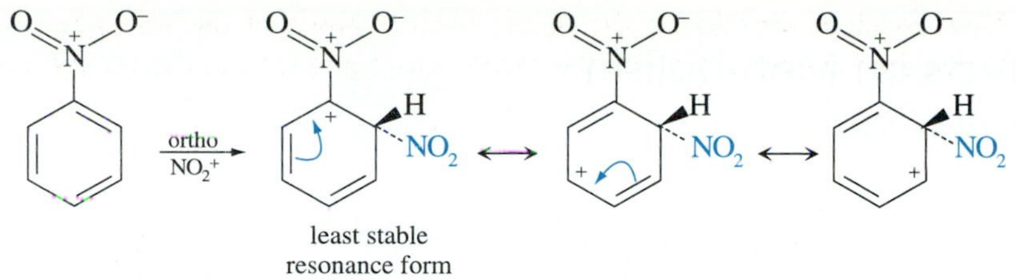

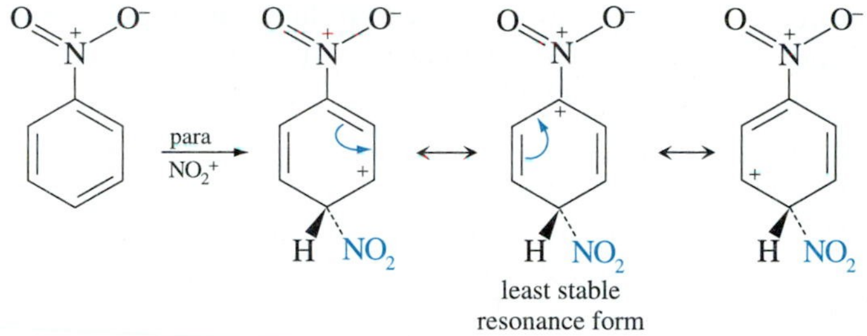

ortho
NO_2^+
NO_2
NO_2
NO_2

least stable resonance form

para
NO_2^+
H NO_2
H NO_2
H NO_2

least stable resonance form

Now consider an attack at the meta position. None of the resonance forms of the intermediate has a positive charge on the carbon atom bonded to a nitro group—whose nitrogen atom, we noted previously, has the formal charge of +1. Thus, the intermediate resulting from attack at the meta position is more stable overall than the intermediates formed from ortho or para substitution. As a consequence, meta substitution is favored over ortho or para substitution.

meta
NO_2^+
NO_2
NO_2
NO_2

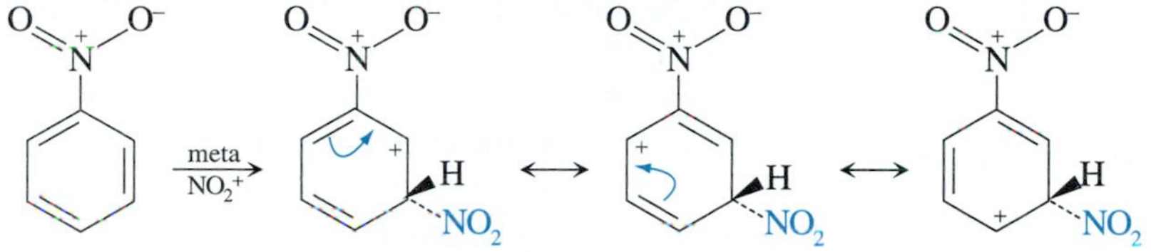

Finally, we consider the halogen substituents that are weakly deactivating but yet are ortho,para directors. Because the halogens are more electronegative than a benzene ring, they withdraw electron density from the ring by an inductive effect. But because halogens have lone pair electrons, they can donate electrons to the carbocation intermediate. However, this resonance effect, which supplies electrons to the aromatic ring, only comes into play if the entering electrophile attacks ortho or para to the halogen atom. Thus, although weakly deactivating, the halogens are ortho,para directors.

Cl: ⟷ Cl: ⟷ Cl:

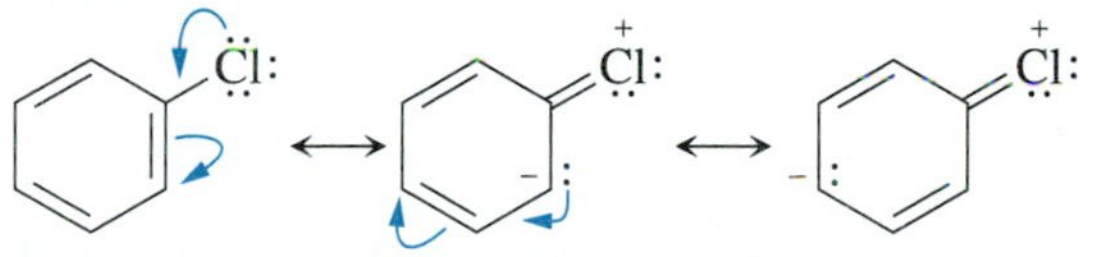

As we noted earlier, the 3*p* orbital of chlorine and the 4*p* orbital of bromine do not overlap effectively with the 2*p* orbital of carbon, so their electrons are not donated as effectively to the ring as the lone pair electrons of either oxygen or nitrogen.

5.9 Reactions of Side Chains

A group of carbon atoms bonded to an aromatic ring constitutes a **side chain.** Side-chain carbon atoms that are separated from the aromatic ring by two or more σ bonds behave independently of the aromatic ring. However, carbon atoms directly bonded to the aromatic ring are influenced by the ring. For

Substituent Effects on Metabolism of Benzene Derivatives

We have seen that benzene is remarkably unreactive even under very strong reaction conditions. We might therefore expect benzene to be inert in living cells at pH 7 and 37°C, and it is. Benzene itself is not metabolized in most human cells; rather it accumulates in the liver, where it does great harm. Benzene is carcinogenic and extremely toxic. Oxidation of benzene and aromatic compounds by cytochrome *P*-450 often yields phenols. Although the process appears to be aromatic hydroxylation, the reaction actually occurs via a three-membered heterocyclic ring called an epoxide (Chapter 9). The epoxide intermediates, called **arene oxides,** rearrange to phenols.

(an arene) → *P*-450, O_2 → (an arene oxide) → rearrangement → (a phenol)

All arene oxide intermediates are very reactive and undergo several types of reactions besides the rearrangement reaction to form phenols. Arene oxides react with proteins, RNA, and DNA. As a consequence, serious cellular disruptions can occur. These processes will be discussed in Chapter 9, when the chemistry of epoxides is presented.

Some drugs containing aromatic rings are hydroxylated at the para position when metabolized. Phenytoin, an anticonvulsant, is an example. The phenolic compounds react further to form water-soluble derivatives.

phenytoin   *p*-hydroxyphenytoin

When some drugs are hydroxylated in the liver, they are also pharmacologically active. The site of hydroxylation of phenylbutazone, an anti-inflammatory agent, is at the para position. This hydroxylated product has been produced in the laboratory and is now marketed under the trade names Tandearil® and Oxalid®.

phenylbutazone   Tandearil

Groups that deactivate aromatic rings toward aromatic substitution affect their metabolic reactions. Dioxin (2,3,7,8-tetrachlorodibenzo-*p*-dioxin) and PCBs (polychlorinated biphenyls) have multiple deactivating groups. Dioxin is an impurity formed in the production of the herbicide 2,4,5-trichlorophen-oxyacetic acid. Approximately one million tons of PCBs were produced between 1930 and 1979. They were used as heat exchange agents in products such as electrical transformers.

2,3,7,8-tetrachlorodibenzo-*p*-dioxin

a polychlorobiphenyl

Because the aromatic ring loses electrons in the oxidation process, the electron-withdrawing groups slow the rate of biological oxidation. These halogenated compounds are nonpolar and quite soluble in fatty tissue. Thus, they tend to persist in the bodies of organisms that inadvertently ingest them. As a consequence they accumulate in the food chain. For example, fish concentrate PCBs from contaminated waters in their tissues. The amount of PCBs increases in the tissues of birds or humans who then eat the fish.

example, a **benzyl carbocation** has a positively charged carbon atom directly attached to a benzene ring and is more stable than primary and secondary carbocations and of comparable stability to a tertiary carbocation. A benzyl carbocation is resonance stabilized in the same manner as the allylic carbocation we discussed in the previous chapter. That is, the positive charge at the benzylic carbon atom is delocalized among the carbon atoms of the benzene ring.

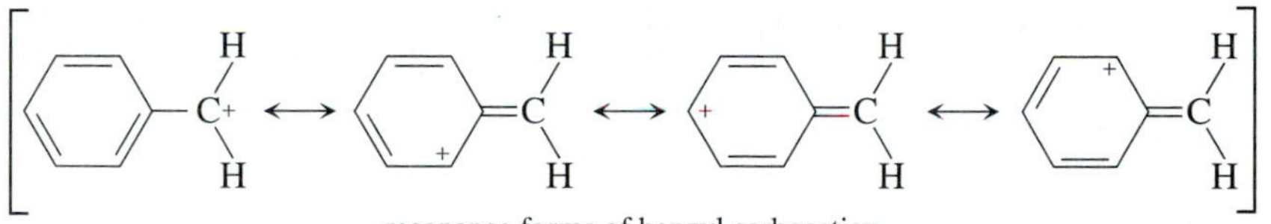

resonance forms of benzyl carbocation

The effect of resonance stabilization of a benzyl carbocation is illustrated in the addition reaction of HBr to indene. Electrophilic attack of H^+ at one of the carbon atoms of the carbon-carbon double bond gives a secondary benzyl carbocation. Attack at the other carbon atom directly would produce a much less stable secondary carbocation.

indene

less stable intermediate
(a secondary carbocation)

more stable intermediate
(a benzylic secondary carbocation)

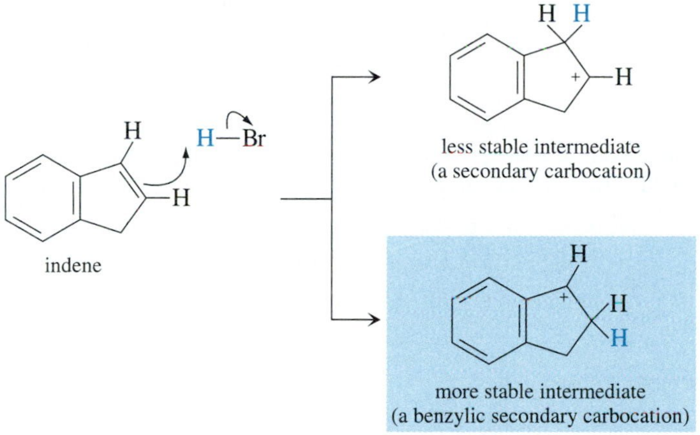

The only product obtained in the reaction is derived from the more stable carbocation, which then reacts with the bromide ion.

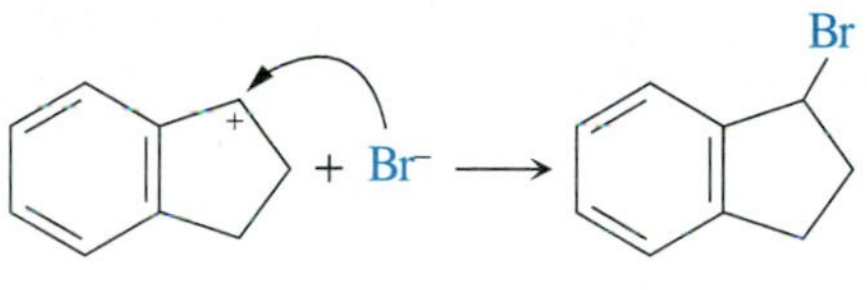

Although the aromatic ring causes special reactivity on the side chain, the ring itself is quite unreactive toward many reagents, and remains intact. We recall that potassium permanganate (Section 4.7) reacts with the π bonds of an alkene under conditions where the σ bonds of the saturated part of the molecule do not react. The benzene ring, in spite of being "unsaturated," is not oxidized by potassium permanganate even under vigorous conditions that totally oxidize side chains to produce a carboxylic acid at the site of the alkyl group.

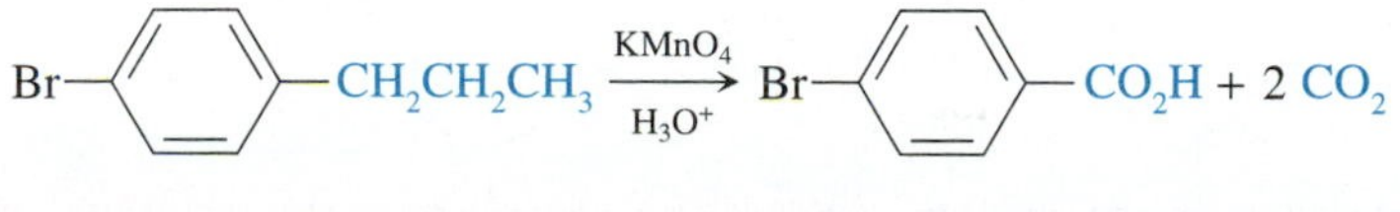

$$\text{Br}-\text{C}_6\text{H}_4-\text{CH}_2\text{CH}_2\text{CH}_3 \xrightarrow[\text{H}_3\text{O}^+]{\text{KMnO}_4} \text{Br}-\text{C}_6\text{H}_4-\text{CO}_2\text{H} + 2\ \text{CO}_2$$

EXAMPLE 5-7

Predict the product of the reaction of the following compound with potassium permanganate.

Br, CH(CH$_3$)CH$_2$CH$_3$, NO$_2$ (1-bromo-3-sec-butyl-5-nitrobenzene)

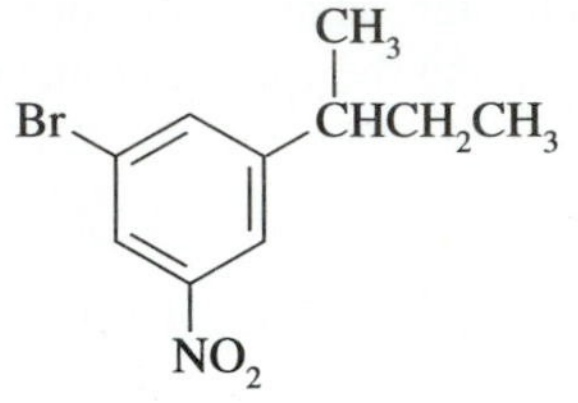

Solution

Potassium permanganate oxidizes the side chain of a substituted aromatic compound completely and forms a carboxylic acid. The *sec*-butyl group is oxidized, and a —CO_2H group results. The product of the reaction is 3-bromo-5-nitrobenzoic acid.

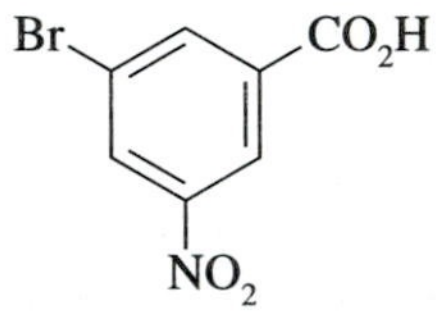

Br, CO_2H, NO$_2$

Problem 5.7

A compound with molecular formula $C_{10}H_{14}$ is oxidized to give the following dicarboxylic acid. What are the structures of three possible compounds that would give this result?

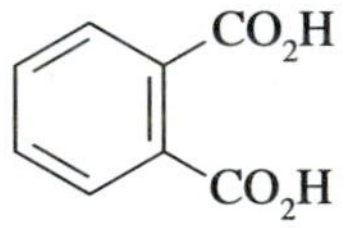

CO_2H, CO_2H (ortho)

5.10 Functional Group Modification

Functional group modifications are important because only a few functional groups can be placed directly on an aromatic ring by electrophilic aromatic substitution. The remaining groups are obtained by modifying a group already bonded to the aromatic ring. One such example is the conversion of a methyl group into a carboxylic acid group.

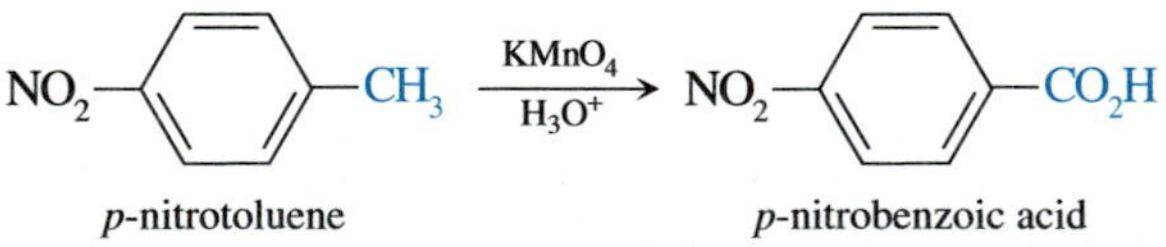

NO_2—C$_6$H$_4$—CH_3 $\xrightarrow[H_3O^+]{KMnO_4}$ NO_2—C$_6$H$_4$—CO_2H

p-nitrotoluene *p*-nitrobenzoic acid

We have also learned that an acyl group bonded to a benzene ring can be converted into an alkyl group by reduction with zinc/mercury amalgam in HCl (Section 5.5).

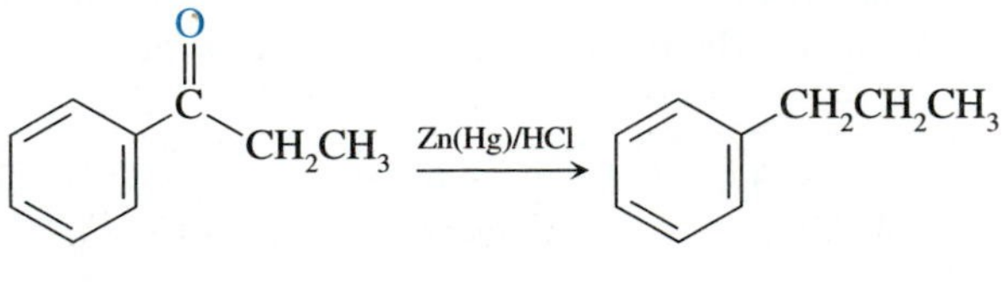

C$_6$H$_5$—C(=O)—CH_2CH_3 $\xrightarrow{Zn(Hg)/HCl}$ C$_6$H$_5$—$CH_2CH_2CH_3$

A nitro group can be attached to a benzene ring by electrophilic aromatic substitution, but an amino group cannot be attached to an aromatic ring in one step. However, after a nitro group is introduced, it can easily be reduced to an amino group; the product is an aniline.

$$\text{(2-nitrotoluene: } NO_2, CH_3) \xrightarrow{\text{Sn/HCl}} \text{(2-methylaniline: } NH_2, CH_3)$$

The amino groups of anilines can be converted into many other groups by first converting the amino group into a diazonium ion, $Ar{-}N_2^+$. A diazonium ion is made by treating an aniline with nitrous acid, HNO_2, prepared by reaction of sodium nitrite with sulfuric acid. This step, which produces a diazonium ion, is called **diazotization.**

$$Ar{-}NH_2 \xrightarrow[H_2SO_4]{NaNO_2} Ar{-}N_2^+$$

Aromatic diazonium ions are extremely reactive. They react with nucleophiles, replacing the diazonium group and liberating nitrogen gas. Thus, the nucleophile occupies the position of the original amino group.

$$Ar{-}N_2^+ + :Nu^- \longrightarrow Ar{-}Nu + N_2$$

In 1884, the German chemist T. Sandmeyer found that diazonium ions react with nucleophiles supplied in the form of a Cu(I) salt. Thus, a solution of an aromatic diazonium ion can be treated with Cu_2Cl_2 or Cu_2Br_2 to yield chlorobenzene or bromobenzene, respectively. These reactions are known collectively as the **Sandmeyer reaction**.

$$\text{(}NH_2, CH_3\text{)} \xrightarrow[H_2SO_4]{NaNO_2} \text{(}N_2^+, CH_3\text{)} \xrightarrow{Cu_2Br_2} \text{(}Br, CH_3\text{)}$$

Cuprous salts of the cyanide ion result in the formation of aryl nitriles.

$$\text{(}Br, NH_2\text{)} \xrightarrow[H_2SO_4]{NaNO_2} \text{(}Br, N_2^+\text{)} \xrightarrow{Cu_2(CN)_2} \text{(}Br, CN\text{)}$$

Phenols can be synthesized by reaction of the aryldiazonium compound with hot aqueous acid. This reaction is one of the few ways to introduce an —OH group onto an aromatic ring.

$$\text{(}NH_2, CH(CH_3)_2\text{)} \xrightarrow[H_2SO_4]{NaNO_2} \text{(}N_2^+, CH(CH_3)_2\text{)} \xrightarrow{H_3O^+} \text{(}OH, CH(CH_3)_2\text{)}$$

A diazonium group can also be replaced by hydrogen using hypophosphorous acid, H_3PO_2. This process removes the amino substituent from the aromatic ring after its role as a directing group in a series of synthetic steps is concluded (see Section 5.11).

$$\text{3,5-dibromoaniline } (NH_2) \xrightarrow[H_2SO_4]{NaNO_2} \text{3,5-dibromobenzenediazonium } (N_2^+) \xrightarrow{H_3PO_2} \text{1,3-dibromobenzene } (H)$$

5.11 Synthesis of Substituted Aromatic Compounds

Chemists often must design benzene derivatives with two or more substituents strategically placed around the ring. A project of this type begins with an analysis of the ortho,para- or meta-directing characteristics of the substituents. For example, consider the problem of synthesis of *m*-chloronitrobenzene. A nitro group is meta-directing; a chloro group is ortho,para-directing. The order in which we add these groups is clearly important. If chlorination precedes nitration, the entering nitro group will be largely directed to form *o*-chloronitrobenzene and *p*-chloronitrobenzene. Very little of the desired meta isomer will be formed.

$$\text{benzene} \xrightarrow[FeCl_3]{Cl_2} \text{chlorobenzene (Cl)} \xrightarrow[H_2SO_4]{HNO_3} \text{o-chloronitrobenzene (Cl, } NO_2\text{)} + \text{p-chloronitrobenzene (Cl, } NO_2\text{)}$$

However, the desired compound results by introducing the nitro group first and the chlorine group second. Because the nitro group is a meta director, the entering chlorine atom is directed to the desired meta position.

$$\text{benzene} \xrightarrow[H_2SO_4]{HNO_3} \text{nitrobenzene } (NO_2) \xrightarrow[FeCl_3]{Cl_2} \text{m-chloronitrobenzene } (NO_2\text{, Cl})$$

It may also be necessary to substitute a group onto the aromatic ring and then modify it. Consider the synthesis of *m*-dibromobenzene, a task that appears at first glance to be impossible. The bromo groups are meta to each other, but bromine is an ortho,para director! Direct bromination of benzene would place on the ring one bromine atom that would then direct the second bromine atom into the ortho or para position.

$$\text{benzene} \xrightarrow[FeBr_3]{Br_2} \text{bromobenzene (Br)} \xrightarrow[FeBr_3]{Br_2} \text{o-dibromobenzene (Br, Br)} + \text{p-dibromobenzene (Br, Br)}$$

However, the nitro group is a meta director. So nitration to form nitrobenzene followed by bromination gives *m*-bromonitrobenzene.

$$\text{C}_6\text{H}_6 \xrightarrow[\text{H}_2\text{SO}_4]{\text{HNO}_3} \text{C}_6\text{H}_5\text{NO}_2 \xrightarrow[\text{FeBr}_3]{\text{Br}_2} m\text{-BrC}_6\text{H}_4\text{NO}_2$$

A nitro group can be converted to a bromo group by (1) reducing the nitro group to an amino group, (2) converting the amino group to a diazonium group, and (3) treating the diazonium compound with copper(I) bromide. The procedure requires several steps, but it accomplishes the apparently "impossible" task of preparing *m*-dibromobenzene.

$$m\text{-BrC}_6\text{H}_4\text{NO}_2 \xrightarrow{\text{Sn/HCl}} m\text{-BrC}_6\text{H}_4\text{NH}_2 \xrightarrow[\text{H}_2\text{SO}_4]{\text{NaNO}_2} m\text{-BrC}_6\text{H}_4\text{N}_2^+ \xrightarrow{\text{Cu}_2\text{Br}_2} m\text{-BrC}_6\text{H}_4\text{Br}$$

EXAMPLE 5-8

Devise a synthesis of *m*-bromoaniline starting from benzene.

Solution

Bromine, which is an ortho,para-director, can be introduced directly onto the benzene ring by reaction with bromine and $FeBr_3$. The amino group, $—NH_2$, of aniline is also an ortho,para-director. It can be introduced indirectly by first nitrating benzene and then reducing the nitro compound. Recall that the nitro group is a meta director.

Bromination of benzene followed by nitration gives a mixture of *ortho*- and *para*-bromonitrobenzene. The desired meta isomer is not formed. Nitration of benzene gives nitrobenzene—a compound that now directs subsequent electrophilic substitution reactions to produce the meta isomer. Thus, bromination of nitrobenzene followed by reduction of the product gives the desired *m*-bromoaniline.

$$\text{C}_6\text{H}_6 \xrightarrow[\text{H}_2\text{SO}_4]{\text{HNO}_3} \text{C}_6\text{H}_5\text{NO}_2 \xrightarrow[\text{FeBr}_3]{\text{Br}_2} m\text{-BrC}_6\text{H}_4\text{NO}_2 \xrightarrow{\text{Sn / HCl}} m\text{-BrC}_6\text{H}_4\text{NH}_2$$

Problem 5.8

Propose a method to synthesize *p*-nitrobenzoic acid starting from benzene.

$$\text{NO}_2\text{—C}_6\text{H}_4\text{—CO}_2\text{H}$$

p-nitrobenzoic acid

EXAMPLE 5-9

Propose a synthesis of *m*-bromophenol starting from benzene.

Solution

The hydroxyl and bromo groups are both ortho,para directors. Thus, the synthesis must proceed via an intermediate compound that has a meta director such as the nitro group. Nitration of benzene followed by bromination gives *m*-bromonitrobenzene.

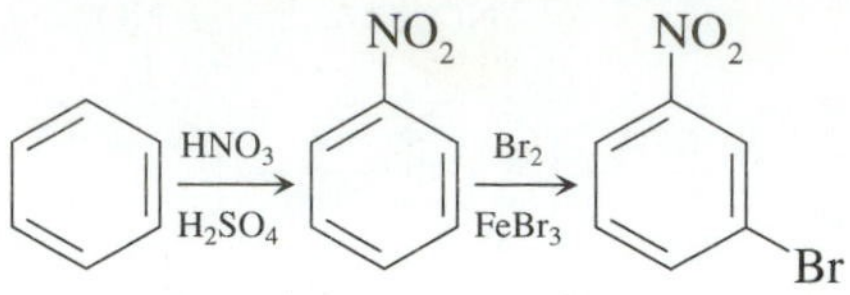

NO_2 NO_2

HNO$_3$ / H$_2$SO$_4$ → ; Br$_2$ / FeBr$_3$ → Br

Reduction of the nitro group followed by conversion of the amino group into a diazonium ion gives an intermediate that can be transformed into a phenol.

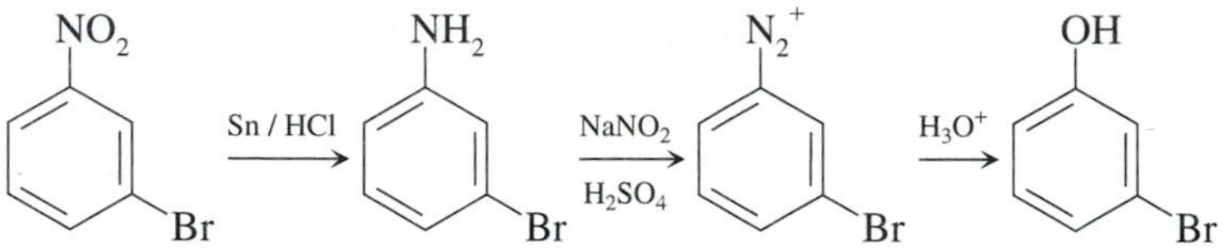

NO_2 Br → Sn / HCl → NH_2 Br → $NaNO_2$ / H_2SO_4 → N_2^+ Br → H_3O^+ → OH Br

Problem 5.9

Propose a synthesis of *p*-bromobenzonitrile starting from bromobenzene.

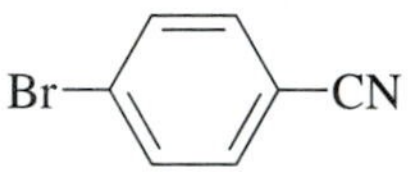

Br—⟨benzene⟩—CN

p-bromobenzonitrile

Summary of Reactions

1. Halogenation (Section 5.5)

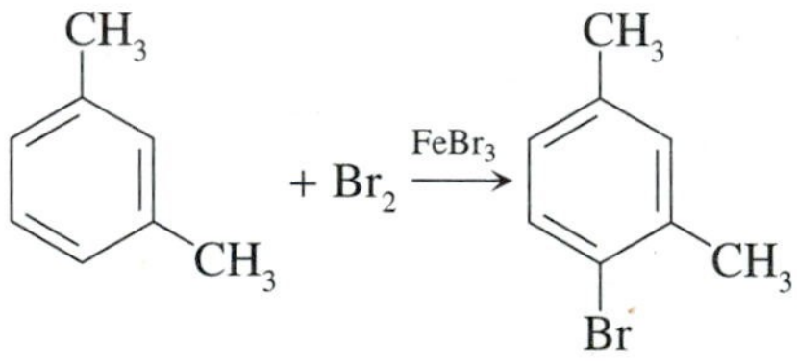

CH_3, CH_3 + Br_2 → FeBr$_3$ → CH_3, CH_3, Br

2. Nitration (Section 5.5)

CH_3, OCH_3 + HNO_3 → H_2SO_4 → CH_3, NO_2, OCH_3

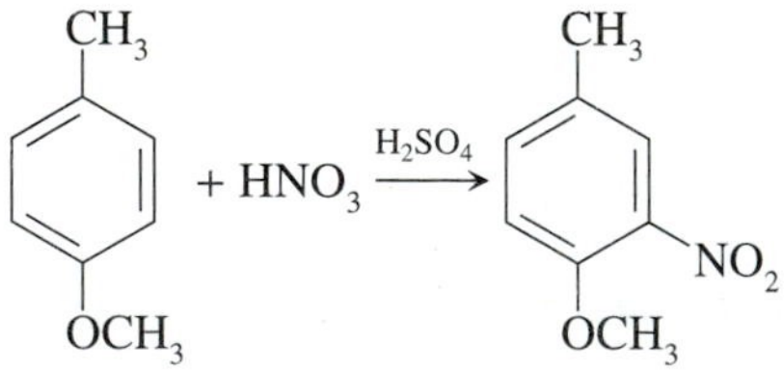

3. Sulfonation (Section 5.5)

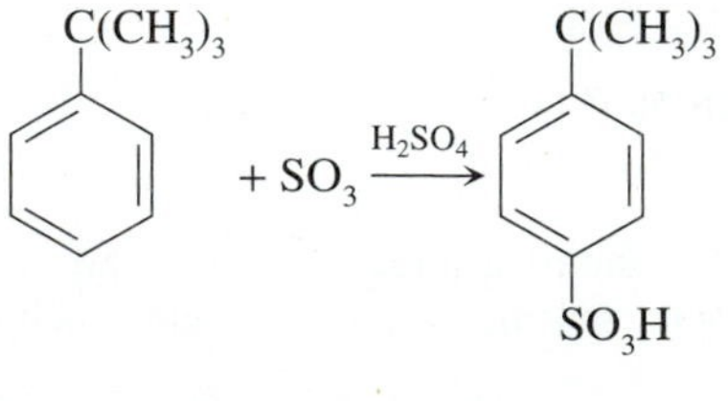

$C(CH_3)_3$ + SO_3 → H_2SO_4 → $C(CH_3)_3$, SO_3H

4. **Alkylation (Friedel-Crafts) (Section 5.5)**

+ Cl— $\xrightarrow{AlCl_3}$

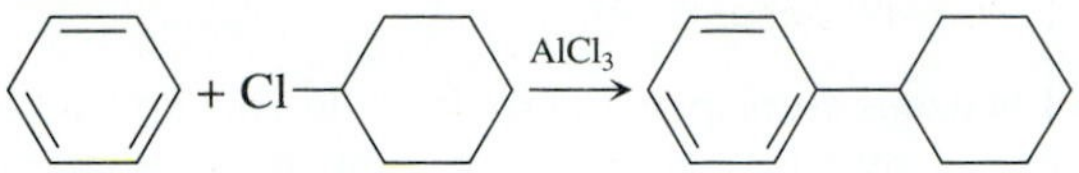

5. **Acylation (Friedel-Crafts) (Section 5.5)**

O

+ Cl $\xrightarrow{AlCl_3}$ O

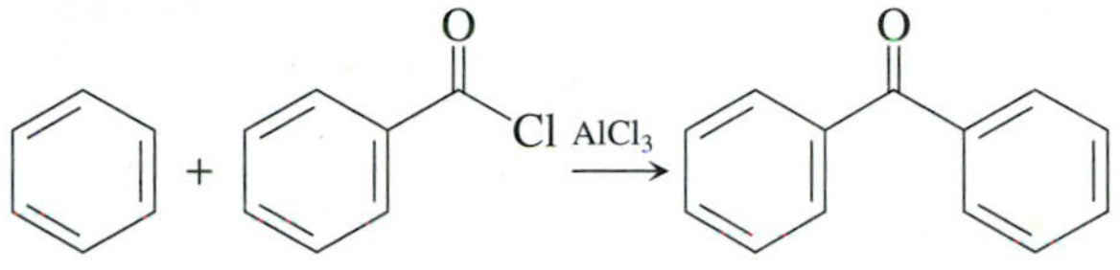

6. **Oxidation of Side Chains (Section 5.9)**

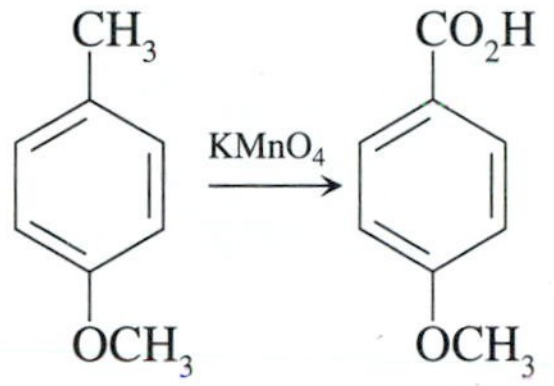

CH_3 $\xrightarrow{KMnO_4}$ CO_2H

OCH_3 OCH_3

7. **Reduction of Acyl Side Chains (Section 5.5)**

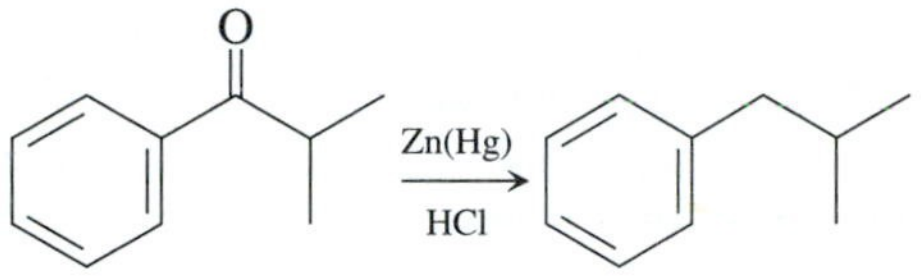

O

$\xrightarrow[HCl]{Zn(Hg)}$

8. **Reduction of Nitro Groups (Section 5.10)**

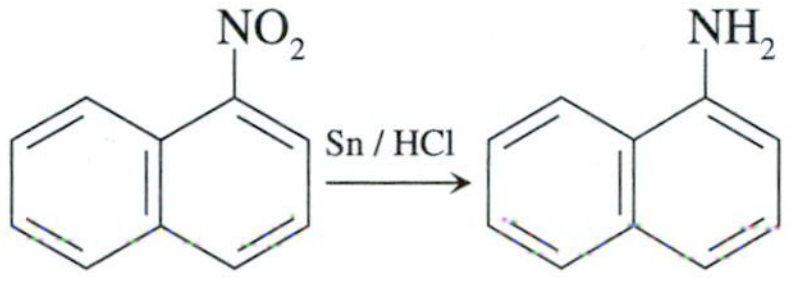

NO_2 $\xrightarrow{Sn / HCl}$ NH_2

9. **Reactions of Amino Groups via Diazonium Ion (Section 5.10)**

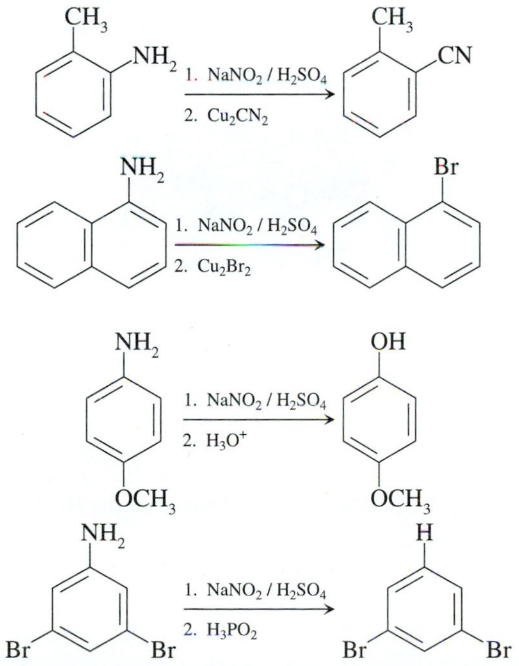

CH_3 NH_2 $\xrightarrow[2.\ Cu_2CN_2]{1.\ NaNO_2 / H_2SO_4}$ CH_3 CN

NH_2 $\xrightarrow[2.\ Cu_2Br_2]{1.\ NaNO_2 / H_2SO_4}$ Br

NH_2 $\xrightarrow[2.\ H_3O^+]{1.\ NaNO_2 / H_2SO_4}$ OH

OCH_3 OCH_3

NH_2 $\xrightarrow[2.\ H_3PO_2]{1.\ NaNO_2 / H_2SO_4}$ H

Br Br Br Br

Explorations with Molecular Models

1. Prepare a ball-and-stick model of benzene corresponding to the following space-filling model. How many sets of equivalent carbon atoms are there? How many sets of equivalent hydrogen atoms are there? Replace one of the hydrogen atoms with a chlorine atom. Now how many sets of equivalent carbon atoms and how many sets of equivalent hydrogen atoms are there? Replace one atom in each of the possible sets of hydrogen atoms with a chlorine atom, one at a time, and determine the number of possible dichlorobenzene isomers. Write the structures of the products and name each compound.

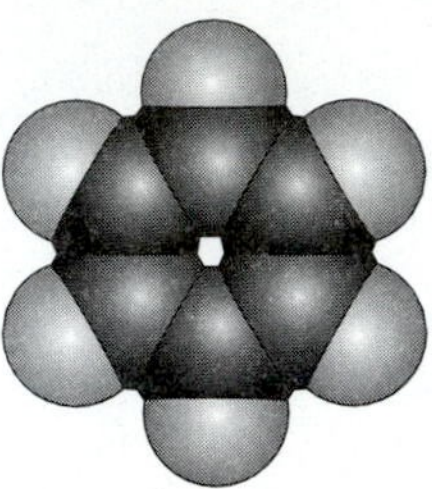

2. Consider each possible isomer prepared in Exploration Exercise 1 and determine the number of sets of equivalent hydrogen atoms in each one. Taking one isomer at a time, replace one member of each set of nonequivalent hydrogen atoms with a chlorine atom. Write the structures of the products and name each compound derived by converting each isomeric dichlorobenzene into isomeric trichlorobenzenes. Examine each set of drawn structures and determine which are are duplicated in two or three of the sets of structures. Eliminating duplicates, how many isomeric trichlorobenzenes are there?

3. Prepare a model of benzene and remove a pair of ortho hydrogen atoms. Connect the two carbon atoms by a chain of methylene units. How many atoms are in the smallest ring that can be formed without straining the model? Draw a standard two-dimensional representation of the molecule as a bond-line structure.

4. Prepare a model of benzene and remove hydrogen atoms from the 1 and 4 carbon atoms. Connect the two carbon atoms by a chain of methylene units. How many atoms are in the smallest ring that can be formed without straining the model? *Hint:* It is necessary to pass the chain over the top of the plane of the structure to reach the other side.

5. Consider the following space-filling model of naphthalene. How many sets of equivalent carbon atoms and how many sets of equivalent hydrogen atoms are there? Replace one atom from each of the possible sets of hydrogen atoms with a chlorine atom and determine the number of possible monosubstituted isomers. Write the structures of the products.

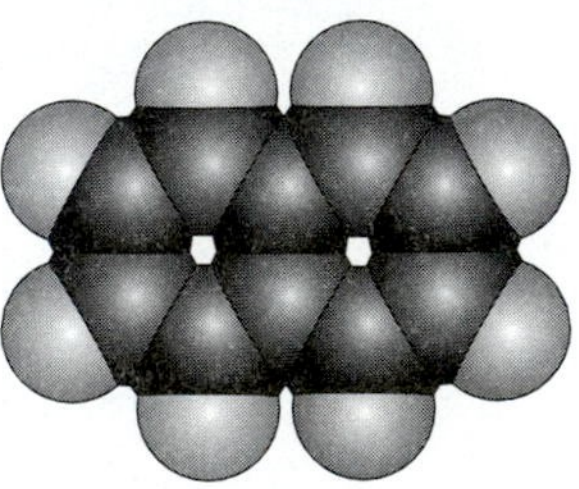

Exercises

Aromaticity

5.1 Determine which of the following is an aromatic compound based on the Hückel rule.

(a) (b) (c)

5.2 Determine which of the following is an aromatic compound based on the Hückel rule.

(a) (b) (c)

Polycyclic and Heterocyclic Aromatic Compounds

5.3 There are two isomeric bromonaphthalenes. Draw their structures.

5.4 There are three isomeric bromoanthracenes. Draw their structures.

5.5 There are three isomeric diazines, $C_4N_2H_4$, that resemble benzene but have two nitrogen atoms in place of carbon in the ring. Draw their structures. Which of the isomers should have no dipole moment?

5.6 There are three isomeric triazines, $C_3N_3H_3$, that resemble benzene but have three nitrogen atoms in place of carbon in the ring. Draw their structures. Which of the isomers should have no dipole moment?

5.7 How many electrons does each heteroatom contribute to the π system in each of the following compounds?

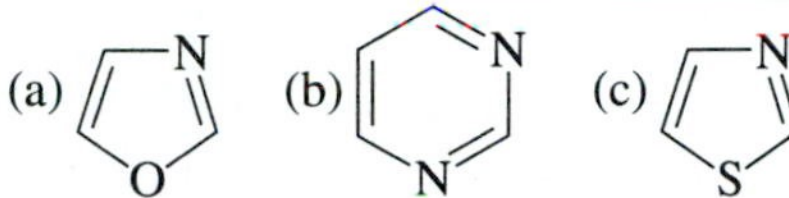

(c) N S

5.8 How many electrons does each heteroatom contribute to the π system in each of the following compounds?

(a) N N H (b) N N H (c) N N

5.9 Identify the heterocyclic ring structure contained in each of the following compounds, which have been investigated as possible male contraceptives.

(a) O_2N—(furan)—CH=N—NHCNH$_2$ (C=O) (b) Cl—(thiophene)—C(=O)CH_3

5.10 Identify the aromatic heterocyclic ring structure contained in each of the following compounds.

(a) tolmetin, a drug used to lower blood sugar levels

CH_3 C O N CH_3 $CH_2CO_2H_2$

(b) ethionamide, an antitubercular agent

S C NH_2 N CH_2CH_3

Isomerism in Aromatic Compounds

5.11 There are three isomeric dichlorobenzenes. One compound is nonpolar. Which one?

5.12 The boiling points of benzyl alcohol ($C_6H_5CH_2OH$) and anisole ($C_6H_5OCH_3$) are 205 and 154°C, respectively. Explain this difference.

5.13 There are four isomeric substituted benzene compounds with the molecular formula C_8H_{10}. Name each compound.

5.14 There are three isomeric substituted benzene compounds with the molecular formula $C_6H_3Br_3$. Name each compound.

Nomenclature of Aromatic Compounds

5.15 Identify each of the following as an ortho-, meta-, or para-substituted compound.

(a) methylparaben, a food preservative (b) DEET, an insect repellant

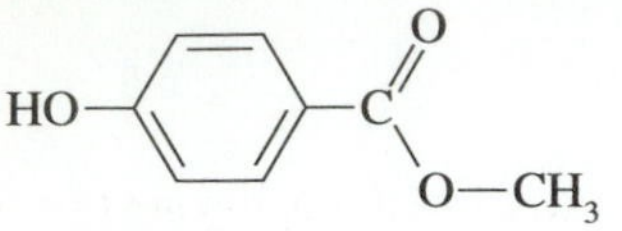

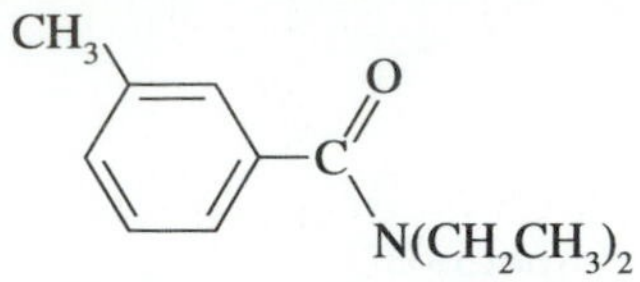

HO O C O—CH_3 CH_3 O C $N(CH_2CH_3)_2$

5.16 Identify each of the following as an ortho-, meta-, or para-substituted compound.

(a) resorcinol monoacetate, a germicide (b) salicylamide, an analgesic

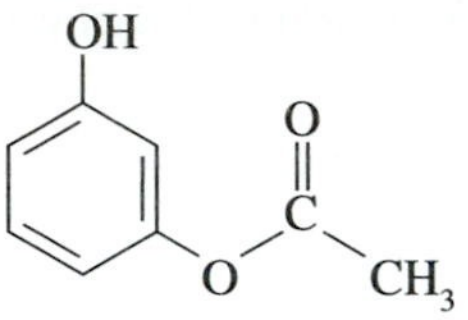

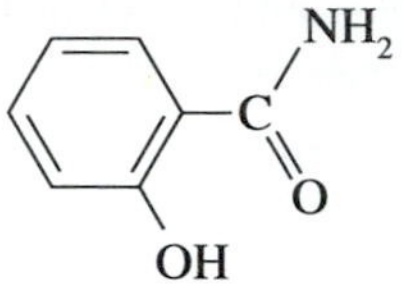

OH O C O CH_3 NH_2 C O OH

5.17 Name each of the following compounds.

(a) CH_2CH_3 CH_2CH_3 (b) $CH(CH_3)_2$ (c) CH_3 CH_3 CH_3

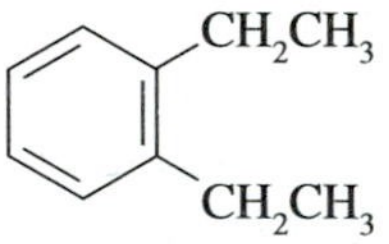

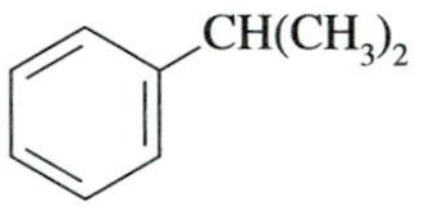

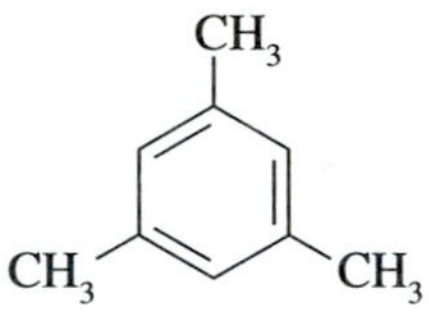

5.18 Name each of the following compounds.

(a) OH Br Br (b) NH_2 Cl (c) CH_3 Br Cl

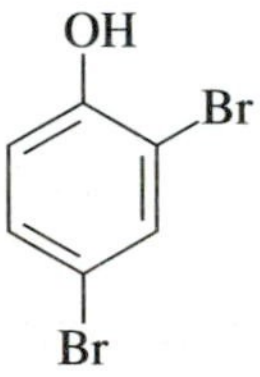

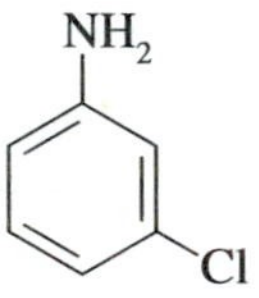

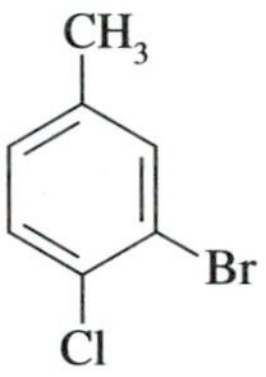

5.19 Name each of the following compounds, which serve the indicated function.

(a) an antiseptic agent used to treat athlete's foot (b) a disinfectant

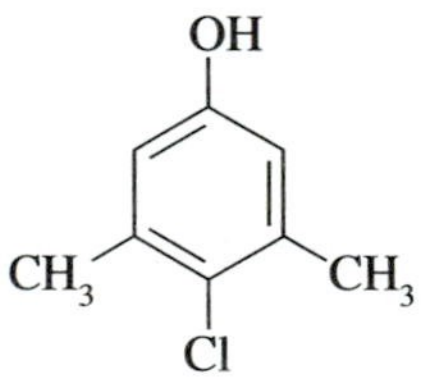

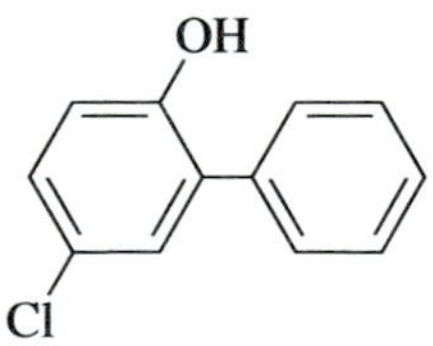

OH CH_3 CH_3 Cl OH Cl

5.20 Draw the structure of each of the following compounds.

(a) 5-isopropyl-2-methylphenol, found in oil of marjoram (b) 2-isopropyl-5-methylphenol, found in oil of thyme

(c) 2-hydroxybenzyl alcohol, found in the bark of the willow tree

5.21 Draw the structure of 3,4,6-trichloro-2-nitrophenol, a lampricide used to control sea lampreys in the Great Lakes.

5.22 *N,N*-Dipropyl-2,6-dinitro-4-trifluoromethylaniline is the IUPAC name for Treflan®, a herbicide. Draw its structure. (The prefix *N* signifies the location of substituents on a nitrogen atom.)

Electrophiles and Ring Substituents

5.23 Consider the thiomethyl group, —SCH_3. Predict whether it is an activating or deactivating group. Will it be an ortho,para-directing group or a meta-directing group?

5.24 The sulfonamide group is found in sulfa drugs. Consider its structure and determine if it is an activating or deactivating group. Will it be an ortho,para-directing or a meta-directing group?

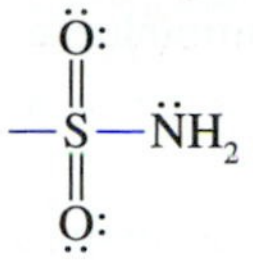

5.25 What product will result from the Friedel-Crafts alkylation of benzene using 2-methyl-1-chloropropane and aluminum trichloride?

5.26 Alkylation of benzene can be accomplished by using an alkene, such as propene, and an acid catalyst. What is the electrophile and what is the product?

5.27 Indicate on which ring and at what position bromination of the following compound will occur.

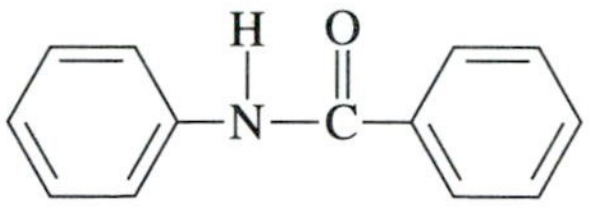

5.28 Indicate on which ring and at what position nitration of the following compound will occur.

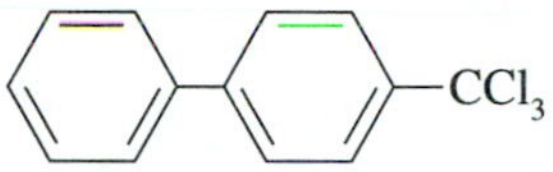

Reactions of Side Chains of Aromatic Compounds

5.29 Draw the oxidation product of each compound in Exercise 5.17 when reacted with potassium permanganate.

5.30 Predict the oxidation product of each of the following compounds when reacted with potassium permanganate.

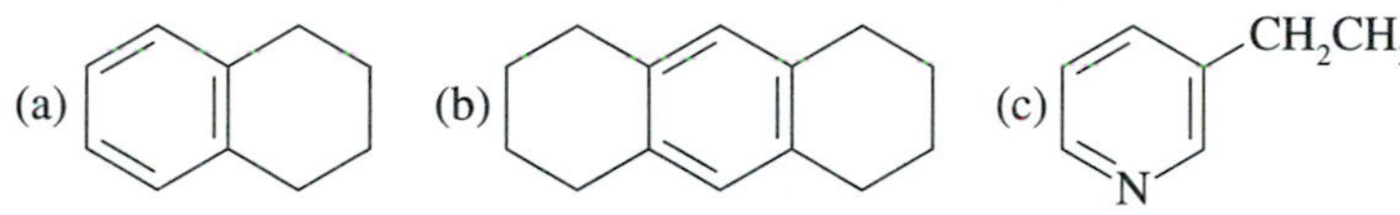

5.31 Free radical bromination of propylbenzene yields essentially one product. What is its structure? Suggest a reason why other isomers are not formed.

5.32 Treatment of allyl benzene with dilute acid causes isomerization to an isomeric compound. Suggest its structure and propose a mechanism for its formation.

Synthesis of Aromatic Compounds

5.33 What reagent is required for each of the following reactions? Will an ortho,para mixture of products or the meta isomer predominate?

(a) nitration of bromobenzene (b) sulfonation of nitrobenzene (c) bromination of ethylbenzene
(d) methylation of anisole

5.34 What reagent is required for each of the following reactions? Will an ortho,para mixture of products or the meta isomer predominate?

(a) bromination of benzoic acid (b) acetylation of isopropylbenzene (c) nitration of acetophenone
(d) nitration of phenol

5.35 Starting with benzene, describe the series of reagents and reactions required to produce each of the following compounds.

(a) *p*-bromonitrobenzene (b) *m*-bromonitrobenzene (c) *p*-bromoethylbenzene (d) *m*-bromoethylbenzene

5.36 Starting with benzene, describe the series of reagents and reactions required to produce each of the following compounds.

(a) *m*-bromobenzenesulfonic acid (b) *p*-bromobenzenesulfonic acid (c) *p*-nitrotoluene (d) *p*-nitrobenzoic acid

5.37 Starting with either benzene or toluene, describe the series of reagents and reactions required to produce each of the following compounds.

(a) 3,5-dinitro-1-chlorobenzene (b) 2,4,6-trinitrotoluene (c) 2,6-dibromo-4-nitrotoluene

5.38 Starting with either benzene or toluene, describe the series of reagents and reactions required to produce each of the following compounds.

(a) 2,4,6-tribromobenzoic acid (b) 2-bromo-4-nitrotoluene (c) 1-bromo-3,5-dinitrobenzene

5.39 Starting with either benzene or toluene, describe the series of reagents and reactions required to produce each of the following compounds.

(a) *m*-bromophenol (b) *m*-bromoaniline (c) *p*-methylphenol

5.40 Starting with either benzene or toluene, describe the series of reagents and reactions required to produce each of the following compounds.

(a) 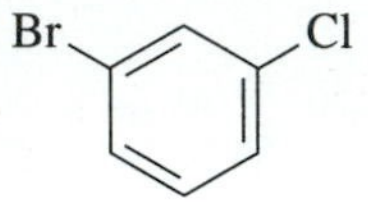 Br, Cl (b) 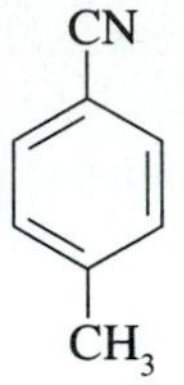 CN, CH_3 (c) 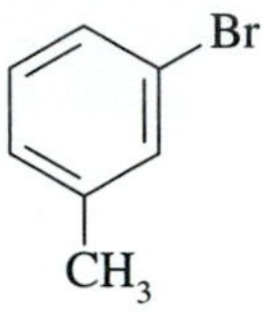 Br, CH_3

Metabolic Oxidation of Aromatic Compounds

5.41 Explain why aromatic hydroxylation of chlorpromazine, an antipsychotic drug, occurs at the indicated position and in that ring.

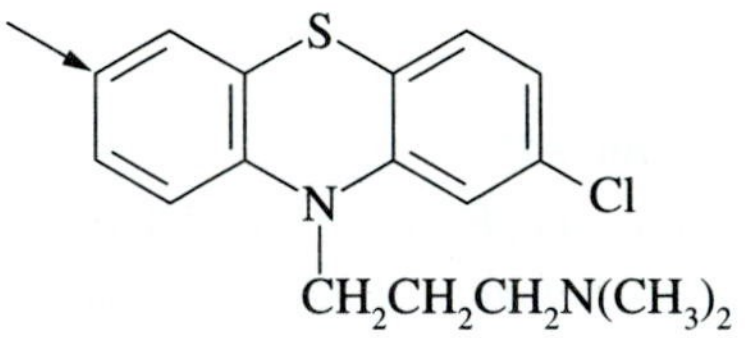

S, N, Cl, $CH_2CH_2CH_2N(CH_3)_2$

5.42 Why doesn't aromatic hydroxylation of probenecid, a drug used to treat chronic gout, occur?

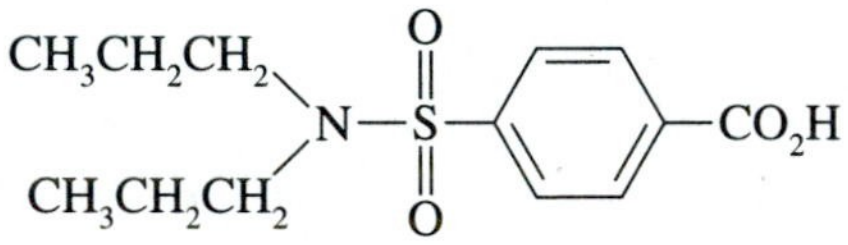

$CH_3CH_2CH_2$, $CH_3CH_2CH_2$, N—S(=O)(=O)—, CO_2H

Spectroscopy

5.43 If you have read Chapter 18 and the material has been assigned by your professor, then Exercises 18.1 and 18.23, which are based on the structure of aromatic hydrocarbons, may be done.

CHAPTER 6

STEREOCHEMISTRY

Overview

6.1 Configuration of Molecules

In Chapters 3 and 4 we considered the structures of geometric isomers, which are one of a general class of stereoisomers. **Stereoisomers** have the same connectivity—the same sequence of bonded atoms—but different arrangements of the atoms in space. The different three-dimensional arrangements of atoms in space determine their **configurations.** Geometric isomers have different configurations.

The configuration of a molecule plays a major role in its biological properties. Stereoisomers often have entirely different biological properties. Geometric isomers invariably elicit different responses in organisms. For example, bombykol, the sex attractant of the male silkworm moth, has a trans-cis arrangement about the double bonds at the C-10 and C-12 positions. It is 10^9 to 10^{13} times more potent than the other three possible geometric isomers. Disparlure, the sex attractant of the female gypsy moth, is biologically active only if the alkyl groups bonded to the three-membered ring are in a cis configuration.

$CH_3(CH_2)_2$–CH=CH–CH=CH–$(CH_2)_9OH$ (cis double bond; trans double bond)

bombykol

$CH_3(CH_2)_9$–CH(O)CH–$(CH_2)_4CH(CH_3)_2$ (cis alkyl groups)

disparlure

Geometric isomerism is only one type of stereoisomerism. Another type of stereoisomerism is the result of the mirror image relationships between molecules, the subject of this chapter. These molecules differ in configuration about a tetrahedral carbon atom bearing four different groups of atoms, which is called a **stereogenic center.** This phenomenon is not as easily visualized as geometric isomers, but its consequences are even more vital to life processes.

6.2 Mirror Images and Chirality

The fact that we live in a three-dimensional world has important personal consequences. In the simple act of looking into a mirror, you see someone who does not actually exist—namely, your mirror image. Every object has a mirror image, but this reflected image need not be identical to the actual object. Let's consider a few common three-dimensional objects. A simple wooden chair looks exactly like its mirror image (Figure 6.1). When an object and its mirror image exactly match, we say that they are **superimposable.** Superimposable objects can be "placed" on each other so that each three-dimensional feature of one object coincides with an equivalent three-dimensional feature in the mirror image.

Now let's consider some objects and their mirror images that are not identical. These are said to be **nonsuperimposable.** One example is the side-arm

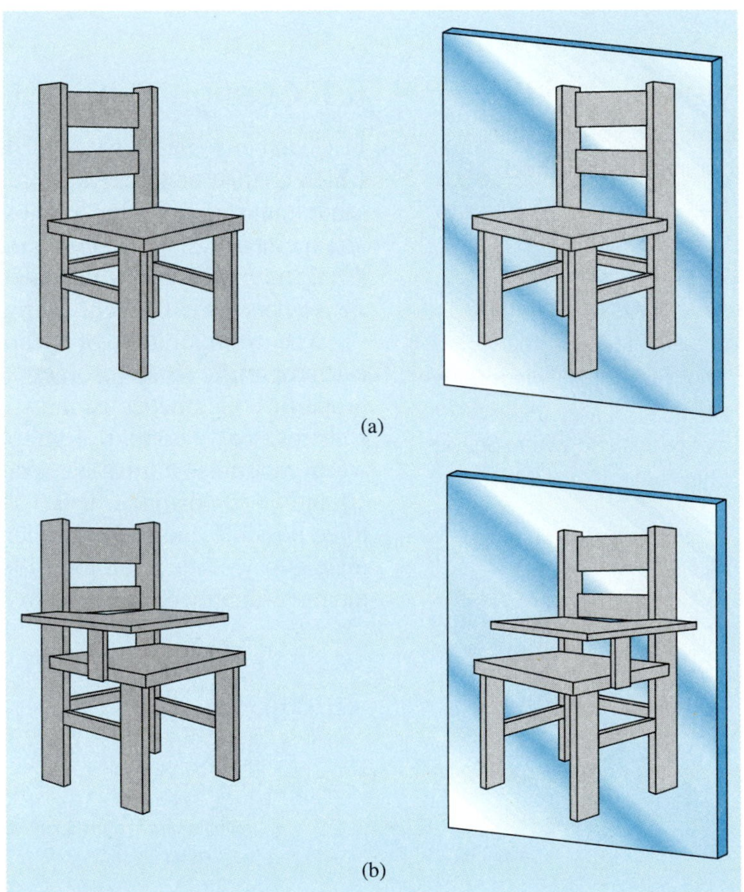

FIGURE 6.1
Objects and Their Mirror Images

In (a) the simple chair and its mirror image are identical. The mirror image can be superimposed on the original chair. In (b) the side arm chair has a mirror image that is different. The chair is a right-handed object; the mirror image is a left-handed object. The mirror image cannot be superimposed on the original chair.

chair found in many classrooms. When a chair with a "right-handed arm" is reflected in a mirror, it becomes a chair with a "left-handed arm" (Figure 6.1). We can convince ourselves of this by imagining sitting in the chair or its mirror image.

Now consider the nonsuperimposability of our hands, which are also related as mirror images. The mirror image of a left hand looks like a right hand. But when we try to superimpose our hands, we find that it cannot be done. Thus, our hands are related as nonsuperimposable mirror images.

An object that is not superimposable on its mirror image is said to be **chiral** (Greek, *chiron,* hand). Thus, objects such as gloves and shoes that have a "handedness" are chiral. An object that is superimposable on its mirror image is **achiral.** Objects such as a cup or a hammer are achiral. We can determine whether an object is chiral or achiral without trying to superimpose its mirror image. An object, such as a cup, that has a plane of symmetry that divides it so that one half of it is the mirror image of the other half, is achiral (Figure 6.2). However, if an object, such as your hand, has no such plane of symmetry, it is chiral. Thus, a cup is achiral, but a hand is chiral.

The Nautilus shell is chiral and exists only in one form.

Molecular Chirality

The concepts of chirality can be extended from macroscopic objects to molecules. Most molecules produced in living organisms are chiral. A molecule is chiral if it contains a stereogenic center, which is a tetrahedral carbon atom attached to four different atoms or groups of atoms. These four atoms or groups of atoms can be arranged in two different ways corresponding to two different stereoisomers. Let's consider the stereoisomers of bromochlorofluoromethane. One stereoisomer of this molecule and its mirror image are illustrated in Figure 6.3. The carbon atom is a stereogenic center, and the two structures cannot be superimposed; therefore, bromochlorofluoromethane is chiral.

In our discussion of macroscopic achiral objects we noted that these contain a plane of symmetry. This generalization applies equally well to molecules.

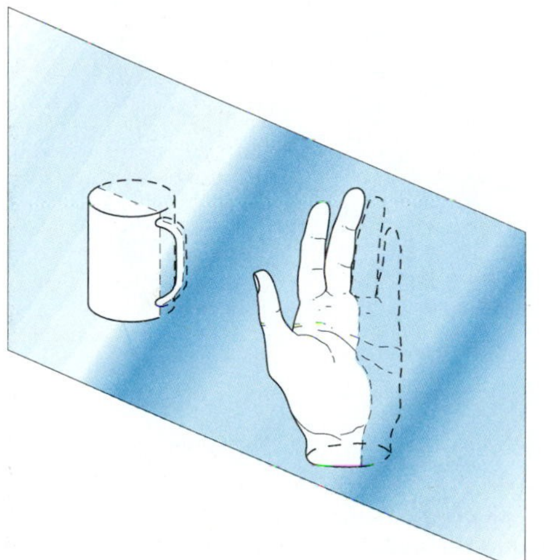

FIGURE 6.2
Plane of Symmetry
The cup shown can be divided into two equal halves by its plane of symmetry. One half is the mirror image of the other half. The cup is achiral. The plane shown does not split a hand into two equal halves. A hand is chiral.

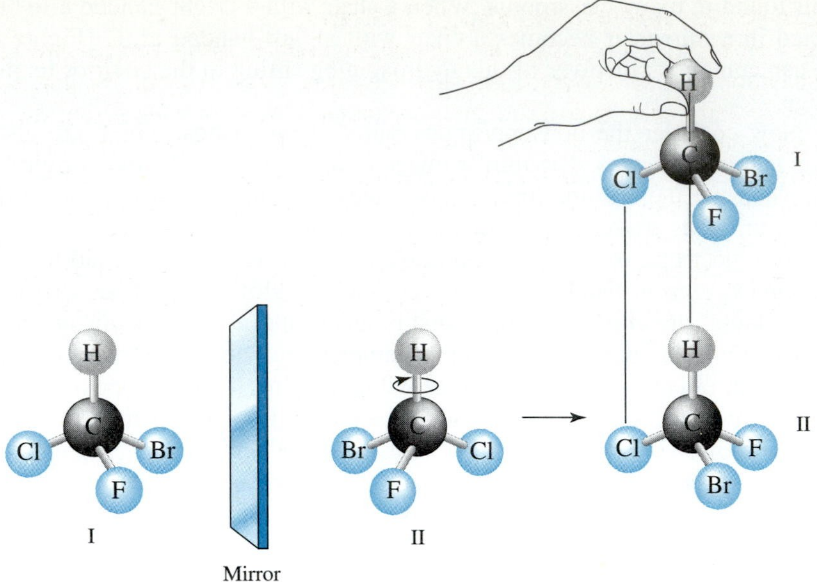

FIGURE 6.3
Criteria for Chirality in Molecules

The spheres of bromochlorofluoromethane, structure I, held in the hand do not line up with those of structure II, which has been rotated to line up the chlorine atoms. Structure I and its mirror image, structure II, cannot be superimposed.

Figure 6.4 shows ball-and-stick models of dichloromethane and bromochloromethane. Neither molecule has a stereogenic center. Dichloromethane has two planes of symmetry; bromochloromethane has one plane of symmetry. Each molecule can be superimposed on its mirror image and is therefore achiral. In contrast, bromochlorofluoromethane does not have a plane of symmetry and is chiral.

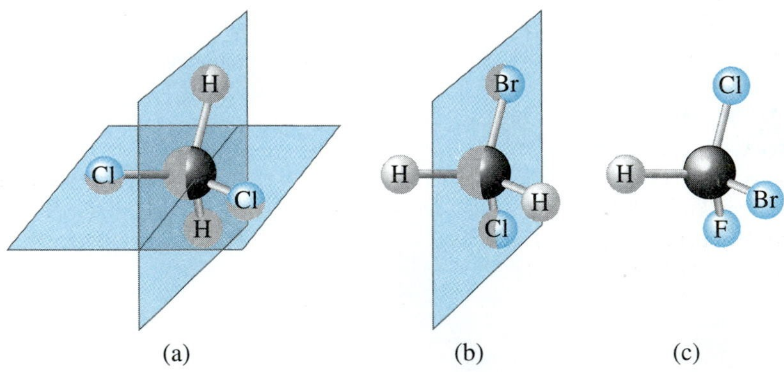

FIGURE 6.4
Planes of Symmetry for Molecules

(a) Dichloromethane is bisected by two planes of symmetry. For each plane, one part of the structure mirrors the part of the molecule lying on the other side of the plane. (b) Bromochloromethane has one plane of symmetry. (c) Bromochlorofluoromethane has no plane of symmetry.

Enantiomers Are Mirror Image Isomers

Stereoisomers that are related as nonsuperimposable mirror images are called **enantiomers** (Greek, *enantios,* opposite + *meros,* part). We can tell that a substance is chiral and predict that two enantiomers exist by identifying the substituents on each carbon atom. A carbon atom with four different substituents is a stereogenic center, and a molecule containing this carbon atom can exist as a pair of enantiomers. For example, 2-butanol is chiral because the C-2 atom, a stereogenic center, is attached to four different groups (CH_3, CH_3CH_2, H, and OH). In contrast, 2-propanol is achiral because it does not contain any carbon atom attached to four different groups. The C-2 atom has two methyl groups attached to it.

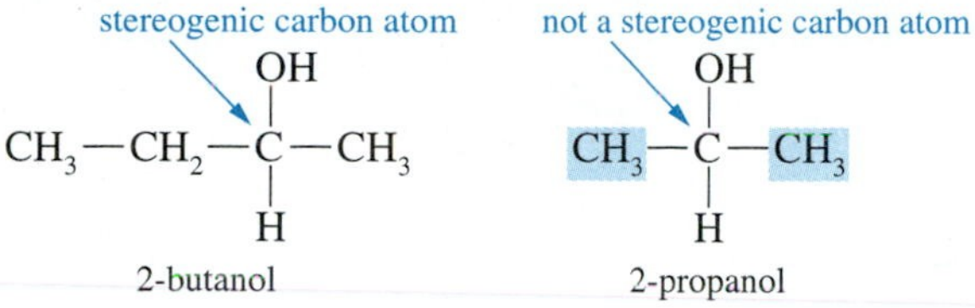

stereogenic carbon atom

$$CH_3-CH_2-\underset{H}{\overset{OH}{C}}-CH_3$$

2-butanol

not a stereogenic carbon atom

$$CH_3-\underset{H}{\overset{OH}{C}}-CH_3$$

2-propanol

Physical Properties of Enantiomers

Enantiomers have many identical physical properties, such as density, melting point, and boiling point. They also have the same chemical properties in a symmetrical environment. However, they can be distinguished in a chiral environment. This difference is important in many reactions in living cells.

We can regard our hands as analogous to the enantiomers of a chiral molecule. Let's consider the interaction of our hands with a symmetrical object such as a pair of tweezers. The tweezers have a plane of symmetry, and they can be used equally well by either a left-handed or right-handed person. The tweezers cannot be used to identify chiral objects. For example, a blindfolded person cannot use tweezers to distinguish right- and left-handed gloves. However, if you were blindfolded you could easily use your hands to distinguish right- and left-handed gloves. Our hands are "a chiral environment," and in this environment, mirror image gloves do not have identical properties. The right glove will fit comfortably only on the right hand. We can distinguish chiral objects only because we, ourselves, are chiral. In much the same way, only one of a pair of enantiomers fits into a specific site in a biological molecule such as an enzyme catalyst, because the enzyme is chiral. The binding of this enantiomer is said to be **stereospecific.**

EXAMPLE 6-1

Consider phenytoin, a compound used as an anticonvulsant drug. Is the molecule chiral or achiral? Determine your answer by identifying the number of different groups bonded to tetrahedral carbon atoms and by determining whether or not the molecule has a plane of symmetry.

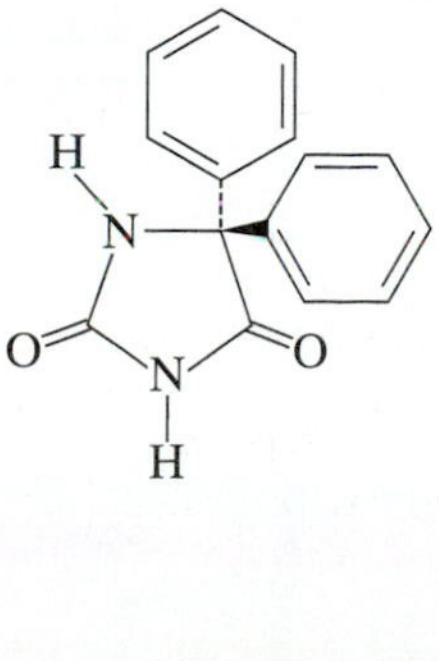

Solution

Phenytoin has only one tetrahedral carbon atom in the entire molecule! That carbon atom is bonded to a nitrogen atom, a carbonyl group, and two phenyl groups. Because the tetrahedral carbon atom is attached to two identical phenyl groups, it is not a stereogenic center, and as a result the molecule is achiral.

Phenytoin has a plane of symmetry that lies in the plane of the page. The phenyl groups of phenytoin lie above and below the symmetry plane. Note that the other atoms of phenytoin are bisected by this plane.

Problem 6.1

Consider the following structural formula for nicotine. Is the molecule chiral?

6.3 Optical Activity

Although enantiomers have many identical physical properties, they behave differently toward plane-polarized light. This one difference in physical properties is used to distinguish a chiral molecule from its enantiomer.

Light consists of waves oscillating in an infinite number of planes that lie at right angles to the direction of propagation of the light. When a beam of "ordinary" light passes through a *polarizing filter,* it is converted to a beam of **plane-polarized light** that oscillates in a single plane. We are familiar with this phenomenon in everyday life: plane-polarized light can be produced by Polaroid® sunglasses, which reduce glare by acting as a polarizing filter.

The interaction of plane-polarized light with chiral molecules can be measured by an instrument called a **polarimeter** (Figure 6.5). In a polarimeter, light having a single vibrational frequency—that is, monochromatic light—passes through a polarizing filter. The polarized light then traverses a tube containing a

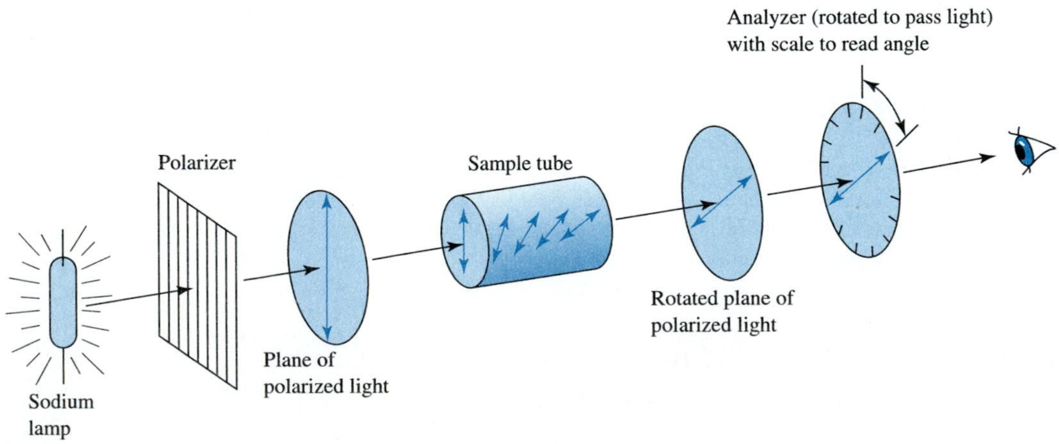

FIGURE 6.5
Representation of a Polarimeter.

6.4 Fischer Projection Formulas

Drawing molecules in three dimensions is difficult, especially for compounds containing several stereogenic carbon atoms (Section 6.6). The enantiomers of chiral substances are conveniently drawn in two dimensions by a convention proposed by the German chemist Emil Fischer over a century ago. Using this convention, the configurations of substances are all related to the configuration of a reference compound called glyceraldehyde.

The enantiomers of glyceraldehyde are written in two dimensions using a projection method proposed by Fischer (Figure 6.6). The carbonyl group, —CHO, the hydroxymethyl group, $—CH_2OH$, and the stereogenic carbon atom are arranged vertically with the most oxidized group, —CHO, at the "top." The stereogenic carbon atom is placed in the plane of the paper. In this arrangement, the —CHO group and the $—CH_2OH$ group extend behind the plane of the page, and the hydrogen atom and the hydroxyl group extend in front of the plane. The projection of these four groups into a plane gives a two-dimensional structure called the **Fischer projection formula.** The stereogenic carbon atom is usually not shown. It is located at the point where the bond lines cross.

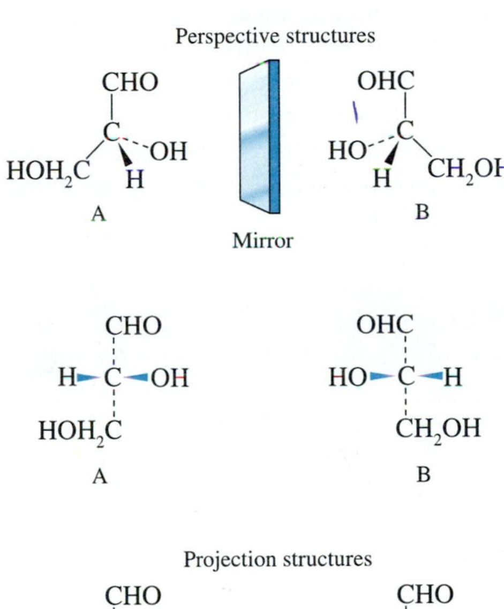

FIGURE 6.6

Projection Formula of Enantiomers of Glyceraldehyde

The Fischer projection formulas of the two enantiomers of glyceraldehyde have crossed lines at a point where the stereogenic carbon atom would be. The vertical lines are assumed to project away from the viewer. The horizontal lines project toward the viewer.

solution of the compound to be examined. During this passage, the plane of polarized light is rotated by an interaction with the chiral compound. After the plane-polarized light leaves the sample tube, it passes through a second polarizing filter called an analyzer. The analyzer is rotated either clockwise or counterclockwise to allow the light to escape. The angle of rotation of the analyzer equals the angle by which the light has been rotated by the chiral compound.

Because chiral molecules rotate plane-polarized light, they are called **optically active.** Achiral molecules do not rotate plane-polarized light and are optically inactive.

Specific Rotation

The amount of rotation observed in a polarimeter depends on the structure of the substance, the number of molecules encountered by the light, and the wavelength of the light. The optical activity of a substance is reported as its **specific rotation,** symbolized by $[\alpha]_D$. It is the number of degrees of rotation of a solution at a concentration of 1 g/mL in a tube 1 dm long. The standard conditions selected for these experiments are 25°C and the yellow light (D line, 589 nm) of the sodium vapor lamp.

If a chiral substance rotates plane-polarized light to the right, in a positive (+), clockwise direction, the substance is **dextrorotatory** (Latin, *dextro,* right). If a chiral substance rotates plane-polarized light to the left, in a negative (−), counterclockwise direction, the substance is **levorotatory** (Latin, *laevus,* left). The dextrorotatory and levorotatory isomers of a given substance rotate polarized light the same number of degrees but in opposite directions. Therefore, they are sometimes called **optical isomers.** The (+) isomer is sometimes called the **d form,** where *d* refers to dextrorotatory. The (−) isomer is called the **l form** where *l* refers to levorotatory. The specific rotations of some bioactive substances are listed in Table 6.1.

Table 6.1 Specific Rotations of Bioactive Compounds

Compound	$[\alpha]_D$
azidothymidine (AZT)	+99
cefotaxine (a cephalosporin)	+55
cholesterol	−31.5
cocaine	−16
codeine	−136
epinephrine (adrenaline)	−5.0
heroin	−107
levodopa	−13.1
monosodium glutamate (MSG)	−25.5
morphine	−132
oxacillin (a penicillin)	+201
progesterone (female sex hormone)	+172
sucrose (table sugar)	+66.5
testosterone (male sex hormone)	+109

It might appear that if one projection formula is lifted out of the plane and flipped over, we would obtain the formula of the enantiomer. However, if the Fischer projection formula of molecule A were flipped over, the carbonyl group and the hydroxymethyl group, originally behind the plane, would be in front of the plane. These groups would not occupy identical positions with respect to the carbonyl group and hydroxymethyl group of molecule B, which are behind the plane. Therefore, to avoid the error of apparently achieving a two-dimensional equivalence of nonequivalent three-dimensional molecules, it is important not to lift two-dimensional representations out of the plane of the paper.

Fischer projection formulas can be drawn to depict any chiral substance if we "know" the configuration at the stereogenic carbon atom. However, the configuration could not be determined in Fischer's era because there was no way to "see" the arrangement of the atoms in space. Therefore, Fischer arbitrarily assigned a configuration to one member of an enantiomeric pair of glyceraldehydes. The dextrorotatory enantiomer of glyceraldehyde, which rotates plane-polarized light in a clockwise direction (+13.5), was assigned to the Fischer projection with the hydroxyl group on the right side and was called **D-glyceraldehyde.** The mirror image compound, (−)-glyceraldehyde, was assigned a structure in which the hydroxyl group is on the left and was called **L-glyceraldehyde.** It rotates plane-polarized light in a counterclockwise direction.

```
   CHO              CHO
    |                |
H—C—OH          HO—C—H
    |                |
  CH2OH            CH2OH
```

D-glyceraldehyde $[\alpha]_D = +13.5°$ L-glyceraldehyde $[\alpha]_D = -13.5°$

Note that the designations D and L refer only to configuration at the stereogenic center; they do not indicate the direction of rotation. The direction of rotation of light depends on configuration and the identity of the atoms bonded to the stereogenic carbon atom (a subject that will not be discussed further). The configurations of other chiral compounds are determined by chemically relating them to either D- or L-glyceraldehyde. If D-(+)-glyceraldehyde is converted into lactic acid by converting the $—CH_2OH$ group into $—CH_3$ and the —CHO group into $—CO_2H$, the resulting lactic acid must have the D configuration (Figure 6.7). The lactic acid produced rotates light in a counterclockwise direction; it is levorotatory and is designated D-(−)-lactic acid. This one transformation illustrates the fact that the D and L notations refer to configuration of the stereogenic carbon atom, not the direction of rotation of plane-polarized light. Both (+)-glyceraldehyde and (−)-lactic acid are designated with a D configuration, because both have the —OH group on the right in the Fischer projection formula.

```
   CHO                            CO2H
    |        several steps          |
H—C—OH     ————————————→     H—C—OH
    |                               |
  CH2OH                            CH3
```

D-(+)-glyceraldehyde $[\alpha]_D = +13.5$ D-(−)-lactic acid $[\alpha]_D = -3.8$

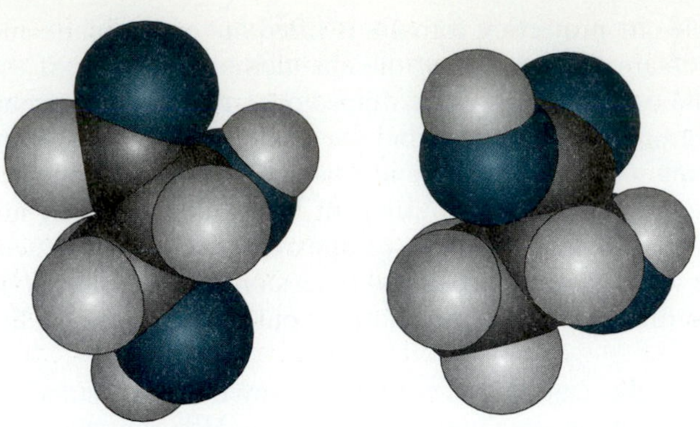

FIGURE 6.7
Relative Configuration of Molecules
Conversion of D-glyceraldehyde to D-lactic acid occurs without changes of any of the bonds to the stereogenic center. As a consequence the configuration is not changed.

The Fischer convention provides the **relative configuration** of molecules, meaning that the configuration is related to the arbitrary selection of the configuration of glyceraldehyde, the reference molecule. To determine the absolute configuration—the actual arrangement of atoms in space—we need to "see" the atoms. This can be done using X-ray crystallography. The absolute configuration of an optically active substance was determined in 1950. The arrangement of its atoms in space corresponds to the arrangement of atoms as related to the configuration of D-(+)-glyceraldehyde arbitrarily assigned by Fischer. As a result, all configurations deduced by using D-(+)-glyceraldehyde as the reference compound are also correct.

6.5 Absolute Configuration

The configuration of the stereogenic center of some molecules, such as amino acids and carbohydrates, can easily be compared to that of D-(+)-glyceraldehyde. But this procedure is not easily applied to molecules whose structures differ considerably from the reference compound. To circumvent this difficulty, R. S. Cahn, K. C. Ingold, and V. Prelog established a set of rules that define the absolute configuration of any stereogenic center. The configuration is designated by placing the symbol ***R*** or ***S*** within parentheses in front of the name of the compound.

R,*S* Configurations

The ***R*,*S* system** of configurational nomenclature for describing absolute configurations is related to the method we introduced in Section 4.3 to describe the location of groups in geometric isomers of alkenes. In the *R*,*S* system, the four groups bonded to each stereogenic carbon atom are arranged from highest to lowest priority. The highest priority group is assigned the number 1, the lowest priority group the number 4. Then the molecule is oriented so that the bond from the carbon atom to the group of lowest priority is arranged directly along our line of sight (Figure 6.8). When this has been done, the three highest priority groups point toward us and lie on the circumference of a circle. (It may help to imagine holding the lowest priority group in your hand like the stem of a flower as you then examine the petals.) Consider the path taken as we trace the groups ranked 1 to 3. In Figure 6.8 this direction is clockwise. Therefore, the configuration is designated *R* (Latin, *rectus,* right). If we trace a counterclockwise path from groups ranked 1 to 3, the configuration is designated *S* (Latin, *sinister,* left).

Priority Rules

The priority rules we defined in Section 4.3 for determining the configuration of geometric isomers also apply for chiral compounds.

1. *Atoms.* We rank the four *atoms* bonded to a stereogenic carbon atom in order of decreasing atomic number; the higher the atomic number, the higher the priority:

$$\mathrm{I > Br > Cl > F > O > N > C > {}^{2}H > {}^{1}H}$$

highest priority — lowest priority

   As indicated by the priority order ^{2}H (deuterium) > ^{1}H, isotopes are ranked in order of decreasing mass.

2. *Groups of atoms.* If a decision cannot be made using rule 1 because two or more of the directly attached atoms are the same, move along the chain until a point of first difference is encountered. Then apply rule 1. For example, the

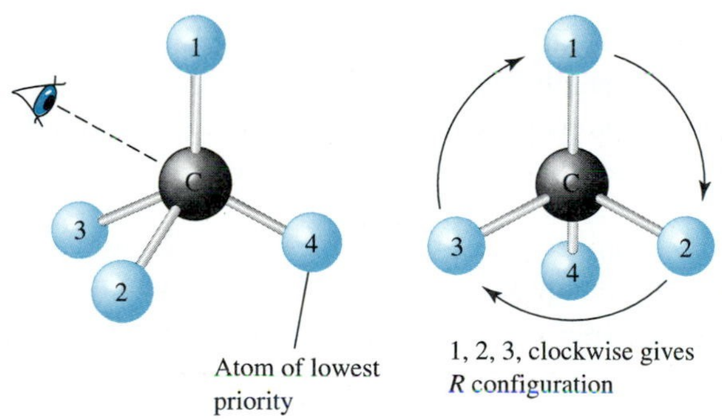

FIGURE 6.8
The Cahn-Ingold-Prelog System
Place the atom of lowest priority away from your eye and view the stereogenic center along the axis of the carbon bond to that atom. Imagine holding the atom of lowest priority as you move the molecule into an alternate position for viewing.

ethyl group has a higher priority than the methyl group because the next atom in the chain is a carbon atom in the ethyl group but a hydrogen atom in the methyl group. Using this rule, the priority of alkyl groups is $(CH_3)_3C—$ $> (CH_3)_2CH— > CH_3CH_2— > CH_3—$.

3. *Multiple bonds*. If a group contains a double bond, both atoms are doubled. That is, a double bond is counted as two single bonds to each of the atoms of the double bond. The same principle is used for a triple bond. Thus, the order $HC{\equiv}C— > CH_2{=}CH— > CH_3CH_2—$ is obtained. The priority order for common functional groups containing oxygen is $—CO_2H$ (carboxylic acid) > —CHO (aldehyde) > $—CH_2OH$ (alcohol).

Let's use the *R*,*S* system to determine the configuration of one of the enantiomers of alanine, an amino acid isolated from proteins. Alanine has one stereogenic carbon atom; it is bonded to a hydrogen atom, a methyl group, a carboxyl group, $—CO_2H$, and an amino group, $—NH_2$. Consider the perspective drawing of this enantiomer of alanine.

O, OH, C, CH_3, C, H, NH_2

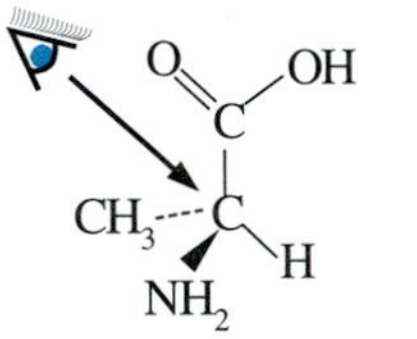

alanine

First, we rank the four groups attached to the stereogenic carbon atom in order of their priority from lowest to highest. The lowest priority (4) is given to the atom with the lowest atomic number *directly attached* to the stereogenic carbon atom; namely, hydrogen. The highest priority (1) is given to the atom with highest atomic number *directly attached* to the stereogenic carbon atom; namely, nitrogen. The stereogenic carbon atom is attached to two other carbon atoms: one is in a methyl group, and the other is a carboxyl group. The carboxyl group has the higher priority (2) because the carbon atom is attached to two oxygen atoms, whereas the carbon atom in the methyl group is attached only to hydrogen atoms; it has priority (3). Having assigned priorities, we next look into the molecule along the C—H bond and align the molecule so that the hydrogen atom points away from us. When this is done, we trace a path from the group with priority 1 to the group with priority 2, to the group with priority 3. The highest priority amino group lies at "4 o'clock", the next highest group ($—CO_2H$) is at the "12 o'clock" position, and the methyl group is at the "8 o'clock" position. Tracing a path from the amino group to the carboxyl group to the methyl group requires a counterclockwise motion. Therefore, the configuration of this enantiomer of alanine in proteins is *S*.

EXAMPLE 6-2

Arrange the groups at the stereogenic center of ethchlorvynol, a sedative-hypnotic, in order from low to high priority.

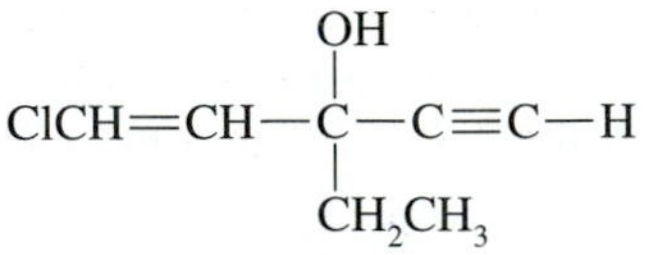

$$ClCH{=}CH{-}\underset{\displaystyle CH_2CH_3}{\overset{\displaystyle OH}{\underset{|}{\overset{|}{C}}}}{-}C{\equiv}C{-}H$$

Solution

The stereogenic center has three bonds to carbon atoms and one to an oxygen atom. Because oxygen has a higher atomic number than carbon, the —OH group has the highest priority. The priorities of the other three groups are determined by going to the next atom in the chain. In each case the next atom is carbon. However, there are different numbers of carbon-carbon bonds. The ethyl group has the lowest priority, followed by the —CH=CHCl group and the —C≡CH group in order of increasing priority. Thus, the groups around the stereogenic carbon atom increase in priority in the order $—CH_2CH_3 < —CH{=}CHCl < —C{\equiv}CH < —OH$.

Problem 6.2

Arrange the groups at the stereogenic center of baclofen, an antispastic, in order from low to high priority.

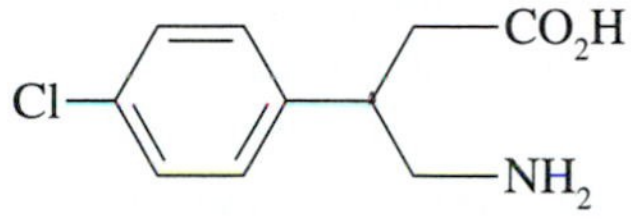

Cl—C_6H_4—CH(CH_2—CO_2H)(CH_2—NH_2)

EXAMPLE 6-3

Warfarin is an anticoagulant drug that is used both to treat thromboembolic disease and, in larger doses, as a rat poison. Assign its configuration.

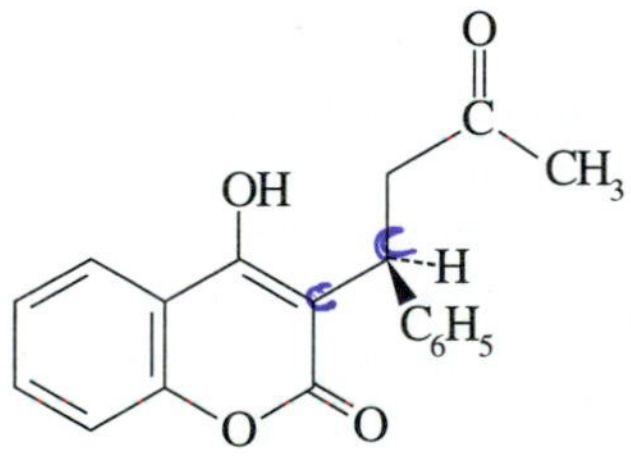

OH, O, C, CH_3, H, C_6H_5, O, O

Solution

Warfarin contains only one tetrahedral carbon atom that is attached to four different groups. That stereogenic carbon atom is attached to a hydrogen atom, a $—C_6H_5$ group (a phenyl group), and two other more complex groups. The lowest priority group is the hydrogen atom. The remaining three groups are linked through carbon atoms. One of them, the methylene group at the 12 o'clock position, has the next lowest priority (3) because it is attached to two hydrogen atoms. Next, we assign the priorities of the phenyl group and the ring system to the left. Both groups are attached to the stereogenic carbon atom by a carbon atom with a single and a double bond. Therefore, we must move to the next atom. When we do this in the complex ring, we find a carbon atom bonded to oxygen, which has a higher priority than the carbon atom bonded to hydrogen at the equivalent position in the phenyl ring. Therefore, the complex ring has a higher priority (1) than phenyl (2). Looking into the carbon-hydrogen bond at the stereogenic carbon atom, so that the hydrogen atom points away from us, we trace a counterclockwise path from group 1 to group 2 to group 3: this enantiomer of warfarin has the *S* configuration.

Problem 6.3

Assign the configuration of epinephrine, commonly known as adrenaline.

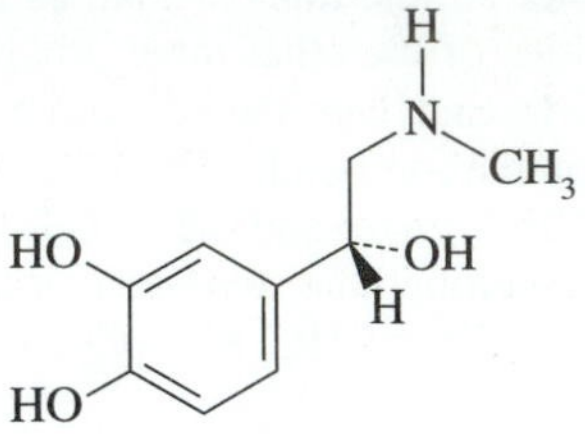

6.6 Multiple Stereogenic Centers

Many compounds contain several stereogenic centers. For example, the antibiotic erythromycin contains 18 stereogenic centers! How is the number of stereoisomers in a molecule with two or more stereogenic centers related to the number of stereogenic carbon atoms? The chirality of a molecule with two or more stereogenic centers depends on whether the centers are equivalent or nonequivalent. The term *nonequivalent* means that the stereogenic carbon atoms are not bonded to identical sets of substituents. We will first consider molecules with nonequivalent stereogenic centers. The number of optically active isomers for a molecule containing n nonequivalent stereogenic carbon atoms is 2^n. Consider 2,3,4-trihydroxybutanal.

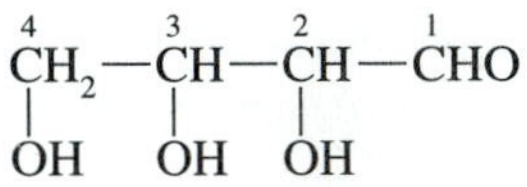

2,3,4-trihydroxybutanal

The C-2 and C-3 atoms are both stereogenic centers. They are also nonequivalent, because they are not bonded to identical sets of substituents. Therefore, the indicated structure has four stereoisomers (Figure 6.9).

Structures I and II are mirror images, are not superimposable on each other (they are chiral), and are enantiomers. This can be verified in two dimensions by imagining a mirror placed between I and II. Structures III and IV are also

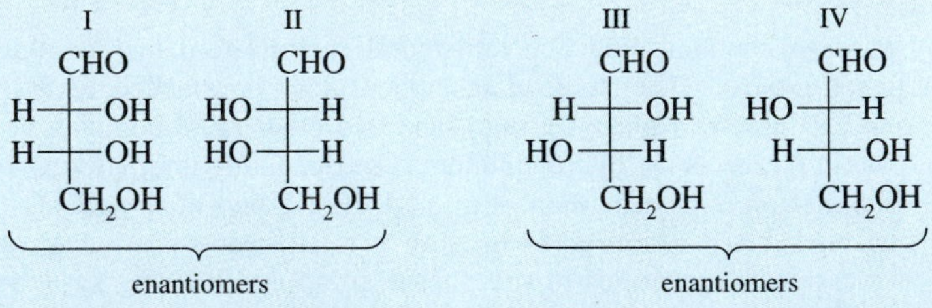

	I	II	III	IV
$[\alpha]_D$	−21.5	+21.5	−29.1	+29.1
Melting point	(liquid)	(liquid)	130 °C	130 °C
Solubility in ethanol	Very soluble	Very soluble	Slightly soluble	Slightly soluble

FIGURE 6.9

Enantiomers and Diastereomers

There are four stereoisomers of a compound containing two nonequivalent stereogenic centers. There are two sets of enantiomers. Any combination of two stereoisomers that are not enantiomers are diastereomers.

mirror images and are nonsuperimposable. Like all enantiomers, they rotate plane-polarized light in opposite directions but by the same absolute value.

Now consider structures I and III. These stereoisomers do not have a mirror-image relationship. Stereoisomers that are not enantiomers are called **diastereomers.** The pairs II and III, I and IV, and II and IV are also diastereomeric pairs. In contrast to enantiomers, which have the same chemical and physical properties, diastereomers have different chemical and physical properties. A comparison of some physical properties of enantiomers and diastereomers is given in Figure 6.9.

Nomenclature of Diastereomers

The name of a compound with two or more stereogenic centers must indicate the configuration of every center. The configuration of each stereogenic carbon atom is indicated by a number, which corresponds to its position in the carbon chain, and the letter *R* or *S*, separated by commas. Figure 6.9 shows the structures of the four stereoisomers of 2,3,4-trihydroxybutanal. Each structure has two stereogenic carbon atoms: C-2 and C-3. Each of these stereogenic carbon atoms can be *R* or *S*. Thus, the four possibilities are (2*R*,3*R*), (2*S*,3*S*), (2*S*,3*R*), and (2*R*,3*S*). The enantiomer of the (2*R*,3*R*) compound is the (2*S*,3*S*) isomer, which has the opposite configuration at each stereogenic center. Compounds whose configurations differ at only one of the two stereogenic centers are diastereomers. For example, the (2*R*,3*R*) compound is a diastereomer of the (2*S*,3*R*) isomer.

Achiral Diastereomers

Compounds that have two or more equivalent stereogenic centers, but are nevertheless achiral, are called **meso compounds** (Greek, *meso,* middle). Meso compounds are not optically active. Let's consider compounds with two equivalently substituted stereogenic centers, as in the tartaric acids given in Figure 6.10. In each structure, the C-2 and C-3 atoms are connected to four different groups. But instead of the four diastereomers that would exist if the stereogenic centers were nonequivalent, only three stereoisomers exist and one is optically inactive. The compounds labeled (2*S*,3*S*) and (2*R*,3*R*) are enantiomers; therefore, they are optically active. But look at the structures labeled (2*R*,3*S*) and (2*S*,3*R*). Although the structures are drawn as "mirror images," these mirror images are, in fact, superimposable and are identical. To show that this is so, rotate one structure 180° in the plane of the paper: the resulting structure superimposes on the original structure. Thus the two structures represent the same molecule, which is not optically active.

The structures labeled (2*R*,3*S*) and (2*S*,3*R*) have two equivalent stereogenic carbon atoms. Each of these structures has a plane of symmetry. We recall from Section 6.2 that a structure with a plane of symmetry is achiral and that it is superimposable on its mirror image. In the case of this tartaric acid, the plane of symmetry is between the C-2 and C-3 atoms, so that the top half of the molecule is the mirror image of the bottom half.

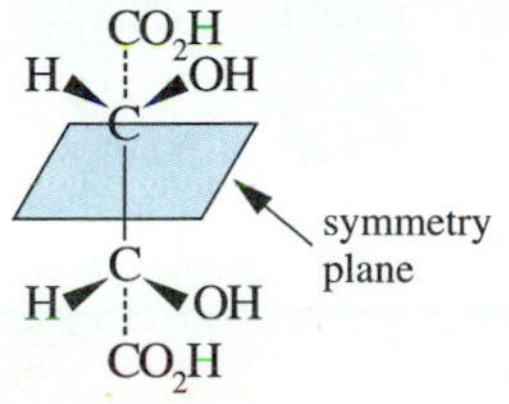

Metabolism May Vary by Species and within Species

The metabolism of drugs is often species dependent—a fact that must be considered because drugs are usually tested on animals prior to human trials. Even within the same species there are often strain differences, which are common among inbred test animals such as mice and rabbits.

Genetic differences in drug metabolism have been clearly established in humans. The differences between Northern Europeans, Eskimos, Mediterraneans, and Asians must be considered in prescribing certain drugs. For example, the antituberculosis agent isoniazid is metabolized at different rates by individuals with different genetic backgrounds. Eskimos metabolize the drug far faster than do Egyptians.

Drug metabolism also varies by sex. Some oxidative processes are controlled by sex hormones, particularly the androgens. This factor has become a matter of some concern recently, because it has been common practice to test drugs on men rather than on women. Metabolism in men is more easily studied because of smaller hormonal changes day to day. Also of concern is the possible effect of drugs on a very early fetus before a woman knows she's pregnant. However, recent protocols proposed by federal agencies stipulate that information about the metabolism of drugs in humans be determined on a more representative sample of the population that might use the drug.

The anticonvulsant phenytoin shows a dramatic difference in metabolism depending on species.

phenytoin

This achiral compound is oxidized to a chiral phenol. In humans, the hydroxylation occurs at the para position of one ring, and the compound has the *S* configuration. In dogs, the hydroxylation occurs at the meta position of the other ring, and the compound has the *R* configuration. (Molecular models of these two isomers are shown at the end of this discussion.)

(*S*)-(−)-*p*-hydroxyphenytoin (human)

(*R*)-(+)-*m*-hydroxyphenytoin (dog)

The processes by which metabolites are converted into water-soluble compounds to be excreted are also species dependent. For this reason, it is not advisable to feed animals certain foods that humans may enjoy. For example, the metabolites from chocolate can accumulate and poison a dog.

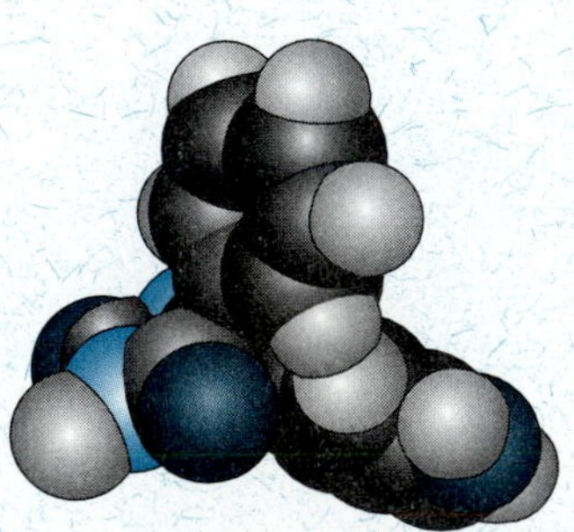

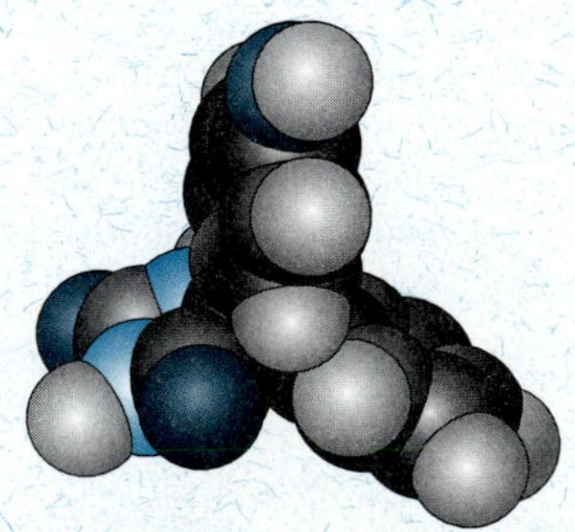

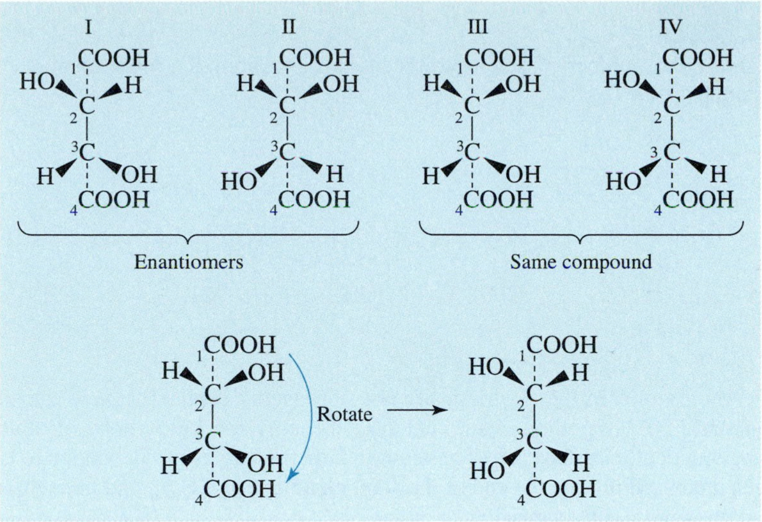

Enantiomers

Same compound

Rotate

FIGURE 6.10

Optically Active and Meso Stereoisomers

There are only three stereoisomers of a compound with two equivalent stereogenic centers. Two stereoisomers are enantiomers. The third compound has a plane of symmetry. It is optically inactive and is a meso compound.

EXAMPLE 6-4

Threonine, an amino acid isolated from proteins, has the following condensed molecular formula. Write the Fischer projections of the possible stereoisomers.

$$\overset{4}{C}H_3\overset{3}{C}H(OH)\overset{2}{C}H(NH_2)\overset{1}{C}O_2H$$

Solution

Both the C-2 and C-3 atoms are attached to four different substituents. Therefore, the compound has two stereogenic centers. Because the stereogenic centers are nonequivalent, four diastereomers are possible. The Fischer projections are written by placing the carboxyl group at the top of the vertical chain. The amino and hydroxyl groups may be on the right or left sides of the projection formulas.

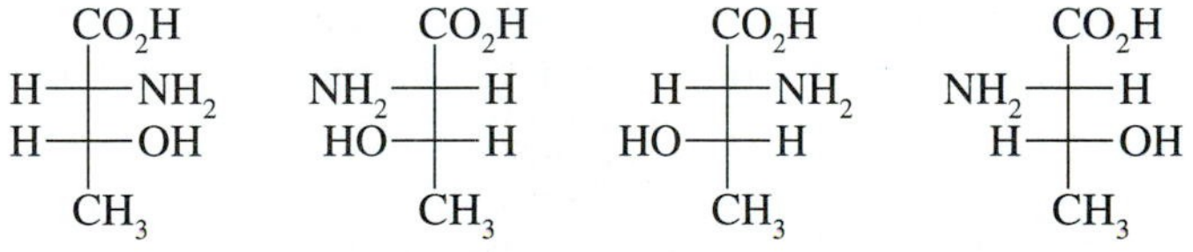

CO_2H / H—NH_2 / H—OH / CH_3

CO_2H / NH_2—H / HO—H / CH_3

CO_2H / H—NH_2 / HO—H / CH_3

CO_2H / NH_2—H / H—OH / CH_3

The structure of threonine isolated from proteins is given by the Fischer projection at the right.

Problem 6.4

Write the Fischer projection formulas of the stereoisomeric 2,3-dibromobutanes. What relationships should exist between the optical activities of these isomers?

EXAMPLE 6-5

Determine the number of stereogenic centers in vitamin K_1. How many stereoisomers are possible?

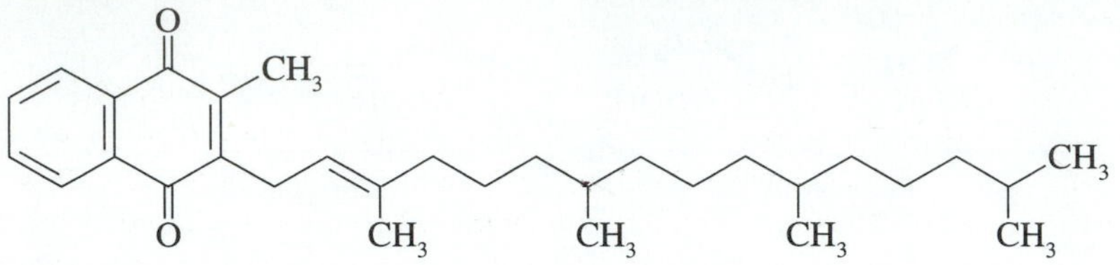

Solution

The carbon atoms in the two rings are not stereogenic centers because they are not tetrahedral. The long alkyl chain contains 10 methylene units, none of which is a stereogenic center because a carbon atom in a methylene group is bonded to two hydrogen atoms. Similarly, the end of the alkyl chain is $CH(CH_3)_2$, and this tertiary carbon atom is not a stereogenic center either.

Next, consider the positions in the middle of the alkyl chain that have methyl group branches. The methyl group on the left is bonded to a double-bonded carbon atom, which does not have four groups bonded to it: it is not a stereogenic center. The next two methyl groups are located on stereogenic centers. Because there are two stereogenic carbon atoms, $2^2 = 4$ isomers are possible.

Problem 6.5

Determine the number of stereogenic centers in nootkatone, found in grapefruit oil. How many stereoisomers are possible?

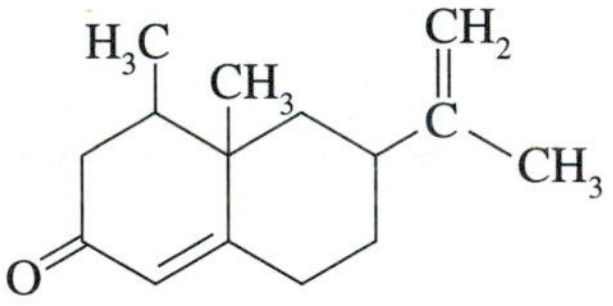

6.7 Synthesis of Stereoisomers

Reaction of an achiral reactant with an achiral reagent to produce a compound with a stereogenic center gives a 50:50 mixture of enantiomers called a **racemic mixture.** Consider the reduction of pyruvic acid with $NaBH_4$. The C-2 atom is a carbonyl carbon atom. The atoms directly bonded to the carbonyl carbon atom are arranged in a trigonal plane. Addition of hydrogen to the carbon atom can occur from either face of the molecular plane. Thus the tetrahedral carbon atom of the lactic acid formed can have two possible configurations.

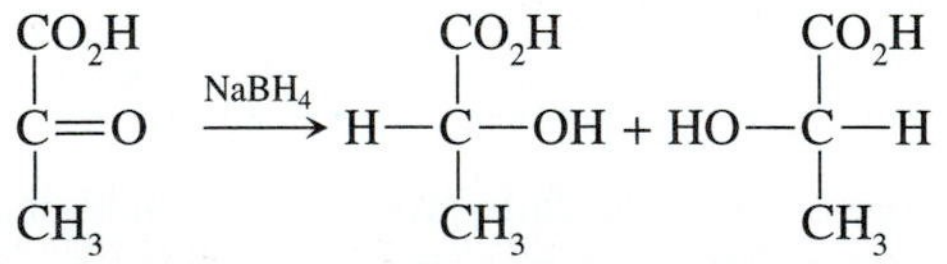

pyruvic acid $\xrightarrow{NaBH_4}$ (*R*)-lactic acid + (*S*)-lactic acid

racemic mixture of lactic acids

The individual lactic acid molecules produced are optically active. But a solution containing the products of the reaction is not optically active because the rotation of plane-polarized light by the (*R*)-lactic acid is canceled by the oppo-

Chirality and Your Senses

Our senses are sensitive to the configurations of molecules. Both the sense of taste and the sense of smell result from changes in specific sensory receptors when a receptor binds a specific molecule. The receptor's conformation changes, and this triggers a sequence of events that transmits a nerve impulse to the brain by sensory neurons. The brain interprets the input from sensory neurons as the "odor" of, say, spearmint.

Both enantiomeric and diastereomeric molecules interact differently in living systems. Differences in biological response to diastereomeric compounds are perhaps easier to understand. After all, diastereomeric compounds have different physical properties. Mannose, a carbohydrate, exists in two diastereomeric forms that differ in configuration at one center. The α form tastes sweet, but the β form tastes bitter.

α-D-mannose

β-D-mannose

Although the relationship of structure to the physiology of taste is not well understood, our response to sweetness is related to hydrogen bonds formed between a compound and our taste buds. Thus, the arrangement of the hydroxyl groups in diastereomeric carbohydrates such as mannose, glucose, and galactose (Chapter 11) determines their relative sweetness.

Sensory receptors associated with taste and smell can distinguish between some pairs of enantiomers. The specificity of response is similar to the relationship between our hands and how they fit into gloves. Because sensory receptors are chiral, they stereospecifically interact with only one of a pair of enantiomers.

The two enantiomeric forms of carvone have very different odors. (+)-Carvone is present in spearmint oil and has the odor of spearmint. In contrast, its enantiomer, (−)-carvone, is present in caraway seed and has the odor associated with rye bread. ■

(+)-carvone (spearmint)

(−)-carvone (caraway)

site optical rotation of (*S*)-lactic acid, which is formed in equal amounts. Note that there is a difference between a racemic mixture and a meso compound. A racemic mixture contains optically active components; the meso compound is a single achiral substance.

If we wish to synthesize a chiral product from an achiral reactant, the reaction must occur in a chiral environment. Protein catalysts called enzymes are examples of chiral reagents. Reduction of pyruvic acid using the liver enzyme lactate dehydrogenase yields exclusively (*S*)-lactic acid. The reducing agent for the reaction is nicotinamide adenine dinucleotide, NADH.

$$\underset{\text{pyruvic acid}}{CH_3-C(=O)-CO_2H} \xrightarrow{\text{NADH}} \underset{(S)\text{-lactic acid}}{HO-CH(CH_3)-CO_2H}$$

6.8 Reactions of Chiral Compounds

As noted in the previous section, chiral compounds produced in reactions catalyzed by enzymes have a single configuration. Enzymes can also distinguish between enantiomers and bind only one of the two isomers to cause a selective reaction. Enzymes themselves contain many stereogenic centers and have a "handedness." They will bind a particular molecule, but not its mirror image (Figure 6.11). For example, the enzyme tryptophan synthetase catalyzes the formation of the amino acid tryptophan from the amino acid serine and indole. Serine contains one stereogenic carbon atom, and can exist as the mirror image pair (*R*)- and (*S*)-serine. Of the two possible enantiomers, tryptophan synthetase binds only (*S*)-serine and produces only (*S*)-tryptophan.

$$\underset{\text{indole}}{C_8H_7N} + \underset{(S)\text{-serine}}{HOCH_2-CH(NH_2)-CO_2H} \xrightarrow{\text{tryptophan synthetase}} \underset{(S)\text{-tryptophan}}{C_8H_6N-CH_2-CH(NH_2)-CO_2H}$$

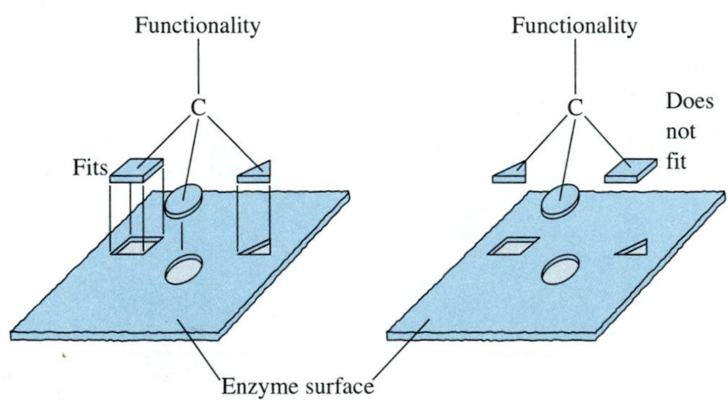

FIGURE 6.11
Stereospecificity of an Enzyme for an Enantiomer

The enantiomer represented on the left fits into the template created by the enzyme and the functionality is available for reaction. The enantiomer on the right does not fit the enzyme and its reactions are not catalyzed.

Reactions in the Chemistry Laboratory

We will see in subsequent chapters that the chirality of the products of reactions provides chemists with information about how bonds are broken and formed. Three types of reactions will be encountered:

1. Reactions that do not involve a stereogenic carbon atom;
2. Reactions that create a new stereogenic carbon atom;
3. Reactions that take place at a stereogenic carbon atom.

Any laboratory synthesis using chiral reactants will give chiral products if the bonds at the stereogenic center are not affected by the reagents. Formation of an ester (esterification) of (*R*)-lactic acid gives exclusively (*R*)-methyl lactate because the stereogenic carbon atom is not involved.

$$CH_3OH + H{-}\underset{\displaystyle CH_3}{\overset{\displaystyle CO_2H}{\underset{|}{\overset{|}{C}}}}{-}OH \longrightarrow H{-}\underset{\displaystyle CH_3}{\overset{\displaystyle CO_2CH_3}{\underset{|}{\overset{|}{C}}}}{-}OH + H_2O$$

(*R*)-lactic acid $[\alpha]_D = -3.8°$ (*R*)-methyl lactate $[\alpha]_D = +7.5°$

Reactions that break and make bonds at a stereogenic carbon atom will be the subject of the next chapter.

Explorations with Molecular Models

1. Prepare a model corresponding to the following structure of 2-bromobutane. Arrange the structure to view the stereogenic center by sighting with the carbon-hydrogen bond directed away from your eye. Determine whether the structure is *R* or *S*.

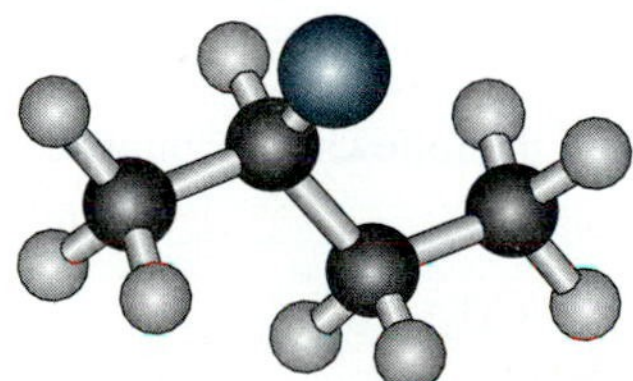

2. Prepare a model that is the mirror image of the structure of 2-bromobutane given in Exploration Exercise 1. Arrange the structure so that the carbon chain is vertical, with the C-1 atom at the top of the structure. Make sure that the hydrogen and bromine atoms bonded to the C-2 atom are directed toward you. Is this structure D or L?

3. Prepare the following model of 2,3-dibromobutane. Rotate the structure about the C-2 to C-3 bond until the bromine atoms are eclipsed. Does this structure have a plane of symmetry?

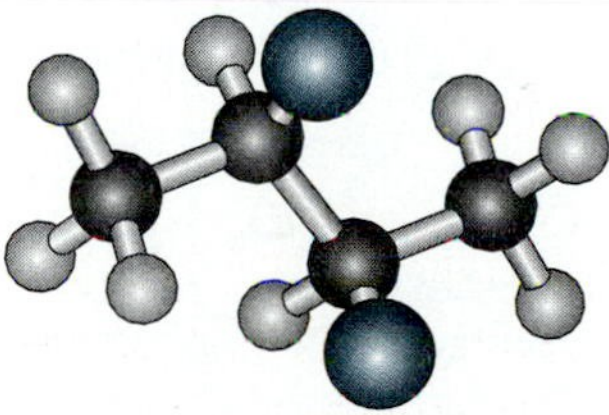

4. Interchange a hydrogen atom and bromine atom bonded to one of the carbon atoms of the model prepared for Exploration Exercise 3. Rotate the structure about the C-2 to C-3 bond until the bromine atoms are eclipsed. Does this structure have a plane of symmetry?

5. Prepare a model of *cis*-1,3-dibromocyclopentane. The structure has a plane of symmetry. Describe the atoms and/or bonds that are intersected by that plane.

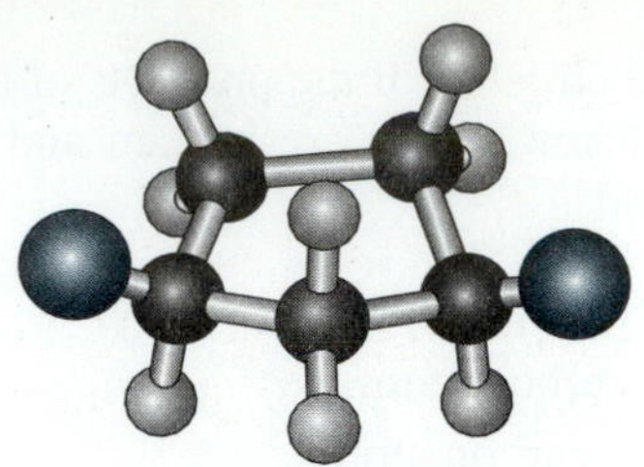

Exercises

Chirality

6.1 Which of the following isomeric methylheptanes has a stereogenic center?
(a) 2-methylheptane (b) 3-methylheptane (c) 4-methylheptane

6.2 Which of the following isomeric bromohexanes has a stereogenic center?
(a) 1-bromohexane (b) 2-bromohexane (c) 3-bromohexane

6.3 Which of the compounds with molecular formula $C_5H_{11}Cl$ has a stereogenic center?

6.4 Which of the compounds with molecular formula $C_3H_6Cl_2$ has a stereogenic center?

6.5 How many stereogenic centers does each of the following alkanes have?

(a) 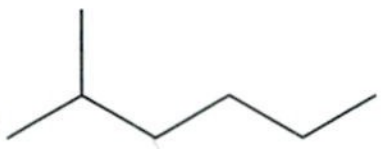 (b) 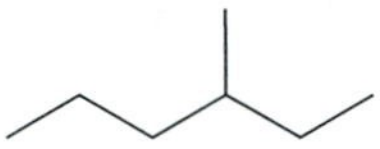 (c)

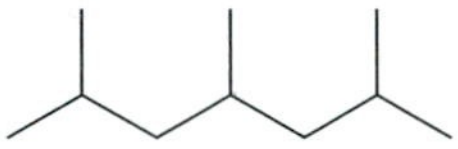

6.6 How many stereogenic centers does each of the following cyclic compounds have?

(a) 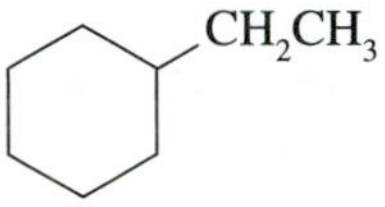 CH_2CH_3 (b) 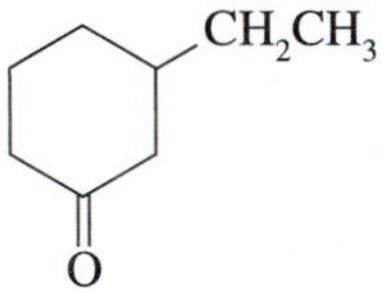 CH_2CH_3, O (c) 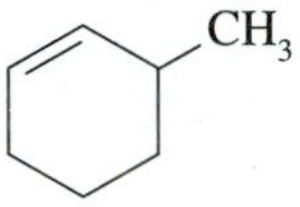 CH_3

6.7 How many stereogenic centers does each of the following barbiturates have?

(a) phenobarbital (b) secobarbital (c) hexobarbital

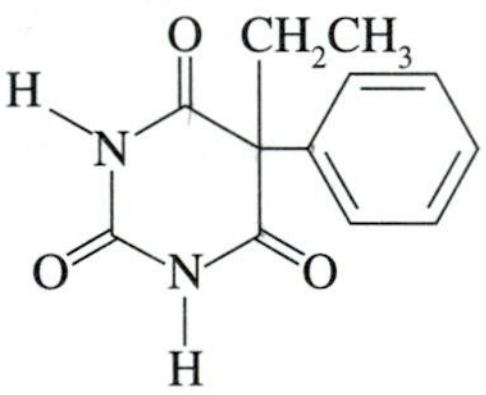

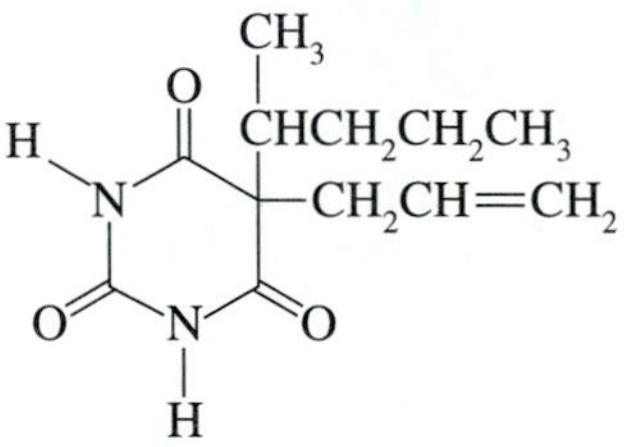

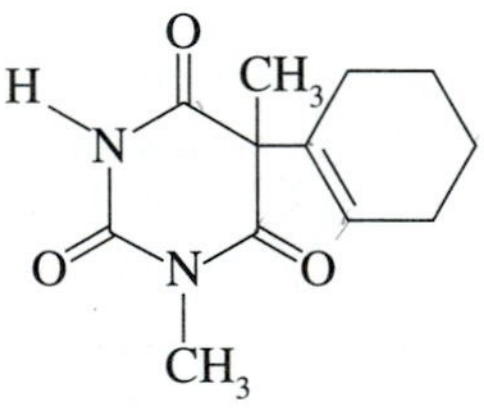

6.8 How many stereogenic centers does each of the following drugs have?

(a) ibuprofen, an analgesic (b) chloramphenicol, an antibiotic

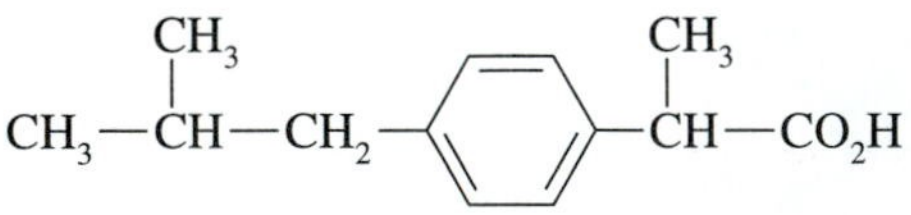

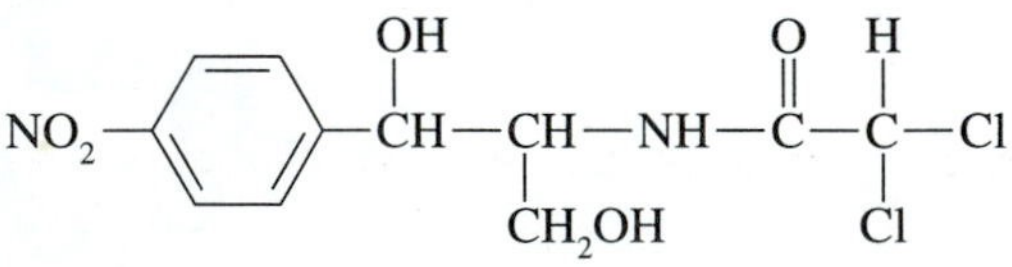

Priority Rules

6.9 Arrange the following groups in order of increasing priority.
(a) $-OH$, $-SH$, $-SCH_3$, $-OCH_3$ (b) $-CH_2Br$, $-CH_2Cl$, $-Cl$, $-Br$
(c) $-CH_2-CH{=}CH_2$, $-CH_2-O-CH_3$, $-CH_2-C{\equiv}CH$, $-C{\equiv}C-CH_3$
(d) $-CH_2CH_3$, $-CH_2OH$, $-CH_2CH_2Cl$, $-OCH_3$

6.10 Arrange the following groups in order of increasing priority.

(a) $-O-\overset{O}{\overset{\|}{C}}-CH_3$ $\quad -\overset{O}{\overset{\|}{C}}-CH_3$ $\quad -\overset{O}{\overset{\|}{C}}-OH$ (b) $-O-\overset{O}{\overset{\|}{C}}-CH_3$ $\quad -NH-\overset{O}{\overset{\|}{C}}-CH_3$ $\quad -\overset{O}{\overset{\|}{C}}-NH_2$

(c) $-S-\overset{O}{\overset{\|}{C}}-CH_3$ $\quad -O-\overset{O}{\overset{\|}{C}}-CH_2Br$ $\quad -\overset{O}{\overset{\|}{C}}-Cl$ (d) $-C{\equiv}CH$ $\quad -C{\equiv}N$ $\quad -N{\equiv}C$

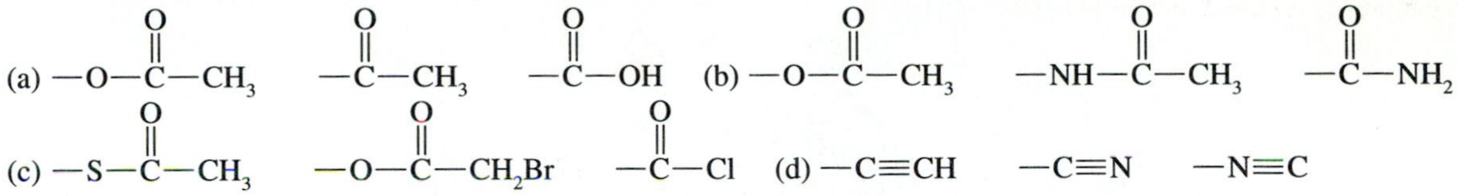

6.11 Consider the stereogenic center in the following drugs and arrange the groups in order from low to high priority.

(a) chlorphenesin carbamate, a muscle relaxant (b) mexiletine, an antiarrhythmic

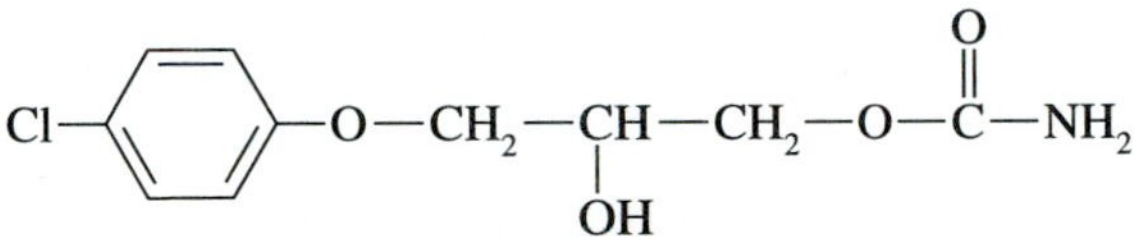

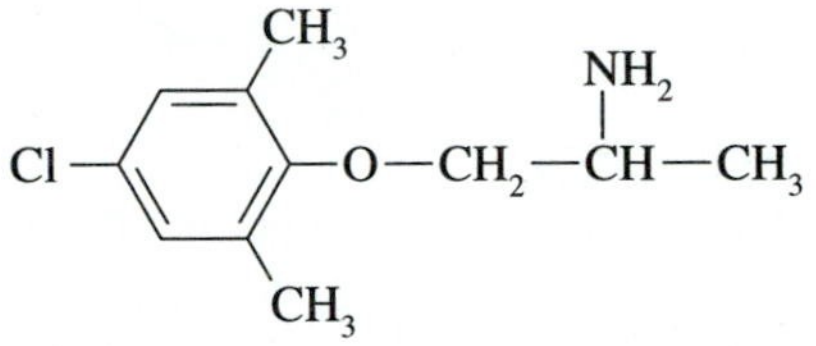

$Cl-C_6H_4-O-CH_2-\underset{OH}{\underset{|}{CH}}-CH_2-O-\overset{O}{\overset{\|}{C}}-NH_2$

$Cl-C_6H_2(CH_3)_2-O-CH_2-\overset{NH_2}{\overset{|}{CH}}-CH_3$

6.12 Consider the stereogenic center in the following drugs and arrange the groups in order of increasing priority.

(a) brompheniramine, an antihistamine (b) fluoxetine, an antidepressant

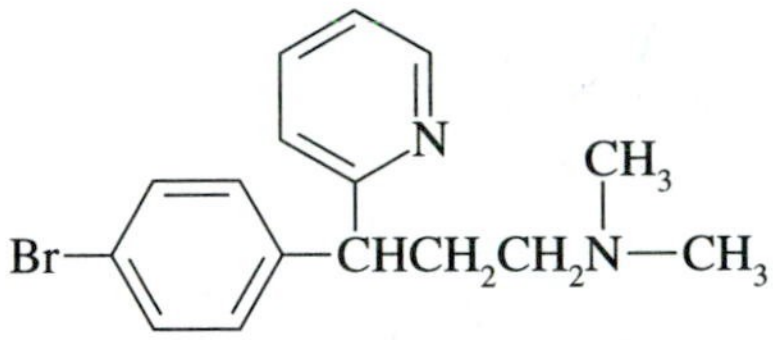

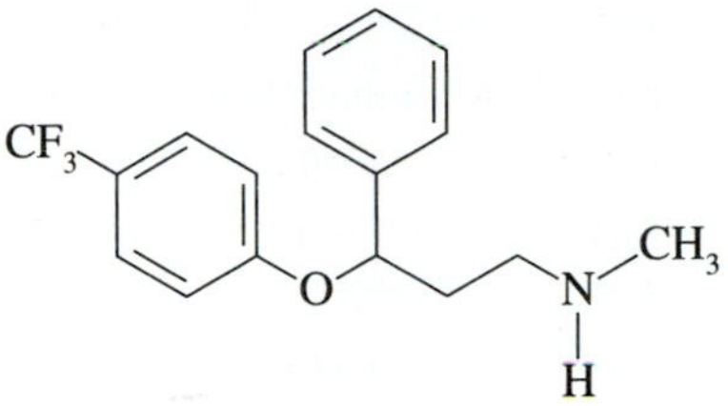

$Br-C_6H_4-\underset{(2\text{-pyridyl})}{CH}CH_2CH_2\overset{CH_3}{\overset{|}{N}}-CH_3$

$CF_3-C_6H_4-O-CH(C_6H_5)-CH_2CH_2-N(H)-CH_3$

R,S Configuration

6.13 Draw the structure of each of the following compounds.

(a) (*R*)-2-chloropentane (b) (*R*)-3-chloro-1-pentene (c) (*S*)-3-chloro-2-methylpentane

6.14 Draw the structure of each of the following compounds.

(a) (*S*)-2-bromo-2-phenylbutane (b) (*S*)-3-bromo-1-hexyne (c) (*R*)-2-bromo-2-chlorobutane

6.15 Assign the configuration of each of the following compounds.

(a) 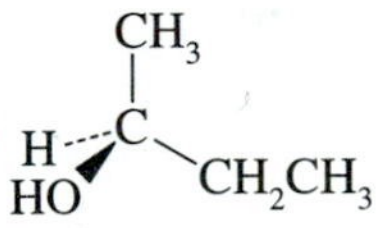 C bearing CH_3 (up), H (hashed), HO (wedged), CH_2CH_3
(b) C bearing CH_2CH_3 (up), H (hashed), Cl (wedged), CH_2CH_2Cl
(c) C bearing Br (up), H, CH_2CH_3 (hashed), CH_3 (wedged)

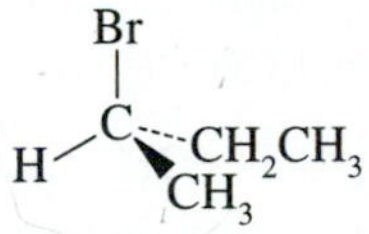

6.16 Assign the configuration of each of the following compounds.

(a) 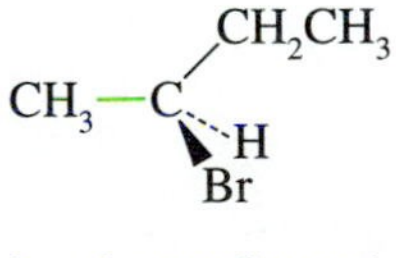 CH_3-C bearing CH_2CH_3, H (hashed), Br (wedged)
(b) CH_3-C bearing OCH_2CH_3, H (hashed), SCH_3 (wedged)
(c) C bearing CH_3, $-SCH_3$, H (hashed), Br (wedged)

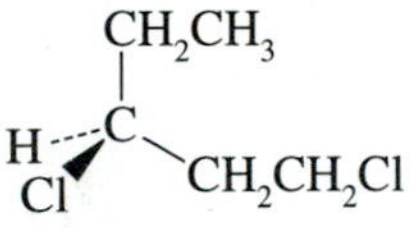

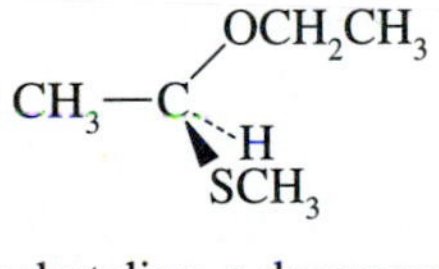

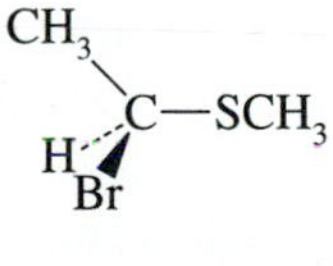

6.17 Assign the configuration of terbutaline, a drug used to treat bronchial asthma.

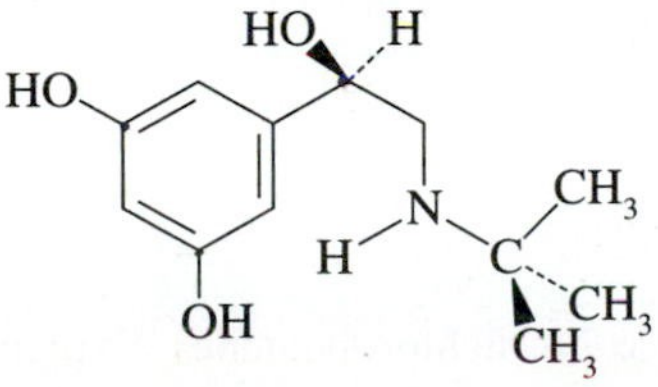

3,5-(HO)$_2C_6H_3$–C(HO)(H)–CH_2–N(H)–C(CH_3)$_3$

6.18 Assign the configuration of the following hydroxylated metabolite of diazepam, a sedative.

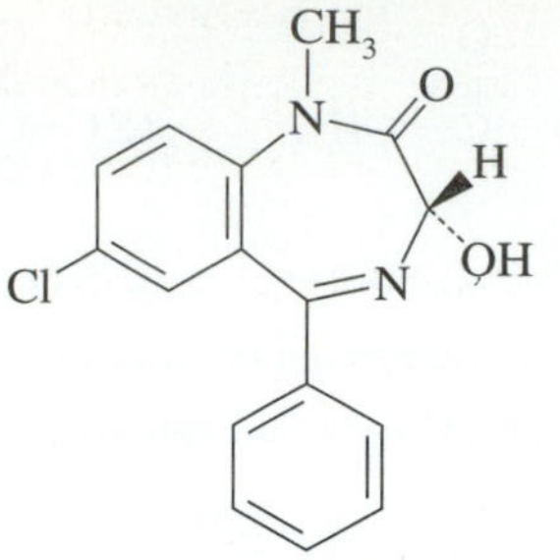

Optical Activity

6.19 The naturally occurring form of glucose has a specific rotation of +53. What is the specific rotation of its enantiomer?

6.20 The naturally occurring form of the amino acid threonine has a specific rotation of +28.3. What is the specific rotation of its enantiomer?

6.21 What do the various prefixes in (*R*)-(+)-glyceraldehyde and (*S*)-(−)-lactic acid mean?

6.22 (*R*)-(−)-lactic acid is converted into a methyl ester when it reacts with methanol. What is the configuration of the ester? Can you predict its sign of rotation?

6.23 Consider the following four projection formulas. Determine the two missing rotations.

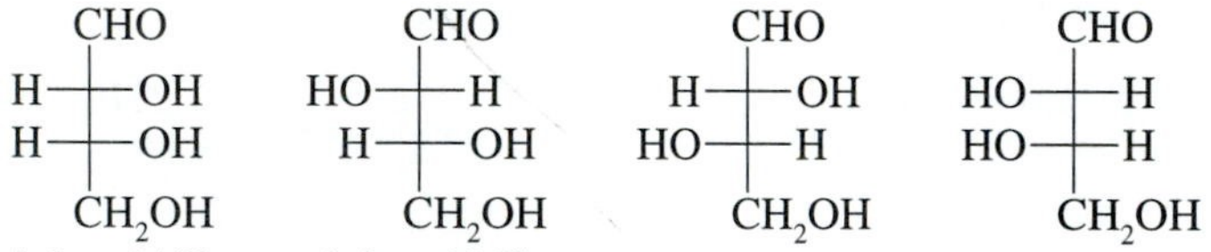

CHO	CHO	CHO	CHO
H—OH	HO—H	H—OH	HO—H
H—OH	H—OH	HO—H	HO—H
CH_2OH	CH_2OH	CH_2OH	CH_2OH
[α] = −14.8°	[α] = +19.6°		

6.24 What relationships between the rotations would exist for the following four compounds?

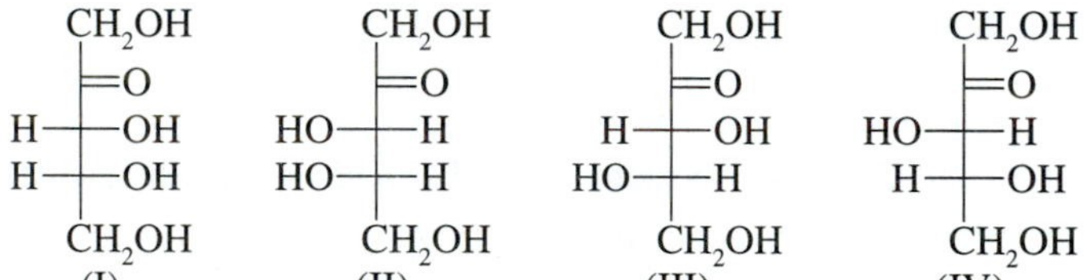

CH_2OH	CH_2OH	CH_2OH	CH_2OH
=O	=O	=O	=O
H—OH	HO—H	H—OH	HO—H
H—OH	HO—H	HO—H	H—OH
CH_2OH	CH_2OH	CH_2OH	CH_2OH
(I)	(II)	(III)	(IV)

Diastereomers

6.25 Consider the structure of 5-hydroxylysine, and determine the number of stereoisomers possible.

$NH_2CH_2CH(OH)CH_2CH_2CH(NH_2)—CO_2H$

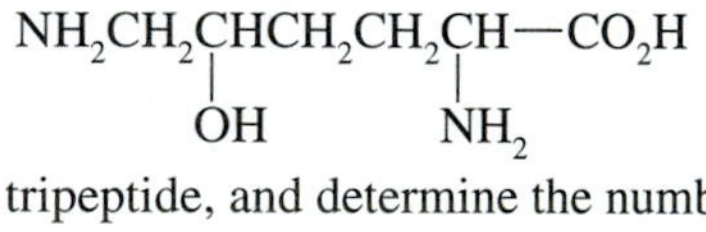

6.26 Consider the structure of the following tripeptide, and determine the number of stereoisomers possible.

$CH_3CH_2CH(CH_3)CH(NH_2)—C(=O)—NH—CH(CH_3)—C(=O)—NH—CH(CH(CH_3)_2)—C(=O)—OH$

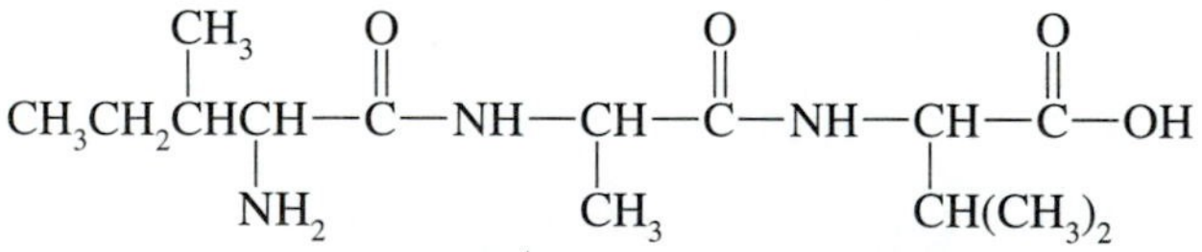

6.27 Ribose is optically active but ribitol, its reduction product, is optically inactive. Why?

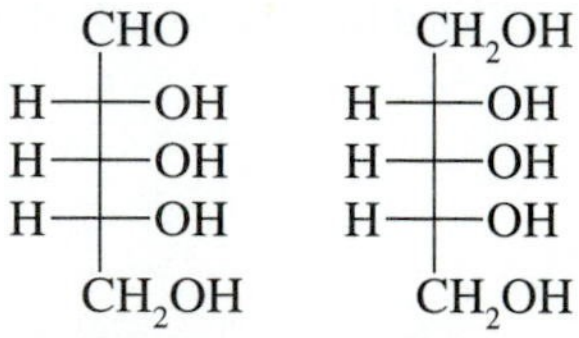

CHO	CH_2OH
H—OH	H—OH
H—OH	H—OH
H—OH	H—OH
CH_2OH	CH_2OH
ribose	ribitol

6.28 There are four isomeric 2,3-dichloropentanes but only three isomeric 2,4-dichloropentanes. Explain why.

Chemical Reactions

6.29 Addition of HBr to 1-butene yields a racemic mixture of 2-bromobutanes. Explain why.

6.30 Reduction of acetophenone with $NaBH_4$ produces a racemic mixture of 1-phenyl-1-ethanols. Explain why.

6.31 How many products are possible when HBr adds to the double bond of (*R*)-3-bromo-1-butene? Which are optically active?

6.32 How many products are possible when HBr adds to the double bond of 4-methylcyclohexene? Which are optically active?

Stereoisomers in Biochemistry

6.33 Natural glucose is a sugar that the body can metabolize. Suggest what would happen if one were to eat its enantiomer.

6.34 The mold *Penicillium glaucum* can metabolize one enantiomer of optically active tartaric acid. Explain what would happen if a racemic mixture of tartaric acid were "fed" to the mold.

6.35 Natural adrenaline is levorotatory. The enantiomer has about 5% of the biological activity of the natural compound. Explain why.

6.36 The following isomer of hydroxycitronellal has the odor of lily of the valley. Its mirror image has a minty odor. Explain why.

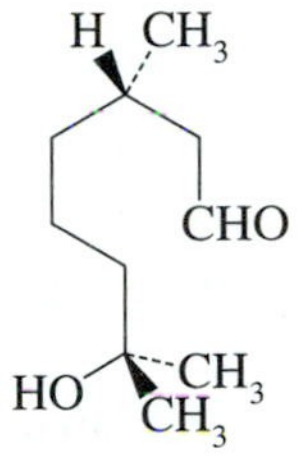

H CH_3

CHO

HO CH_3 CH_3

CHAPTER 7

Nucleophilic Substitution and Elimination Reactions

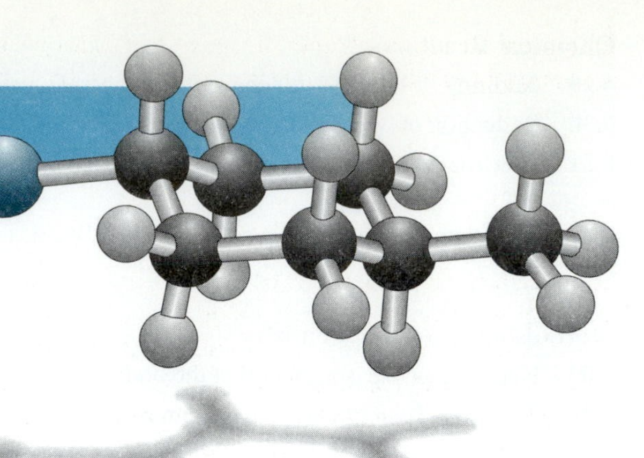

Overview

7.1 Reaction Mechanisms and Haloalkanes

The concept of functional groups and their role in the organization of the structures of organic molecules was introduced in Section 1.9. The use of reaction mechanisms as an organizational device to classify chemical reactions was described in Section 2.9. The details of the electrophilic addition reactions of alkenes (Section 4.9) and electrophilic substitution reactions of aromatic compounds (Section 5.5) are examples of two important reaction mechanisms. In this chapter we examine two additional types of reactions mechanisms—nucleophilic substitution and elimination reactions. These two mechanisms often occur in competition with one another and are used to describe the reactions of several classes of compounds, such as haloalkanes (also called alkyl halides) and alcohols. In this chapter we focus on the reactions of haloalkanes to outline the principles required to understand the role of structure in determining the degree to which the two reaction mechanisms compete.

Reactivity of Haloalkanes

Haloalkanes have a halogen atom bonded to an sp^3-hybridized carbon atom. As a result of the greater electronegativity of the halogens, the carbon atom of the carbon-halogen bond bears a partial positive charge and the halogen atom has a partial negative charge.

$$\overset{\delta+}{\text{C}}-\overset{\delta-}{\text{X}} \qquad \text{where X} = \text{F, Cl, Br, I}$$

As a result, a haloalkane has two sites of reactivity. One is at the carbon atom bonded to the halogen atom. This carbon atom is electropositive and reacts with nucleophiles. The second site of reactivity in a haloalkane is the hydrogen atom bonded to the carbon atom adjacent to the carbon atom bonded to the halogen atom. This hydrogen atom is more acidic than the hydrogen atoms in alkanes because the halogen atom on the adjacent carbon atom withdraws electron density by an inductive effect.

$$-\overset{|}{\underset{\uparrow \atop H^{\delta+}}{C}}\rightarrow\overset{|}{\underset{|}{C}}{}^{\delta+}\rightarrow X^{\delta-}$$

First, let's consider the reaction of the hydroxide ion with a carbon atom bonded to a halogen atom. Hydroxide ion can displace the halide ion in a substitution reaction to produce an alcohol. However, the hydroxide ion is not only a nucleophile but also a strong base that can remove a proton from the carbon atom adjacent to the one bonded to the halogen atom. Abstraction of a proton and the departure of a halide ion in an elimination reaction give an alkene.

$$H-\ddot{\underset{..}{O}}:^{-} \quad -\underset{H}{\overset{|}{\underset{|}{C}}}-\overset{|}{\underset{|}{C}}-\ddot{\underset{..}{Br}}: \longrightarrow -\underset{H}{\overset{|}{\underset{|}{C}}}-\overset{|}{\underset{|}{C}}-\ddot{\underset{..}{O}}-H + :\ddot{\underset{..}{Br}}:^{-}$$

$$H-\ddot{\underset{..}{O}}:^{-} \quad -\underset{H}{\overset{|}{\underset{|}{C}}}-\overset{|}{\underset{|}{C}}-\ddot{\underset{..}{Br}}: \longrightarrow \;\rangle C=C\langle\; + H_2O + :\ddot{\underset{..}{Br}}:^{-}$$

The substitution and elimination reactions usually occur concurrently, and mixtures of products result. In the following sections, we will learn how one reaction may be favored over the other.

Nomenclature of Haloalkanes

Low molecular weight haloalkanes are often named using the name of the alkyl group followed by the name of the halide.

CH_3-CH_2-Br — ethyl bromide

$CH_3-CH(CH_3)-F$ — isopropyl fluoride

$CH_3-C(CH_3)_2-Cl$ — *tert*-butyl chloride

Allyl and benzyl halides also have a halogen atom bonded to an sp^3-hybridized carbon atom and have reactivities similar to those of haloalkanes.

an allyl halide (Cl) — a benzyl halide (Br)

Compact disks are made of a polymer of vinyl chloride.

The IUPAC rules for naming haloalkanes were decribed in Section 3.10. The halogen atoms have a lower priority than all other functional groups, such as double and triple bonds. Thus, the double bond in 4-chlorocyclohexene is used to number the cyclohexene ring, which then determines the number of the carbon atom bonded to the chlorine atom.

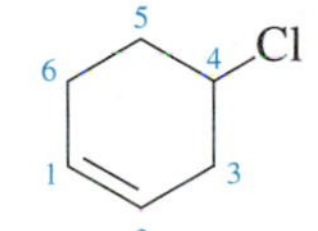

4-chlorocyclohexene

EXAMPLE 7-1

(*E*)-8-Bromo-3,7-dichloro-2,6-dimethyl-1,5-octadiene is produced by a species of red algae. Draw its structure.

Solution

Draw the eight-carbon-atom parent chain, and select a direction for numbering it. Place double bonds between the C-1 and C-2 atoms and between the C-5 and C-6 atoms.

$$\underset{1}{C}{=}\underset{2}{C}{-}\underset{3}{C}{-}\underset{4}{C}{-}\underset{5}{C}{=}\underset{6}{C}{-}\underset{7}{C}{-}\underset{8}{C}$$

Next, place a bromine atom at the C-8 atom, chlorine atoms at the C-3 and C-7 atoms, and methyl groups at the C-2 and C-6 atoms.

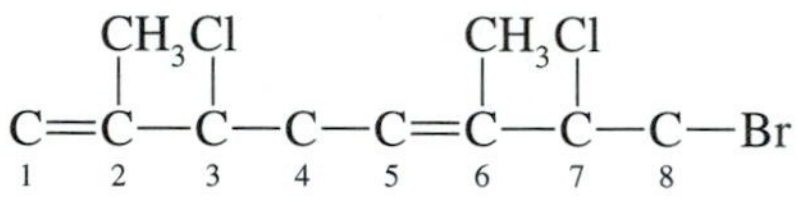

$$\underset{1}{C}{=}\underset{2}{\overset{CH_3}{C}}{-}\underset{3}{\overset{Cl}{C}}{-}\underset{4}{C}{-}\underset{5}{C}{=}\underset{6}{\overset{CH_3}{C}}{-}\underset{7}{\overset{Cl}{C}}{-}\underset{8}{C}{-}Br$$

Finally, arrange the groups about the double bond between the C-5 and C-6 atoms to give the *E* configuration. The higher priority groups at both the C-5 and C-6 atoms are the alkyl groups that are part of the parent carbon chain. Indicate the (*E*) configuration by placing the atoms of the carbon chain on the opposite sides of the double bond. Finally, fill in the requisite hydrogen atoms.

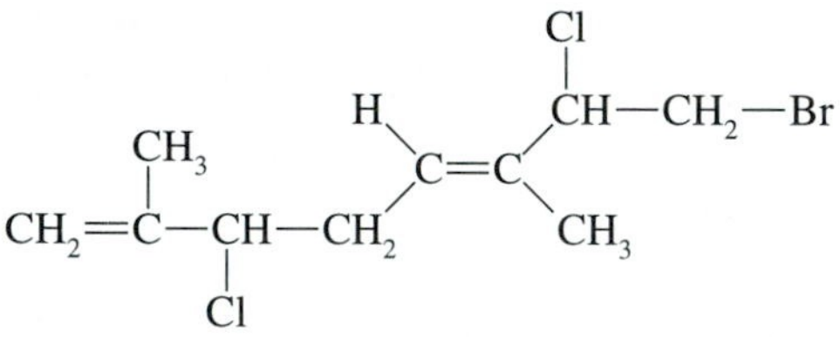

$$CH_2{=}\overset{CH_3}{C}{-}\underset{Cl}{CH}{-}CH_2{-}C(H){=}C(CH_3){-}\overset{Cl}{CH}{-}CH_2{-}Br$$

Problem 7.1

Name each of the following compounds.

(a) 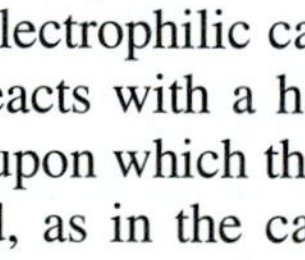 cyclohexane ring with CH_3 (wedge) and F (dash) on adjacent carbons

(b) 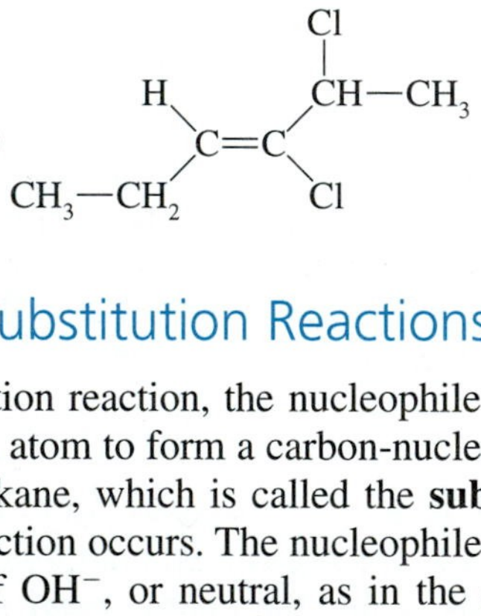 $CH_3{-}CH_2(H)C{=}C(Cl){-}\overset{Cl}{CH}{-}CH_3$

(c) benzene ring with CH_2Br and Cl on adjacent carbons

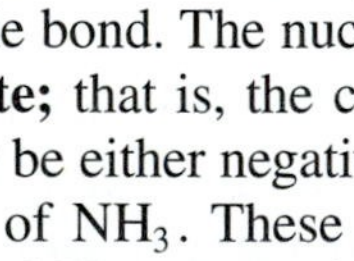

7.2 Nucleophilic Substitution Reactions

In a nucleophilic substitution reaction, the nucleophile donates an electron pair to the electrophilic carbon atom to form a carbon-nucleophile bond. The nucleophile reacts with a haloalkane, which is called the **substrate;** that is, the compound upon which the reaction occurs. The nucleophile may be either negatively charged, as in the case of OH^-, or neutral, as in the case of NH_3. These two types of nucleophiles are commonly represented as $Nu{:}^-$ and Nu:, respectively. If the nucleophile is negatively charged, the product has no net charge. If the nucleophile is neutral, the product is positively charged.

$$Nu{:}^- + R{-}X \longrightarrow R{-}Nu + X{:}^-$$

$$Nu{:} + R{-}X \longrightarrow R{-}Nu^+ + X{:}^-$$

The group displaced by the nucleophile is called the **leaving group.** It has an electron pair that was originally in the C—X bond. The leaving group may

Halogen Compounds in Ocean Organisms

Many pharmaceuticals in use today were developed initially from naturally occurring compounds found in terrestrial plants. It has been only recently that the chemistry of the organisms of the ocean has been examined in the search for new compounds that might be useful in the design of drugs. Given the diversity of species in the ocean, there may be a wealth of potentially useful compounds to be discovered.

As in terrestrial plants and animals, most of the heterocyclic compounds in the ocean have five- and six-membered rings. In contrast to compounds obtained from terrestrial sources, a high percentage of heterocyclic compounds from ocean organisms are often halogenated. For example, a Bahamian sponge produces several chlorinated indoles or indole-derived compounds, such as batzelline.

batzelline

In addition to sponges, halogen-containing compounds occur widely in other marine organisms, such as algae and mollusks. The compounds made by these species have unusual structures, and some of them are clinically useful as antimicrobial, antifungal, and antitumor agents.

The question of why various marine species produce these halogen-containing compounds, many of which are cytotoxic, is an interesting one. The answer may be that they are used in "chemical warfare" in an environment where every organism must find a food source. On the coral reef, many invertebrates are limited to a confined space in which to live and eat. Organisms that are essentially stationary or even those that move, albeit slowly, apparently use halogen-containing compounds as part of a chemical defense mechanism to avoid predators and make survival easier. Thus, they can better survive in an environment where virtually every organism is simultaneously predator and prey. For example, red algae produce halogen compounds as a chemical shield that repels most herbivores. However, the sea hare, a soft-bodied, shell-less mollusk, is not repelled by compounds from the red algae, which are a source of food for the sea hare.

(from red algae) (from sea hare)

Most mollusks are protected from predators by a hard shell, so the shell-less sea hare might seem to have little prospect of survival in the face of large carnivores. However, the sea hare converts the halogen-containing compounds in red algae into closely related substances and secretes them in a mucous coating. This coating protects its soft body against carnivorous fish, which are repelled by the compounds. ■

be negatively charged, as in the case of halide ions. Some neutral leaving groups, such as H_2O, will be examined in later chapters.

Haloalkanes react with nucleophilic anions derived from the halogens, oxygen, sulfur, and even carbon. Haloalkanes also react with neutral nucleophiles that contain nitrogen, such as NH_3 or amines (Chapter 14). At this point let's consider some simple examples of nucleophilic substitution reactions using bromomethane as the substrate. Haloalkanes can react with a halide anion, as in the case of the nucleophilic substitution of iodide ion for chloride or bromide ion.

$$:\ddot{\underset{..}{I}}:^- \; CH_3{-}\ddot{\underset{..}{Br}}: \longrightarrow CH_3{-}\ddot{\underset{..}{I}}: + :\ddot{\underset{..}{Br}}:^-$$

A similar reaction occurs when the hydroxide ion replaces the halide ion to produce an alcohol. When the oxygen-containing nucleophile is an alkoxide ion (RO^-), the product is an ether. These two reactions will be discussed in Chapters 8 and 9.

$$H{-}\ddot{\underset{..}{O}}:^- \; CH_3{-}\ddot{\underset{..}{Br}}: \longrightarrow CH_3{-}\ddot{\underset{..}{O}}{-}H + :\ddot{\underset{..}{Br}}:^-$$

an alcohol

$$CH_3CH_2{-}\ddot{\underset{..}{O}}:^- \; CH_3{-}\ddot{\underset{..}{Br}}: \longrightarrow CH_3{-}\ddot{\underset{..}{O}}{-}CH_2CH_3 + :\ddot{\underset{..}{Br}}:^-$$

an ether

Nucleophilic substitution reactions by sulfur-containing nucleophiles, such as hydrogen sulfide ion, HS^-, and thiolate ions, RS^-, also occur. These reactions yield sulfur analogs of alcohols and ethers—namely, thiols and thioethers (Chapter 8).

$$H{-}\ddot{\underset{..}{S}}:^- \; CH_3{-}\ddot{\underset{..}{Br}}: \longrightarrow CH_3{-}\ddot{\underset{..}{S}}{-}H + :\ddot{\underset{..}{Br}}:^-$$

a thiol

$$CH_3CH_2{-}\ddot{\underset{..}{S}}:^- \; CH_3{-}\ddot{\underset{..}{Br}}: \longrightarrow CH_3{-}\ddot{\underset{..}{S}}{-}CH_2CH_3 + :\ddot{\underset{..}{Br}}:^-$$

a thioether

Haloalkanes also react with carbon-containing nucleophiles to form carbon-carbon bonds, which increase the length of the carbon chain. The cyanide ion, CN^-, is one carbon-containing nucleophile. It produces a nitrile with the formula RCN. Nitriles can be transformed into carboxylic acids (Chapter 12) and amines (Chapter 14). Carbon-containing nucleophiles derived from alkynes are called **alkynide** ions. These nucleophiles, the conjugate bases of alkynes (Chapter 4), react to form alkynes containing the carbon atoms of both the haloalkane and the alkynide.

$$:N{\equiv}C:^- \; CH_3{-}\ddot{\underset{..}{Br}}: \longrightarrow CH_3{-}C{\equiv}N: + :\ddot{\underset{..}{Br}}:^-$$

cyanide ion — a nitrile

$$R{-}C{\equiv}C:^- \; CH_3{-}\ddot{\underset{..}{Br}}: \longrightarrow CH_3{-}C{\equiv}C{-}R + :\ddot{\underset{..}{Br}}:^-$$

an alkynide ion — an alkyne

EXAMPLE 7-2

Using compounds containing no more than three carbon atoms, propose two ways to prepare the thioether $CH_3CH_2{-}S{-}CH_2CH_2CH_3$.

Solution

A thioether can be prepared by reaction of a thiolate ion with a haloalkane. This thioether has two different alkyl groups bonded to sulfur. One alkyl group can be bonded to the sulfur atom in the thiolate ion and the other in the haloalkane. Thus, two possible combinations of reactants can yield the product.

$$\underset{\text{ethylthiolate}}{CH_3CH_2{-}S^-} + \underset{\text{1-bromopropane}}{Br{-}CH_2CH_2CH_3} \longrightarrow CH_3CH_2{-}S{-}CH_2CH_2CH_3$$

$$\underset{\text{bromoethane}}{CH_3CH_2{-}Br} + \underset{\text{propylthiolate}}{{}^-S{-}CH_2CH_2CH_3} \longrightarrow CH_3CH_2{-}S{-}CH_2CH_2CH_3$$

Problem 7.2

Propose two ways to prepare 2-pentyne using nucleophilic substitution reactions.

7.3 Nucleophilicity versus Basicity

Haloalkanes undergo substitution reactions, in which a halide ion is displaced by a nucleophile, or elimination reactions, in which a halide ion and a hydrogen ion are removed from adjacent carbon atoms to give an alkene. The type of reaction that occurs depends on two properties of the nucleophile, termed nucleophilicity and basicity.

The property of the nucleophile that gives it the ability to displace the leaving group is called **nucleophilicity.** These same nucleophiles that supply an electron pair to carbon are also bases. In an elimination reaction, a nucleophile acts as a base and abstracts a proton from the carbon atom adjacent to the one bonded to the halogen atom. Therefore, the elimination reaction depends upon the Brønsted basicity of the nucleophile. Hence, the terms *nucleophilicity* and *basicity* describe different phenomena. Nucleophilicity affects the rate of a substitution reaction at a carbon center. Basicity affects the equilibrium constant for an acid-base reaction between the hydrogen atom of the haloalkane and the nucleophile in an elimination reaction.

The nucleophilicities and basicities of a series of structurally or chemically related nucleophiles such as halide ions, oxygen-containing anions, and sulfur-containing anions are not always related in a simple way. However, trends are evident based on periodic properties of the elements. The relative rates of the reaction of various nucleophiles with iodomethane are given in Table 7.1.

Table 7.1 Relative Rates of Reaction of Nucleophiles with Iodomethane

Nucleophile	Relative rate
CH_3OH	1
NO_3^-	30
F^-	5×10^2
SO_4^{2-}	3×10^3
$CH_3CO_2^-$	2×10^4
Cl^-	2.5×10^4
NH_3	3.2×10^5
N_3^-	6×10^5
Br^-	6×10^5
CH_3O^-	2×10^6
CN^-	5×10^6
I^-	2.5×10^7
CH_3S^-	1×10^9

Trends within a Period

For a group of nucleophilic ions derived from elements in the same period of the periodic table, the nucleophilicity and basicity parallel each other and decrease from left to right in the period. Thus, the methoxide ion is both more basic and more nucleophilic than the fluoride ion (Table 7.1). The oxygen atom of the methoxide ion is less electronegative and holds its electrons less firmly than the fluoride ion. As a consequence, the nonbonding electrons of the oxygen atom can be more easily donated to carbon in a nucleophilic substitution reaction. The periodic trend of nucleophilicities parallels that of the basicities for organic anions.

RCH_2^-	RNH^-	RO^-	F^-
carbanion	amide ion	alkoxide ion	fluoride ion

← increasing basicity

← increasing nucleophilicity

Trends within a Group

For a group of nucleophiles containing atoms in the same group of the periodic table, the order of nucleophilicity runs opposite to the order of basicity. First, consider the nucleophilicities of thiolate and alkoxide ions. Thiolate is more nucleophilic but less basic than alkoxide ion (Table 7.1).

$R—\ddot{S}:^-$	$R—\ddot{O}:^-$
more nucleophilic less basic	less nucleophilic more basic

Similarly, when we compare the nucleophilicities and basicities of the halides, we find that the least basic one, iodide, is the most nucleophilic, whereas the most basic one, fluoride, is the least nucleophilic.

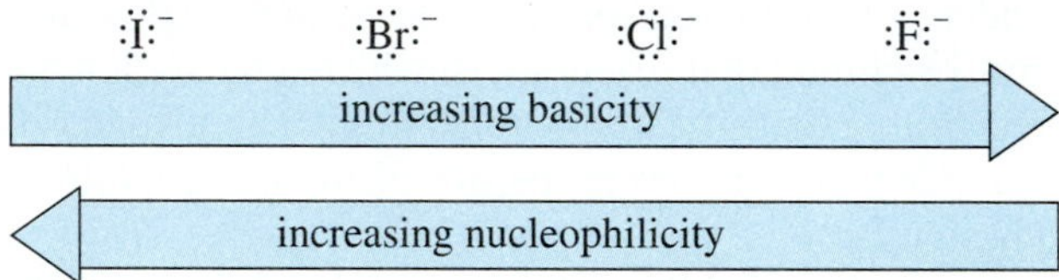

The order of nucleophilicities reflects the polarizability of the atoms. We recall that the atomic radii of elements increase going down a group in the periodic table. As a consequence, the electrons are more polarizable. Thus, iodide ion is more polarizable and therefore more nucleophilic than bromide ion. The polarizability of a nucleophile is important in a nucleophilic substitution reaction because an electron pair of the nucleophile forms a bond to the electrophilic carbon atom during the reaction. Basicity, which is a measure of the ability to bond to a proton, a center of highly concentrated positive charge, is not as sensitive to polarizability.

Effect of Charge

When a nucleophile can exist either as an anion or as its uncharged conjugate acid, the anion is more nucleophilic (Table 7.1). Thus, alkoxide ions (RO^-) are better nucleophiles than alcohols (ROH), and hydroxide ion is a better nucleophile than water. Similarly, thiolate ions (RS^-) are better nucleophiles than thiols (RSH).

$R—\ddot{O}:^-$	>	$R—\ddot{O}:—H$		$R—\ddot{S}:^-$	>	$R—\ddot{S}:—H$
better nucleophile		poorer nucleophile		better nucleophile		poorer nucleophile

7.4 Mechanisms of Substitution Reactions

Nucleophilic substitution reactions can occur by either of two mechanisms. These mechanisms are described by the symbols S_N2 and S_N1, where the term S_N means *substitution, nucleophilic.* The numbers 2 and 1 refer to the number of reactants that are present in the transition state for the rate-determining step.

The S_N2 Mechanism

In the S_N2 mechanism, a nucleophile attacks the substrate and the leaving group, L, departs simultaneously. The reaction occurs in one step and is therefore **concerted.** The substrate and the nucleophile are both present in the transi-

Biological Substitution Reactions with Sulfur Compounds

Many cellular molecules possess a nucleophilic sulfur atom. Of these, one of the most important is glutathione, which contains a sulfhydryl group (—SH). It participates in several enzyme-catalyzed reactions in which the sulfhydryl group reacts as a nucleophile with various toxic intermediates that are produced when drugs are metabolized in liver cells

$$^{-}O_2C-\overset{NH_3^+}{\overset{|}{C}}HCH_2CH_2-\overset{O}{\overset{\|}{C}}-NH\underset{\underset{CH_2-SH}{|}}{C}H-\overset{O}{\overset{\|}{C}}-NHCH_2-CO_2^-$$

glutathione — nucleophilic sulfur atom

The sulfhydryl group of glutathione, often represented as GSH, is a nucleophile that displaces substituents bonded to carbon. The various leaving groups of reactive metabolites are each represented as L. They are all strongly electron-withdrawing groups, and they make the carbon atom to which they are bonded partially positive and susceptible to nucleophilic attack by an essential macromolecule with a nucleophilic center (M—NuC) or by glutathione (GSH). Glutathione protects cells by reacting with toxic metabolites, represented here by R—L, before they react with cellular macromolecules (M—NuC).

$$R-L \xrightarrow{M-Nu:} R-\overset{+}{Nu}-M + L^-$$

$$R-L \xrightarrow{GS-H} R-S-G \quad + HL$$

Glutathione not only protects cells against toxic byproducts of drug metabolism but also provides some degree of protection against toxic industrial chemicals, such as benzyl, allyl, and methyl halides. However, long-term exposure to these chemicals eventually overwhelms the protection provided by glutathione and damages the organism.

Sulfur also plays an important role in substitution reactions in all living cells requiring transfer of a methyl group. The sulfur atom in *S*-adenosylmethionine (SAM) has three carbon atoms bonded to it, forming a positively charged sulfur atom, known as a sulfonium ion.

S-adenosylmethionine

The sulfur atom is part of a large leaving group called *S*-adenosylhomocysteine, which results from transfer of the methyl group to a substrate. This nucleophilic substitution reaction is shown here with a generic nucleophile (NuC^-) and an abbreviated representation of SAM.

$$Nu: + CH_3-\overset{R}{\overset{|}{\overset{+}{S}}}-R \longrightarrow CH_3-Nu + :\overset{R}{\overset{|}{S}}-R$$

methylated substrate

An important example of methyl group transfer from SAM to a nucleophile occurs in the biosynthesis of the neurotransmitter epinephrine. In this reaction, an amino group of norepinephrine attacks the methyl group carbon atom of *S*-adenosylmethionine in a displacement reaction to produce epinephrine. ■

norepinephrine: $(HO)_2C_6H_3-CHCH_2NH_2$ (with OH on CH)

epinephrine: $(HO)_2C_6H_3-CHCH_2NH-CH_3$ (with OH on CH)

tion state for this step. That is, two molecules are present, and the reaction is **bimolecular,** as indicated by the number 2 in the S_N2 symbol. As a consequence, the reaction rate depends on the concentrations of both the nucleophile and the substrate. If the substrate concentration is doubled, the reaction rate is doubled. Similarly, if the concentration of the nucleophile is doubled, the rate again doubles. This relationship between the rate and the concentration of the reactants exists because the reactants must collide in the rate-determining step for the reaction to occur. The probability that the nucleophile will collide with the substrate increases if the concentration of either or both species is increased.

Let's consider the S_N2 reaction of hydroxide ion with chloromethane to give methanol and chloride ion. This reaction is shown with the energy diagram in Figure 7.1. In this plot, the transition state, which occurs at the point of highest energy, contains the hydroxide ion and the substrate. As the reaction proceeds through the transition state, a bond between carbon and hydroxide ion forms and the bond between carbon and chlorine breaks.

The rates of reaction for haloalkanes via the S_N2 mechanism decrease in the order primary >secondary > tertiary. This trend is observed because alkyl groups block the approach of the nucleophile and slow the rate of the reaction (Figure 7.2). This crowding of alkyl groups around the reactive carbon atom is called **steric hindrance.**

When (*S*)-2-bromobutane reacts with sodium hydroxide, the substitution product is (*R*)-2-butanol. The reaction occurs with inversion of configuration. Thus, the nucleophile approaches the electrophilic carbon atom from the back

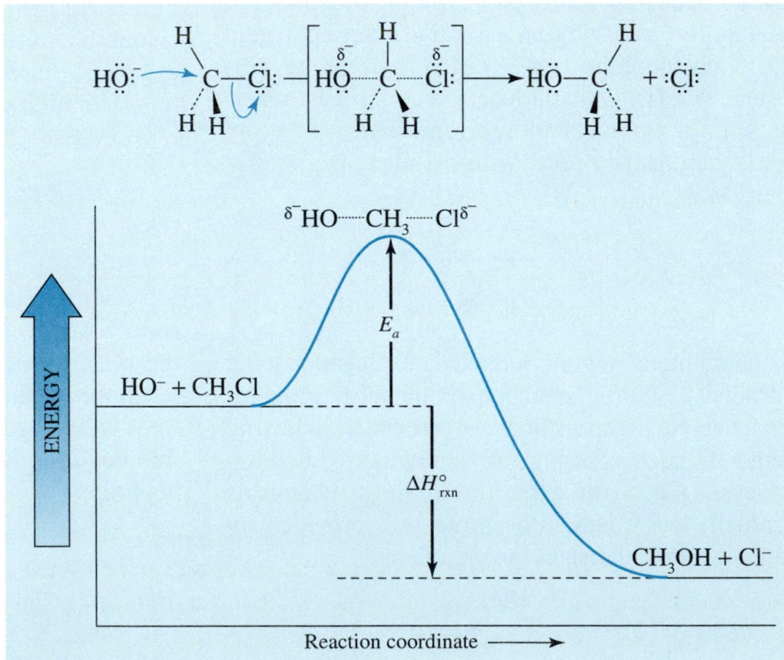

FIGURE 7.1
Activation Energy and the S_N2 Reaction Mechanism

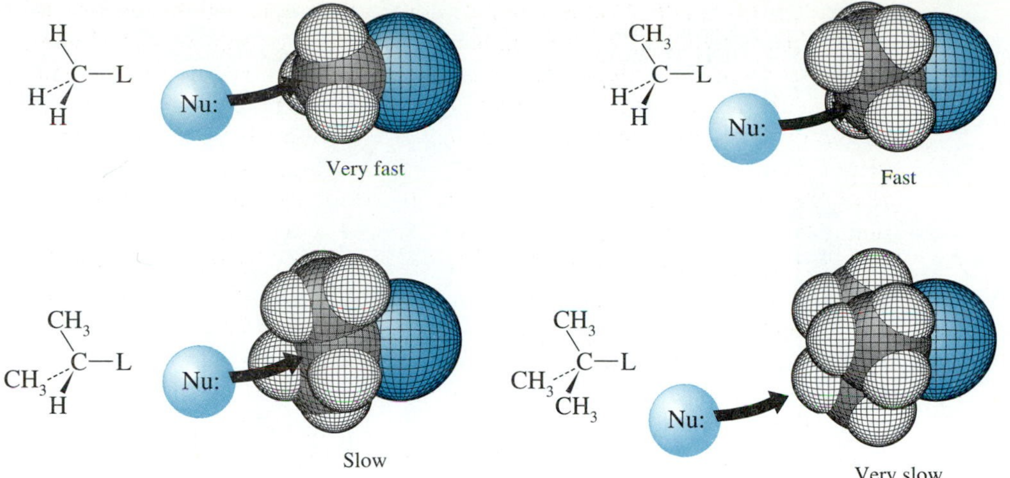

FIGURE 7.2
Effect of Steric Hindrance on the S_N2 Reaction

and the leaving group simultaneously departs from the front of the substrate in the S_N2 mechanism.

$$HO:^- + (CH_3)(CH_3CH_2)(H)C{-}Br \longrightarrow HO{-}C(CH_3)(CH_2CH_3)(H) + :Br:^-$$

(*S*)-2-bromobutane (*R*)-2-butanol

The S_N1 Mechanism

A nucleophilic substitution reaction that occurs by an S_N1 mechanism proceeds in two steps. In the first step, the bond between the carbon atom and the leaving group breaks to produce a carbocation and, most commonly, an anionic leaving group. In the second step, the carbocation reacts with the nucleophile to form the substitution product.

$$C{-}L \xrightarrow{\text{slow}} C^+ + L:^-$$

substrate carbocation

$$Nu:^- + C^+ \xrightarrow{\text{fast}} C{-}Nu$$

nucleophile carbocation

The formation of a carbocation in the first step in an S_N1 reaction is the slow or rate-determining step. The formation of a bond between the nucleophile and the carbocation in the second step occurs rapidly. Because the slow step of the reaction involves only the substrate, the reaction is **unimolecular.** As a result, the rate of the reaction depends only on the concentration of the substrate and not on the concentration of the nucleophile.

An energy diagram tracing the progress of a reaction that occurs by an S_N1 mechanism is shown in Figure 7.3. The rate of the reaction reflects the activation energy required to form the carbocation intermediate. The activation en-

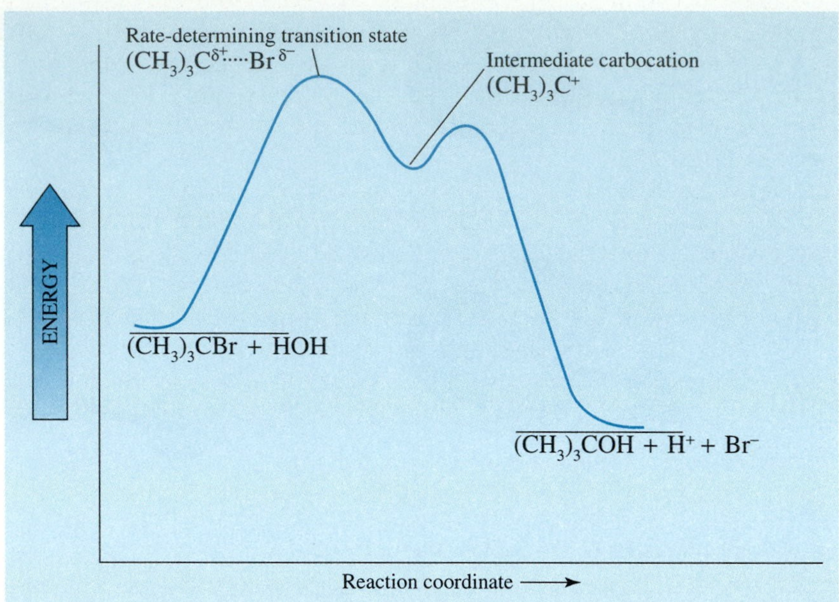

FIGURE 7.3
Activation Energy and the S_N1 Reaction Mechanism

ergy required for the second step, addition of the nucleophile to the carbocation, is much smaller, so step 2 is very fast. The rate of the second step has no effect on the overall rate of the reaction.

The rates of S_N1 reactions decrease in the order $3° > 2° > 1°$, which is the reverse of the order observed in S_N2 reactions. The relative reactivity of haloalkanes in S_N1 reactions corresponds to the relative stability of carbocation

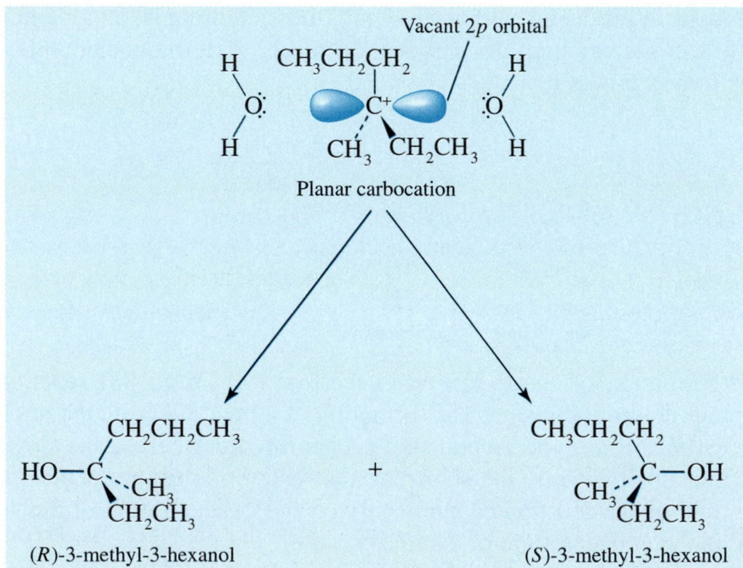

FIGURE 7.4
Stereochemistry of an S_N1 Reaction

intermediates that form during the reaction. We recall from Chapter 4 that the order of stability of carbocations is tertiary > secondary > primary. A tertiary carbocation forms faster than a less stable secondary carbocation, which in turn forms very much faster than a highly unstable primary carbocation. However, S_N1 mechanisms are also favored by resonance-stabilized primary carbocations such as benzyl and allyl.

In contrast to S_N2 reactions at stereogenic centers, which occur with inversion of configuration, an S_N1 reaction gives a racemic mixture of enantiomers that has no optical rotation. For example, (*S*)-3-bromo-3-methylhexane reacts with water to give a racemic mixture of 3-methyl-3-hexanols. The reaction occurs via an achiral carbocation intermediate with a plane of symmetry (Figure 7.4). The carbocation intermediate's plane of symmetry allows the nucleophile to attack equally well from either side. The product is then a racemic mixture of enantiomers. Thus, a chiral substrate loses chirality in a reaction that occurs by an S_N1 mechanism.

7.5 S_N2 versus S_N1 Reactions

Now that we have outlined the general properties of S_N2 and S_N1 reactions, let's see if we can predict which one is likely to occur. We will consider (1) the structure of the substrate, (2) the nucleophile, and (3) the nature of the solvent.

Structure of the Substrate

Primary haloalkanes almost always react in nucleophilic substitution reactions by the S_N2 mechanism, whereas tertiary haloalkanes react by the S_N1 mechanism. Secondary haloalkanes may react by either mechanism depending on the nature of the nucleophile and the solvent.

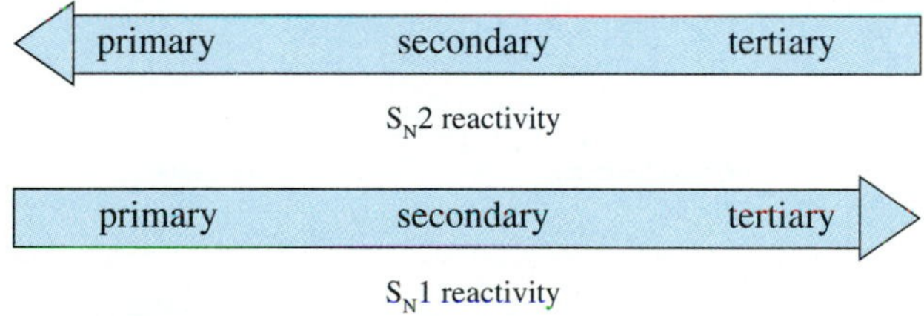

Effect of the Nucleophile

The nature of the nucleophile sometimes determines the mechanism of a nucleophilic substitution reaction. If the nucleophile is a highly polarizable species such as thiolate ion, RS^-, it tends to react by an S_N2 mechanism. If the nucleophile is an uncharged species such as H_2O or CH_3OH, an S_N1 mechanism is more likely.

Effect of Solvent

Until now, we have neglected the role of solvent in nucleophilic substitution reactions, but the choice of solvent can tip the balance in favor of a particular substitution mechanism. We noted that secondary haloalkanes can react by either an S_N1 or an S_N2 mechanism. In these cases, the polarity of the solvent plays an important role. The S_N1 process forms a carbocation intermediate. Because a polar solvent stabilizes charged species better than a nonpolar solvent, it increases the rate of S_N1 reactions much more than it affects the rate of S_N2 reactions.

The solvent also affects nucleophilicity. Solvents that have proton-donating ability, such as alcohols, are called **protic solvents.** A protic solvent interacts strongly with nucleophilic anions by forming hydrogen bonds with the unshared pairs of electrons on the nucleophiles. When the nucleophile is hydrogen-bonded to the solvent, its nucleophilicity decreases, which decreases the likelihood of an S_N2 reaction.

Solvents that do not have protons available for hydrogen bonding to nucleophiles are called **aprotic solvents.** Examples of polar aprotic solvents include dimethylformamide (DMF) and dimethyl sulfoxide (DMSO).

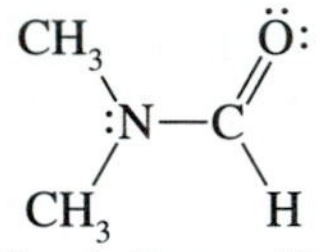

CH_3 Ö: CH_3
:N—C :S=Ö:
CH_3 H CH_3

dimethylformamide 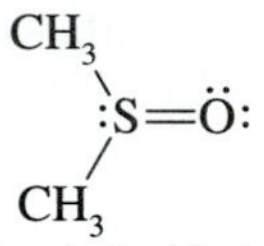 dimethyl sulfoxide

The electron pairs of the oxygen atoms of these aprotic solvents cause them to solvate cations but not anions. For example, these solvents tie up the cation of KCN, but leave the CN^- ion free. Consequently, the nucleophilicity of CN^- is greater in dimethyl sulfoxide than in ethanol (CH_3CH_2OH). Thus, an aprotic solvent such as dimethyl sulfoxide favors an S_N2 reaction.

EXAMPLE 7-3

Explain why the reaction of 3-bromocyclohexene with methanol (CH_3OH) is faster than the reaction of bromocyclohexane with methanol.

Br Br

bromocyclohexane 3-bromocyclohexene

Solution

Both substrates are secondary bromides. The reaction with methanol, a neutral nucleophile, will tend to occur by an S_N1 process. Although both carbocations are secondary, the one derived from 3-bromocyclohexene is also a resonance-stabilized allylic carbocation. Its resonance stabilization enhances the rate of its formation. No such stabilization occurs in the reaction of bromocyclohexane.

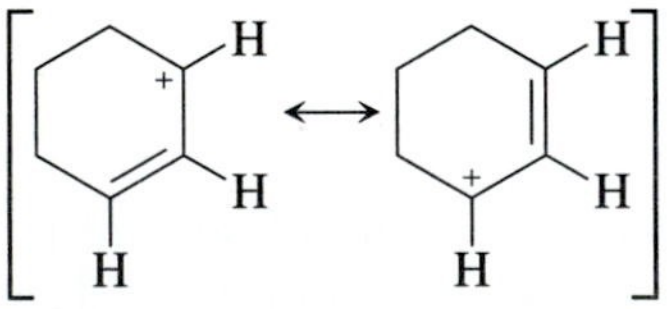

H H
+ ⟷ +
H H
H H

resonance-stabilized allylic carbocation

Problem 7.3

The relative rates of reaction of 1-iodobutane with chloride ion in methanol, formamide, and dimethylformamide are 1, 12, and 1.2×10^6, respectively. Explain the relatively small difference in rate between methanol and formamide and the large difference in rate between formamide and dimethylformamide.

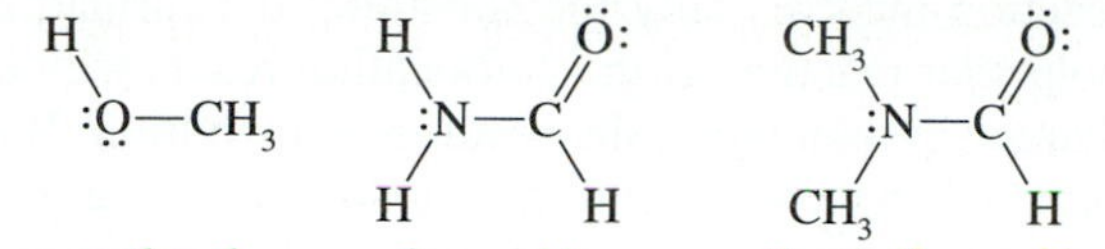

methanol formamide dimethylformamide

7.6 Mechanisms of Elimination Reactions

We noted in Section 7.1 that when a haloalkane undergoes a nucleophilic substitution reaction, a competing elimination reaction may also occur because a nucleophile is also a base. In a substitution reaction, the nucleophile attacks a carbon atom in the substrate. When a haloalkane undergoes an elimination reaction, the nucleophile acts as a base and removes a proton from a carbon atom adjacent to the carbon atom bearing the halogen. Both the proton and the leaving group are eliminated and a carbon-carbon π bond is formed. When a proton and a halogen are lost by a substrate, the elimination reaction is called **dehydrohalogenation.**

An elimination reaction is denoted by the symbol E. Elimination reactions can occur by two mechanisms, designated E2 and E1. An E1 or an E2 process may compete with S_N1 and S_N2 reactions.

The E2 Mechanism

Like the S_N2 reaction mechanism, the E2 mechanism is a concerted process. In an E2 dehydrohalogenation reaction, the base (nucleophile) removes a proton on the carbon atom adjacent to the carbon atom containing the leaving group. As the proton is removed, the leaving group departs and a double bond forms.

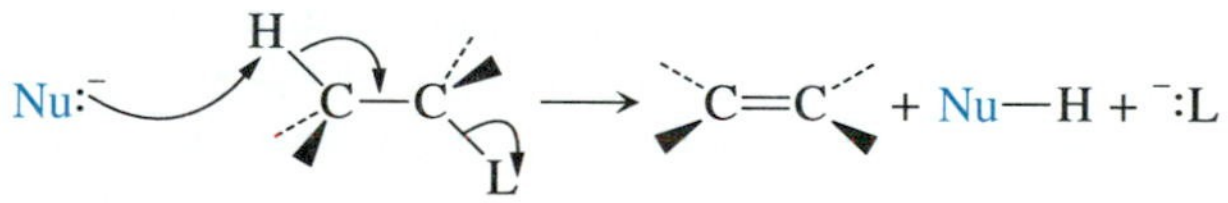

Nu:$^-$ H—C—C—L ⟶ C=C + Nu—H + $^-$:L

Like the S_N2 reaction, an E2 reaction requires a precise molecular arrangement. The anti conformation of the hydrogen and halogen atoms is preferred for the reaction because it aligns the orbitals properly for the formation of the π bond. In terms of "electron pushing," we can visualize the process as the removal of the proton to provide an electron pair that attacks the neighboring carbon atom from the back to displace the leaving group.

An E2 reaction occurs at a rate that depends on the concentrations of both the substrate and the base. If the substrate concentration is doubled, the reaction rate also doubles, as in S_N2 processes. Thus, both E2 and S_N2 mechanisms are affected in the same way, and the two mechanisms compete with each other.

The E1 Mechanism

We recall that an S_N1 reaction proceeds in two steps, and that the rate-determining step is formation of a carbocation intermediate. Similarly, an E1 mechanism occurs in two steps, and the rate-determining step is the formation of a carbocation. Thus, just as E2 and S_N2 mechanisms compete with each other, an E1 mechanism competes with an S_N1 mechanism. Because the rate-determining step of an E1 reaction involves only the substrate, the formation of the carbocation is a unimolecular reaction. If the carbocation reacts with a nucleophile at the positively charged carbon atom, the result is substitution. But if the nucleophile acts as a base and removes a proton from the carbon atom adjacent to the cationic center, the net result is the loss of HL and the formation of a π bond—that is, an elimination reaction.

step 1 $$\text{H}{-}\text{C}{-}\text{C}{-}\text{L} \xrightarrow{\text{slow}} \text{H}{-}\text{C}{-}\text{C}^{+} + :\text{L}^{-}$$

step 2 $$\text{H}{-}\text{C}{-}\text{C}^{+} \xrightarrow[\text{Nu}:^{-}]{\text{fast}} \text{H}{-}\text{C}{-}\text{C}{-}\text{Nu}$$

$$\text{H}{-}\text{C}{-}\text{C}^{+} \xrightarrow[\text{Nu}:^{-}]{\text{fast}} \text{C}{=}\text{C} + \text{H}{-}\text{Nu}$$

7.7 Effect of Structure on Competing Reactions

Let's now examine the variety of product mixtures that result from competing substitution and elimination processes. We will divide our discussion according to the type of haloalkane. These results are summarized in Table 7.2.

Tertiary Haloalkanes

Tertiary haloalkanes can undergo substitution reactions only by an S_N1 mechanism because there is too much steric hindrance for an S_N2 reaction to occur. A tertiary haloalkane can undergo an elimination reaction by either an E2 or E1 process. The mechanism depends on the basicity of the nucleophile and the polarity of the solvent. If the nucleophile is a weak base, S_N1 and E1 processes compete, and the amount of the two types of products depends only on the carbocation formed. For example, 2-bromo-2-methylbutane reacts in ethanol to give about 64% of the product from an S_N1 process.

Table 7.2 Summary of Substitution and Elimination Reactions

Structure	S_N1	S_N2	E1	E2
RCH_2X	Does not occur	Very favored	Does not occur	Occurs readily with strong bases
R_2CHX	Occurs for allylic and benzylic compounds	E2 reactions occur in competition with substitution	Occurs for allylic and benzylic compounds	Occurs readily with strong bases
R_3CX	Occurs readily in polar solvents	Does not occur	Occurs in competition with S_N1 reaction	Occurs readily with bases

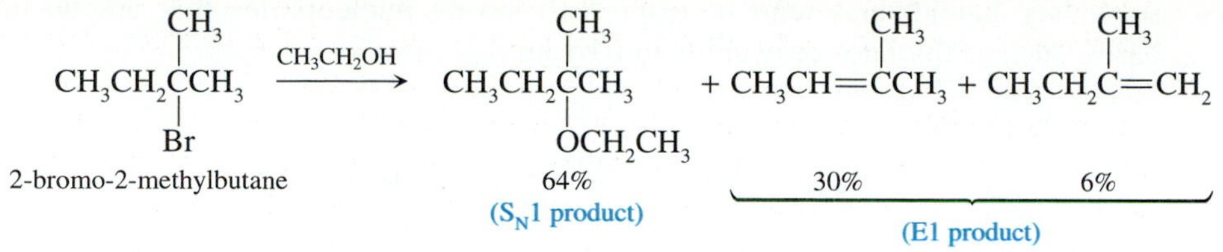

$$\underset{\text{2-bromo-2-methylbutane}}{CH_3CH_2C(CH_3)(Br)CH_3} \xrightarrow{CH_3CH_2OH} \underset{\substack{64\% \\ (S_N1\ \text{product})}}{CH_3CH_2C(CH_3)(OCH_2CH_3)CH_3} + \underset{30\%}{CH_3CH{=}C(CH_3)CH_3} + \underset{6\%}{CH_3CH_2C(CH_3){=}CH_2}$$

(30% + 6%: E1 product)

However, if sodium ethoxide, a strongly basic nucleophile, is added to the ethanol, an E2 process competes with the substitution reaction. The amount of elimination product is increased to a total of about 93% of the product mixture; only 7% of the S_N1 product is formed.

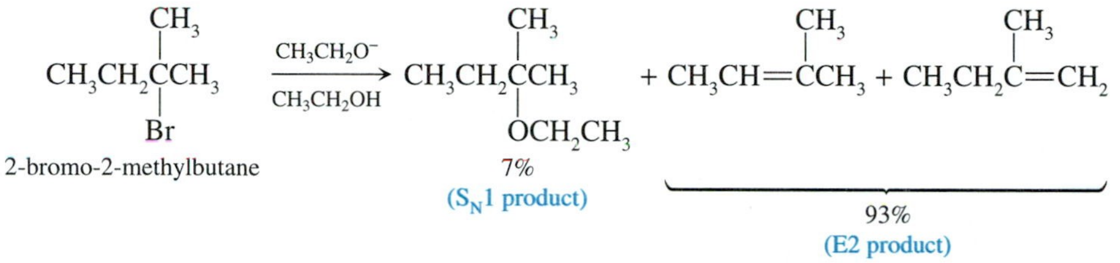

$$\underset{\text{2-bromo-2-methylbutane}}{CH_3CH_2C(CH_3)(Br)CH_3} \xrightarrow[CH_3CH_2OH]{CH_3CH_2O^-} \underset{\substack{7\% \\ (S_N1\ \text{product})}}{CH_3CH_2C(CH_3)(OCH_2CH_3)CH_3} + CH_3CH{=}C(CH_3)CH_3 + CH_3CH_2C(CH_3){=}CH_2$$

(93%: E2 product)

Primary Haloalkanes

Primary haloalkanes can undergo either S_N2 or E2 reactions. They do not undergo S_N1 or E1 reactions because a primary carbocation is very unstable. Primary haloalkanes react with strongly nucleophilic, weakly basic reactants such as ethylthiolate, $CH_3CH_2S^-$, exclusively by an S_N2 process. However, a primary haloalkane reacts with ethoxide ion, which is a weaker nucleophile but a stronger base than ethylthiolate, to give some elimination product, but mostly the S_N2 product.

$$\underset{\text{1-bromobutane}}{CH_3CH_2CH_2CH_2{-}Br} \xrightarrow{CH_3CH_2S^-} \underset{\substack{\text{exclusive product} \\ (S_N2\ \text{mechanism})}}{CH_3CH_2CH_2CH_2{-}S{-}CH_2CH_3}$$

$$\underset{\text{1-bromobutane}}{CH_3CH_2CH_2CH_2{-}Br} \xrightarrow{CH_3CH_2O^-} \underset{\substack{90\% \\ (S_N2\ \text{product})}}{CH_3CH_2CH_2CH_2{-}O{-}CH_2CH_3} + \underset{\substack{10\% \\ (\text{E2 product})}}{CH_3CH_2CH{=}CH_2}$$

If a primary haloalkane is treated with *tert*-butoxide ion, $(CH_3)_3CO^-$, instead of ethoxide, the amount of elimination product increases significantly. The *tert*-butoxide ion is not only more basic than the ethoxide ion but also much more sterically hindered. The combination of these two factors favors elimination by an E2 process over substitution by an S_N2 process.

$$\underset{\text{1-bromobutane}}{CH_3CH_2CH_2CH_2{-}Br} \xrightarrow{(CH_3)_3CO^-} \underset{\substack{15\% \\ (S_N2\ \text{product})}}{CH_3CH_2CH_2CH_2{-}O{-}C(CH_3)_3} + \underset{\substack{85\% \\ (\text{E2 product})}}{CH_3CH_2CH{=}CH_2}$$

Secondary Haloalkanes

Secondary haloalkanes can react by S_N2, E2, S_N1, and E1 mechanisms, and it is sometimes difficult to predict which will occur in a given reaction. However, secondary haloalkanes tend to react with strong nucleophiles that are weak bases, such as thiolates or cyanide ion, by an S_N2 process.

$$\underset{\text{2-bromobutane}}{CH_3CH_2\overset{\overset{Br}{|}}{C}HCH_3} \xrightarrow{CH_3S^-} \underset{(S_N2 \text{ product})}{CH_3CH_2\overset{\overset{SCH_3}{|}}{C}HCH_3}$$

$$\underset{\text{2-bromobutane}}{CH_3CH_2\overset{\overset{Br}{|}}{C}HCH_3} \xrightarrow{CN^-} \underset{(S_N2 \text{ product})}{CH_3CH_2\overset{\overset{CN}{|}}{C}HCH_3}$$

On the other hand, a secondary haloalkane tends to react with weak nucleophiles that are also weak bases, such as ethanol, by an S_N1 process with some accompanying E1 product. In contrast, we can tip the scales in the other direction by adding sodium ethoxide to ethanol. By adding a strong base, the product of the S_N1 reaction drops to 18% of the total, and E2 products account for the rest.

$$\underset{\text{2-bromobutane}}{CH_3CH_2\overset{\overset{Br}{|}}{C}HCH_3} \xrightarrow{CH_3CH_2OH} \underset{\substack{95\% \\ (S_N1 \text{ product})}}{CH_3CH_2\overset{\overset{OCH_2CH_3}{|}}{C}HCH_3} + \underbrace{\underset{4\%}{CH_3CH{=}CHCH_3} + \underset{1\%}{CH_3CH_2CH{=}CH_2}}_{\text{(E1 products)}}$$

$$\underset{\text{2-bromobutane}}{CH_3CH_2\overset{\overset{Br}{|}}{C}HCH_3} \xrightarrow[CH_3CH_2OH]{CH_3CH_2O^-} \underset{\substack{18\% \\ (S_N1 \text{ product})}}{CH_3CH_2\overset{\overset{OCH_2CH_3}{|}}{C}HCH_3} + \underbrace{\underset{66\%}{CH_3CH{=}CHCH_3} + \underset{16\%}{CH_3CH_2CH{=}CH_2}}_{\text{(E2 products)}}$$

EXAMPLE 7-4

Which of the following two methods of preparing $CH_3CH_2OCH(CH_3)_2$ will give the better yield?

$$\underset{\text{ethoxide}}{CH_3CH_2{-}O^-} + \underset{\text{2-bromopropane}}{Br{-}CH(CH_3)_2} \longrightarrow CH_3CH_2{-}O{-}CH(CH_3)_2$$

$$\underset{\text{bromoethane}}{CH_3CH_2{-}Br} + \underset{\text{isopropoxide}}{{}^-O{-}CH(CH_3)_2} \longrightarrow CH_3CH_2{-}O{-}CH(CH_3)_2$$

Solution

An ether can be prepared by treating a haloalkane with an alkoxide in an S_N2 reaction. This ether has two different alkyl groups bonded to oxygen: one from the alkoxide and the other from the haloalkane.

The first reaction is a nucleophilic displacement at a secondary center by a nucleophile that is also a strong base. A competing elimination reaction to yield propene will also occur. The second reaction occurs by an S_N2 mechanism at a primary center, which tends to occur with little competition from an elimination reaction. Therefore, the better way to make the desired product is by the second reaction.

Problem 7.4

The amount of elimination product for the reaction of 1-bromodecane with an alkoxide in the corresponding alcohol solvent is about 1% for methoxide ion and 85% for *tert*-butoxide ion. Explain these data.

Summary of Reactions

1. Nucleophilic Substitution of Haloalkanes (Section 7.2)

$$CH_3CH_2CH_2\underset{|}{\overset{CH_3}{C}}HCH_2CH_2Cl + CH_3O^- \longrightarrow CH_3CH_2CH_2\underset{|}{\overset{CH_3}{C}}HCH_2CH_2OCH_3$$

$$C_6H_5CH_2CH_2Br + CH_3CH_2S^- \longrightarrow C_6H_5CH_2CH_2SCH_2CH_3$$

$$C_5H_9{-}CH_2CH_2Br + CN^- \longrightarrow C_5H_9{-}CH_2CH_2CN$$

$$C_5H_9{-}CH_2CH_2Br + CH_3CH_2C{\equiv}C^- \longrightarrow C_5H_9{-}CH_2CH_2C{\equiv}CCH_2CH_3$$

2. Dehydrohalogenation of Haloalkanes (Section 7.6)

$$\text{Br} \xrightarrow[CH_3CH_2OH]{CH_3CH_2O^-}$$

Explorations with Molecular Models

1. Construct the following model of 1,2-dichloroethane. Arrange the structure to sight along the carbon-carbon bond. What conformational term is used to describe the structure? Does this conformation have a dipole moment? Draw its Newman projection formula.

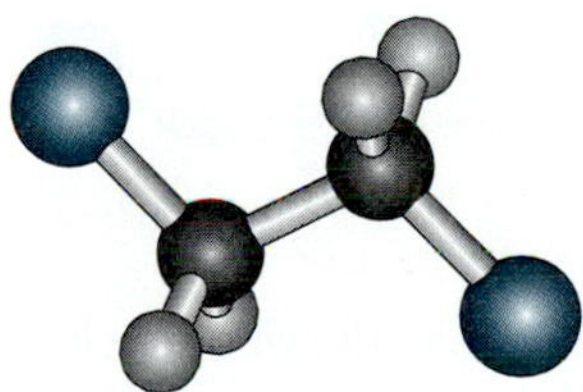

2. Construct the following model of 1-chloro-4-methylcyclohexane. What stereochemical term is used to describe this isomer? Construct a model of the substitution product that results from reaction with hydroxide ion. What sterochemical term is used to describe this isomer?

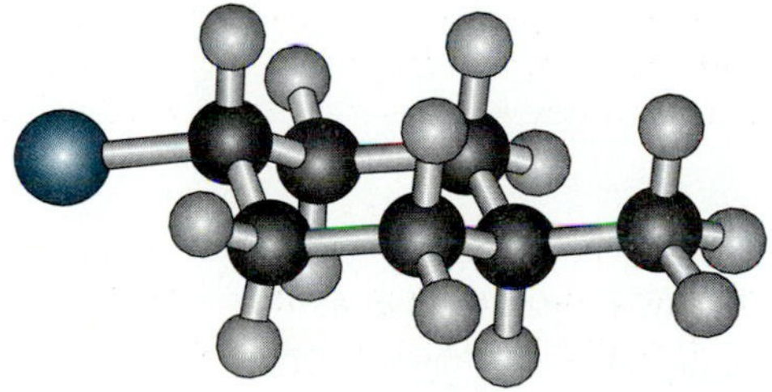

3. Construct the following model of 2-chloro-3-methylpentane. Rotate the structure about the C-2 to C-3 bond to arrange the C-3 hydrogen atom and the chlorine atom in an anti arrangement. Write the structure of the E2 elimination product derived from this conformation. What sterochemical term is used to describe this isomer?

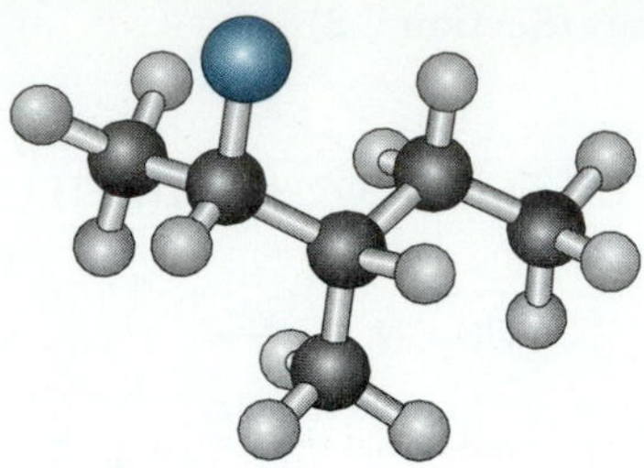

4. Construct the following model of chloroethane. Arrange the structure to sight along the carbon-chlorine bond with the carbon atom closest to your eye, and view the atoms that might interfere with an S_N2 process. Successively replace hydrogen atoms at C-1 of the original chloroethane molecule with methyl groups. How does the steric environment change?

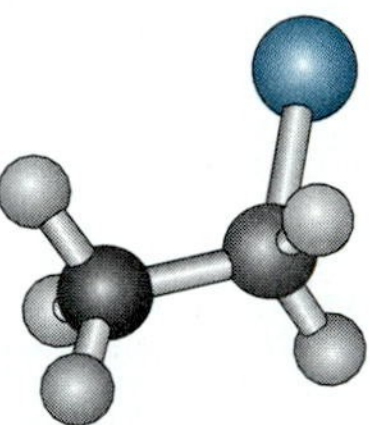

5. Prepare the following model of 1-chloropropane in its gauche conformation. Arrange the structure to sight along the carbon-chlorine bond with the carbon atom closest to your eye. Rotate the C-1 and C-2 atoms about the bond between them to obtain all possible staggered conformations. How many of the conformations have atoms or groups of atoms that might interfere with an S_N2 process? Successively replace hydrogen atoms at C-2 of the structure with methyl groups to obtain 1-chloro-2-methylpropane and 1-chloro-2,2-dimethylpropane. Repeat the rotation process described for 1-chloropropane for each compound. How many of the conformations have atoms or groups of atoms that might interfere with an S_N2 process for each compound?

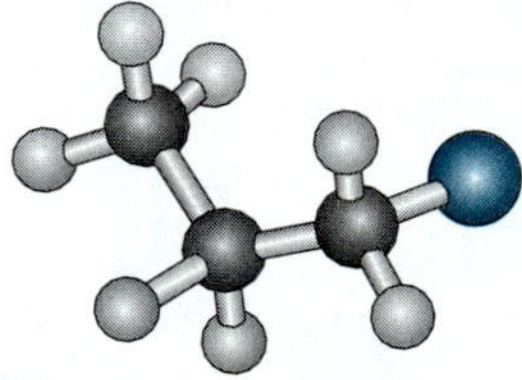

6. Prepare the model of 1-chloro-4-methylcyclohexane shown in Exploration Exercise 2. What bonds at the C-2 and C-6 atoms are anti to the carbon-chlorine bond? Can this compound easily undergo an elimination reaction?

7. Interchange the hydrogen atom and the chlorine atom on the C-1 atom of the model prepared in Exploration Exercise 2. What stereochemical term is used to describe this isomer? What bonds at the C-2 and C-6 atoms are anti to the carbon-chlorine bond? Can this compound easily undergo an elimination reaction?

Exercises

Nomenclature

7.1 What is the IUPAC name for each of the following compounds?

(a) vinyl fluoride (b) allyl chloride (c) propargyl bromide

7.2 What is the IUPAC name for each of the following compounds?

(a) $(CH_3)_3CCH_2Cl$ (b) $(CH_3)_2CHCH_2CH_2Br$ (c) $C_6H_5CH_2CH_2F$

7.3 Draw the structure of each of the following compounds.

(a) *cis*-1-bromo-2-methylcyclopentane (b) 3-chlorocyclobutene (c) (*E*)-1-fluoro-2-butene (d) (*Z*)-1-bromo-1-propene

7.4 What is the IUPAC name for each of the following compounds?

(a) CH_3 Br (b) Br Br (c) Br

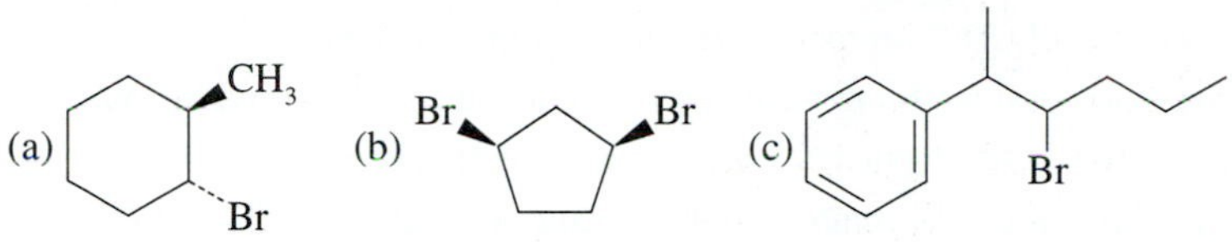

Nucleophilic Substitution Reactions

7.5 Write the structure of the product obtained for each of the following combinations of reactants.

(a) 1-chloropentane and sodium iodide (b) 1,3-dibromopropane and excess sodium cyanide
(c) *p*-methylbenzylchloride and sodium acetylide (d) 2-bromobutane and sodium hydrosulfide (NaSH)

7.6 What haloalkane and nucleophile are required to produce each of the following compounds?

(a) $CH_3CH_2CH_2C{\equiv}CH$ (b) $(CH_3)_2CHCH_2CN$ (c) $CH_3CH_2SCH_2CH_3$

7.7 Alcohols (ROH) are converted into alkoxides (RO^-) by reaction with NaH. Treatment of the following compound with sodium hydride yields C_4H_8O. What is the structure of the product? How is it formed?

$$HOCH_2CH_2CH_2CH_2{-}Br + NaH \longrightarrow C_4H_8O + H_2 + NaBr$$

7.8 Treatment of the following compound with sodium sulfide yields C_4H_8S. What is the structure of the product? How is it formed?

$$ClCH_2CH_2CH_2CH_2Cl + Na_2S \longrightarrow C_4H_8S + 2\ NaCl$$

Structure and Rates of Substitution Reactions

7.9 Which compound in each of the following pairs reacts with sodium iodide at the faster rate in an S_N2 reaction?

(a) 1-bromobutane or 2-bromobutane
(b) 1-chloropentane or chlorocyclopentane
(c) 2-bromo-2-methylpentane or 2-bromo-4-methylpentane

7.10 The rate of reaction of *cis*-1-bromo-4-*tert*-butylcyclohexane with methylthiolate (CH_3S^-) is faster than for the trans isomer. Suggest a reason for this difference.

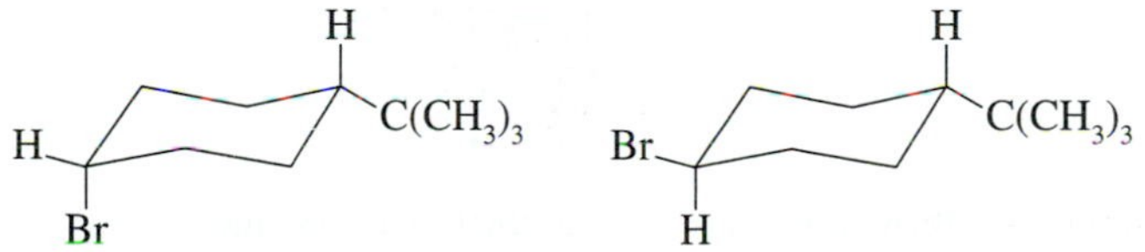

H H $C(CH_3)_3$ Br *cis* isomer

H Br $C(CH_3)_3$ H *trans* isomer

7.11 Predict which compound in each of the following pairs reacts with methanol (CH_3OH) at the faster rate in an S_N1 reaction. Explain why.

(a) bromocyclohexane or 1-bromo-1-methylcyclohexane
(b) isopropyl iodide or isobutyl iodide
(c) 3-bromo-1-pentene or 4-bromo-1-pentene

7.12 Predict which compound in each of the following pairs reacts with ethanol (CH_3CH_2OH) at the faster rate in an S_N1 reaction. Explain why.

(a) 1-bromo-1-phenylpropane or 2-bromo-1-phenylpropane
(b) 3-chlorocyclopentene or 4-chlorocyclopentene
(c) benzyl bromide or *p*-methylbenzyl bromide

7.13 *p*-Methylbenzyl bromide reacts faster than *p*-nitrobenzyl bromide with ethanol to form an ether product. Suggest a reason for this observation.

7.14 Although 1-bromo-2,2-dimethylpropane is a primary halide, S_N2 reactions of this compound are about 10,000 times slower than those of 1-bromopropane. Suggest a reason.

Stereochemistry of Substitution Reactions

7.15 Write the structure of the product of the reaction of (*R*)-2-bromobutane with sodium cyanide.

7.16 Write the structure of the product of the reaction of *cis*-1-bromo-2-methylcyclopentane with methylthiolate.

7.17 Write the structure of the product of the reaction of (*R*)-1-phenyl-1-bromobutane with water.

7.18 Write the structure of the product of the reaction of (*S*)-3-bromo-3-methylhexane with ethanol (CH_3CH_2OH).

7.19 Optically active 2-iodooctane slowly becomes racemic when treated with sodium iodide in an inert polar solvent. Explain why.

7.20 Optically active 2-butanol, $CH_3CH_2CH(OH)CH_3$, slowly becomes racemic when treated with dilute acid. Explain why.

Elimination Reactions

7.21 What alkene will be formed in the E2 reaction of the following compound with sodium methoxide ($CH_3O^-Na^+$)?

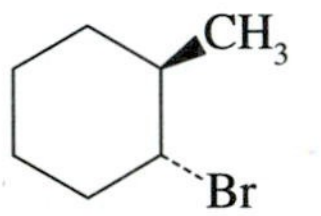

CH_3

Br

7.22 How many isomeric alkenes can be formed by the elimination of HBr from the following compound by an E1 process? Which one will predominate?

$$CH_3-\underset{\displaystyle |}{\overset{\displaystyle CH_3}{\overset{|}{C}H}}-\underset{\underset{\displaystyle CH_3}{|}}{\overset{\overset{\displaystyle Br}{|}}{C}}-CH_2-CH_3$$

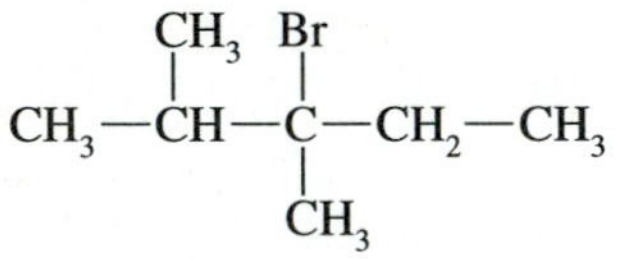

7.23 The rate of elimination of hydrogen bromide in the reaction of *cis*-1-bromo-4-*tert*-butylcyclohexane with hydroxide ion is faster than for the trans isomer. Suggest a reason for this difference.

7.24 Eight diastereomers of 1,2,3,4,5,6-hexachlorocyclohexane are possible. The following isomer undergoes an E2 reaction about 1000 times slower than any of the others. Why?

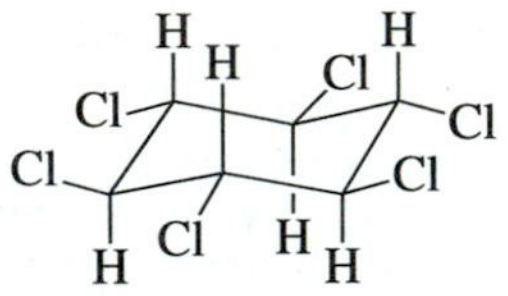

H H H Cl Cl Cl Cl Cl Cl Cl H H H H

7.25 What is the configuration of the alkene formed by the elimination of one molar equivalent of HBr from the following compound?

H H CH_3 C—C Br CH_3 Br

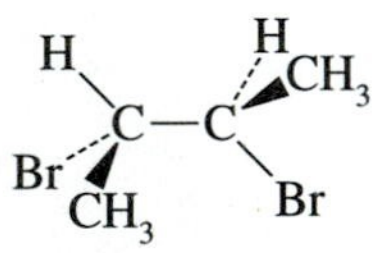

7.26 How many isomers can result from the elimination of HBr from the following compound? Which one will predominate?

Br

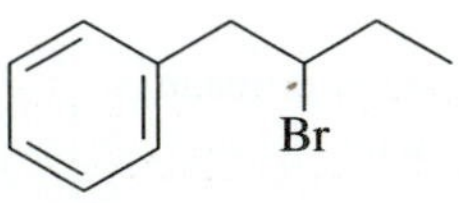

CHAPTER 8

Alcohols and Phenols

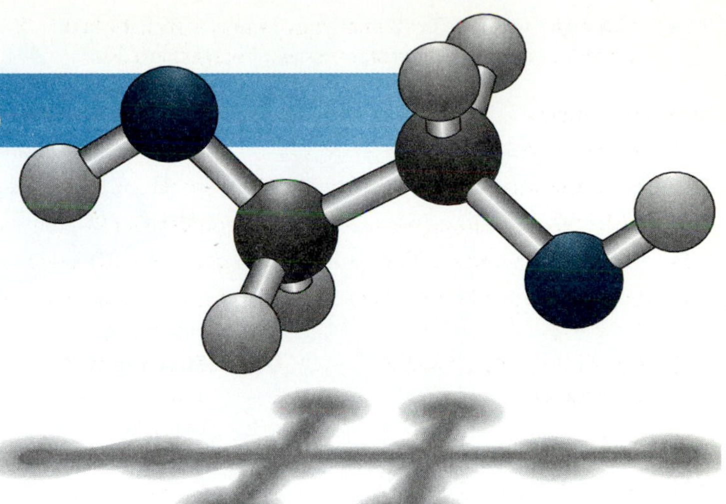

Overview

8.1 The Hydroxyl Group

Families of organic compounds that have functional groups containing oxygen include alcohols, phenols, ethers, aldehydes, ketones, acids, esters, and amides. Alcohols and phenols both contain a hydroxyl group (—OH). A hydroxyl group is also present in carboxylic acids, but it is bonded to a carbonyl carbon atom. As a result, the chemistry of carboxylic acids, the subject of Chapter 12, is substantially different from the chemistry of alcohols and phenols.

Alcohols and phenols can be viewed as organic "relatives" of water in which one hydrogen atom is replaced by an alkyl group or an aryl group. **Alcohols** contain a hydroxyl group bonded to an sp^3-hybridized carbon atom. **Phenols** have a hydroxyl group bonded to an sp^2-hybridized carbon atom of an aromatic ring.

This hydroxyl group is bonded to an sp^3-hybridized carbon atom. The compound is an alcohol.

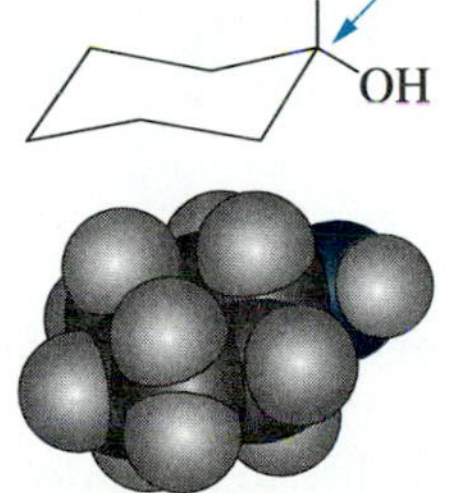

This hydroxyl group is bonded to an sp^2-hybridized carbon atom. The compound is a phenol.

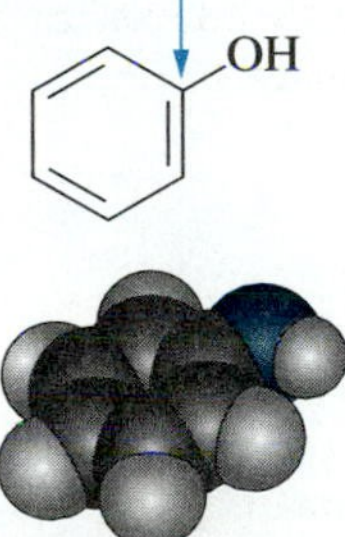

The C—O—H bond angle in methyl alcohol is 108.9° (Figure 8.1). The lone pair electrons in the remaining two sp^3 hybrid orbitals are viewed as directed to the remaining "corners" of a tetrahedron. The radius of the oxygen atom is smaller than that of a carbon atom. As a result, the O—H bond length (96 pm) is shorter than the C—H bond length

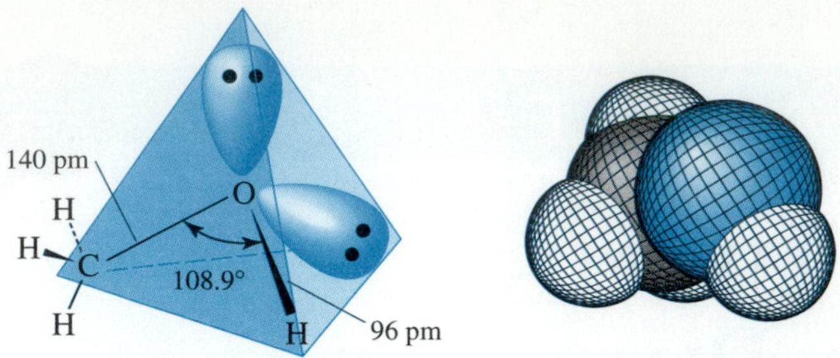

FIGURE 8.1
Structure of Methanol

The C—O—H bond angle is close to the tetrahedral angle. The two sets of lone pair electrons are in sp^3 hybrid orbitals that are directed to two of the corners of a tetrahedron.

(110 pm), and the C—O bond length (140 pm) is shorter than the C—C bond length (154 pm).

Alcohols and phenols react differently at the carbon atom bearing the hydroxyl group because the hybridization of the carbon atoms differs. However, alcohols and phenols both have oxygen-hydrogen bonds and undergo reactions similar to those of water. For example, alcohols and phenols react with alkali metals to produce hydrogen gas, alkali metal ions, and an oxyanion derived from the alcohol or phenol. The oxyanions are called **alkoxides** and **phenoxides,** respectively.

$$2\ Na + 2\ H_2O \longrightarrow H_2 + 2\ \underset{\text{hydroxide}}{HO^-} + 2\ Na^+$$

$$2\ Na + 2\ ROH \longrightarrow H_2 + 2\ \underset{\text{alkoxide}}{RO^-} + 2\ Na^+$$

$$2\ Na + 2\ ArOH \longrightarrow H_2 + 2\ \underset{\text{phenoxide}}{ArO^-} + 2\ Na^+$$

Some alcohols contain two or more hydroxyl groups. For example, ethylene glycol has two carbon atoms and two hydroxyl groups; glycerol has three carbon atoms and three hydroxyl groups.

$CH_2—CH_2$ with OH on each carbon (OH OH)
ethylene glycol

$CH_2—CH—CH_2$ with OH on each carbon (OH OH OH)
glycerol

Ethylene glycol is the major component of antifreeze.

Ethylene glycol is the major ingredient in automobile antifreeze. It is very soluble in water and lowers the freezing point of water considerably. Ethylene glycol is highly toxic; a dose of 50 mL is fatal. Ethylene glycol has important industrial uses in the manufacture of the synthetic fiber Dacron® and Mylar® film, which is used in cassette recorder tapes.

Glycerol, also known as glycerin, provides the backbone of glycerophospholipids, which are the major components of cell membranes. Glycerol is also a component of triacylglycerols (fats), which are a major source of cellular energy (Chapter 13).

8.2 Classification and Nomenclature of Alcohols

Alcohols are classified by the number of alkyl groups bonded to the carbon atom bearing the hydroxyl group. You may wish to review the classification of carbon atoms in Chapter 3. Alcohols are classified as primary (1°), secondary (2°), or tertiary (3°) based on the number of alkyl or aryl groups bonded to the carbon atom to which the hydroxyl group is attached.

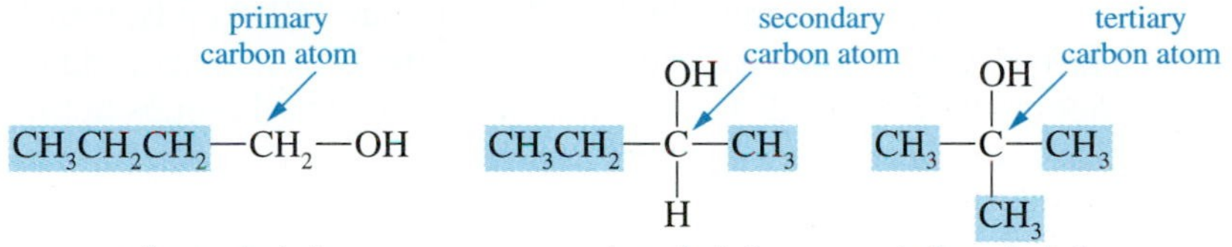

primary carbon atom — $CH_3CH_2CH_2—CH_2—OH$ — primary alcohol

secondary carbon atom — $CH_3CH_2—C(OH)(H)—CH_3$ — secondary alcohol

tertiary carbon atom — $CH_3—C(OH)(CH_3)—CH_3$ — tertiary alcohol

Common Names of Alcohols

The common names of alcohols consist of the name of the alkyl group (Section 3.3) followed by the term *alcohol.* For example, CH_3CH_2OH is ethyl alcohol and $CH_3CH(OH)CH_3$ is isopropyl alcohol. Other common names are allyl alcohol and benzyl alcohol.

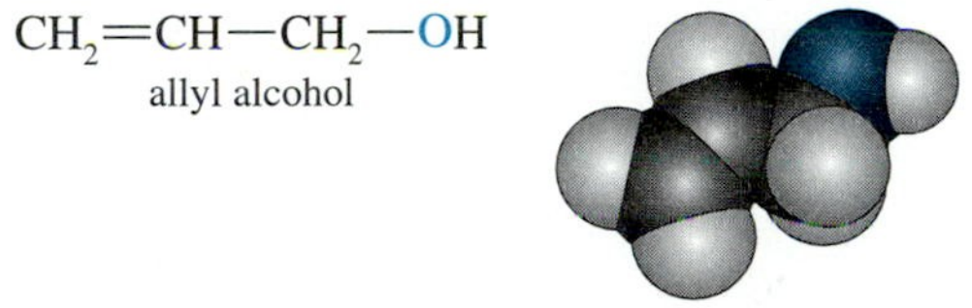

$CH_2{=}CH—CH_2—OH$
allyl alcohol

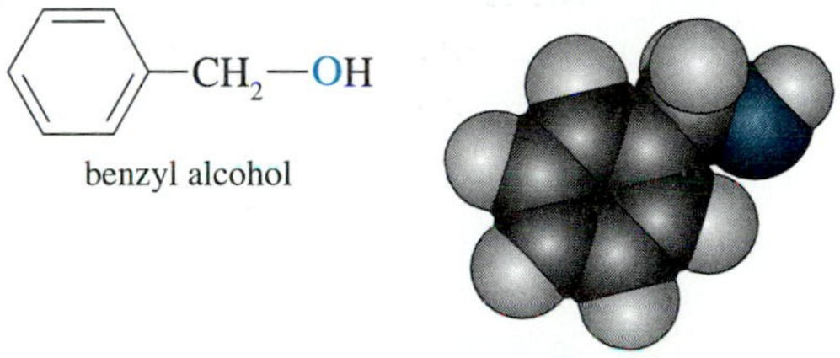

$C_6H_5—CH_2—OH$
benzyl alcohol

IUPAC Names

The IUPAC system of naming alcohols is based on the longest chain of carbon atoms that includes the hydroxyl group as the parent chain. The parent name is obtained by substituting the suffix *-ol* for the final *-e* of the corresponding alkane. The IUPAC rules are as follows:

1. The position of the hydroxyl group is indicated by the number of the carbon atom to which it is attached. The chain is numbered so that the carbon atom bearing the hydroxyl group has the lower number.

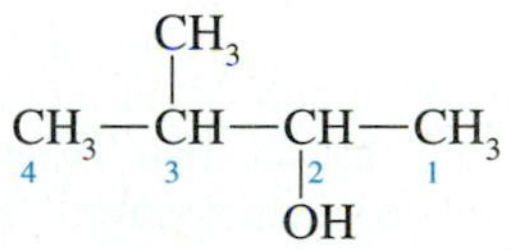

   $\underset{4}{CH_3}—\underset{3}{CH}(CH_3)—\underset{2}{CH}(OH)—\underset{1}{CH_3}$

   This is 3-methyl-2-butanol, not 2-methyl-3-butanol.

2. When the hydroxyl group is attached to a ring, the ring is numbered starting with the carbon atom bearing the hydroxyl group. Numbering continues in the direction that gives the lowest numbers to carbon atoms with substituents

such as alkyl groups. The number 1 is not used in the name to indicate the position of the hydroxyl group.

trans-2-methylcyclobutanol 3,3-dimethylcyclohexanol

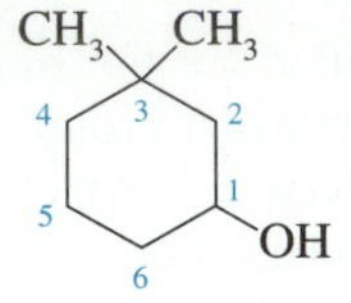

3. Alcohols containing two or more hydroxyl groups are called diols, triols, and so on. The terminal *-e* in the name of the parent alkane is retained, and the suffix *-diol* or *-triol* is added. The positions of the hydroxyl groups in the parent chain are indicated by numbers as a prefix.

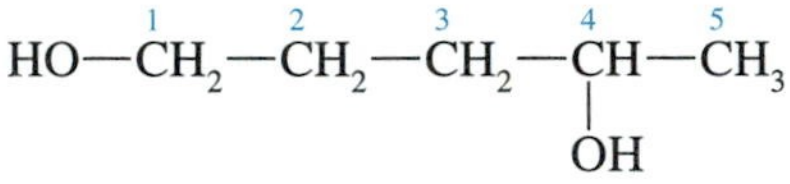

$$HO{-}CH_2{-}CH_2{-}CH_2{-}CH(OH){-}CH_3$$

1,4-pentanediol

4. The hydroxyl group has precedence in numbering the carbon chain of alcohols containing a double or triple bond. The number that indicates the position of the multiple bond is located in the prefix of the name of the alkene (or alkyne). The number that indicates the position of the hydroxyl group is appended to the name of the alkene (or alkyne) along with the suffix *-ol*.

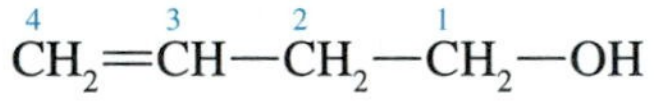

$$CH_2{=}CH{-}CH_2{-}CH_2{-}OH \qquad CH_3{-}C{\equiv}C{-}CH(OH){-}CH_3$$

3-buten-1-ol 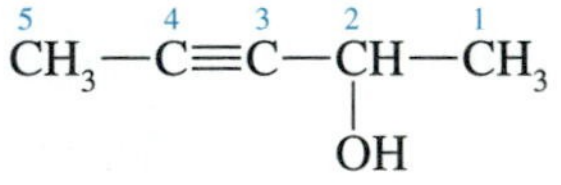 3-pentyn-2-ol

EXAMPLE 8-1

Classify the alcohol functional groups in the broad-spectrum antibiotic chloramphenicol.

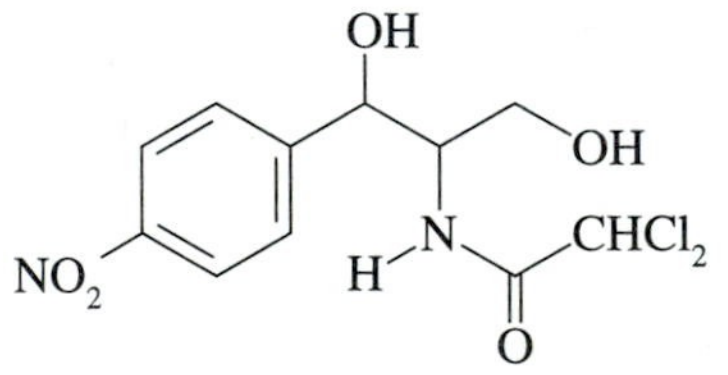

Solution

First, locate the oxygen atoms in the structure. The five oxygen atoms appear at four sites. Two oxygen atoms are bonded to a nitrogen atom in the nitro group; the double-bonded oxygen atom is part of a carbonyl group of an amide. The hydroxyl group on the right is bonded to a carbon atom that has two hydrogen atoms and a carbon atom bonded to it; this is a primary alcohol. The hydroxyl group in the middle of the molecule is bonded to a carbon atom with two other carbon atoms bonded to it; one carbon atom is part of a substituted alkyl group, the other carbon atom is part of an aryl group. This alcohol is secondary.

Problem 8.1

Classify the alcohol functional groups in riboflavin (vitamin B_2)

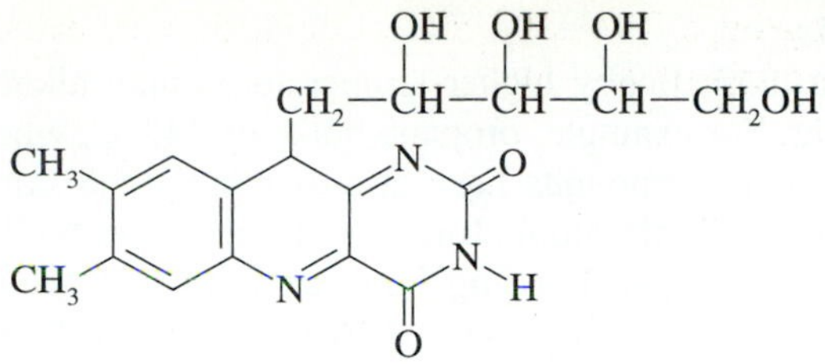

OH OH OH
$CH_2-CH-CH-CH-CH_2OH$
CH_3 N O
N H
CH_3 N
O

EXAMPLE 8-2

(*E*)-1-Chloro-3-ethyl-1-penten-4-yn-3-ol is a sedative-hypnotic known as ethchlorvynol. Draw its structure.

Solution

The term *1-penten* tells us that the parent chain contains five carbon atoms and has a double bond between the C-1 and C-2 atoms. The term *4-yn* informs us that there is a triple bond between the C-4 and C-5 atoms. Using this information, we draw the five-carbon-atom chain and select a direction for numbering it. Then we place the multiple bonds in the proper places.

$$\underset{5}{C}\equiv\underset{4}{C}-\underset{3}{C}-\underset{2}{C}=\underset{1}{C}$$

The IUPAC name also tells us that a chlorine atom is located at the C-1 atom, an ethyl group at the C-3 atom, and a hydroxyl group at the C-3 atom.

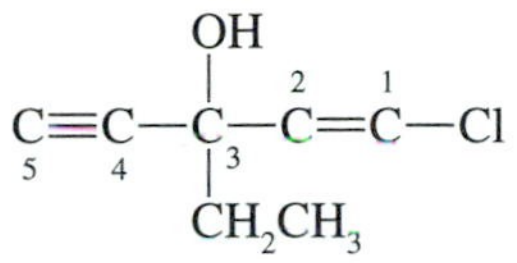

OH
$\underset{5}{C}\equiv\underset{4}{C}-\underset{3}{C}-\overset{2}{C}=\overset{1}{C}-Cl$
CH_2CH_3

Next, we draw the (*E*) configuration by placing the chlorine atom and the C-3 atom on opposite sides of the double bond. Finally, we add the necessary hydrogen atoms.

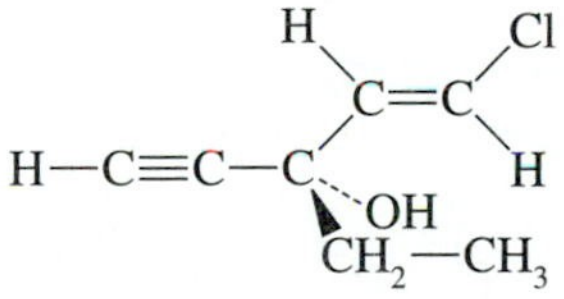

H Cl
C=C
$H-C\equiv C-C$ H
OH
CH_2-CH_3

Problem 8.2

Assign the IUPAC name for citronellol, a compound found in geranium oil, which is used in perfumes.

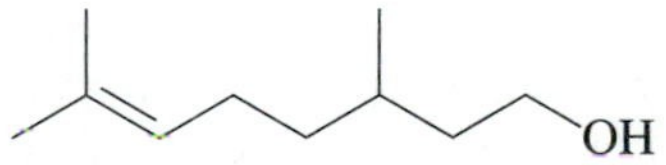

OH

8.3 Physical Properties of Alcohols

The dipole moments of ethanol and propane are 1.69 and 0.08 D, respectively. Alcohols are much more polar than alkanes because they have both a polar C—O bond and a polar O—H bond. However, the strong intermolecular forces in alcohols are the result of hydrogen bonds, which have an enormous influence on the physical properties of alcohols.

Boiling Points

Alcohols boil at dramatically higher temperatures than alkanes of comparable molecular weight. For example, propane boils at −42°C, whereas ethanol boils at 78°C. These two compounds have approximately the same London forces, but ethanol also has dipole-dipole forces of attraction as well as large forces of attraction due to hydrogen bonding between hydroxyl groups. The hydroxyl group of an alcohol can serve as both a hydrogen bond donor and a hydrogen bond acceptor. As a result, much more energy is needed to separate hydrogen-bonded alcohol molecules than is required to disrupt the relatively weak London forces in alkanes.

hydrogen bond

$H_3C{-}\ddot{O}{-}H \cdots H{-}\ddot{O}{-}CH_3$

The differences in the boiling points of the 1-alkanols and alkanes of approximately the same molecular weight are illustrated in Figure 8.2. As the molecular weights of alkanes and alcohols increase, the two curves approach each other. In alcohols with high molecular weights, hydrogen bonding is still possible, but interactions due to London forces increase due to the longer carbon chain. As a consequence, the difference in boiling point between an alcohol and an alkane of comparable molecular weight decreases.

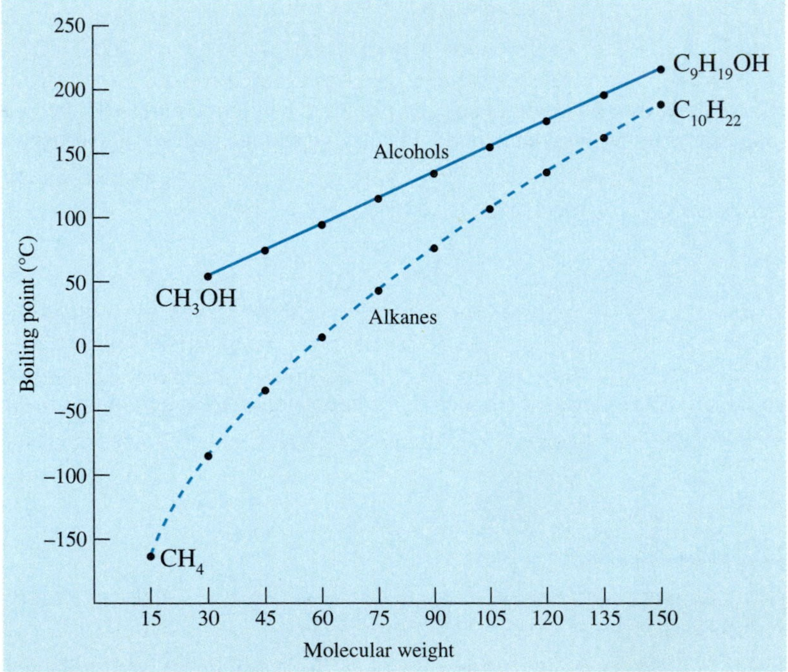

FIGURE 8.2

Boiling Points of Alkanes and Alcohols

The boiling points of both alkanes and alcohols increase with increasing chain length. Alcohols have higher boiling points than alkanes of comparable molecular weight.

Table 8.1 Boiling Points and Solubilities of Alcohols

Name	Formula	Boiling Point (°C)	Solubility (g/100 mL water)
methanol	CH_3OH	65	miscible
ethanol	CH_3CH_2OH	78	miscible
1-propanol	$CH_3CH_2CH_2OH$	97	miscible
1-butanol	$CH_3CH_2CH_2CH_2OH$	117	7.9
1-pentanol	$CH_3CH_2CH_2CH_2CH_2OH$	137	2.7
1-hexanol	$CH_3(CH_2)_4CH_2OH$	158	0.59
1-heptanol	$CH_3(CH_2)_5CH_2OH$	176	0.09
1-octanol	$CH_3(CH_2)_6CH_2OH$	194	insoluble
1-nonanol	$CH_3(CH_2)_7CH_2OH$	213	insoluble
1-decanol	$CH_3(CH_2)_8CH_2OH$	229	insoluble

Solubility of Alcohols in Water

The lower molecular weight alcohols are miscible with water (Table 8.1). These molecules, like water, are highly polar, and we know that "like dissolves like." Furthermore, the hydroxyl group can form three hydrogen bonds to water. Two can form between the two sets of lone pair electrons of the alcohol oxygen atom, which are hydrogen bond acceptors, and the hydrogen atoms of water. The third can form between the hydrogen atom of the hydroxyl group, which is a hydrogen bond donor, and the lone pair electrons on the oxygen atom in water. As the size of the alkyl group increases, alcohols more closely resemble alkanes, and the hydroxyl group has a smaller effect on their physical properties. Water can still can form hydrogen bonds to the hydroxyl group. However, the long chain interferes with other water molecules and prevents them from hydrogen bonding to each other. As a result, the solubility of alcohols decreases with increasing size of the alkyl group.

Alcohols as Solvents

Ethanol is an excellent solvent for many organic compounds, especially those with lone pair electrons that are hydrogen bond acceptors. Polar compounds dissolve readily in the "like" polar solvent. Nonpolar compounds dissolve in alcohols to some extent, but the solubility is often limited because the extensive hydrogen bonding network of the alcohol must be broken to accommodate the solute.

8.4 Acid-Base Reactions of Alcohols

We know that water can act as a proton donor (a Brønsted acid) in some reactions and as a proton acceptor (a Brønsted base) in other reactions depending on conditions. Alcohols can also act as acids or bases. That is, alcohols are amphoteric substances.

Alcohols are slightly weaker acids than water; the K_a of ethanol is 1.3×10^{-16} (pK_a = 15.9). The pK_a values of some common alcohols are listed in Table 8.2. We recall that a strong acid has a large K_a and a small pK_a.

$$CH_3CH_2OH + H_2O \rightleftharpoons CH_3CH_2O^- + H_3O^+ \qquad K_a = 1.3 \times 10^{-16}$$

Table 8.2 Effects of Structure on Acidity of Alcohols

Compound	K_a	pK_a
CH_3OH	3.2×10^{-16}	15.5
CH_3CH_2OH	1.3×10^{-16}	15.9
$(CH_3)_2CHOH$	1×10^{-18}	18.0
$(CH_3)_3COH$	1×10^{-19}	19.0
$ClCH_2CH_2OH$	5×10^{-15}	14.3
CF_3CH_2OH	4×10^{-13}	12.4
$CF_3CH_2CH_2OH$	2.5×10^{-15}	14.6
$CF_3CH_2CH_2CH_2OH$	4×10^{-16}	15.4

The acidity of alcohols increases when electronegative substituents are added to the carbon atoms near the hydroxyl group. Such substituents withdraw electron density from the oxygen atom by an inductive effect that acts through the network of σ bonds. This inductive effect weakens the O—H bond, making the alcohol more acidic. The inductive effect stabilizes the negative charge of the conjugate base. Table 8.2 shows this effect. Replacing a hydrogen atom at the C-2 atom of ethanol with a chlorine atom decreases the pK_a from 15.9 to 14.3, which means that K_a increases by a factor of 40. Replacing all three hydrogen atoms at the C-2 atom of ethanol with the more electronegative fluorine atoms substantially decreases the pK_a. The inductive effect of the electron withdrawing —CF_3 group decreases with distance from the oxygen atom. The pK_a of 4,4,4-trifluorobutanol is similar to the pK_a of a primary alcohol such as ethanol.

When an alcohol loses a proton, a conjugate base called an **alkoxide ion** is produced. Because alcohols are weaker acids than water, alkoxides are somewhat stronger bases than hydroxide ion. Alkoxides are used as bases in organic solvents because they are more soluble than hydroxide salts. Alkoxide anions can easily be prepared by reaction with sodium hydride, a strong base.

$$NaH + CH_3OH \longrightarrow H_2 + CH_3O^- + Na^+$$

Alcohols can act as bases because they have two sets of lone pair electrons on the oxygen atom. But alcohols are very weak bases and can only be protonated by strong acids to form the conjugate acid, an **oxonium ion.** The formation of an oxonium ion is analogous to the reaction of water with a strong acid to give the hydronium ion. Oxonium ions are intermediates in many reactions catalyzed by strong acids.

$$CH_3CH_2OH + HA \rightleftharpoons \underset{\text{an oxonium ion}}{CH_3CH_2OH_2^+} + A^-$$

$$H_2O + HA \rightleftharpoons \underset{\text{hydronium ion}}{H_3O^+} + A^-$$

In summary, alcohols are amphoteric. They exist as alkoxide ions in strongly basic solution, as alcohols in neutral solution, and as oxonium ions in strongly acidic solution.

8.5 Substitution Reactions of Alcohols

The hydroxyl group of an alcohol can be replaced by a halogen atom in either an S_N2 or S_N1 reaction (Chapter 7). For example, treating a primary alcohol with hydrogen bromide, HBr, produces an alkyl bromide. Similarly, treating a primary alcohol with HCl in the presence of $ZnCl_2$, which is required as a catalyst, produces an alkyl chloride.

$$CH_3CH_2CH_2CH_2OH + HBr \longrightarrow CH_3CH_2CH_2CH_2Br + H_2O$$

$$CH_3CH_2CH_2OH + HCl \xrightarrow{ZnCl_2} CH_3CH_2CH_2Cl + H_2O$$

These reactions also occur when secondary and tertiary alcohols are the substrates. The relative reaction rates depend on the type of alcohol and decrease in the order tertiary > secondary > primary alcohols.

The reaction mechanism depends on the structure of the alkyl group. Primary alcohols react by an S_N2 mechanism, whereas secondary and tertiary alcohols react by an S_N1 mechanism. However, in each case the leaving group is not hydroxide ion but a water molecule. The acid catalyst is required to form the conjugate acid of the alcohol. The departure of water, a neutral leaving group, from a developing carbocation center requires less energy than the departure of hydroxide ion, a negatively charged leaving group.

$$R{-}\ddot{O}{:}{-}H \overset{H^+}{\rightleftharpoons} R{-}\ddot{O}^+{-}H_2$$

$$\xrightarrow[S_N1]{-H_2O} R^+ \xrightarrow{Nu:^-} R{-}Nu$$

$$\xrightarrow[S_N2]{Nu:^-} R{-}Nu + H_2O$$

Water is a better leaving group than hydroxide ion because water is the weaker base. There is a general correlation between Lewis basicity and leaving-group ability. A weak base is a better leaving group than a stronger base in both S_N1 and S_N2 reactions.

Primary and secondary alcohols, which react only slowly with HBr and HCl, react readily with thionyl chloride and phosphorus trihalides, such as phosphorus tribromide, to give the corresponding alkyl halides. The products of these reactions are easily separated from the inorganic byproducts. Thionyl chloride produces hydrogen chloride and sulfur dioxide, which are released from the reaction as gases; the chloroalkane remains in solution. The reaction with phosphorus tribromide produces phosphorous acid, which has a high boiling point and is water-soluble. Thus, the bromoalkane can be separated from the reaction mixture by distillation or by adding water.

$$R{-}OH + SOCl_2 \longrightarrow R{-}Cl + HCl(g) + SO_2(g)$$

$$3\,R{-}OH + PBr_3 \longrightarrow 3\,R{-}Br + H_3PO_3$$

Both reactions are commonly written as unbalanced equations that don't show the inorganic products and contain the reagent placed over the reaction arrow.

$$CH_3CH_2CH(CH_3)CH_2CH_2OH \xrightarrow{SOCl_2} CH_3CH_2CH(CH_3)CH_2CH_2Cl$$

$$C_6H_{11}CH_2OH \xrightarrow{PBr_3} C_6H_{11}CH_2Br$$

8.6 Dehydration of Alcohols

The removal of a water molecule from an alcohol is a **dehydration reaction,** which is an example of an elimination reaction. This reaction requires an acid catalyst, such as sulfuric acid or phosphoric acid, and is illustrated by the formation of ethylene from ethanol.

These atoms are eliminated.

$$\mathrm{H{-}C(H)(OH){-}C(H)(H){-}H \xrightarrow{H_2SO_4} H_2C{=}CH_2 + H_2O}$$

In a dehydration reaction, the carbon-oxygen bond of one carbon atom and a carbon-hydrogen bond of an adjacent carbon atom both break. Because there is often more than one adjacent carbon atom, such as in 2-butanol, mixtures of alkenes can result.

Elimination of water can occur in two possible ways.

$$\mathrm{H{-}C(H)(H){-}C(H)(OH){-}C(H)(H){-}C(H)(H){-}H \qquad H{-}C(H)(H){-}C(H)(OH){-}C(H)(H){-}C(H)(H){-}H}$$

In general, when two or more products are possible in a dehydration reaction, the product formed in higher yield is the isomer that contains the greater number of alkyl groups attached to the double bond (the more substituted alkene). This more stable alkene is often called the **Zaitsev product** (Section 4.12). If the products are a mixture of geometric isomers, the more stable trans isomer predominates.

$$\mathrm{CH_3CH_2CH(OH)CH_2CH_3 \xrightarrow{H_2SO_4} \underset{\text{cis-2-pentene (25\%)}}{CH_3CH_2(H)C{=}C(H)CH_3} + \underset{\text{trans-2-pentene (75\%)}}{CH_3CH_2(H)C{=}C(CH_3)H}}$$

cis-2-pentene (25%) *trans*-2-pentene (75%)

Dehydration reactions (and the reverse hydration reactions) occur in metabolic processes. These reactions are catalyzed by enzymes and are rapid, even though high concentrations of acids are not present and the reaction temperature is only 37°C. One example is the dehydration of citric acid, catalyzed by the enzyme aconitase, to give *cis*-aconitic acid.

$$\mathrm{\underset{\text{citric acid}}{HO_2C{-}CH_2{-}C(OH)(CO_2H){-}CH_2{-}CO_2H} \xrightarrow{aconitase} \underset{\text{cis-aconitic acid}}{(HO_2CCH_2)(CO_2H)C{=}C(H)(CO_2H)} + H_2O}$$

citric acid *cis*-aconitic acid

Mechanism of Dehydration

The dehydration of alcohols occurs by mechanisms that depend on the structure of the alcohol. Tertiary alcohols undergo acid-catalyzed dehydration by an E1 mechanism; primary alcohols are dehydrated by an E2 mechanism. In either mechanism, the first step is the protonation of the oxygen atom (an acid-base reaction) to produce an oxonium ion. The acid is represented as HA.

$$(CH_3)_3C{-}\ddot{O}{-}H + H{-}A \longrightarrow (CH_3)_3C{-}\overset{+}{O}H_2 + {:}A^-$$

an oxonium ion

A tertiary alcohol loses water by an S_N1 process to produce a tertiary carbocation. The tertiary carbocation then loses a proton from the carbon atom adjacent to the carbon atom bearing the positive charge, and an alkene is produced.

$$(CH_3)_3C{-}\overset{+}{O}H_2 \longrightarrow CH_3{-}\overset{+}{C}(CH_3)_2 + H_2O$$

a tertiary carbocation

$$H_3C{-}\overset{+}{C}(CH_3)_2 \longrightarrow H_2C{=}C(CH_3)_2 + H^+$$

Primary alcohols are dehydrated by an E2 mechanism. The first step again is protonation of the —OH group to give an oxonium ion. However, in the second step, a base removes a proton from the carbon atom adjacent to the carbon atom bearing the oxygen atom. An electron pair in the C—H bond "moves" to form a carbon-carbon double bond and the electron pair of the C—O bond is retained by the oxygen atom. Formation of the double bond and loss of water occur in a single concerted step.

$$H{-}\ddot{O}{-}H + CH_3{-}CH_2{-}CH_2{-}\overset{+}{O}H_2 \longrightarrow CH_3CH{=}CH_2 + H_3O^+ + H_2O$$

Note that in both the E1 process and the E2 process, the acid HA serves only as a catalyst; it is regenerated in the last step of the reaction.

EXAMPLE 8-3

Predict the product(s) of the dehydration of 1-methylcyclohexanol.

HO CH_3

Solution

This tertiary alcohol has three carbon atoms adjacent to the carbon atom bearing the hydroxyl group. Each carbon atom can lose a hydrogen atom in the dehydration reaction. However, loss of a hydrogen atom from either the C-2 or C-6 atoms results in the same product. Thus, only two isomers are formed.

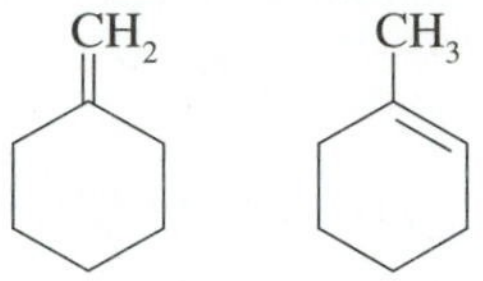

The second structure, with the double bond within the six-membered ring, predominates because it is the more highly substituted alkene.

Problem 8.3

Write the structures of the products of dehydration of 4-methyl-2-pentanol. Which of the isomeric alkenes should be the major product?

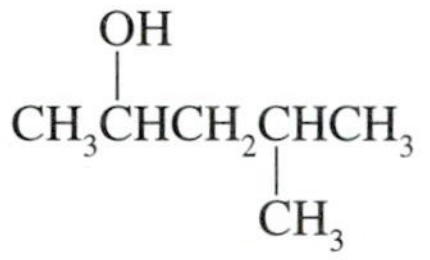

$$CH_3CH(OH)CH_2CH(CH_3)CH_3$$

8.7 Oxidation of Alcohols

Primary and secondary alcohols react differently with oxidizing agents. Primary alcohols, which have the general formula RCH_2OH, can be oxidized to aldehydes, which have the general formula RCHO. Note that this oxidation occurs with the loss of two hydrogen atoms. Aldehydes are easily oxidized and may react further to produce carboxylic acids, RCOOH. When an aldehyde is oxidized to a carboxylic acid, the oxidized carbon atom gains an oxygen atom.

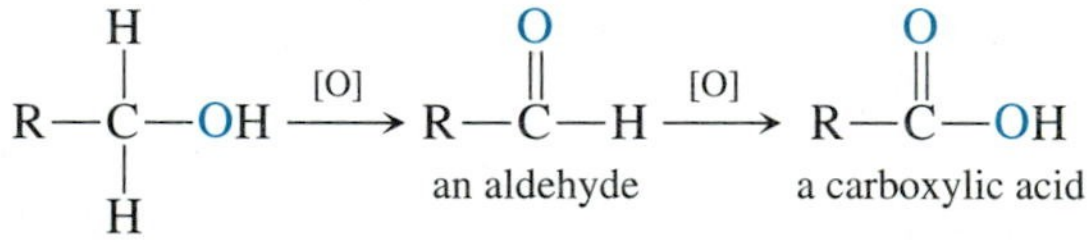

$$R{-}CH_2{-}OH \xrightarrow{[O]} \underset{\text{an aldehyde}}{R{-}C({=}O){-}H} \xrightarrow{[O]} \underset{\text{a carboxylic acid}}{R{-}C({=}O){-}OH}$$

Secondary alcohols are oxidized to form ketones, RCOR, which cannot be further oxidized because there is no hydrogen atom on the oxygen-bearing carbon atom of the ketone. Tertiary alcohols are not oxidized because the carbon atom bearing the —OH group has no hydrogen atom.

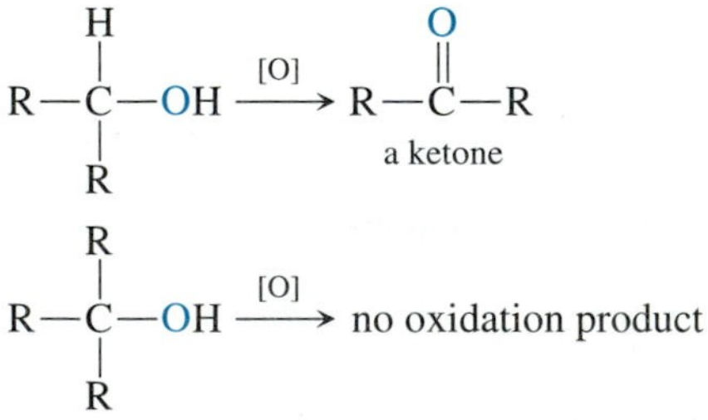

$$R{-}CH(R){-}OH \xrightarrow{[O]} \underset{\text{a ketone}}{R{-}C({=}O){-}R}$$

$$R{-}C(R)_2{-}OH \xrightarrow{[O]} \text{no oxidation product}$$

Alcohols are oxidized by the Jones reagent, which consists of chromium trioxide in aqueous sulfuric acid and acetone. The Jones reagent oxidizes primary alcohols to carboxylic acids. This reagent also converts secondary alcohols to ketones.

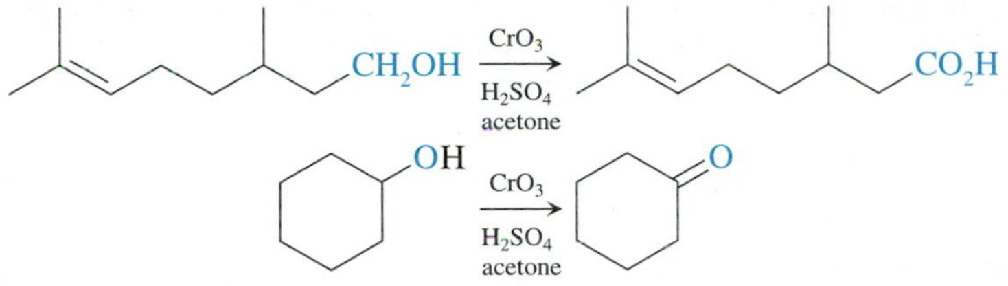

$$\text{(geraniol-type)}\ CH_2OH \xrightarrow[\text{H}_2\text{SO}_4,\ \text{acetone}]{\text{CrO}_3} CO_2H$$

$$\text{cyclohexanol (OH)} \xrightarrow[\text{H}_2\text{SO}_4,\ \text{acetone}]{\text{CrO}_3} \text{cyclohexanone (=O)}$$

Alcohols are also oxidized by pyridinium chlorochromate (PCC) in methylene chloride (CH_2Cl_2) as solvent. Secondary alcohols are oxidized to ketones. However, in contrast to the Jones reagent, oxidation of primary alcohols by PCC yields aldehydes without continued oxidation to carboxylic acids.

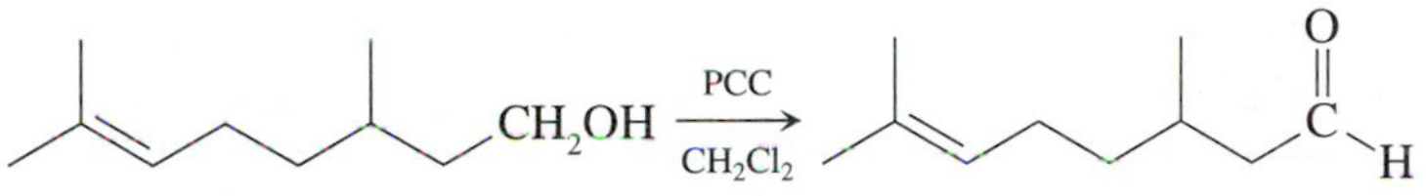

$$CH_2OH \xrightarrow[\text{CH}_2\text{Cl}_2]{\text{PCC}} \overset{O}{\overset{\|}{C}}-H$$

PCC is prepared by dissolving CrO_3 in hydrochloric acid and then adding pyridine. The isolated PCC is used in CH_2Cl_2 as solvent.

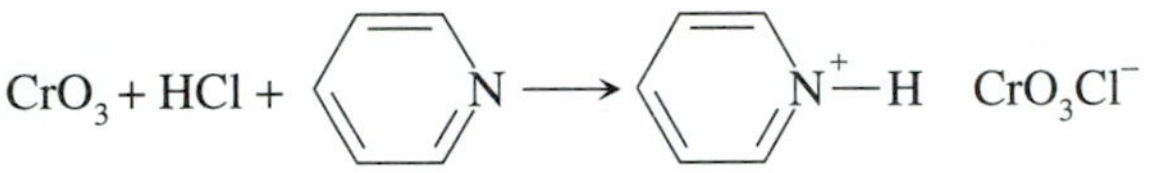

$$CrO_3 + HCl + C_5H_5N \longrightarrow C_5H_5\overset{+}{N}-H\ \ CrO_3Cl^-$$

pyridinium chlorochromate (PCC)

EXAMPLE 8-4

Which of the isomeric $C_4H_{10}O$ alcohols react with Jones reagent to produce a ketone, C_4H_8O?

Solution

There are four isomeric alcohols because there are four C_4H_9— alkyl groups (Section 3.3). The *n*-butyl and isobutyl groups are primary; the *tert*-butyl group is tertiary. Only the *sec*-butyl group provides a secondary alcohol, and only secondary alcohols yield ketones when oxidized. Thus, only 2-butanol (*sec*-butyl alcohol) yields a ketone.

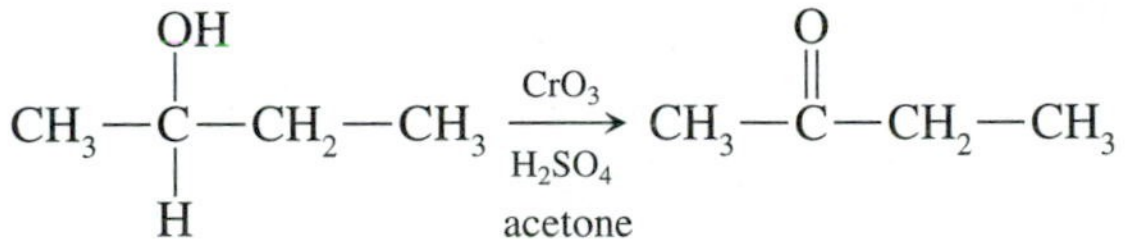

$$CH_3-\underset{H}{\overset{OH}{C}}-CH_2-CH_3 \xrightarrow[\text{H}_2\text{SO}_4,\ \text{acetone}]{\text{CrO}_3} CH_3-\overset{O}{\overset{\|}{C}}-CH_2-CH_3$$

Problem 8.4

Which of the isomeric $C_5H_{12}O$ alcohols react with PCC to produce an aldehyde, $C_5H_{10}O$?

8.8 Synthesis of Alcohols

We have already encountered two methods of preparing alcohols: (1) substitution of a halide by hydroxide and (2) addition of water to an alkene. These reactions often have low yields. The yield in the substitution reaction is diminished

Toxicity of Alcohols

Many alcohols are poisonous substances. Methanol is highly toxic. Drinking as little as 15 mL (1 tablespoon) of pure methanol can cause blindness; 30 mL will cause death. Prolonged breathing of methanol vapors is also a serious health hazard.

Although ethanol is the least toxic of the simple alcohols, it is still a poisonous substance that must be oxidized in the body to prevent high blood alcohol levels, which can "poison" the brain. Alcohol is oxidized in the liver by the enzyme alcohol dehydrogenase (ADH). ADH requires a coenzyme, nicotinamide adenine dinucleotide (NAD^+), as an oxidizing agent. The coenzyme can exist in an oxidized form, NAD^+, and a reduced form, NADH. Ethanol is oxidized to ethanal (acetaldehyde) by NAD^+-dependent liver ADH. Subsequent oxidation of ethanal yields ethanoic acid (acetic acid), which is nontoxic.

$$CH_3-CH_2-OH \xrightarrow[NAD^+]{ADH} CH_3-\overset{O}{\overset{\|}{C}}-H$$

ethanal (acetaldehyde)

$$CH_3-\overset{O}{\overset{\|}{C}}-H \xrightarrow[NAD^+]{ADH} CH_3-\overset{O}{\overset{\|}{C}}-OH$$

ethanoic acid (acetic acid)

Acetic acid is not toxic; it is a source of acetyl groups for the Krebs cycle—an important source of biological energy. Excess acetyl groups are converted into fat in a series of fatty acid biosynthesis reactions. Nevertheless, ingestion of alcohol as a food source is not recommended because it is a source of "empty" calories—it lacks minerals, vitamins, and proteins. Furthermore, the liver—the major site of alcohol metabolism—is damaged when excessive amounts of ethyl alcohol are consumed.

The oxidation products of some other alcohols are toxic. In the case of methanol, oxidation catalyzed by ADH gives methanal (formaldehyde) and then methanoic acid (formic acid). Methanol is not normally ingested. However, it is contained in illicit sources of ethyl alcohol such as moonshine.

$$H-\underset{H}{\overset{H}{C}}-OH \xrightarrow[NAD^+]{ADH} H-\overset{O}{\overset{\|}{C}}-H$$

methanal (formaldehyde)

$$H-\overset{O}{\overset{\|}{C}}-H \xrightarrow[NAD^+]{ADH} H-\overset{O}{\overset{\|}{C}}-OH$$

methanoic acid (formic acid)

Formaldehyde is transported in the blood throughout the body and reacts very rapidly with proteins, destroying their biological function. For example, it reacts with an amine functional group of lysine contained in rhodopsin, a protein required for vision (Chapter 10), which is why the ingestion of methanol causes blindness. Formaldehyde also reacts with amino groups in other proteins, including many enzymes, and the loss of the function of these biological catalysts causes death.

Ethylene glycol is toxic. This sweet-tasting substance is the major component of antifreeze. Dogs are poisoned when they ingest antifreeze left in open containers. Oxidation occurs to give oxalic acid, which causes kidney failure.

$$H-\underset{H}{\overset{OH}{C}}-\underset{H}{\overset{OH}{C}}-H \xrightarrow{[O]} HO-\overset{O}{\overset{\|}{C}}-\overset{O}{\overset{\|}{C}}-OH$$

ethylene glycol oxalic acid

Intravenous injections of ethanol are used to treat methanol or ethylene glycol poisoning before substantial oxidation has occurred. ADH binds more tightly to ethanol than to methanol or ethylene glycol, and the rate of oxidation of ethanol is about six times faster than that of methanol. The ethanol is also present in higher concentration because it is directly injected. As a result, neither methanol nor ethylene glycol are oxidized to toxic products. These "toxic" alcohols are then slowly excreted by the kidneys. ■

by the competing elimination reaction (Chapter 7). The yield in the hydration of an alkene is somewhat limited because the reaction is reversible (Chapter 4).

In this section we will discuss two types of reactions that give excellent yields of alcohols. The first type is the reduction of carbonyl compounds; the second is an "indirect" hydration of alkenes. In both reactions, the functional group converted to the hydroxyl group is located on the proper hydrocarbon skeleton. In Chapter 10 we will examine reactions to simultaneously form alcohols and build new hydrocarbon skeletons.

Although grapes are the source of ethanol in wine, most ethanol for other products is produced by hydration of ethylene.

Reduction of Carbonyl Compounds

Alcohols can be produced by reducing the carbonyl group of aldehydes and ketones with hydrogen gas at approximately 100 atm pressure in the presence of a metal catalyst such as palladium, platinum, or a special reactive form of nickel called Raney nickel. Aldehydes yield primary alcohols; ketones yield secondary alcohols.

$$\underset{\text{an aldehyde}}{R{-}\overset{\overset{\displaystyle O}{\|}}{C}{-}H} \xrightarrow{[H]} \underset{\text{a primary alcohol}}{R{-}\overset{H}{\underset{H}{C}}{-}OH}$$

$$\underset{\text{a ketone}}{R{-}\overset{\overset{\displaystyle O}{\|}}{C}{-}R} \xrightarrow{[H]} \underset{\text{a secondary alcohol}}{R{-}\overset{H}{\underset{R}{C}}{-}OH}$$

The reduction reaction occurs by the transfer of hydrogen atoms bound to the surface of the metal catalyst to the carbonyl oxygen and carbon atoms. Note that the same catalysts are used for the hydrogenation of alkenes, which is a much faster reaction that occurs at 1 atm pressure. Thus, any carbon-carbon double bonds in a molecule are reduced under the conditions required to reduce a carbonyl group.

$$\text{cyclohexenyl}{-}\overset{\overset{\displaystyle O}{\|}}{C}{-}H \xrightarrow[\text{Raney Ni}]{H_2} \text{cyclohexyl}{-}CH_2{-}OH$$

A carbonyl group can be reduced to an alcohol selectively by reagents attracted to the highly polarized carbonyl group. The carbonyl carbon atom has a partial positive charge and tends to react with nucleophiles.

$$\text{nucleophile attacks here} \longrightarrow \overset{\delta^+}{>C}{=}\overset{\delta^-}{\ddot{O}}:$$

The carbon-carbon double bond of alkenes is not polar, so it does not react with nucleophiles. This difference in reactivity is the basis for the reduction of carbonyl compounds by metal hydrides, such as sodium borohydride, $NaBH_4$, and lithium aluminum hydride, $LiAlH_4$, neither of which reacts with alkenes. Both reagents are a source of a nucleophilic hydride ion. Sodium borohydride is less reactive than lithium aluminum hydride, but both easily reduce both aldehydes and ketones. Sodium borohydride can be used in ethanol as the solvent.

$$C_6H_5\text{-}CHO \xrightarrow[CH_3CH_2OH]{NaBH_4} C_6H_5\text{-}CH_2\text{-}OH$$

benzaldehyde benzyl alcohol

In reduction by sodium borohydride, a hydride ion of the borohydride ion, BH_4^-, is transferred to the carbonyl carbon atom from boron, and the carbonyl oxygen atom is protonated by the ethanol.

$$H_3\overset{-}{B}\text{—}H + \,\rangle C{=}\ddot{O}: + H\text{—}\ddot{O}CH_2CH_3 \longrightarrow H\text{—}\overset{|}{\underset{|}{C}}\text{—}\ddot{O}\text{—}H + CH_3CH_2\ddot{O}BH_3^-$$

ethoxyborohydride

The ethoxyborohydride product in the preceding reaction has three remaining hydride ions available for further reduction reactions, and the ultimate boron product is tetraethoxyboride, $(RO)_4B^-$. Thus, one mole of $NaBH_4$ reduces four moles of a carbonyl compound.

When lithium aluminum hydride is used to reduce carbonyl compounds, an ether such as diethyl ether, $(CH_3CH_2)_2O$, is used as the solvent.

$$\text{cyclohexanone} \xrightarrow[2.\ H_3O^+]{1.\ LiAlH_4/\text{ether}} \text{cyclohexanol}$$

cyclohexanone cyclohexanol

The reduction of a carbonyl group by lithium aluminum hydride occurs by the transfer of a hydride anion from AlH_4^- to the carbonyl carbon atom. The carbonyl oxygen atom forms a salt with aluminum. The salt is an alkoxyaluminate.

$$H_3\overset{-}{Al}\text{—}H + \,\rangle C{=}\ddot{O}: \longrightarrow H\text{—}\overset{|}{\underset{|}{C}}\text{—}\ddot{O}\text{—}AlH_3^-$$

an alkoxyaluminate

The initial alkoxyaluminate has three remaining hydride ions available for further reduction reactions, and the ultimate aluminum product is tetraalkoxyaluminate, $(RO)_4Al^-$. Thus, one mole of $LiAlH_4$ reduces four moles of a carbonyl compound. The tetraalkoxyaluminate is hydrolyzed with aqueous acid in a separate, second step.

$$(RO)_4Al^-Li^+ + 4\ H_2O \longrightarrow 4\ ROH + Li^+Al(OH)_4^-$$

Reduction of carbonyl compounds in metabolic processes occurs via NADH, the reduced form of the coenzyme NAD^+, nicotinamide adenine dinucleotide. The structures of the two forms of the coenzyme are shown with an R group representing a portion of the molecule that is not directly involved in its action.

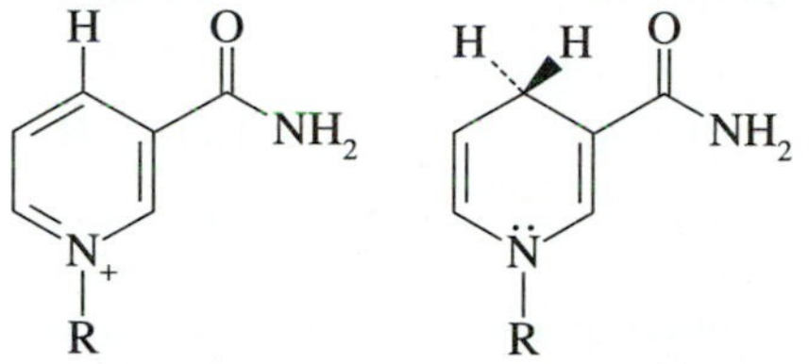

NAD$^+$
(oxidized form)

NADH
(reduced form)

The reduced form formally behaves as a hydride source by transferring a hydrogen atom and a bonding pair of electrons to a carbonyl carbon atom.

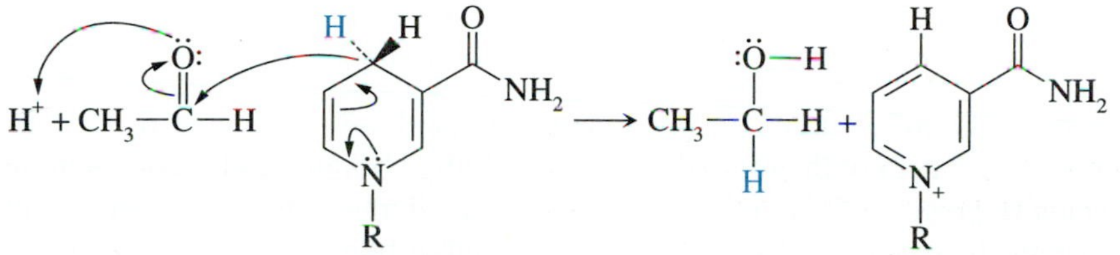

$$H^+ + CH_3-\overset{\displaystyle O}{\overset{\|}{C}}-H + \text{NADH} \longrightarrow CH_3-\underset{\displaystyle H}{\overset{\displaystyle OH}{C}}-H + \text{NAD}^+$$

NADH
(reduced form)

NAD$^+$
(oxidized form)

Indirect Hydration of Alkenes

In Chapter 4 we discussed the "direct" electrophilic addition of water to alkenes to give alcohols. In this section we consider two "indirect" ways to add the elements of water to a double bond. These methods are indirect because the hydroxyl group, the hydrogen atom, or both originate in reagents other than water. One such reaction, oxymercuration-demercuration of an alkene, gives a product that corresponds to Markovnikov addition of water. Hydroboration-oxidation, another indirect reaction, gives the equivalent of an anti-Markovnikov addition of water to the alkene.

In an oxymercuration-demercuration reaction, an alkene is treated with mercuric acetate, $Hg(OAc)_2$, and the product is treated with sodium borohydride. The net result is a **Markovnikov addition product** in which the —OH group is bonded to the more substituted carbon atom of the alkene.

$$\underset{\text{1-hexene}}{CH_3(CH_2)_3CH{=}CH_2} \xrightarrow[\text{2. NaBH}_4]{\text{1. Hg(OAc)}_2/\text{H}_2\text{O}} \underset{\text{2-hexanol}}{CH_3(CH_2)_3\overset{\displaystyle HO}{\overset{|}{C}}H-\overset{\displaystyle H}{\overset{|}{C}}H_2}$$

In the first step, an electrophilic $HgOAc^+$ ion adds to the double bond followed by attack by a nucleophilic water molecule. The net result is the bonding of —HgOAc and a hydroxyl group on adjacent carbon atoms. The product corresponds to a Markovnikov addition because the electrophile attacks the less substituted carbon atom.

$$CH_3(CH_2)_3CH{=}CH_2 + {}^+HgOAc \xrightarrow{H_2O} CH_3(CH_2)_3\overset{\displaystyle HO}{\overset{|}{C}}H-\overset{\displaystyle HgOAc}{\overset{|}{C}}H_2 + H^+$$

The organomercury compound is then reduced with sodium borohydride, and the —HgOAc group is replaced by a hydrogen atom. Thus, oxymercuration-demercuration gives the product that would result from direct hydration of an alkene.

$$CH_3(CH_2)_3CH(OH)-CH_2(HgOAc) \xrightarrow{NaBH_4} CH_3(CH_2)_3CH(OH)-CH_2(H) + Hg + OAc^-$$

Hydroboration-oxidation of alkenes, which was developed by the American chemist H. C. Brown, also requires two steps. The sequence of reactions adds the elements of water to a double bond to give a product that corresponds to **anti-Markovnikov addition.**

$$\underset{\text{1-hexene}}{CH_3(CH_2)_3CH{=}CH_2} \xrightarrow[\text{2. } H_2O_2\,/\,OH^-]{\text{1. } B_2H_6} \underset{\text{1-hexanol}}{CH_3(CH_2)_3CH(H)-CH_2(OH)}$$

In the hydroboration step, an alkene is treated with diborane, B_2H_6. Diborane acts as if it were the monomeric species called borane, BH_3. The compound is usually prepared in an ether solvent such as diethyl ether or tetrahydrofuran (Chapter 9). It adds to the carbon-carbon double bond of one alkene and then adds successively to two additional alkenes to produce a trialkylborane, R_3B. These steps are hydroboration reactions.

$$>C{=}C< + H-BH_2 \longrightarrow H-\overset{|}{\underset{|}{C}}-\overset{|}{\underset{|}{C}}-BH_2 \longrightarrow \longrightarrow \left[H-\overset{|}{\underset{|}{C}}-\overset{|}{\underset{|}{C}}\right]_3 B$$

In the oxidation step, the trialkylborane is treated with hydrogen peroxide and base to oxidize the organoborane to form an alcohol.

$$\left[H-\overset{|}{\underset{|}{C}}-\overset{|}{\underset{|}{C}}\right]_3 B \xrightarrow[OH^-]{H_2O_2} 3\,H-\overset{|}{\underset{|}{C}}-\overset{|}{\underset{|}{C}}-OH$$

The hydroboration-oxidation of 1-methylcyclohexene gives the anti-Markovnikov product; the hydrogen atom is added to the more substituted carbon atom, and the hydroxyl group is on the less substituted carbon atom. (Only a single addition of BH_3 is shown for the sake of simplicity.)

$$\text{1-methylcyclohexene} \xrightarrow{B_2H_6} \text{(H, } CH_3\text{; H, } BH_2\text{)} \xrightarrow[OH^-]{H_2O_2} \text{(H, } CH_3\text{; H, OH)}$$

Note that the hydrogen atom and BH_2 unit are introduced from the same side of the double bond. This mode of addition is observed because hydroboration is a concerted process. That is, the carbon-boron and carbon-hydrogen bonds are formed at the same time that the boron-hydrogen bond is broken. In the oxidation step a hydroxyl group replaces the boron with retention of configuration. (The mechanism of this process is beyond the scope of this text.)

Borane reacts with alkenes for two reasons. First, the boron atom in borane is an electron-deficient species—it has only six electrons. Thus, the boron atom has a vacant 2*p* orbital and is an electrophile. It bonds to the least substituted carbon atom—much like a proton. Second, boron is more electropositive than hydrogen. Therefore, the hydrogen atom of the boron-hydrogen bond has a partial negative charge. This hydrogen atom behaves like a hydride ion, not like a proton. In summation, two properties of BH_3-the electrophilic character of the boron atom and the hydride character of the hydrogen atom—account for anti-Markovnikov addition of BH_3 to alkenes.

EXAMPLE 8-5

What product is formed from the following alkene by oxymercuration-demercuration? What product is formed by hydroboration-oxidation?

(cyclopentane)$=CH_2$

Solution

The alkene is disubstituted, and both alkyl groups are bonded to the same carbon atom. The double-bonded CH_2 is the less substituted carbon atom; the ring carbon atom is the more substituted carbon atom. An oxymercuration-demercuration reaction places a hydrogen atom at the CH_2 site and a hydroxyl group on the ring carbon atom. This is equivalent to Markovnikov addition of water to the alkene.

$$\text{(cyclopentane)}{=}CH_2 \xrightarrow[\text{2. NaBH}_4]{\text{1. Hg(OAc)}_2/\text{H}_2\text{O}} \text{(cyclopentane)}(OH)(CH_3)$$

The hydroboration-oxidation product has a hydroxyl group at the CH_2 site and a hydrogen atom at the ring carbon atom. This process is equivalent to anti-Markovnikov addition of water to the double bond.

$$\text{(cyclopentane)}{=}CH_2 \xrightarrow[\text{2. H}_2\text{O}_2/\text{OH}^-]{\text{1. B}_2\text{H}_6} \text{(cyclopentane)}(H)(CH_2OH)$$

Problem 8.5

Write the product of the oxymercuration-demercuration of 3,3-dimethyl-1-butene. Do the same for the hydroboration-oxidation of this compound.

8.9 Phenols

The chemistry of phenols is quite different from the chemistry of alcohols because it is very difficult to break the C—O bond in phenols. The carbon atom is sp^2-hybridized and the C—O bond in phenols is shorter and stronger than the C—O bond of alcohols, where the carbon atom is sp^3-hybridized. As a result, neither S_N2 nor S_N1 processes occur.

Acidity of Phenols

Phenols are stronger acids than alcohols, but they are still quite weak acids. A typical alcohol has a pK_a of 16–17. In contrast, phenol is 10 million times more acidic: its pK_a is 10.

$$R{-}\ddot{\underset{..}{O}}{-}H + H_2O \rightleftharpoons R{-}\ddot{\underset{..}{O}}{:}^- + H_3O^+ \qquad pK_a = 17$$

$$Ar{-}\ddot{\underset{..}{O}}{-}H + H_2O \rightleftharpoons Ar{-}\ddot{\underset{..}{O}}{:}^- + H_3O^+ \qquad pK_a = 10$$

Industrial Synthesis of Alcohols

About 10 billion pounds of methanol are produced annually in the United States by reaction of carbon monoxide and hydrogen in a 1:2 mixture known as synthesis gas.

$$CO + 2\,H_2 \xrightarrow[400°C,\ 600\ atm]{Cr/Zn} CH_3OH$$

Synthesis gas is obtained by adjusting the ratio of carbon monoxide to hydrogen of the water gas mixture obtained by the reaction of water with methane.

$$CH_4 + H_2O \xrightarrow[700°C]{NiO} CO + 3\,H_2$$

Methanol is used as a solvent and as an antifreeze in windshield wiper fluid. It is also used as a fuel in Indianapolis-class cars because it is a clean-burning fuel and has an octane number of 116. However, the majority of the methanol produced is converted to other compounds. About 50% is catalytically oxidized by oxygen in air to methanal (formaldehyde) for use in the synthesis of resins and plastics such as Bakelite®.

$$2\,CH_3OH + O_2 \xrightarrow[400°C]{ZnO/Cr_2O_3} 2\,H_2C{=}O + 2\,H_2O$$

The gasoline additive MTBE (*tert*-butyl methyl ether) is produced by the addition of methanol to 2-methylpropene by passing a mixture of the two reactants as gases at high temperature over an acidic catalyst. The addition product is that predicted by Markovnikovs rule.

$$CH_3\ddot{O}H + CH_2{=}C(CH_3)_2 \xrightarrow{\text{acidic catalyst}} CH_3{-}C(\ddot{O}CH_3)(CH_3){-}CH_3$$

MTBE
(*tert*-butyl methyl ether)

MTBE increases the octane number of gasoline by about 5 units. Thus, this additive is being used as one of the replacements for earlier additives such as tetraethyllead, which has been eliminated due to long-term environmental effects of lead emissions. Current production levels of MTBE are 1.7 billion gallons annually in the United States.

Ethanol has been prepared by distillation of the mixture obtained by fermentation of grains and sugar. However, currently only 5% of ethanol is produced by this industrial method. Most ethanol is obtained by acid-catalyzed hydration of ethylene. About 100 million gallons of ethanol are produced annually in the United States by this method as a chemical intermediate for the production of varied industrial products.

$$H{-}\ddot{O}{-}H + CH_2{=}CH_2 \xrightarrow[300°C]{H_3PO_4} CH_3{-}CH_2{-}\ddot{O}H$$

Isopropyl alcohol is also produced by an acid-catalyzed hydration reaction. It is used as a solvent and as an intermediate for other industrial syntheses, such as the production of propanone (acetone), which is an important solvent.

$$H{-}\ddot{O}{-}H + CH_2{=}CH{-}CH_3 \xrightarrow[300°C]{H_3PO_4} CH_3{-}CH(\ddot{O}H){-}CH_3$$

$$CH_3{-}CH(\ddot{O}H){-}CH_3 \xrightarrow[400°C]{ZnO} CH_3{-}C({=}\ddot{O}){-}CH_3 + H_2$$

acetone

Isopropyl alcohol is used as an aqueous solution called rubbing alcohol. It evaporates from skin and produces a cooling effect. Isopropyl alcohol is also a weak bactericide and can be used to sterilize medical instruments.

About 3 million tons of ethylene oxide are produced annually from ethylene by direct air oxidation over a silver oxide catalyst. Most of this product is used to produce ethylene glycol in an acid-catalyzed hydrolysis reaction.

$$H_2C{=}CH_2 \xrightarrow{O_2/Ag_2O} \text{ethylene oxide} \xrightarrow{H_3O^+} HOCH_2{-}CH_2OH$$

Phenol is more acidic than cyclohexanol and acyclic alcohols because the phenoxide ion is more stable than the alkoxide ion. In an alkoxide ion, such as derived from cyclohexanol, the negative charge is localized at the oxygen atom. However, in a phenoxide ion, the negative charge is stabilized by resonance, in which the charge is delocalized over the benzene ring.

localized negative charge

delocalized resonance contributors

Listerine® contains several substituted phenols that are germicides.

Phenols are more acidic when the ring is substituted with electron-withdrawing groups. These substituents stabilize the phenoxide ion by further delocalizing the negative charge. Phenols substituted with electron-donating groups are less acidic than phenol.

p-bromophenol	phenol	*p*-methylphenol
($pK_a = 9.35$)	($pK_a = 10.00$)	($pK_a = 10.26$)

Phenols are much more acidic than alcohols, and as a result they dissolve in basic solutions. However, phenols are not sufficiently acidic to react with aqueous sodium bicarbonate, in contrast to organic acids such as carboxylic acids (pK_a = 5), which react with aqueous sodium bicarbonate to form carbon dioxide gas and a carboxylate anion.

Oxidation of Phenols

Phenols are oxidized to give conjugated 1,4-diketones called quinones. Hydroquinone, a phenol with two hydroxyl groups, is very easily oxidized. It will reduce silver bromide that has been activated by exposure to light in photographic film emulsions.

hydroquinone + 2 AgBr ⟶ 1,4-benzoquinone + 2 Ag + 2 HBr

Hydroquinone is used as a developer. It reacts faster with the light-activated silver bromide than with unexposed grains of the salt. As a consequence, silver deposits in the film at points where exposure to light occurred. The result is a "negative" image. Areas in the film negative that are clear correspond to unlit areas in the real world; areas that are very dark in the negative correspond to well-lit areas in the scene photographed.

Quinones are widely distributed in nature, where along with the reduced hydroquinone form they serve as reducing and oxidizing agents. Coenzyme Q, also called ubiquinone, is a quinone found within the mitochondria of cells. Coenzyme Q, in its oxidized form, oxidizes NADH to regenerate NAD^+, a common oxidizing agent in biological reactions. The side chain represented by R in the following structure is a polyunsaturated unit that consists of isoprene units.

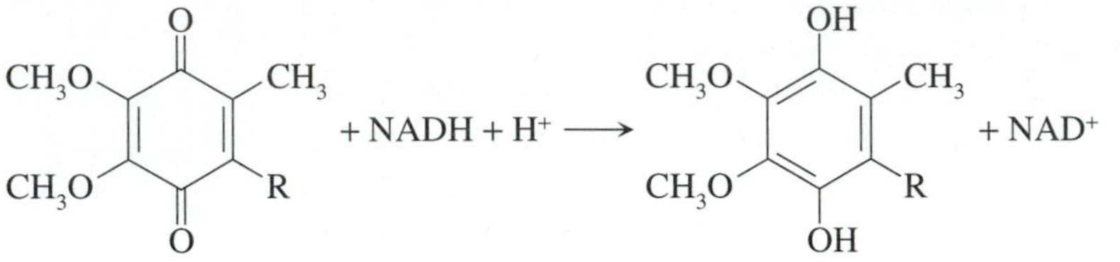

$$\text{coenzyme Q (oxidized form)} + NADH + H^+ \longrightarrow \text{coenzyme Q (reduced form)} + NAD^+$$

coenzyme Q (oxidized form)

coenzyme Q (reduced form)

8.10 Sulfur Compounds

Sulfur is in the same group of the periodic table as oxygen. It forms compounds that are structurally similar to alcohols. The —SH group is called the **sulfhydryl group.** Compounds containing an —SH group are called **mercaptans** or **thiols.** The nomenclature of these compounds resembles that of alcohols, except that the suffix *-thiol* replaces the suffix *-ol*. The *-e* of the alkane name is retained.

$$CH_3{-}CH_2{-}\underset{}{\overset{SH}{\overset{|}{C}}}H{-}CH_2{-}CH_3$$

3-pentanethiol

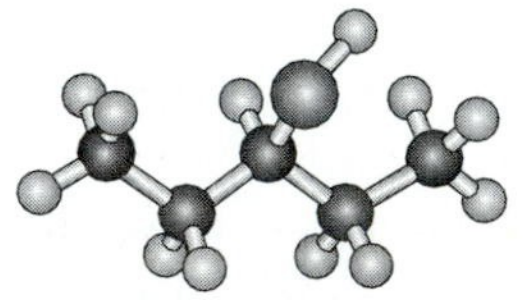

Alcohols and thiols differ in some significant respects. For example, thiols have lower boiling points than the corresponding alcohols because sulfur does not form hydrogen bonds. We recall from Chapter 2 that only nitrogen, oxygen, and fluorine form hydrogen bonds.

$$\underset{\text{bp} = 98^\circ\text{C}}{CH_3CH_2CH_2CH_2SH} \qquad \underset{\text{bp} = 117^\circ\text{C}}{CH_3CH_2CH_2CH_2OH}$$

Some alcohols have rather sweet odors, but one of the distinguishing properties of thiols is their strong, disagreeable odor. The odor of the striped skunk is due to 3-methyl-1-butanethiol. Thiols can be detected by the human nose at the level of parts per billion in air. Small amounts of thiols are added to natural gas so that leaks can easily be detected. A skunk takes more drastic measures to fend off predators.

Although thiols are weak acids, they are far stronger than alcohols.

$$R{-}\ddot{\underset{..}{S}}{-}H + H_2O \rightleftharpoons R{-}\ddot{\underset{..}{S}}{:}^- + H_3O^+ \quad pK_a = 8$$

$$R{-}\ddot{\underset{..}{O}}{-}H + H_2O \rightleftharpoons R{-}\ddot{\underset{..}{O}}{:}^- + H_3O^+ \quad pK_a = 17$$

The sulfhydryl group is acidic enough to react with hydroxide ions to form thiolate salts.

$$R{-}\ddot{\underset{..}{S}}{-}H + OH^- \longrightarrow \underset{\text{a thiolate}}{R{-}\ddot{\underset{..}{S}}{:}^-} + H_2O$$

Phenols are Germicides

Germicides are classified as disinfectants and antiseptics. Disinfectants are compounds that decrease the bacterial count on objects such as medical instruments. Antiseptics also inhibit bacterial growth but are used on living tissue. The English surgeon Joseph Lister used phenol itself as a hospital disinfectant in the late nineteenth century. But phenol is no longer used as an antiseptic, because it causes severe burns to skin. A 2% solution was formerly used to decontaminate medical instruments. However, this use has also been largely discontinued, because substituted phenols and other compounds are more effective.

The efficiency of a germicide is measured in terms of its **phenol coefficient (PC).** Phenol itself has a PC value of 1. If a 1% solution of a germicide is as effective as a 10% solution of phenol, it has a PC of 10. If a phenol is substituted with alkyl groups, its germicidal action increases. For example, the methyl-substituted phenols called *ortho-*, *meta-*, and *para*-cresol are used in the commercial disinfectant Lysol®. Another phenol, called thymol, is used by dentists to sterilize a tooth before filling it.

OH, CH_3 — *p*-methylphenol (*p*-cresol)

OH, CH_3, $CH(CH_3)_2$ — 5-isopropyl-2-methylphenol (thymol)

The phenol coefficient increases when the phenol is halogenated, particularly if the halogen is para to the hydroxyl group. The structures of a few halogenated phenols are shown in the following drawings. Chlorophene is a more effective germicide than *o*-phenylphenol. 4-Chloro-3,5-dimethylphenol is more effective than the cresols; it is used in topical preparations for athlete's foot and jock itch. Hexachlorophene (PC = 120) has been used in some toothpastes, deodorants, and soaps. However, it is toxic to infants and is no longer used in commercial products, although it is still used as a surgical scrub.

OH, Cl — chlorophene

OH, CH_3, CH_3, Cl — 4-chloro-3,5-dimethylphenol

Cl, OH, HO, Cl, CH_2, Cl, Cl, Cl, Cl — hexachlorophene

Phenols containing two hydroxyl groups are also germicides. When the two hydroxyl groups are meta, the phenols are called resorcinols. Resorcinol and hexylresorcinol are effective germicides. Resorcinol has a PC of only 0.4, but it is useful in the treatment of psoriasis and seborrhea. As in the case of phenols, alkyl substitution increases the PC. Hexylresorcinol has a PC of 98; it is used in throat lozenges.

OH, OH — resorcinol

OH, OH, $CH_2(CH_2)_4CH_3$ — hexylresorcinol

The sensitivity of biological function to differences in chemical structure is illustrated by the properties of phenols having hydroxyl groups in a 1,2 arrangement (pyrocatechols). A series of compounds where the R group at the 3-position is a polyunsaturated chain of 15 carbon atoms are urushiols. They are the skin irritant found in poison ivy and poison oak! ■

OH, OH — pyrocatechol

OH, OH, R — various urushiols

Thiols can be obtained from haloalkanes by nucleophilic displacement of halide ion with the sulfhydryl ion (HS^-). Thioethers, the sulfur analogs of ethers, are obtained from haloalkanes by nucleophilic displacement with thiolates.

$$R-\ddot{\underset{..}{Br}}: + H-\ddot{\underset{..}{S}}:^- \longrightarrow R-\ddot{\underset{..}{S}}-H + :\ddot{\underset{..}{Br}}:^-$$

a thiol

$$R'-\ddot{\underset{..}{Br}}: + R-\ddot{\underset{..}{S}}:^- \longrightarrow R'-\ddot{\underset{..}{S}}-R + :\ddot{\underset{..}{Br}}:^-$$

a thioether

Thiols are easily oxidized, but yield disulfides rather than the structural analogs of aldehydes and ketones. In the following equation, the symbol [O] represents an unspecified oxidizing agent that removes the hydrogen atoms.

$$2\,R-\ddot{\underset{..}{S}}-H \xrightarrow{[O]} R-\ddot{\underset{..}{S}}-\ddot{\underset{..}{S}}-R$$

a disulfide

This reaction is of great biological importance because many proteins contain the amino acid cysteine (Chapter 15), which contains a sulfhydryl group. Oxidation of the —SH group of cysteine gives a disulfide bond in cystine.

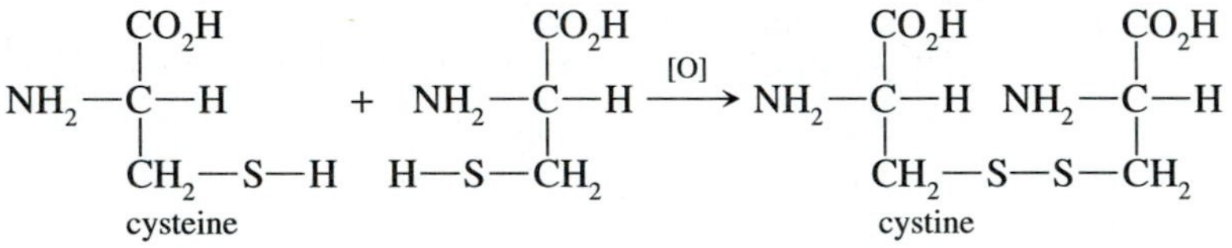

$$NH_2-\underset{CH_2-S-H}{\overset{CO_2H}{C}}-H \quad + \quad NH_2-\underset{H-S-CH_2}{\overset{CO_2H}{C}}-H \xrightarrow{[O]} NH_2-\overset{CO_2H}{\underset{|}{C}}-H \quad NH_2-\overset{CO_2H}{\underset{|}{C}}-H$$

$$CH_2-S-S-CH_2$$

cysteine — cystine

In some cases the sulfhydryl groups in an enzyme must be maintained in the reduced state for proper biological function. If an essential cysteine sulfhydryl group is oxidized, the enzyme becomes inactive.

Sulfhydryl groups in enzymes (E—S—H) also react with salts of lead and mercury. Because these reactions render the enzymes inactive, lead and mercury salts are highly toxic.

$$2\,E-S-H + Hg^{2+} \longrightarrow E-S-Hg-S-E + 2\,H^+$$

Summary of Reactions

1. Synthesis of Haloalkanes from Alcohols (Section 8.5)

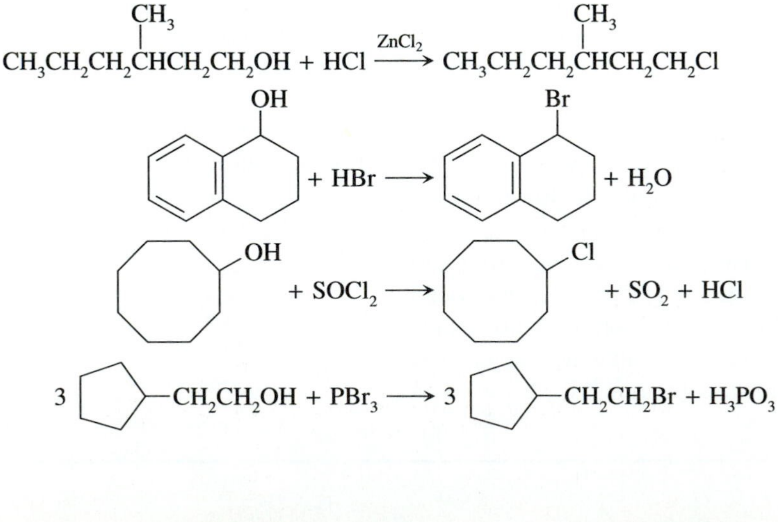

$$CH_3CH_2CH_2\overset{CH_3}{\overset{|}{C}}HCH_2CH_2OH + HCl \xrightarrow{ZnCl_2} CH_3CH_2CH_2\overset{CH_3}{\overset{|}{C}}HCH_2CH_2Cl$$

1-Tetralol (OH) + HBr ⟶ 1-bromotetralin (Br) + H_2O

Cyclooctanol (OH) + $SOCl_2$ ⟶ chlorocyclooctane (Cl) + SO_2 + HCl

3 cyclopentyl—$CH_2CH_2OH + PBr_3 \longrightarrow$ 3 cyclopentyl—$CH_2CH_2Br + H_3PO_3$

2. Dehydration of Alcohols (Section 8.6)

$$CH_3CH_2C(OH)(CH_2CH_3)CH_2CH_3 \xrightarrow{H_2SO_4} (CH_3CH_2)_2C{=}CHCH_3$$

3. Oxidation of Alcohols (Section 8.7)

$$C_5H_9\text{—}CH_2OH \xrightarrow{\text{Jones reagent}} C_5H_9\text{—}CO_2H$$

$$C_5H_9\text{—}CH_2OH \xrightarrow{PCC} C_5H_9\text{—}CHO$$

$$\text{cyclooctanol (—OH)} \xrightarrow{\text{Jones reagent}} \text{cyclooctanone (=O)}$$

4. Synthesis of Alcohols by Reduction of Carbonyl Compounds (Section 8.8)

$$\text{cyclohexenyl—C(=O)CH}_3 \xrightarrow[(100\ atm)]{H_2/Ni} \text{cyclohexyl—CH(OH)CH}_3$$

$$C_6H_5C(=O)CH_3 \xrightarrow[2.\ H_3O^+]{1.\ LiAlH_4} C_6H_5CH(OH)CH_3$$

$$CH_3CH_2CH_2CH(CH_3)CH_2CHO \xrightarrow[CH_3CH_2OH]{NaBH_4} CH_3CH_2CH_2CH(CH_3)CH_2CH_2OH$$

5. Synthesis of Alcohols from Alkenes (Section 8.8)

$$\text{cyclohexyl—CH=CH}_2 \xrightarrow[2.\ NaBH_4]{1.\ Hg(OAc)_2/H_2O} \text{cyclohexyl—CH(OH)CH}_3$$

$$\text{cyclohexyl—CH=CH}_2 \xrightarrow[2.\ H_2O_2/OH^-]{1.\ B_2H_6/THF} \text{cyclohexyl—CH}_2CH_2OH$$

6. Synthesis of Sulfides (Section 8.10)

$$CH_3CH_2CH(CH_3)CH_2SH \xrightarrow[2.\ CH_3CH_2I]{1.\ NaOH} CH_3CH_2CH(CH_3)CH_2SCH_2CH_3$$

Explorations with Molecular Models

1. Prepare a model corresponding to the following structure of 3-methyl-2-butanol. Arrange the structure to view the stereogenic center with the carbon-hydrogen bond directed away from your eye. Determine whether the structure is *R* or *S*.

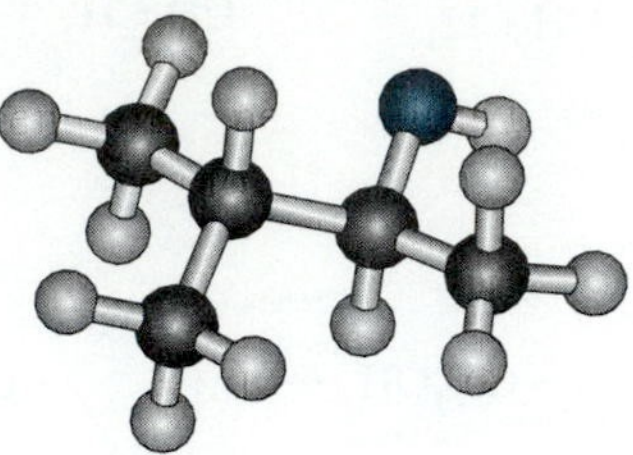

2. Prepare a model corresponding to the following structure. Rotate about the C-2 to C-3 bond to obtain a conformation that has the hydroxyl groups eclipsed. Is the compound optically active?

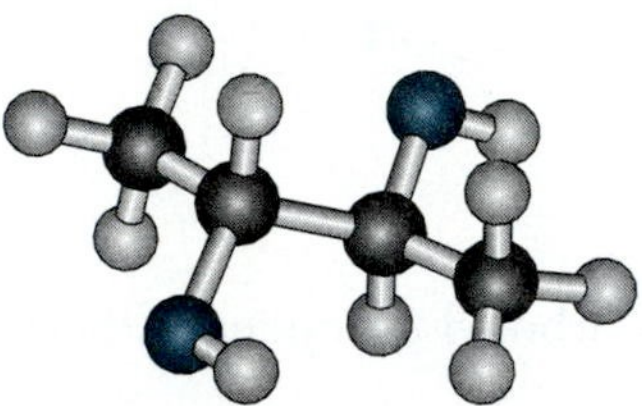

3. Prepare a model of the following conformation of 2-butanol. Assume that a derivative of the alcohol can be prepared to allow an elimination reaction that occurs via an E2 mechanism. As formation of the carbon-carbon double bond occurs, what changes result in the proximity of the methyl groups? What is the structure of the product?

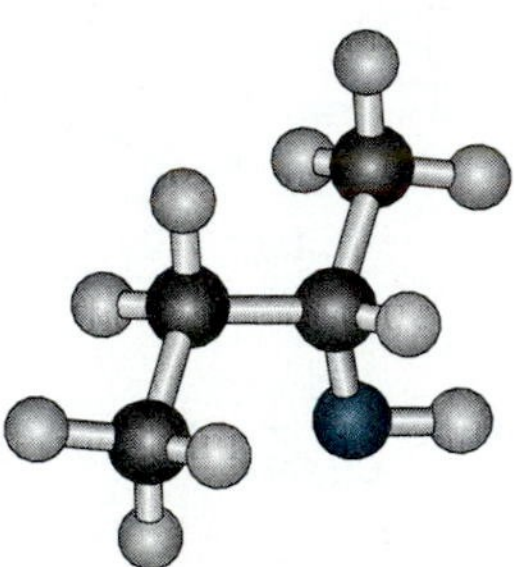

4. Using the model constructed in Exploration Exercise 3, rotate the structure about the C-2 to C-3 bond until the "back" hydrogen atom at C-3 is moved to the "top" position originally occupied by the other hydrogen atom. Assume that a derivative of the alcohol can be prepared to allow an elimination reaction that occurs via an E2 mechanism. As formation of the carbon-carbon double bond occurs, what changes result in the proximity of the methyl groups? What is the structure of the product? Considering steric effects, which product is obtained via the lowest energy transition state—that derived from Exploration Exercise 3 or that derived from this Exercise?

Exercises

Nomenclature of Alcohols

8.1 Write the structural formula of each of the following.

(a) 2-methyl-2-pentanol (b) 2-methyl-1-butanol (c) 2,3-dimethyl-1-butanol

8.2 Write the structural formula of each of the following.

(a) 2-methyl-3-pentanol (b) 3-ethyl-3-pentanol (c) 4-methyl-2-pentanol

8.3 Write the structural formula of each of the following.

(a) 1-methylcyclohexanol (b) *trans*-2-methylcyclohexanol (c) *cis*-3-ethylcyclopentanol

8.4 Write the structural formula of each of the following.

(a) 1,2-hexanediol (b) 1,3-propanediol (c) 1,2,4-butanetriol

8.5 Name each of the following compounds.

(a) 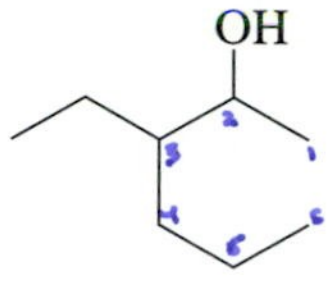 (b) 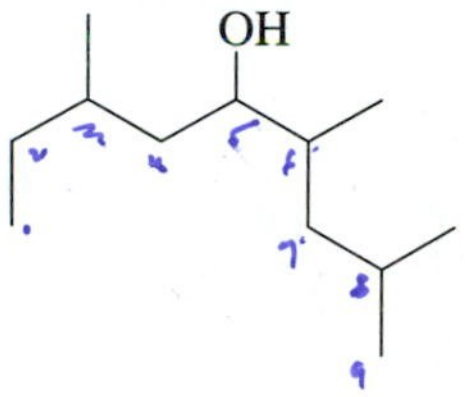 (c)

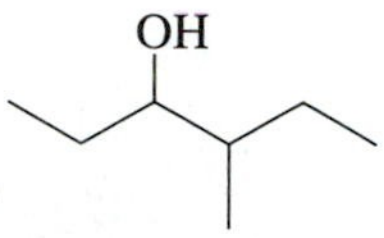

8.6 Name each of the following compounds.

(a)  CH_3 ... OH (b) 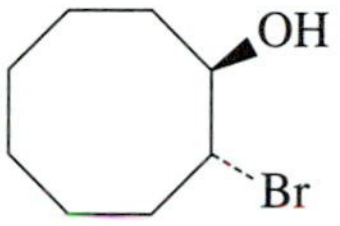 OH, Br (c) 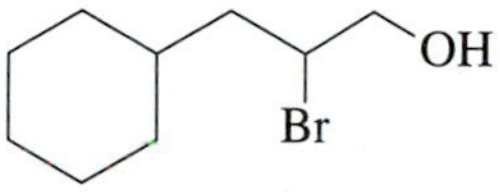 OH, Br

8.7 Name the following compound, which is the sex attractant of the Mediterranean fruit fly.

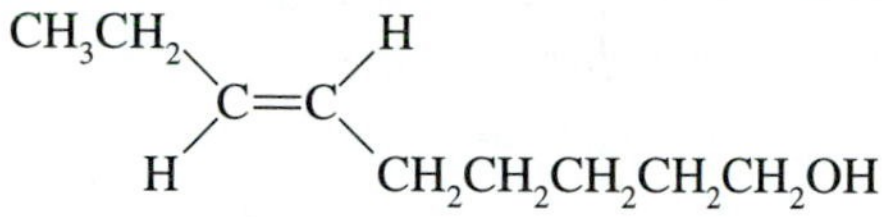

$$CH_3CH_2(H)C{=}C(H)CH_2CH_2CH_2CH_2CH_2OH$$

8.8 Name the following compound, which is used as a mosquito repellant.

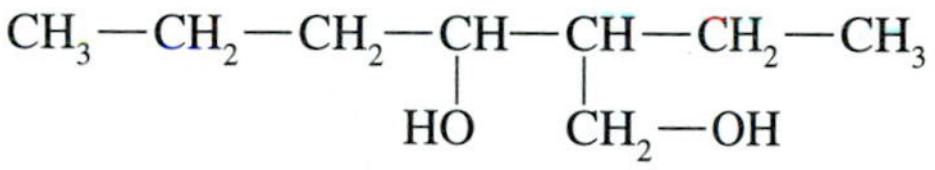

$$CH_3{-}CH_2{-}CH_2{-}CH(OH){-}CH(CH_2{-}OH){-}CH_2{-}CH_3$$

Classification of Alcohols

8.9 Classify each of the following alcohols.

(a) 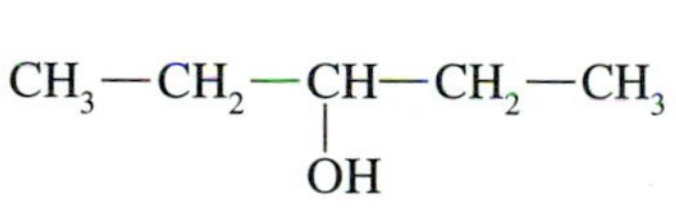 $CH_3{-}CH_2{-}CH(OH){-}CH_2{-}CH_3$ (b) 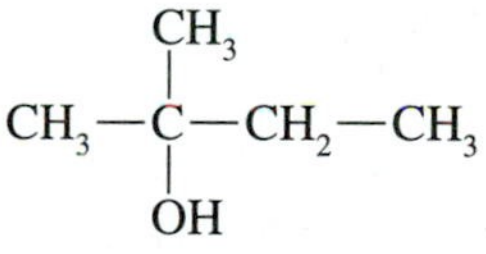 $CH_3{-}C(CH_3)(OH){-}CH_2{-}CH_3$ (c) $CH_3{-}C(CH_3)_2{-}OH$

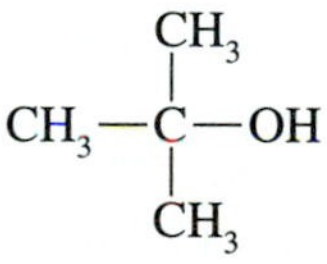

8.10 Classify each of the following alcohols.

(a)  CH_3 ... CH_3, OH (b) 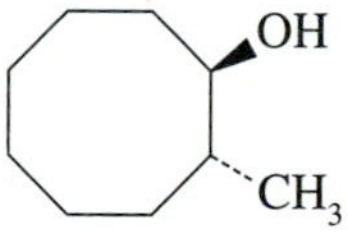 OH, CH_3 (c) 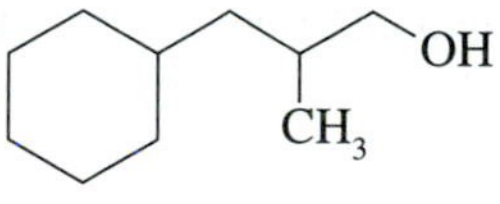 OH, CH_3

8.11 Classify each of the hydroxyl groups in the following vitamins.

(a) pyridoxal (vitamin B_6) (b) thiamine (vitamin B_1)

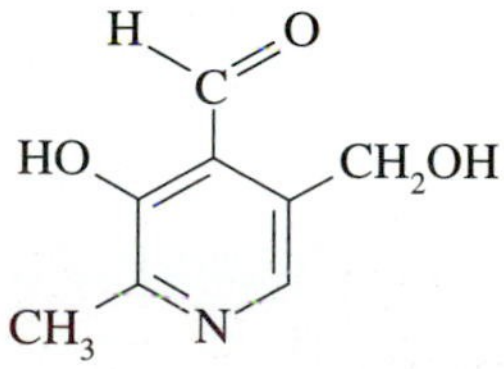

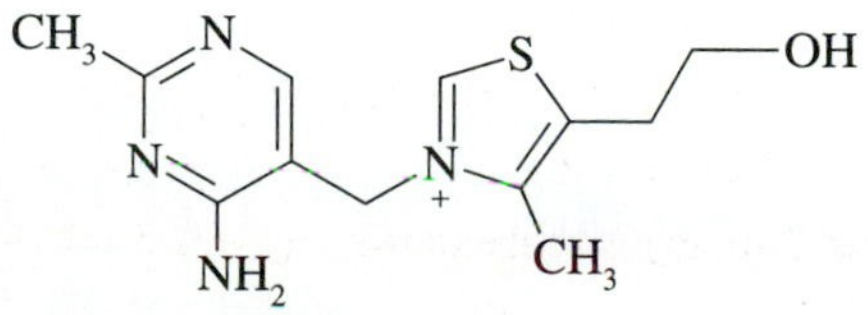

H, C=O, HO, CH_2OH, CH_3, N

CH_3, N, N, NH_2, N^+, S, OH, CH_3

8.12 Classify each of the hydroxyl groups in the following steroids.

(a) digitoxigenin, a cardiac glycoside (b) hydrocortisone, an anti-inflammatory drug

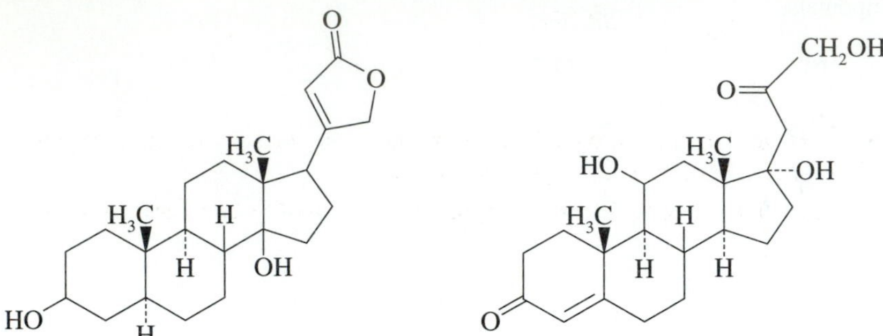

Physical Properties of Alcohols

8.13 1,2-Hexanediol is very soluble in water but 1-heptanol is not. Explain why these two compounds with similar molecular weights have different solubilities.

8.14 Ethylene glycol and 1-propanol boil at 198 and 97°C, respectively. Explain why these two compounds with similar molecular weights have different boiling points.

8.15 Explain why 1-butanol is less soluble in water than 1-propanol.

8.16 Suggest a reason why 2-methyl-1-propanol is much more soluble in water than 1-butanol.

Acid-Base Properties of Alcohols

8.17 1,1,1-Trichloro-2-methyl-2-propanol is used as a bacteriostatic agent. Compare its K_a to that of 2-methyl-2-propanol.

8.18 Based on the data in Table 8.2, estimate the K_a of 2-bromoethanol.

8.19 Based on the data in Table 8.2, estimate the K_a of cyclohexanol.

8.20 Which base is the stronger, methoxide ion or *tert*-butoxide ion?

Substitution Reactions

8.21 Rank the following isomeric compounds according to reactivity with HBr.

(I) cyclohexyl–$CH(OH)CH_3$ (OH, $CHCH_3$)

(II) cyclohexane ring bearing OH and CH_2CH_3 on the same carbon

(III) cyclohexyl–CH_2CH_2OH

8.22 Rank the following isomeric compounds according to reactivity with HCl and $ZnCl_2$.

(I) $CH_3-CH_2-C(CH_3)_2-OH$

(II) $CH_3-CH_2-CH(CH_3)-CH_2-OH$

(III) $CH_3-CH(CH_3)-CH(OH)-CH_3$

8.23 Write the structure of the product of the reaction of each of the following compounds with PBr_3.

(a) $CH_3-C(CH_3)_2-CH_2-CH_2-OH$ (b) $CH_3-C(OH)(CH_2-CH_3)-CH_2-CH_3$ (c) $CH_3-CH(CH_3)-CH_2-CH(OH)-CH_3$

8.24 Write the structure of the product of the reaction of each of the following compounds with $SOCl_2$.

(a) 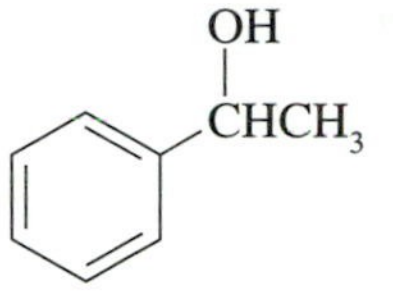 OH | $CHCH_3$ (b) 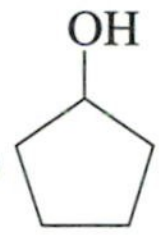 OH (c) 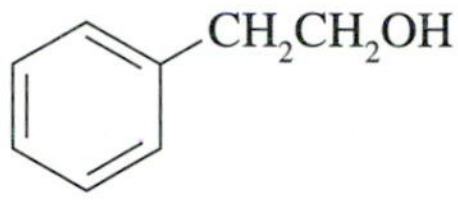 CH_2CH_2OH

8.25 Reaction of 3-buten-2-ol with HBr yields a mixture of two products: 3-bromo-1-butene and 1-bromo-2-butene. Explain why. (*Hint:* The reaction of this allylic alcohol occurs via an S_N1 process.)

8.26 The rate of reaction of the unsaturated alcohol (I) with HBr is faster than the the rate of reaction of the saturated alcohol (II). Explain why.

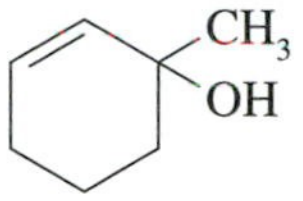

CH_3 OH CH_3 OH

(I) 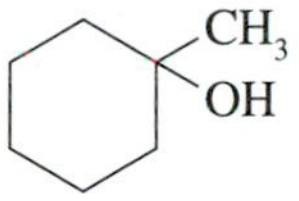 (II)

Dehydration of Alcohols

8.27 Draw the structure of the dehydration product(s) when each of the following compounds reacts with sulfuric acid. If more than one product is formed, predict the major isomer.

(a) $CH_3C(OH)(CH_3)CH_2CH_3$ (b) $CH_3CH_2CH(CH_3)CH_2CH_2OH$ (c) $CH_3CH_2CH(OH)CH_2CH_3$

8.28 Draw the structure of the dehydration product(s) when each of the following compounds reacts with sulfuric acid. If more than one product is formed, predict the major isomer.

(a) 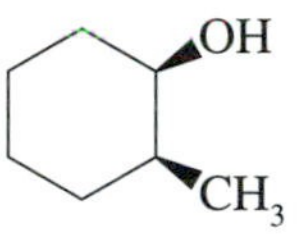 OH CH_3 (b) 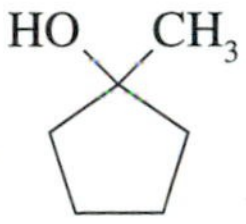 HO CH_3 (c) 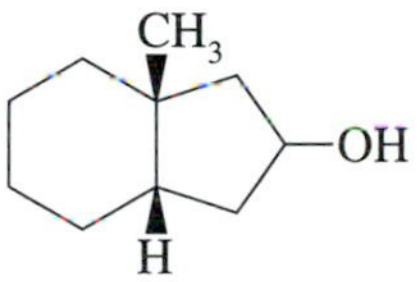 CH_3 OH H

8.29 Write the expected product of the acid-catalyzed dehydration of 1-phenyl-2-propanol. The reaction is more rapid than the rate of dehydration of 2-propanol. Explain why.

8.30 The dehydration of *trans*-2-methylcyclopentanol occurs via an E2 process to give predominantly 3-methylcyclopentene rather than 1-methylcyclopentene. What does this information indicate about the stereochemistry of the elimination reaction?

Oxidation of Alcohols

8.31 Write the product formed from the oxidation of each of the compounds in Exercise 8.21 with the Jones reagent.

8.32 Write the product formed from the oxidation of each of the compounds in Exercise 8.22 with the Jones reagent.

8.33 Write the product formed from the oxidation of the sex attractant of the Mediterranean fruit fly (see Exercise 8.7) with PCC.

8.34 Write the product formed by oxidation with PCC of the mosquito repellant in Exercise 8.8.

8.35 Which of the compounds in Exercises 8.23 and 8.24 will give a ketone when oxidized by Jones reagent?

8.36 Which of the compounds in Exercises 8.23 and 8.24 will give an acid product when oxidized by Jones reagent?

Preparation of Alcohols

8.37 Name the final product of oxymercuration-demercuration of each of the following compounds.

(a) 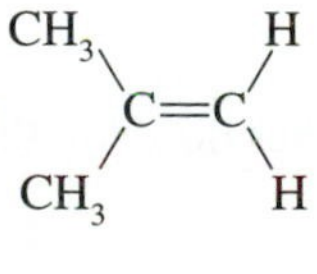 $(CH_3)_2C{=}CH_2$ (b) 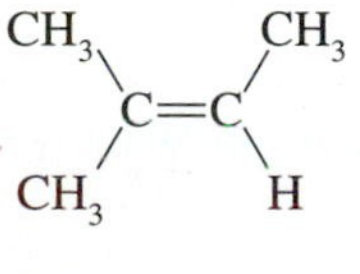 $(CH_3)_2C{=}CHCH_3$ (c) 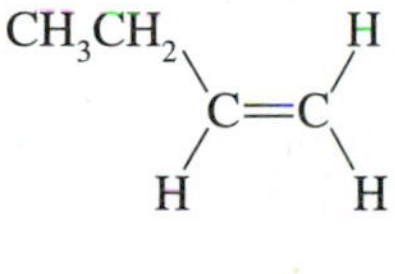 $CH_3CH_2CH{=}CH_2$

8.38 Name the final product of hydroboration-oxidation of each of the compounds in Exercise 8.37.

8.39 Draw the structure of the hydroboration-oxidation product of each of the following compounds.

(a) 1-methylcyclodecene (CH_3) (b) cyclobutyl–$CH{=}CH_2$ (c) cyclohexylidene$=C(H)CH_2CH_3$

8.40 Draw the structure of the oxymercuration-demercuration product of each compound in Exercise 8.39.

Synthesis Sequences

8.41 Write the structure of the final product of each of the following sequences of reactions.

(a) cyclopentyl–$CH_2CH{=}CH_2 \xrightarrow[2.\ NaBH_4]{1.\ Hg(OAc)_2} \xrightarrow[H_2SO_4\ acetone]{CrO_3}$

(b) phenyl–$CH{=}CH_2 \xrightarrow[2.\ H_2O_2/OH^-]{1.\ B_2H_6} \xrightarrow{PCC}$

(c) $CH_3CH_2CH_2CH{=}CH_2 \xrightarrow[2.\ H_2O_2/OH^-]{1.\ B_2H_6} \xrightarrow[H_2SO_4\ acetone]{CrO_3}$

8.42 Write the structure of the final product of each of the following sequences of reactions.

(a) cyclohexyl–$C(=O)H \xrightarrow[CH_3CH_2OH]{NaBH_4} \xrightarrow{SOCl_2}$ (b) cyclopentanone ($=O$) $\xrightarrow[2.\ H_3O^+]{1.\ LiAlH_4} \xrightarrow{PBr_3}$ (c) phenyl–$C(=O)CH_3 \xrightarrow[CH_3CH_2OH]{NaBH_4} \xrightarrow{HBr}$

Phenols

8.43 *p*-Nitrophenol is a much stronger acid than phenol. Explain why.

8.44 Which phenoxide is the stronger base, *p*-ethylphenoxide or *p*-chlorophenoxide? Explain your choice.

8.45 Draw the structure of the quinone obtained from the oxidation of the following substituted naphthalene.

(2-chloronaphthalene-1,4-diol: OH, Cl, OH)

8.46 2-Methylhydroquinone is more easily oxidized to a quinone than is 2-chlorohydroquinone. Explain why.

Sulfur Compounds

8.47 There are four isomeric $C_4H_{10}S$ compounds with an —SH group. Draw the structures of the compounds.

8.48 There are three isomeric C_3H_8S compounds. Draw their structures.

8.49 Draw the structure of each of the following compounds.

(a) 1-propanethiol (b) 2-methyl-3-pentanethiol (c) cyclopentanethiol

8.50 Draw the structure of each of the following compounds.

(a) 2-propanethiol (b) 2-methyl-1-propanethiol (c) cyclobutanethiol

8.51 Addition of sodium hydroxide to an aqueous solution of $CH_3CH_2CH_2SH$ substantially eliminates the odor. Explain why.

8.52 The boiling points of ethanethiol and dimethyl sulfide are 35 and 37°C, respectively. Why are the boiling points similar? What types of intermolecular forces are responsible for this similarity?

8.53 Indicate two methods to produce the scent marker of the red fox using a thiol as one of the reactants.

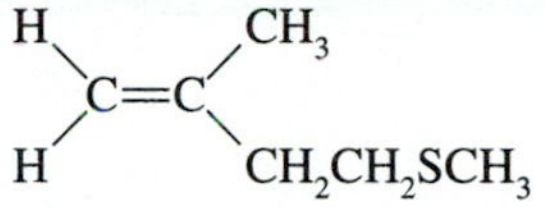

$$\begin{array}{ccc} H & & CH_3 \\ & C{=}C & \\ H & & CH_2CH_2SCH_3 \end{array}$$

8.54 Outline a series of reactions to produce the compound used for defense by the skunk, starting with 3-methyl-1-butene.

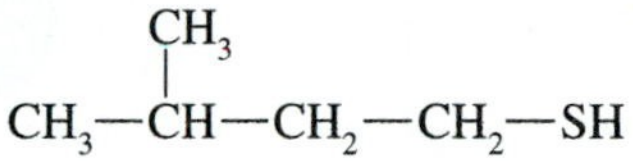

$$\begin{array}{c} \quad CH_3 \\ \quad | \\ CH_3{-}CH{-}CH_2{-}CH_2{-}SH \end{array}$$

8.55 Dimethyl disulfide is emitted by female hamsters as an attractant for male hamsters. How might this compound be produced?

8.56 Minks secrete the following disulfide. How could this compound be prepared in a laboratory synthesis starting from 1-bromo-3-methylbutane?

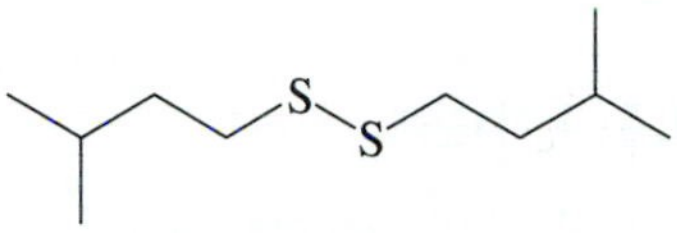

S S

Spectroscopy

8.57 If you have read Chapter 18 and the material has been assigned by your professor, then Exercises 18.26, 18.27, and 18.32, which are based on the structure of alcohols, may be done.

CHAPTER 9

ETHERS AND EPOXIDES

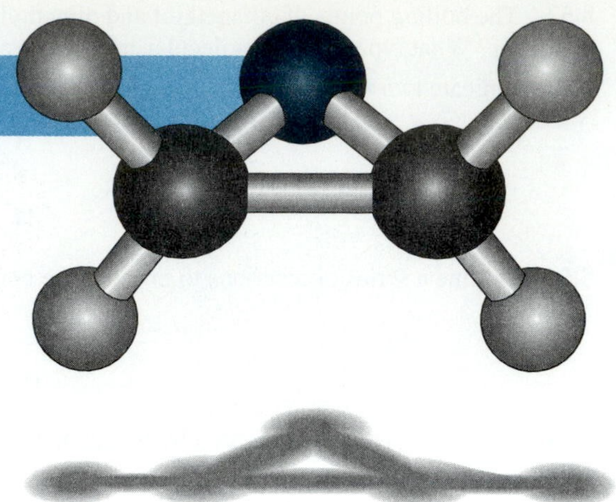

Overview

9.1 Ethers

Like alcohols, ethers can be viewed as organic "cousins" of water. **Ethers** contain two groups, which may be alkyl or aryl groups, bonded to an oxygen atom. The groups are the same in a **symmetrical ether** and different in an **unsymmetrical ether.**

$CH_3CH_2—O—CH_2CH_3$ diethyl ether (symmetrical ether)

$CH_3CH_2CH_2—O—C_6H_5$ phenyl propyl ether (unsymmetrical ether)

$C_6H_5—O—C_6H_5$ diphenyl ether (symmetrical ether)

The oxygen atom of an ether is sp^3-hybridized, and the C—O—C bond angle is approximately the tetrahedral bond angle. The bond angle in dimethyl ether is 112° (Figure 9.1). The two O—C bonds are directed to two of the corners of a tetrahedron. Based on the hybridization of oxygen, the lone pair electrons in the remaining two sp^3 hybrid orbitals are viewed as directed to the remaining "corners" of a tetrahedron.

Predictions can be made about the most stable conformations of ethers by comparing them to the structures of similar hydrocarbons. Imagine creating an ether by replacing a $—CH_2—$ group of an alkane with an oxygen atom. For example, replacing the C-3 methylene group of pentane with an oxygen atom would give diethyl ether. Diethyl ether has an anti arrangement of all carbon and oxygen atoms in its most stable conformation.

anti conformation of diethyl ether

A similar situation prevails when we compare conformations of cyclic ethers and cycloalkanes. For example, tetrahydropyran, the ether analog of cyclohexane, also exists in a chair conformation. The tetrahedral oxy-

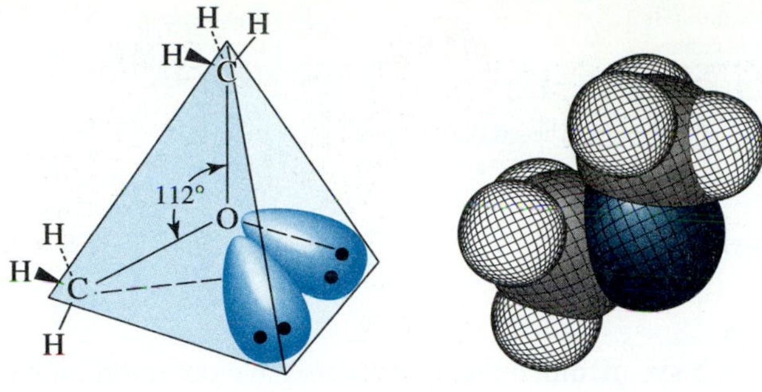

FIGURE 9.1
Structure of Dimethyl Ether

gen atom has two lone pairs that occupy positions in space corresponding to the axial and equatorial C—H bonds of cyclohexane. The conformation of tetrahydropyran is particularly important because many carbohydrates contain tetrahydropyran rings (Chapter 11).

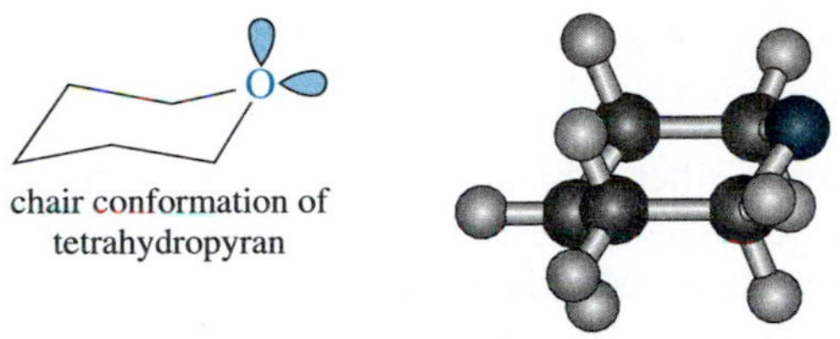

9.2 Nomenclature of Ethers

Simple ethers are named as alkyl alkyl ethers. The name is constructed by listing the alkyl (or aryl) groups in alphabetical order and appending the name *ether*. For example, an unsymmetrical ether that has a butyl group and a methyl group bonded to an oxygen atom is called butyl methyl ether. Symmetrical ethers are named by using the prefix *di-* in conjunction with the name of the alkyl group. For example, an ether with two isopropyl groups bonded to an oxygen atom is called diisopropyl ether.

$$CH_3{-}CH_2{-}CH_2{-}CH_2{-}O{-}CH_3$$

butyl methyl ether

$$CH_3{-}\underset{}{\overset{CH_3}{\overset{|}{CH}}}{-}O{-}\overset{CH_3}{\overset{|}{CH}}{-}CH_3$$

diisopropyl ether

IUPAC Names of Ethers

Ethers are named according to IUPAC nomenclature as alkoxyalkanes, where the smaller alkyl group and the oxygen atom constitute an **alkoxy group.** An alkoxy group is treated as a substituent on the larger parent alkane chain or cycloalkane ring. For example, a five-carbon chain with an $—OCH_3$ group at the C-2 atom is named 2-methoxypentane and a cyclohexane ring with an $—OCH_2CH_3$ group is named ethoxycyclohexane.

The larger group is the parent chain.

$CH_3CH_2CH_2CHCH_3$ (with $O{-}CH_3$ on C-2) — The smaller group is the alkoxy substituent.

2-methoxypentane

The larger group is the cyclohexane ring.

(cyclohexane)$-OCH_2CH_3$ — The smaller group is the alkoxy substituent.

ethoxycyclohexane

Cyclic Ethers

The three- through six-membered cyclic ethers have common names. In all ring systems, the oxygen atom is assigned the number 1, and the rings are numbered in the direction that gives the first substituent the lower number. Cyclic ethers with three-atom rings are called epoxides. Because these compounds are formed from the oxidation of an alkene, the common name of an epoxide is derived by adding oxide to the name of the alkene.

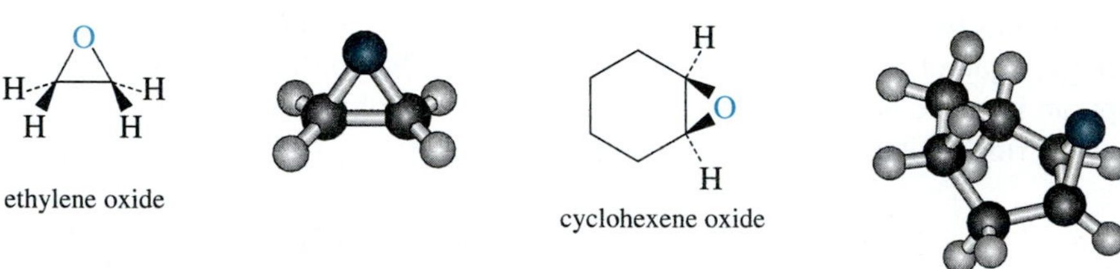

ethylene oxide

cyclohexene oxide

The four-membered ether ring compounds, called trimethylene oxides, are not common substances. The name of the five-membered ring ether is tetrahydrofuran (THF), based on its relationship to the aromatic compound furan. Similarly, tetrahydropyran (THP), a six-membered ring ether, is related to pyran, an unsaturated ether.

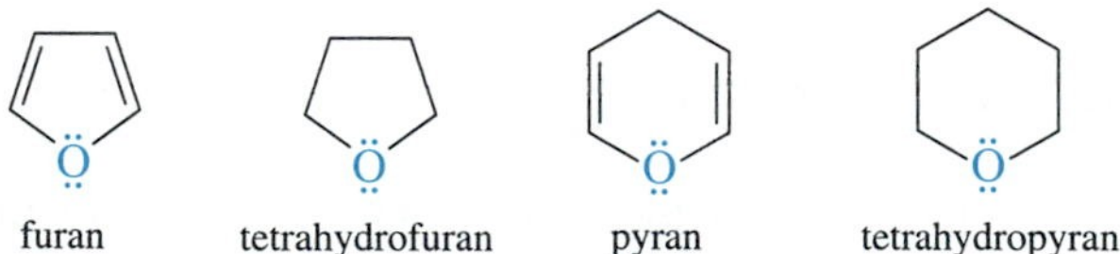

furan tetrahydrofuran pyran tetrahydropyran

The IUPAC names for cyclic ethers having three-, four-, five-, and six-membered rings are oxirane, oxetane, oxolane, and oxane, respectively. The oxygen atom in each of these rings receives the number 1. The ring is numbered in the direction that gives the lowest numbers to substituents.

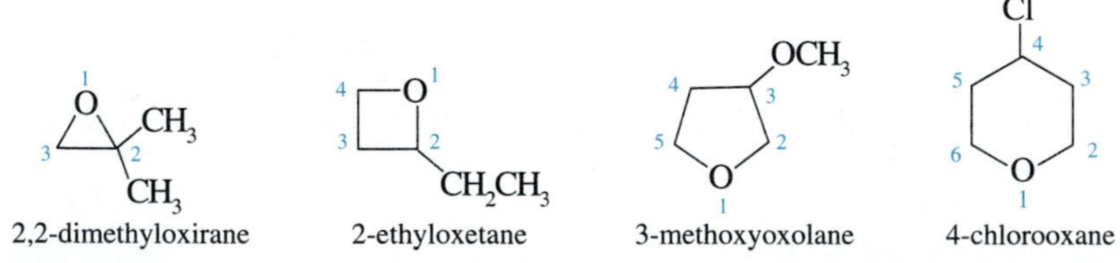

2,2-dimethyloxirane 2-ethyloxetane 3-methoxyoxolane 4-chlorooxane

EXAMPLE 9-1

What are the common and IUPAC names of the following compound?

(benzene ring)$-O-CH_2-CH_3$

Solution

The common name of an unsymmetrical ether is based on the names of the two alkyl (or aryl) groups. In this case they are phenyl and ethyl. The names of the two groups are arranged alphabetically, and the word ether is added to give the name ethyl phenyl ether.

The IUPAC name is obtained by first identifying the smaller of the two groups bonded to the oxygen atom and then selecting it to constitute the alkoxy group on the larger parent alkane chain. In this case, the benzene ring is selected as the parent and the ethoxy group is the substituent. The IUPAC name is ethoxybenzene.

Problem 9.1

What are the IUPAC names of the following ethers?

(a) OCH_2CH_3 (b) OCH_3, OCH_3 (c) OCH_2CH_3, OCH_2CH_3

9.3 Physical Properties of Ethers

Ethers have two polar C—O bonds at a tetrahedral angle and, as a result, ethers have substantial dipole moments. Ethers are more polar than alkanes and somewhat less polar than alcohols, as evidenced by their respective dipole moments (Table 9.1).

Ethers do not have an O—H bond and therefore cannot serve as hydrogen bond donors. As a consequence, ether molecules do not hydrogen bond to each other. Thus, ethers have boiling points substantially lower than those of alcohols of comparable molecular weight (Table 9.1). The boiling points of ethers are very close to the boiling points of alkanes of similar molecular weight.

Because ethers are polar, they are more soluble in water than alkanes of similar molecular weight. The solubility of ethers in water results from hydrogen bonds between the water molecules and the nonbonding electron pairs of ether molecules. The solubility of diethyl ether is about 7.5 g per 100 mL of water, much higher than that of pentane, which dissolves only slightly in water

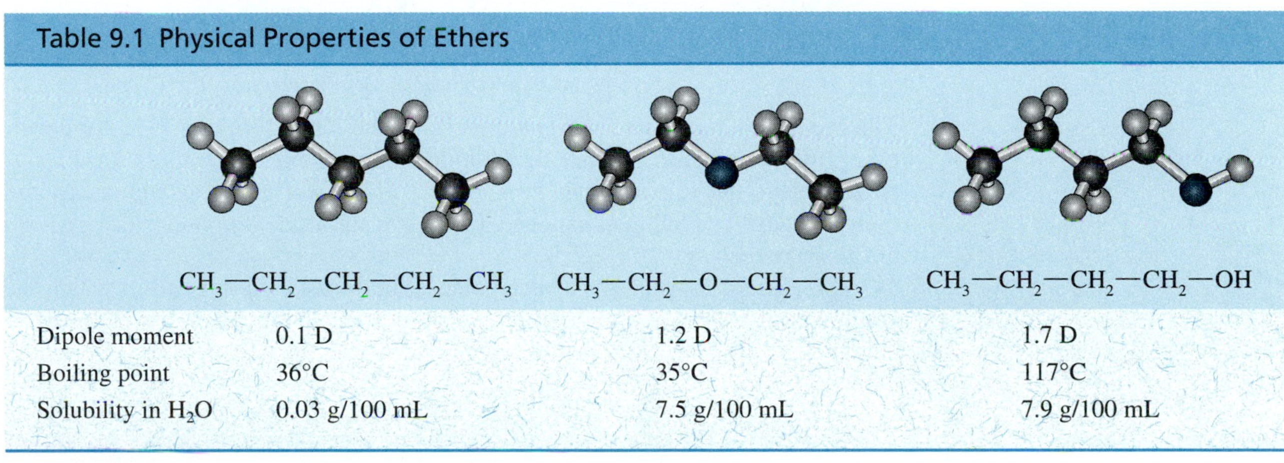

Table 9.1 Physical Properties of Ethers

	$CH_3—CH_2—CH_2—CH_2—CH_3$	$CH_3—CH_2—O—CH_2—CH_3$	$CH_3—CH_2—CH_2—CH_2—OH$
Dipole moment	0.1 D	1.2 D	1.7 D
Boiling point	36°C	35°C	117°C
Solubility in H_2O	0.03 g/100 mL	7.5 g/100 mL	7.9 g/100 mL

(Table 9.1). The solubility of diethyl ether in water is similar to that of 1-butanol, because water can form hydrogen bonds to each compound.

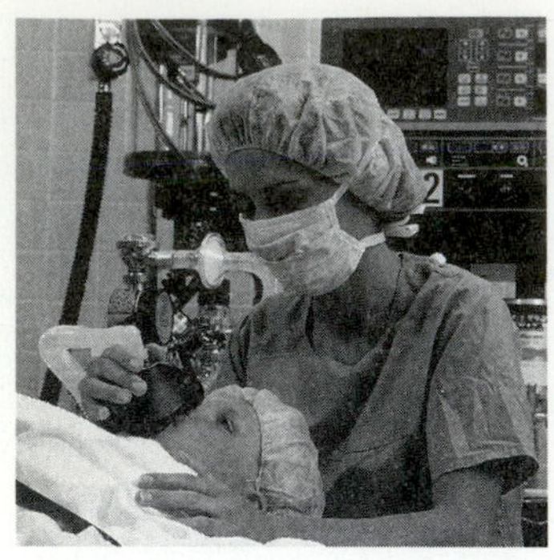

Diethyl ether as an anesthetic has been largely replaced by other gases.

Ethers as Solvents

Ethers such as diethyl ether dissolve a variety of nonpolar and polar compounds. Nonpolar compounds are generally more soluble in diethyl ether than in alcohols such as ethanol, because ethers do not have a hydrogen bonding network that would have to be broken up to dissolve the solute. Because diethyl ether has a dipole moment, polar substances readily dissolve in it. Polar compounds that can serve as hydrogen bond donors dissolve in diethyl ether because they can form hydrogen bonds to the nonbonding electron pairs of the ether oxygen atoms.

The nonbonding electron pairs of ethers can stabilize electron-deficient species such as BF_3 and borane, BH_3. For example, the borane—THF complex is used in the hydroboration of alkenes (Section 8.8).

$$\text{THF}\ \ddot{O}^{+}\!-\!B^{-}H_3$$

borane—THF complex

9.4 The Grignard Reagent and Ethers

Ethers such as diethyl ether or tetrahydrofuran are excellent solvents for certain reagents that would otherwise react with protons supplied by protic solvents. One such example is the **Grignard reagent,** represented as RMgX, which can be prepared from haloalkanes as well as from aryl halides.

$$R—X \xrightarrow[\text{ether}]{\text{Mg}} R—Mg—X$$

The oxygen atom of diethyl ether (or THF) forms a complex with the magnesium atom of the Grignard reagent. These reagents in ether solution are very useful in organic synthesis.

$$(CH_3CH_2)_2\ddot{O}\cdots Mg(R)(X)\cdots\ddot{O}(CH_2CH_3)_2$$

Grignard reagent solvated by diethyl ether

The French chemist Victor Grignard received the Nobel Prize in 1912 for developing the methods to prepare these organomagnesium compounds. In a Grignard reagent, the R group may be a 1°, 2°, or 3° alkyl group as well as a vinyl or aryl group. The halogen may be Cl, Br, or I; fluorine compounds do not form Grignard reagents.

A Grignard reagent has a very polar carbon-magnesium bond in which the carbon atom has a partial negative charge and the metal a partial positive charge.

$$\overset{\delta-}{C}—\overset{\delta+}{MgX}$$

Polyethers in the Laboratory and in Nature

Compounds such as 1,2-dimethoxyethane (glyme) and 1,4-dioxane readily dissolve polar compounds because they contain two polar sites.

$CH_3{-}O{-}CH_2{-}CH_2{-}O{-}CH_3$

glyme

1,4-dioxane

Some polyethers can even solvate cations. You may recall from an earlier chemistry course that cations are solvated by several water molecules in aqueous solution. For example, Co^{2+} coordinates with six water molecules in a species represented as $Co(H_2O)_6{}^{2+}$. Cyclic polyethers can similarly solvate cations and increase the solubility of ionic compounds in organic solvents. These compounds, called **crown ethers,** are named *x*-crown-*y*, where *x* is the total number of atoms in the ring and *y* is the number of oxygen atoms. The 18-crown-6 ether chelates the potassium ion in a cavity within the ring.

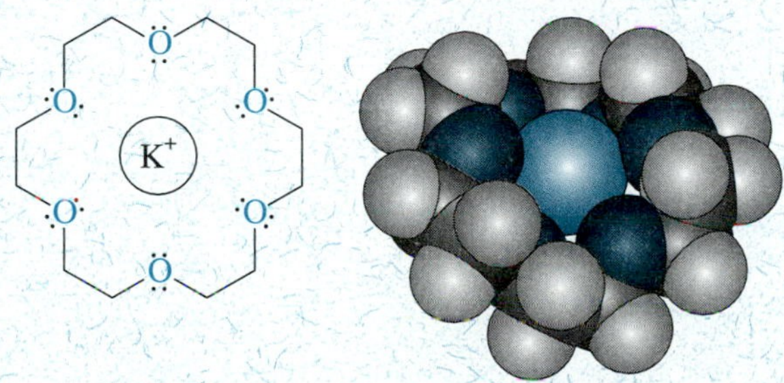

The chelating characteristics of crown ethers depend on the match between the size of the cavity and the radius of the ion. The diameter of the internal cavity of 18-crown-6 is close to the diameter of the potassium ion. Not only does the potassium ion fit within the cavity, but the conformation of the crown ether allows all six oxygen atoms to coordinate to it effectively. As a result, potassium salts are soluble in this crown ether.

Several cyclic and acyclic polyethers act as antibiotics by transporting ions across certain biological membranes. These ethers, called **ionophores** (ion carriers), function by disrupting the electrolyte balance between the interior and exterior of cells that is necessary for normal maintenance of the cell. The cyclic ether nonactin selectively transports potassium out of bacterial cells. This compound, which contains four five-membered ring ether units linked by ester units, binds potassium about 10 times better than it binds sodium. Because a cell must maintain a higher internal concentration of potassium ions than of sodium ions, the selective removal of potassium ions kills bacteria. ■

nonactin

This bond polarity is opposite that of the carbon-halogen bond of haloalkanes. Because the carbon atom in a Grignard reagent has a partial negative charge, it resembles a carbanion and reacts with electrophiles. Grignard reagents are very reactive reactants that are used synthetically to form new carbon-carbon bonds. Examples of these reactions will be discussed in Section 10.6.

Grignard reagents react rapidly with acidic hydrogen atoms in molecules such as alcohols and water to produce alkanes. Thus, formation of the Grignard reagent followed by reaction with water provides a way to convert a haloalkane to an alkane in two steps.

$$R{-}X \xrightarrow[\text{ether}]{Mg} R{-}Mg{-}X \xrightarrow{H_2O} R{-}H$$

EXAMPLE 9-2

Devise a synthesis of $CH_3CH_2CHDCH_3$ starting from 1-butene and heavy water (D_2O).

Solution

Reaction of a Grignard reagent R—MgBr with D_2O will yield R—D. The necessary Grignard reagent is obtained from the corresponding bromoalkane R—Br.

$$CH_3CH_2\underset{}{\overset{Br}{C}}HCH_3 \xrightarrow[\text{ether}]{Mg} CH_3CH_2\overset{MgBr}{C}HCH_3 \xrightarrow{D_2O} CH_3CH_2\overset{D}{C}HCH_3$$

The required 2-bromobutane can be prepared from 1-butene by adding HBr. This reaction occurs according to Markovnikov's rule, so that a hydrogen atom adds to the less substituted carbon atom of the double bond.

$$CH_3CH_2CH{=}CH_2 \xrightarrow{HBr} CH_3CH_2\overset{Br}{C}HCH_3$$

Problem 9.2

Devise a synthesis of the following compound starting from benzene.

$$(CH_3)_2CH{-}C_6H_4{-}D$$

9.5 Synthesis of Ethers

Ethers can be prepared by a method called the **Williamson synthesis.** In this reaction, a halide ion is displaced from an alkyl halide by an alkoxide ion in an S_N2 reaction. The alkoxide ion is prepared by the reaction of an alcohol with a strong base such as sodium hydride.

$$C_5H_9{-}\ddot{O}{-}H + NaH \longrightarrow C_5H_9{-}\ddot{O}{:}^- Na^+ + H_2$$

$$C_5H_9{-}\ddot{O}{:}^- Na^+ + CH_3{-}Br \longrightarrow C_5H_9{-}\ddot{O}{-}CH_3 + NaBr$$

The Williamson synthesis gives the best yields with primary halides, because the reaction occurs by an S_N2 displacement in which a halide ion is the leaving group. The yield is lower for secondary alkyl halides because they can also react with the alkoxide ion in a competing elimination reaction. The Williamson synthesis cannot be used with tertiary alkyl halides because they undergo elimination reactions instead. Thus, to make an unsymmetrical ether that has a primary and a tertiary alkyl group, a primary alkyl halide and a tertiary alkoxide ion must be used. For example, *tert*-butyl methyl ether can be prepared by the reaction of sodium *tert*-butoxide with methyl iodide, but not by the reaction of sodium methoxide with 2-chloro-2-methylpropane.

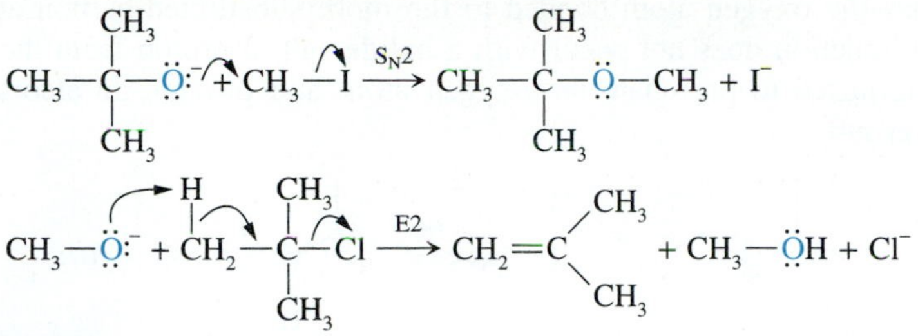

$$(CH_3)_3C{-}\ddot{O}{:}^- + CH_3{-}I \xrightarrow{S_N2} (CH_3)_3C{-}\ddot{O}{-}CH_3 + I^-$$

$$CH_3{-}\ddot{O}{:}^- + H{-}CH_2{-}C(CH_3)_2{-}Cl \xrightarrow{E2} CH_2{=}C(CH_3)_2 + CH_3{-}\ddot{O}H + Cl^-$$

EXAMPLE 9-3

Propose a synthesis of 2-ethoxynapthalene, known by its tradename Nerolin II. It is used in perfumery for its orange blossom odor.

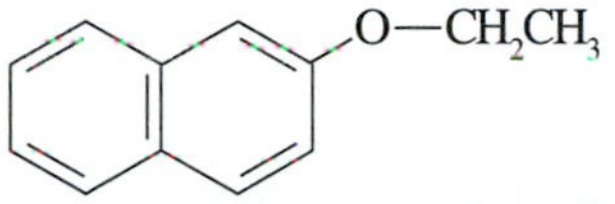

(2-naphthyl)$-O-CH_2CH_3$

Solution

Consider reactions of the following two possible combinations of reagents.

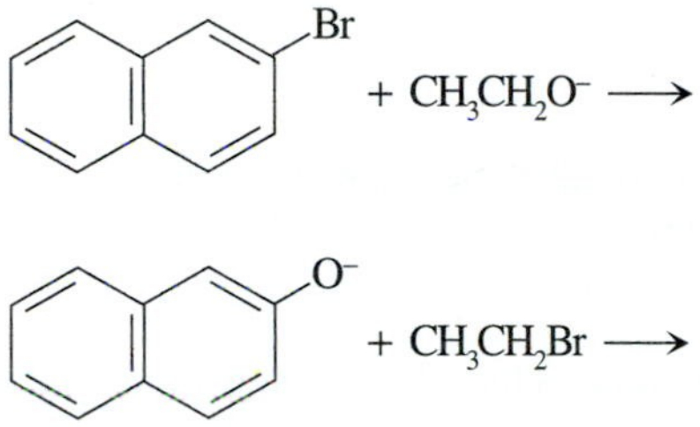

2-bromonaphthalene $+ CH_3CH_2O^- \longrightarrow$

2-naphthoxide O^- $+ CH_3CH_2Br \longrightarrow$

The first combination will not give the ether product because S_N2 reactions cannot occur at the sp^2-hybridized carbon atom of this aromatic compound. However, the second reaction of a phenoxide type ion with bromoethane occurs readily because bromoethane is a primary alkyl halide. The required nucleophile is formed by reacting the following phenol with sodium hydride.

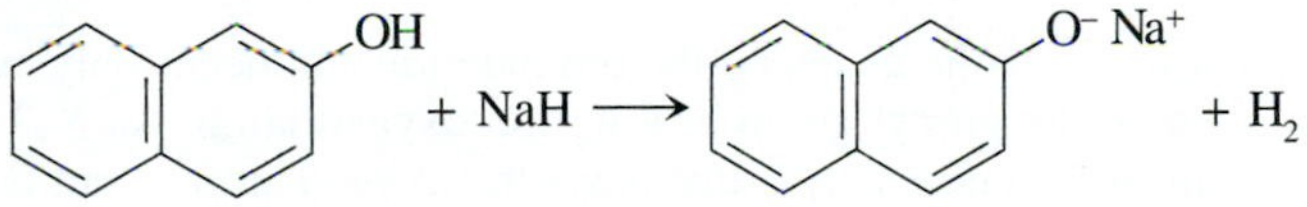

2-naphthol (OH) $+ NaH \longrightarrow$ (2-naphthyl)$O^- Na^+ + H_2$

Problem 9.3

Propose a synthesis of benzyl *tert*-butyl ether using the Williamson synthesis.

9.6 Reactions of Ethers

Ethers are stable compounds that do not react with most common reagents. They do not react with bases because they are aprotic, but do react with strong acids whose conjugate bases are good nucleophiles. For example, ethers react with HI (or with HBr), and the carbon-oxygen bond is cleaved to produce alkyl iodides (or bromides).

$$R{-}O{-}R' \xrightarrow{HX} R{-}O{-}H + X{-}R'$$

$$R{-}O{-}H \xrightarrow{HX} R{-}X$$

In general, the less substituted alkyl halide is formed by an S_N2 reaction. The halide ion attacks the less hindered carbon atom, and the displaced alkoxy group has the oxygen atom bonded to the more substituted carbon atom. The cleavage reaction does not occur with a halide salt: a proton from the halogen acid is required to protonate the oxygen atom and provide an alcohol as the leaving group.

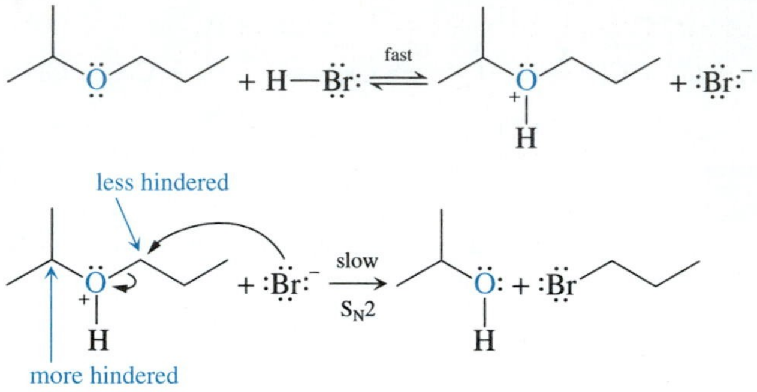

fast

less hindered

slow

S_N2

more hindered

In a subsequent reaction, the alcohol product reacts with the hydrogen halide to give a second mole of an alkyl halide. Both alkyl groups of the ether are eventually converted into alkyl halides if sufficient hydrogen halide is used.

EXAMPLE 9-4

Based on the mechanism of ether cleavage, write the products of the reaction of HBr with phenyl propyl ether.

Solution

First, the strong acid protonates the ether oxygen atom. The resulting cation is an oxonium ion.

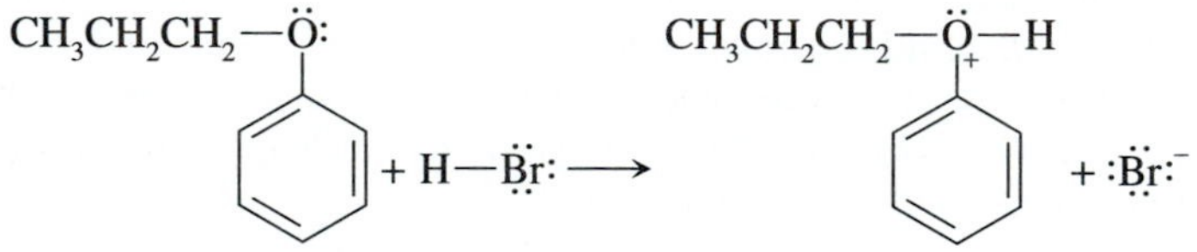

$CH_3CH_2CH_2$—Ö: + H—Br: ⟶ $CH_3CH_2CH_2$—$\overset{+}{O}$—H + :Br:$^-$

Subsequent nucleophilic attack by the bromide ion can occur only at the methylene carbon atom of the propyl group bearing the oxygen atom. An S_N2 reaction at the carbon atom of the benzene ring that bears the oxygen atom is not possible, so phenol will not react further with HBr.

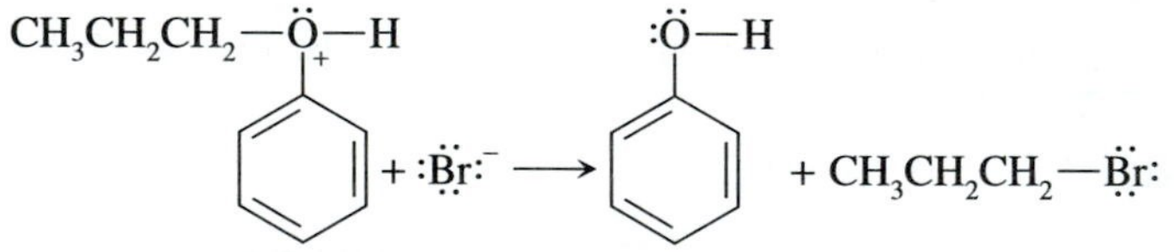

$CH_3CH_2CH_2$—$\overset{+}{O}$—H + :Br:$^-$ ⟶ :Ö—H + $CH_3CH_2CH_2$—Br:

Problem 9.4

Based on the mechanism of the cleavage of ethers, predict the structure of the bromoalcohol formed in the first step of the reaction of the following ether with HBr.

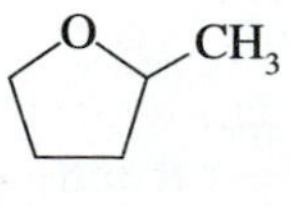

O CH$_3$

9.7 Synthesis of Epoxides

The synthesis and reactions of cyclic ethers containing four or more atoms are similar to those of acyclic ethers. The three-membered cyclic ethers, which are important intermediates in synthesis (Section 9.8), can be synthesized by oxidizing an alkene with a peroxyacid (RCO_3H). Peroxyacetic acid, CH_3CO_3H, is used in industry, but *m*-chloroperoxybenzoic acid (MCPBA) may be used to prepare smaller amounts of epoxides in the laboratory.

$$\text{cyclohexene} + m\text{-}ClC_6H_4C(=O)OOH \xrightarrow[25°C]{CH_2Cl_2} \text{cyclohexene oxide} + m\text{-}ClC_6H_4C(=O)OH$$

In the epoxidation of alkenes with *m*-chloroperoxybenzoic acid, the stereochemistry of the groups in the alkene is retained. That is, groups that are cis in the alkene are cis in the epoxide, and groups that are trans in the alkene remain trans in the epoxide.

$$CH_3(H)C{=}C(H)CH_3 \xrightarrow{MCPBA} \text{cis-}CH_3CH{-}O{-}CHCH_3$$

9.8 Reactions of Epoxides

The cyclic ethers tetrahydrofuran and tetrahydropyran are as unreactive as acyclic ethers and are often used as solvents. In contrast, epoxides are highly reactive because the three-membered ring has considerable bond angle strain. The products of the ring-opening reactions have normal tetrahedral bond angles and are not strained.

Acid-Catalyzed Ring Opening

Consider the reaction of water with ethylene oxide to form ethylene glycol.

$$\text{ethylene oxide} \xrightarrow{H_3O^+} HOCH_2{-}CH_2OH$$

The acid-catalyzed ring opening of epoxides occurs by an S_N2 process in which water is the nucleophile and the "leaving group" is the protonated oxygen atom of the epoxide.

$$H{-}\ddot{O}(H): + \text{protonated epoxide} \longrightarrow H{-}\overset{+}{\ddot{O}}(H){-}CH_2CH_2{-}\ddot{O}H \overset{H_2O}{\rightleftharpoons} H{-}\ddot{O}{-}CH_2CH_2{-}\ddot{O}H + H_3O^+$$

Ring Opening by Nucleophiles

Ethers do not generally react with nucleophiles under neutral or basic conditions. However, epoxides are so reactive that the ring is opened even by nucleophiles such as OH^-, SH^-, or NH_3 or the related organic species RO^-, RS^-, and

RNH_2. The reaction of ethylene oxide with ammonia gives 2-amino-ethanol, a compound used commercially as a corrosion inhibitor.

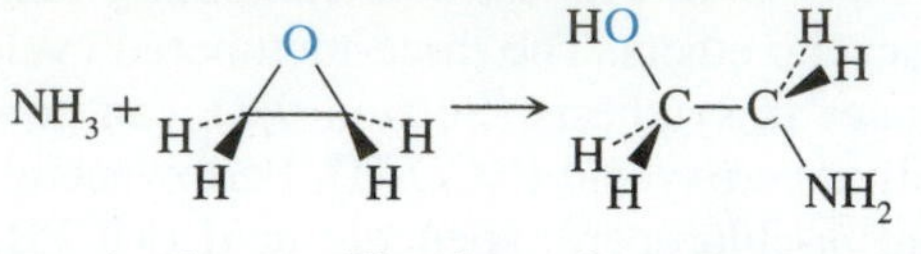

$$NH_3 + \text{ethylene oxide} \longrightarrow HOCH_2CH_2NH_2$$

Similar reactions occur in the sterilization of temperature-sensitive equipment when it is exposed to ethylene oxide gas. The epoxide ring reacts with a variety of nucleophilic functional groups in bacterial macromolecules, and the bacteria die as a result.

Epoxides react with Grignard reagents to produce alcohols with two more carbon atoms than the starting alkyl halide. The sequence of reactions is

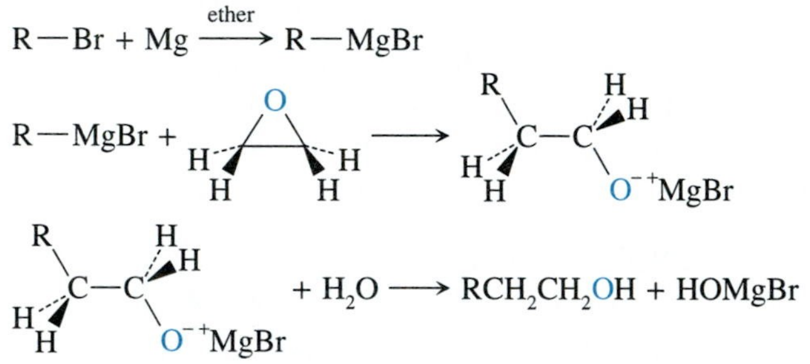

$$R{-}Br + Mg \xrightarrow{\text{ether}} R{-}MgBr$$

$$R{-}MgBr + \text{ethylene oxide} \longrightarrow RCH_2CH_2O^{-}\,{}^{+}MgBr$$

$$RCH_2CH_2O^{-}\,{}^{+}MgBr + H_2O \longrightarrow RCH_2CH_2OH + HOMgBr$$

Direction of Ring Opening

Unsymmetrical epoxides give different products under acid-catalyzed and base-catalyzed conditions.

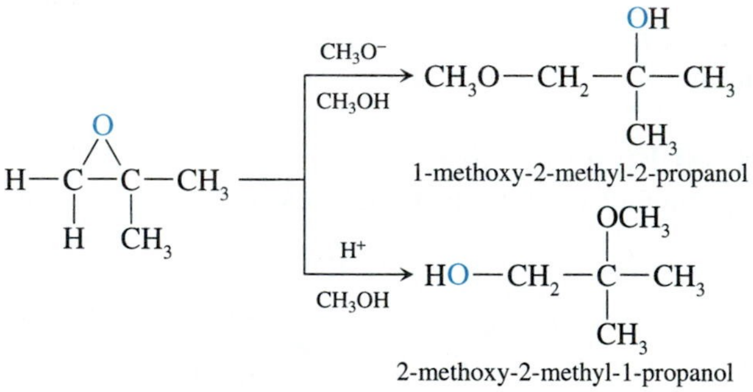

$$H{-}C(H){-}C(CH_3){-}CH_3 \text{ (epoxide)} \xrightarrow[CH_3OH]{CH_3O^-} CH_3O{-}CH_2{-}C(OH)(CH_3){-}CH_3$$

1-methoxy-2-methyl-2-propanol

$$\xrightarrow[CH_3OH]{H^+} HO{-}CH_2{-}C(OCH_3)(CH_3){-}CH_3$$

2-methoxy-2-methyl-1-propanol

In the case of the ring opening by a nucleophile under basic conditions, the reaction is controlled by the same features as the S_N2 displacement reactions we considered in Chapter 7. The nucleophile attacks the less hindered primary carbon atom instead of the tertiary carbon atom. The resulting alkoxide ion then exchanges a proton with the solvent, and the methoxide base is regenerated.

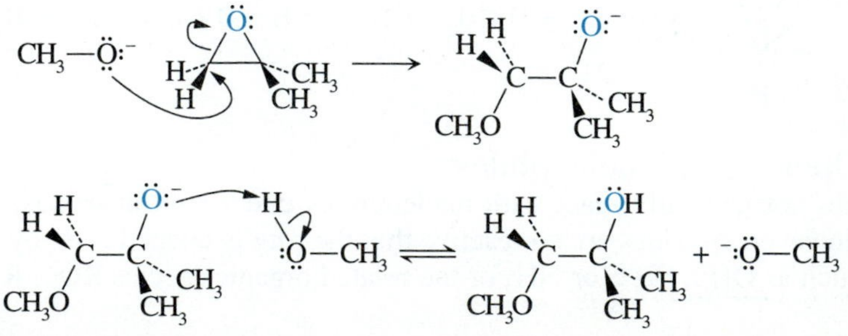

$$CH_3{-}\ddot{O}{:}^- + \text{2,2-dimethyloxirane} \longrightarrow CH_3OCH_2C(CH_3)_2{-}{:}\ddot{O}{:}^-$$

$$CH_3OCH_2C(CH_3)_2{-}{:}\ddot{O}{:}^- + H{-}\ddot{O}{-}CH_3 \rightleftharpoons CH_3OCH_2C(CH_3)_2{-}\ddot{O}H + {}^{-}{:}\ddot{O}{-}CH_3$$

Under acid-catalyzed conditions, the protonated epoxide has some positive charge on each carbon atom in the ring. Consider the following resonance forms:

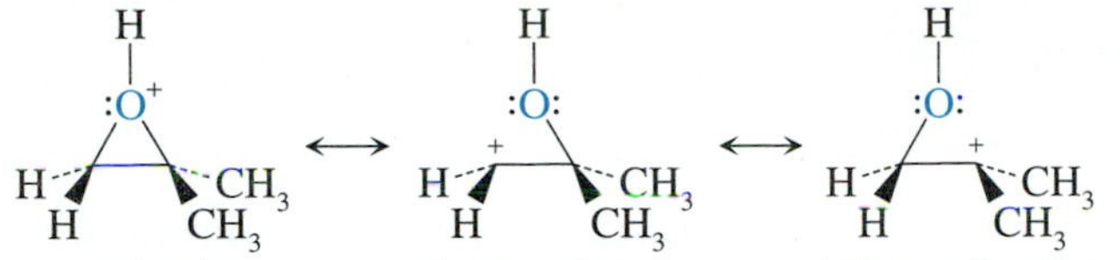

oxonium ion ⟷ primary carbocation ⟷ tertiary carbocation

The positive charge is more stable on the tertiary carbon atom than on the primary carbon atom. This tertiary carbocation resonance form is a more important contributor than the primary one. Thus, the tertiary carbocation center combines with the nucleophile to give the conjugate acid of the observed product of the acid-catalyzed ring opening. The conjugate acid then exchanges a proton with the solvent.

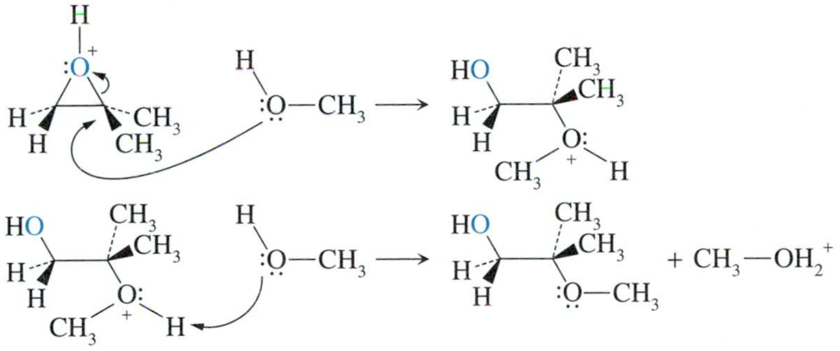

$$+\ CH_3{-}OH_2^+$$

EXAMPLE 9-5

Predict the product of the reaction of 2-methyloxirane with the Grignard reagent prepared from bromoethane.

Solution

The Grignard reagent reacts as a nucleophile. Thus, the Grignard reagent of bromoethane behaves as an ethyl carbanion. Nucleophilic attack of the ethyl carbanion on methyloxirane occurs at the primary rather than the secondary carbon atom. Subsequent hydrolysis of the magnesium alkoxide yields 2-pentanol.

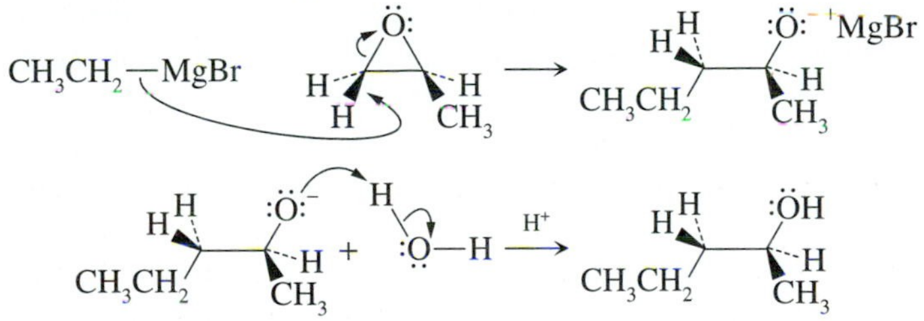

$CH_3CH_2{-}MgBr$

Problem 9.5

Predict the product of the reaction of styrene oxide in an acid-catalyzed reaction with methanol.

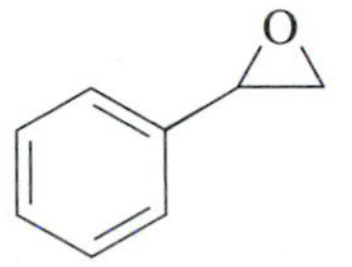

styrene oxide

Ring Opening of Biological Epoxides

Epoxides are produced biologically as oxidation products of alkenes and aromatic compounds. These epoxides are formed in the liver by cytochrome *P*-450 and then they undergo ring-opening reactions. The oxygen of the epoxide comes from molecular oxygen. If the epoxide reacts with a biological macromolecule, the result is potentially devastating. The epoxides made from aromatic compounds are called **arene oxides.** These molecules can undergo four kinds of reactions. With one exception, the reactions give products that are not harmful to the organism.

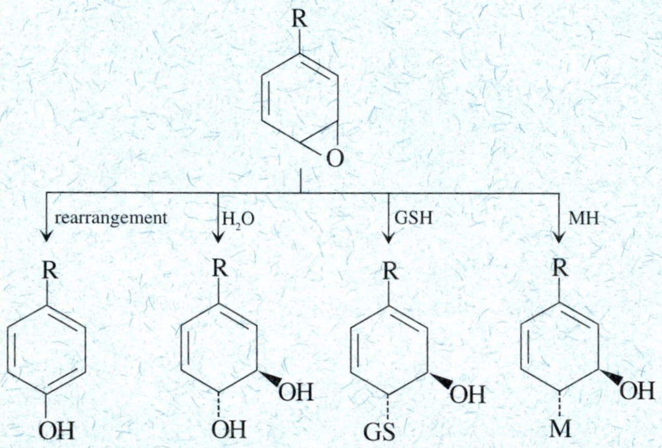

The rearrangement of an arene oxide gives a water-soluble phenol that is easily eliminated from the body as polar derivatives. Both sulfate esters or acetals of glucuronic acid (Chapter 11) are typical derivatives. Hence, this pathway does not lead to the accumulation of toxic byproducts.

Ring opening of the arene oxide by water gives a trans diol by an S_N2 process. These diols tend to be water-soluble and easily eliminated from the body. In the case of naphthalene, the epoxide forms at the C-1 to C-2 bond rather than the C-2 to C-3 bond.

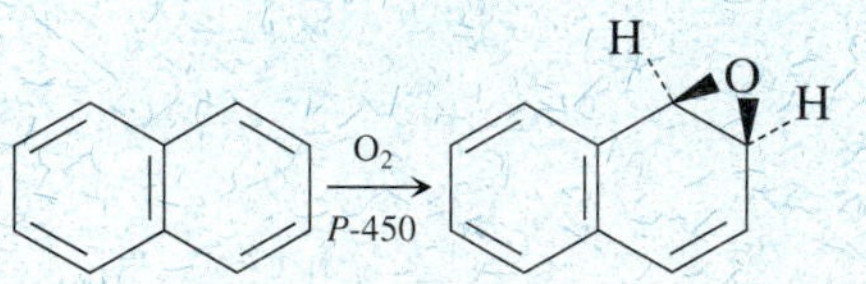

Isotopic labeling studies have shown that the oxygen atom of the benzylic hydroxyl group in the diol formed by ring opening of the epoxide is the original oxygen atom of the epoxide, and the other hydroxyl group comes from water.

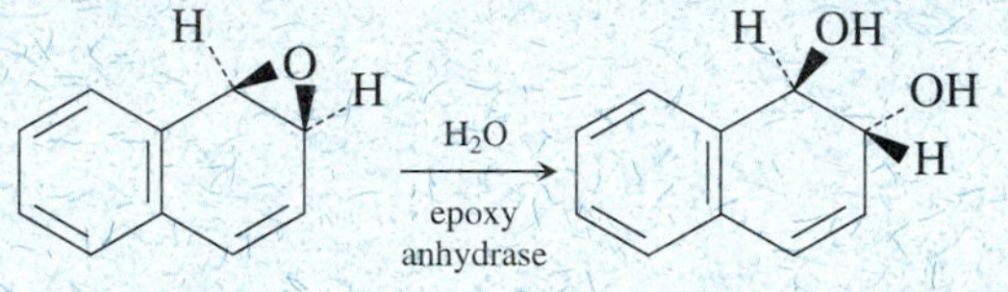

Glutathione contains a nucleophilic sulfhydryl group and acts as a scavenger that reacts with toxic metabolites. Glutathione (GSH) reacts with arene oxides in a ring-opening reaction. The product contains many polar functional groups, so it is water soluble and easily eliminated from the body.

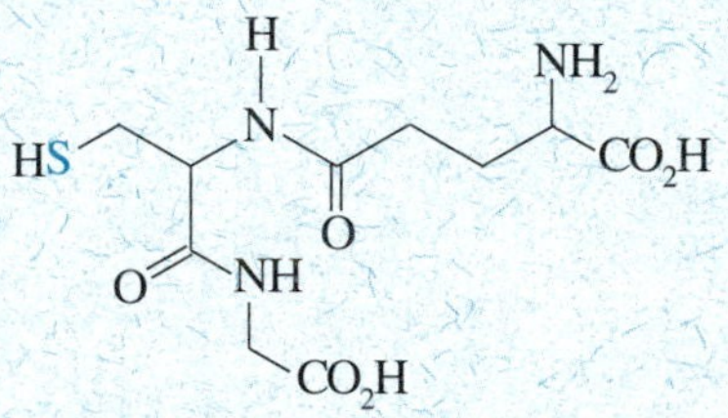

glutathione (GSH)

Arene oxides react with the nucleophilic functional groups present in most macromolecules (represented by MH), including enzymes and the polynucleotides RNA and DNA. These reactions can cause significant alterations in biological functions. A particularly dangerous arene oxide is the epoxide of benzo[*a*]pyrene. Epoxidation followed by ring opening of the epoxide to give a diol and further epoxidation ultimately produces an epoxydiol.

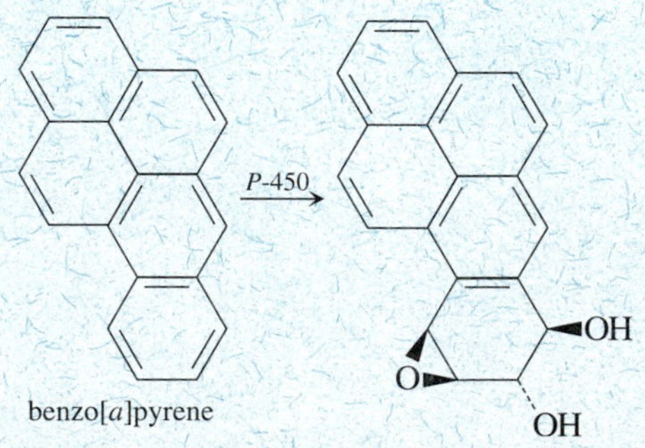

benzo[*a*]pyrene

This epoxydiol reacts with amino groups in DNA and the altered DNA can cause cancer. Benzo[*a*]pyrene is a combustion product found in tobacco smoke as well as in smoke formed by fat dripping in charcoal grills.

The epoxide metabolites of alkenes tend to be more stable than arene oxides. They undergo ring opening with water to give diols as in the case of the ring opening of the epoxide formed in the metabolism of the anticonvulsant drug carbamazepine.

It is not easy to predict whether an epoxide will react with water or glutathione and thus be nontoxic or whether it will react harmfully with macromolecules. However, it appears that relatively stable epoxides tend to undergo ring opening by water or glutathione. Also, epoxides that have sterically hindered oxirane rings—for example, benzo[*a*]pyrene—tend to react with nucleophilic groups of macromolecules.

Like arene oxides, epoxides formed from alkenes also react with glutathione to produce water-soluble compounds that are easily eliminated from the body. Epoxides derived from alkenes also undergo ring-opening reactions with macromolecules. ■

carbamazepine $\xrightarrow{P\text{-}450}$ epoxide $\xrightarrow{\text{epoxy anhydrase}}$ diol

Summary of Reactions

1. Synthesis of Ethers (Section 9.5)

$$CH_3CH_2CH_2CH(CH_3)CH_2OH \xrightarrow[2.\ CH_3CH_2Br]{1.\ NaH} CH_3CH_2CH_2CH(CH_3)CH_2OCH_2CH_3$$

2. Cleavage of Ethers (Section 9.6)

$$\text{cyclohexyl-CH(OCH}_3\text{)CH}_3 \xrightarrow[-\ CH_3Br]{HBr} \text{cyclohexyl-CH(OH)CH}_3 \xrightarrow{HBr} \text{cyclohexyl-CH(Br)CH}_3$$

3. Synthesis of Epoxides (Section 9.7)

$$(CH_3CH_2)HC{=}CH(CH_3) \xrightarrow[CH_2Cl_2]{MCPBA} \text{epoxide}$$

4. Ring Cleavage of Epoxides (Section 9.8)

$$\text{CH}_3\text{CH}_2\text{-epoxide} \xrightarrow[2.\ H_3O^+]{1.\ CH_3O^-/CH_3OH} CH_3CH_2CH(OH)CH_2OCH_3$$

$$\text{CH}_3\text{CH}_2\text{-epoxide} \xrightarrow{CH_3OH/H^+} CH_3CH_2CH(OCH_3)CH_2OH$$

Explorations with Molecular Models

1. Prepare the following model of methyl ethyl ether. View the molecule along the oxygen to methylene carbon bond. What conformational term describes this structure? Draw a Newman projection formula of the structure.

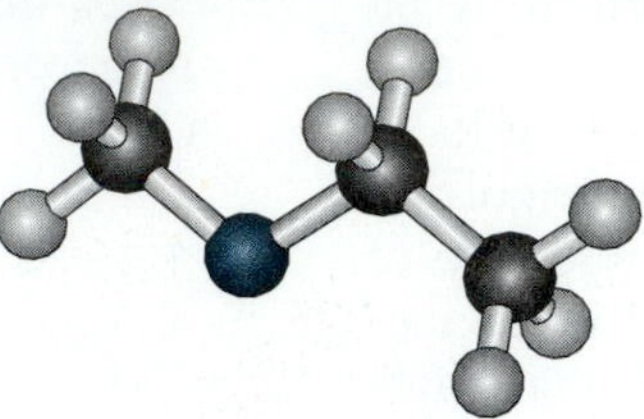

2. Prepare a model of dimethyl ether. Remove one hydrogen atom that is pointed toward you from each methyl group and determine the minimum number of additional methylene groups necessary to form a cyclic ether without any undue strain on the molecular model. It may be necessary to rotate atoms about carbon-carbon bonds to close the ring. Draw the structure and name the compound.

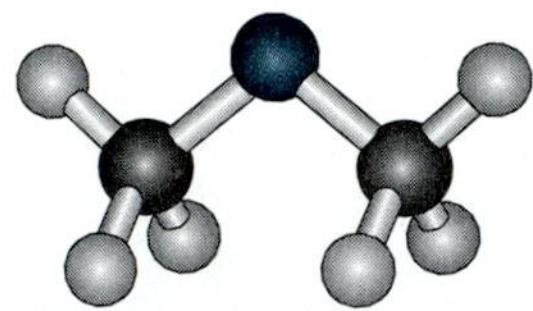

3. Prepare the following model of ethyl isopropyl ether. Consider the cleavage reaction with HBr in which the oxygen atom is protonated and the bromide ion serves as a nucleophile to displace an alcohol. First sight along the bond from the methylene group to the oxygen atom and evaluate the steric environment that would be encountered in an S_N2 displacement. Then sight along the bond from the secondary carbon atom of the isopropyl group to the oxygen atom. Evaluate the steric environment that would be encountered in an S_N2 displacement. Which of the two paths is the least sterically hindered?

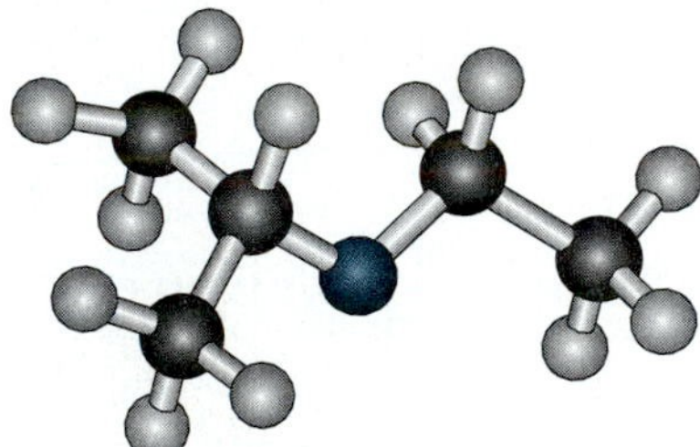

4. Consider the following model of ethylene oxide. How many planes of symmetry are there? Describe the atoms and/or bonds that are intersected by each plane.

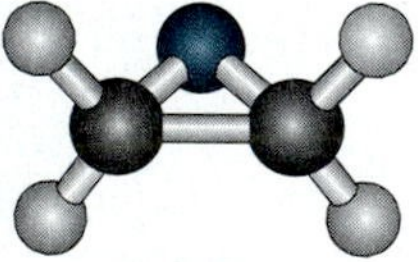

5. Consider the following model of 2-methyloxirane. Determine whether the stereogenic center is *R* or *S*.

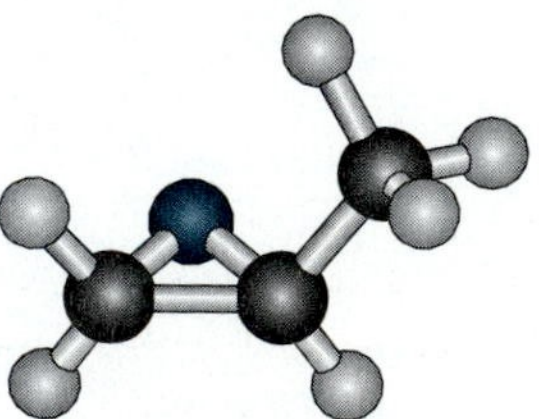

6. Consider the 2-methyloxirane shown in Exploration Exercise 5. The cleavage reaction of this compound using hydroxide ion occurs by a nucleophilic displacement reaction. Consider two possible reactions. First sight along the bond from the methylene group to the oxygen atom and evaluate the steric environment that would be encountered in an S_N2 displacement.

Then sight along the bond from the secondary carbon atom to the ring oxygen atom. Evaluate the steric environment that would be encountered in an S_N2 displacement. Which of the two paths is the least sterically hindered?

7. Consider the following model of 2,3-dimethyloxirane. How many planes of symmetry are there? Describe the atoms and/or bonds that are intersected by each plane. Assign the *R*/*S* configurations of both the ring carbon atom on the right and the ring carbon atom on the left.

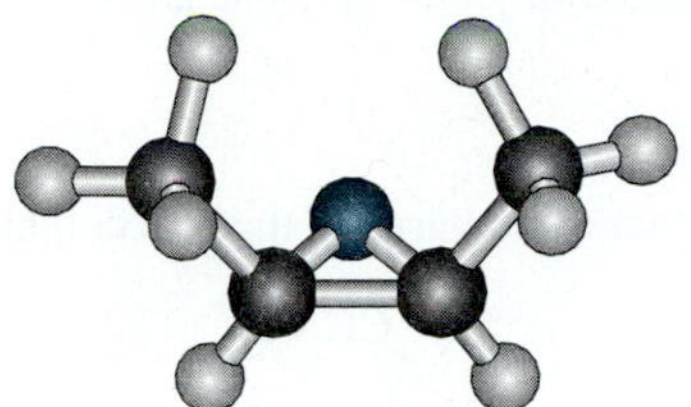

Exercises

Ether Isomers

9.1 Draw the three isomeric ethers with the molecular formula $C_4H_{10}O$.

9.2 Draw the four isomeric methyl ethers with the molecular formula $C_5H_{12}O$.

9.3 Draw the isomeric saturated ethers with the molecular formula C_3H_6O.

9.4 Draw the isomeric unsaturated ethers with the molecular formula C_4H_8O.

Nomenclature of Ethers

9.5 Give the common name each of the following compounds.

(a) cyclopentyl–O–cyclopentyl (b) phenyl–$O—CH_2CH_2CH_3$ (c) cyclopentyl–$O—CH_2CH_2CH_3$

9.6 Give the common name each of the following compounds.

(a) phenyl–$O—CH(CH_3)_2$ (b) phenyl–$CH_2—O—C(CH_3)_3$ (c) cyclohexyl–$O—CH{=}CH_2$

9.7 Assign the IUPAC name of each of the following compounds.

(a) $CH_3CH_2CH_2CHCH_3$ with OCH_3 on C-4 (b) $CH_3CHCH_2CHCH_3$ with CH_3 and OCH_3 substituents (c) $CH_3CH_2CH_2CHCH_2CH_3$ with OCH_2CH_3 substituent

9.8 Assign the IUPAC name of each of the following compounds.

(a) $CH_3CH_2CH_2CHCH_2OCH_3$ with OCH_3 substituent (b) $CH_3CHCH_2CHCH_3$ with CH_3O and OCH_3 substituents (c) $CH_3CH_2CHCH_2CHCH_3$ with CH_3CH_2O and OCH_2CH_3 substituents

9.9 What is the common name of the following compound, which is used as an anesthetic?

$$CH_2{=}CH—O—CH{=}CH_2$$

9.10 2-Chloro-1,1,2-trifluoro-1-difluoromethoxyethane is the IUPAC name of enflurane, a general anesthetic. Draw its structure.

Properties of Ethers

9.11 Dioxane is miscible with water. Why?

9.12 *p*-Ethylphenol is more soluble in aqueous solution than the isomeric ethoxybenzene. Explain why.

9.13 The boiling points of dipropyl ether and diisopropyl ether are 91 and 68°C, respectively. Explain why the boiling points of these isomeric ethers differ.

9.14 The boiling points of 1-ethoxypropane and 1,2-dimethoxyethane are 64 and 83°C, respectively. Explain why.

9.15 Explain why dipropyl ether is soluble in concentrated sulfuric acid whereas heptane is insoluble.

9.16 Aluminum trichloride dissolves in tetrahydropyran and heat is liberated. Explain why.

Grignard Reagents

9.17 Devise a synthesis of 1-deutero-1-methylcyclohexane starting from 1-methylcyclohexene.

9.18 Devise a synthesis of 1,4-dideuterobutane starting from any organic compound that does not contain deuterium.

9.19 Why can't a Grignard reagent be formed using ethanol (CH_3CH_2OH) as a solvent?

9.20 Reaction of 1,2-dibromoethane with magnesium yields ethylene and magnesium bromide. Write a mechanism that accounts for this reaction.

Synthesis of Ethers

9.21 Consider the structure of the following ether and determine the best combination of reactants to synthesize it.

CH_2CH_3

OCH_2CH_3

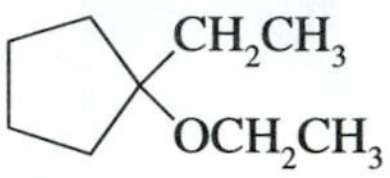

9.22 Consider the structure of the following ether and determine the best combination of reactants to synthesize it.

$-CH_2CH_2-O-$

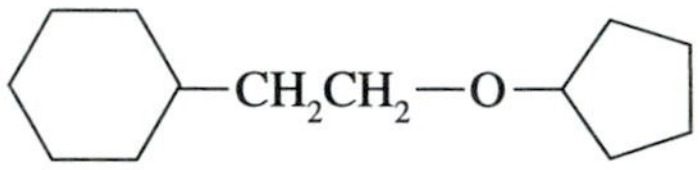

9.23 How could the local anesthetic dibucaine be prepared using the Williamson synthesis?

H

O N

C $CH_2CH_2N(CH_2CH_3)_2$

N $O-CH_2CH_2CH_2CH_3$

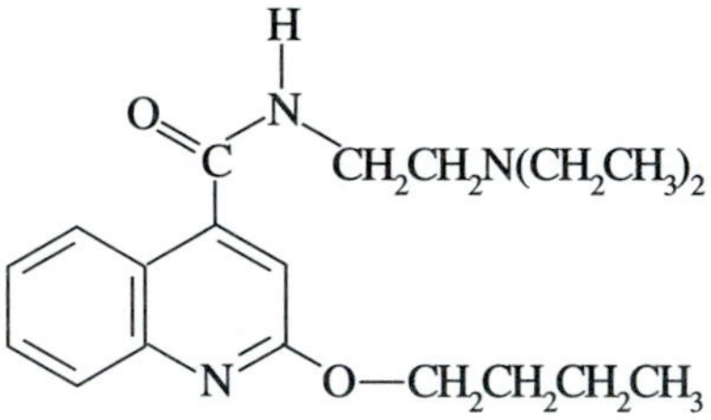

9.24 Consider synthetic methods to produce the antihistamine diphenylpyraline using the Williamson synthesis. What difficulties might be encountered in this synthesis?

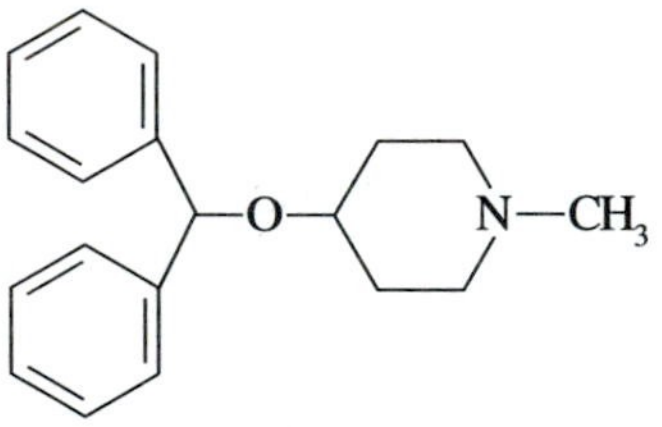

-O- $N-CH_3$

9.25 Reaction of 1-hexene with mercuric acetate in methanol as solvent followed by reduction of the intermediate product with sodium borohydride yields 2-methoxyhexane. What is the structure of the intermediate product? How is it formed?

9.26 Reaction of 5-chloro-2-pentanol with sodium hydride yields 2-methyltetrahydrofuran. How is the product formed?

Synthetic Sequences

9.27 Write the structure of the final product of each of the following sequences of reactions.

(a) 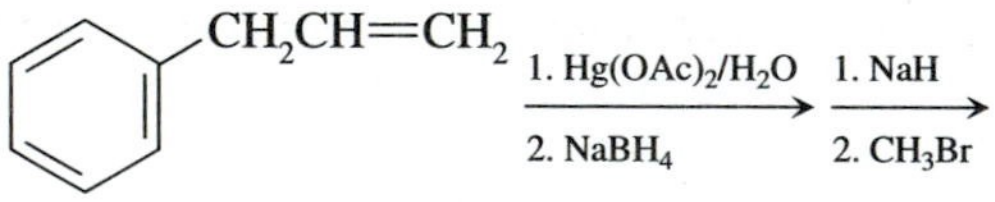 $CH_2CH{=}CH_2$ $\xrightarrow[\text{2. NaBH}_4]{\text{1. Hg(OAc)}_2/\text{H}_2\text{O}}$ $\xrightarrow[\text{2. CH}_3\text{Br}]{\text{1. NaH}}$

(b) 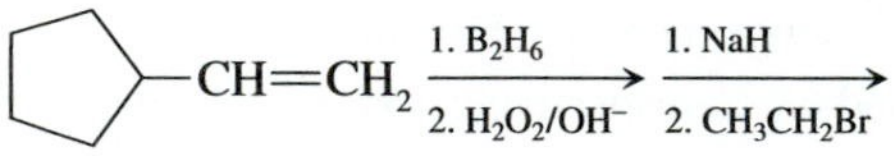 $-CH{=}CH_2$ $\xrightarrow[\text{2. H}_2\text{O}_2/\text{OH}^-]{\text{1. B}_2\text{H}_6}$ $\xrightarrow[\text{2. CH}_3\text{CH}_2\text{Br}]{\text{1. NaH}}$

(c) 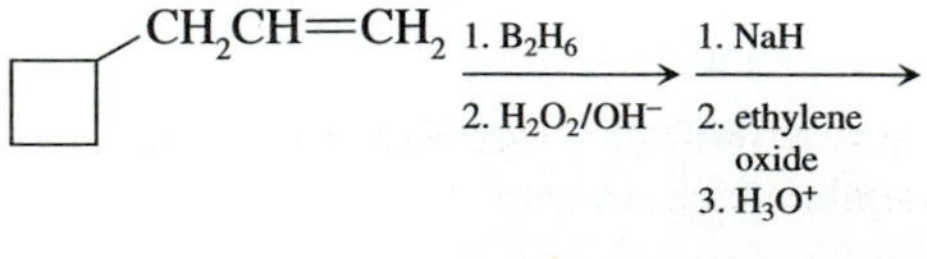 $CH_2CH{=}CH_2$ $\xrightarrow[\text{2. H}_2\text{O}_2/\text{OH}^-]{\text{1. B}_2\text{H}_6}$ $\xrightarrow[\text{2. ethylene oxide}]{\text{1. NaH}}$ 3. H_3O^+

9.28 Write the structure of the final product of each of the following sequences of reactions.

(a) 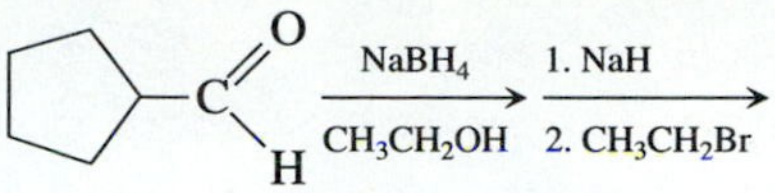 cyclopentyl–CHO $\xrightarrow[CH_3CH_2OH]{NaBH_4}$ $\xrightarrow[2.\ CH_3CH_2Br]{1.\ NaH}$

(b) 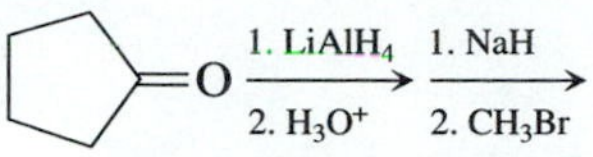 cyclopentanone $\xrightarrow[2.\ H_3O^+]{1.\ LiAlH_4}$ $\xrightarrow[2.\ CH_3Br]{1.\ NaH}$

(c) 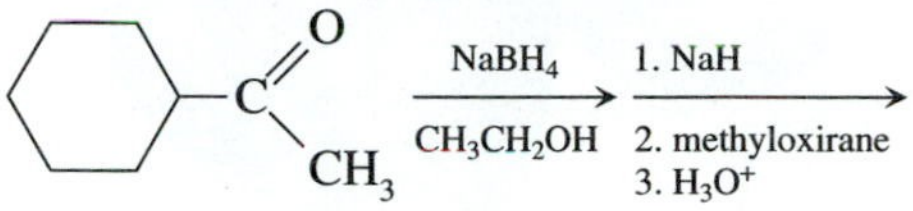 cyclohexyl–C(=O)CH$_3$ $\xrightarrow[CH_3CH_2OH]{NaBH_4}$ $\xrightarrow[2.\ methyloxirane\ 3.\ H_3O^+]{1.\ NaH}$

Reactions of Ethers

9.29 A compound of formula $C_5H_{12}O_2$ reacts with HI to give a mixture of iodomethane, iodoethane, and 1,2-diiodoethane. What is the structure of the compound?

9.30 A compound of formula $C_5H_{12}O_2$ reacts with HI to give a mixture of iodomethane and 1,3-diiodopropane. What is the structure of the compound?

9.31 A compound of formula $C_5H_{10}O$ reacts with HI to give only 1,5-diiodopentane. What is the structure of the compound?

9.32 A compound of formula $C_4H_8O_2$ reacts with HI to give only 1,2-diiodoethane. What is the structure of the compound?

Synthesis of Epoxides

9.33 Two products can be obtained by the epoxidation of the following bicycloalkene with MCPBA. Draw their structures.

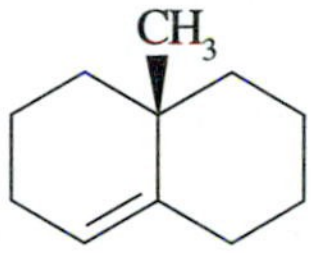

CH_3

9.34 What alkene is required to synthesize the following compound, which is the sex attractant of the Gypsy moth, using MCPBA?

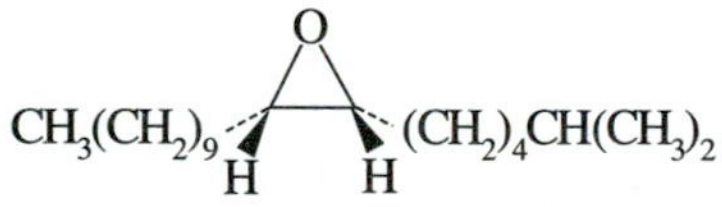

O

$CH_3(CH_2)_9$ ... $(CH_2)_4CH(CH_3)_2$

H H

Reactions of Epoxides

9.35 Ethyl cellosolve, $CH_3CH_2—O—CH_2CH_2—OH$, is an industrial solvent. Suggest a synthesis of this compound using ethylene oxide as one of the reactants.

9.36 2-Phenylethanol is used in some perfumes. How can this compound be prepared starting from bromobenzene?

9.37 What is the product of the reaction of 2-methyloxirane with methanol in the presence of an acid catalyst?

9.38 Epoxide rings can be cleaved by metal hydrides. Write the product of the reaction of cyclohexene oxide and $LiAlD_4$.

9.39 A mixture of 2,2-dimethyloxirane and ethanethiol is treated with sodium hydroxide. Write the structure of the expected product.

9.40 Epoxide rings can be cleaved by phenoxides. Propose a synthesis of the muscle relaxant methocarbamol using this fact.

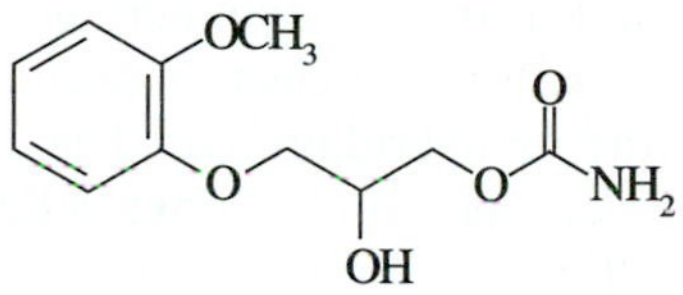

OCH_3 O

O O NH_2

OH

Spectroscopy

9.41 If you have read Chapter 18 and the material has been assigned by your professor, then Exercises 18.7, 18.17, and 18.18, which are based on the structure of ethers and epoxides, may be done.

CHAPTER 10

ALDEHYDES AND KETONES

Overview

10.1 The Carbonyl Group

A **carbonyl group** consists of a **carbonyl carbon atom** and a **carbonyl oxygen atom** linked by a double bond. Formaldehyde, CH_2O, is the simplest compound with a carbonyl group. Note that the lone pairs on the oxygen atom of the carbonyl group are not always shown in drawings of molecular structures.

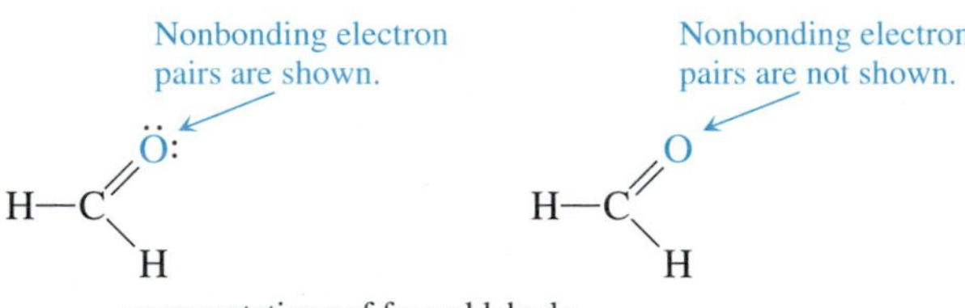

representations of formaldehyde

The carbonyl carbon atom, which is sp^2-hybridized, contributes one electron to each of the three hybrid orbitals, forming three σ bonds. Formaldehyde has two σ bonds to hydrogen atoms and one σ bond to the carbonyl oxygen atom. These coplanar bonds lie at approximately 120° to each other. The fourth electron of the carbonyl carbon atom occupies a $2p$ orbital perpendicular to the plane of the three sp^2 hybrid orbitals. The carbonyl oxygen atom, also sp^2-hybridized, contributes one of its six valence electrons to the sp^2 hybrid orbital that forms a σ bond with the carbonyl carbon atom. Four valence electrons remain as two sets of nonbonding electron pairs in the other two sp^2 hybrid orbitals. They lie at approximately 120° to each other and to the carbon-oxygen bond (Figure 10.1). The last valence electron occupies a $2p$ orbital perpendicular to the plane of the sp^2 hybrid orbitals. The $2p$ orbitals of the carbon and oxygen atoms overlap sideways to form a π bond.

Because oxygen is more electronegative than carbon, the oxygen atom attracts the electrons in the carbon-oxygen double bond, making the carbonyl bond polar. In addition, the carbonyl group is resonance-stabilized, as shown by the charged contributing structure (2).

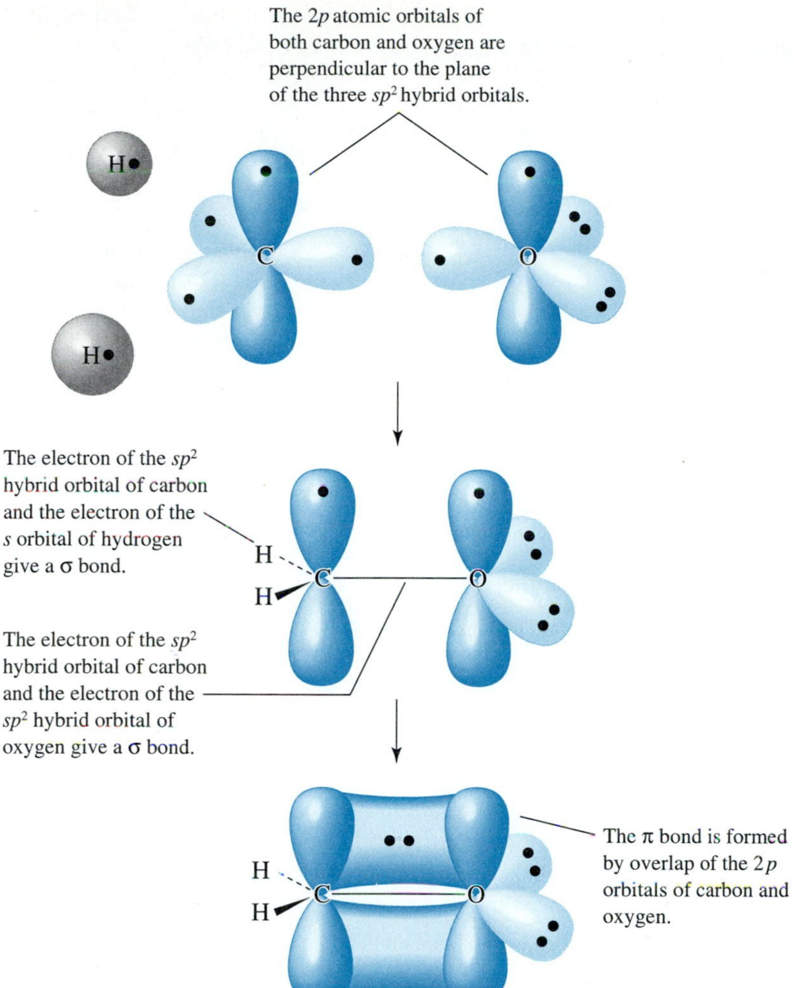

The 2*p* atomic orbitals of both carbon and oxygen are perpendicular to the plane of the three sp^2 hybrid orbitals.
H•
H•
C
O
The electron of the sp^2 hybrid orbital of carbon and the electron of the *s* orbital of hydrogen give a σ bond.
H
H
The electron of the sp^2 hybrid orbital of carbon and the electron of the sp^2 hybrid orbital of oxygen give a σ bond.
The π bond is formed by overlap of the 2*p* orbitals of carbon and oxygen.

FIGURE 10.1
Bonding in the Carbonyl Group

$$\underset{(1)}{>C^{\delta+}{=}\ddot{\underset{\cdot\cdot}{O}}{:}^{\delta-}} \longleftrightarrow \underset{(2)}{>\overset{+}{C}{-}\ddot{\underset{\cdot\cdot}{O}}{:}^{-}}$$

Contributing structure (1) is more important because each atom has a Lewis octet. However, the dipolar structure (2) is often used to account for the chemical reactions of aldehydes and ketones. We shall see that the carbonyl carbon atom reacts with nucleophiles, and the carbonyl oxygen atom reacts with electrophiles.

Carbonyl Compounds

When a carbonyl carbon atom is bonded to at least one hydrogen atom, the resulting compound is an **aldehyde.** Formaldehyde, the simplest aldehyde, has two hydrogen atoms bonded to the carbonyl carbon atom. The carbonyl group in other aldehydes is bonded to one hydrogen atom and either an alkyl group (R) or an aromatic group (Ar). Although the bond angles around the carbonyl carbon atom are approximately 120°, structures are often written with the carbonyl oxygen atom at 90° to a linear arrangement of atoms.

$$R{-}C(=\ddot{O}:){-}H \quad \text{or} \quad R{-}\overset{\overset{\ddot{O}:}{\|}}{C}{-}H$$

general structural formulas of aldehydes

A carbonyl carbon atom is bonded to two other carbon atoms of either alkyl or aryl groups in a **ketone.** As with aldehydes, a ketone has 120° bond angles at the carbonyl carbon atom, but structures are often written with the carbonyl oxygen atom at 90° to a linear arrangement of atoms.

$$R{-}C(=\ddot{O}:){-}R \quad \text{or} \quad R{-}\overset{\overset{\ddot{O}:}{\|}}{C}{-}R$$

general structural formulas of ketones

An aldehyde can be written with the condensed formula RCHO or ArCHO, where the symbol CHO indicates that both the hydrogen and oxygen atoms are bonded to the carbonyl carbon atom. A ketone has the condensed formula RCOR. In this condensed formula the symbol CO represents the carbonyl group, and the two R groups flanking the CO group are bonded to the carbonyl carbon atom.

Occurrence of Aldehydes and Ketones

The carbonyl group is prevalent in compounds isolated from biological sources. Aldehydes and ketones that are sufficiently volatile often have pleasant odors. For that reason, they are used in commercial products such as air fresheners, soaps, and perfumes. For example, α-ionone and jasmone are fragrant ketones responsible for the scent of irises and jasmine, respectively (Figure 10.2). The relationship between structure and the physiological response of odor is not well understood.

At one time the extraction of fragrant compounds from flowers and other plants was the sole source of materials for products such as perfumes. However, it is now more economical to synthesize these compounds in the laboratory. In addition, synthesis allows for the production of compounds with "new" odors.

10.2 Nomenclature of Aldehydes and Ketones

Aldehydes and ketones with low molecular weights are often referred to by their common names. The names of aldehydes are derived from the common names of related acids (Chapter 12).

$$H{-}\overset{\overset{\ddot{O}:}{\|}}{C}{-}H \qquad CH_3{-}\overset{\overset{\ddot{O}:}{\|}}{C}{-}H \qquad CH_3{-}\overset{\overset{\ddot{O}:}{\|}}{C}{-}CH_3$$

formaldehyde, acetaldehyde, acetone

$$C_6H_5{-}\overset{\overset{\ddot{O}:}{\|}}{C}{-}H \qquad C_6H_5{-}\overset{\overset{\ddot{O}:}{\|}}{C}{-}CH_3 \qquad C_6H_5{-}\overset{\overset{\ddot{O}:}{\|}}{C}{-}C_6H_5$$

benzaldehyde, acetophenone, benzophenone

α-ionone

jasmone

space-filling models

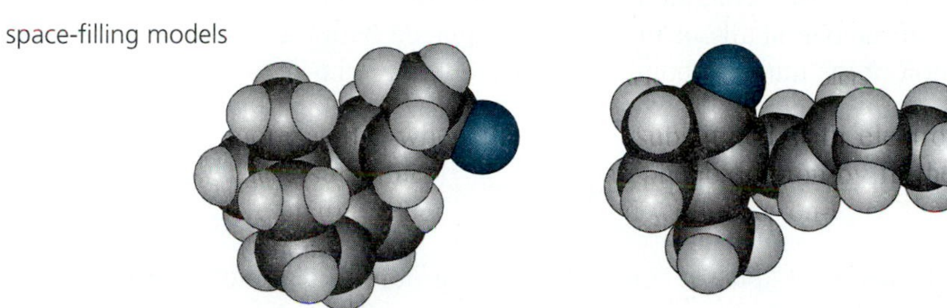

Ball and stick models

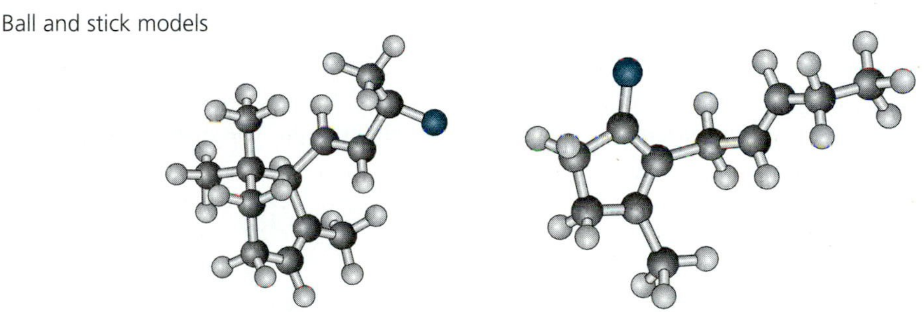

FIGURE 10.2
Structures of Some Fragrant Ketones

IUPAC Names of Aldehydes

Aldehydes are named by IUPAC rules similar to those outlined for alcohols. The final *-e* of the parent hydrocarbon corresponding to the aldehyde is replaced by the ending *-al*. The IUPAC rules are as follows:

1. The parent chain is numbered so that the carbonyl carbon atom is C-1. The number 1 is not used in the name because it is understood that the carbonyl carbon atom is located at the end of the chain. The name of each substituent and the number of the carbon atom to which it is attached is added to the parent name as a prefix.

$$\underset{4}{CH_3}-\underset{3}{\overset{CH_3}{\overset{|}{CH}}}-\underset{\underset{CH_3}{|}}{\overset{2}{CH}}-\underset{1}{\overset{O}{\overset{||}{C}}}-H$$

This is 2,3-dimethylbutanal, not 2,3-dimethyl-1-butanal.

2. The aldehyde functional group has a higher priority than alkyl, halogen, hydroxyl, and alkoxy groups. The names and positions of these groups are indicated as prefixes to the name of the parent aldehyde.

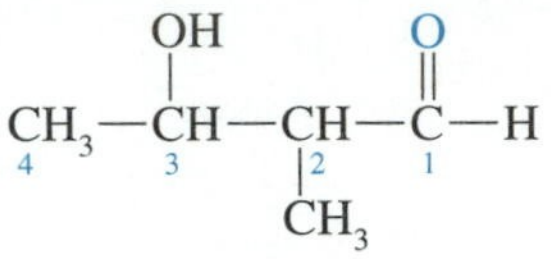

$$\underset{4}{CH_3}-\underset{3}{\overset{OH}{\overset{|}{CH}}}-\underset{\underset{CH_3}{|2}}{CH}-\underset{1}{\overset{O}{\overset{\|}{C}}}-H$$

This is 3-hydroxy-2-methylbutanal.

3. The aldehyde functional group has a higher priority than double or triple bonds. For a chain containing a double or triple bond, the final *-e* of the name of the parent alkene or alkyne is replaced with the suffix *-al*. The position of the multiple bond is indicated by a prefix.

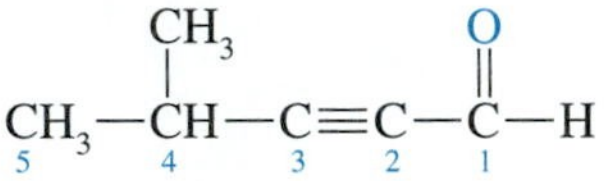

$$\underset{5}{CH_3}-\underset{4}{\overset{CH_3}{\overset{|}{CH}}}-\underset{3}{C}\equiv\underset{2}{C}-\underset{1}{\overset{O}{\overset{\|}{C}}}-H$$

This is 4-methyl-2-pentynal.

4. For aldehydes or ketones containing other groups with a higher priority, such as carboxylic acids, the carbonyl group is given the prefix *oxo-*. The position of the oxo group is given by a number. The priority order is carboxylic acid > aldehyde > ketone.

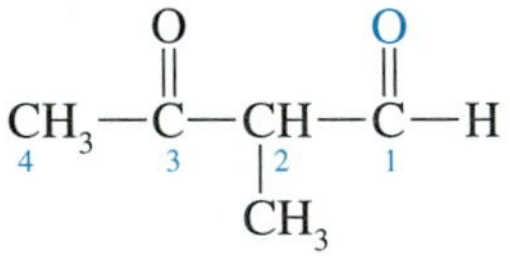

$$\underset{4}{CH_3}-\underset{3}{\overset{O}{\overset{\|}{C}}}-\underset{\underset{CH_3}{|2}}{CH}-\underset{1}{\overset{O}{\overset{\|}{C}}}-H$$

This is 2-methyl-3-oxobutanal.

5. The suffix *carbaldehyde* is used to designate an aldehyde group attached to a ring.

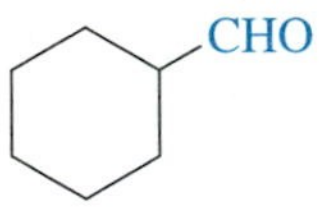

CHO

cyclohexanecarbaldehyde

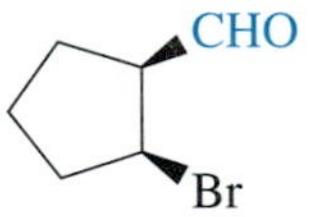

CHO

Br

cis-2-bromocyclopentanecarbaldehyde

IUPAC Names of Ketones

The IUPAC rules for naming ketones are similar to those used for aldehydes. The final *-e* of the parent hydrocarbon is replaced with the ending *-one*. However, the carbonyl group in a ketone is not on a terminal carbon atom and its position must be indicated.

1. The carbon chain is numbered so that the carbonyl carbon atom has the lower number. This number appears as a prefix to the parent name. The identity and location of substituents are indicated with a prefix to the parent name.

$$\underset{5}{CH_3}-\underset{4}{\overset{CH_3}{\overset{|}{CH}}}-\underset{3}{CH_2}-\underset{2}{\overset{O}{\overset{\|}{C}}}-\underset{1}{CH_3}$$

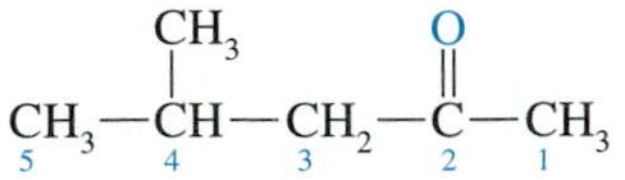

This is 4-methyl-2-pentanone, not 2-methyl-4-pentanone.

2. Cyclic ketones are named as cycloalkanones. The carbonyl carbon atom receives the number 1. The ring is then numbered in the direction that gives the lower number to the first substituent encountered.

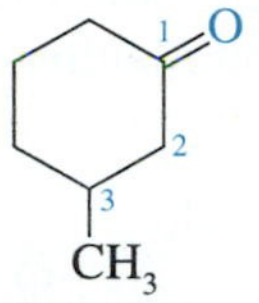

3-methylcyclohexanone 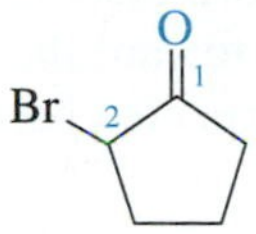 2-bromocyclopentanone

3. Halogens, the hydroxyl group, alkoxy groups, and multiple bonds have lower priorities than the ketone group. The name of a substituted ketone is determined by the method described previously for aldehydes.

EXAMPLE 10-1

1-Phenyl-2,4-hexadiyn-1-one is the IUPAC name for Capillin®, which is used against skin fungi. Draw its structure.

Solution

When we dissect the name, we see that it has the suffix 1-one and the stem name hexa, indicating that the parent chain is a ketone containing six carbon atoms. We write the carbon skeleton and number the chain. Place the carbonyl oxygen atom on the C-1 atom.

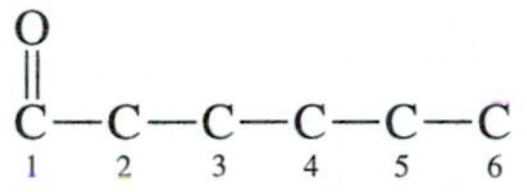

The name has the prefix 1-phenyl. Therefore, we add a phenyl group at the C-1 atom. Note that the presence of the phenyl group makes the compound a ketone. A carbonyl carbon atom at the end of a chain would otherwise be part of an aldehyde.

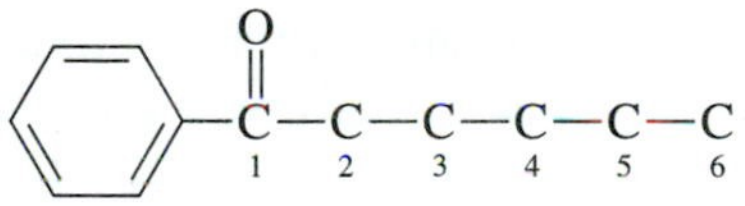

The diyn tells us that there are two triple bonds; they are located at the C-2 and C-4 atoms. Fill in the requisite hydrogen atoms.

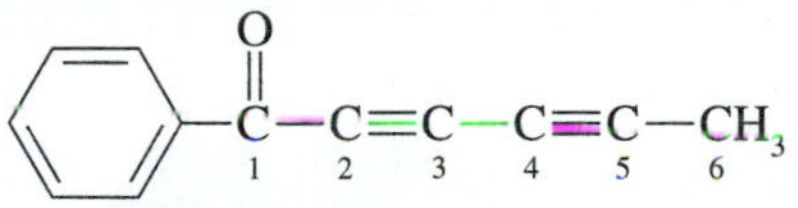

Problem 10.1

Give the IUPAC name for the following compound. It is an alarm pheromone in some species of ants.

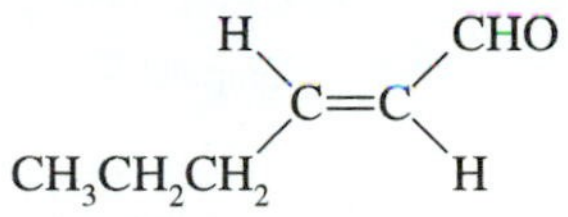

10.3 Physical Properties of Aldehydes and Ketones

Because oxygen is more electronegative than carbon, the electrons in the carbonyl bond are pulled toward the oxygen atom, and the carbonyl group is polar. This polarity is also expected based on the contributing charged structure of the resonance hybrid.

$$CH_3CH_2(H)\overset{\delta+}{C}{=}\overset{\delta-}{\ddot{O}}: \longleftrightarrow CH_3CH_2(H)\overset{+}{C}{-}\ddot{\underset{..}{O}}:^{-}$$

As a result, the dipole moment for propanal, a typical aldehyde, is larger than the dipole moment of butane and 1-propanol (Table 10.1). The dipole moment reflects the polarity of the carbonyl group.

Boiling Points

Aldehydes and ketones have higher boiling points than the alkanes of similar molecular weight (Table 10.1) because of dipole-dipole intermolecular forces due to the carbonyl group. However, alcohols have higher boiling points than aldehydes and ketones even though alcohols have smaller dipole moments than carbonyl compounds. This order of boiling points is the result of hydrogen bonding in alcohols that is not possible in carbonyl compounds. As the molecular weights of the carbonyl compounds increase, their dipole-dipole attractive forces become less important compared to London forces of the hydrocarbon skeleton. As a result, the physical properties of aldehydes and ketones become more like those of hydrocarbons as chain length increases. The boiling point differences become smaller, although the order of boiling points is still alcohol > carbonyl compound > alkane.

Solubility

Aldehydes and ketones cannot form hydrogen bonds with one another because they cannot function as hydrogen bond donors. However, the carbonyl oxygen atom has lone pair electrons that can serve as hydrogen bond acceptors. Thus, carbonyl groups can form hydrogen bonds with water. Hence, the lower molec-

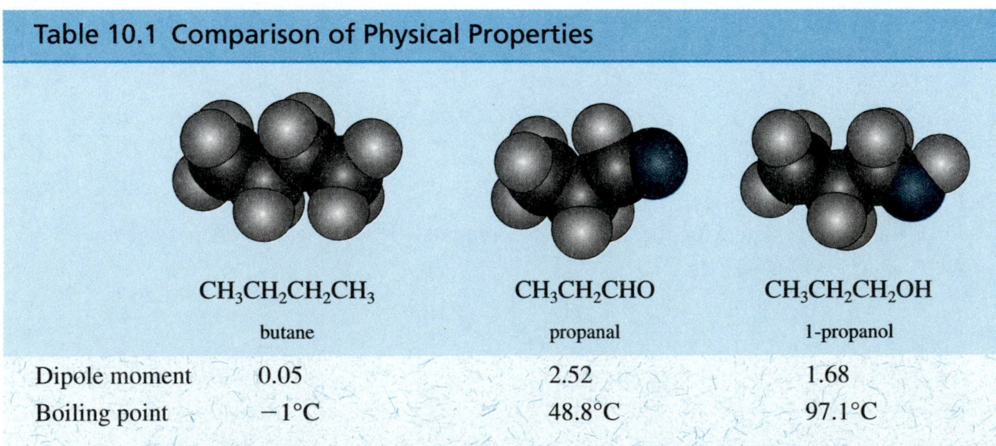

Table 10.1 Comparison of Physical Properties

	$CH_3CH_2CH_2CH_3$ butane	CH_3CH_2CHO propanal	$CH_3CH_2CH_2OH$ 1-propanol
Dipole moment	0.05	2.52	1.68
Boiling point	−1°C	48.8°C	97.1°C

ular weight compounds formaldehyde, acetaldehyde, and acetone are soluble in water in all proportions.

$$\mathrm{H_2O \cdots O{=}C(CH_3)_2 \cdots H_2O}$$

The lone pair electrons of the carbonyl oxygen atom serve as a hydrogen bond acceptor.

However, the solubility of carbonyl compounds in water decreases as the chain length increases, and their solubilities become more like those of hydrocarbons.

Both acetone and 2-butanone (known in industry as methyl ethyl ketone, or MEK) are excellent solvents for a variety of organic compounds. These polar solvents dissolve polar solutes because "like dissolves like." These solvents also readily dissolve protic solutes such as alcohols and carboxylic acids because the carbonyl group acts as a hydrogen bond acceptor for these compounds.

10.4 Redox Reactions of Carbonyl Compounds

The carbonyl group is in an oxidation state between that of an alcohol and a carboxylic acid. Thus, a carbonyl group can be reduced to an alcohol or oxidized to a carboxylic acid.

$$-\overset{\overset{\large OH}{|}}{\underset{|}{C}}-H \xleftarrow{\text{reduction}} -\overset{\overset{\large O}{\|}}{C}- \xrightarrow{\text{oxidation}} -\overset{\overset{\large O}{\|}}{C}-OH$$

Oxidation Reactions

In Chapter 8 we saw that primary alcohols are oxidized to aldehydes, which are subsequently easily oxidized to acids. Under the same conditions, secondary alcohols are oxidized to ketones, which are not oxidized further. This difference in reactivity distinguishes these classes of compounds. For example, Tollens's reagent and Fehling's solution are mild oxidizing reagents. Each of these converts aldehydes to carboxylic acids; neither of them oxidizes ketones. These reagents therefore provide a simple qualitative way of distinguishing aldehydes from ketones.

Tollens's reagent is a basic solution of a silver ammonia complex ion. When Tollens's reagent is added to a test tube that contains an aldehyde, the aldehyde is oxidized and metallic silver is deposited as a mirror on the wall of the test tube.

The reflection of an image in a mirror is due to a metallic silver deposit.

$$R-\overset{\overset{\large O}{\|}}{C}-H + 2\,Ag(NH_3)_2^{\,+} + 3\,OH^- \longrightarrow R-\overset{\overset{\large O}{\|}}{C}-O^- + 2\,Ag(s) + 2\,H_2O + 2\,NH_3$$

Fehling's solution contains cupric ion, Cu^{2+}, as a complex ion in a basic solution. It oxidizes aldehydes to carboxylic acids as the Cu^{2+} is reduced to Cu^+, which forms a brick-red precipitate, Cu_2O. Fehling's solution has the characteristic blue color of Cu^{2+}, which fades as the red precipitate of Cu_2O

forms. Because Fehling's solution is basic, the carboxylic acid product is formed as its conjugate base.

$$R{-}\overset{\overset{\displaystyle O}{\|}}{C}{-}H + \underset{\substack{\text{blue}\\ \text{solution}}}{2\,Cu^{2+}} + 5\,OH^{-} \longrightarrow R{-}\overset{\overset{\displaystyle O}{\|}}{C}{-}O^{-} + \underset{\substack{\text{red}\\ \text{precipitate}}}{Cu_2O(s)} + 3\,H_2O$$

Reduction to Alcohols

In Chapter 4 we saw that carbon-carbon double bonds can be reduced by hydrogen gas with nickel, palladium, or platinum as catalysts. In Chapter 8 we found that aldehydes and ketones can be catalytically reduced to alcohols by hydrogen gas and catalysts such as palladium and Raney nickel. However, the reduction of aldehydes and ketones with hydrogen gas requires more severe conditions than are required to reduce alkenes. We also learned that both lithium aluminum hydride, $LiAlH_4$, and sodium borohydride, $NaBH_4$, reduce carbonyl groups, but neither reagent reduces carbon-carbon double or triple bonds (Section 8.8). Thus, either of these reagents may be used to reduce a carbonyl group selectively in compounds with carbon-carbon multiple bonds.

$$\text{cyclohexenyl}{-}\overset{\overset{\displaystyle O}{\|}}{C}{-}H \xrightarrow[CH_3CH_2OH]{NaBH_4} \text{cyclohexenyl}{-}CH_2OH$$

$$\text{(bicyclic alkene)}{-}\overset{\overset{\displaystyle O}{\|}}{C}{-}CH_3 \xrightarrow[\text{2. } H_3O^+]{\text{1. } LiAlH_4} \text{(bicyclic alkene)}{-}\overset{\overset{\displaystyle OH}{|}}{C}H{-}CH_3$$

Reduction to a Methylene Group

A carbonyl group can be reduced directly to a methylene group by using either the Clemmensen reduction or the Wolff-Kishner reduction. The former uses a zinc amalgam (Zn/Hg) and HCl, and the latter uses hydrazine (NH_2NH_2) and base.

$$\underset{\text{cyclohexanecarbaldehyde}}{C_6H_{11}{-}\overset{\overset{\displaystyle O}{\|}}{C}{-}H} \xrightarrow[HCl]{Zn/Hg} \underset{\text{methylcyclohexane}}{C_6H_{11}{-}CH_3}$$

$$\underset{\text{acetophenone}}{C_6H_5{-}\overset{\overset{\displaystyle O}{\|}}{C}{-}CH_3} \xrightarrow[KOH]{NH_2NH_2} \underset{\text{ethylbenzene}}{C_6H_5{-}CH_2CH_3}$$

We introduced the reduction of a carbonyl group to a methylene group in Chapter 5 as a method of converting the product of a Friedel-Crafts acylation to an alkyl group that could not be produced by direct Friedel-Crafts alkylation.

EXAMPLE 10-2

Can the isomeric carbonyl-containing compounds of molecular formula C_4H_8O be distinguished from each other by Tollens's reagent?

Solution

There are three isomeric carbonyl-containing compounds—two aldehydes and one ketone.

$CH_3CH(CH_3)CHO$ $\quad$ $CH_3CH_2CH_2CHO$ $\quad$ $CH_3CH_2C(=O)CH_3$

Both aldehydes react with Tollens's reagent to produce a silver mirror. Therefore, these two compounds cannot be distinguished from each other with this reagent. However, the ketone does not react with Tollens's reagent. Therefore, the compound that does not yield a silver mirror is 2-butanone.

Problem 10.2

Explain how the following two isomeric compounds could be distinguished using Fehling's solution.

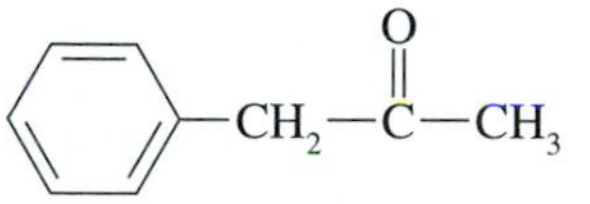

C_6H_5—CH_2—C(=O)—CH_3 (1) $\quad$ C_6H_5—CH_2—CH_2—C(=O)—H (2)

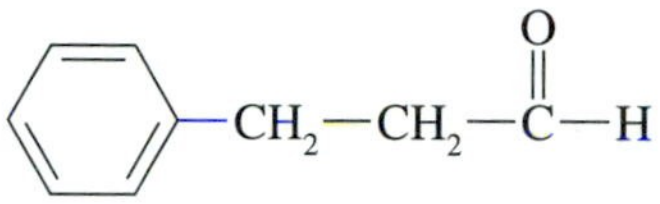

EXAMPLE 10-3

Reduction of the following ketone using sodium borohydride gives a mixture of two isomeric compounds. Draw their structures.

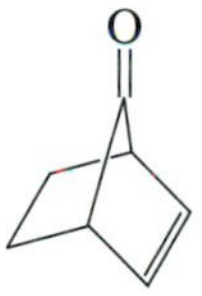

Solution

The hydride ion can attack from either the right or left of the planar carbonyl group. Attack from the left places the hydroxyl group on the same side of the ring as the carbon-carbon double bond. Attack from the right gives an isomer with the hydroxyl group on the side opposite that of the carbon-carbon double bond.

H OH $\quad$ HO H

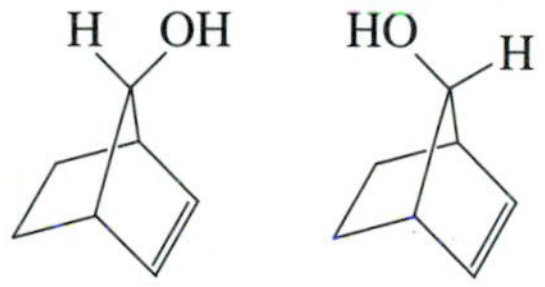

Problem 10.3

Draw the structure of the product of the reaction of the following compound with each of the following reducing agents: (a) hydrazine and base, (b) palladium and hydrogen at 1 atm, and (c) Raney nickel and hydrogen at 100 atm.

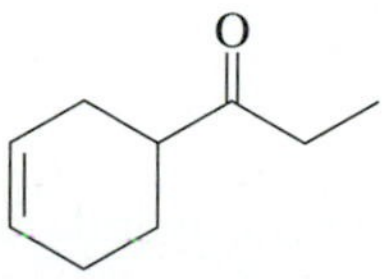

O

10.5 Addition Reactions of Carbonyl Compounds

Aldehydes and ketones contain a π bond that undergoes addition reactions. Because the carbonyl bond is polar, an unsymmetrical reagent reacts with it so that the electrophilic part bonds to the carbonyl oxygen atom and the nucleophilic part to the carbonyl carbon atom.

Nucleophiles attack here. $\overset{\delta+}{C}=\overset{\delta-}{\ddot{O}}:$ Electrophiles attack here.

Many reagents that add to carbonyl compounds can be represented as H—Nu. The electrophilic part of the reagent is H^+; the nucleophilic part is $Nu:^-$. The addition reaction occurs in several steps. The order of the steps depends on whether the reaction is acid- or base-catalyzed.

For the base-catalyzed reaction, the nucleophile attacks the carbonyl group, giving a tetrahedral intermediate that is subsequently protonated.

$$Nu:^- + C=\ddot{O}: \rightleftharpoons Nu-C-\ddot{O}:^- \overset{H_2O}{\rightleftharpoons} Nu-C-\ddot{O}-H + OH^-$$

trigonal planar carbonyl compound — tetrahedral intermediate — tetrahedral product

The net reaction is the addition of a nucleophile and a proton across the π bond of the carbonyl group. The nucleophile $Nu:^-$ is formed by an acid-base reaction of H—Nu with OH^-. Note that OH^- is regenerated as a result of protonation of the oxygen atom of the tetrahedral intermediate.

The acid-catalyzed reaction occurs by initial protonation of the carbonyl oxygen atom to produce a resonance-stabilized intermediate.

$$C=\ddot{O}: + H^+ \rightleftharpoons \left[\overset{+}{C}-\ddot{O}-H \longleftrightarrow C=\ddot{O}^+-H\right]$$

trigonal planar carbonyl compound — resonance-stabilized oxocarbocation

As a result, the carbonyl carbon atom is more susceptible to attack by even weak nucleophiles. This reaction can be shown using either of the two contributing resonance forms. Subsequent loss of the proton bonded to the original nucleophile in an acid-base reaction with solvent regenerates the proton required in the first step.

$$H-Nu: + C=\ddot{O}^+-H \rightleftharpoons H-Nu^+-C-\ddot{O}-H \overset{-H^+}{\rightleftharpoons} Nu-C-\ddot{O}-H$$

tetrahedral intermediate

Equilibria in Addition Reactions

Nucleophiles that add to carbonyl compounds are classed according to the reversibility of the reaction. Nucleophiles that add irreversibly include hydride ion derived from lithium aluminum hydride or sodium borohydride (Section

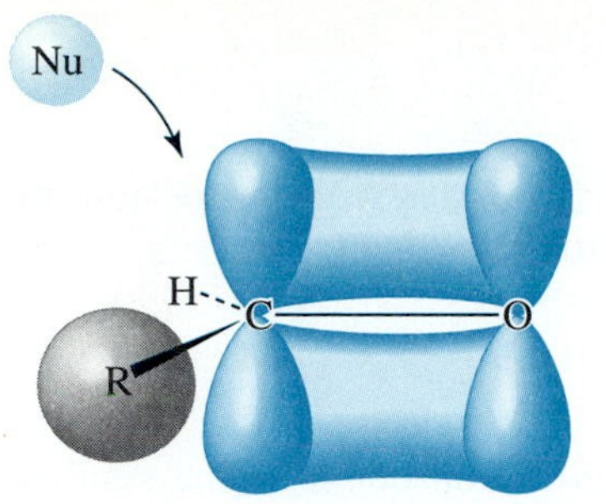

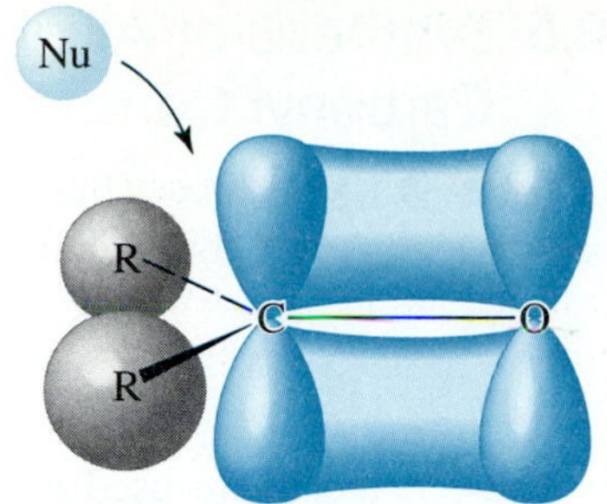

FIGURE 10.3
Effect of Steric Hindrance on Addition Reactions

10.4) and the carbanion derived from a Grignard reagent (Section 10.6). These nucleophiles are conjugate bases of weak acids and are thus poor leaving groups. As a result, once they add to the carbonyl carbon atom the reverse reaction is unfavorable. Nucleophiles that add reversibly are conjugate bases of strong acids and are good leaving groups. We will consider examples of these reactions in Sections 10.7 through 10.9. The important point to note is that these reversible reactions are usually driven to completion by adjusting reaction conditions based on Le Châtelier's principle.

Relative Reactivities of Aldehydes and Ketones

Nucleophiles react faster with aldehydes than with ketones because of both electronic and steric effects. First, we will consider electronic effects. A ketone has two alkyl groups attached to the carbonyl carbon atom. These groups donate electron density to the carbonyl carbon atom and stabilize its partial positive charge. (We recall that alkyl groups stabilize carbocations in the same way.) The carbonyl carbon atom of an aldehyde is attached to only one alkyl group. Therefore, the carbonyl carbon atom of an aldehyde has a larger partial positive charge than that of a ketone. As a consequence, nucleophiles react faster with aldehydes than with ketones.

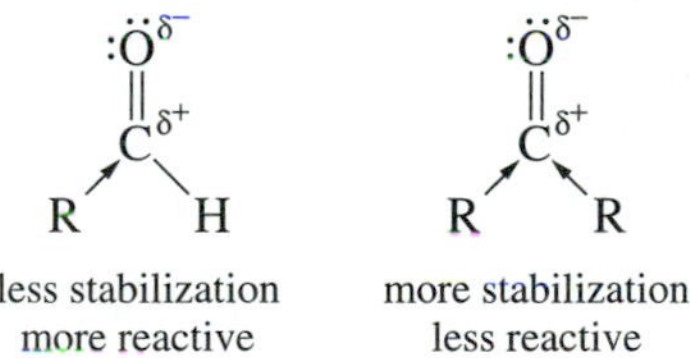

The sizes of groups also affect the reactivity of aldehydes and ketones. Because a ketone has two alkyl groups attached to the carbonyl carbon atom, it is sterically hindered relative to the carbonyl carbon atom of an aldehyde, which has only a hydrogen atom and one alkyl group bonded to it. Thus, a nucleophile can approach the carbonyl group of an aldehyde more readily, so the reaction occurs faster with an aldehyde (Figure 10.3). Therefore, both electronic and steric effects reinforce each other to make aldehydes more reactive than ketones.

10.6 Synthesis of Alcohols from Carbonyl Compounds

In Chapter 9 we discussed the reaction of haloalkanes with magnesium to produce organometallic compounds called **Grignard reagents.**

$$R—X \xrightarrow[\text{ether}]{\text{Mg}} R—MgX$$

These reagents are used to form carbon-carbon bonds in the synthesis of complex molecules from simpler molecules. Grignard reagents contain a very strongly polarized carbon-magnesium bond in which the carbon atom has a partial negative charge.

$$\overset{\delta^-}{C}—\overset{\delta^+}{MgX}$$

The carbon atom of the Grignard reagent resembles a carbanion. It reacts as a nucleophile and adds to the electrophilic carbon atom of a carbonyl group in an aldehyde or ketone. The magnesium ion forms a salt with the negatively charged oxygen atom. The resulting product is a magnesium alkoxide, which is hydrolyzed to obtain an alcohol.

$$R'—MgX \quad C{=}\ddot{O}: \xrightarrow{\text{ether}} R'—C—\ddot{O}—MgX \xrightarrow{H_2O} R'—C—\ddot{O}—H + HOMgX$$

a magnesium alkoxide

A Grignard reagent adds to various types of carbonyl compounds to give primary, secondary, and tertiary alcohols. Primary alcohols are synthesized by reacting the Grignard reagent, R′—MgX, with formaldehyde.

$$R'—MgX + H_2C{=}\ddot{O}: \xrightarrow{\text{ether}} R'—CH_2—\ddot{O}—MgX \xrightarrow{H_2O} R'—CH_2—\ddot{O}—H + HOMgX$$

a magnesium alkoxide — a primary alcohol

$$C_6H_{11}—MgCl + H_2C{=}O \xrightarrow[\text{2. } H_2O]{\text{1. ether}} C_6H_{11}—CH_2OH$$

Secondary alcohols are obtained by reacting the Grignard reagent, R′—MgX, with an aldehyde, RCHO. Note that the carbon atom bearing the hydroxyl group is bonded to the alkyl groups from both the Grignard reagent and the aldehyde.

$$R'—MgX + R—\overset{O}{\overset{\|}{C}}—H \xrightarrow{\text{ether}} R—\underset{R'}{CH}—O—MgX \xrightarrow{H_2O} R—\underset{R'}{CH}—O—H + HOMgX$$

a magnesium alkoxide — a secondary alcohol

$$CH_3{-}MgBr + C_6H_5{-}\overset{O}{\overset{\|}{C}}{-}H \xrightarrow[2.\ H_2O]{1.\ ether} C_6H_5{-}\underset{}{\overset{OH}{\overset{|}{C}H}}{-}CH_3$$

Tertiary alcohols are made by reacting the Grignard reagent, R′—MgX, with a ketone. Two of the alkyl groups bonded to the carbon atom bearing the hydroxyl group were part of the ketone; one alkyl group is provided from the Grignard reagent.

$$R'{-}MgX + R{-}\overset{O}{\overset{\|}{C}}{-}R \xrightarrow{ether} R{-}\underset{R'}{\underset{|}{\overset{R}{\overset{|}{C}}}}{-}O{-}MgX \xrightarrow{H_2O} R{-}\underset{R'}{\underset{|}{\overset{R}{\overset{|}{C}}}}{-}O{-}H + HOMgX$$

a magnesium alkoxide — a tertiary alcohol

$$CH_3CH_2{-}MgBr + CH_3CH_2CH_2\overset{O}{\overset{\|}{C}}CH_3 \xrightarrow[2.\ H_2O]{1.\ ether} CH_3CH_2CH_2\underset{CH_2CH_3}{\underset{|}{\overset{OH}{\overset{|}{C}}}}CH_3$$

Acetylenic Alcohols

Alkynide ions, the conjugate bases of alkynes, react with carbonyl groups in much the same way as Grignard reagents do. The alkynides are prepared in an acid-base reaction with acetylene or a terminal alkyne using sodium amide in ammonia. If a carbonyl compound is then added to the reaction mixture, an alcohol forms after acid workup. If the alkynide is derived from acetylene, an acetylenic alcohol forms.

$$H{-}C{\equiv}C{-}H + NaNH_2 \xrightarrow{NH_3} H{-}C{\equiv}C{-}Na + NH_3$$

$$\text{cyclohexanone} + H{-}C{\equiv}C{-}Na \xrightarrow[2.\ H_3O^+]{1.\ NH_3} \text{1-ethynylcyclohexanol (cyclohexane ring bearing OH and } C{\equiv}C{-}H\text{)}$$

1-ethynylcyclohexanol

EXAMPLE 10-4

The European bark beetle produces a pheromone that causes beetles to aggregate. Describe two ways that the compound could be synthesized in the laboratory by a Grignard reagent.

$$CH_3{-}CH_2{-}CH_2{-}\overset{CH_3}{\overset{|}{C}H}{-}\underset{OH}{\underset{|}{C}H}{-}CH_2{-}CH_3$$

Solution

The compound is a secondary alcohol that can be made from an aldehyde and a Grignard reagent. The carbon atom bonded to the hydroxyl group must be chosen as the carbonyl carbon atom. One of the two alkyl groups bonded to this carbon atom must be part of the aldehyde. The other must be the alkyl group of the Grignard reagent. The two possible components that could be introduced by a Grignard reagent are an ethyl group or a 1-methylbutyl group.

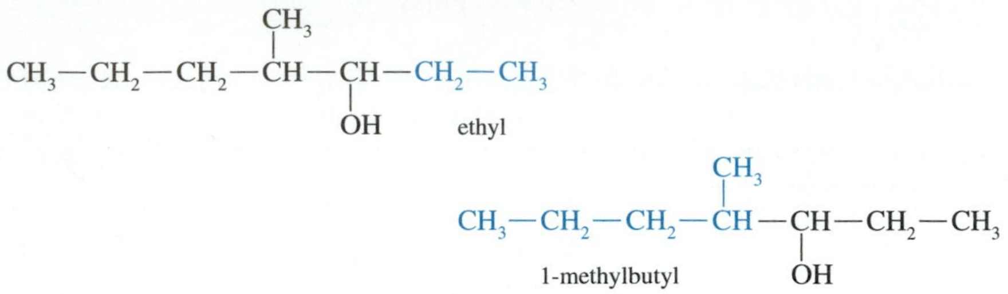

$$CH_3-CH_2-CH_2-\underset{}{\overset{CH_3}{\overset{|}{C}}}H-\underset{\underset{OH}{|}}{C}H-CH_2-CH_3$$

ethyl

$$CH_3-CH_2-CH_2-\overset{CH_3}{\overset{|}{C}}H-\underset{\underset{OH}{|}}{C}H-CH_2-CH_3$$

1-methylbutyl

If ethyl magnesium bromide is the Grignard reagent, the required aldehyde for the reaction is 2-methylpentanal. If the Grignard reagent is 1-methylbutyl magnesium bromide, propanal is required as the aldehyde.

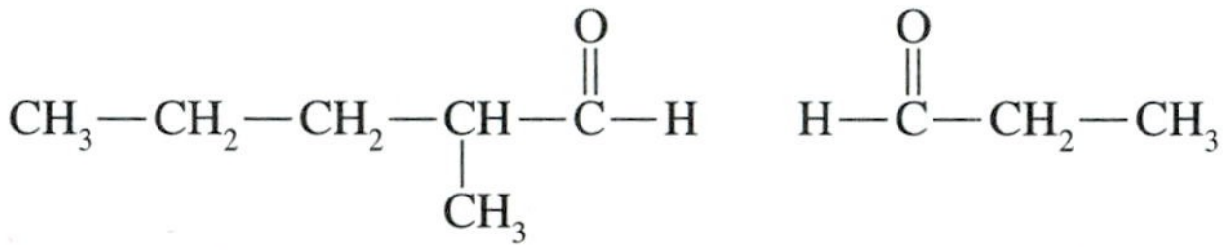

$$CH_3-CH_2-CH_2-\underset{\underset{CH_3}{|}}{C}H-\overset{\overset{O}{\|}}{C}-H \qquad H-\overset{\overset{O}{\|}}{C}-CH_2-CH_3$$

2-methylpentanal propanal

Problem 10.4

The methyl Grignard reagent reacts with 4-*tert*-butylcyclohexanone to give a mixture of two isomeric products. Draw their structures.

10.7 Addition of Oxygen Compounds

The nucleophilic oxygen atom of both water and alcohols can attack the carbonyl carbon atom of aldehydes and ketones to give addition products. Addition of water yields a hydrate; addition of an alcohol yields a hemiacetal or hemiketal. Because water and alcohols are weak nucleophiles, each of these reactions is reversible.

Addition of Water

Water adds to aldehydes and ketones to form **hydrates.** The proton of water bonds to the oxygen atom of the carbonyl group; the hydroxide ion adds to the carbon atom. Formaldehyde is over 99% hydrated. The hydrate of formaldehyde, called formalin, was formerly used to preserve biological specimens. It is no longer used because formaldehyde is a suspected carcinogen.

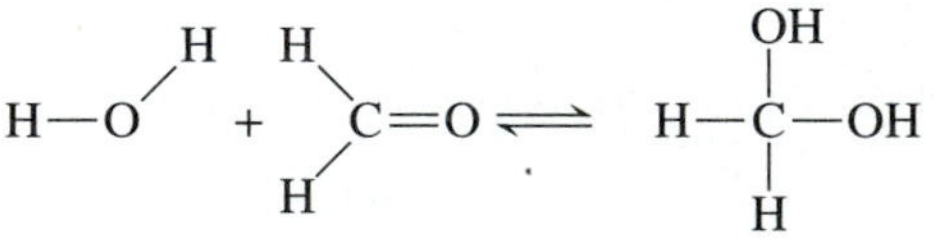

$$H-O-H + H_2C{=}O \rightleftharpoons H-\underset{\underset{H}{|}}{\overset{\overset{OH}{|}}{C}}-OH$$

formaldehyde hydrate

Other aldehydes are substantially less hydrated. Ketones are hydrated to an even smaller extent, usually less than 1%. The hydrates of aldehydes and ketones usually cannot be isolated. They exist only in solution, where the large amount of water forces the equilibrium position to the right.

Oral Contraceptives

The female sex hormones are collectively called estrogens. Estrogens, such as estradiol, are released during pregnancy and inhibit further ovulation. Oral contraceptives are designed to mimic this effect of pregnancy in that they inhibit ovulation.

estradiol

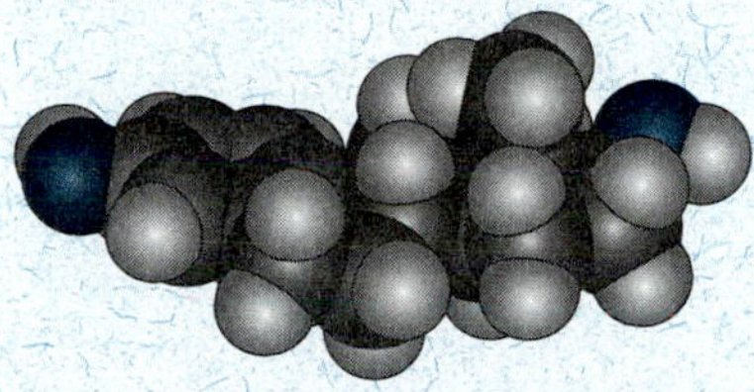

Estradiol itself is not an effective oral contraceptive because its C-17 hydroxyl group is rapidly oxidized to estrone in metabolic reactions. This oxidation product has greatly reduced estrogenic activity.

estrone

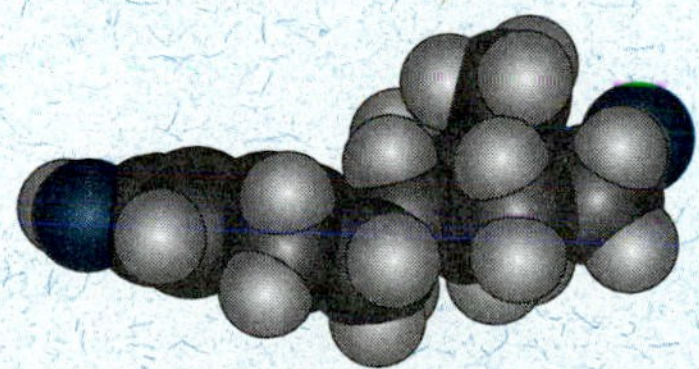

The hormonal action of estradiol is related to the C-17 hydroxyl group, which is located above the plane of the five-membered ring. Thus, it was decided to synthesize a compound structurally related to estradiol that is a tertiary alcohol with the correct stereochemistry. Such a compound would survive metabolic oxidation because tertiary alcohols cannot be oxidized. Based on the reactivity of Grignard reagents, one might propose to add methyl Grignard reagent to estrone to produce a tertiary alcohol. However, estrone has a phenolic hydroxyl group that would react with the Grignard reagent. This difficulty has been cleverly bypassed. Instead of making a tertiary alcohol by adding an alkyl group derived from a Grignard reagent, a tertiary alcohol is made by adding acetylide anion—the conjugate base of acetylene (Section 4.5). The reaction of the acetylide ion with a carbonyl compound produces an acetylenic alcohol. Sodium acetylide reacts with estrone to give ethynyl estradiol. Although the acetylide ion could potentially attack from either side of the C-17 carbonyl group of estrone, a methyl group extends above the plane of the ring in the vicinity of the carbonyl group. The acetylide anion thus approaches the "bottom" of the ring, so the product has its —OH group "up," which is the stereochemistry required for hormonal activity. This tertiary alcohol cannot be oxidized and is an effective oral contraceptive. ■

ethynyl estradiol

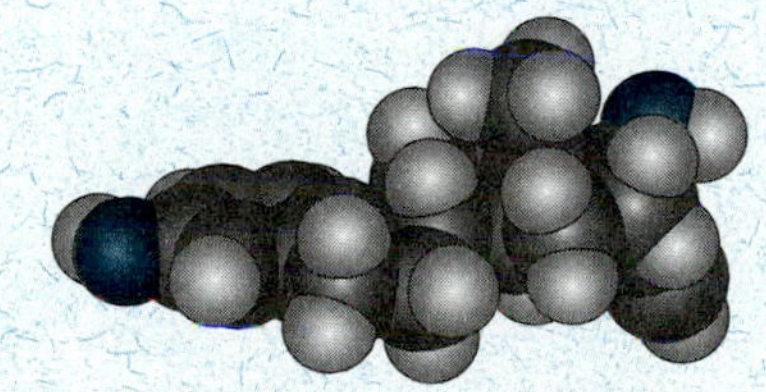

Addition of Alcohols

Alcohols add to carbonyl compounds in an acid-catalyzed reaction. The reaction does not occur in basic or neutral solution. The hydrogen atom of the alcohol adds to the carbonyl oxygen atom, and the —OR′ portion (alkoxy group) adds to the carbon atom. The product is called a **hemiacetal** if the carbonyl compound is an aldehyde and a **hemiketal** if the carbonyl compound is a ketone. These molecules have both an —OH group and an —OR′ group attached to the same carbon atom. The hemiacetal has a hydrogen atom and an alkyl group attached to the original carbonyl carbon atom, whereas the hemiketal has two alkyl groups attached.

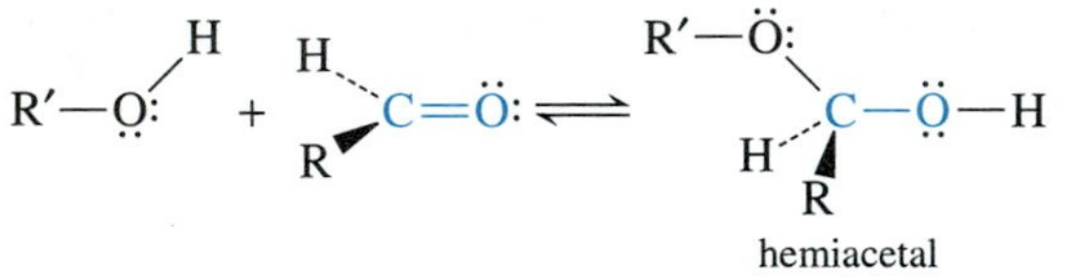

hemiacetal

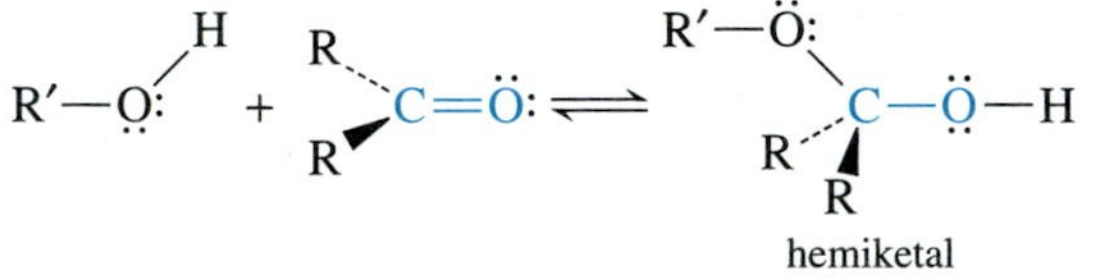

hemiketal

Hemiacetals and hemiketals are usually unstable compounds. That is, the equilibrium constant for formation of either a hemiacetal or a hemiketal is less than 1, and the equilibrium position for the preceeding reactions lies to the left. However, when both the carbonyl group and the alcohol are part of the same molecule, the equilibrium constant is larger because a stable cyclic product forms. Cyclization is favorable because the two functional groups are close to each other.

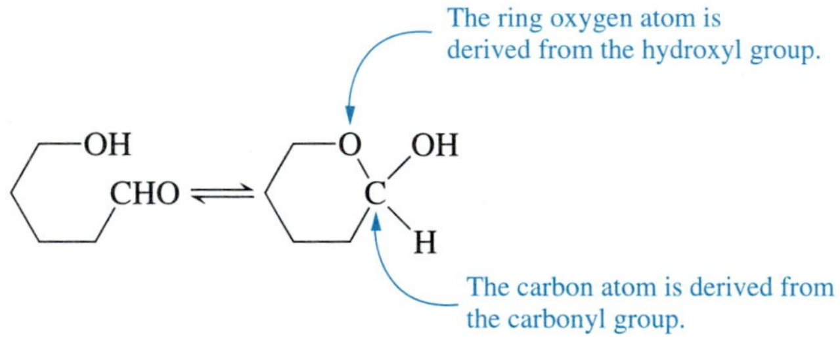

The ring oxygen atom is derived from the hydroxyl group.

The carbon atom is derived from the carbonyl group.

Carbohydrates (Chapter 11), which contain both carbonyl and hydroxyl groups, exist to only a small extent as open-chain molecules. They exist largely as cyclic hemiacetals or hemiketals.

Mechanism of Addition of Alcohols

The first step in the acid-catalyzed addition of an alcohol to an aldehyde or ketone is protonation of the carbonyl oxygen atom (an electron pair donor, or Lewis base) to produce a resonance-stabilized carbocation.

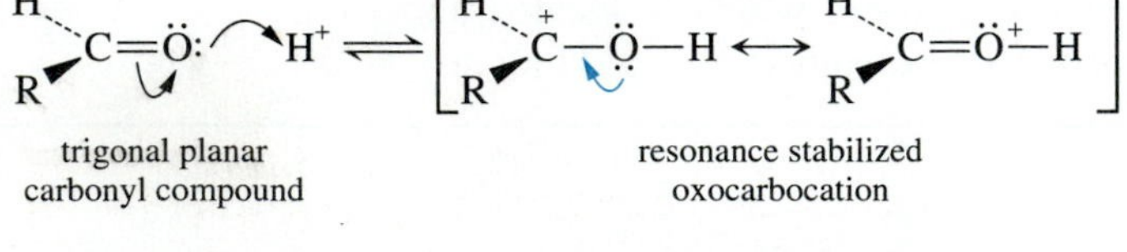

trigonal planar carbonyl compound

resonance stabilized oxocarbocation

Then the carbocation, which acts as a Lewis acid, reacts with the lone pair electrons of the oxygen atom of the alcohol R′—OH.

$$R'\text{—}\ddot{O}\text{—}H + H(R)C{=}\ddot{O}^{+}\text{—}H \rightleftharpoons R'\text{—}\ddot{O}^{+}(H)\text{—}C(H)(R)\text{—}\ddot{O}\text{—}H$$

conjugate acid of a hemiacetal

A proton is transferred from the conjugate acid of the hemiacetal in an acid-base reaction and becomes available again for the first step of the reaction sequence. Thus, the reaction is acid catalyzed.

10.8 Formation of Acetals and Ketals

The —OH group in either a hemiacetal or a hemiketal can be replaced by substitution of the —OH by another alkoxy group, —OR′. This reversible reaction occurs readily in acidic solution to produce an acetal or ketal.

$$R'\text{—}\ddot{O}\text{—}H + R'\text{—}\ddot{O}\text{—}C(H)(R)\text{—}\ddot{O}\text{—}H \overset{H^+}{\rightleftharpoons} R'\text{—}\ddot{O}\text{—}C(H)(R)\text{—}\ddot{O}\text{—}R' + H_2O$$

hemiacetal acetal

$$R'\text{—}\ddot{O}\text{—}H + R'\text{—}\ddot{O}\text{—}C(R)(R)\text{—}\ddot{O}\text{—}H \overset{H^+}{\rightleftharpoons} R'\text{—}\ddot{O}\text{—}C(R)(R)\text{—}\ddot{O}\text{—}R' + H_2O$$

hemiketal ketal

Note that both **acetals** and **ketals** have two alkoxy groups (—OR′) attached to the same carbon atom. An acetal also has a hydrogen atom and an alkyl group attached to the carbon atom, whereas the ketal has two alkyl groups attached. The formation of an acetal or a ketal requires two molar equivalents of alcohol per mole of the original carbonyl compound.

$$2\ CH_3\text{—}\ddot{O}\text{—}H + CH_3CH_2CH_2(H)C{=}\ddot{O}: \overset{H^+}{\rightleftharpoons} CH_3\text{—}\ddot{O}\text{—}C(H)(CH_3CH_2CH_2)\text{—}\ddot{O}\text{—}CH_3 + H_2O$$

Cyclic hemiacetals or hemiketals react with alcohols to produce cyclic acetals or ketals. Consider the cyclic hemiacetal of 5-hydroxypentanal. Its ring oxygen atom was originally the 5-hydroxyl oxygen atom. When this cyclic hemiacetal reacts with an alcohol, R′—OH, the product is a cyclic acetal. The oxygen atom in the —OR′ group of the acetal originated in the alcohol R′—OH. We will see this reaction again when we consider carbohydrates in Chapter 11.

$$HO(CH_2)_4CHO \overset{H^+}{\rightleftharpoons} \text{cyclic C(OH)(H)} \underset{H^+}{\overset{CH_3OH}{\rightleftharpoons}} \text{cyclic C(OCH}_3\text{)(H)} + H_2O$$

cyclic hemiacetal cyclic acetal

Reactivity of Acetals and Ketals

The conversion of a hemiacetal to an acetal and the conversion of a hemiketal to a ketal are reversible in acid solution. The position of the equilibrium can be shifted toward formation of an acetal or ketal by removing the water formed in the reaction or by increasing the concentration of the alcohol.

$$\text{hemiacetal + alcohol} \rightleftharpoons \text{acetal + water}$$

Adding alcohol "pushes" equilibrium to the right.

Removing water "pulls" equilibrium to the right.

The reverse reaction, the acid-catalyzed hydrolysis of acetals or ketals, is favored when water is added. Acetals and ketals react with water in a hydrolysis reaction to give a carbonyl compound and the alcohol. However, acetals and ketals do not react in neutral or basic solution.

$$CH_3CH_2-C(OCH_2CH_3)_2-CH_3 + H_2O \overset{H^+}{\rightleftharpoons} CH_3CH_2-C(=O)-CH_3 + 2\ CH_3CH_2OH$$

Mechanism of Acetal and Ketal Formation

The conversion of hemiacetals and hemiketals to acetals and ketals occurs in four reversible, acid-catalyzed steps. These steps are shown here for the conversion of a hemiacetal to an acetal. In step 1, the acid protonates the oxygen atom of the hydroxyl group. In step 2, water leaves, and a resonance-stabilized oxocarbocation forms.

$$R'-\ddot{O}-CH(R)-\ddot{O}-H + H^+ \rightleftharpoons R'-\ddot{O}-CH(R)-\overset{+}{O}H_2 \overset{-H_2O}{\rightleftharpoons} \left[H(R)\overset{+}{C}-\ddot{O}-R' \longleftrightarrow H(R)C=\overset{+}{\ddot{O}}-R' \right]$$

resonance-stabilized oxocarbocation

In step 3, the carbocation (a Lewis acid) combines with the alcohol (a Lewis base). In step 4, the proton bonded to the oxygen atom is lost to give an acetal. Note that H^+ is a catalyst: it starts the reaction by protonating the hemiacetal and is regenerated in the last step when the acetal forms.

$$R'-\ddot{O}-H + H(R)C=\overset{+}{\ddot{O}}-R' \rightleftharpoons R'-\overset{+}{O}(H)-CH(R)-\ddot{O}-R' \overset{-H^+}{\rightleftharpoons} R'-\ddot{O}-CH(R)-\ddot{O}-R'$$

EXAMPLE 10-5

Identify the class to which each of the following compounds belongs.

(a) $CH_3CH_2-C(H)(OCH_2CH_3)_2$ (b) $CH_3-C(OH)(OCH_3)-CH_2CH_3$

Solution

A carbon atom in (a) is bonded to two $-OCH_2CH_3$ groups and a hydrogen atom as well as one alkyl group. Compound (a) is therefore an acetal made from propanal and ethanol. In (b) a carbon atom is linked to an $—OCH_3$ group, an —OH group, and two alkyl groups. It is therefore a hemiketal that can be formed in equilibrium with 2-butanone and methanol.

Problem 10.5

Identify the class to which each of the following compounds belongs.

(a) $CH_3CH_2CH_2—C(OH)(OCH_2CH_3)—H$ (b) $CH_3CH_2—C(OCH_3)(OCH_3)—CH_2CH_3$

10.9 Addition of Nitrogen Compounds

The carbonyl groups of aldehydes and ketones react with nucleophiles that contain nitrogen; that is, they react with ammonia, NH_3, and with amines of the general formula RNH_2 (Chapter 14). The reaction gives imines, which are compounds with the carbonyl carbon atom bonded to the nitrogen atom by a double bond. The reaction occurs in two steps called an **addition-elimination reaction.** In the addition step, the nitrogen atom bonds to the carbonyl carbon atom and a hydrogen atom bonds to the carbonyl oxygen atom. This step to give a hemiaminal resembles the addition of an alcohol to a carbonyl compound. The initial addition product loses a molecule of water in an elimination reaction to give an **imine.** The net result of reacting an aldehyde or ketone with GNH_2, where G represents any group, is the replacement of the carbonyl oxygen atom by G—N=.

$$G—NH_2 + C{=}O \rightleftharpoons [G—NH—C—OH] \rightleftharpoons C{=}N—G + H_2O$$

a hemiaminal an imine

Note that the formation of an imine is accompanied by the release of water as a product. Imines can be isolated, but they react with water in the reverse reaction to produce the original carbonyl compound and the nitrogen compound. Thus, one way to isolate an imine is to remove water from the solution as the imine forms.

EXAMPLE 10-6

Write the structure of the intermediate hemiaminal produced by reaction of benzaldehyde and methylamine. Write the structure of the imine that results from dehydration of the hemiaminal.

Solution

Bond the nitrogen atom of the amine to the carbonyl carbon atom and add a hydrogen atom to the oxygen atom. This hemiaminal addition product is the result of nucleophilic attack of the nitrogen atom on the carbonyl carbon atom followed by addition of a proton to the oxygen atom.

$$C_6H_5—CH{=}O + CH_3—NH_2 \rightleftharpoons C_6H_5—CH(OH)—NH—CH_3$$

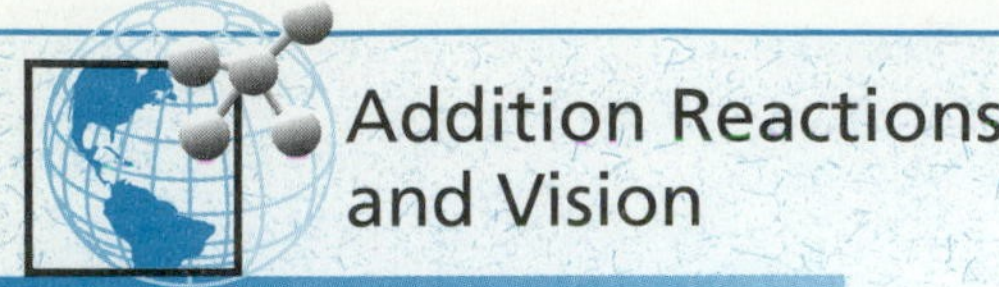

Addition Reactions and Vision

We learned at our mother's knee that "carrots are good for us." This homely injunction is true because carrots contain β-carotene, which mammals require for vision. β-Carotene is a pigment that is largely responsible for the color of carrots. It is also available in egg yolk, liver, and various fruits and vegetables. β-Carotene has an all-trans configuration.

Persons who do not have adequate β-carotene in their diets suffer from *night blindness*. Mammals have a liver enzyme system that splits β-carotene in half to give two molecules of an aldehyde named retinal.

There are several isomeric retinals because geometric isomers can exist about each of the double bonds. The all-trans compound and the isomer with a cis orientation about one of the double bonds, known as *cis*-11-retinal, play an important role in vision.

cis-11-Retinal undergoes an addition reaction with a protein in the retina called opsin to form a substance called rhodopsin. The aldehyde group of *cis*-11-retinal reacts with a specific amino group in the protein to form an imine. The shape of the imine adduct of *cis*-11-retinal allows it to "fit" into the protein.

Rhodopsin is a visual receptor in the retina that absorbs visible light. When light strikes rhodopsin, the cis double bond is isomerized to a trans double bond, a process called photoisomerization. The resulting all-trans isomer no longer fits into the opsin, the imine spontaneously hydrolyzes, and the all-trans retinal is released from opsin. This process occurs in about one millisecond. During that time a nerve impulse is generated and travels to the brain, where it is translated into a visual image.

If *cis*-11-retinal cannot bind opsin to give rhodopsin, vision is impaired. We recall from earlier discussions that formaldehyde, which is produced by the oxidation of methyl alcohol, can cause blindness. Blindness occurs because formaldehyde competes with *cis*-11-retinal for the reactive amine group of opsin. If no rhodopsin is formed, then no "light-induced" messages will get to the brain. ■

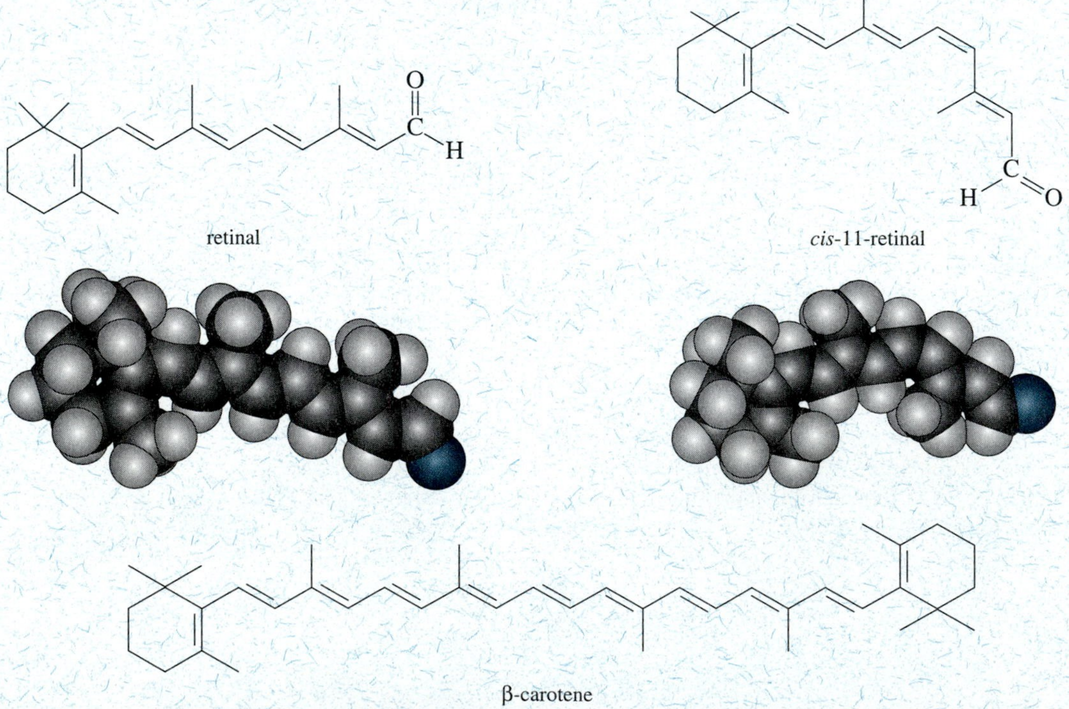

Dehydration occurs by elimination of the —OH group bonded to the carbon atom and the hydrogen atom bonded to the nitrogen atom to give the imine.

H :Ö—H, C, :N—CH$_3$, H ⇌ H, C=N̤, CH$_3$ + H_2O

Note that two geometric isomers can result, because the nitrogen atom is sp^2-hybridized. The trans isomer is shown. The cis isomer is less stable due to steric hindrance of the methyl and phenyl groups.

H, C=N̈, CH$_3$

cis isomer

The vision of an owl, as for all animals, depends on a chemical reaction to give rhodopsin, an imine.

Problem 10.6

Write the structure of the reactants required to produce the following compound:

CH$_3$, C=N̤, H, N—C, O, NH$_2$

10.10 Reactivity of the α-Carbon Atom

The carbonyl group itself is not the only reactive site of carbonyl compounds. Many important reactions occur at the carbon atom directly attached to the carbonyl carbon atom. This carbon atom is called the **α-carbon atom.**

O H ← α-hydrogen atom

—C—C—C—

α-carbon atoms

Acidity of α-Hydrogen Atoms

The carbonyl carbon atom has a partial positive charge and it attracts electrons in neighboring bonds by an inductive effect. As a result, the α-carbon atom loses electron density and acquires a partial positive charge. This effect is transmitted in turn to the bonds holding hydrogen atoms to the α-carbon atom. Thus, an **α-hydrogen atom** is more acidic than a hydrogen atom in a C—H bond of a hydrocarbon.

$$CH_3CH_3 + H_2O \rightleftharpoons CH_3CH_2^- + H_3O^+ \qquad pK_a = 50$$

$$CH_3\overset{O}{\overset{\|}{C}}CH_3 + H_2O \rightleftharpoons CH_3\overset{O}{\overset{\|}{C}}CH_2^- + H_3O^+ \qquad pK_a = 20$$

However, this tremendous difference of 30 powers of 10 in K_a is due to more than just the inductive effect of the carbonyl group. When a carbonyl compound loses its α-hydrogen atom, the resulting anion, called an **enolate anion,** is stabilized by resonance. One of its contributing resonance structures has a negative charge on the oxygen atom, and the other has a negative charge on the carbon atom. Because the charge on the anion is delocalized, an enolate anion is more stable than a carbanion, such as $CH_3CH_2^-$, in which no such resonance stabilization is possible.

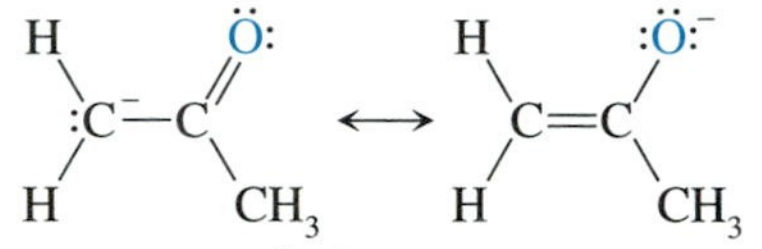

$$H_2\ddot{C}^- - C(=\ddot{O}:)CH_3 \longleftrightarrow H_2C=C(-\ddot{O}:^-)CH_3$$

contributing resonance structures of enolate ion

Keto-Enol Equilibria

The acidity of the α-hydrogen atoms of carbonyl compounds has another consequence: both aldehydes and ketones exist as an equilibrium mixture of isomeric compounds called the **keto** and the **enol** forms. Simple aldehydes and ketones exist predominantly in the keto form; acetone has less than 0.01% of the enol form.

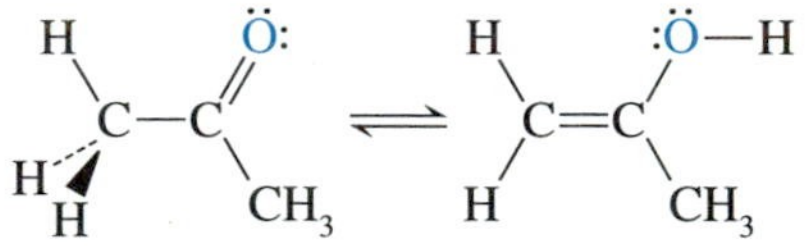

$$H_3C - C(=\ddot{O}:)CH_3 \rightleftharpoons H_2C=C(-\ddot{O}-H)CH_3$$

acetone (keto form)   acetone (enol form)

This type of isomerism is called **tautomerism,** and the two isomeric forms are called **tautomers.** Note that tautomers differ in the location of a hydrogen atom and a double bond. Thus, they are structural isomers, not contributing resonance forms, as in the case of the enolate ion, where only the locations of the electrons differ. The tautomers equilibrate in a rearrangement reaction in which a hydrogen atom bonded to an α-carbon atom is transferred to the carbonyl oxygen atom. The carbon-oxygen double bond becomes a single bond, and the carbon-carbon single bond becomes a double bond.

Although the enol form is present only in small amounts at equilibrium, it is often responsible for the reactivity of carbonyl compounds. Tautomerism is important in the chemistry of carbohydrates (Chapter 11) and in the metabolism of these compounds. For example, one of the intermediates in carbohydrate metabolism is a three-carbon compound called dihydroxyacetone phosphate. It undergoes an enzyme-catalyzed isomerization reaction to form another three-carbon molecule called D-glyceraldehyde 3-phosphate. The isomerization reaction occurs by transfer of a hydrogen atom from the α-carbon atom of dihydroxyacetone phosphate to the carbonyl oxygen atom. Tautomerization in the first step of the reaction yields an **enediol intermediate.** The second tautomerization step yields D-glyceraldehyde 3-phosphate.

$$\underset{\text{dihydroxyacetone phosphate}}{HOCH_2-C(=O)-CH_2OPO_3^{2-}} \rightleftharpoons \underset{\text{enediol intermediate}}{H(OH)C{=}C(OH)CH_2OPO_3^{2-}} \rightleftharpoons \underset{\text{D-glyceraldehyde 3-phosphate}}{HC(=O)-CH(OH)-CH_2OPO_3^{2-}}$$

Isomerization occurs because the enediol intermediate is in equilibrium with dihydroxyacetone phosphate and D-glyceraldehyde 3-phosphate. Similar isomerization reactions occur in many enzyme-catalyzed reactions of carbohydrates.

10.11 The Aldol Condensation

As a result of the acidity of the α-hydrogen atoms of carbonyl compounds, two carbonyl compounds can react with one another to give a condensation product that is both an aldehyde and an alcohol. Hence, the product is called an **aldol,** and this base-catalyzed reaction is called an **aldol condensation.**

In an aldol condensation, a new carbon-carbon bond is formed between the α-carbon atom of one carbonyl compound and the carbonyl carbon atom of the other one. Note that the product has just one carbon atom between the aldehyde and alcohol carbon atoms.

$$R-CH_2-\overset{O}{\overset{\|}{C}}-H + \underset{R}{\underset{|}{CH_2}}-\overset{O}{\overset{\|}{C}}-H \overset{OH^-}{\rightleftharpoons} \underset{\text{an aldol}}{R-CH_2-\overset{OH}{\overset{|}{CH}}-\underset{R}{\underset{|}{CH}}-\overset{O}{\overset{\|}{C}}-H}$$

The base-catalyzed aldol condensation occurs in a three-step mechanism. The reaction mechanism is illustrated with acetaldehyde.

1. One aldehyde molecule reacts with base (OH^-) at its α C—H bond to give a nucleophilic enolate anion.

$$H\ddot{O}:^- + H_3C-CH(=\ddot{O}:) \rightleftharpoons \underset{\text{enolate anion}}{H_2\ddot{C}^--CH(=\ddot{O}:)} + H_2O$$

2. The nucleophilic enolate anion reacts with the carbonyl carbon atom of another aldehyde molecule. The product is the alkoxide anion of an aldol.

$$H_3C-CH(=\ddot{O}:) + \underset{\text{nucleophile}}{H_2\ddot{C}^--CH(=\ddot{O}:)} \rightleftharpoons CH_3-\overset{:\ddot{O}:^-}{\overset{|}{\underset{H}{\underset{|}{C}}}}-CH_2-CH(=\ddot{O}:)$$

3. A proton from water combines with the alkoxide anion and regenerates a hydroxide anion.

$$H{-}\ddot{O}{:}\;\; H{-}\ddot{O}^{-}{-}CH(CH_3){-}CH_2{-}CHO \rightleftharpoons CH_3{-}CH(OH){-}CH_2{-}CHO + H{-}\ddot{O}{:}^{-}$$

Aldol products easily dehydrate to give a double bond in conjugation with the carbonyl group. Dehydration of the aldol shown followed by reduction gives 1-butanol in a commercial process.

$$CH_3{-}CH(OH){-}CH_2{-}CHO \xrightarrow[-H_2O]{H^+} CH_3{-}CH{=}CH{-}CHO \xrightarrow[Ni]{H_2} CH_3CH_2CH_2CH_2OH$$

Mixed Aldol Condensation

In the aldol condensation reaction, the same aldehyde provides the enolate and the substrate attacked by the enolate anion. But if two different aldehydes are mixed in a basic solution, the enolate anion of one can react with the carbonyl form of the other. Mixtures of products can result because any two aldehydes can react with each other. Thus, if the aldehydes are A_1 and A_2, aldol condensations can produce $A_1{-}A_1$, $A_2{-}A_2$, $A_1{-}A_2$, and $A_2{-}A_1$. In short, a dreadful mixture is produced. This unhappy outcome can be avoided if one of the aldehydes does not have any α-hydrogen atoms and one of the aldehydes is less reactive toward nucleophiles than the other. For example, an aldol condensation occurs between benzaldehyde and acetaldehyde to give a high yield of a single aldol product. Reaction between two benzaldehyde molecules cannot occur because benzaldehyde does not have any α-hydrogen atoms. Therefore, benzaldehyde is mixed with a base, and acetaldehyde is then slowly added. The acetaldehyde is converted to an enolate anion. Because the concentration of free acetaldehyde in the mixture is much smaller than that of benzaldehyde, the enolate of acetaldehyde reacts with benzaldehyde. Only one addition product is obtained. The condensation reaction of two different aldehydes yields a product called a **mixed aldol.**

$$C_6H_5{-}CHO + CH_3{-}CHO \xrightarrow{OH^-} C_6H_5{-}CH(OH){-}CH_2{-}CHO$$

Subsequent dehydration of this mixed aldol product gives cinnamaldehyde, the component of cinnamon responsible for its characteristic odor.

$$C_6H_5{-}CH(OH){-}CH_2{-}CHO \xrightarrow{-H_2O} C_6H_5{-}CH{=}CH{-}CHO$$

cinnamaldehyde

When an aldol condensation occurs, a new carbon-carbon bond forms. Mixed aldol condensations are important in metabolism. For example, an aldol condensation occurs in the synthesis of glucose from three-carbon precursors. In this reaction, which is catalyzed by the enzyme aldolase, the enolate anion derived from dihydroxyacetone phosphate reacts with D-glyceraldehyde to give fructose 1,6-bisphosphate (Figure 10.4). This molecule is subsequently converted to glucose.

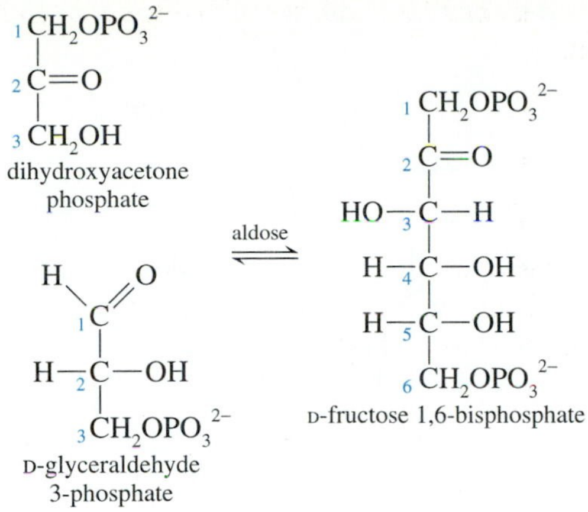

$$\begin{array}{l} {}_1CH_2OPO_3^{2-} \\ | \\ {}_2C{=}O \\ | \\ {}_3CH_2OH \end{array}$$

dihydroxyacetone phosphate

$$\begin{array}{c} H \quad\;\; O \\ \diagdown \;\; /\!/ \\ {}_1C \\ | \\ H{-}{}_2C{-}OH \\ | \\ {}_3CH_2OPO_3^{2-} \end{array}$$

D-glyceraldehyde 3-phosphate

$\xrightleftharpoons{\text{aldose}}$

$$\begin{array}{c} {}_1CH_2OPO_3^{2-} \\ | \\ {}_2C{=}O \\ | \\ HO{-}{}_3C{-}H \\ | \\ H{-}{}_4C{-}OH \\ | \\ H{-}{}_5C{-}OH \\ | \\ {}_6CH_2OPO_3^{2-} \end{array}$$

D-fructose 1,6-bisphosphate

FIGURE 10.4

Biological Aldol Condensation to Form Fructose

The enzyme aldolase catalyzes the aldol condensation of dihydroxyacetone phosphate and D-glyceraldehyde 3-phosphate.

EXAMPLE 10-7

Without writing the individual steps of the mechanism, draw the product of the aldol condensation of propanal.

Solution

First draw the structural formula of propanal with the carbonyl group on the right.

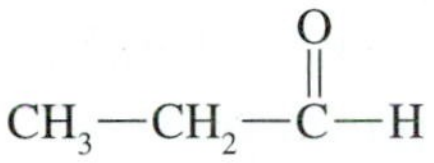

$$CH_3{-}CH_2{-}\overset{\overset{\displaystyle O}{\|}}{C}{-}H$$

Next, draw a second structural formula of propanal with the α-carbon atom near the carbonyl carbon atom of the first structure.

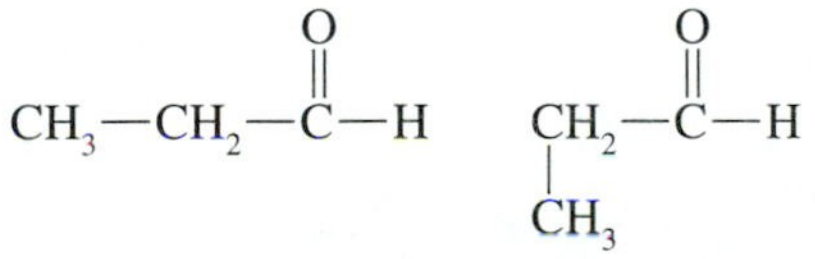

$$CH_3{-}CH_2{-}\overset{\overset{\displaystyle O}{\|}}{C}{-}H \qquad \underset{\underset{\displaystyle CH_3}{|}}{CH_2}{-}\overset{\overset{\displaystyle O}{\|}}{C}{-}H$$

Remove one of the α-hydrogen atoms of the structure on the right and add it to the carbonyl oxygen atom of the structure on the left. Now form the carbon-carbon bond between the α-carbon atom of the structure on the right and the carbonyl carbon atom of the structure on the left. The resulting structure is the aldol product.

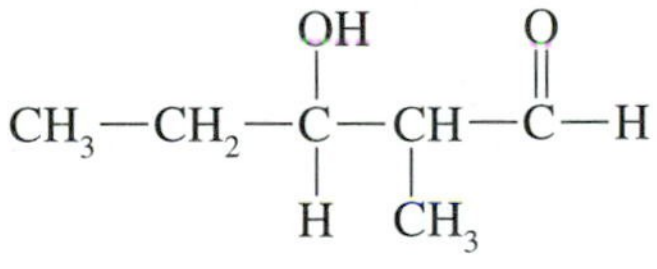

$$CH_3{-}CH_2{-}\underset{\underset{\displaystyle H}{|}}{\overset{\overset{\displaystyle OH}{|}}{C}}{-}\underset{\underset{\displaystyle CH_3}{|}}{CH}{-}\overset{\overset{\displaystyle O}{\|}}{C}{-}H$$

Problem 10.7

What compound is required to form the following unsaturated compound using an aldol condensation?

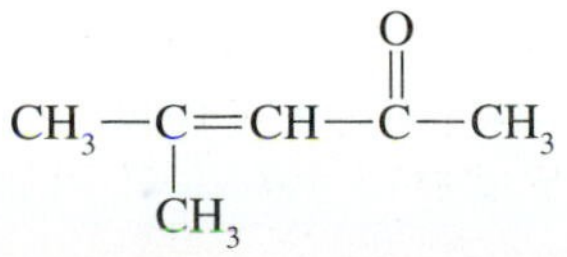

$$CH_3{-}\underset{\underset{\displaystyle CH_3}{|}}{C}{=}CH{-}\overset{\overset{\displaystyle O}{\|}}{C}{-}CH_3$$

Summary of Reactions

1. Oxidation of Aldehydes (Section 10.4)

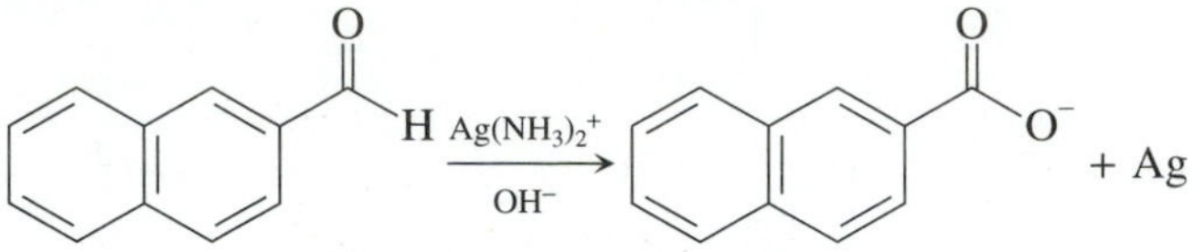

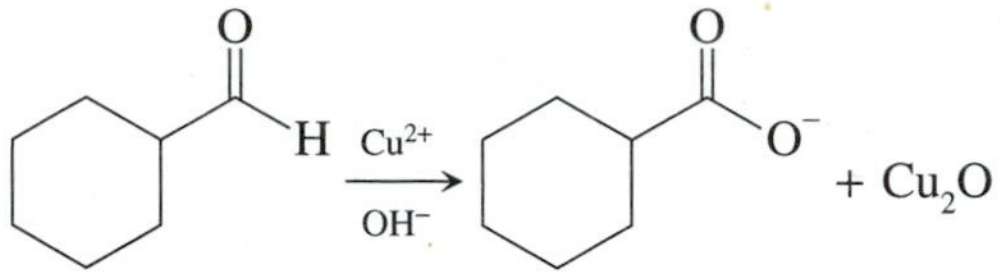

O ‖ H $\xrightarrow[OH^-]{Ag(NH_3)_2^+}$ O ‖ O^- + Ag

O ‖ H $\xrightarrow[OH^-]{Cu^{2+}}$ O ‖ O^- + Cu_2O

2. Reduction of Aldehydes and Ketones to Alcohols (Section 10.4)

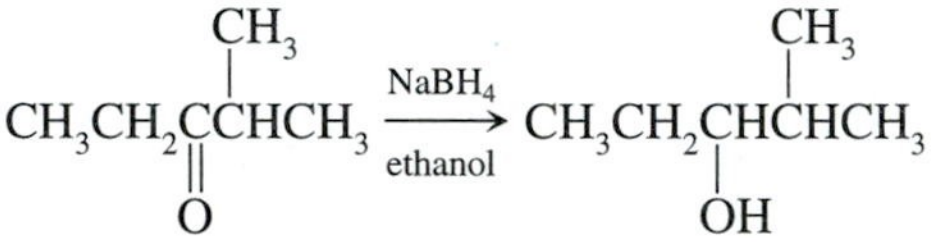

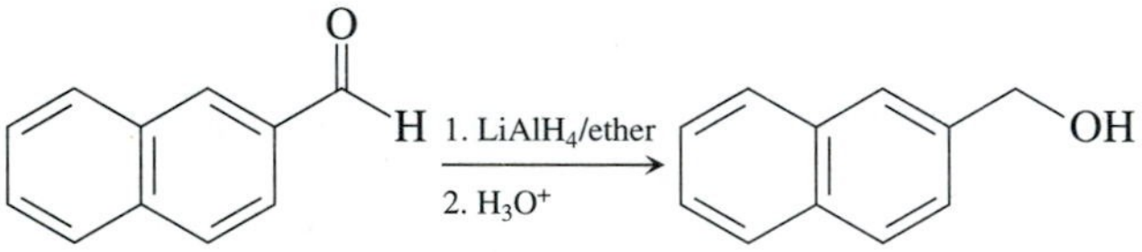

$$CH_3CH_2\underset{\underset{O}{\|}}{C}\overset{\overset{CH_3}{|}}{C}HCH_3 \xrightarrow[\text{ethanol}]{NaBH_4} CH_3CH_2\underset{\underset{OH}{|}}{C}H\overset{\overset{CH_3}{|}}{C}HCH_3$$

O ‖ H $\xrightarrow[\text{2. } H_3O^+]{\text{1. LiAlH}_4\text{/ether}}$ OH

3. Reduction of Aldehydes and Ketones to Methylene Groups (Section 10.4)

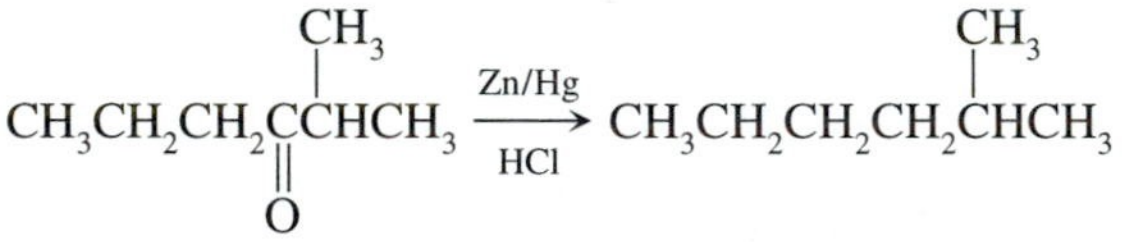

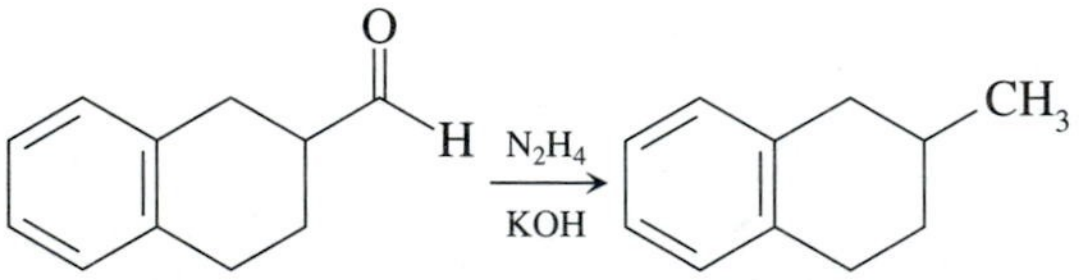

$$CH_3CH_2CH_2\underset{\underset{O}{\|}}{C}\overset{\overset{CH_3}{|}}{C}HCH_3 \xrightarrow[HCl]{Zn/Hg} CH_3CH_2CH_2CH_2\overset{\overset{CH_3}{|}}{C}HCH_3$$

O ‖ H $\xrightarrow[KOH]{N_2H_4}$ CH_3

4. Synthesis of Alcohols Using Grignard Reagents (Section 10.6)

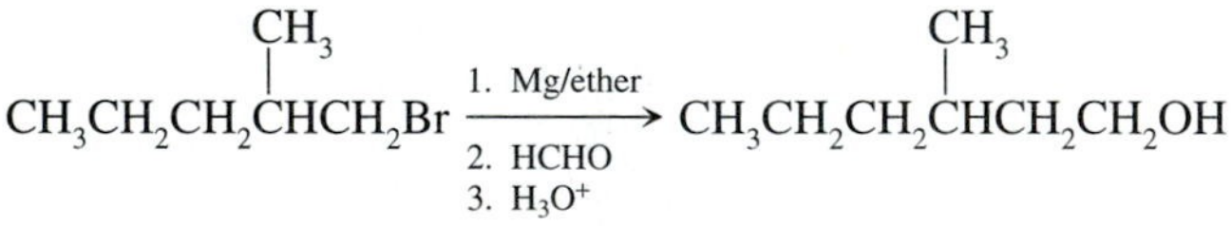

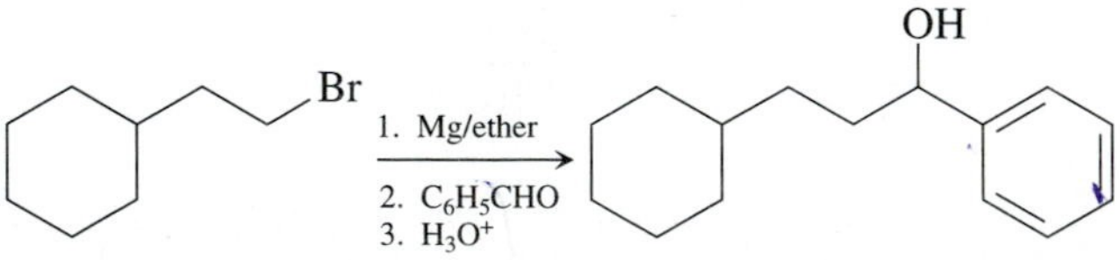

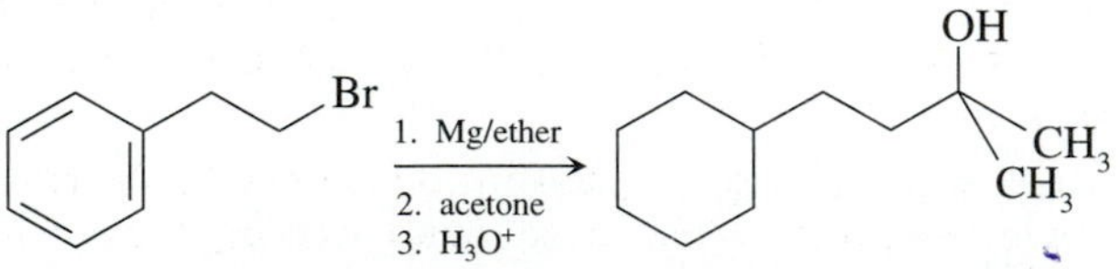

$$CH_3CH_2CH_2\overset{\overset{CH_3}{|}}{C}HCH_2Br \xrightarrow[\text{2. HCHO; 3. } H_3O^+]{\text{1. Mg/ether}} CH_3CH_2CH_2\overset{\overset{CH_3}{|}}{C}HCH_2CH_2OH$$

Br $\xrightarrow[\text{2. } C_6H_5CHO\text{; 3. } H_3O^+]{\text{1. Mg/ether}}$ OH

Br $\xrightarrow[\text{2. acetone; 3. } H_3O^+]{\text{1. Mg/ether}}$ OH, CH_3, CH_3

5. **Formation of Acetals and Ketals (Section 10.8)**

$$CH_3CH_2CH_2CH_2\overset{\overset{\displaystyle O}{\|}}{C}CH_3 \underset{H_3O^+}{\overset{CH_3OH}{\rightleftharpoons}} CH_3CH_2CH_2CH_2\overset{\overset{\displaystyle OCH_3}{|}}{\underset{\underset{\displaystyle OCH_3}{|}}{C}}CH_3 + H_2O$$

$$C_6H_5\overset{\overset{\displaystyle O}{\|}}{C}H \underset{H_3O^+}{\overset{CH_3CH_2OH}{\rightleftharpoons}} C_6H_5\overset{\overset{\displaystyle OCH_2CH_3}{|}}{\underset{\underset{\displaystyle OCH_2CH_3}{|}}{C}}H + H_2O$$

6. **Addition of Nitrogen Compounds to Carbonyl Compounds (Section 10.9)**

$$CH_3CH_2\overset{\overset{\displaystyle CH_3}{|}}{C}HCH_2\overset{\overset{\displaystyle O}{\|}}{C}{-}H \underset{H_3O^+}{\overset{CH_3CH_2NH_2}{\rightleftharpoons}} CH_3CH_2\overset{\overset{\displaystyle CH_3}{|}}{C}HCH_2\overset{\overset{\displaystyle NCH_2CH_3}{\|}}{C}{-}H + H_2O$$

7. **Aldol Condensation (Section 10.10)**

$$2\ C_6H_{11}{-}CH_2\overset{\overset{\displaystyle O}{\|}}{C}{-}H \overset{OH^-}{\rightleftharpoons} C_6H_{11}{-}CH_2\overset{\overset{\displaystyle OH}{|}}{C}H{-}\underset{\underset{\displaystyle C_6H_{11}}{|}}{C}H{-}\overset{\overset{\displaystyle O}{\|}}{C}{-}H \xrightarrow{-H_2O} C_6H_{11}{-}CH_2CH{=}\underset{\underset{\displaystyle C_6H_{11}}{|}}{C}{-}\overset{\overset{\displaystyle O}{\|}}{C}{-}H$$

$$(m\text{-}ClC_6H_4){-}\overset{\overset{\displaystyle O}{\|}}{C}{-}H + CH_3CH_2\overset{\overset{\displaystyle O}{\|}}{C}{-}H \overset{OH^-}{\rightleftharpoons} (m\text{-}ClC_6H_4){-}\overset{\overset{\displaystyle OH}{|}}{C}H{-}\underset{\underset{\displaystyle CH_3}{|}}{C}H\overset{\overset{\displaystyle O}{\|}}{C}{-}H$$

Explorations with Molecular Models

1. The lowest energy conformations of carbonyl compounds have either a C—H or a C—C bond from the carbon atom bonded to the carbonyl carbon atom in an eclipsed conformation with respect to the carbonyl oxygen atom. Construct a ball-and-stick model that corresponds to the following space-filling model of 2-butanone in an eclipsed conformation. View the carbonyl carbon atom along an axis perpendicular to the plane containing the carbonyl carbon atom and its three adjacent atoms. Do the same thing from the other side of the plane. Would the path of a nucleophile be hindered as it approached the carbonyl carbon atom from either of the two directions?

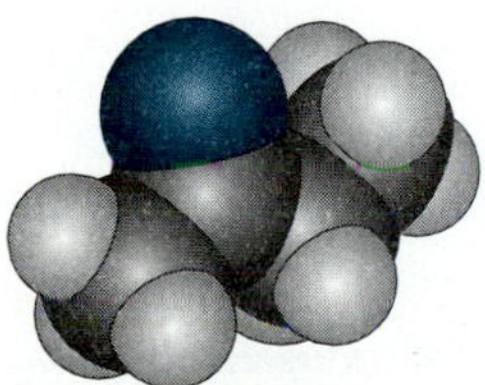

2. Use the model prepared in Exploration Exercise 1. Rotate the structure about the C-2 to C-3 bond until a hydrogen atom of the C-3 atom eclipses the carbonyl oxygen atom. View the carbonyl carbon atom along an axis perpendicular to the plane containing the carbonyl carbon atom and its three adjacent atoms. Do this in both possible directions. Is the steric hindrance to nucleophilic attack affected in either direction?

3. Construct a ball-and-stick model corresponding to the space-filling model of 3,3-dimethyl-2-butanone on the next page. View the carbonyl carbon atom along an axis perpendicular to the plane containing the carbonyl carbon atom and its three ad-

jacent atoms. Do this in both possible directions. Would the path of a nucleophile be hindered as it approaches the carbonyl carbon atom in either of the two directions?

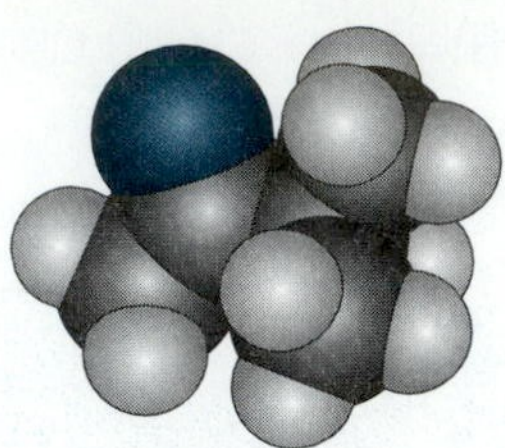

4. Construct a ball-and-stick model corresponding to the following space-filling model of 4-methylcyclohexanone. From the equatorial direction, view the carbonyl carbon atom along an axis perpendicular to the plane containing the carbonyl carbon atom and its three adjacent atoms. Do the same thing from the axial side of the plane. Would the path of a nucleophile be hindered as it approaches the carbonyl carbon atom from either direction?

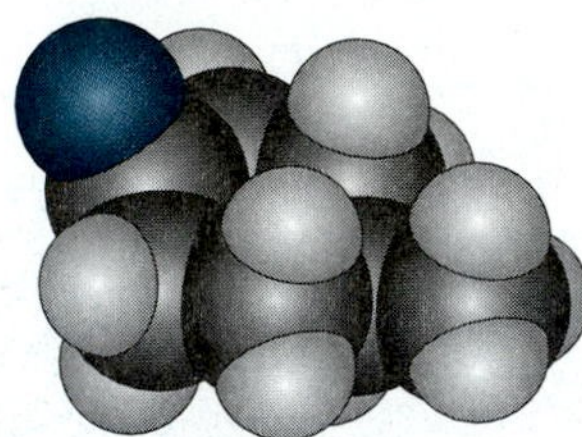

5. Construct the following ball-and-stick model. What class of compounds does this structure represent? What compounds could be used to synthesize the structure?

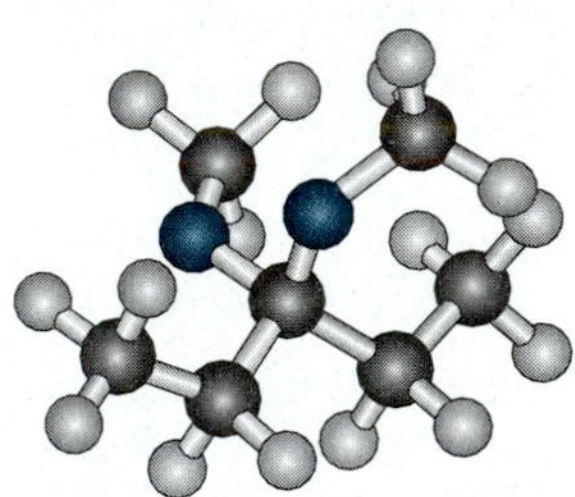

Exercises

Nomenclature of Aldehydes and Ketones

10.1 Write the structural formula for each of the following compounds.

(a) 2-methylbutanal (b) 3-ethylpentanal (c) 2-bromopentanal

10.2 Write the structural formula for each of the following compounds.

(a) 3-bromo-2-pentanone (b) 2,4-dimethyl-3-pentanone (c) 4-methyl-2-pentanone

10.3 Give the IUPAC name for each of the following compounds.

$$\text{(a) } \mathrm{CH_3CH_2CH_2CHO} \qquad \text{(b) } \mathrm{CH_3\overset{\displaystyle CH_3}{\underset{\displaystyle CH_3}{C}}CH_2CHO} \qquad \text{(c) } \mathrm{CH_3CH_2\overset{\displaystyle CH_3}{C}H\underset{\displaystyle CH_2CH_3}{C}HCHO}$$

10.4 Give the IUPAC name for each of the following compounds.

(a) $CH_3CH_2\overset{O}{\overset{\|}{C}}CH_2CH_3$ (b) $CH_3CH(CH_3)\overset{O}{\overset{\|}{C}}CH_2CH_3$ (c) $CH_3CH(CH_3)CH_2\overset{O}{\overset{\|}{C}}CH_2CH_3$

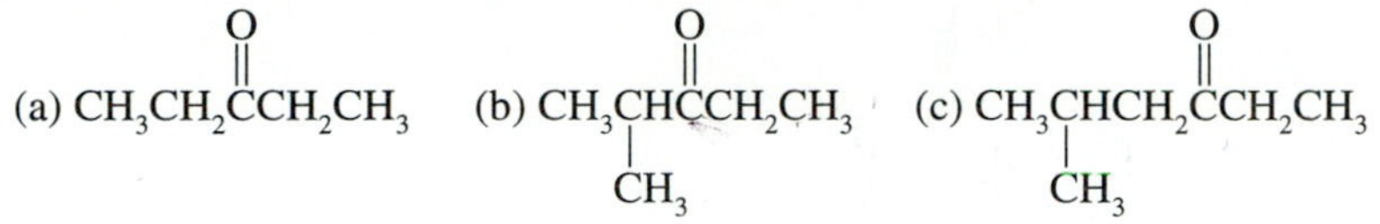

10.5 Give the IUPAC name for each of the following compounds.

(a) Cl CH₃ CH₃ H O (b) CH₃ O CH₃ (c) CH₃ CH₃ CH₃ CH₃ Br O H

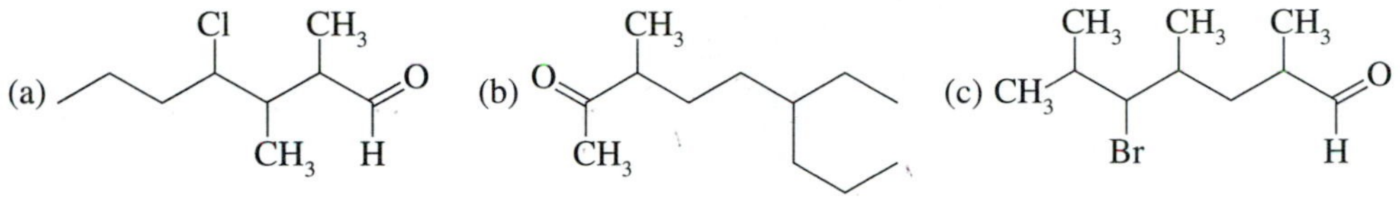

10.6 Give the IUPAC name for each of the following compounds.

(a) O (b) O CH₃ (c) O

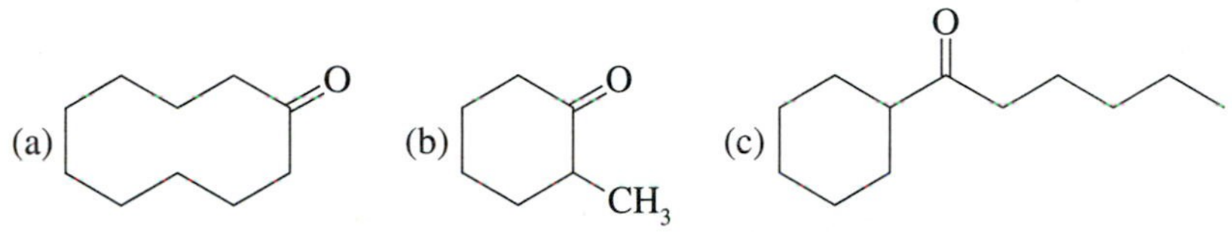

10.7 Draw the product of ozonolysis (Section 4.7) obtained from 1-methylcyclohexene. Name the compound.

10.8 Draw the products of ozonolysis of vitamin K_1 assuming that the double bond in the quinone ring is unaffected. Name the carbonyl compound derived from the side chain.

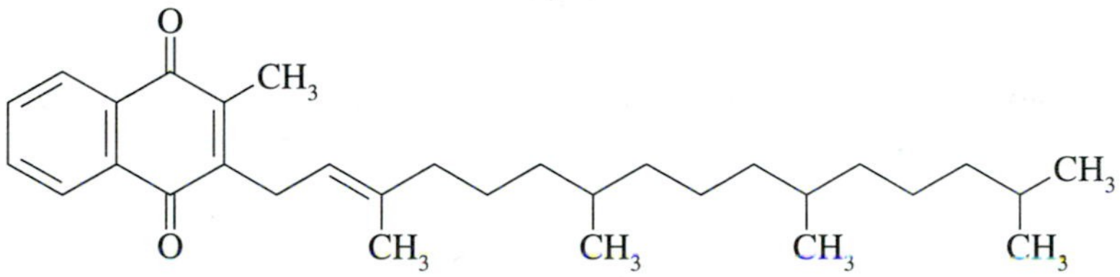

O CH₃ O CH₃ CH₃ CH₃ CH₃ CH₃

Properties of Aldehydes and Ketones

10.9 The dipole moments of 1-butene and butanal are 0.34 and 2.52 D, respectively. Explain this difference.

10.10 The dipole moments of acetone and isopropyl alcohol are 2.7 and 1.7 D, respectively. Explain this difference.

10.11 The boiling points of butanal and 2-methylpropanal are 75 and 61°C, respectively. Explain this difference.

10.12 The boiling points of 2-heptanone, 3-heptanone, and 4-heptanone are 151, 147, and 144°C, respectively. What is responsible for this trend?

10.13 The solubilities of butanal and 1-butanol in water are 7 and 9 g/100 mL, respectively. Explain this difference.

10.14 The solubilities of butanal and 2-methylpropanal in water are 7 and 11 g/100 mL, respectively. Explain this difference.

Oxidation and Reduction

10.15 What is observed when an aldehyde reacts with Fehling's solution? What is observed when an aldehyde reacts with Tollens's reagent?

10.16 Draw the structure of the product of each of the following reactions.

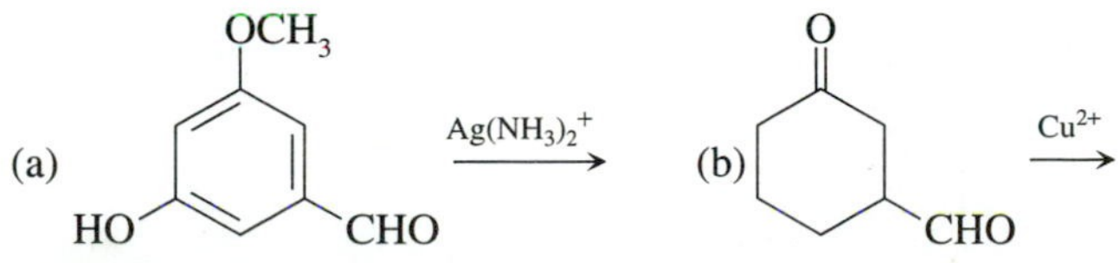

(a) OCH_3, HO, CHO $\xrightarrow{Ag(NH_3)_2^+}$ (b) O, CHO $\xrightarrow{Cu^{2+}}$

10.17 What is the product when each of the following reacts with lithium aluminum hydride?

(a) O, C, CH_3 (b) O (c) O, C, H

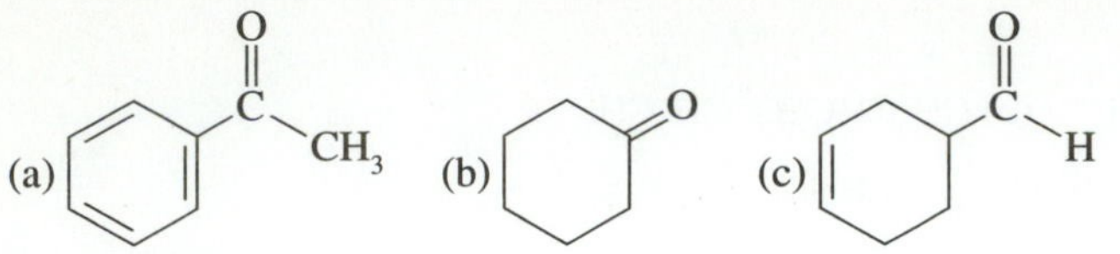

10.18 What is the product when each of the following reacts with sodium borohydride?

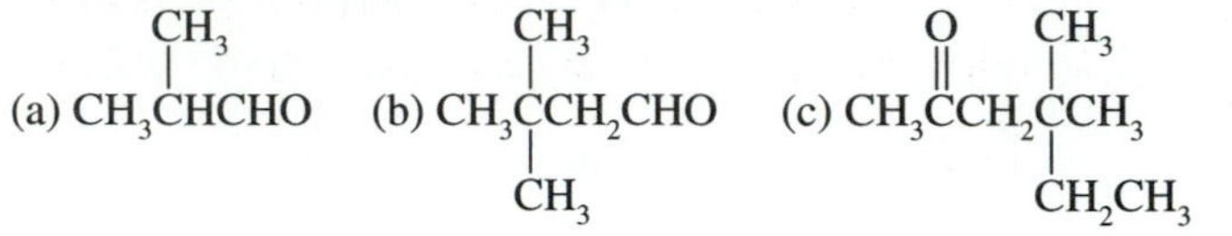

(a) $CH_3CH(CH_3)CHO$ (b) $CH_3C(CH_3)_2CH_2CHO$ (c) $CH_3C(=O)CH_2C(CH_3)(CH_2CH_3)CH_3$

10.19 The reduction of carvone by lithium aluminum hydride yields two products. Explain why.

CH_3, O, C, CH_3, CH_2

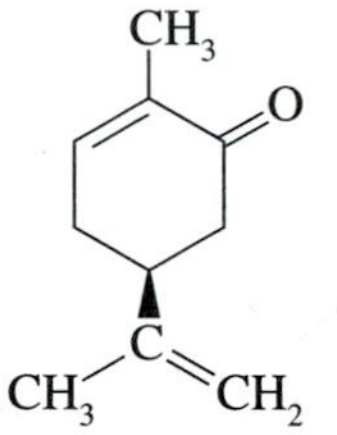

10.20 The reduction of the following compound by sodium borohydride yields two products. Explain why.

CH_3, O

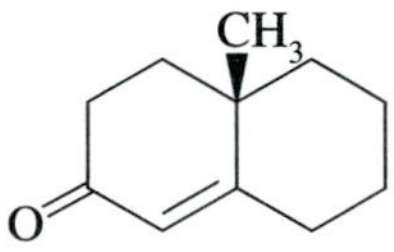

Addition Reactions

10.21 Formaldehyde has been used to disinfect rooms and surgical instruments. Why is this compound so effective compared to other carbonyl compounds?

10.22 Glutaraldehyde is used as a sterilizing solution for instruments that cannot be heated in an autoclave. Explain its action in sterilizing objects.

$$H-\overset{O}{\overset{\|}{C}}-CH_2CH_2CH_2CH_2-\overset{O}{\overset{\|}{C}}-H$$

10.23 The equilibrium constants for the hydration of ethanal and 2,2-dimethylpropanal are 1.8×10^{-2} and 4.1×10^{-3}, respectively. Explain this difference.

10.24 The equilibrium constants for the hydration of acetone and hexafluoroacetone are 2.5×10^{-5} and $2.2 \times 10^{+4}$, respectively. Explain this difference.

10.25 Hydrogen cyanide, $H-C\equiv N$, reacts with aldehydes under basic conditions to give an addition product. Write the structure of the addition product of hydrogen cyanide and benzaldehyde.

10.26 Hydrogen cyanide, $H-C\equiv N$, reacts with 2-propanone to give a good yield of an addition product, but 2,2,4,4-tetramethyl-3-pentanone gives a poor yield in the same reaction. Why?

Grignard Reactions

10.27 What carbonyl compound and Grignard reagent are required to produce each of the following compounds?

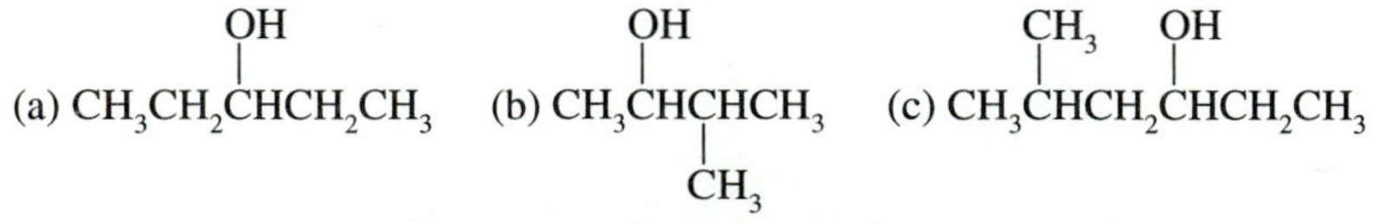

(a) $CH_3CH_2CH(OH)CH_2CH_3$ (b) $CH_3CH(OH)CH(CH_3)CH_3$ (c) $CH_3CH(CH_3)CH_2CH(OH)CH_2CH_3$

10.28 What combination(s) of a carbonyl compound and Grignard reagent are required to produce each of the following compounds?

(a) 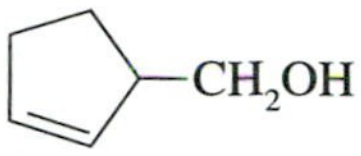 $-CH_2OH$ (b) 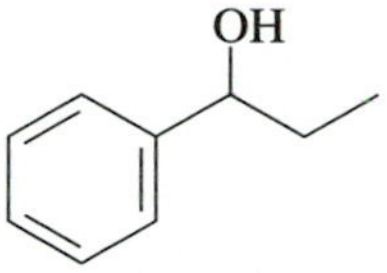 OH (c) 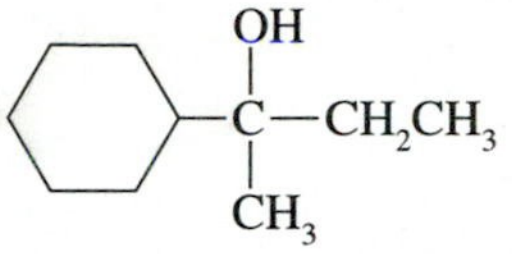 OH, $-C-CH_2CH_3$, CH_3

10.29 Propose a two-step sequence of reactions that could be used to convert cyclohexanone into 1-methylcyclohexene.

10.30 Propose a three-step sequence of reactions that could be used to convert 2-bromopentane into the following carboxylic acid.

$$CH_3CH_2CH_2CH(CH_3)CO_2H$$

Addition of Alcohols

10.31 Identify each of the following as a hemiacetal, hemiketal, acetal, or ketal.

(a) $CH_3CH_2CH(OCH_3)_2$ (b) $CH_3CH_2C(OCH_3)_2CH_3$ (c) $CH_3CH(OH)OCH_2CH_3$

10.32 Identify each of the following as a hemiacetal, hemiketal, acetal, or ketal.

(a) 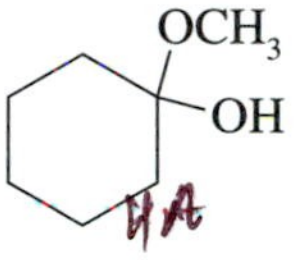 OCH_3, OH (b) 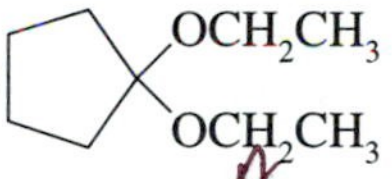 OCH_2CH_3, OCH_2CH_3 (c) 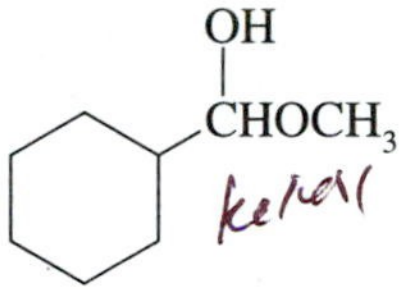 OH, $CHOCH_3$

10.33 Identify each of the following as a hemiacetal, hemiketal, acetal, or ketal.

(a) 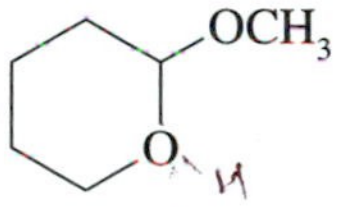 OCH_3, O (b) 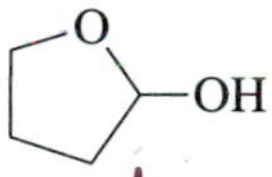 O, OH (c) 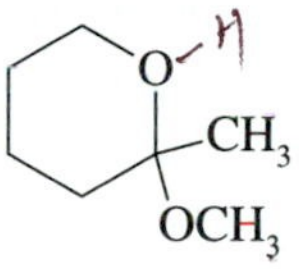 O, CH_3, OCH_3

10.34 Identify each of the following as a hemiacetal, hemiketal, acetal, or ketal.

(a) 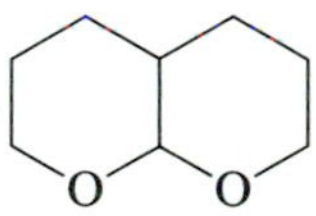 O, O (b) 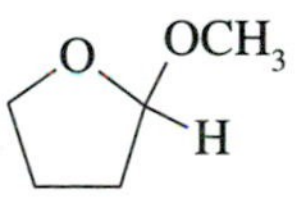 O, OCH_3, H (c) 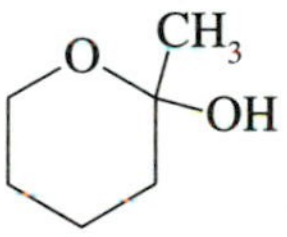 O, CH_3, OH

10.35 What carbonyl compound and alcohol are required to form each of the following isomeric compounds?

(a) 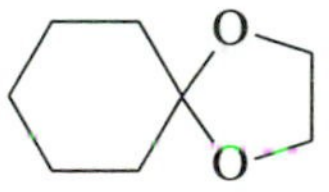 O, O (b) 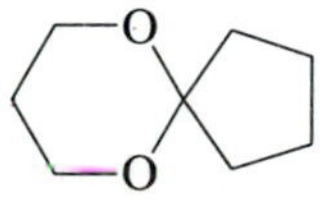 O, O

10.36 Reduction of the following compound by the Wolff-Kishner method gives $C_{11}H_{20}O_2$, but reduction by the Clemmensen method gives C_9H_{18}. Explain the difference in the products formed.

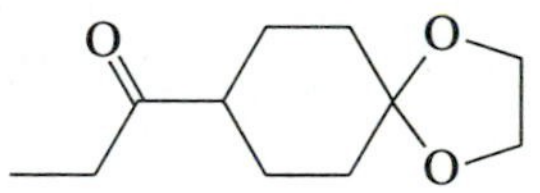

O, O, O

Addition of Nitrogen Compounds

10.37 Write the structure of the product of reaction of each of the following combinations of reactants.

(a) ethanal and CH_3NH_2

(b) acetone and $CH_3CH_2NH_2$

(c) cyclohexanone and $(CH_3)_2CHNH_2$

10.38 What reactants are required to produce the following isomeric imines?

(a) 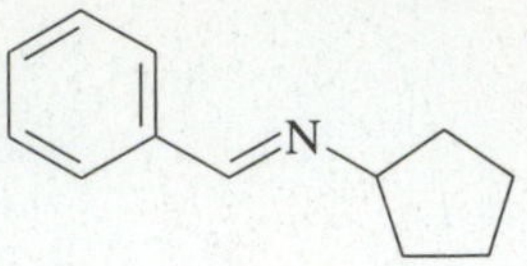 (b)

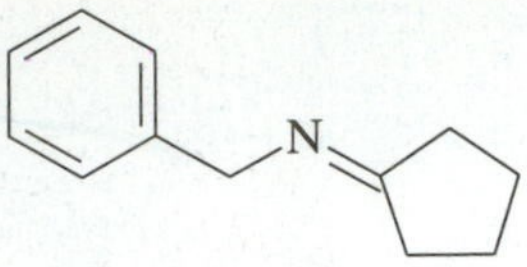

10.39 Hydroxylamine ($HONH_2$) reacts with cyclohexanone to yield a single product. However, cyclopentanecarbaldehyde yields two isomeric products. Explain why.

10.40 Draw the structure of the product of reaction of hydrazine (NH_2NH_2) with 2 moles of benzaldehyde.

Chemistry of the α-Carbon Atom

10.41 Formaldehyde cannot exist as an enol. Why?

10.42 2,2-Dimethylpropanal cannot exist as an enol. Why?

10.43 Write the enol form of each of the following compounds.

(a) acetone (b) butanal (c) cyclohexanone

10.44 There are three possible enols of 2-butanone. Explain why and write the structures.

10.45 Draw the keto structure for each of the following enols.

(a) 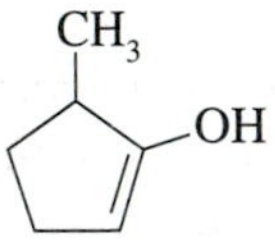 CH_3, OH (b) 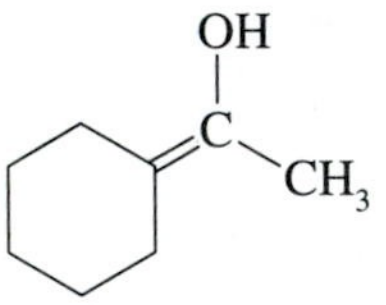 OH, C, CH_3

10.46 There are two possible enols of 2-methylcyclohexanone. Draw both structures. Predict which is the more stable.

Aldol Condensation

10.47 Draw the structure of the aldol product made by reacting butanal with a base.

10.48 Draw the structure of the aldol product made by reacting phenylethanal with a base.

10.49 Acetone will form a mixed aldol with benzaldehyde. Name the product.

10.50 Acetophenone will form a mixed aldol with formaldehyde. Name the product.

10.51 Jasmone can be synthesized from the following unsaturated diketone by an intramolecular aldol reaction followed by dehydration of the product. Which α-carbon atom serves as the nucleophile and which carbonyl group is attacked by the nucleophile?

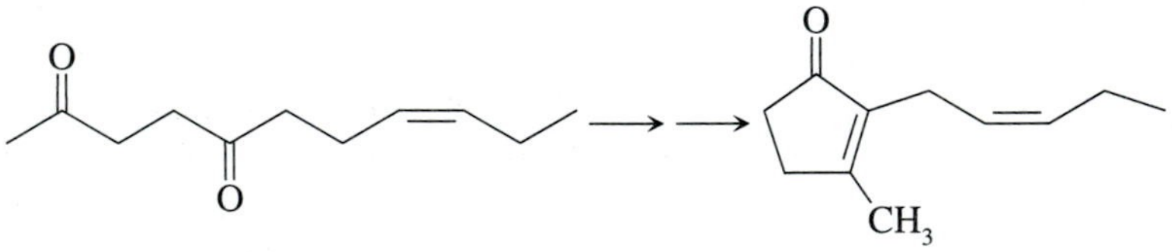

O, O → → O, CH_3

Jasmone

10.52 The following compound has been synthesized for use in perfumes. It is made by a mixed aldol condensation followed by dehydration and reduction of the double bond. What two carbonyl compounds are required? What type of reducing agent must be used?

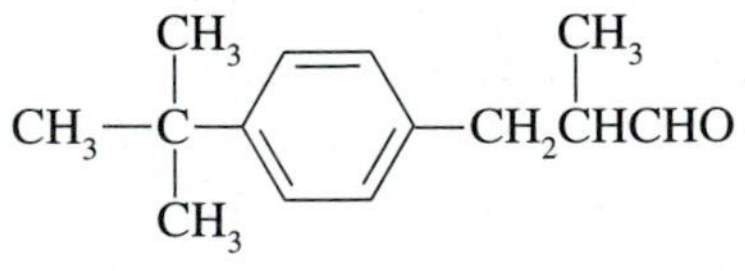

$(CH_3)_3C-C_6H_4-CH_2CH(CH_3)CHO$

Spectroscopy

10.53 If you have read Chapter 18 and the material has been assigned by your professor, then Exercises 18.4, 18.5, 18.8, and 18.28, which are based on the structure of carbonyl compounds, may be done.

CHAPTER 11

Carbohydrates

Overview

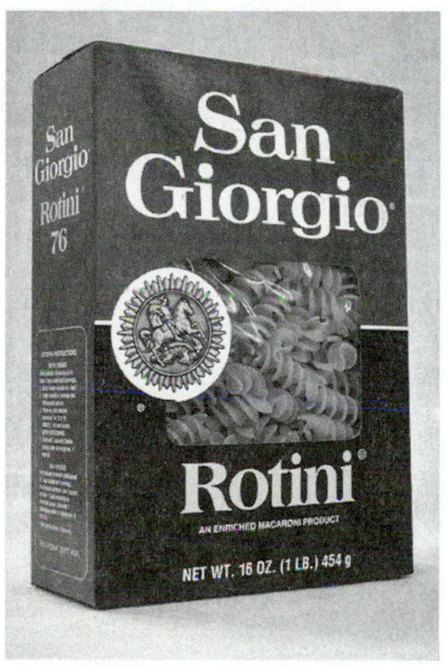

Numerous forms of pasta are a ready source of carbohydrates.

11.1 Classification of Carbohydrates

If importance were measured by abundance, the carbohydrates would hold first prize, for they are far and away the most abundant molecules in the biological world (excepting only water). The functions of carbohydrates are as varied as their structures. They are a major source of metabolic energy and are important structural components in plants as well as animal cells. **Carbohydrates** are polyhydroxy aldehydes or ketones or compounds that can be hydrolyzed to form them. They span a wide range of structures, from molecules containing three carbon atoms to giant molecules containing thousands of carbon atoms.

Carbohydrates are divided into three large structural classes—monosaccharides, oligosaccharides, and polysaccharides. Carbohydrates such as glucose and fructose that cannot be hydrolyzed into smaller molecules are **monosaccharides.**

Carbohydrates consisting of 2 to 10 or so monosaccharides are **oligosaccharides.** Hydrolysis of oligosaccharides may yield identical monosaccharides or two or more different monosaccharides. Oligosaccharides are called **disaccharides, trisaccharides,** and so forth, depending on the number of linked monosaccharide units. The disaccharide lactose, also called "milk sugar," contains one molecule of glucose and one of galactose. Maltose, another disaccharide, contains two glucose units.

Polysaccharides contain thousands of covalently linked monosaccharides. Those that contain only one type of monosaccharide are **homopolysaccharides.** Examples include starch and cellulose that are made by plants and contain only glucose. Glycogen, found in animals, is a homopolysaccharide containing glucose. Polysaccharides that contain more than one type of monosaccharide are called **heteropolysaccharides.**

The monosaccharides in oligo- and polysaccharides are linked by acetal or ketal bonds, which are called **glycosidic bonds** in carbohydrate chemistry. These bonds link the aldehyde or ketone site of one monosaccharide and a hydroxyl group of another monosaccharide. Hydrolysis of the glycosidic bonds yields the component monosaccharides.

Monosaccharides are further classified by their most highly oxidized functional group. Monosaccharides are called **aldoses** if their most highly oxidized functional group is an aldehyde. They are **ketoses** if their most highly oxidized functional group is a ketone. The suffix *-ose* indicates that a compound is a carbohydrate. The prefix *aldo-* or *keto-* indicates that the compound is an aldehyde or ketone. The number of carbon atoms in an aldose or ketose is indicated by the prefix *tri-*, *tetr-*, *pent-*, and *hex-*. Aldoses are numbered from the carbonyl carbon atom, whereas ketoses are numbered from the end of the carbon chain closest to the carbonyl carbon atom.

```
                              1
                               CHO
      1                       2|
       CH2OH           H—C—OH
      2|                      3|
       C=O             H—C—OH
      3|                      4|
  H—C—OH              H—C—OH
      4|                      5|
       CH2OH                CH2OH
  a ketotetrose        an aldopentose
```

11.2 Chirality of Monosaccharides

Monosaccharides are conveniently represented by their Fischer projection formulas (Section 6.4). We recall that in a Fischer projection formula the carbon chain is represented by a vertical line. Groups attached at the top and bottom of the vertical line represent bonds going into the page, and horizontal lines represent bonds coming out of the page. By convention, the carbonyl carbon atom, the most oxidized carbon atom in these compounds, is placed near the "top" in the Fischer projection formula. The simplest aldose, glyceraldehyde, has three carbon atoms, one of which is a stereogenic center. This aldotriose can exist in two enantiomeric forms.

```
      CHO               CHO              CHO              CHO
  H►C◄OH           H——+——OH        HO►C◄H          HO——+——H
     CH2OH             CH2OH            CH2OH            CH2OH
D-glyceraldehyde  Fischer projection  L-glyceraldehyde  Fischer projection
```

Monosaccharides with multiple stereogenic centers are represented with the carbon backbone continually pushed back behind the plane of the page. A curve of atoms in a C shape results, with the attached hydrogen atoms and hydroxyl groups pointing out from the backbone of the carbon chain.

```
      CHO               CHO              CH2OH             CH2OH
  H►C◄OH           H——+——OH            C=O              ══O
 HO►C◄H           HO——+——H         HO►C◄H          HO——+——H
 HO►C◄H           HO——+——H          H►C◄OH          H——+——OH
  H►C◄OH           H——+——OH          H►C◄OH          H——+——OH
     CH2OH             CH2OH            CH2OH            CH2OH
  D-galactose    Fischer projection    D-fructose     Fischer projection
```

D and L Series of Monosaccharides

The stereochemistry of each stereogenic center of a monosaccharide can be assigned by the *R,S* notation. However, late in the nineteenth century, the German chemist Emil Fischer devised a stereochemical nomenclature that preceded the *R,S* notation. It is still in common use for carbohydrates and amino acids (Chapter 15). In the Fischer stereochemical system, the configurations of all stereogenic centers are based on their relationship to the naturally occurring stereoisomer of glyceraldehyde. We recall that when this aldotriose has the hydroxyl group on its C-2 atom located on the right in the projection formula, its configuration is symbolized as D. Its enantiomer, called L-glyceraldehyde, has the hydroxyl group on the left at the C-2 atom in the Fischer projection formula.

Monosaccharides are made in cells from the "building block" D-glyceraldehyde, so nearly all naturally occurring monosaccharides have the same configuration as D-glyceraldehyde at the stereogenic carbon atom farthest from the carbonyl group. For example, the hydroxyl groups on the C-4 atom of ribose and on the C-5 atoms of glucose and fructose are on the right side in the following projection formulas. Each compound is a member of the D series of aldoses.

CHO
H—C—OH
CH_2OH
D-glyceraldehyde

CHO
H—C—OH
H—C—OH
H—C—OH
CH_2OH
D-ribose

CHO
H—C—OH
HO—C—H
H—C—OH
H—C—OH
CH_2OH
D-glucose

CH_2OH
C=O
HO—C—H
H—C—OH
H—C—OH
CH_2OH
D-fructose

The configuration of the hydroxyl groups of a monosaccharide relative to one another determines its chemical identity. Thus, in ribose the hydroxyl groups at the stereogenic carbon atoms are all on the same side in the Fischer projection formula. The name D-ribose has two components: the term D defines the configuration at the stereogenic carbon atom farthest from the aldehyde; the term *ribo* by itself defines the relative configuration of the three stereogenic centers at the C-2, C-3, and C-4 atoms. Taken together, the name D-ribose defines the absolute configuration at every stereogenic center in the molecule. Thus, in D-ribose, the hydroxyl groups at the C-2, C-3, and C-4 atoms are all on the right in the Fischer projection formula. In the enantiomer, L-ribose, the hydroxyl groups at the C-2, C-3, and C-4 atoms are all on the left because the entire structure is the mirror image of the D isomer.

mirror

CHO
H—C—OH
H—C—OH
H—C—OH
CH_2OH
D-ribose

CHO
HO—C—H
HO—C—H
HO—C—H
CH_2OH
L-ribose

determines D-configuration

determines L-configuration

The Fischer projection formulas of the aldotetroses, aldopentoses, and aldohexoses of the D series are shown in Figure 11.1. D-Glyceraldehyde, at the top of the "tree," is the parent aldose. When we insert a new stereogenic center (H—C—OH) between the carbonyl carbon atom and the stereogenic center below it, the resulting molecules are D-aldotetroses. Because the new CHOH group can have its —OH group on the right or left, two aldotetroses, D-erythrose and D-threose, are possible. Note that aldotetroses contain two nonequivalent stereogenic centers, so $2^2 = 4$ stereoisomers are possible. The two L-aldotetroses are not shown in Figure 11.1. D-Erythrose and L-erythrose are enantiomers, as are D-threose and L-threose.

Inserting an additional stereogenic center (H—C—OH), which can have either of two configurations, between the carbonyl carbon atom and the chiral center at the C-2 atom in D-erythrose leads to two D-aldopentoses: D-ribose and D-arabinose. Similarly, inserting a new stereogenic center (H—C—OH) between the carbonyl carbon atom and the stereogenic center at the C-2 atom in D-threose gives D-xylose and D-lyxose. Repeating the process one more time in each of the four D-aldopentoses gives a total of eight D-aldohexoses. D-Glucose and D-galactose are the most widely found in nature; D-mannose and D-talose occur in smaller amounts. The others are extremely rare.

Any of the members of a group of isomeric monosaccharides shown in Figure 11.1 are diastereomers of each other. They are not enantiomers because they are not mirror images. The enantiomer of each compound is the L-monosaccharide of the same name.

Epimers

We recall that diastereomers are stereoisomers, but not enantiomers. Diastereomers that contain two or more stereogenic carbon atoms but differ in configuration at only one stereogenic center are called **epimers.** Thus, the diastereomers D-glucose and D-galactose are epimers because they differ in configuration only at the C-4 atom. D-Glucose and D-mannose are epimers that differ in configuration at the C-2 atom.

	D-galactose		D-glucose		D-mannose
C-1	CHO		CHO		CHO
C-2	H—OH		H—OH	← C-2 epimers →	HO—H
C-3	HO—H		HO—H		HO—H
C-4	HO—H	← C-4 epimers →	H—OH		H—OH
C-5	H—OH		H—OH		H—OH
C-6	CH_2OH		CH_2OH		CH_2OH

The interconversion of epimers such as D-glucose and D-mannose at the C-2 atom illustrates the chemical reactions described in Chapter 10. The α-hydrogen atom of an aldehyde is involved in a tautomerization process and is responsible for an equilibrium between keto and enol forms of carbonyl compounds.

$$\text{H}-\text{C}(=\text{O})-\text{CH}_2-\text{R} \rightleftharpoons \text{H}(\text{HO})\text{C}=\text{CH}-\text{R}$$

keto form ⇌ enol form

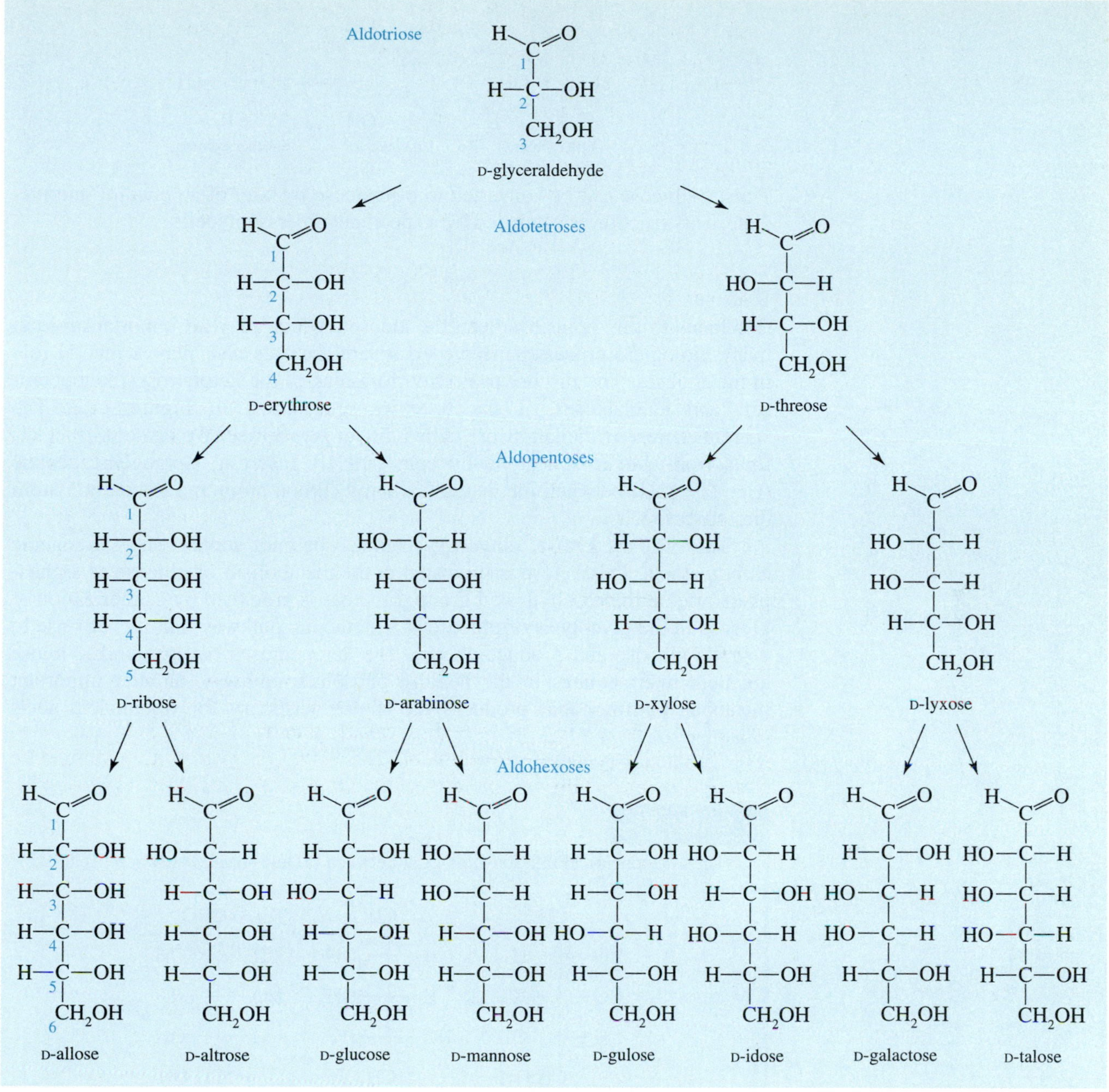

FIGURE 11.1
Structures of the D-Aldoses

In an aldose the α-carbon atom is a stereogenic center with a hydroxyl group bonded to it. Tautomerization yields an **enediol** in which the α-carbon atom is not a stereogenic center. In the reverse reaction to regenerate the aldose, a stereogenic center is formed again at the α-carbon atom, which can have either of two configurations. The resultant compounds are C-2 epimers.

H–C=O		H–C–OH		H–C=O
H—C—OH	⇌	‖	⇌	HO—C—H
R		R–C–OH		R
one epimer		an enediol		another epimer

Thus, D-glucose can be converted to D-mannose by way of an enediol intermediate. This reaction is catalyzed by a specific epimerase in cells.

Ketoses

Our focus to this point has been the aldoses, which play an important role in many biological processes. However, several ketoses also play a pivotal role in metabolism. The Fischer projection formulas of the ketotetroses, ketopentoses, and ketohexoses of the D series are shown in Figure 11.2. The "parent" ketose is the ketotriose called dihydroxyacetone. We can construct ketoses from dihydroxyacetone by consecutively inserting stereogenic centers (H—C—OH) between the ketone carbonyl carbon atom and the carbon atom directly below it.

The simplest ketose, dihydroxyacetone, does not contain any stereogenic carbon atoms. This ketose is produced in the metabolism of glucose as a phosphate ester at the C-3 hydroxyl group. Fructose is produced by isomerization of glucose in the glycolysis of glucose—a metabolic pathway that all cells use to degrade glucose and produce energy. The ketopentoses ribulose and xylulose are both intermediates in the pentose phosphate pathway, another important metabolic pathway that produces the ribose necessary for ribonucleic acids (Chapter 16).

EXAMPLE 11-1

What stereochemical relationships exist between (a) and (b) and between (a) and (c)?

(a)	(b)	(c)
CHO	CHO	CHO
HO—H	H—OH	H—OH
HO—H	H—OH	HO—H
H—OH	HO—H	H—OH
CH_2OH	CH_2OH	CH_2OH

Solution

Structures (a) and (b) are enantiomers because the configuration at every center in (b) is opposite that of corresponding centers in (a). Structures (a) and (c) are diastereomers. Both are D-monosaccharides and differ in configuration only at the C-2 atom, so they are epimers.

Problem 11.1

What relationship exists between D-allose and D-talose (see Figure 11.1)?

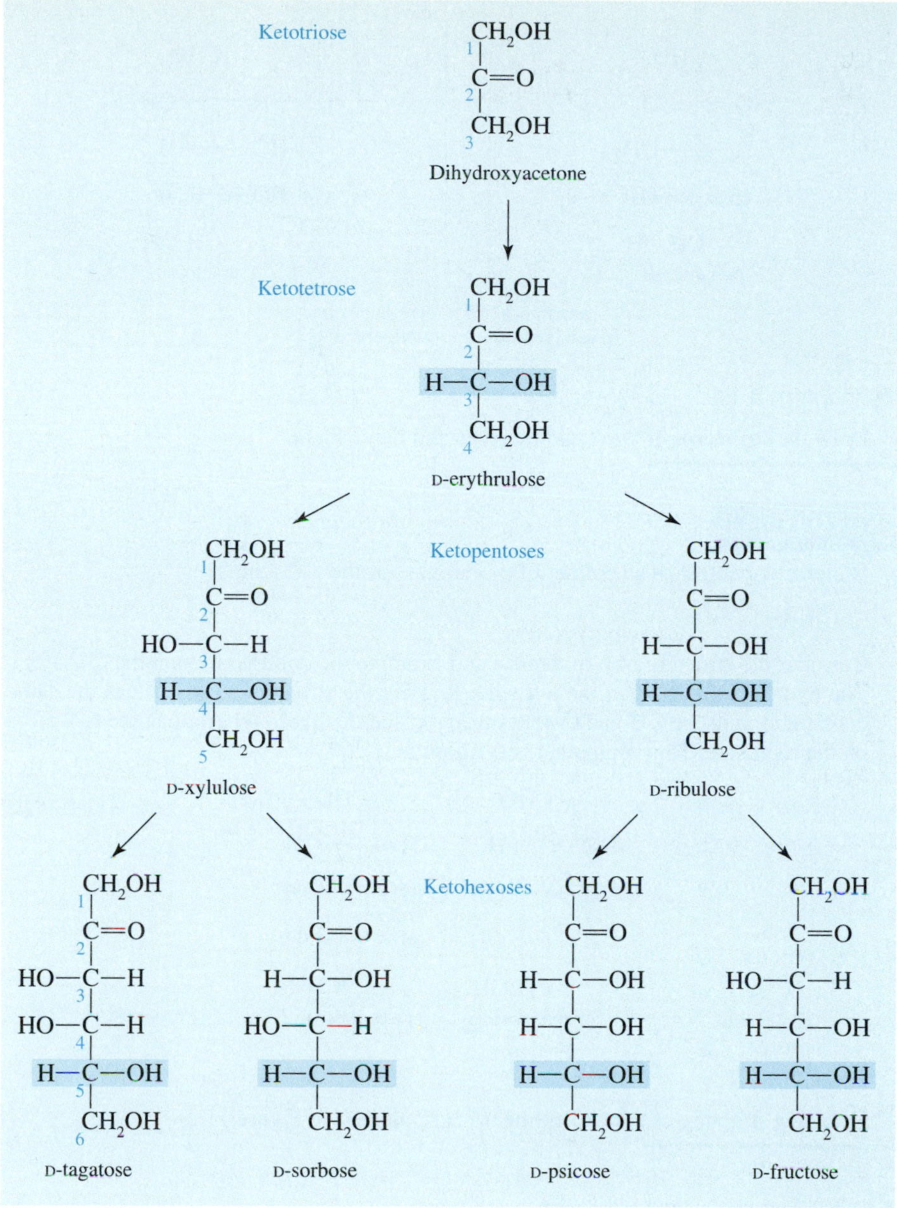

Ketotriose

CH_2OH (1)
C=O (2)
CH_2OH (3)

Dihydroxyacetone

Ketotetrose

CH_2OH (1)
C=O (2)
H—C—OH (3)
CH_2OH (4)

D-erythrulose

Ketopentoses

CH_2OH (1)
C=O (2)
HO—C—H (3)
H—C—OH (4)
CH_2OH (5)

D-xylulose

CH_2OH
C=O
H—C—OH
H—C—OH
CH_2OH

D-ribulose

Ketohexoses

CH_2OH (1)
C=O (2)
HO—C—H (3)
HO—C—H (4)
H—C—OH (5)
CH_2OH (6)

D-tagatose

CH_2OH
C=O
H—C—OH
HO—C—H
H—C—OH
CH_2OH

D-sorbose

CH_2OH
C=O
H—C—OH
H—C—OH
H—C—OH
CH_2OH

D-psicose

CH_2OH
C=O
HO—C—H
H—C—OH
H—C—OH
CH_2OH

D-fructose

FIGURE 11.2
Structures of the D-2-Ketoses

EXAMPLE 11-2

What is the structure of L-arabinose, which is present in some antiviral drugs?

Solution

L-Arabinose is the enantiomer of D-arabinose (see Figure 11.1). Therefore, the L-isomer can be drawn by reflecting the planar projection formula in an imagined mirror perpendicular to the plane of the page and parallel to the carbon chain. Each hydroxyl group that is on the right in D-arabinose is on the left in L-arabinose, and vice versa.

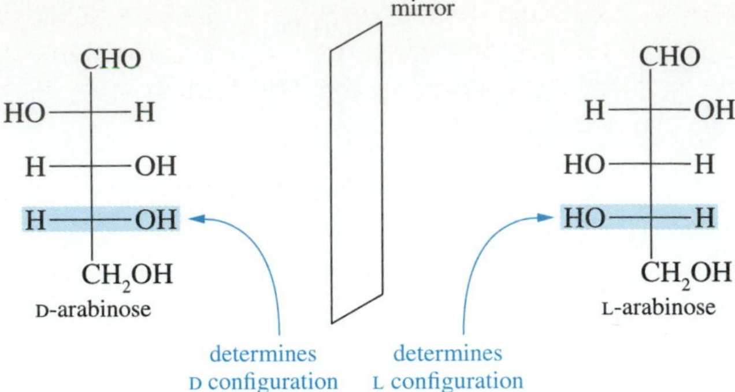

Problem 11.2

Draw the structures of two L-aldotetroses and name them.

EXAMPLE 11-3

Which aldopentose is an epimer of D-arabinose at the C-2 atom?

Solution

Consider the structure of D-arabinose and examine the configuration at the C-2 atom. The hydroxyl group is on the left side. Rewrite the structure so that it has the same configuration at the C-3 and C-4 atoms, but place the hydroxyl group at the C-2 atom on the right side. This compound is D-ribose.

D-arabinose	D-ribose
CHO	CHO
HO—C—H	H—C—OH
H—C—OH	H—C—OH
H—C—OH	H—C—OH
CH_2OH	CH_2OH

Problem 11.3

Draw the structure of the C-3 epimer of D-ribulose (see Figure 11.2).

11.3 Hemiacetals and Hemiketals

We recall that aldehydes and ketones react reversibly with alcohols to form hemiacetals and hemiketals, respectively (Chapter 10).

$$R'{-}\ddot{O}{-}H + (H)(R)C{=}\ddot{O}: \rightleftharpoons (R'{-}\ddot{O}{-})(H)(R)C{-}\ddot{O}{-}H$$

hemiacetal

$$R'{-}\ddot{O}{-}H + (R)(R)C{=}\ddot{O}: \rightleftharpoons (R'{-}\ddot{O}{-})(R)(R)C{-}\ddot{O}{-}H$$

hemiketal

When the hydroxyl group and the carbonyl group are part of the same molecule, a cyclic hemiacetal is formed in an intramolecular reaction. Cyclic hemiacetals containing five or six atoms in the ring form readily because of the proximity of the two functional groups.

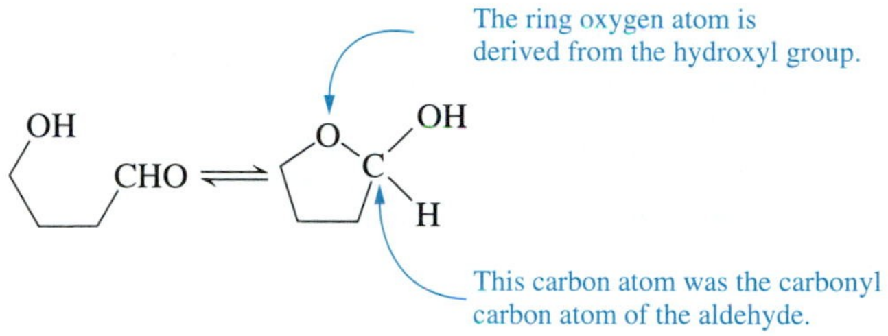

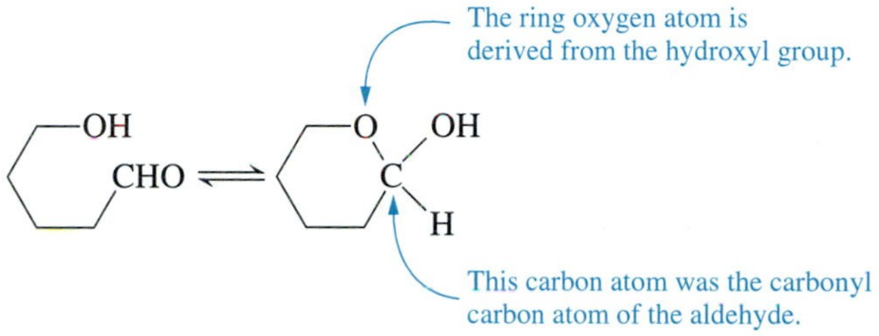

The ring oxygen atom is derived from the hydroxyl group.

This carbon atom was the carbonyl carbon atom of the aldehyde.

The ring oxygen atom is derived from the hydroxyl group.

This carbon atom was the carbonyl carbon atom of the aldehyde.

The cyclic hemiacetal or hemiketal forms of aldo- and ketohexoses and pentoses are the predominant forms of monosaccharides rather than the open-chain structures we have discussed to this point. Cyclic hemiacetals and hemiketals of carbohydrates that contain five-membered rings are called **furanoses;** cyclic hemiacetals and hemiketals that contain six-membered rings are called **pyranoses.** These structures are usually represented by planar structures called Haworth projection formulas.

Haworth Projection Formulas

In a Haworth projection formula, a cyclic hemiacetal or hemiketal is represented as a planar structure and viewed edge-on. Bond lines representing atoms toward the viewer are written as heavy wedges; bond lines away from the viewer are written as unaccentuated lines. The carbon atoms are arranged clockwise with the C-1 atom of the pyranose of an aldohexose on the right. For the furanose of the ketohexose shown, the C-2 atom is placed on the right side of the structure.

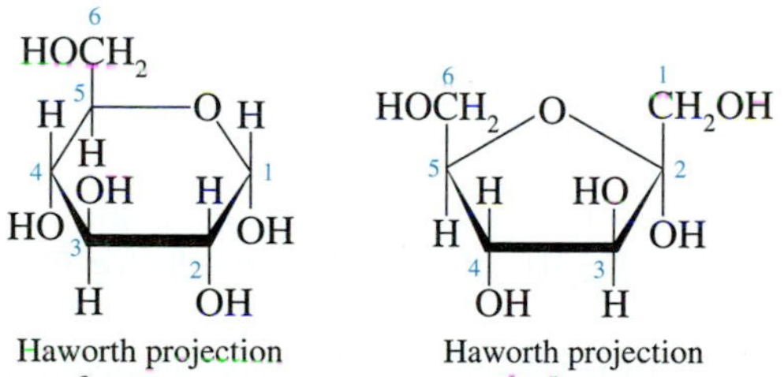

Haworth projection of a pyranose

Haworth projection of a furanose

Let's see how the Fischer projection formula of D-glucose can be converted into a hemiacetal written as a Haworth projection formula. The open-chain form of D-glucose is shaped like a "C" that is arranged vertically on the page. Now tilt this curved chain to the right so that it is horizontal. Groups on the right in the Fischer projection are then directed downward, whereas groups on the left are directed upward.

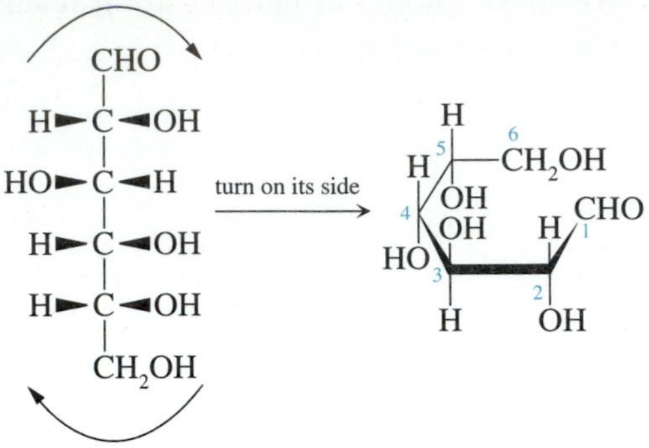

In this conformation, the C-5 —OH group is not near enough to the carbonyl carbon atom to form a ring. To bring the C-5 —OH group near the carbonyl carbon atom, rotate the structure about the bond between the C-4 and C-5 atoms. The —CH_2OH group is now above the plane of the curved carbon chain, and the C-5 hydrogen atom is below the plane.

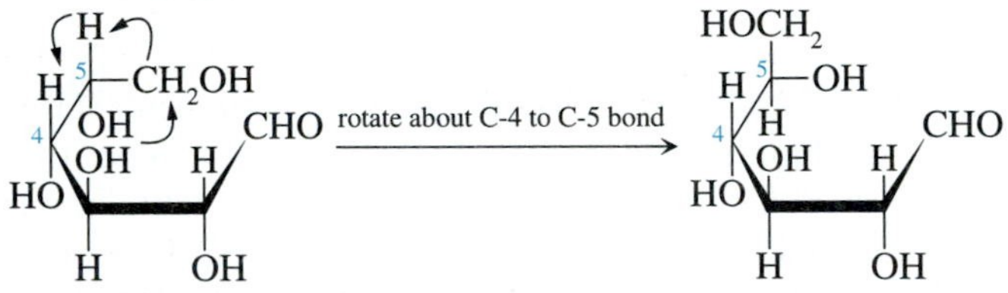

Now add the oxygen atom of the C-5 —OH group to the carbonyl carbon atom, and add a hydrogen atom to the carbonyl oxygen atom. A six-membered ring that contains five carbon atoms and one oxygen atom results. There are now four different groups bonded to the C-1 atom in this cyclic hemiacetal. Thus, a new stereogenic center is formed at the original carbonyl carbon atom, and two configurations are possible at the C-1 atom. If the hydroxyl group of the hemiacetal is directed below the plane, the compound is α-D-glucopyranose; if it is above the plane, the compound is β-D-glucopyranose.

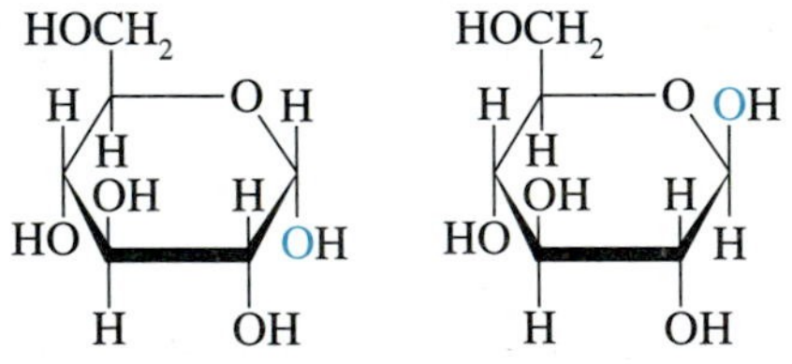

α-D-glucopyranose β-D-glucopyranose

Note that in both forms of glucose, as well as in all carbohydrates with a D configuration, the —CH_2OH group is located above the ring in a Haworth projection.

The α and β forms of D-glucose are diastereomers that differ in configuration at one center. Hence, they are epimers. Compounds whose configurations differ only at the hemiacetal center are a special type of epimer called **anomers.** The chiral carbon atom at the hemiacetal center that forms in the cyclization reaction is called the **anomeric carbon atom.**

Now let's consider the cyclic form of the ketohexose D-fructose. D-Fructose cyclizes in aqueous solution to give a mixture that contains 20% α- and β-D-fructofuranose and 80% α- and β-D-fructopyranose. The furanose isomers are formed when the C-5 —OH group adds to the carbonyl carbon atom of the C-2 keto group. A ring of four carbon atoms and one oxygen atom results. The pyranose isomers form when the C-6 hydroxyl group adds to the C-2 carbonyl carbon atom. Again, α and β designate the configuration at the anomeric carbon atom.

6 $HOCH_2$ O 1 CH_2OH 5 H HO 2 anomeric center H OH 4 3 OH H

α-D-fructofuranose
(a hemiketal)

6 $HOCH_2$ O OH 5 H HO 2 anomeric center H CH_2OH 1 4 3 OH H

β-D-fructofuranose
(a hemiketal)

Mutarotation

Evidence for the reversible closure of an open-chain monosaccharide to form two anomeric cyclic forms is provided by a phenomenon known as mutarotation. When D-glucose is crystallized from methanol, α-D-glucopyranose, which melts at 146°C, is obtained. It has $[\alpha]_D = +112.2$. When D-glucose is crystallized from acetic acid, the β-anomer, which melts at 150°C, is obtained. It has $[\alpha]_D = +18.7$. We recall that diastereomers have different chemical and physical properties, so these data are not surprising.

When α-D-glucopyranose is dissolved in water, the rotation of the solution slowly changes from the initial value of +112.2 to an equilibrium value of +54. When β-D-glucopyranose is dissolved in water, the rotation of the solution slowly changes from the initial value of +18.7 to the same equilibrium value of +54. The gradual changes in rotation to an equilibrium point for either of a set of anomeric carbohydrates is known as **mutarotation.** Mutarotation results from the interconversion of the cyclic hemiacetals with the open-chain form in solution.

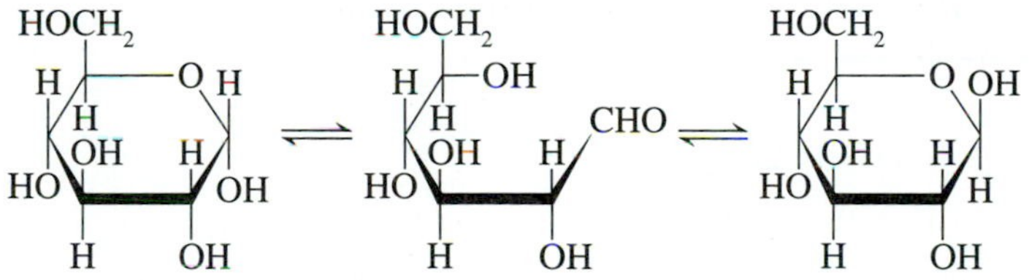

$HOCH_2$ H O H H OH H HO OH H OH ⇌ $HOCH_2$ H OH H CHO OH H HO H OH ⇌ $HOCH_2$ H O OH H OH H HO H H OH

α-D-glucopyranose β-D-glucopyranose

Ring opening followed by ring closure can form either the α- or β-anomer. At equilibrium there is 36% of the α-anomer and 64% of the β-anomer for glucose; less than 0.01% of the open-chain form is present. The mutarotation of glucose in cells is catalyzed by an enzyme called mutarotase.

EXAMPLE 11-4

Draw the Haworth projection of the α-anomer of the pyranose form of D-galactose; that is, α-D-galactopyranose.

Solution

First write galactose in the Fischer projection. Because the pyranose form is a six-membered ring, draw the ring consisting of five carbon atoms and one oxygen atom. Place the C-6 —CH_2OH group above the plane of the ring.

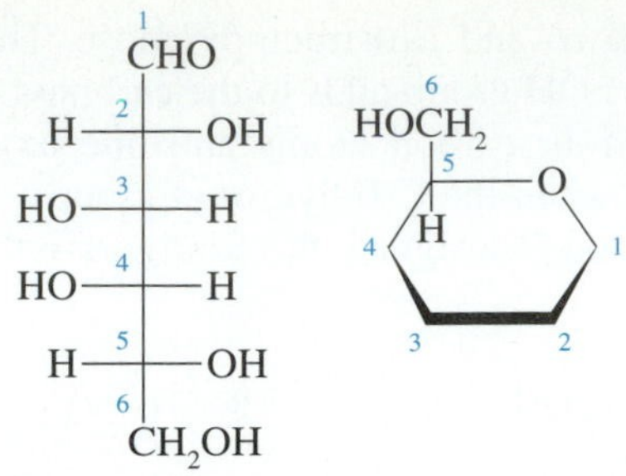

Now enter the hydroxyl groups and hydrogen atoms at C-2, C-3, and C-4. An atom or group on the right in the Fischer projection is below the ring of the Haworth projection, and an atom or group on the left is above the ring.

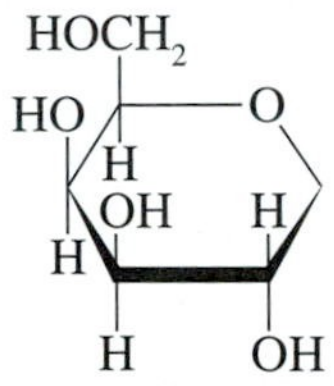

Finally, the α-anomer must have a hydroxyl group below the plane of the ring at the anomeric carbon atom, C-1. The hydrogen atom at C-1 is above the plane.

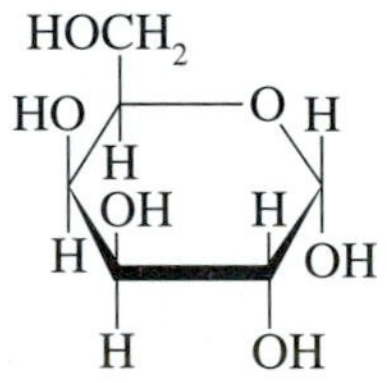

α-D-galactopyranose

Problem 11.4

Draw the Haworth projection formula for the pyranose form of D-mannose with the β configuration at the anomeric center; that is, β-D-mannopyranose.

11.4 Conformations of Monosaccharides

Haworth projection formulas do not give an accurate three-dimensional representation of carbohydrates, because the six atoms of pyranose rings exist in chair conformations just like cyclohexane (Section 3.6). Any hydroxyl group (or other group) that is up in the Haworth projection is also up in the chair conformation. However, "up" and "down" do not correspond to axial and equatorial, respectively. Each carbon atom must be individually examined. On one set of alternating carbon atoms, an "up" substituent is axial; on the intervening carbon atoms, an "up" substituent is equatorial.

Haworth projection formulas are converted into chair representations by "moving" two carbon atoms. The anomeric carbon atom is lowered below the plane of the ring, and the C-4 atom is raised above the plane of the ring. The remaining four atoms—three carbon atoms and the ring oxygen atom—are unchanged. This process is shown in Figure 11.3 for both α-D-glucopyranose and β-D-glucopyranose. Both the hydrogen atoms and the hydroxyl groups can be

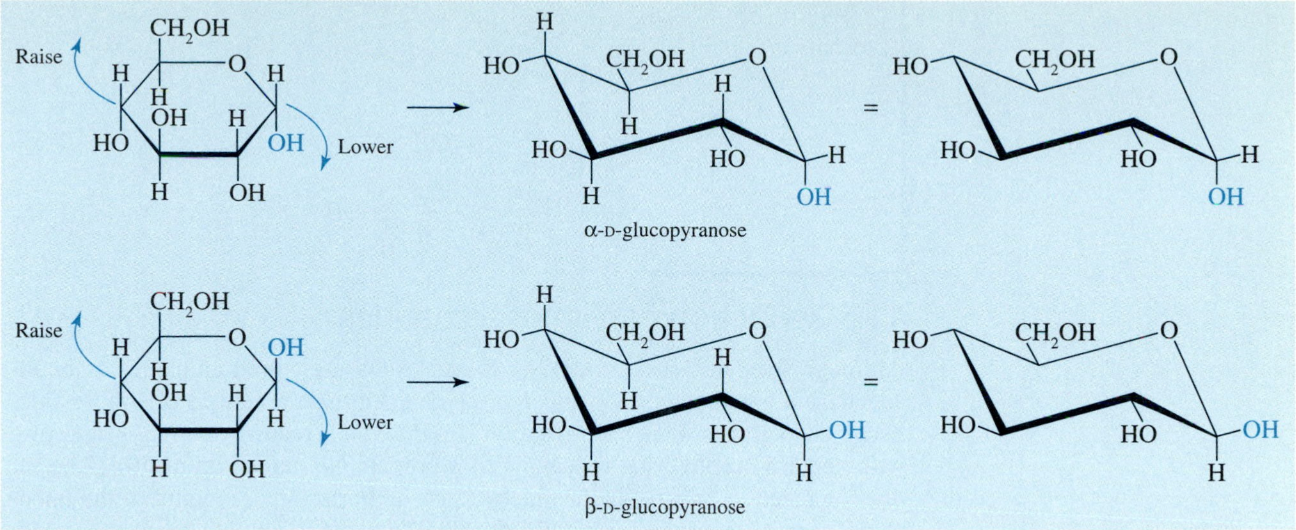

α-D-glucopyranose

β-D-glucopyranose

FIGURE 11.3
Conversion of Haworth Projection into Chair Representation

shown. However, a more condensed form that eliminates the C—H bonds is often used.

Note the changes in the locations of the hydroxyl groups in the Haworth projection compared to those in the chair conformation. Although the hydroxyl groups were both up and down in the Haworth projection formula, all hydroxyl groups are equatorial in β-D-glucopyranose. This anomer is the more stable based on the position of equilibrium observed in mutarotation experiments. β-D-Glucopyranose, the most abundant aldohexose, is the only aldohexose that has all of its hydroxyl groups in equatorial positions.

EXAMPLE 11-5

Draw the chair conformation of α-D-galactopyranose.

Solution

Although the chair conformation could be derived from the Haworth projection—which can, in turn, be derived from the open-chain formula—there is an easier way to obtain the structure. We need only recall that the β-anomer of glucose has all of its hydroxyl groups in equatorial positions. The α-anomer of galactose must therefore have the hydroxyl group at the C-1 atom in an axial position. We also recall that galactose is the C-4 epimer of glucose. Therefore, the hydroxyl group at the C-4 atom must also be axial.

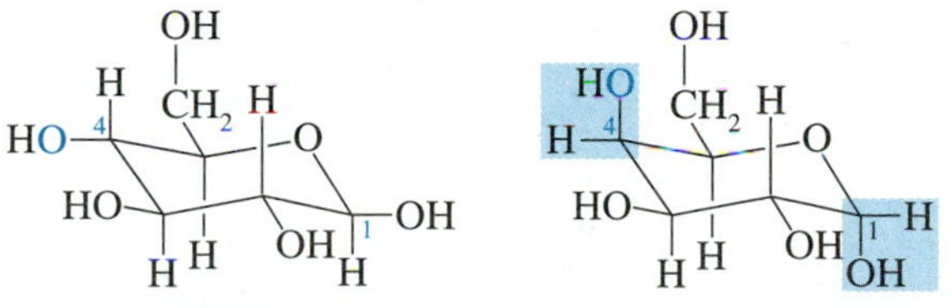

β-D-glucopyranose

α-D-galactopyranose

Problem 11.5

Identify the following monosaccharide.

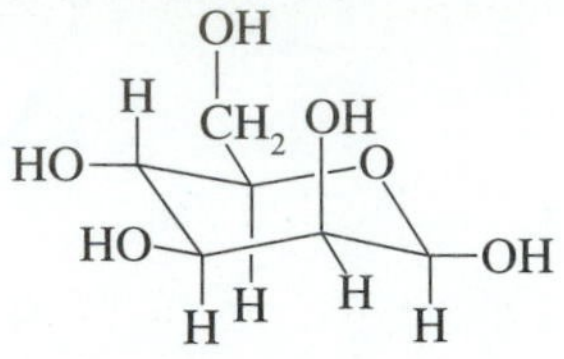

11.5 Reduction of Monosaccharides

Although five- and six-carbon monosaccharides exist predominantly as hemiacetals and hemiketals, they undergo the characteristic reactions of simple aldehydes and ketones. One such reaction is reduction. Treating an aldose or ketose with sodium borohydride reduces it to a polyalcohol called an **alditol.** The reduction reaction occurs via the aldehyde group in the small amount of the open-chain form of the aldose in equilibrium with its cyclic hemiacetal. As the aldehyde is reduced, the equilibrium shifts to produce more aldehyde until eventually all of the monosaccharide is reduced. The alditol derived from D-glucose is called D-glucitol. D-Glucitol occurs in some fruits and berries. Produced and sold commercially as a sugar substitute, it is also called sorbitol.

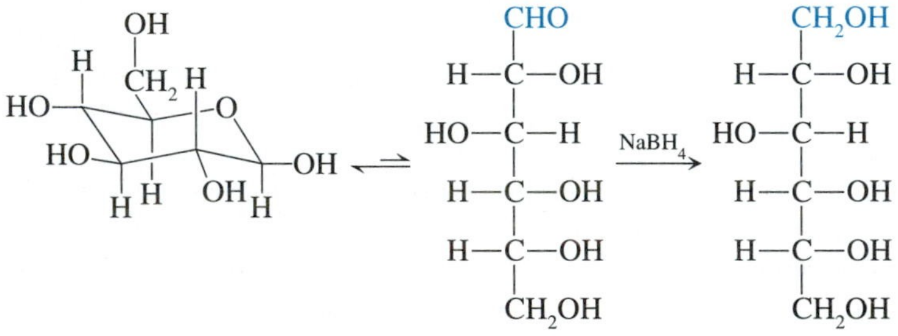

D-glucose $\xrightarrow{NaBH_4}$ D-glucitol (an alditol)

EXAMPLE 11-6

D-Xylitol is used as a sweetener in some chewing gums that are said to have a lower probability of causing cavities in teeth compared to those containing glucose and fructose. Deduce the structure of D-xylitol from its name.

Solution

The name resembles D-xylose, which can be reduced by sodium borohydride to an alditol. The configurations of the carbon atoms bearing the hydroxyl groups in D-xylitol are the same as those in D-xylose.

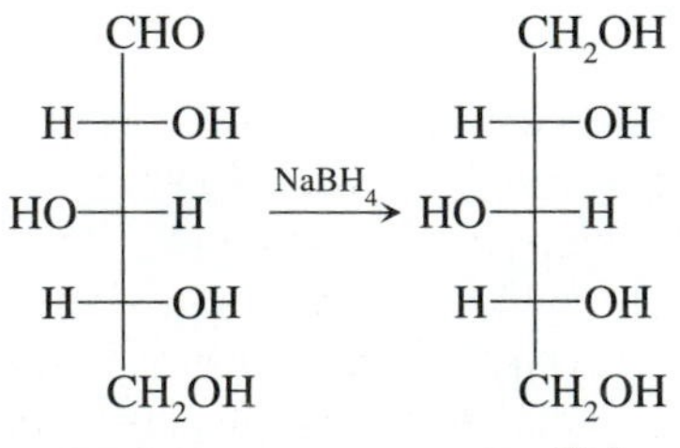

D-xylose $\xrightarrow{NaBH_4}$ D-xylitol

Problem 11.6

Reduction of ribulose (see Figure 11.2) by sodium borohydride gives a mixture of two isomeric compounds with different physical properties. Explain why.

11.6 Oxidation of Carbohydrates

In Chapter 10 we saw that aldehydes are oxidized by Tollens's reagent and Fehling's solution. These reagents also oxidize open-chain aldoses that exist in equilibrium with the cyclic hemiacetal form. When some of the open-chain form reacts, the equilibrium shifts to form more compound for subsequent oxidation, and eventually all of the aldose is oxidized. Oxidation yields a product with a carboxylic acid at the original C-1 atom. This product is called an **aldonic acid.**

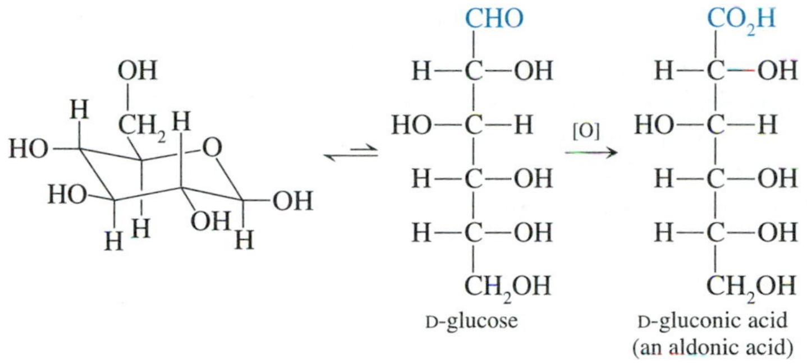

D-glucose

D-gluconic acid (an aldonic acid)

If Tollens's reagent is used as the oxidizing agent, metallic silver forms a mirror on the walls of the test tube. If Fehling's solution is used, a red precipitate of Cu_2O indicates that a reaction has occurred. Carbohydrates that react with Fehling's solution are called **reducing sugars.** The term *reducing* refers to the effect of the carbohydrate on the reagent that is reduced in the reaction.

Fehling's solution also oxidizes ketoses. We certainly do not expect this, because ketones are not oxidized by Fehling's solution. However, we recall that α-hydroxy ketones tautomerize in basic solution. Because Fehling's solution is basic, the tautomer of a ketose is an enediol that not only reverts to the α-hydroxy ketone, but also forms an isomeric α-hydroxy aldehyde.

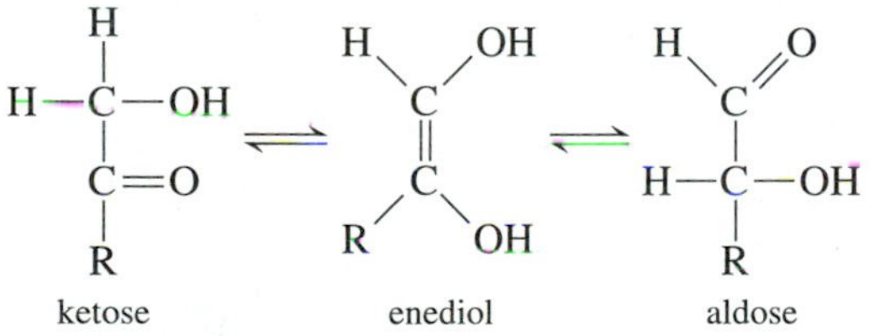

ketose

enediol

aldose

Shifting a hydrogen atom from the C-2 hydroxyl group to the C-1 atom in the enediol regenerates the original ketose. However, shifting a hydrogen atom from the C-1 hydroxyl group to the C-2 atom forms an aldose. In basic solution, then, a ketose is in equilibrium with an aldose. The aldose reacts with Fehling's solution, and more ketose is converted into aldose. The equilibrium shifts as predicted by Le Châtelier's principle, and eventually all of the ketose is oxidized.

Other hydroxyl groups of aldoses may be oxidized if stronger oxidizing agents are used. For example, dilute nitric acid oxidizes both the aldehyde group and the primary alcohol of aldoses to give **aldaric acids.**

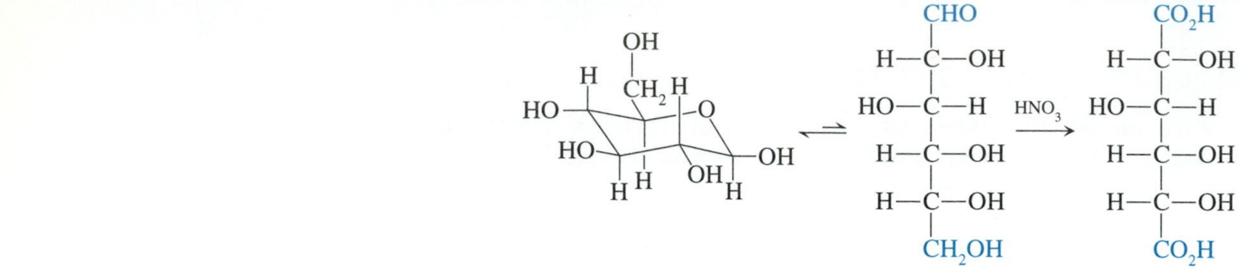

D-glucose

D-glucaric acid (an aldaric acid)

The terminal $—CH_2OH$ group of an aldose can be oxidized enzymatically in cells without oxidation of the aldehyde group. The product is a **uronic acid.** The enzyme responsible for this reaction uses $NADP^+$ (nicotinamide adenine dinucleotide phosphate, which has a structure similar to NAD^+) as the oxidizing agent. An example of this reaction is the oxidation of D-glucose to give D-glucuronic acid, a component in the polysaccharide hyaluronic acid, which is found in the vitreous humor of the eye.

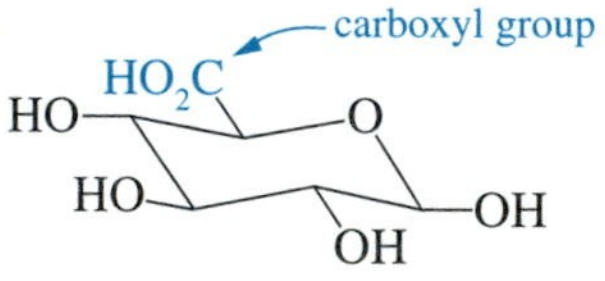

carboxyl group

D-glucuronic acid

11.7 Glycosides

In Chapter 10 we saw that hemiacetals and hemiketals react with alcohols to yield acetals and ketals, respectively. The equilibrium of this acid-catalyzed reaction is shifted to the right by using excess alcohol, or by removal of the water formed. In this substitution reaction an $—OR'$ group replaces the $—OH$ group.

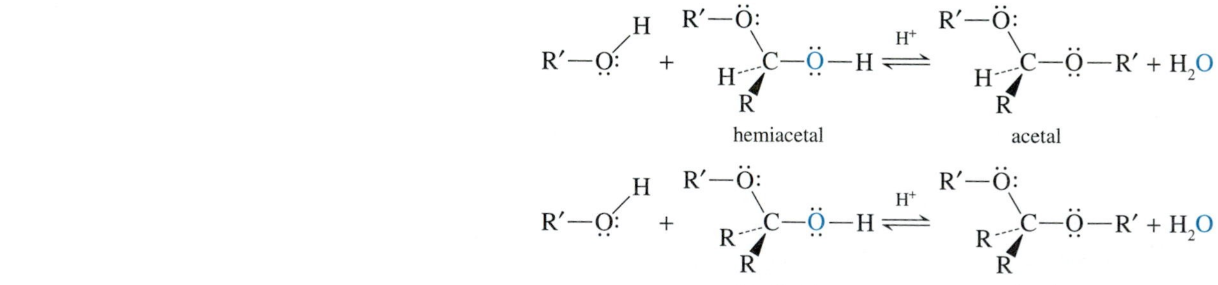

hemiacetal

acetal

hemiketal

ketal

The cyclic hemiacetal and hemiketal forms of monosaccharides also react with alcohols to form acetals and ketals. These acetals and ketals are called **glycosides,** and the carbon-oxygen bond formed is called a **glycosidic bond.** The group bonded to the anomeric carbon atom of a glycoside is an **aglycone.** In aglycones, an oxygen atom from an alcohol or phenol is linked to the anomeric carbon atom.

Glycosides are named by citing the aglycone group first and then replacing the *-ose* ending of the carbohydrate with *-oside*. The configuration at the glycosidic carbon atom must be indicated.

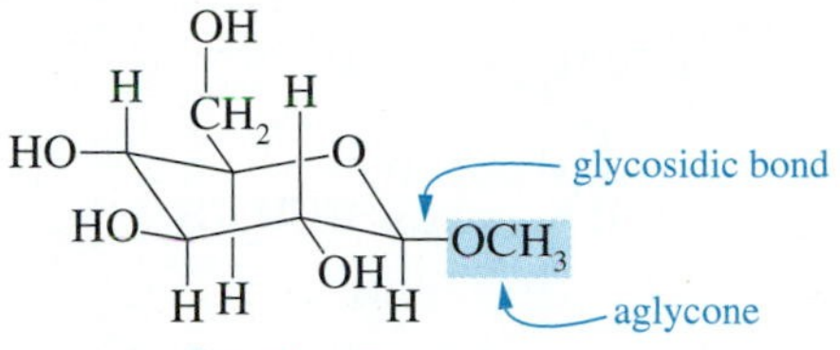

methyl β-D-glucopyranoside

Recall that the conversion of hemiacetals or hemiketals into acetals or ketals, respectively, occurs via a resonance-stabilized oxocarbocation (Section 10.8) formed by protonation of the —OH group followed by loss of water.

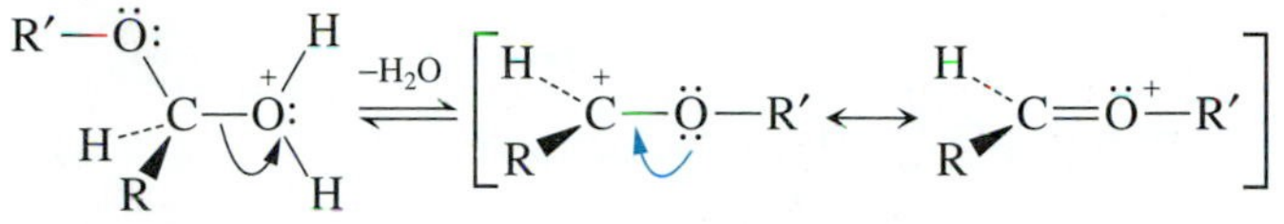

resonance-stabilized oxocarbocation

Subsequently, the intermediate oxocarbocation reacts with an alcohol serving as a nucleophile.

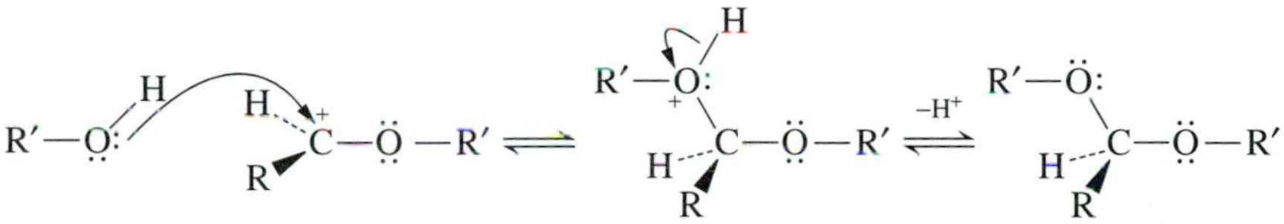

Attack of the alcohol in the case of the intermediate formed from a monosaccharide can occur from either of two directions. As a result, a mixture of α- and β-anomers forms.

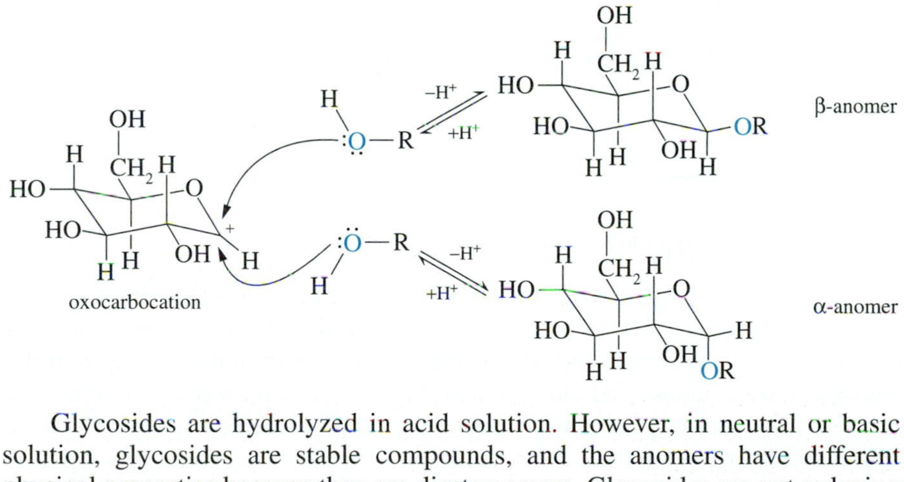

oxocarbocation

β-anomer

α-anomer

Glycosides are hydrolyzed in acid solution. However, in neutral or basic solution, glycosides are stable compounds, and the anomers have different physical properties because they are diastereomers. Glycosides are not reducing sugars because they do not hydrolyze to form a free aldehyde group in Fehling's solution, which is basic.

EXAMPLE 11-7

Examine the following molecule to determine its component functional groups. From what compounds can the substance be formed?

Solution

The compound is a furanose form of a monosaccharide, because there are four carbon atoms and one oxygen atom in a five-membered ring. The ring carbon atom on the right is an acetal center because there are two —OR groups and one hydrogen atom bonded to it.

one C—OR bond
hydrogen atom from an aldehyde
acetal center
one C—OR bond

The acetal has the α configuration, and the alcohol used to form the acetal is ethanol.

The carbohydrate has the D configuration because the $—CH_2OH$ group is "up" in the Haworth projection. The other two stereogenic carbon atoms of the pentose have hydroxyl groups "down," which corresponds to the right in the Fischer projection. The carbohydrate component of the compound is ribose. The name of the compound is ethyl α-D-ribofuranoside.

Problem 11.7

Examine the following molecule to determine its component functional groups. From what compounds can the substance be formed?

Lactose is a common disaccharide found in milk.

11.8 Disaccharides

Disaccharides have glycosidic linkages between the anomeric center of one monosaccharide unit and the hydroxyl oxygen atom of the second monosaccharide unit, which is the aglycone. A glycosidic bond between the C-1 atom of the hemiacetal of an aldose and the C-4 atom of the second monosaccharide is very common. Such bonds are designated (1,4′). The prime superscript indicates which carbon atom of the monosaccharide that is the aglycone provides the oxygen atom. Maltose, cellobiose, and lactose all have (1,4′) glycosidic bonds (Figure 11.4).

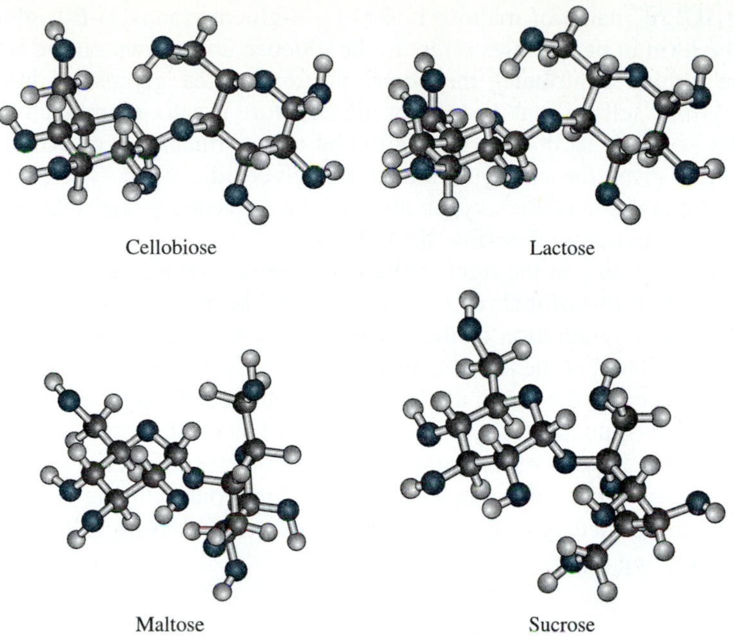

FIGURE 11.4
Molecular Models of Disaccharides

In principle, any of the carbon atoms of a monosaccharide could provide the oxygen atom of the aglycone. In fact, (1,1′), (1,2′), (1,3′), (1,4′), and (1,6′) glycosidic bonds have all been found in naturally occurring disaccharides containing aldohexoses. Note that a (1,1′) glycosidic bond connects both anomeric carbon atoms. Sucrose has a (1,2′) glycosidic bond connecting the anomeric carbon atoms of glucose and fructose (Figure 11.4).

Maltose

In maltose the glycosidic oxygen atom of one glucose is α and is bonded to the C-4 atom of another glucose unit that is the aglycone. Therefore, maltose is an α-1,4′-glycoside. Maltose is produced by the enzymatic hydrolysis of starch (a homopolysaccharide) catalyzed by the enzyme **amylase.** Maltose is further hydrolyzed by the enzyme maltase to produce two molecules of D-glucose.

α-1,4′-glycosidic bond

4-*O*-(α-D-glucopyranosyl)-β-D-glucopyranose
(maltose)

The IUPAC name of maltose is 4-*O*-(α-D-glucopyranosyl)-β-D-glucopyranose. The term in parentheses refers to the glucose unit shown on the left in the structure, which contributes the acetal portion of the glycosidic bond. The suffix *-pyrano-* tells us that this part of the structure is a six-membered ring, and the suffix *-syl* tells us that the ring is linked to a partner by a glycosidic bond. The term α gives the configuration of the glycosidic bond. The prefix *4-O-* refers to the position of the oxygen atom of the aglycone—the right-hand ring. The β-D-glucopyranose describes the aglycone.

The glucose ring at the right of the structure is shown as the β-anomer, but both anomeric forms of maltose can exist in equilibrium in solution. This designation for the configuration of the aglycone ring should not be confused with the glycosidic bond at the acetal center, which is always α in maltose. If we do not want to specify the configuration of the hemiacetal center of maltose, the name 4-*O*-(α-D-glucopyranosyl)-D-glucopyranose is used.

Maltose undergoes mutarotation at its hemiacetal anomeric center. Recall that the process occurs via an open-chain structure containing an aldehyde. The free aldehyde formed by ring opening can react with Fehling's solution, so maltose is a reducing sugar.

Cellobiose

Cellobiose is a disaccharide that has two molecules of D-glucose linked by a β-1,4′-glycosidic bond. Thus, cellulose differs from maltose in the configuration of its glycosidic bond.

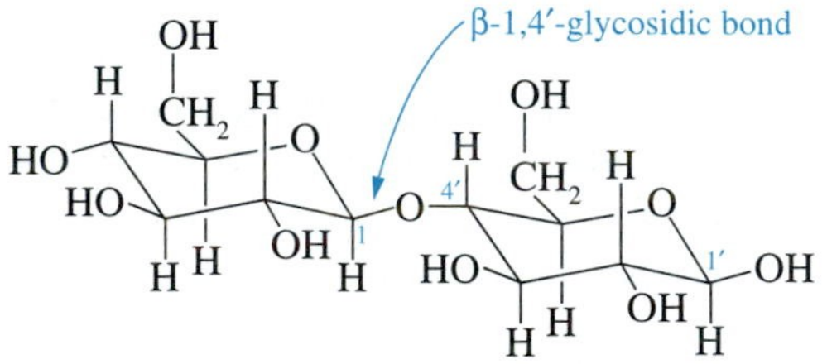

β-1,4′-glycosidic bond

4-*O*-(β-D-glucopyranosyl)-β-D-glucopyranose
(cellobiose)

As in maltose, the aglycone of cellobiose is a hemiacetal, which can be either α or β. In solution, the two forms of cellobiose exist in equilibrium. Thus, cellobiose mutarotates and is a reducing sugar. Again, do not confuse the configuration of the hemiacetal center with that of the glycosidic bond, which is β in cellobiose.

Cellobiose is produced by hydrolysis of cellulose, a homopolysaccharide of glucose in which all units are linked by β-1,4′-glycosidic bonds. Humans do not have an enzyme to hydrolyze cellobiose. Small differences in configuration at the (1,4′) linkage result in remarkable differences in the chemical reactivity of these biomolecules. Glycosidic bonds are hydrolyzed by enzymes called glycosidases. A glycosidase that hydrolyzes α-1,4′-glycosidic bonds does not hydrolyze molecules that have β-1,4′-glycosidic bonds (and vice versa).

Lactose

Lactose, a disaccharide found in different concentrations in the milk of mammals, including both humans and cows, is often called milk sugar. The IUPAC name of lactose is 4-*O*-(β-D-galactopyranosyl)-D-glucopyranose.

axial 4-hydroxyl group of galactose

β-1,4′-glycosidic bond

4-*O*-(β-D-galactopyranosyl)-β-D-glucopyranose
(lactose)

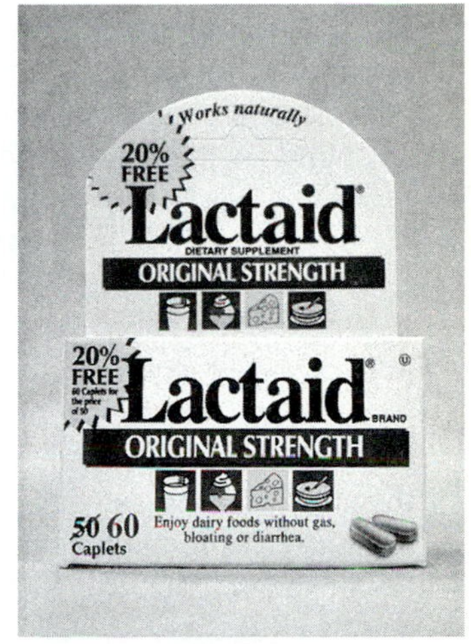

Commercial products provide a source of lactase for use by lactose intolerant individuals.

Compare this structure to that of cellobiose. The monosaccharide shown on the left in the structure has an axial hydroxyl group at the C-4 atom, so it is galactose. The oxygen atom at the C-4 atom of the ring on the right is equatorial, so it is glucose. Both lactose and cellobiose are linked by a β-glycosidic linkage to the C-4 atom of a D-glucopyranose ring on the right. In both lactose and cellobiose, the glycosidic bond is β-1,4′. Humans have an enzyme called β-galactosidase (also known as lactase) that catalyzes the hydrolysis of the β-1,4′-galactosidic linkage. However, β-galactosidase does not catalyze the hydrolysis of the β-1,4′-glucosidic linkage of cellobiose.

As in the case of cellobiose and maltose, the aglycone component of lactose is a hemiacetal, which can be either α or β. In solution, the two forms of lactose exist in equilibrium. Thus, lactose undergoes mutarotation and is a reducing sugar.

The lactose content of milk varies with species; cow's milk contains about 5% lactose, whereas human milk contains about 7%. The enzyme lactase, which is present in the small intestine, catalyzes hydrolysis of lactose to form glucose and galactose. Galactose is then isomerized into glucose in a reaction catalyzed by the enzyme UDP-galactose-4-epimerase.

Sucrose

Some disaccharides have a glycosidic linkage between both anomeric centers. Sucrose, common table sugar, is a disaccharide of α-glucose and β-fructose in which the anomeric centers are linked 1,2′.

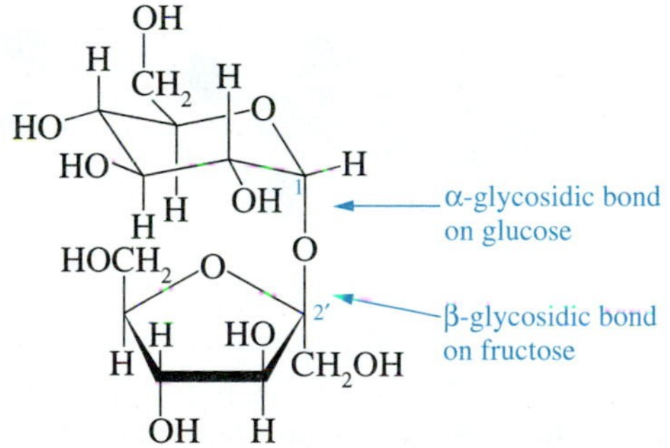

α-D-glucopyranosyl-β-D-fructofuranoside
(sucrose)

Sucrose has both an acetal and a ketal functional group. Neither ring can exist in equilibrium with either an aldehyde or ketone. As a result, sucrose cannot mutarotate, nor is it a reducing sugar. The systematic name, α-D-glucopyranosyl-β-D-fructofuranoside, ends in the suffix *-oside,* which indicates that sucrose exists as a glycoside and so is not a reducing sugar.

Metabolism of Lactose

Lactose is an important source of metabolic energy in infants. It is hydrolyzed to glucose and galactose, which is subsequently epimerized to glucose. However, there are genetic variations in people that can adversely affect one or the other of these processes. These conditions are known as lactose intolerance and galactosemia.

Individuals with lactose intolerance lack the enzyme lactase needed to hydrolyze lactose. If they ingest food that contains lactose, the high level of unhydrolyzed lactose in their intestinal fluids draws water from tissues by osmosis, and the osmotic pressure balance is upset. Unmetabolized lactose in the small intestine is slowly metabolized by intestinal bacterial fermentation, which produces a variety of carboxylic acids and carbon dioxide. The result is abdominal distention, cramping, and diarrhea. Although lactose intolerance is not life threatening, it is quite unpleasant.

The level of the enzyme lactase in humans varies with both age and race. Most humans have sufficient lactase for the early years of life, when milk is a major part of their diet. However, in adulthood the lactase level decreases, and lactose intolerance results. This trait shows remarkable genetic variations. For example, most northern Europeans have high lactase levels, as do several nomadic pastoral tribes in Africa (see the accompanying figure). The ability to digest milk as adults may be the result of an evolutionary process in societies that consume large amounts of milk and milk products, such as cheese. Those individuals with the enzyme necessary to digest milk may have had an adaptational advantage.

Some peoples, such as the Thai and the Chinese, have a high lactose intolerance. Similarly, the Ibo and Yoruba of Nigeria cannot tolerate lactose as adults. The Fula and Hausa of the Sudan differ in the extent of their lactose intolerance. The Fula raise a breed of cattle called fulani, whereas the Hausa, who show lactose intolerance, do not raise cattle. The Tussi, a cattle-owning class of the Rundi of eastern Africa, also can digest lactose.

The inability to metabolize galactose is the result of a more serious genetic disease termed galactosemia. Epimers are interconverted in cells by enzymes called **epimerases.** The necessary epimerase

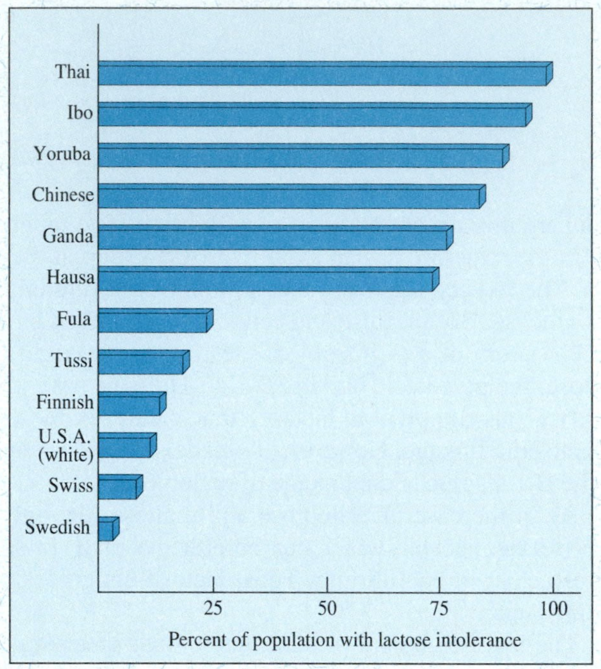

to catalyze the conversion of D-galactose into D-glucose is not present in some newborn children. When the epimerase that converts galactose to glucose is missing, D-galactose accumulates in the body, causing liver damage that usually is fatal.

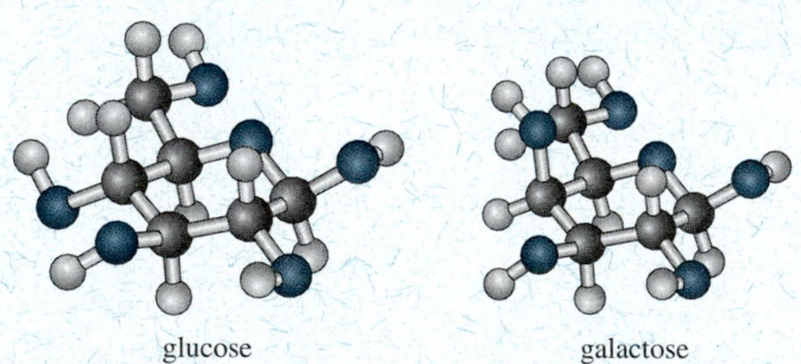

If the individual survives, mental retardation and/or blindness due to severe cataracts is likely. In the absence of the epimerase, galactose is reduced to an alditol called galactitol. Galactitol accumulates in the lens of the eye and causes cataracts.

Blood tests of cells in the umbilical cord can be used to check for the absence of the epimerase in the newborn. The consequences of galactosemia can be prevented if the diet of affected newborns does not contain milk or milk products.

EXAMPLE 11-8

Describe the structure of the following disaccharide.

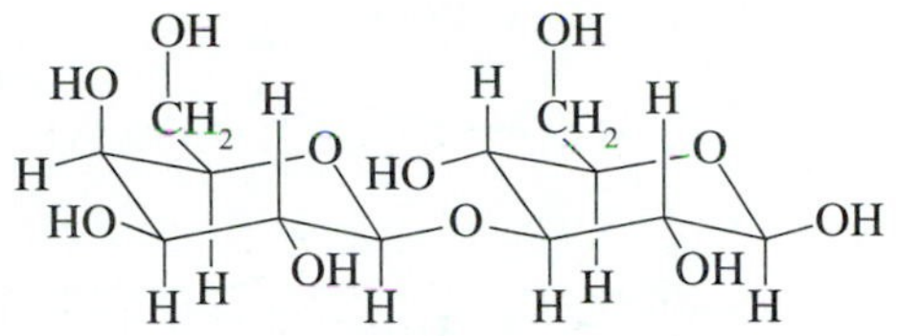

Solution

The hemiacetal center located on the aglycone ring (at the right) has a hydroxyl group in the β-configuration. The glycosidic bond is from the C-1 atom of the acetal ring (on the left) to the C-3 atom of the aglycone ring (on the right). Furthermore, the oxygen bridge is formed through a β-glycosidic bond. Thus, the bridge is β-1,3′.

Next, we examine both rings to determine the identity of the monosaccharides. The ring on the left is galactose: all of its hydroxyl groups are equatorial except the one at the C-4 atom, which is axial. The ring on the right is glucose. The compound is 3-*O*-(β-D-galactopyranosyl)-β-D-glucopyranose.

Problem 11.8

Describe the structure of the following disaccharide.

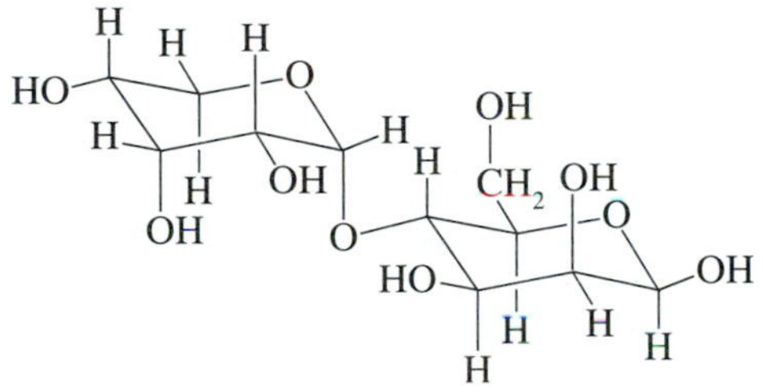

11.9 Polysaccharides

Polysaccharides are high molecular weight substances consisting of monosaccharides linked to one another by glycosidic bonds. Because the structures of heteropolysaccharides are more complex than those of homopolysaccharides, only homopolysaccharides will be considered in this section.

The homopolysaccharides starch and cellulose, found in plants, contain only glucose. About 20% of starch is **amylose,** which is soluble in cold water; the remaining 80%, called **amylopectin,** is insoluble in water. Starch is present in potatoes, rice, wheat, and other cereal grains. The amount of amylose and amylopectin in starch is variable and depends on its source.

Starch and cellulose differ by one structural feature, but this difference has great biological importance. Starch, whose glucosyl units are linked α-1,4′, can be digested by most animals. Cellulose, whose glucosyl units are linked β-1,4′, can be digested only by cattle and other herbivores, because microorganisms present in their digestive tracts have enzymes that hydrolyze β-glucosides. Termites can also digest cellulose.

Amylose is a linear polymer with 200 to 2000 α-linked glucose units that serves as a major source of food for some animals. The molecular weight of amylose ranges from 40,000 to 400,000 amu. Cellulose is a β-linked polymer

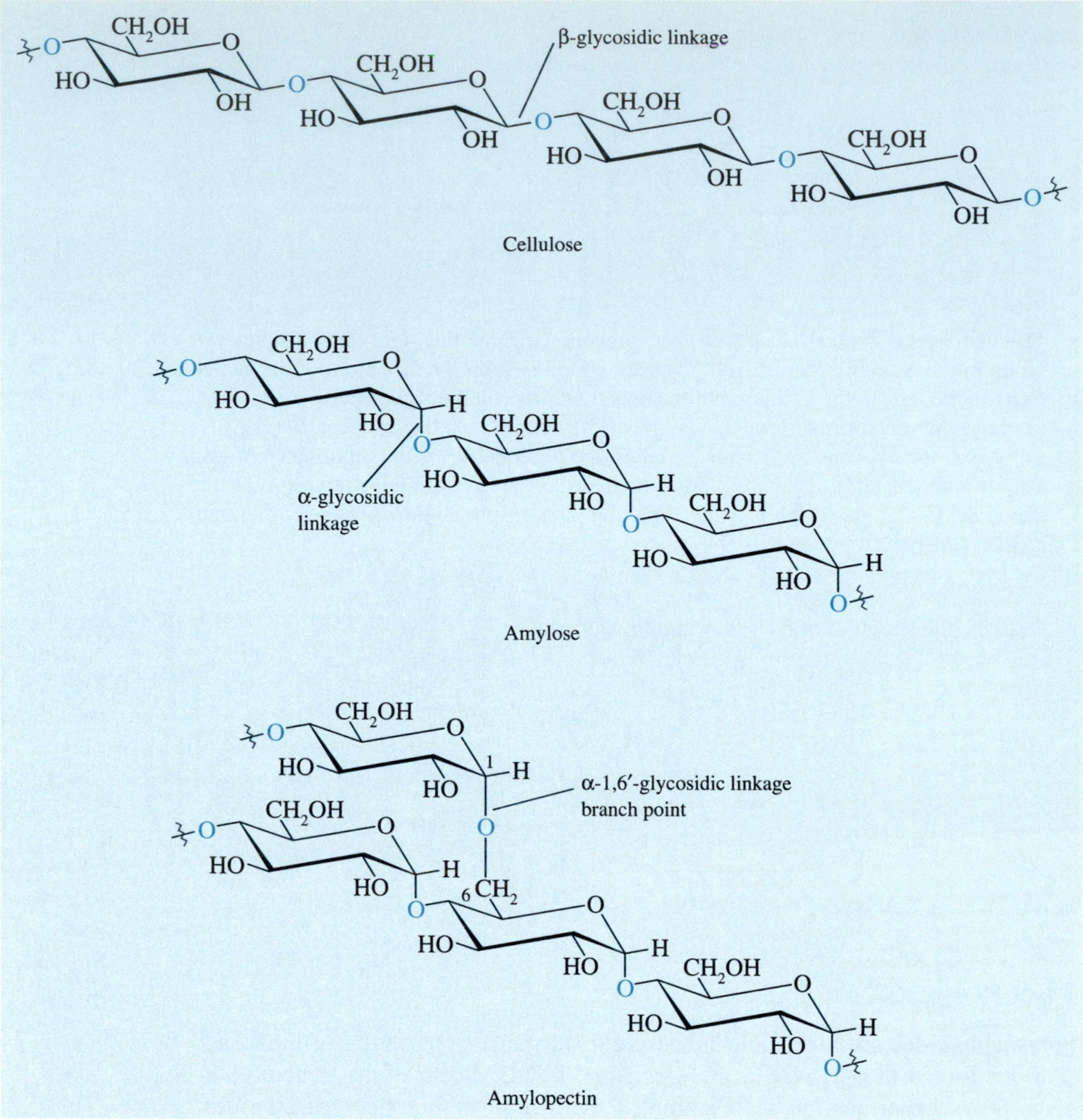

FIGURE 11.5
Structures of Polysaccharides

of glucose (Figure 11.5) that can contain 5,000 to 10,000 glucose units. Certain algae produce cellulose molecules that contain more than 20,000 glucose units.

Amylopectin contains chains similar to those in amylose, but only about 25 glucose units occur per chain. Amylopectin has branches of glucose-containing chains interconnected by a glycosidic linkage between the C-6 hydroxyl group of one chain and the C-1 atom of another glucose chain (Figure 11.5). The molecular weight of amylopectin may be as high as 1 million. Because each chain has an average molecular weight of 3000, there may be as many as 300 interconnected chains.

Human Blood Groups

Complex carbohydrates coat the surfaces of nearly all human cells, acting as markers that identify the cell. Human blood cells contain surface markers that divide blood into three major classes, designated A, B, and O. The classification of blood groups relies on differences in the structures of oligosaccharides bonded to a protein called glycophorin, which is embedded in the membrane of red blood cells.

The blood-group oligosaccharides contain several different monomers: galactose (Gal), *N*-acetylgalactosamine (GalNAc), and *N*-acetylglucosamine (GlcNAc).

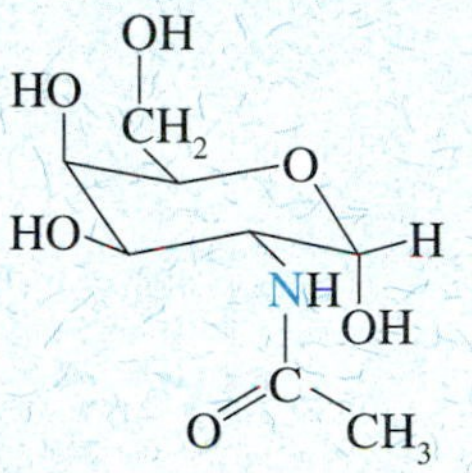

α-*N*-acetylgalactosamine (GalNAc)

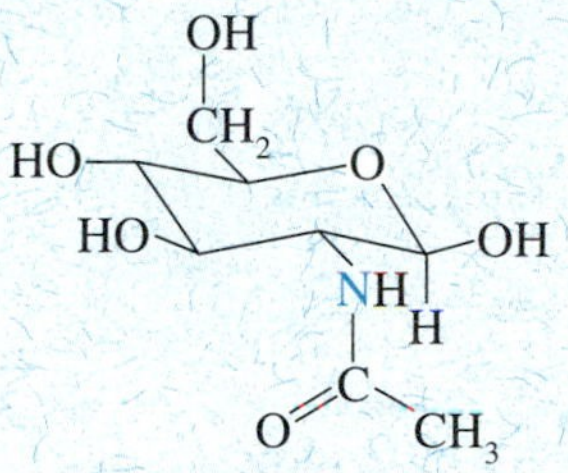

β-*N*-acetylglucosamine (GlcNAc)

They also contain the rather unusual sugar 6-deoxy-α-L-galactose. This sugar has the common name α-L-fucose.

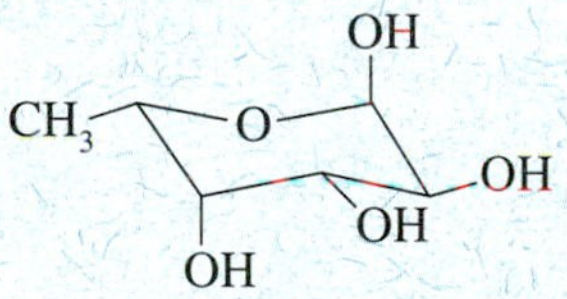

6-deoxy-α-l-galactose (l-fucose)

The α-L-fucose moiety of each blood group is attached to a trisaccharide in blood groups A and B and to a disaccharide in blood group O (see structures in the drawings below). Type B, which is not shown, differs from type A by a galactose on the left side of the structure rather than *N*-acetylgalactosamine. In each oligosaccharide the β-galactose residue is attached to α-L-fucose by a 1,2′-glycosidic bond. The β-*N*-acetylglucosamine portion of the oligosaccharide shown is bonded to a "spacer unit" that consists of an oligosaccharide chain that may contain from 2 to as many as 50 monosaccharide units. This chain is bonded to a protein in a variety of possible ways including a glycosidic bond to the hydroxyl group of a serine residue in the protein.

Each blood group is further subdivided into two types of chains that differ in their glycosidic linkages. In a type 1 chain, the central β-Gal moiety is linked to β-GlcNAc by a 1,4′-glycosidic bond. In a type 2 chain, the β-Gal moiety is linked to β-GlcNAc by a 1,3′-glycosidic bond. The structures below show a type 1 linkage.

These carbohydrates are the *antigenic determinants* of their groups. A person with type A blood makes antibodies that "attack" type B blood, forming clumps of type B cells. Similarly, a person with type B blood makes antibodies that "attack" type A blood. However, persons who are type A or type B do not make antibodies against type O blood, so type O persons are called universal donors. They are not, however, universal acceptors, because they produce antibodies against both type A and type B blood. ■

Type A

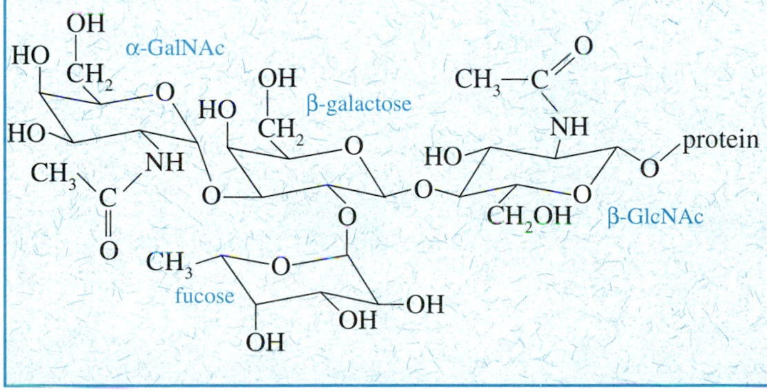

α-GalNAc · β-galactose · β-GlcNAc · fucose · protein

Type O

β-galactose · β-GlcNAc · fucose · protein

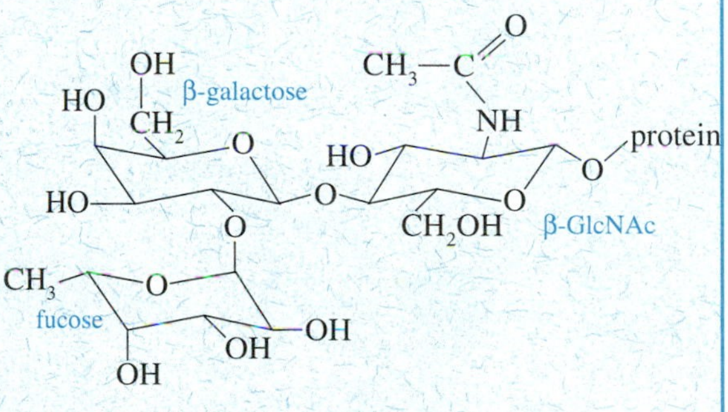

The structure of glycogen is similar to that of amylopectin, but glycogen has more branches, and the branches are shorter than those of amylopectin. The average chain length in glycogen is 12 glucose units. Glycogen has a molecular weight greater than 3 million. Glycogen is synthesized by animals as a storage form of glucose. Although glycogen is found throughout the body, the largest amounts are in the liver. An average human adult has enough glycogen for about 15 hours of normal activity.

Summary of Reactions

1. Reduction of Monosaccharides (Section 11.5)

$$\begin{array}{c} CHO \\ H-\!\!\!-OH \\ H-\!\!\!-OH \\ CH_2OH \end{array} \xrightarrow{NaBH_4} \begin{array}{c} CH_2OH \\ H-\!\!\!-OH \\ H-\!\!\!-OH \\ CH_2OH \end{array}$$

2. Oxidation of Monosaccharides (Section 11.6)

$$\begin{array}{c} CHO \\ H-\!\!\!-OH \\ H-\!\!\!-OH \\ CH_2OH \end{array} \xrightarrow{Ag(NH_3)_2^+} \begin{array}{c} CO_2H \\ H-\!\!\!-OH \\ H-\!\!\!-OH \\ CH_2OH \end{array}$$

$$\begin{array}{c} CHO \\ H-\!\!\!-OH \\ HO-\!\!\!-H \\ H-\!\!\!-OH \\ CH_2OH \end{array} \xrightarrow{HNO_3} \begin{array}{c} CO_2H \\ H-\!\!\!-OH \\ HO-\!\!\!-H \\ H-\!\!\!-OH \\ CO_2H \end{array}$$

3. Isomerization of Monosaccharides (Section 11.2)

$$\begin{array}{c} CHO \\ H-\!\!\!-OH \\ H-\!\!\!-OH \\ H-\!\!\!-OH \\ CH_2OH \end{array} \rightleftharpoons \begin{array}{c} CH_2OH \\ =O \\ H-\!\!\!-OH \\ H-\!\!\!-OH \\ CH_2OH \end{array} \rightleftharpoons \begin{array}{c} CHO \\ HO-\!\!\!-H \\ H-\!\!\!-OH \\ H-\!\!\!-OH \\ CH_2OH \end{array}$$

4. Formation of Glycosides (Section 11.7)

$$\text{glucopyranose} \underset{H^+}{\overset{CH_3CH_2OH}{\rightleftharpoons}} \text{ethyl glucoside } (\beta, \text{ anomeric } OCH_2CH_3 \text{ equatorial}) + \text{ethyl glucoside } (\alpha, \text{ anomeric } OCH_2CH_3 \text{ axial})$$

Explorations with Molecular Models

1. Construct a ball-and-stick model corresponding to the following space-filling model of one of the enantiomers of glyceraldehyde. Arrange the chain of atoms in the orientation required to generate a Fischer projection formula. Does the structure correspond to D- or L-glyceraldehyde?

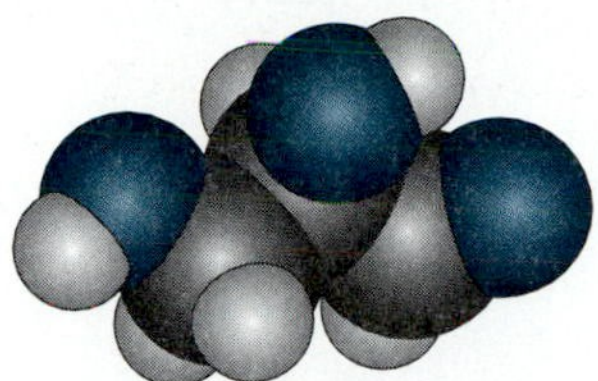

2. Construct a ball-and-stick model corresponding to the following space-filling model. Rotate the structure about the C-2 to C-3 bond to obtain a conformation that has eclipsed carbon atoms. Arrange this conformation in the orientation required to generate a Fischer projection formula. Does the compound have the D or L configuration? Write the Fischer projection formula. Name the compound.

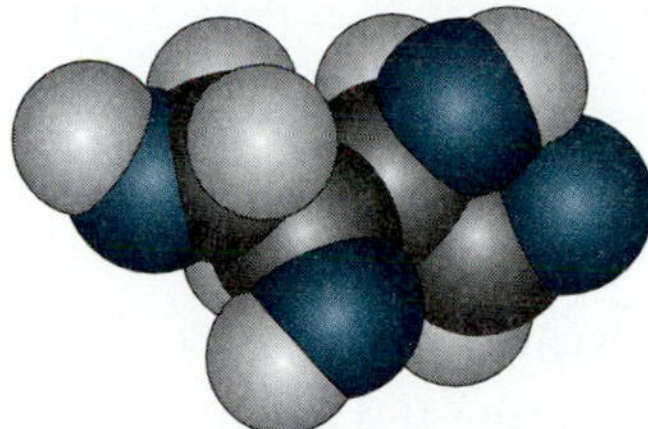

3. Construct the following model of a monosaccharide in its cyclic form. Is the configuration at the anomeric center α or β? Identify the monosaccharide.

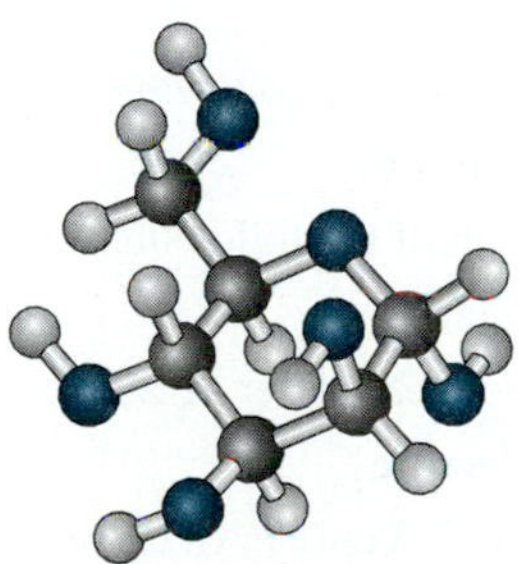

4. Construct the following model of a glycoside of a monosaccharide. Rotate the entire structure to place it in the conventional orientation required for writing a Haworth projection formula. Is the glycoside an acetal or a ketal? What is the configuration at the glycosidic center? Identify the monosaccharide that formed the glycoside.

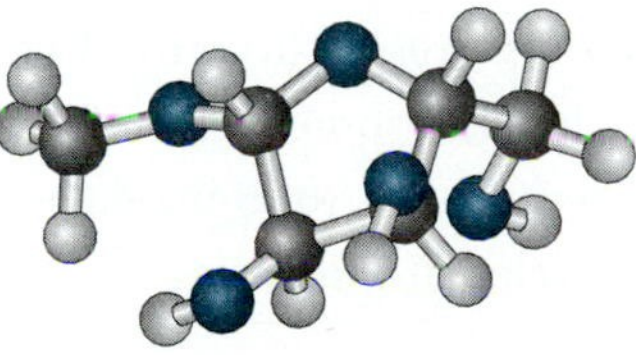

Exercises

Classification of Monosaccharides

11.1 What is an aldose? How does it differ from a ketose?

11.2 To what carbon atom do the letters D and L refer in monosaccharides?

11.3 Classify each of the following monosaccharides based on the number of carbon atoms and the carbonyl group.

(a)
CHO
HO—H
HO—H
H—OH
CH_2OH

(b)
CH_2OH
=O
HO—H
H—OH
HO—H
CH_2OH

(c)
CH_2OH
=O
HO—H
HO—H
CH_2OH

(d)
CHO
HO—H
HO—H
H—OH
H—OH
CH_2OH

11.4 Classify each of the following monosaccharides based on the number of carbon atoms and the carbonyl group.

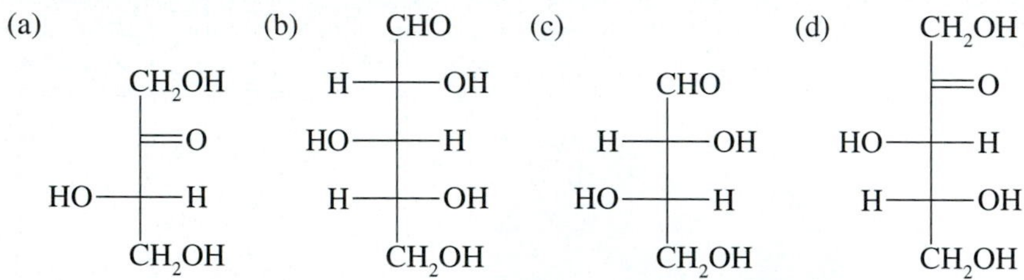

11.5 Classify each of the monosaccharides in Exercise 11.3 as D or L.

11.6 Classify each of the monosaccharides in Exercise 11.4 as D or L.

11.7 Draw the Fischer projection formulas of the isomeric D-ketopentoses with the carbonyl group at the C-2 atom.

11.8 Draw the Fischer projection formulas of the isomeric 3-ketopentoses. Indicate whether each isomer is chiral or achiral.

11.9 Draw the Fischer projection formula of each of the following.
(a) L-xylose (b) L-erythrose (c) L-galactose

11.10 Draw the Fischer projection formula of each of the following.
(a) L-ribose (b) L-threose (c) L-mannose

11.11 Draw the Fischer projection formula of the C-2 epimer of each of the following.
(a) D-xylose (b) D-erythrose (c) D-galactose

11.12 Draw the Fischer projection formula of the C-3 epimer of each of the following.
(a) D-ribose (b) D-glucose (c) D-mannose

Haworth Projection Formulas

11.13 Draw the Haworth projection formula of the pyranose form of each of the following compounds.
(a) α-D-mannose (b) β-D-galactose (c) α-D-glucose (d) α-D-galactose

11.14 Draw the Haworth projection formula of the furanose form of each of the following compounds.
(a) α-D-fructose (b) β-D-fructose (c) α-D-ribulose (d) β-D-xylulose

11.15 Identify the monosaccharide represented by each of the following structures.

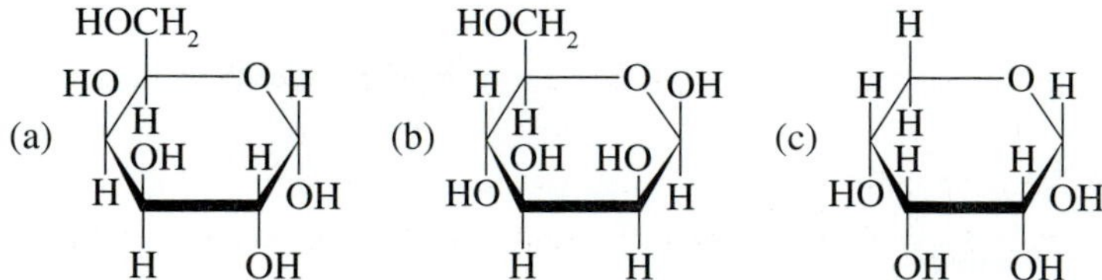

11.16 Identify the monosaccharide represented by each of the following structures.

(a) 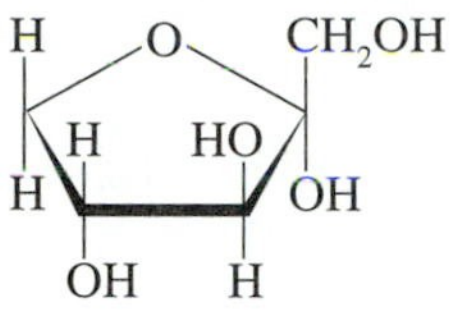 (b) 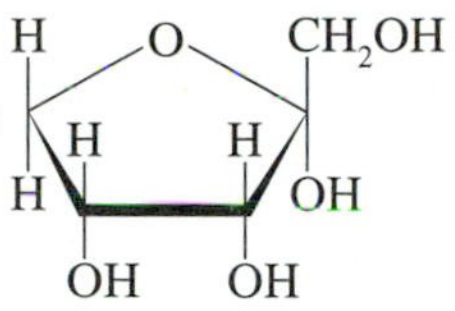 (c)

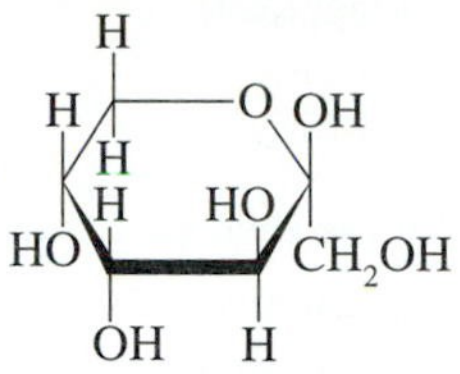

11.17 Name each compound in Exercise 11.15, indicating the type of ring and configuration of the anomeric center.

11.18 Name each compound in Exercise 11.16, indicating the type of ring and configuration of the anomeric center.

Conformations of Monosaccharides

11.19 Draw the chair conformations of β-galactopyranose and β-mannopyranose and compare the number of axial hydroxyl groups in each compound.

11.20 Draw the chair conformations of β-talopyranose and β-allopyranose and compare the number of axial hydroxyl groups in each compound.

Mutarotation

11.21 Can all aldopentoses mutarotate? Why?

11.22 Can L-glucose mutarotate?

11.23 Which of the following compounds can mutarotate?

(a) 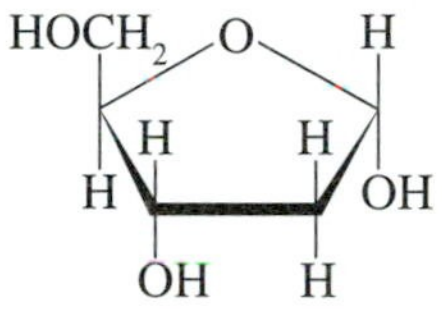 (b) 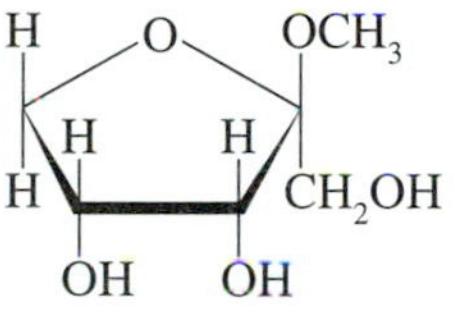 (c)

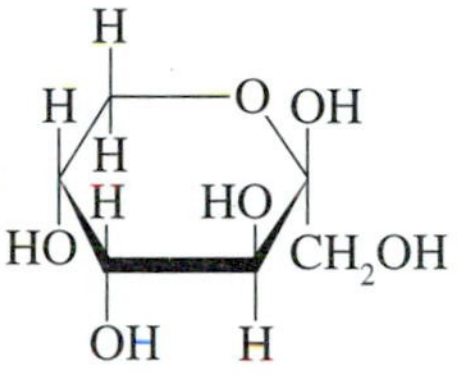

11.24 Which of the following compounds can mutarotate?

(a) 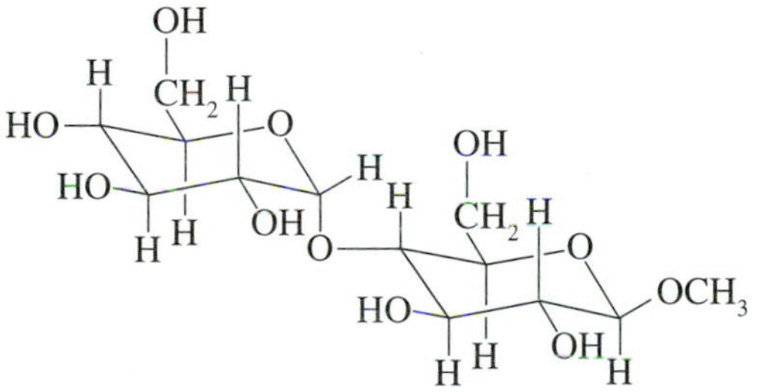 (b)

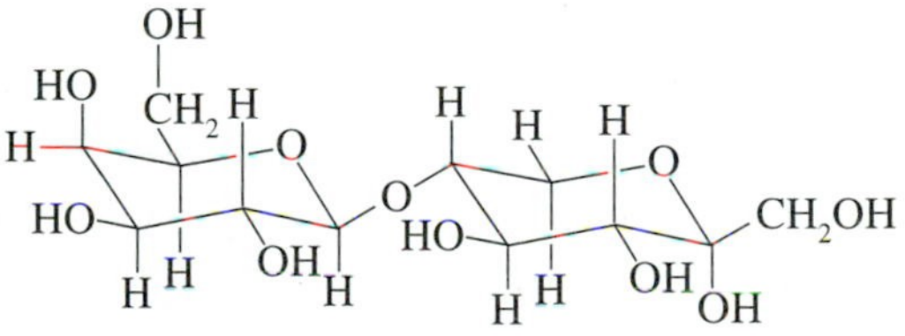

11.25 The $[\alpha]_D$ of the α- and β-anomers of D-galactose are +150.7 and +52.8, respectively. In water, mutarotation of D-galactose results in a specific rotation of +80.2. Which anomer predominates?

11.26 In solution, D-ribose forms an equilibrium mixture containing 6% α-furanose, 18% β-furanose, 20% α-pyranose, and 56% β-pyranose. Explain why the β-pyranose form predominates at equilibrium.

Reduction of Monosaccharides

11.27 Draw the Fischer projections of the alditols of D-erythrose and D-threose. One compound is optically active and the other is a meso compound. Explain why.

11.28 Which of the alditols of the D-aldopentoses are optically inactive? Explain why.

11.29 Reduction of D-fructose with sodium borohydride yields a mixture of two alditols. Explain why. Name the two alditols.

11.30 Reduction of D-tagatose with sodium borohydride yields a mixture of galactitol and talitol. Without refering to Figure 11.2, what is the structure of D-tagatose?

Oxidation of Monosaccharides

11.31 Draw the structures for each of the following compounds.

(a) D-mannonic acid (b) D-galactonic acid (c) D-ribonic acid

11.32 Draw the structures for each of the following compounds.

(a) D-allonic acid (b) D-talonic acid (c) D-xylonic acid

11.33 Oxidation of D-erythrose and D-threose with nitric acid yields aldaric acids, one of which is optically inactive. Which one? Explain why.

11.34 Which of the D-aldopentoses will yield optically inactive aldaric acids when oxidized with nitric acid?

Isomerization of Monosaccharides

11.35 Draw the structure of the aldose and ketose that can exist in equilibrium with D-allose in basic solution.

11.36 Draw the structure of two aldoses that can exist in equilibrium with D-xylulose in basic solution.

Glycosides

11.37 Draw the Haworth projection formulas of the two glycosides formed from the pyranose forms of mannose and methyl alcohol.

11.38 Draw the Haworth projection formulas of the two glycosides formed from the furanose forms of fructose and ethyl alcohol.

11.39 Vanillin is found as a β-glycoside of D-glucose. Draw the structure of the glycoside containing a pyranose ring.

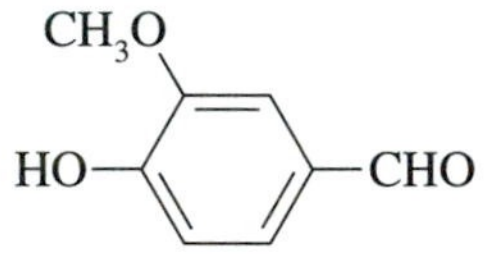

vanillin

11.40 Salicin is found in the bark of several species of fruit trees. What are the hydrolysis products of salicin?

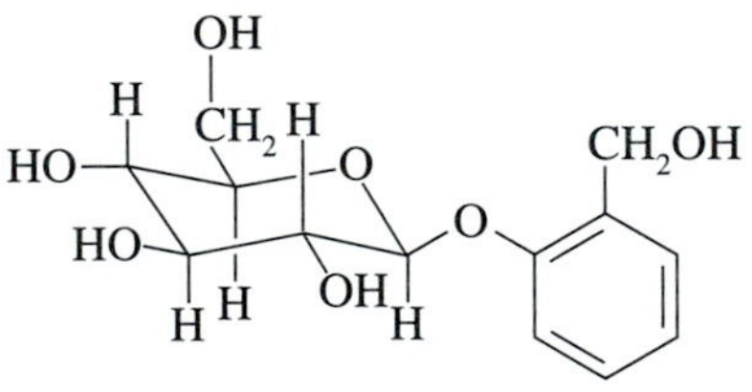

salicin

Disaccharides

11.41 Determine the component monosaccharides of each of the following compounds and describe the type of glycosidic linkage in each.

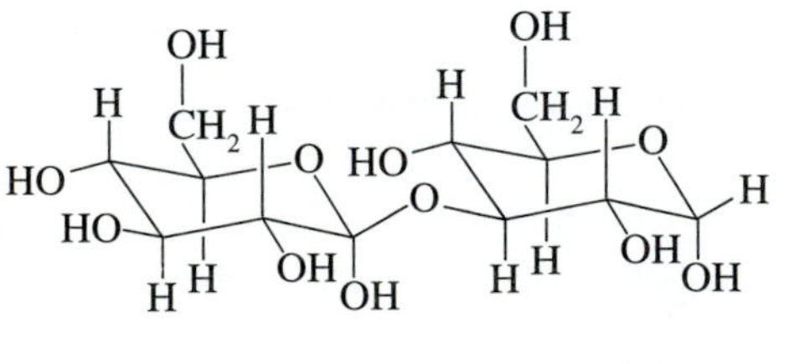

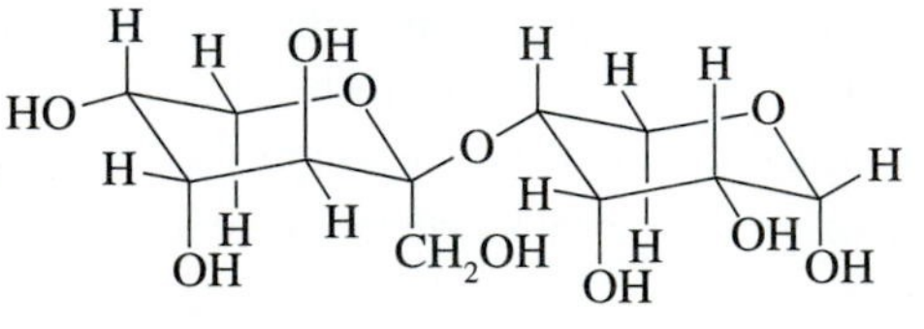

11.42 Determine the component monosaccharides of each of the following compounds and describe the type of glycosidic linkage in each.

(a)

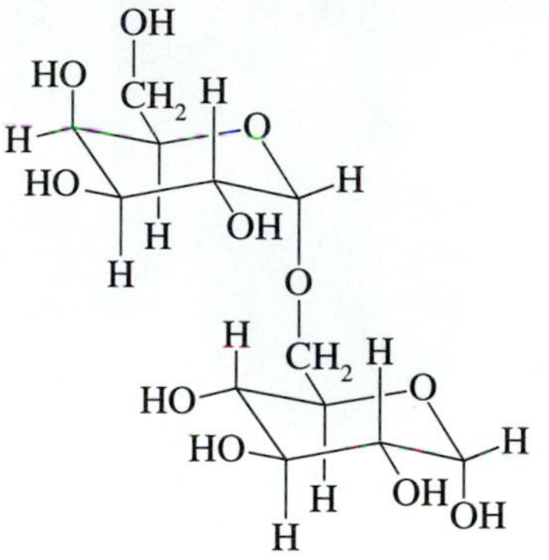

(b)

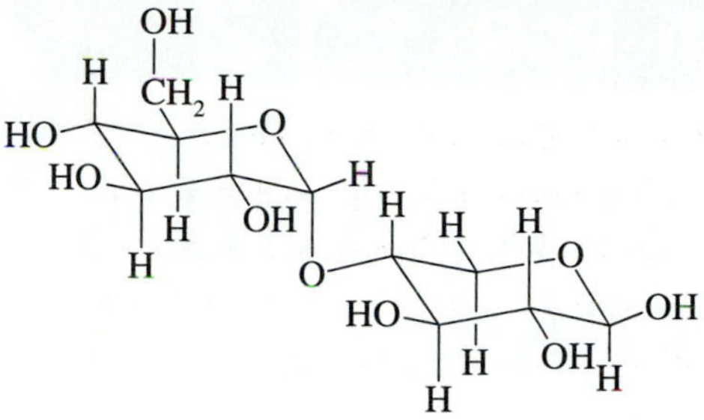

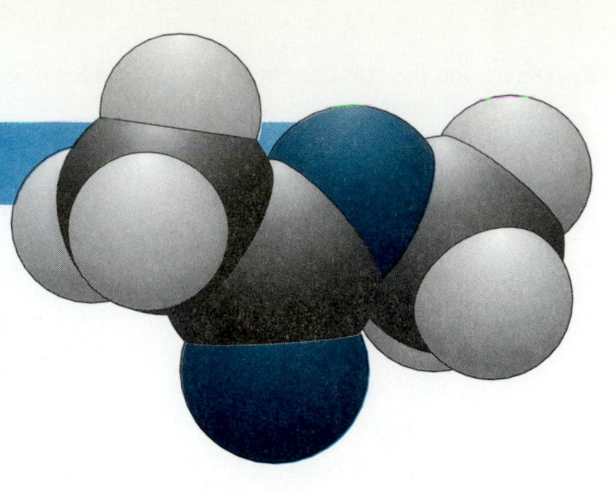

CHAPTER 12

CARBOXYLIC ACIDS AND ESTERS

Overview

12.1 The Carboxyl and Acyl Groups

In this chapter we consider the structure, properties, and reactions of carboxylic acids. These compounds have a **carboxyl group** bonded to a hydrocarbon unit, which may be saturated, unsaturated, aromatic, or heterocyclic.

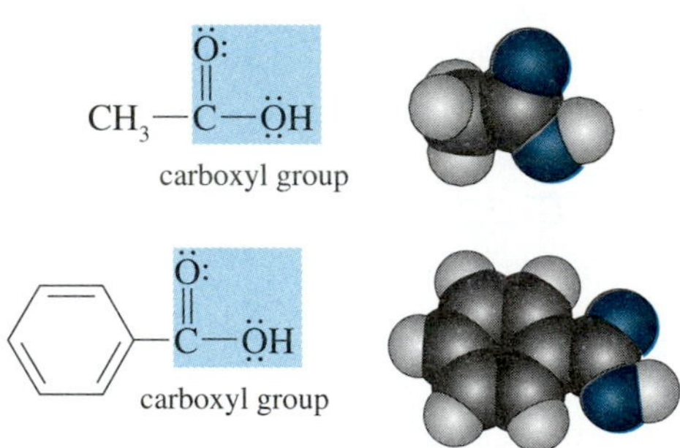

The carboxyl carbon atom is sp^2-hybridized, and three of its valence electrons form three σ bonds at 120° angles to one another (Figure 12.1). One of the σ bonds is to a hydrogen atom or a carbon atom of an alkyl, aromatic, or heterocyclic group. The other two σ bonds are to oxygen atoms: one to the hydroxyl oxygen atom and the other to the carbonyl oxygen atom. The carbonyl carbon atom of the carboxyl group also has one electron in a $2p$ orbital forming a π bond with an electron in a $2p$ orbital of the carbonyl oxygen atom.

Although the bond angles at the carboxyl carbon atom are all approximately 120°, the carboxyl group is often represented with only vertical and horizontal lines. To save space, two condensed representations of the carboxyl group are commonly used. Unless required to account for the mechanism of a reaction, the nonbonding electrons are not shown.

$$R{-}\overset{\displaystyle O}{\overset{\|}{C}}{-}OH \qquad R{-}COOH \qquad R{-}CO_2H$$

equivalent representations of a carboxylic acid

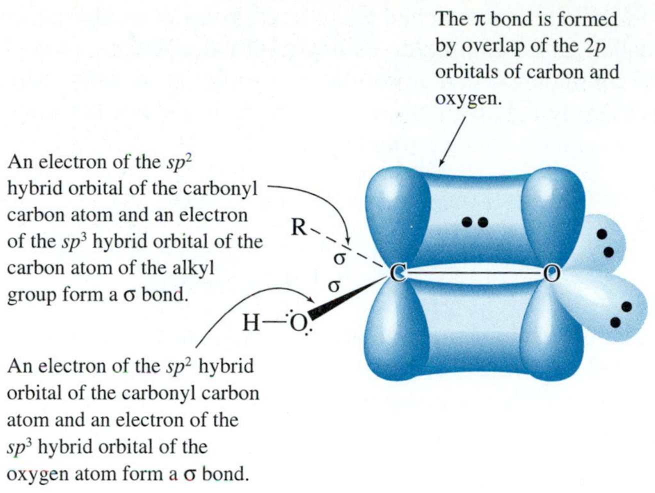

The π bond is formed by overlap of the 2*p* orbitals of carbon and oxygen.
An electron of the sp^2 hybrid orbital of the carbonyl carbon atom and an electron of the sp^3 hybrid orbital of the carbon atom of the alkyl group form a σ bond.
R
σ
σ
C
O
H—O
An electron of the sp^2 hybrid orbital of the carbonyl carbon atom and an electron of the sp^3 hybrid orbital of the oxygen atom form a σ bond.

FIGURE 12.1
Bonding in Carboxylic Acids

The Acyl Group and Acid Derivatives

The RCO unit contained in a carboxylic acid is called an **acyl group.** Several families of acid derivatives have oxygen-containing groups or electronegative atoms, such as nitrogen or a halogen, in place of the OH group of a carboxylic acid. This group, represented by L, can be replaced in a nucleophilic acyl substitution reaction (Section 12.6).

$$\mathrm{R{-}C({=}O){-}L + Nu{:}^- \longrightarrow R{-}C({=}O){-}Nu + L{:}^-}$$

If an alkoxy (—OR) or phenoxy (—OAr) group is bonded to the acyl group, the derivative is an **ester.** Esters are formed from a carboxylic acid and an alcohol in a condensation reaction (Section 12.8). Esters are moderately reactive toward water in hydrolysis reactions and produce a carboxylic acid and an alcohol (Section 12.9).

$$\mathrm{R{-}C({=}\ddot{O}{:}){-}\ddot{O}{-}R' \qquad R{-}C({=}\ddot{O}{:}){-}\ddot{O}{-}Ar \qquad Ar{-}C({=}\ddot{O}{:}){-}\ddot{O}{-}Ar}$$

general structures of esters

If the substituent is linked to the acyl group through a nitrogen atom, the compound is called an **amide.** The classification of amides depends on the number of carbon groups, including the acyl group, bonded to the nitrogen atom. These compounds are much less reactive than esters in hydrolysis reactions. The amide functional group is responsible for the structural stability of proteins. Amides are discussed in Chapter 14.

$$\mathrm{R{-}C({=}\ddot{O}{:}){-}\ddot{N}(H){-}H \qquad R{-}C({=}\ddot{O}{:}){-}\ddot{N}(H){-}R' \qquad R{-}C({=}\ddot{O}{:}){-}\ddot{N}(R'){-}R'}$$

a primary amide — a secondary amide — a tertiary amide

When the substituent attached to an acyl group is a chlorine atom, the derivative is called an **acid chloride** or **acyl chloride.** When two acyl groups are bonded to a common oxygen atom, the compound is an **acid anhydride.** Neither of these highly reactive compounds occurs in nature, but they are used in the laboratory synthesis of esters and amides.

$$\mathrm{R{-}\overset{\overset{\Large ..}{O}:}{\overset{\|}{C}}{-}\ddot{\underset{..}{Cl}}:}$$
an acid chloride

$$\mathrm{R{-}\overset{\overset{\Large ..}{O}:}{\overset{\|}{C}}{-}\ddot{\underset{..}{O}}{-}\overset{\overset{\Large ..}{O}:}{\overset{\|}{C}}{-}R}$$
an acid anhydride

When a substituent is linked to an acyl group through a sulfur atom, the derivative is called a **thioester.** Thioesters are less reactive than acid chlorides and acid anhydrides, but they are sufficiently reactive to participate in many biochemical acyl transfer reactions.

$$\mathrm{R{-}\overset{\overset{\Large ..}{O}:}{\overset{\|}{C}}{-}\ddot{\underset{..}{S}}{-}R'}$$
a thioester

Esters, amides, anhydrides, and thioesters may make up part of a cyclic structure. Cyclic esters are called **lactones.** Cyclic amides are called **lactams.** Cyclic acyl derivatives behave chemically like acyclic acyl compounds.

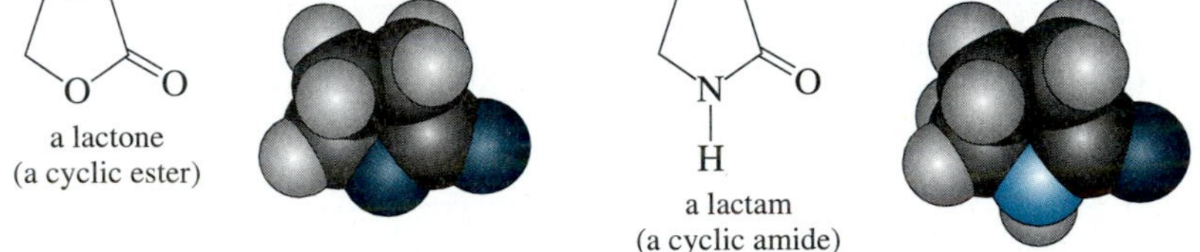

a lactone (a cyclic ester)

a lactam (a cyclic amide)

Runners may tire due to buildup of lactic acid (2-hydroxypropanoic acid) in the muscles.

12.2 Nomenclature of Carboxylic Acid Derivatives

Common Names

The carboxylic acids and their derivatives are abundant in nature and were among the first organic substances to be isolated. Because they have been known for so long, the common acids, esters, and other acyl compounds are often referred to by their common names. Some common acids are formic acid (HCO_2H), acetic acid (CH_3CO_2H), and benzoic acid ($C_6H_5CO_2H$). Both the common and the IUPAC names of a few commonly encountered carboxylic acids are given in Table 12.1.

In the common names, the positions of groups attached to the parent chain are designated alpha (α), beta (β), gamma (γ), delta (δ), and so forth. The —COOH group itself is not designated by a Greek letter.

$$\mathrm{\underset{\varepsilon}{CH_3}{-}\underset{\delta}{CH_2}{-}\underset{\gamma}{\overset{\overset{\textstyle Br}{|}}{CH}}{-}\underset{\beta}{\overset{\overset{\textstyle CH_2CH_3}{|}}{CH}}{-}\underset{\alpha}{CH_2}{-}CO_2H}$$
γ-bromo-β-ethylcaproic acid

Table 12.1 Nomenclature of Carboxylic Acids

Formula	Common Name	IUPAC Name
HCO_2H	formic acid	methanoic acid
CH_3CO_2H	acetic acid	ethanoic acid
$CH_3CH_2CO_2H$	propionic acid	propanoic acid
$CH_3(CH_2)_2CO_2H$	butyric acid	butanoic acid
$CH_3(CH_2)_3CO_2H$	valeric acid	pentanoic acid
$CH_3(CH_2)_4CO_2H$	caproic acid	hexanoic acid
$CH_3(CH_2)_6CO_2H$	caprylic acid	octanoic acid
$CH_3(CH_2)_8CO_2H$	capric acid	decanoic acid
$CH_3(CH_2)_{10}CO_2H$	lauric acid	dodecanoic acid
$CH_3(CH_2)_{12}CO_2H$	myristic acid	tetradecanoic acid
$CH_3(CH_2)_{14}CO_2H$	palmitic acid	hexadecanoic acid
$CH_3(CH_2)_{16}CO_2H$	stearic acid	octadecanoic acid

Some unbranched carboxylic acids contain a —COOH group at each end of the chain. The common names and IUPAC names of some of these dicarboxylic acids are listed in Table 12.2.

IUPAC Names

The IUPAC rules to name carboxylic acids are similar to those outlined for aldehydes. The final *-e* of the parent hydrocarbon is replaced by the ending *-oic acid.* The carboxyl group has a higher priority than aldehyde, ketone, halogen, hydroxyl, and alkoxy groups. The priority order for carbonyl groups is carboxylic acid > aldehyde > ketone, and the carbonyl group of an aldehyde or ketone is indicated by the prefix *oxo-*. Compounds that have a $—CO_2H$ group bonded to a cycloalkane ring are named as derivatives of the cycloalkane, and the suffix *-carboxylic acid* is added. The ring carbon atom to which the carboxyl carbon atom is bonded is assigned the number 1, but this number is not

Table 12.2 Nomenclature of Dicarboxylic Acids

Structural Formula	IUPAC Name	Common Name
$HO_2C—CO_2H$	ethanedioic acid	oxalic acid
$HO_2C—CH_2—CO_2H$	propanedioic acid	malonic acid
$HO_2C—(CH_2)_2—CO_2H$	butanedioic acid	succinic acid
$HO_2C—(CH_2)_3—CO_2H$	pentanedioic acid	glutaric acid
$HO_2C—(CH_2)_4—CO_2H$	hexanedioic acid	adipic acid

included in the name. The following examples illustrate the proper use of IUPAC rules in naming carboxylic acids.

$CH_3-CH(OH)-CH(CH_3)-C(=O)-OH$

3-hydroxy-2-methylbutanoic acid

$CH_3-CH(CH_3)-C\equiv C-C(=O)-OH$

4-methyl-2-pentynoic acid

$CH_3-C(=O)-CH(CH_3)-C(=O)-OH$

2-methyl-3-oxobutanoic acid

cis-2-chlorocyclohexanecarboxylic acid

Bread products often contain salts such as sodium propionate which act as preservatives.

Names of Carboxylic Acid Derivatives

The name of the conjugate base of a carboxylic acid, a **carboxylate** anion, is obtained by changing the *-oic acid* ending to *-oate.* For a salt of a carboxylic acid, the name of the carboxylate anion is preceded by the name of the metal ion.

$CH_3CH_2CH_2-C(=O)-O^- Na^+$

sodium butanoate
(sodium butyrate)

$C_6H_5-CH_2CH_2-C(=O)-O^- K^+$

potassium 3-phenylpropanoate
(potassium β-phenylpropionate)

An ester is named by first writing the name of the alkyl or aryl group bonded to the bridging oxygen atom, followed by the name of the acyl portion of the ester, which is derived from a carboxylic acid and named as a carboxylate.

$CH_3CH_2CH_2-C(=O)-O-CH_2CH_3$

ethyl butanoate
(ethyl butyrate)

$C_6H_{11}-CH_2CH_2-C(=O)-O-CH_2CH_2CH_3$

propyl 3-cyclohexylpropanoate
(propyl β-cyclohexylpropionate)

An acid halide is named by changing the ending *-oic acid* of the carboxylic acid to the ending *-oyl halide,* where the name of the halide is fluoride, chloride, bromide, or iodide. An acid halide functional group bonded to a cycloalkane ring is named as a *carbonyl halide.*

$CH_3CH(CH_3)CH_2-C(=O)-Br$

3-methylbutanoyl bromide

$CH_3-C_6H_{10}-C(=O)-Cl$

trans-4-methylcyclohexanecarbonyl chloride

An acid anhydride consists of two acyl groups bonded through a bridging oxygen atom. Although acid anhydrides can have two different acyl groups, compounds containing identical acyl groups are more common. They are named by replacing the suffix *-oic acid* with *-oic anhydride.*

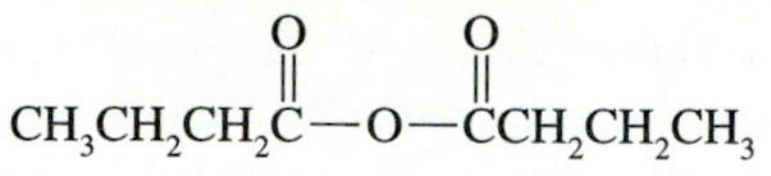

$CH_3CH_2CH_2C(=O)-O-C(=O)CH_2CH_2CH_3$

butanoic anhydride

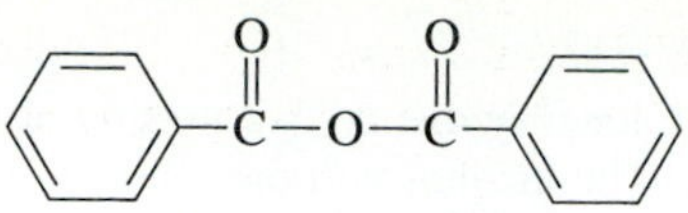

$C_6H_5-C(=O)-O-C(=O)-C_6H_5$

benzoic anhydride

Amides are named by replacing the suffix for the acid (*-oic acid*) with the suffix *-amide.* An amide functional group bonded to a cycloalkane ring is named as a carboxamide.

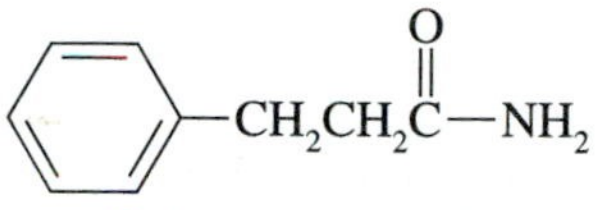

$C_6H_5-CH_2CH_2C(=O)-NH_2$

3-phenylpropanamide

$C_6H_{11}-C(=O)-NH_2$

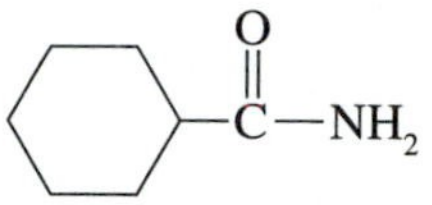

cyclohexanecarboxamide

In secondary and tertiary amides, the nitrogen atom is bonded to one or more alkyl or aryl groups instead of hydrogen atoms. The names of amides with groups bonded to nitrogen will be discussed along with names of amines in Chapter 14.

EXAMPLE 12-1

The structure of oleic acid, an unsaturated carboxylic acid present as an ester in vegetable oils, is shown here. What is the IUPAC name of oleic acid?

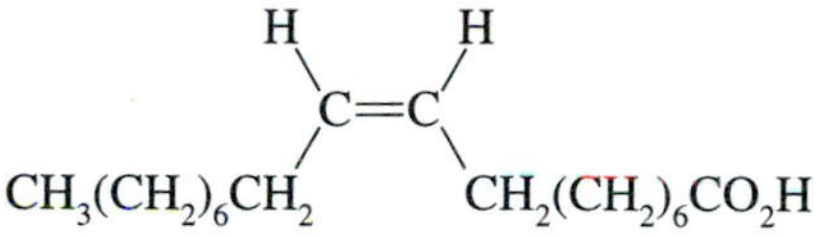

H H
C=C
$CH_3(CH_2)_6CH_2$ $CH_2(CH_2)_6CO_2H$

Solution

First, we determine the length of the longest continuous chain that contains the —COOH group: it contains 18 carbon atoms. The parent name in the IUPAC system is octadecanoic acid. The double bond is located at carbon atom 9 in the chain, numbering from the carboxyl group on the right. Thus, the compound is a 9-octadecenoic acid. The configuration about the double bond is Z, so the complete name is (*Z*)-9-octadecenoic acid.

Problem 12.1

Mevalonic acid is required to form isopentenyl pyrophosphate in terpene biosynthesis. Assign its IUPAC name.

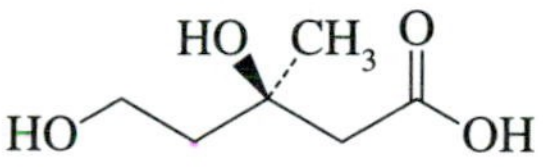

HO CH₃ O
HO OH

EXAMPLE 12-2

Assign the IUPAC name to clofibrate, a drug used to lower the concentration of blood triacylglycerols and cholesterol.

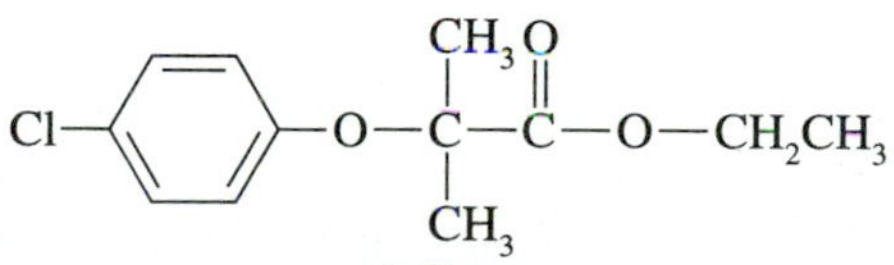

$Cl-C_6H_4-O-C(CH_3)_2-C(=O)-O-CH_2CH_3$

clofibrate

Solution

First, identify the alcohol portion of the ester; it is located at the right of the molecule. The alcohol portion contains two carbon atoms, so the compound is an ethyl ester.

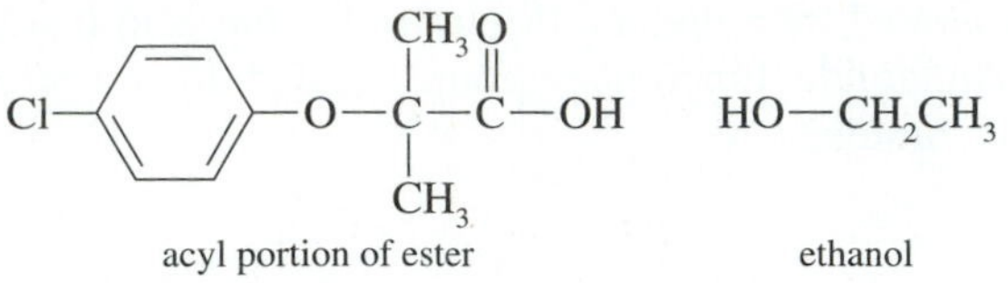

$$\text{Cl}-\text{C}_6\text{H}_4-\text{O}-\text{C}(\text{CH}_3)_2-\overset{\text{O}}{\overset{\|}{\text{C}}}-\text{OH} \qquad \text{HO}-\text{CH}_2\text{CH}_3$$

acyl portion of ester — ethanol

The acyl portion is a substituted propanoic acid with a methyl group and an aryl-containing group at the C-2 atom. Imagine removing the aryl-containing group from the acid and adding a hydrogen atom to its oxygen atom. The resulting compound is *p*-chlorophenol. The original group is *p*-chlorophenoxy.

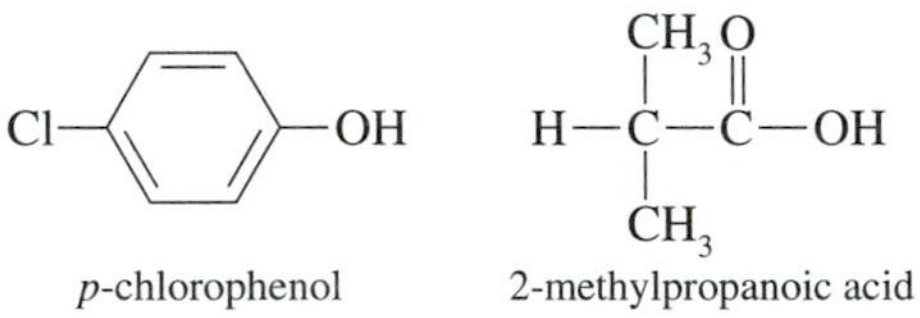

$$\text{Cl}-\text{C}_6\text{H}_4-\text{OH} \qquad \text{H}-\text{C}(\text{CH}_3)_2-\overset{\text{O}}{\overset{\|}{\text{C}}}-\text{OH}$$

p-chlorophenol — 2-methylpropanoic acid

The name of the acid is 2-(*p*-chlorophenoxy)-2-methylpropanoic acid. Now change the *-oic* ending of the acid to *-oate* and write the name of the alkyl group of the alcohol as a separate word in front of the modified acid name. The ester is named ethyl 2-(*p*-chlorophenoxy)-2-methylpropanoate.

Problem 12.2

Isobutyl formate has the odor of raspberries. Based on this common name, draw its structural formula. What is the IUPAC name?

12.3 Physical Properties of Carboxylic Acids and Esters

The physical properties of carboxylic acids and esters, such as boiling point and solubility, are distinctly different because of the different types of intermolecular interactions that are possible in each class of compounds. Differences in biological properties—such as odor—that depend on physiological responses are not well understood.

Boiling Points and Solubilities

The boiling points of carboxylic acids are high (Table 12.3). Carboxylic acids interact very strongly by forming hydrogen-bonded dimers. These dimers have higher boiling points than substances of comparable molecular weight that do not form dimers.

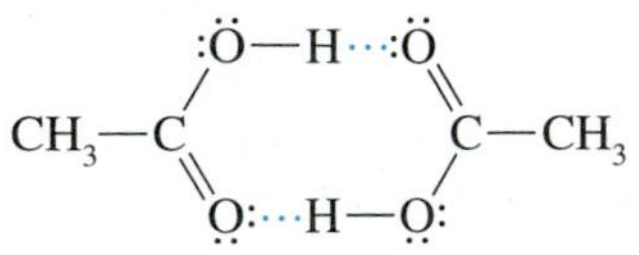

$$\text{CH}_3-\text{C}(=\ddot{\text{O}}\text{:}\cdots\text{H}-\ddot{\text{O}}\text{:})-\ddot{\text{O}}-\text{H}\cdots\text{:}\ddot{\text{O}}=\text{C}-\text{CH}_3$$

Industrial Synthesis of Acetic Acid

The production of acetic acid by chemical industry in 1995 was 4.7 billion pounds. This amount ranked acetic acid thirty-third overall for all chemicals produced and twenty-first among organic compounds. For almost 50 years, starting in 1915, the commercial synthesis of acetic acid was based on acetylene as a "feedstock." This substance was produced by reaction of water with calcium carbide, which had to be prepared by heating calcium oxide with coke at 2500°C in an electric furnace.

$$CaO + 3\,C \xrightarrow{2500°C} CaC_2 + CO$$

$$CaC_2 + 2\,H_2O \longrightarrow H{-}C{\equiv}C{-}H + Ca(OH)_2$$

The conversion of acetylene into acetic acid involved two steps, a hydration reaction followed by an oxidation reaction. Hydration of acetylene using mercuric sulfate and sulfuric acid gives acetaldehyde via an enol intermediate (Section 4.10). Oxidation of acetaldehyde by oxygen gas using cobalt(III) acetate as a catalyst gives acetic acid.

$$H_2O + H{-}C{\equiv}C{-}H \xrightarrow[H_2SO_4]{HgSO_4} CH_3{-}\overset{O}{\overset{\|}{C}}{-}H$$

$$CH_3{-}\overset{O}{\overset{\|}{C}}{-}H \xrightarrow[Co^{3+}]{O_2} CH_3{-}\overset{O}{\overset{\|}{C}}{-}OH$$

Large-scale industrial syntheses are developed using low-cost feedstocks in efficient reactions that have low energy requirements. Although the reactants in the described synthesis of acetic acid are relatively inexpensive and the reactions occur in high yield, the process requires large amounts of energy. Over a half-century, the cost of energy increased so much that the process eventually became uneconomical.

Ethylene is inexpensive and is readily available from the refining of petroleum and natural gas. Its production ranked fifth overall in 1995 among all industrial chemicals and first among organic compounds. Chemical engineers devised a process with much lower energy requirements that catalytically oxidizes ethylene to acetaldehyde using molecular oxygen and a catalyst of palladium(II) and copper(II). This process, which became the primary source of acetaldehyde, was devised by Wacker-Chemie of Germany in 1959 and is known as the **Wacker process.**

$$O_2 + 2\,CH_2{=}CH_2 \xrightarrow[Cu^{2+}]{Pd^{2+}} 2\,CH_3{-}\overset{O}{\overset{\|}{C}}{-}H$$

Subsequent catalytic oxidation of acetaldehyde using molecular oxygen with cobalt(III) acetate as a catalyst is used to produce acetic acid.

This process dominated the market for only 15 years, because in 1973 the Monsanto process was developed using methanol as a feedstock and reacting it with carbon monoxide. The formation of the carbon-carbon bond is catalyzed by rhodium(III), HI, and water.

$$CO + CH_3OH \xrightarrow[HI,\,H_2O]{Rh^{3+}} CH_3{-}\overset{O}{\overset{\|}{C}}{-}OH$$

Methanol ranked twenty-first among the commercial chemicals produced and tenth among organic compounds in 1995 when 11.3 billions pounds were produced. Using various catalysts, methane and water react to produce hydrogen and carbon monoxide in a 3 : 1 molar ratio, known as water gas.

$$CH_4 + H_2O \longrightarrow CO + 3\,H_2$$

The components of the water gas mixture can be adjusted to a 2 : 1 mixture of hydrogen gas to carbon monoxide, known as synthesis gas, and that mixture is converted to methanol in a catalyzed process.

$$CO + 2\,H_2 \longrightarrow CH_3OH$$

As a consequence, the water gas mixture is the source of both compounds required in the Monsanto process. Thus, methane, obtained from natural gas, and water are the starting materials for the synthesis of acetic acid. Catalysts have been developed for all of the necessary reactions to make the various steps efficient and economical. ■

Table 12.3 Physical Properties of Carboxylic Acids

IUPAC Name	Melting Point (°C)	Boiling Point (°C)	Solubility (g/100 g H_2O, at 20°C)
methanoic acid		101	miscible
ethanoic acid		118	miscible
propanoic acid		141	miscible
butanoic acid		164	miscible
pentanoic acid		186	4.97
hexanoic acid		205	0.96
octanoic acid		239	0.068
decanoic acid	32	270	0.015
dodecanoic acid	44	299	0.0055

Carboxylic acids with low molecular weights are soluble in water because the carboxyl group forms several hydrogen bonds with water. A carboxylic acid serves both as a hydrogen bond donor through its hydroxyl hydrogen atom and as a hydrogen bond acceptor through the lone pair electrons of both oxygen atoms. The solubility of carboxylic acids, like that of alcohols, decreases with increasing chain length because the long, nonpolar hydrocarbon chain dominates the physical properties of the acid.

Esters are polar molecules, but their boiling points are lower than those of carboxylic acids and alcohols of similar molecular weight because there is no intermolecular hydrogen bonding between ester molecules.

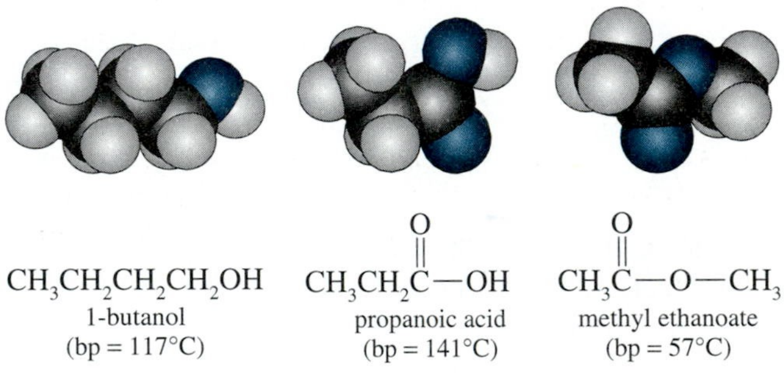

The characteristic fruity smell of preserves is due to esters.

Esters can form hydrogen bonds through their oxygen atoms to the hydrogen atoms of water molecules. Thus, esters are slightly soluble in water. However, because esters do not have a hydrogen atom to form a hydrogen bond to an oxygen atom of water, they are less soluble than carboxylic acids. The solubility and boiling points of some esters are listed in Table 12.4.

Odors of Acids and Esters

Liquid carboxylic acids have sharp, unpleasant odors. For example, butanoic acid occurs in rancid butter and aged cheese. Caproic, caprylic, and capric acids have the smell of goats. (The Latin word for goat, *caper,* is the source of the common names of these acids.)

Table 12.4 Physical Properties of Esters		
Name	**Boiling Point (°C)**	**Solubility (g/100 g H_2O)**
methyl methanoate	32	miscible
methyl ethanoate	57	24.4
methyl propanoate	80	1.8
methyl butanoate	102	0.5
methyl pentanoate	126	0.2
methyl hexanoate	151	0.06
ethyl methanoate	54	miscible
ethyl ethanoate	77	7.4
ethyl propanoate	99	1.7
ethyl butanoate	120	0.5
ethyl pentanoate	145	0.2
propyl ethanoate	102	1.9
butyl ethanoate	125	1.0
methyl benzoate	199	0.1
ethyl benzoate	213	0.08

In contrast to carboxylic acids, esters have pleasant fruity smells. In fact, the odors of many fruits are due to esters. For example, ethyl ethanoate is found in pineapples, 3-methylbutyl ethanoate in apples and bananas, 3-methylbutyl 3-methylbutanoate in apples, and octyl ethanoate in oranges.

Esters have low boiling points, and some fraction of them is driven off in the preparation of processed foods. Thus, esters are added back to the food to make them smell "natural." Nevertheless, government regulations require that the added esters be identified as additives on the label. The esters used in some products need not be the same as those in natural fruits, but of course they are selected to produce the same odor or taste. The choice of esters may be dictated by their cost and availability. Some of these flavoring agents are listed in Table 12.5.

Table 12.5 Esters Used as Flavoring Agents		
Name	**Formula**	**Flavor**
methyl butanoate	$CH_3CH_2CH_2CO_2CH_3$	apple
pentyl butanoate	$CH_3CH_2CH_2CO_2CH_2(CH_2)_3CH_3$	apricot
pentyl ethanoate	$CH_3CO_2CH_2(CH_2)_3CH_3$	bananna
octyl ethanoate	$CH_3CO_2CH_2(CH_2)_6CH_3$	orange
ethyl butanoate	$CH_3CH_2CH_2CO_2CH_2CH_3$	pinapple
ethyl methanoate	$HCO_2CH_2CH_3$	rum

12.4 Acidity of Carboxylic Acids

Although acetic acid and other carboxylic acids are weak acids, they are much more acidic than alcohols or phenols. The K_a for acetic acid is about 10^{11} times larger than the K_a for ethanol.

$$CH_3CH_2OH + H_2O \rightleftharpoons CH_3CH_2O^- + H_3O^+ \qquad K_a = 10^{-16}$$
$$CH_3CO_2H + H_2O \rightleftharpoons CH_3CO_2^- + H_3O^+ \qquad K_a = 1.8 \times 10^{-5}$$

The greater acidity of acetic acid is the result of resonance stabilization of the negative charge in the conjugate base, acetate ion. In the ethoxide ion ($CH_3CH_2O^-$) the negative charge is concentrated on a single oxygen atom.

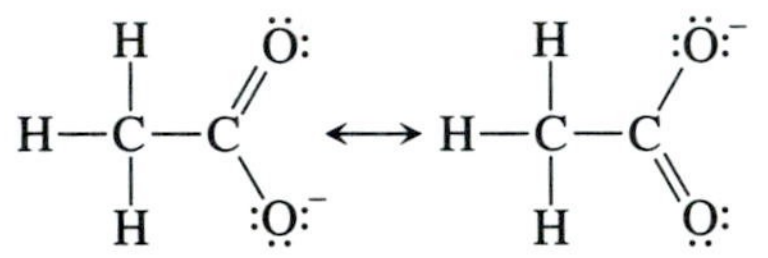

$$H_3C{-}C(=\ddot{O}{:}){-}\ddot{O}{:}^- \longleftrightarrow H_3C{-}C(-\ddot{O}{:}^-){=}\ddot{O}{:}$$

The acidity of carboxylic acids is also partly the result of an inductive effect (Section 2.8). That is, the carbonyl group polarizes the H—O bond by attracting electron density through the σ bonds. The withdrawal of electron density from the H—O bond weakens it and thus increases the acidity of the ionizable hydrogen atom.

The inductive effect of an alkyl or aryl group attached to the carbonyl carbon atom affects K_a. An alkyl group is electron-releasing with respect to hydrogen. This release of electron density to the carboxyl group stabilizes the acid and slightly destabilizes the conjugate base. Thus, acetic acid (pK_a = 4.74) is a weaker acid than formic acid (pK_a = 3.75). In contrast, the sp^2-hybridized carbon atom of an aryl group is electron-withdrawing relative to an alkyl group. Thus, benzoic acid (pK_a = 4.19) is a stronger acid than acetic acid.

Electron-withdrawing groups bonded to the α carbon atom increase the acidity of carboxylic acids. For example, chloroacetic acid is a stronger acid than acetic acid. Other examples of the effect of electron-withdrawing groups on the acidity of carboxylic acids are given in Table 12.6.

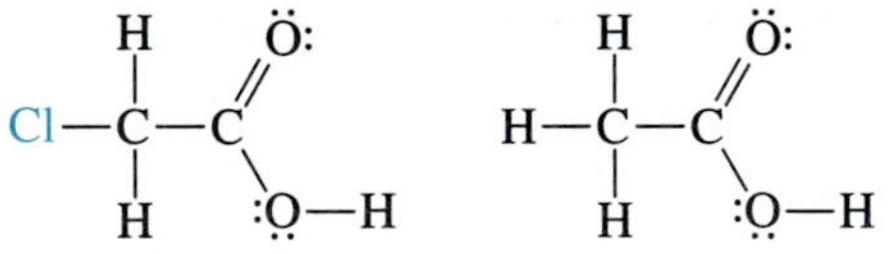

$$ClCH_2{-}C(=\ddot{O}{:}){-}\ddot{O}{-}H \qquad H_3C{-}C(=\ddot{O}{:}){-}\ddot{O}{-}H$$

pK_a = 2.9 pK_a = 4.7

The electrons in the C—Cl bond are "pulled" toward the more electronegative chlorine atom and away from the carbon skeleton. As a result, the electrons of the oxygen atom are drawn away from the O—H bond, and the proton can therefore ionize more easily.

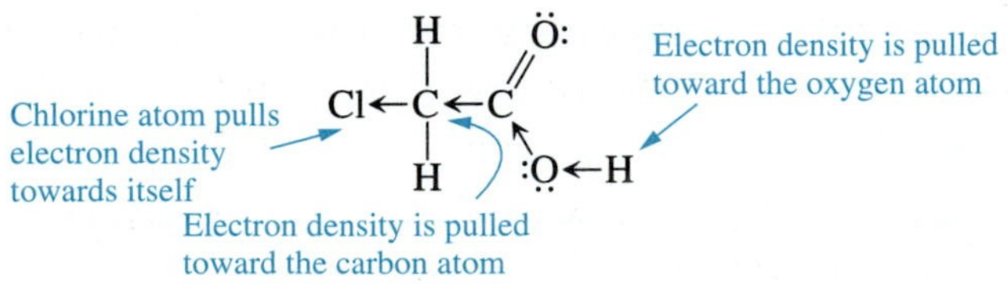

$$Cl \leftarrow CH_2 \leftarrow C(=\ddot{O}{:}) \leftarrow \ddot{O} \leftarrow H$$

Chlorine atom pulls electron density towards itself

Electron density is pulled toward the oxygen atom

Electron density is pulled toward the carbon atom

Table 12.6 pK_a Values of Carboxylic Acids

Name	Formula	pK_a
methanoic acid	HCO_2H	3.75
ethanoic acid	CH_3CO_2H	4.72
propanoic acid	$CH_3CH_2CO_2H$	4.87
butanoic acid	$CH_3(CH_2)_2CO_2H$	4.82
2-methylpropanoic acid	$(CH_3)_2CHCO_2H$	4.84
pentanoic acid	$CH_3(CH_2)_3CO_2H$	4.81
2,2-dimethylpropanoic acid	$(CH_3)_3CCO_2H$	5.03
fluoroethanoic acid	FCH_2CO_2H	2.59
chloroethanoic acid	$ClCH_2CO_2H$	2.86
bromoethanoic acid	$BrCH_2CO_2H$	2.90
iodoethanoic acid	ICH_2CO_2H	3.18
dichloroethanoic acid	Cl_2CHCO_2H	1.26
trichloroethanoic acid	Cl_3CCO_2H	0.64
trifluoroethanoic acid	F_3CCO_2H	0.23
methoxyethanoic acid	$CH_3OCH_2CO_2H$	3.55
cyanoethanoic acid	$NCCH_2CO_2H$	2.46
nitroethanoic acid	$NO_2CH_2CO_2H$	1.70

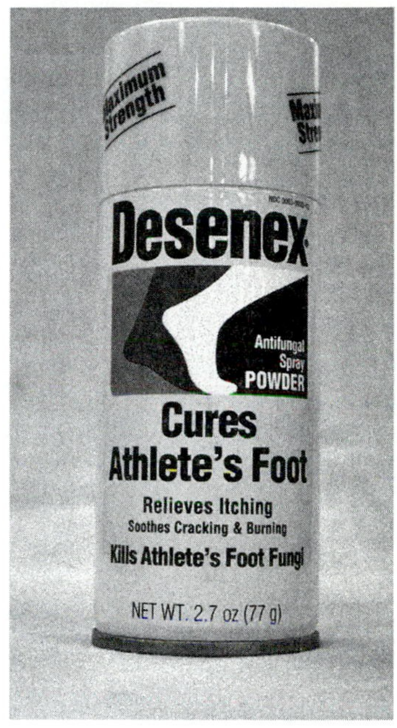

A salt of 10-undecenoic acid is the active ingredient in this antifungal product.

As the distance between the halogen atom and the carboxyl group increases, the inductive effect falls off dramatically. For β- and γ-substituted acids, the K_a values approach that of an unsubstituted carboxylic acid.

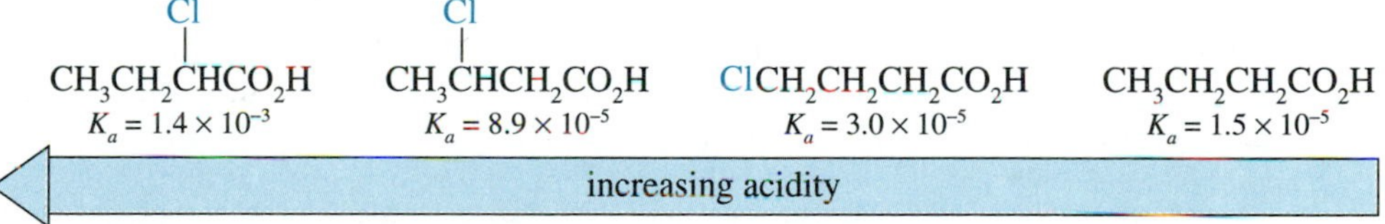

Salts of Carboxylic Acids

The reaction of carboxylic acids with hydroxide ions and the reaction of carboxylate salts with hydronium ions have some practical applications in separating carboxylic acids from mixtures. Because they are ionic, carboxylate salts are more soluble in water than their corresponding carboxylic acids. Carboxylic acids are often separated from other nonpolar organic compounds in the laboratory by adding a solution of sodium hydroxide to form the more soluble carboxylate salt. Consider, for example, a mixture of decanol and decanoic acid. Decanol is not soluble in water and does not react with sodium hydroxide. However, decanoic acid reacts with sodium hydroxide and thus dissolves in the basic solution, whereas the decanol remains undissolved.

$$\underset{\text{(insoluble in water)}}{CH_3(CH_2)_8CO_2H} + OH^- \longrightarrow \underset{\text{(soluble in water)}}{CH_3(CH_2)_8CO_2^-} + H_2O$$

Undissolved decanol is physically separated from the basic solution. Then HCl is added to neutralize the basic solution, and insoluble decanoic acid separates from the aqueous solution.

$$\underset{\text{(soluble in water)}}{CH_3(CH_2)_8CO_2^-} + H_3O^+ \longrightarrow \underset{\text{(insoluble in water)}}{CH_3(CH_2)_8CO_2H} + H_2O$$

This procedure is very useful in isolating acids from complex mixtures in nature. It is also used to purify acids produced by chemical synthesis in the laboratory.

EXAMPLE 12-3

Pyruvic acid is a key metabolic intermediate in oxidative processes that provide energy for the growth and maintenance of cells. Its pK_a is 2.5, indicating that it is about 100 times more acidic than propanoic acid (pK_a = 4.7). Explain why.

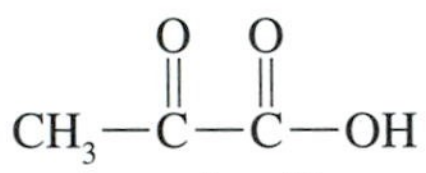

$$CH_3-\overset{\overset{\displaystyle O}{\|}}{C}-\overset{\overset{\displaystyle O}{\|}}{C}-OH$$

pyruvic acid

Solution

Because the pK_a of pyruvic acid is smaller than that of propanoic acid, pyruvic acid is a stronger acid. Thus the group attached to the carbonyl carbon atom pulls electron density away from the O—H bond. We recall that the carbonyl group is very polar and that the carbonyl carbon atom has a partial positive charge. As a consequence, the carbonyl carbon atom of the ketone group inductively pulls electron density from the carboxylic acid group, the polarity of the O—H bond increases, and the acidity increases.

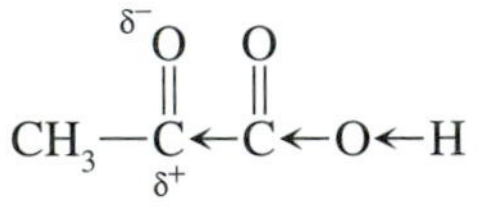

$$CH_3-\underset{\delta^+}{\overset{\overset{\delta^-\displaystyle O}{\|}}{C}}\leftarrow\overset{\overset{\displaystyle O}{\|}}{C}\leftarrow O\leftarrow H$$

Problem 12.3

The pK_a values for the dissociation of the first of the two carboxyl groups of malic acid and oxaloacetic acid are 3.41 and 1.70, respectively. Which compound is the stronger acid? Which of the two carboxyl groups in each compound is the more acidic?

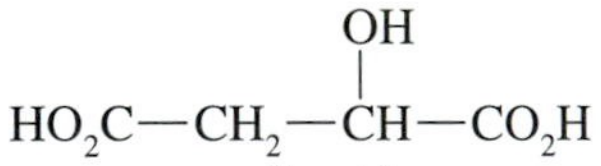

$$\underset{\text{malic acid}}{HO_2C-CH_2-\overset{\overset{\displaystyle OH}{|}}{C}H-CO_2H} \qquad \underset{\text{oxaloacetic acid}}{HO_2C-CH_2-\overset{\overset{\displaystyle O}{\|}}{C}-CO_2H}$$

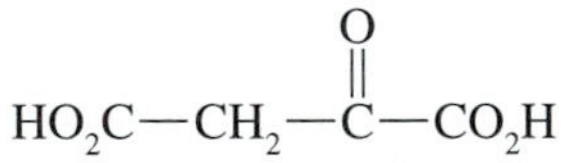

12.5 Synthesis of Carboxylic Acids

Carboxylic acids can be prepared by several oxidation reactions described in earlier chapters. Both aldehydes and alcohols are oxidized by Jones reagent (Section 8.7) to produce carboxylic acids. Aldehydes can be oxidized by Tollens's reagent or Fehling's solution (Section 10.4).

$$\underset{\text{3-methyl-1-hexanol}}{CH_3CH_2CH_2\overset{CH_3}{\overset{|}{C}}HCH_2CH_2OH} \xrightarrow[H_2SO_4/\text{acetone}]{CrO_3} \underset{\text{3-methylhexanoic acid}}{CH_3CH_2CH_2\overset{CH_3}{\overset{|}{C}}HCH_2CO_2H}$$

$$\text{3-cyclohexenecarbaldehyde (CHO)} \xrightarrow{Cu^{2+}} \text{3-cyclohexenecarboxylic acid } (CO_2H)$$

3-cyclohexenecarbaldehyde 3-cyclohexenecarboxylic acid

Alkylbenzenes are oxidized by potassium permanganate to give benzoic acids. The entire side chain is oxidized in this reaction (Section 5.9). The oxidation of tetralin to produce phthalic acid is an example.

$$\text{tetralin} \xrightarrow{KMnO_4} \text{phthalic acid } (CO_2H,\ CO_2H) + 2\ CO_2$$

tetralin phthalic acid

Carboxylic acids can also be made by two more general methods starting from haloalkanes. These methods provide structures that have one more carbon atom than the reactant haloalkane.

In Chapter 10 we saw that the Grignard reagent acts as a nucleophile, and that it reacts with the carbonyl group of aldehydes or ketones. A similar reaction occurs between a Grignard reagent and the carbon-oxygen double bond of carbon dioxide to yield the magnesium salt of a carboxylic acid. Adding aqueous acid to the solution of the conjugate base gives the carboxylic acid.

$$R{-}MgBr \quad :\ddot{O}{=}C{=}\ddot{O}: \longrightarrow R{-}\overset{\overset{\ddot{O}:}{\|}}{C}{-}\ddot{O}:^{-}\ {}^{+}MgBr \xrightarrow{H_3O^+} R{-}\overset{\overset{\ddot{O}:}{\|}}{C}{-}\ddot{O}{-}H$$

Starting from the haloalkane, the reaction sequence requires three steps. First, the haloalkane is converted to a Grignard reagent. Second, the ether solution is poured over solid carbon dioxide (dry ice). Finally, the reaction mixture is acidified. The reaction sequence can be presented using a single arrow with the three steps listed.

$$\text{o-bromotoluene (Br, CH}_3) \xrightarrow[2.\ CO_2\quad 3.\ H_3O^+]{1.\ Mg/\text{ether}} \text{o-methylbenzoic acid } (CO_2H,\ CH_3)$$

o-bromotoluene *o*-methylbenzoic acid

The second synthesis that adds one carbon atom to the parent chain of the reacting haloalkane involves an S_N2 displacement of halide ion by cyanide ion (Chapter 7). The resulting product, called a nitrile (RCN), can be hydrolyzed to produce a carboxylic acid.

$$R{-}\ddot{B}r: + {}^{-}:C{\equiv}N: \longrightarrow R{-}C{\equiv}N: + :\ddot{B}r:^{-}$$

$$R{-}C{\equiv}N: \xrightarrow{H_3O^+} R{-}\overset{\overset{\ddot{O}:}{\|}}{C}{-}\ddot{O}{-}H + NH_4^+$$

The reaction sequence can be presented using a single arrow with the two steps listed.

$$\underset{\text{1-bromo-2-methylpentane}}{CH_3CH_2CH_2\overset{\displaystyle CH_3}{\overset{|}{C}}HCH_2-\ddot{\underset{..}{Br}}:} \xrightarrow[\text{2. } H_3O^+]{\text{1. } CN^-/DMF} \underset{\text{3-methylhexanoic acid}}{CH_3CH_2CH_2\overset{\displaystyle CH_3}{\overset{|}{C}}HCH_2CO_2H}$$

We recall that substitution reactions of the S_N2 type are most effective with primary haloalkanes; elimination reactions decrease the yield for secondary haloalkanes.

EXAMPLE 12-4

Suggest a synthesis that accomplishes the following transformation.

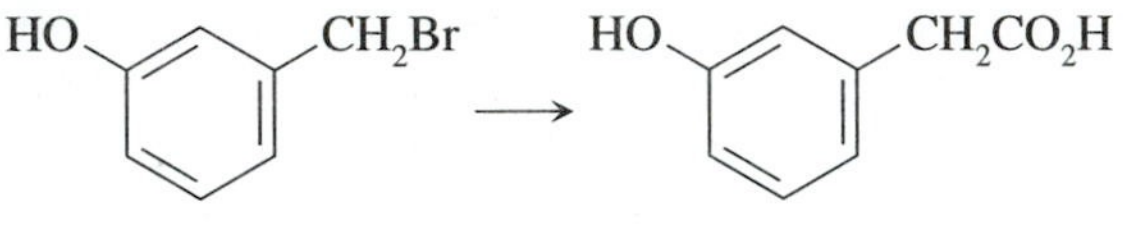

m-$HOC_6H_4CH_2Br \longrightarrow$ m-$HOC_6H_4CH_2CO_2H$

Solution

To carry out this synthesis, one carbon atom must be added to the side chain of the aromatic compound. Which of the two methods given in this section should be selected? A Grignard reagent cannot be made from the *m*-hydroxybenzyl bromide, because the phenolic hydroxyl group has an acidic proton that would destroy the Grignard reagent.

Consider the substitution reaction of the starting material with cyanide ion to give the nitrile. Because benzyl bromide is a primary haloalkane, it readily reacts in an S_N2 reaction. The resulting nitrile is hydrolyzed to form the desired acid.

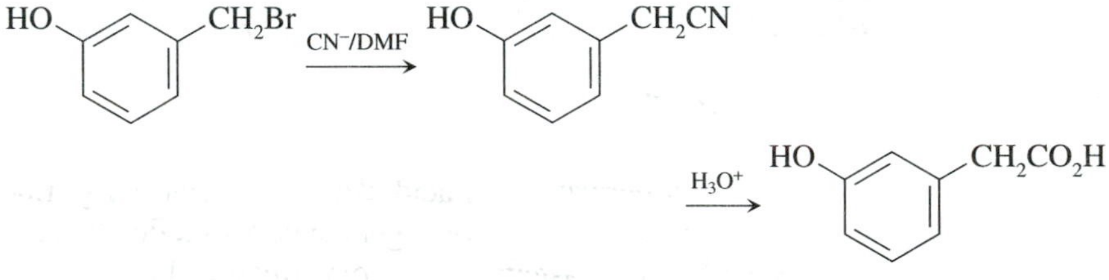

m-$HOC_6H_4CH_2Br \xrightarrow{CN^-/DMF}$ m-$HOC_6H_4CH_2CN \xrightarrow{H_3O^+}$ m-$HOC_6H_4CH_2CO_2H$

Problem 12.4

Outline a series of steps to prepare 2,2-dimethylheptanoic acid starting from 2-methyl-1-heptene.

$$CH_3CH_2CH_2CH_2CH_2\overset{\displaystyle CH_3}{\overset{|}{C}}H{=}CH_2 \longrightarrow CH_3CH_2CH_2CH_2CH_2\underset{\displaystyle CH_3}{\underset{|}{\overset{\displaystyle CH_3}{\overset{|}{C}}}}-CO_2H$$

12.6 Nucleophilic Acyl Substitution

In Chapter 10 we discussed nucleophilic addition reactions in which the nucleophile attacks the electrophilic carbonyl carbon atom of aldehydes and ketones to give a tetrahedral product.

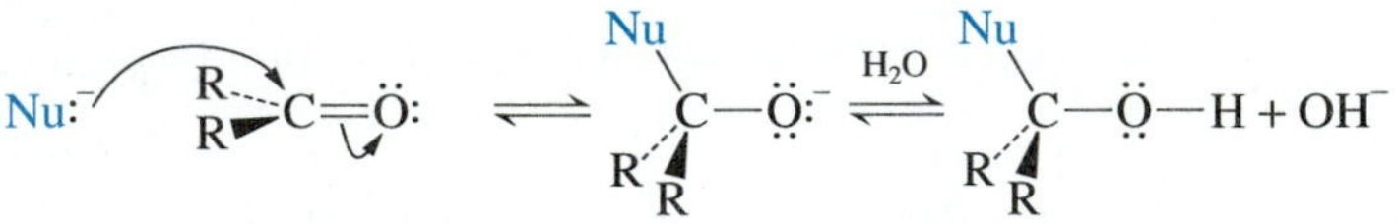

$$Nu:^- + R_2C{=}\ddot{O}: \rightleftharpoons Nu{-}CR_2{-}\ddot{O}:^- \overset{H_2O}{\rightleftharpoons} Nu{-}CR_2{-}\ddot{O}{-}H + OH^-$$

trigonal planar carbonyl compound — tetrahedral product

Acyl derivatives also react with nucleophiles at the carbonyl carbon atom to generate a tetrahedral intermediate. However, the tetrahedral intermediate is unstable, and a leaving group departs to form a different acyl derivative. The overall process is called **nucleophilic acyl substitution.** The process is also called an **acyl transfer reaction** because it transfers an acyl group from one group (the leaving group) to another (the nucleophile).

$$\mathrm{Nu}:^- + \mathrm{R(L)C{=}\ddot{O}}: \longrightarrow \mathrm{Nu(R)(L)C{-}\ddot{O}}:^- \longrightarrow \mathrm{R(Nu)C{=}\ddot{O}}: + \mathrm{L}^-$$

acyl derivative — tetrahedral intermediate

Why don't acyl derivatives behave like aldehydes and ketones and form stable tetrahedral products? The answer is that the intermediate formed from an acyl derivative has a good leaving group. In the case of an acid chloride, the leaving group is the chloride ion, which is a weak base. We recall that leaving group abilities are inversely related to base strength. An intermediate derived from a ketone does not have a good leaving group. A carbanion, the conjugate acid of a hydrocarbon, is an extremely strong base and therefore a very poor leaving group.

The stoichiometry of the nucleophilic acyl substitution reaction resembles that of an S_N2 substitution reaction of haloalkanes. However, the resemblance is only superficial. An S_N2 reaction is a single-step process in which the nucleophile bonds to the carbon atom as the leaving group leaves. Nucleophilic acyl substitution occurs in two steps (Figure 12.2). The rate-determining step is most commonly nucleophilic attack at the carbonyl carbon atom to form a tetrahedral intermediate. The loss of the leaving group occurs in a second faster step.

Relative Reactivity of Acyl Derivatives

The order of reactivity of acyl derivatives is acid chloride > acid anhydride > ester = acid > amide. This order of reactivity might appear to reflect the leaving group abilities, as reflected by their basicities. We know that HCl is a strong acid

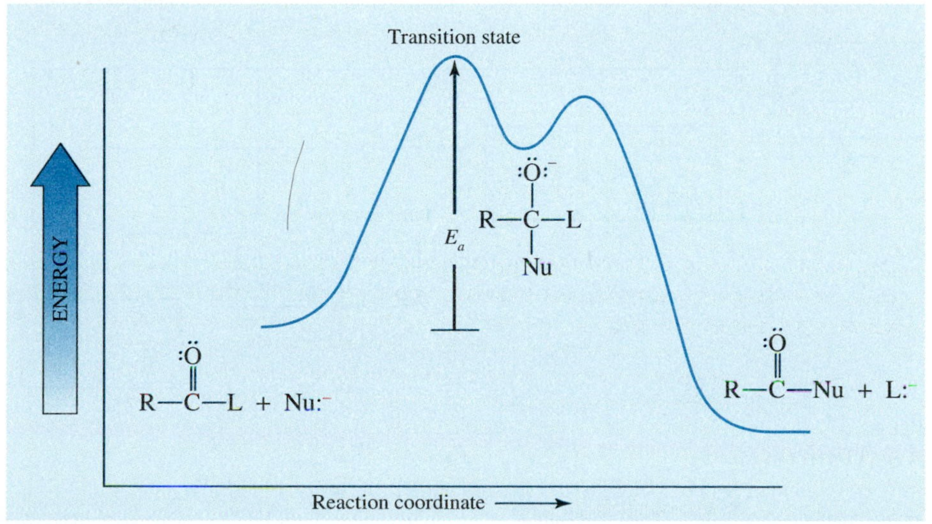

FIGURE 12.2
Mechanism of Nucleophilic Acyl Substitution

and NH_3 is a very weak acid. Thus, Cl^- is a weak base and NH_2^- is a strong base, meaning that Cl^- is a better leaving group than NH_2^-. However, the rate-determining step is not loss of the leaving group but rather attack of the nucleophile.

The accepted explanation of the order of reactivities is resonance stabilization of the reactant. Donation of an electron pair of the atom bonded to the acyl carbon atom decreases the partial positive charge on the carbon atom and decreases its electrophilicity. In the tetrahedral intermediate, no resonance stabilization is involved. Consider the general equation for the first step of the reaction.

$$\left[R{-}C(={\ddot{O}}{:}){-}\ddot{L}{:} \longleftrightarrow R{-}C(-{:}\ddot{O}{:}^-){=}L^+ \right] \xrightarrow{Nu:^-} R{-}C(Nu)(L){-}\ddot{O}{:}^-$$

resonance-stabilized reactant — no resonance stabilization

We recall from the discussion of substituent effects on electrophilic aromatic substitution that second row elements such as oxygen and nitrogen effectively donate electrons by resonance. Furthermore, nitrogen is a better donor of electrons by resonance than oxygen because nitrogen is less electronegative. Thus, a nitrogen atom stabilizes amides more effectively than an oxygen atom stabilizes esters. Finally, we recall that chlorine, a third row element, is not effective in donating electrons by resonance. Thus, acid chlorides have little resonance stabilization by the chlorine atom.

The order of reactivity means that a more reactive acyl derivative can be converted into a less reactive acyl derivative (Figure 12.3). The relative reactivity of acyl derivatives enables us to understand most of the chemical reactions presented in this chapter. Note that acids and esters both have an oxygen atom bonded to the carbonyl carbon atom. Thus, these two classes of compounds have similar reactivities and can be readily interconverted in equilibrium processes. These important reactions will be discussed in later sections.

$$R{-}\overset{O}{\overset{\|}{C}}{-}OH + R'O{-}H \rightleftharpoons R{-}\overset{O}{\overset{\|}{C}}{-}OR' + H_2O$$

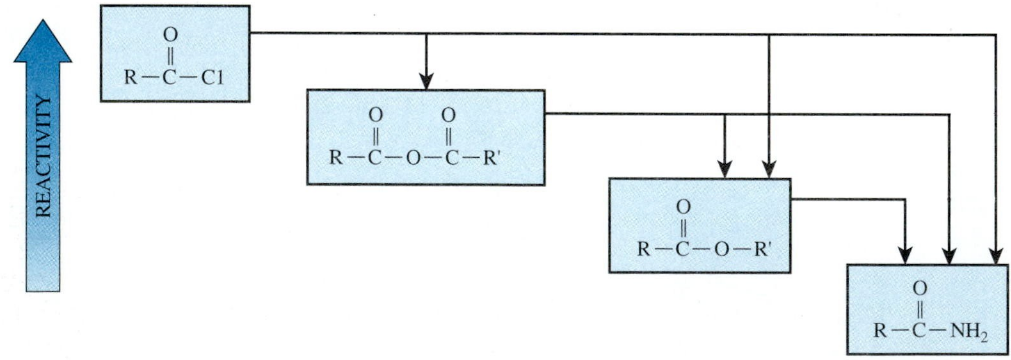

FIGURE 12.3
Reactivity of Acyl Derivatives
The more reactive acyl derivatives can be converted spontaneously to the less reactive acyl derivatives by nucleophilic acyl substitution reactions.

Acid Chlorides

Acid chlorides, the most reactive acyl derivatives, are prepared in the laboratory by treating carboxylic acids with thionyl chloride, $SOCl_2$. This reaction takes advantage of Le Châtelier's principle. Thionyl chloride reacts with carboxylic acids to form acid chlorides because the reaction is driven to the right by the formation of HCl and SO_2, which are released as gases and escape from the reaction mixture. Phosphorus trichloride, PCl_3, can also be used to prepare acyl chlorides.

$$\text{C}_6\text{H}_{11}\text{-C(=O)-OH} \xrightarrow{SOCl_2} \text{C}_6\text{H}_{11}\text{-C(=O)-Cl}$$

cyclohexanecarboxylic acid — cyclohexanecarbonyl chloride

Acid chlorides react with most nucleophiles and are hydrolyzed by the moisture in air. Reaction of an acid chloride with an alcohol gives an ester. Amines are easily converted into amides by reaction with acid chlorides.

$$\text{R-C(=O)-Cl} \xrightarrow{\text{H-O-H}} \text{R-C(=O)-OH} + \text{HCl}$$

$$\text{R-C(=O)-Cl} \xrightarrow{\text{R'-O-H}} \text{R-C(=O)-OR'} + \text{HCl}$$

$$\text{R-C(=O)-Cl} \xrightarrow{\text{R'-NH-H}} \text{R-C(=O)-NHR'} + \text{HCl}$$

Esters are easily produced from acids by first converting the acid to an acid chloride. The acid chloride reacts with an alcohol to give excellent yields of an ester. A base such as pyridine is used to react with the HCl formed in the reaction.

$$\text{C}_3\text{H}_5\text{-C(=O)-OH} \xrightarrow{SOCl_2} \text{C}_3\text{H}_5\text{-C(=O)-Cl} \xrightarrow[\text{pyridine}]{CH_3CH_2OH} \text{C}_3\text{H}_5\text{-C(=O)-O-}CH_2CH_3$$

Acid Anhydrides

Acid anhydrides are less reactive than acid chlorides, but they are still very active acylating agents. Water hydrolyzes acid anhydrides to acids, alcohols react to give esters, and amines give amides. Note that the byproduct in each case is one molar equivalent of a carboxylic acid.

$$\text{R-C(=O)-O-C(=O)-R} \xrightarrow{\text{H-O-H}} \text{R-C(=O)-OH} + \text{R-C(=O)-OH}$$

$$\text{R-C(=O)-O-C(=O)-R} \xrightarrow{\text{R'-O-H}} \text{R-C(=O)-OR'} + \text{R-C(=O)-OH}$$

$$\text{R-C(=O)-O-C(=O)-R} \xrightarrow{\text{R'-NH-H}} \text{R-C(=O)-NHR'} + \text{R-C(=O)-OH}$$

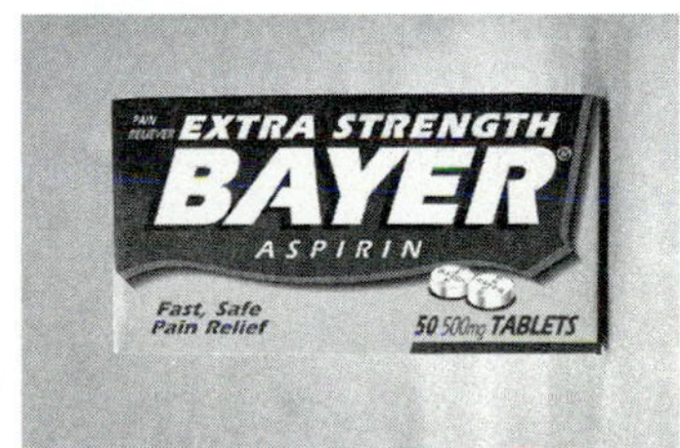

Aspirin is an ester that is produced in large quantities by esterification of a phenol.

The most common acid anhydride is acetic anhydride; over a million tons are produced each year. Reaction of acetic anhydride with the phenol functional

Thioesters Are Nature's Active Acyl Compounds

The interconversion of acyl compounds in cells occurs by transferring acyl groups from one molecule to another. These processes do not involve acid halides or acid anhydrides, because both types of compounds would be hydrolyzed rapidly in the aqueous environment of the body. Esters are widely distributed in cells but generally react too slowly to transfer acyl groups efficiently to other substrates.

Nature uses an acyl transfer reagent that is sufficiently stable toward hydrolysis but yet reactive enough toward nucleophiles to react with them at speeds sufficient to maintain life processes. These "just right" reagents are thioesters. Thioesters are more reactive than esters, and the thiol group is easily replaced by an alkoxy group from an alcohol. The increased reactivity of thioesters over esters is due to the smaller ability of sulfur, a third period element, to donate electrons by resonance to the carbonyl carbon atom and stabilize the thioester. Because thioesters are less stable than esters, the transfer of an acyl group from a thioester to an alcohol is a spontaneous process.

$$CH_3\overset{O}{\overset{\|}{C}}-SR + H-OR' \longrightarrow CH_3\overset{O}{\overset{\|}{C}}-OR' + H-SR$$

The most important thioester is acetyl coenzyme A, which is formed from the thiol group of coenzyme A, a complex thiol that we will abbreviate as CoA—SH. Coenzyme A consists of three substructures—adenosine diphosphate (ADP), the vitamin pantothenic acid, and 2-aminoethanethiol. The thiol group of CoA—SH is bonded to an acyl group in acyl—S—CoA derivatives.

When an acetyl group is linked to CoA—SH, the adduct is acetyl coenzyme A. This extremely important metabolite is produced by the degradation of long-chain carboxylic acids contained in fats. Acetyl—CoA is also produced by metabolic degradation of many amino acids and carbohydrates. Acetyl—CoA is a donor of the two-carbon acetyl group in the biosynthesis of long-chain carboxylic acids.

group of salicylic acid yields an ester called acetylsalicylic acid, or aspirin. The annual production of aspirin in the United States is 12,000 tons.

$$\text{salicylic acid} + CH_3-\overset{O}{\overset{\|}{C}}-O-\overset{O}{\overset{\|}{C}}-CH_3 \longrightarrow \text{acetylsalicylic acid } (C_6H_4(CO_2H)-O-\overset{O}{\overset{\|}{C}}-CH_3) + CH_3-\overset{O}{\overset{\|}{C}}-OH$$

salicylic acid (OH, CO_2H) — acetic anhydride — acetylsalicylic acid — acetic acid

12.7 Reduction of Acyl Derivatives

In Chapter 10 we saw that aldehydes (or ketones) can be reduced by either sodium borohydride or lithium aluminum hydride. These reactions involve nucleophilic attack of hydride ion on the carbonyl carbon atom. A similar process occurs with acyl derivatives. However, there are subsequent steps in the reaction mechanisms that occur after nucleophilic attack by the hydride ion. In this section, the reduction of esters, carboxylic acids, and acid chlorides is considered. The reduction of amides is presented in Chapter 14.

Acetyl coenzyme A reacts with nucleophiles in enzyme-catalyzed biological reactions to give new acyl compounds.

$$CH_3\overset{O}{\overset{\|}{C}}{-}SCoA + HNu{:} \longrightarrow CH_3\overset{O}{\overset{\|}{C}}{-}Nu + CoA{-}SH$$

For example, acetyl coenzyme A provides the acetyl group in the biosynthesis of the neurotransmitter acetylcholine. Choline contains a hydroxyl group that is acetylated by acetyl coenzyme A to make acetylcholine. ■

$$CH_3\overset{O}{\overset{\|}{C}}{-}SCoA + \underset{\text{choline}}{(CH_3)_3\overset{+}{N}CH_2CH_2OH} \longrightarrow \underset{\text{acetylcholine}}{CH_3\overset{O}{\overset{\|}{C}}{-}OCH_2CH_2\overset{+}{N}(CH_3)_3} + CoA{-}SH$$

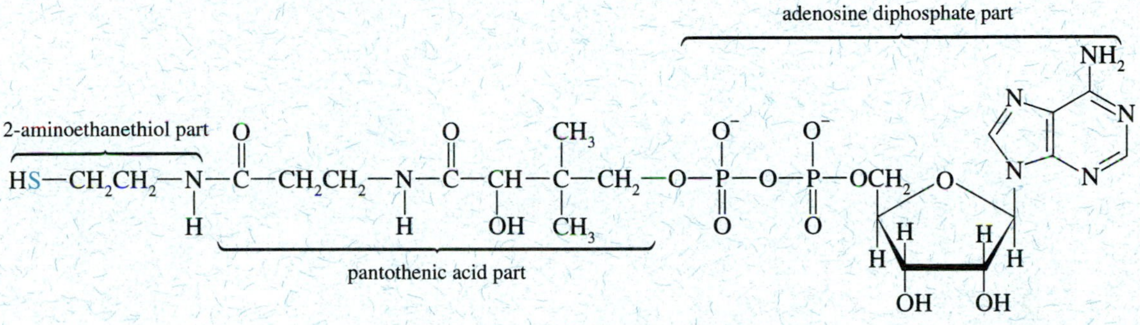

adenosine diphosphate part

2-aminoethanethiol part

$$HS{-}CH_2CH_2{-}\underset{H}{N}{-}\overset{O}{\overset{\|}{C}}{-}CH_2CH_2{-}\underset{H}{N}{-}\overset{O}{\overset{\|}{C}}{-}\underset{OH}{CH}{-}\overset{CH_3}{\underset{CH_3}{C}}{-}CH_2{-}O{-}\overset{O^-}{\underset{O}{P}}{-}O{-}\overset{O^-}{\underset{O}{P}}{-}OCH_2\text{-(ribose-adenine)}$$

pantothenic acid part

coenzyme A

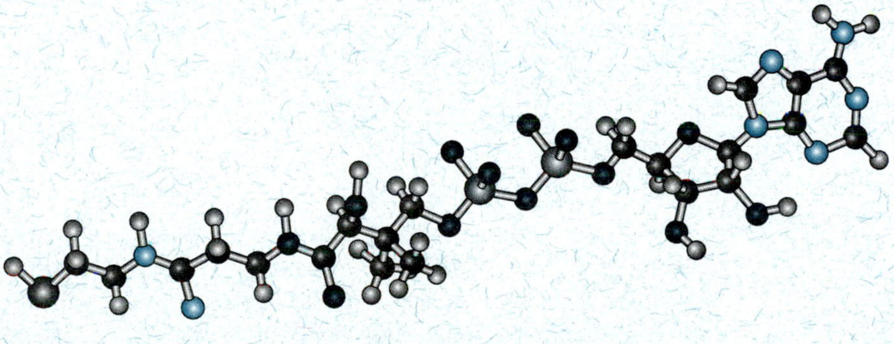

Reduction of Esters

The reduction of esters requires the strong reducing agent lithium aluminum hydride. The milder reagent sodium borohydride does not reduce esters. Remember that lithium aluminum hydride does not reduce carbon-carbon double bonds.

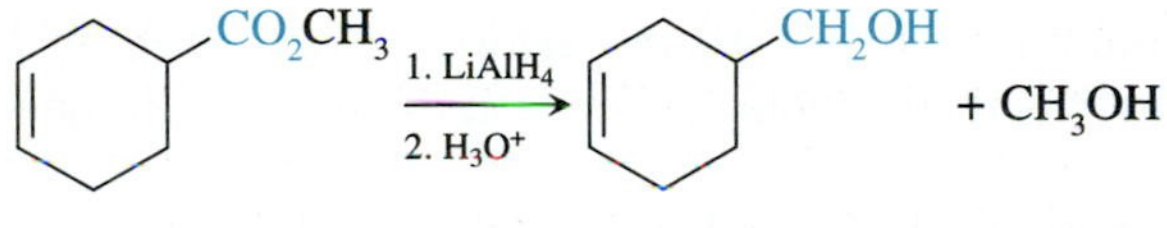

$$\text{(cyclohexenyl)}{-}CO_2CH_3 \xrightarrow[2.\ H_3O^+]{1.\ LiAlH_4} \text{(cyclohexenyl)}{-}CH_2OH + CH_3OH$$

Note that the alcohol portion of the ester is a byproduct of the reaction. The esters that are typically reduced by lithium aluminum hydride contain a low molecular weight alkyl group introduced in the conversion of an acid to an ester.

The alcohol obtained by reduction of the acid portion of the ester is easily separated from the low molecular weight, water-soluble alcohol.

The mechanism of the reduction of an ester is pictured as simply a nucleophilic attack of a one molar equivalent of a "hydride" ion on the carbonyl carbon atom. However, the aluminum atom participates in the reaction and is bonded to the oxygen atom. For simplicity, the structures shown have eliminated the aluminum group. Attack by the hydride ion produces a tetrahedral intermediate that loses an alkoxide ion. The resulting aldehyde is reduced even more rapidly than the original ester by a second molar equivalent of hydride ion. In total, two molar equivalents of hydride ion or one-half molar equivalent of lithium aluminum hydride are required for the reduction.

$$H:^- + R(R'O)C{=}\ddot{O}: \longrightarrow \left[R(R'O)(H)C{-}\ddot{O}:^- \right] \longrightarrow R(H)C{=}\ddot{O}: + R'O^-$$

tetrahedral intermediate

$$H:^- + R(H)C{=}\ddot{O}: \longrightarrow R(H)(H)C{-}\ddot{O}:^- \xrightarrow{H_3O^+} R(H)(H)C{-}\ddot{O}{-}H$$

Reduction of Acids

Carboxylic acids are reduced by lithium aluminum hydride but not by the milder reagent sodium borohydride. As in the case of the reduction of esters, carboxylic acids are always completely reduced to alcohols; no intermediate aldehyde can be isolated.

$$C_6H_5{-}CH_2CH_2CO_2H \xrightarrow[2.\ H_3O^+]{1.\ LiAlH_4} C_6H_5{-}CH_2CH_2CH_2OH$$

3-phenylpropanoic acid → 3-phenyl-1-propanol

Lithium aluminum hydride is a strong base, and it reacts with the acidic proton of the carboxylic acid to give hydrogen gas. One molar equivalent of hydride ion is used in the reaction. The carboxylate salt is then reduced, requiring an additional two molar equivalents of hydride ion. Therefore, reduction of a carboxylic acid requires a total of three molar equivalents of hydride ion.

Reduction of Acid Chlorides

Esters or carboxylic acids cannot be reduced to aldehydes by hydride reducing agents because aldehydes are more reactive than esters or carboxylic acids. Acid chlorides are more reactive than esters toward nucleophilic hydride ion. As a consequence, acid chlorides are more rapidly reduced than esters. However, lithium aluminum hydride is such a strong reducing agent that acid chlorides are still reduced all the way to primary alcohols. The milder reducing agent lithium aluminum tri(*tert*-butoxy)hydride reacts with acid chlorides but much more slowly with aldehydes, which can therefore be isolated.

$$C_6H_{11}{-}CH_2CH_2C({=}O){-}Cl \xrightarrow[2.\ H_3O^+]{1.\ LiAlH[OC(CH_3)_3]_3} C_6H_{11}{-}CH_2CH_2C({=}O){-}H$$

3-cyclohexylpropanoyl chloride → 3-cyclohexylpropanal

12.8 Synthesis of Esters

Carboxylic acids are converted to esters in good yield by a two-step method (Section 12.6). The acid is converted to an acid chloride in one step, which reacts with an alcohol in the second step.

$$R-\overset{O}{\overset{\|}{C}}-OH \xrightarrow{SOCl_2} R-\overset{O}{\overset{\|}{C}}-Cl \xrightarrow[\text{pyridine}]{R'OH} R-\overset{O}{\overset{\|}{C}}-O-R'$$

Carboxylic acids can also be directly converted to esters in a single step called the **Fischer esterification** reaction. This reaction is an example of a condensation reaction; it entails joining two reactants into one larger product with the simultaneous formation of a second smaller product, such as water. A carboxylic acid reacts with an alcohol in a condensation reaction to give an ester and water. Thus, the ester contains a part of both the alcohol and the acid.

Water comes from here.

This C—O bond is formed.

part of the acid

$$\underset{\text{a carboxylic acid}}{R-\overset{O}{\overset{\|}{C}}-OH} + \underset{\text{an alcohol}}{H-O-R'} \longrightarrow \underset{\text{an ester}}{R-\overset{O}{\overset{\|}{C}}-O-R'} + H_2O$$

part of the alcohol

The reaction is catalyzed by inorganic acids, and both the carboxylic acid and its ester are present in substantial amounts at equilibrium. The yield of ester can be increased by distilling the water out of the reaction mixture. The ester yield is also increased by using an excess of the alcohol. Ethyl esters of acids are obtained by using ethanol as a solvent. Under such conditions, the high concentration of ethanol favors a high conversion of the acid to the ester. Thus, removing water or adding alcohol shifts the equilibrium to the side of the product, as predicted by Le Châtelier's principle.

$$CH_3-\overset{O}{\overset{\|}{C}}-OH + CH_3CH_2OH \rightleftharpoons CH_3-\overset{O}{\overset{\|}{C}}-O-CH_2CH_3 + H_2O$$

Adding alcohol "pushes" the reaction to the right.

Removing water "pulls" the reaction to the right.

Mechanism of Esterification

Does the oxygen atom linking the acyl carbon atom and the alkyl carbon atom of an ester "come from" the oxygen of the acid or the alcohol? From a different perspective, does the water formed come from the hydroxyl group of the alcohol and the hydrogen of the acid or from the hydroxyl group of the acid and the hydrogen of the alcohol? These related questions have been answered from studies on the mechanism of the acid-catalyzed esterification reaction using ^{18}O-labeled methanol. When methanol reacts with benzoic acid, the oxygen-18 is contained in the ester.

$$C_6H_5-\overset{O}{\overset{\|}{C}}-OH + CH_3-{}^{18}O-H \longrightarrow C_6H_5-\overset{O}{\overset{\|}{C}}-{}^{18}O-CH_3 + H_2O$$

Based on this information, the mechanism involves nucleophilic attack of the oxygen atom of the alcohol on the carbonyl carbon atom. The mechanism that

Polyesters

Many commercial products, called **condensation polymers,** are made by reacting monomers to give large molecules and some small molecule, such as water, as a byproduct. This polymerization process differs from addition polymerization in which the entire monomer is included in the polymer.

In condensation polymerization reactions, each monomer has two functional groups. An example of condensation polymerization is the reaction of terephthalic acid and ethylene glycol.

terephthalic acid: HO—C(=O)—C₆H₄—C(=O)—OH

ethylene glycol: $HOCH_2CH_2OH$

Terephthalic acid is produced by the oxidation of the methyl groups of *p*-dimethylbenzene (*p*-xylene) using oxygen and V_2O_5 as catalyst. Ethylene glycol is produced by hydration of ethylene oxide (Section 9.8). Reaction of one of the carboxyl groups of terephthalic acid with one of the hydroxyl groups of ethylene glycol gives an ester that also contains a carboxyl group and a hydroxyl group. Both of these groups can react again.

HO—C(=O)—C₆H₄—C(=O)—OCH_2CH_2OH

This carboxyl group can react with ethylene glycol.

This hydroxyl group can react with terephthalic acid.

Reaction at each end produces a larger molecule that continues to react to form the high molecular weight polyester poly(ethylene terephthalate) also known as PET. PET is produced as a fiber called Dacron®. This fiber has very high strength and is resistant to creasing and wrinkling. However, the first Dacron fibers were somewhat unpleasant to touch, and for this reason were used in woven and knitted fabrics in combination with cotton or wool. New production methods have lessened the harshness of this fiber. Dacron fabric is physiologically inert, so it can be used in the form of a mesh to replace diseased sections of blood vessels.

PET is also used to form a film called Mylar®, which can be produced in sheets. It is used to make magnetic recording tape. Thicker versions of Mylar are used in compact disks.

$HOCH_2CH_2O$—[C(=O)—C₆H₄—C(=O)—OCH_2CH_2O]$_n$—C(=O)—C₆H₄—C(=O)—OH

PET

allows for this central reaction involves a series of acid-base reactions. Protonation of the carbonyl oxygen atom gives a reactive oxocarbocation, which is then attacked by the alcohol.

R(HO)C=Ö: + H^+ ⇌ [R(HO)C^+—Ö—H ⟷ R(HO)C=$Ö^+$—H]

R′—Ö(H): + R(HO)C=$Ö^+$—H ⇌ R′—O^+(H)—C(R)(HO)—Ö—H

Subsequent loss of a proton from the oxygen atom of the original alcohol and gain of a proton at the hydroxyl oxygen atom of the original acid occurs in rapid successive equilibrium reactions.

$$R'\text{—}\overset{+}{\ddot{O}}(H)\text{—}C(R)(HO)\text{—}\ddot{O}\text{—}H \overset{-H^+}{\rightleftharpoons} R'\text{—}\ddot{O}\text{—}C(R)(HO)\text{—}\ddot{O}\text{—}H \overset{+H^+}{\rightleftharpoons} R'\text{—}\ddot{O}\text{—}C(R)(HO)\text{—}\overset{+}{\ddot{O}}(H)H$$

Loss of water as a leaving group gives an oxocarbocation that subsequently loses a proton to give the final ester product.

$$R'\text{—}\ddot{O}\text{—}C(R)(HO)\text{—}\overset{+}{\ddot{O}}(H)H \overset{-H_2O}{\rightleftharpoons} \left[R(R'O)\overset{+}{C}\text{—}\ddot{O}\text{—}H \longleftrightarrow R(R'O)C\text{=}\overset{+}{\ddot{O}}\text{—}H \right] \overset{-H^+}{\rightleftharpoons} R(R'O)C\text{=}\ddot{O}\text{:}$$

12.9 Hydrolysis of Esters

The sail of this windsurfer is made of the polyester Dacron.

Hydrolysis reactions split a reactant into two smaller products with the net incorporation of water into the two products (Section 2.5). The hydrolysis of an ester produces an acid and an alcohol. Ester hydrolysis, then, is just the reverse of the esterification reaction. The hydrolysis of an ester is catalyzed by strong acids. The reaction is favored by a large excess of water.

$$R\text{—}\overset{O}{\overset{\|}{C}}\text{—}O\text{—}R' + H_2O \overset{H^+}{\rightleftharpoons} R\text{—}\overset{O}{\overset{\|}{C}}\text{—}OH + H\text{—}O\text{—}R'$$

This C—O bond is cleaved.

Adding water "pushes" the reaction to the right.

The hydrolysis of an ester by a strong base is called **saponification** (from Latin, *sapon,* soap) because this reaction is used to make soaps from esters of long-chain carboxylic acids (Section 13.6). Because the reaction mixture is basic, the conjugate base of the carboxylic acid is formed in the reaction, compared to the carboxylic acid itself in the acid-catalyzed hydrolysis reaction.

$$R\text{—}\overset{O}{\overset{\|}{C}}\text{—}O\text{—}R' + HO^- \longrightarrow R\text{—}\overset{O}{\overset{\|}{C}}\text{—}O^- + H\text{—}O\text{—}R'$$

There is another important distinction between hydrolysis and saponification: hydrolysis is catalyzed by an acid, whereas in the saponification reaction, hydroxide is a reagent. It is not a catalyst, because equal numbers of moles of hydroxide and ester are required. Hydroxide is consumed in the reaction, and, because it is a strong base, the position of equilibrium lies overwhelmingly to the right, so the reaction is essentially irreversible.

EXAMPLE 12-5

The antibiotic chloramphenicol tastes bitter. Its palatability for children is improved by using a suspension of the palmitate ester. Once orally administered, the ester is

Esters and Anhydrides of Phosphoric Acid

Phosphoric acid, pyrophosphoric acid, and triphosphoric acid occur in living cells as esters. Just as alcohols react with carboxylic acids, they also react with phosphoric acid, pyrophosphoric acid, and triphosphoric acid.

ester bond

R—O—P(=O)(OH)—OH

alkyl phosphate
(an ester of phosphoric acid)

ester bond

R—O—P(=O)(OH)—O—P(=O)(OH)—OH

alkyl diphosphate
(an ester of pyrophosphoric acid)

ester bond

R—O—P(=O)(OH)—O—P(=O)(OH)—O—P(=O)(OH)—OH

alkyl triphosphate
(an ester of triphosphoric acid)

Note that these esters are also acids. At physiological pH, the protons in the —OH groups of these esters are ionized. It is for this reason that phosphoric acid derivatives are soluble in the aqueous environment of living systems.

The esters are also acid anhydrides of phosphoric acid. The oxygen atom between the phosphorus atoms is part of the anhydride bond. These compounds are analogs of acid anhydrides of carboxylic acids.

anhydride bond

R—O—P(=O)(OH)—O—P(=O)(OH)—OH

anhydride bond

R—O—P(=O)(OH)—O—P(=O)(OH)—O—P(=O)(OH)—OH

The anhydride bonds of diphosphates and triphosphates are hydrolyzed in many biological reactions. Adenosine triphosphate (ATP), a triphosphate ester of adenosine, stores some of the energy released in the degradation of carbohydrates, fats, and amino acids. When the terminal anhydride bond of ATP is hydrolyzed, adenosine diphosphate (ADP) is produced and 7.3 kcal of energy are released per mole of ADP produced. When the anhydride bond near the adenosine portion of the molecule is hydrolyzed, adenosine monophosphate (AMP) and pyrophosphoric acid are produced, and 8.6 kcal of energy are released per mole of AMP produced.

triphosphate — adenosine

$^{-}$O—P(=O)(O^{-})—O—P(=O)(O^{-})—O—P(=O)(O^{-})—OCH$_2$ (ribose with H, H, H, H, OH, OH; adenine with NH$_2$, N, N, N, N)

ATP transfers phosphate groups to alcohols in many metabolic reactions. For example, as part of the metabolism of glucose the hydroxyl group at C-6 is phosphorylated in a reaction with ATP to give glucose 6-phosphate and ADP. ■

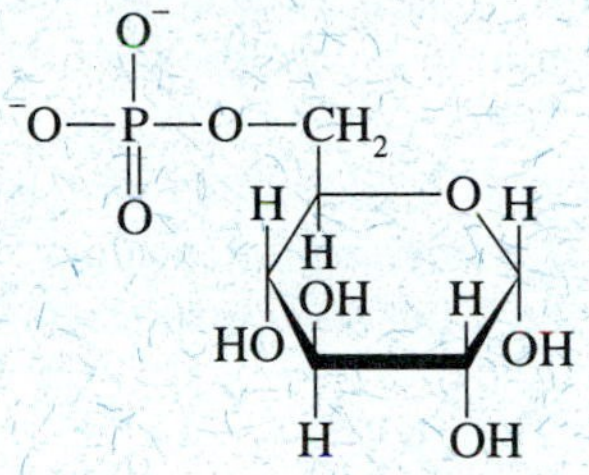

glucose 6-phosphate

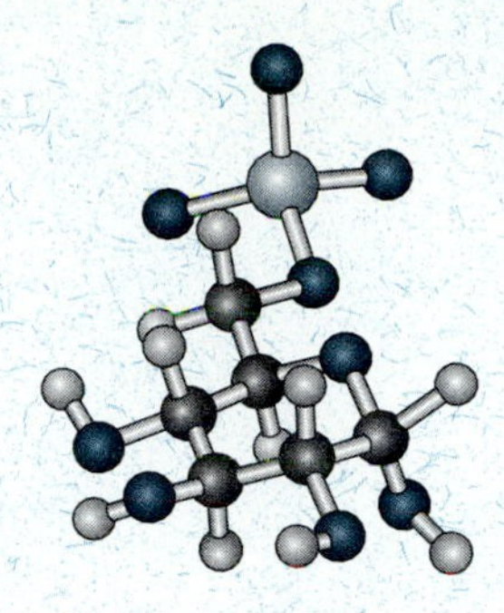

enzymatically hydrolyzed in the intestine. Given the structure of the ester, write the structure of the antibiotic.

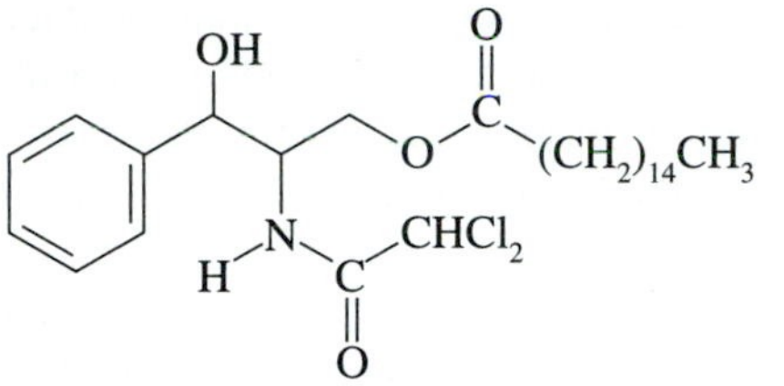

Solution

First, locate the ester functional group by examining the carbonyl carbon atoms. The carbonyl group at the bottom of the structure is bonded to a nitrogen atom. This is an amide group, a very stable acyl functional group. The carbonyl carbon atom at the top of the structure is bonded to an oxygen atom. This is the ester functional group. The carbon atom chain to the right of the carbonyl group is part of the acid. A total of 16 carbon atoms are contained in palmitic acid. Chloramphenicol is bonded in the ester through its primary hydroxyl group.

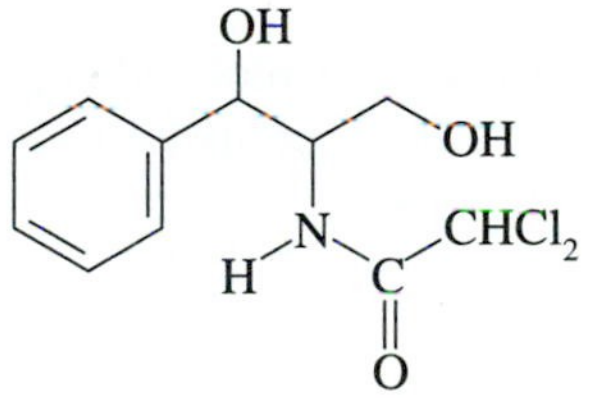

chloramphenicol

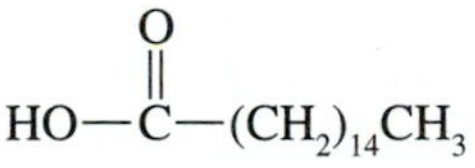

$HO{-}C({=}O){-}(CH_2)_{14}CH_3$

palmitic acid

Problem 12.5

Hydrolysis of diloxanide furanoate in the body is required for it to be effective against intestinal amebiasis. What is the acid component of the drug? Based on the name of the drug, what is the name of the acid?

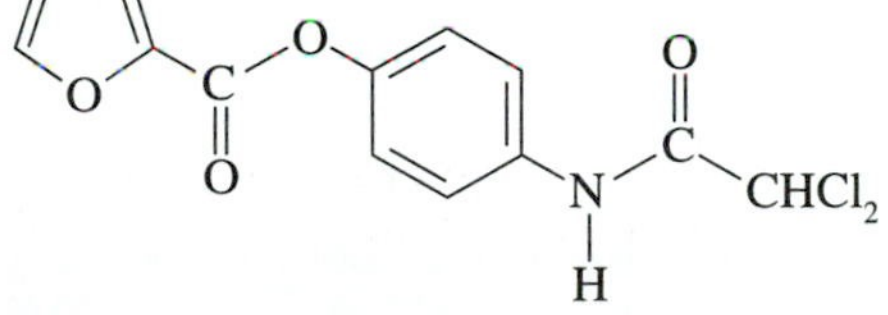

12.10 The Claisen Condensation

We recall that the hydrogen atom bonded to the α-carbon atom in an aldehyde or ketone is acidic (pK_a = 20). As a consequence, carbonyl compounds with an α-carbon atom undergo condensation reactions in which the α-carbon atom of one molecule bonds to the carbonyl carbon atom of another molecule. The result is an aldol (Section 10.11).

Two molecules of an ester also react with each other in the presence of a base such as ethoxide ion to produce a condensation product. The reaction, which produces a β-keto ester, is called the **Claisen condensation.**

$$2\ CH_3-\overset{\overset{\displaystyle O}{\|}}{C}-OCH_2CH_3 \xrightarrow[HOCH_2CH_3]{NaOCH_2CH_3} CH_3-\overset{\overset{\displaystyle O}{\|}}{C}-CH_2-\overset{\overset{\displaystyle O}{\|}}{C}-OCH_2CH_3 + CH_3CH_2OH$$

ethyl acetate → ethyl acetoacetate

The mechanism of the Claisen condensation resembles that of the aldol condensation, but there are important differences. A full equivalent of base is required for the Claisen condensation rather than the catalytic amount required for the aldol condensation. There are several equilibria in the sequence of reactions, and some are not favorable. The entire sequence is made favorable by driving the final step to completion by an acid-base reaction that requires a molar equivalent of base.

Like aldehydes, esters have an acidic α-hydrogen atom. The acid dissociation constants of esters (pK_a = 25) are about 10^5 times smaller than those of aldehydes. The pK_a of ethanol is 16. Thus, the reaction of sodium ethoxide with an ester ($K = 10^{-9}$) produces only a small amount of the enolate at equilibrium.

$$CH_3CH_2\ddot{O}:^- + H_3C-C(=\ddot{O}:)OCH_2CH_3 \rightleftharpoons H_2\overset{-}{C}:-C(=\ddot{O}:)OCH_2CH_3 + CH_3CH_2OH$$

weaker base, weaker acid, stronger base, stronger acid

The conjugate base of the ester reacts with another molecule of ester to form a carbon-carbon bond. The addition product is the conjugate base of a hemiacetal, which loses an alkoxide to give a β-keto ester.

$$CH_3-C(=\ddot{O}:)-\ddot{O}CH_2CH_3 + H_2\overset{-}{C}:-C(=\ddot{O}:)OCH_2CH_3 \rightleftharpoons CH_3-\underset{\underset{\displaystyle :\ddot{O}-CH_2CH_3}{|}}{\overset{\overset{\displaystyle :\ddot{O}:^-}{|}}{C}}-CH_2-C(=\ddot{O}:)OCH_2CH_3$$

$$CH_3-\underset{\underset{\displaystyle :\ddot{O}-CH_2CH_3}{|}}{\overset{\overset{\displaystyle :\ddot{O}:^-}{|}}{C}}-CH_2-C(=\ddot{O}:)OCH_2CH_3 \rightleftharpoons CH_3-\overset{\overset{\displaystyle \ddot{O}:}{\|}}{C}-CH_2-C(=\ddot{O}:)OCH_2CH_3 + CH_3CH_2\ddot{O}:^-$$

β-Keto esters have pK_a values around 11 because the negative charge of the conjugate base can be delocalized over both carbonyl groups. Ethanol is a weaker acid than β-keto esters, and, therefore, ethoxide ion essentially completely removes a proton from the product of the Claisen condensation. This final step drives the overall sequence of reactions to completion.

$$CH_3CH_2\ddot{O}:^- + CH_3-\overset{\overset{\large\ddot{O}:}{\|}}{C}-\underset{\underset{\large H}{|}}{CH}-\overset{\overset{\large\ddot{O}:}{/\!/}}{C}\diagdown OCH_2CH_3 \rightleftharpoons CH_3-\overset{\overset{\large\ddot{O}:}{\|}}{C}-\ddot{C}H^--\overset{\overset{\large\ddot{O}:}{/\!/}}{C}\diagdown OCH_2CH_3 + CH_3CH_2\ddot{O}H$$

Finally, dilute acid added at the end of the reaction converts the conjugate base of the β-keto ester to the product.

$$H-\overset{\overset{\large H}{|}}{\overset{+}{O}}-H \quad CH_3-\overset{\overset{\large\ddot{O}:}{\|}}{C}-\ddot{C}H^--\overset{\overset{\large\ddot{O}:}{/\!/}}{C}\diagdown OCH_2CH_3 \longrightarrow CH_3-\overset{\overset{\large\ddot{O}:}{\|}}{C}-CH_2-\overset{\overset{\large\ddot{O}:}{/\!/}}{C}\diagdown OCH_2CH_3 + H_2\ddot{O}:$$

Biochemical Condensation Reactions

A condensation of acetyl coenzyme A and oxaloacetic acid occurs in the first step of the citric acid cycle, which involves a reaction between an ester and a ketone. The α-carbon atom of acetyl coenzyme A bonds to the carbonyl carbon atom of oxaloacetic acid in a reaction that resembles the aldol condensation. In this reaction, an acetyl group is transferred to oxaloacetic acid to form a compound that is subsequently hydrolyzed to citric acid.

$$\underset{\text{oxaloacetic acid}}{HO_2CCH_2\overset{\overset{O}{\|}}{C}-CO_2H} + \underset{\text{acetyl coenzyme A}}{CH_3-\overset{\overset{O}{\|}}{C}-S-CoA} \longrightarrow HO_2CCH_2-\overset{\overset{OH}{|}}{\underset{\underset{CO_2H}{|}}{C}}-CH_2-\overset{\overset{O}{\|}}{C}-S-CoA$$

$$HO_2CCH_2-\overset{\overset{OH}{|}}{\underset{\underset{CO_2H}{|}}{C}}-CH_2-\overset{\overset{O}{\|}}{C}-S-CoA \xrightarrow{H_2O} \underset{\text{citric acid}}{HO_2CCH_2-\overset{\overset{OH}{|}}{\underset{\underset{CO_2H}{|}}{C}}-CH_2-\overset{\overset{O}{\|}}{C}-OH} + CoA-SH$$

Most of the acetyl coenzyme A produced in this and other metabolic reactions reacts with oxaloacetic acid to form citric acid. However, in certain illnesses, such as diabetes, the metabolism of fats predominates over the metabolism of carbohydrates. When there is not enough oxaloacetic acid to react with all of the acetyl coenzyme A being produced, acetyl coenzyme A reacts with itself in a Claisen condensation.

$$CH_3-\overset{\overset{O}{\|}}{C}-S-CoA + CH_3-\overset{\overset{O}{\|}}{C}-S-CoA \longrightarrow CH_3-\overset{\overset{O}{\|}}{C}-CH_2-\overset{\overset{O}{\|}}{C}-S-CoA + CoA-SH$$

Hydrolysis of the resulting β-keto thioester yields acetoacetic acid (3-ketobutanoic acid). Subsequent reactions produce 3-hydroxybutanoic acid and acetone, which are collectively called ketone bodies. Detection of these compounds in the urine is indicative of diabetes.

Summary of Reactions

1. Synthesis of Carboxylic Acids by Oxidative Methods (Section 12.5)

Cyclopentyl–CH_2OH $\xrightarrow{\text{Jones reagent}}$ Cyclopentyl–CO_2H

4-(CH_3)-1-(OCH_2CH_3)benzene $\xrightarrow{KMnO_4}$ 4-(CO_2H)-1-(OCH_2CH_3)benzene

2. Synthesis of Carboxylic Acids from Haloalkanes (Section 12.5)

3-Bromocyclohexene (Br) $\xrightarrow[\text{2. } CO_2 \text{ 3. } H_3O^+]{\text{1. Mg/ether}}$ Cyclohex-2-ene–CO_2H

$CH_3CH_2CH(CH_3)CH_2—Br \xrightarrow[\text{2. } H_3O^+]{\text{1. NaCN/DMF}} CH_3CH_2CH(CH_3)CH_2C(=O)—OH$

3. Synthesis of Acyl Halides (Section 12.6)

Tetrahydronaphthalene–$C(=O)OH$ $\xrightarrow{SOCl_2}$ Tetrahydronaphthalene–$C(=O)Cl$

4. Reduction of Carboxylic Acids and Acid Derivatives (Section 12.7)

Cyclohexyl–$C(=O)OH$ $\xrightarrow[\text{2. } H_3O^+]{\text{1. } LiAlH_4}$ Cyclohexyl–CH_2OH

Cyclohexyl–$C(=O)OCH_3$ $\xrightarrow[\text{2. } H_3O^+]{\text{1. } LiAlH_4}$ Cyclohexyl–CH_2OH + CH_3OH

$CH_3CH_2CH(CH_3)CH_2C(=O)—Cl \xrightarrow[\text{2. } H_3O^+]{\text{1. } LiAlH[OC(CH_3)_3]_3} CH_3CH_2CH(CH_3)CH_2C(=O)—H$

5. Synthesis of Esters (Sections 12.6 and 12.8)

$$CH_3-CH(CH_3)-CH_2-OH + HO-C(=O)-CH(CH_3)CH_3 \xrightleftharpoons{H^+} CH_3-CH(CH_3)-CH_2-O-C(=O)-CH(CH_3)CH_3$$

$$C_6H_5-CH_2-OH + Cl-C(=O)-CH(CH_3)CH_3 \xrightarrow{\text{pyridine}} C_6H_5-CH_2-O-C(=O)-CH(CH_3)CH_3$$

$$\text{(2-methylcyclohexanol)} + CH_3-C(=O)-O-C(=O)-CH_3 \xrightarrow{OH^-} \text{(2-methylcyclohexyl)}-O-C(=O)-CH_3$$

6. Hydrolysis of Esters (Section 12.9)

$$CH_3-CH(CH_3)-CH_2-O-C(=O)-CH(CH_3)CH_3 \xrightleftharpoons{H_3O^+} CH_3-CH(CH_3)-CH_2-OH + HO-C(=O)-CH(CH_3)CH_3$$

$$\text{(2-furyl)}-C(=O)-OCH_2CH_3 + OH^- \longrightarrow \text{(2-furyl)}-C(=O)-O^- + HOCH_2CH_3$$

7. Claisen Condensation (Section 12.10)

$$C_6H_5-CH_2C(=O)-OCH_2CH_3 \xrightarrow[2.\ H_3O^+]{1.\ CH_3CH_2O^-} C_6H_5-CH_2C(=O)-CH(C_6H_5)-C(=O)-OCH_2CH_3$$

Explorations with Molecular Models

1. Construct a ball-and-stick model corresponding to the space-filling model of the following structure. Count the number of atoms and determine the values of *x* and *y* that satisfy the general molecular formula $C_nH_{2n-x}O_y$ that represents acids.

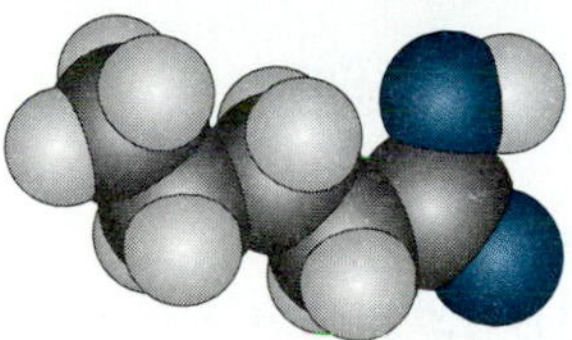

2. Construct a ball-and-stick model corresponding to the space-filling model of the following structure. Does the structure have a stereogenic center? Draw a standard two-dimensional representation of the molecule using 90° and 180° angles. Assign the IUPAC name of this carboxylic acid.

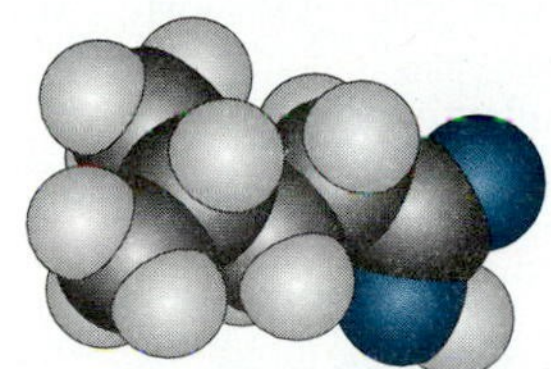

3. Construct a ball-and-stick model corresponding to the space-filling model of the following structure. Assign the IUPAC name of this carboxylic acid. Draw a standard two-dimensional representation of this structure.

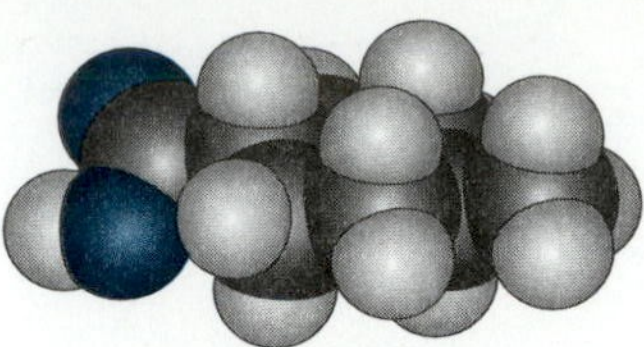

4. Construct a ball-and-stick model corresponding to the space-filling model of methyl acetate. Remove one hydrogen atom from each methyl group and determine the minumum number of additional methylene ($—CH_2—$) units required to form a cyclic ester. (*Hint:* It is necessary to rotate the structure about some of its bonds.) Write a standard two-dimensional representation of the cyclic structure.

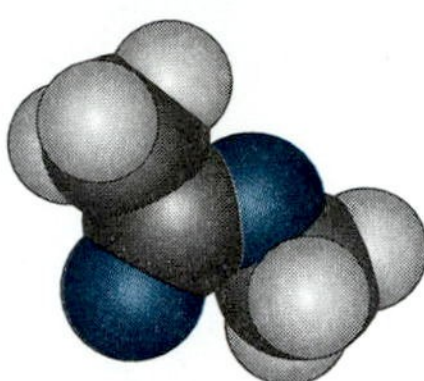

5. Using the original model constructed in Exploration Exercise 4, determine the values of *x* and *y* in the general molecular formula $C_nH_{2n-x}O_y$ that represents esters.

6. Construct a ball-and-stick model corresponding to the following space-filling model of succinic acid. Rotate the structure about its C-2 to C-3 bond until the two carboxyl groups are eclipsed. Remove an —OH group from one carboxyl group and a hydrogen atom from the other carboxyl group. Join the two carbonyl carbon atoms using the remaining oxygen atom. What type of functional group results? Write a standard two-dimensional representation of the structure.

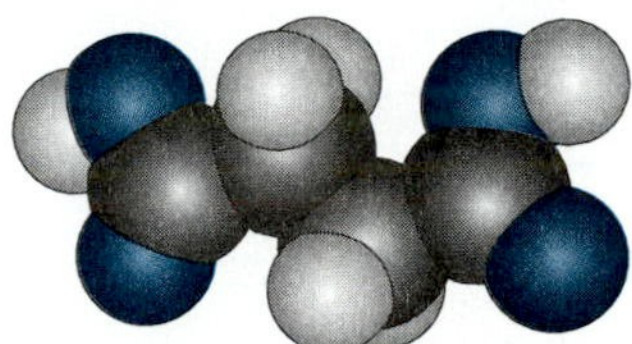

Exercises

Nomenclature

12.1 Give the common name for each of the following acids.

(a) $CH_3CH_2CO_2H$ (b) $CH_3(CH_2)_4CO_2H$ (c) $CH_3(CH_2)_{16}CO_2H$

12.2 Draw the structure of each of the following esters.

(a) octyl acetate (b) ethyl butyrate (c) propyl valerate

12.3 Give the common name for each of the following esters.

(a) $H—\overset{\overset{\large O}{\|}}{C}—O—CH_2CH_3$ (b) $CH_3CH_2CH_2—\overset{\overset{\large O}{\|}}{C}—O—CH_3$ (c) $CH_3—\overset{\overset{\large O}{\|}}{C}—O—CH_2(CH_2)_6CH_3$

12.4 Give the common name for each of the following acids.

(a) $CH_3\underset{\underset{\large Cl}{|}}{C}HCH_2CO_2H$ (b) $Br\underset{\underset{\large CH_3}{|}}{C}HCO_2H$ (c) $CH_3\underset{\underset{\large CH_3}{|}}{C}H\overset{\overset{\large Br}{|}}{C}HCH_2CO_2H$

12.5 Give the IUPAC name for each of the following acids.

(a) 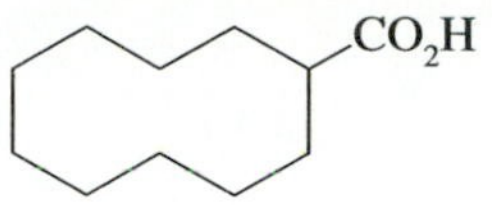 [cyclodecane ring]–CO_2H (b) 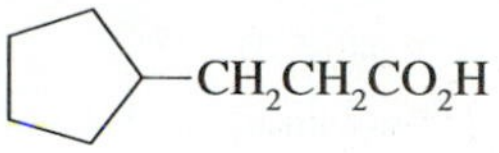 [cyclopentane ring]–$CH_2CH_2CO_2H$ (c)  CH_3O–[cyclohexane ring]–CO_2H

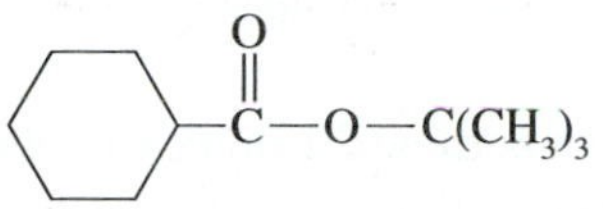

12.6 Give the IUPAC name for each of the following esters.

(a) 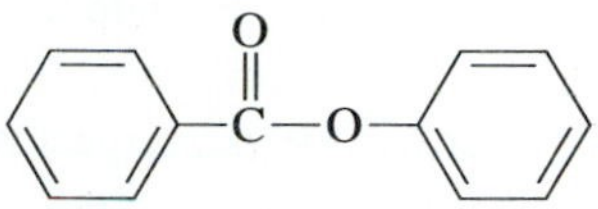 C_6H_5–C(=O)–O–C_6H_5 (b) [cyclohexane ring]–C(=O)–O–$C(CH_3)_3$ (c) [cyclodecane ring]–$CH_2CH_2CO_2CH(CH_3)_2$

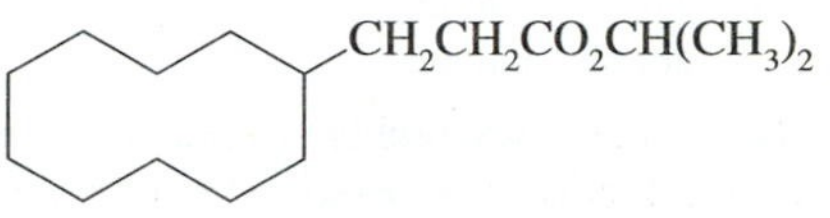

12.7 The IUPAC name of ibuprofen, the analgesic in Motrin®, Advil®, and Nuprin®, is 2-(4-isobutylphenyl)propanoic acid. Draw the structure.

12.8 10-Undecenoic acid is the antifungal agent contained in Desenex® and Cruex®. Write the structure.

Cyclic Acyl Derivatives

12.9 The IUPAC names of lactones are derived by adding the term *lactone* at the end of the name of the parent hydroxy acid. Name each of the following compounds.

(a) 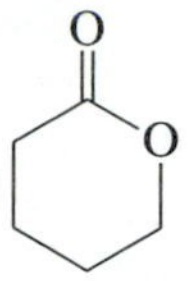 (b) 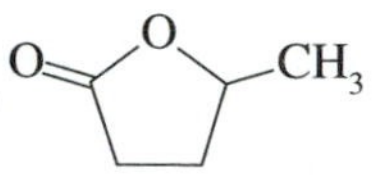 CH_3 (c)

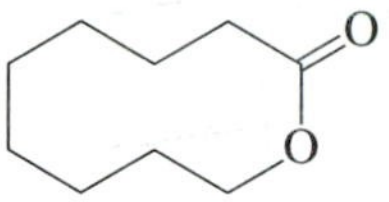

12.10 The IUPAC names of lactams are derived by adding the term *lactam* at the end of the name of the parent amino acid. Write the structure of each of the following lactams.

(a) 3-aminopropanoic acid lactam

(b) 4-aminopentanoic acid lactam

(c) 5-aminopentanoic acid lactam

12.11 Which of the following compounds are lactones?

(a) 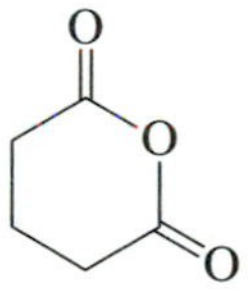 (b) 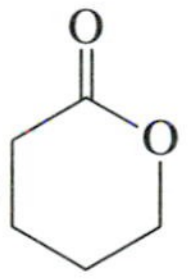 (c)

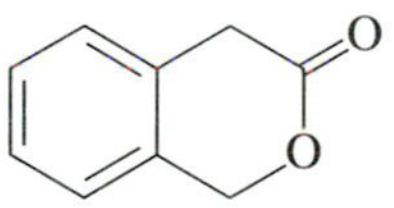

12.12 Which of the following compounds are lactams?

(a) 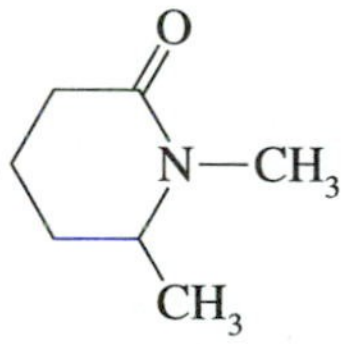 N–CH_3, CH_3 (b) 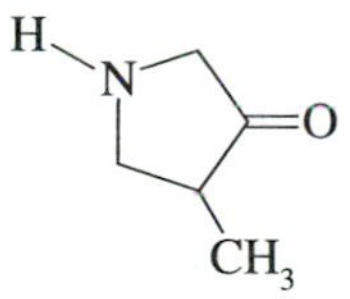 H–N, =O, CH_3 (c) 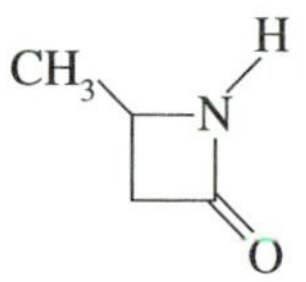 CH_3, N–H, O

Molecular Formulas

12.13 What is the general molecular formula for a saturated carboxylic acid?

12.14 What is the general molecular formula for a saturated dicarboxylic acid?

12.15 How many isomeric acids have the molecular formula $C_4H_8O_2$?

12.16 How many isomeric esters have the molecular formula $C_4H_8O_2$?

Properties of Acids

12.17 Why is 1-butanol less soluble in water than butanoic acid?

12.18 Adipic acid is much more soluble in water than hexanoic acid. Why?

12.19 The boiling point of decanoic acid is higher than that of nonanoic acid. Explain why.

12.20 The boiling points of 2,2-dimethylpropanoic acid and pentanoic acid are 164 and 186°C, respectively. Explain why these values differ as they do.

12.21 The boiling points of methyl pentanoate and butyl ethanoate are 126 and 125°C, respectively. Explain why the values are similar.

12.22 The boiling points of methyl pentanoate and methyl 2,2-dimethylpropanoate are 126 and 102°C, respectively. Explain why these values differ as they do.

Acidity of Carboxylic Acids

12.23 The K_a values of formic acid and acetic acid are 1.8×10^{-4} and 1.8×10^{-5}, respectively. Which compound is the stronger acid?

12.24 The pK_a values of acetic acid and benzoic acid are 4.74 and 4.19, respectively. Which acid is stronger?

12.25 The K_a of methoxyacetic acid is 2.7×10^{-4}. Explain why this value differs from the K_a of acetic acid (1.8×10^{-5}).

12.26 The K_a values of benzoic acid and *p*-nitrobenzoic acid are 6.3×10^{-5} and 3.8×10^{-4}, respectively. Explain why these values differ as they do.

12.27 The pK_a of benzoic acid is 4.2. The pK_a of probenecid, a drug used to treat gout, is 3.4. Explain why these values differ as they do.

$$(CH_3CH_2CH_2)_2N{-}\overset{\overset{\displaystyle O}{\|}}{\underset{\underset{\displaystyle O}{\|}}{S}}{-}C_6H_4{-}CO_2H$$

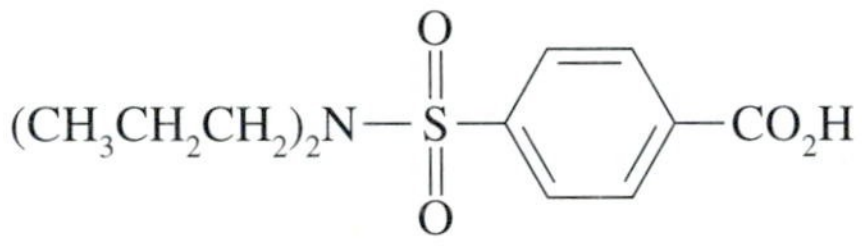

12.28 Predict the pK_a of indomethacin, an anti-inflammatory agent.

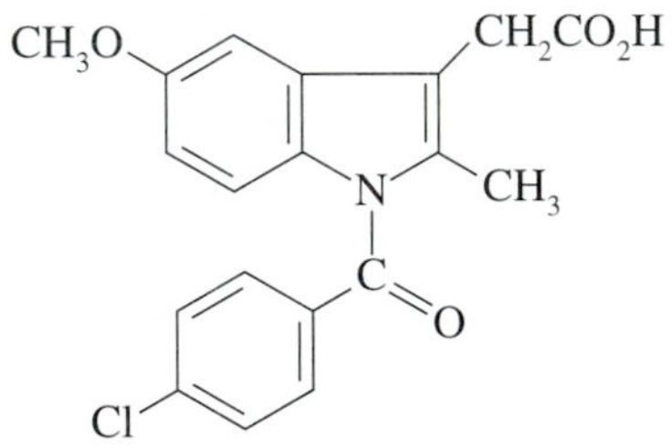

CH_3O CH_2CO_2H N CH_3 C O Cl

Nucleophilic Acyl Substitution

12.29 Indicate whether each of the following reactions will occur.

(a) $CH_3{-}\overset{\overset{\displaystyle O}{\|}}{C}{-}Cl + CH_3OH \longrightarrow CH_3{-}\overset{\overset{\displaystyle O}{\|}}{C}{-}OCH_3 + HCl$

(b) $CH_3{-}\overset{\overset{\displaystyle O}{\|}}{C}{-}NH_2 + CH_3OH \longrightarrow CH_3{-}\overset{\overset{\displaystyle O}{\|}}{C}{-}OCH_3 + NH_3$

(c) $CH_3{-}\overset{\overset{\displaystyle O}{\|}}{C}{-}OCH_3 + CH_3NH_2 \longrightarrow CH_3{-}\overset{\overset{\displaystyle O}{\|}}{C}{-}NHCH_3 + CH_3OH$

12.30 Indicate whether each of the following reactions will occur.

$$\text{(a) } CH_3{-}\overset{\overset{\displaystyle O}{\|}}{C}{-}O{-}\overset{\overset{\displaystyle O}{\|}}{C}{-}CH_3 + NH_3 \longrightarrow CH_3{-}\overset{\overset{\displaystyle O}{\|}}{C}{-}NH_2 + HO{-}\overset{\overset{\displaystyle O}{\|}}{C}{-}CH_3$$

$$\text{(b) } CH_3{-}\overset{\overset{\displaystyle O}{\|}}{C}{-}O{-}\overset{\overset{\displaystyle O}{\|}}{C}{-}CH_3 + HCl \longrightarrow CH_3{-}\overset{\overset{\displaystyle O}{\|}}{C}{-}Cl + HO{-}\overset{\overset{\displaystyle O}{\|}}{C}{-}CH_3$$

$$\text{(c) } CH_3{-}\overset{\overset{\displaystyle O}{\|}}{C}{-}O{-}\overset{\overset{\displaystyle O}{\|}}{C}{-}CH_3 + CH_3OH \longrightarrow CH_3{-}\overset{\overset{\displaystyle O}{\|}}{C}{-}OCH_3 + HO{-}\overset{\overset{\displaystyle O}{\|}}{C}{-}CH_3$$

Reduction of Acyl Derivatives

12.31 Draw the structure of the product of each of the following reactions.

(a) 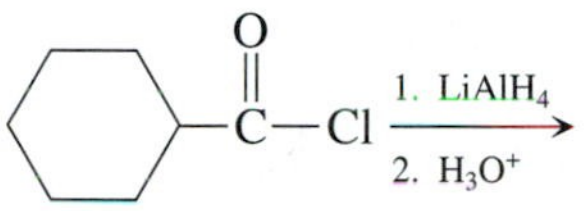 cyclohexyl–$\overset{\overset{\displaystyle O}{\|}}{C}{-}Cl$ $\xrightarrow[\text{2. } H_3O^+]{\text{1. } LiAlH_4}$

$$\text{(b) } CH_3CH_2\overset{\overset{\displaystyle CH_3}{|}}{C}HCH_2(CH_2)_4CH_2{-}\overset{\overset{\displaystyle O}{\|}}{C}{-}O{-}CH_3 \xrightarrow[\text{2. } H_3O^+]{\text{1. } LiAlH_4}$$

(c) 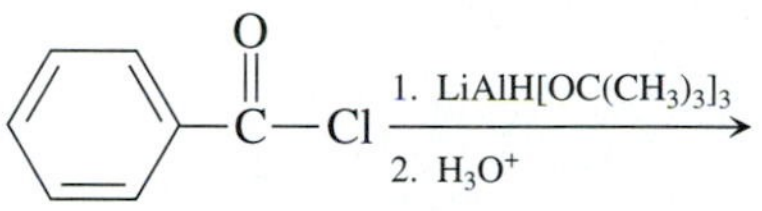 phenyl–$\overset{\overset{\displaystyle O}{\|}}{C}{-}Cl$ $\xrightarrow[\text{2. } H_3O^+]{\text{1. } LiAlH[OC(CH_3)_3]_3}$

12.32 Draw the structure of the product of each of the following reactions.

(a) 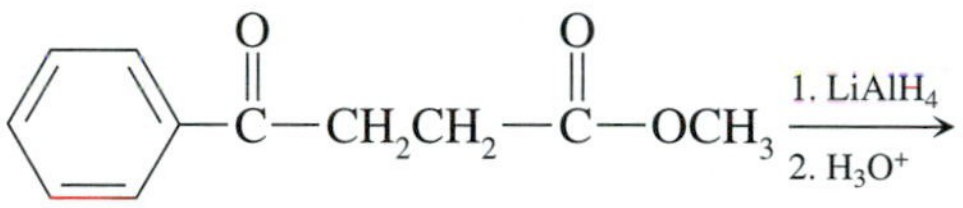 phenyl–$\overset{\overset{\displaystyle O}{\|}}{C}{-}CH_2CH_2{-}\overset{\overset{\displaystyle O}{\|}}{C}{-}OCH_3$ $\xrightarrow[\text{2. } H_3O^+]{\text{1. } LiAlH_4}$

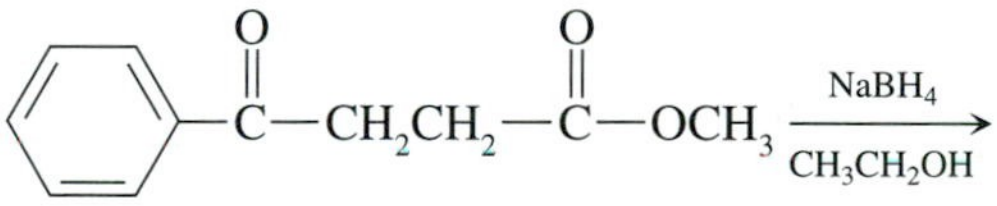

(b) phenyl–$\overset{\overset{\displaystyle O}{\|}}{C}{-}CH_2CH_2{-}\overset{\overset{\displaystyle O}{\|}}{C}{-}OCH_3$ $\xrightarrow[CH_3CH_2OH]{NaBH_4}$

(c) six-membered lactone (ring O adjacent to C=O) $\xrightarrow[\text{2. } H_3O^+]{\text{1. } LiAlH_4}$

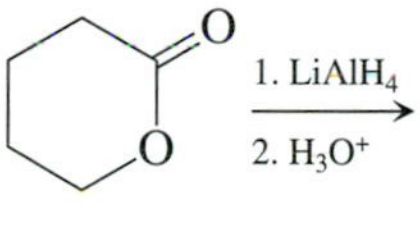

Chemical Reactions

12.33 Write the product of reaction of hexanoic acid with each of the following reagents.
(a) thionyl chloride (b) methanol (c) lithium aluminum hydride

12.34 Write the product of reaction of propanoyl chloride with each of the following compounds.
(a) 1-propanol (b) 2-butanethiol (c) benzylamine ($C_6H_5CH_2NH_2$)

12.35 Write the product of the reaction of butyrolactone with methylamine, CH_3NH_2.

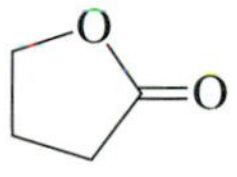

butyrolactone

12.36 Write the product of the reaction of phthalic anhydride with methanol.

phthalic anyhydride

12.37 When succinic acid ($HO_2CCH_2CH_2CO_2H$) is heated, a compound with the formula $C_4H_4O_3$ is formed. Suggest a structure for the compound.

12.38 When 3-(*o*-hydroxyphenyl)-propanoic acid is heated in the presence of an acid catalyst, a compound with molecular formula $C_9H_8O_2$ forms. Draw its structure.

Multistep Synthesis

12.39 Write the structure of the final product of each of the following sequences of reactions.

(a) cyclopentyl–CH_2CH_2OH $\xrightarrow[H_2SO_4/\text{acetone}]{CrO_3}$ $\xrightarrow{SOCl_2}$

(b) phenyl–CO_2CH_3 $\xrightarrow[2.\ H_3O^+]{1.\ LiAlH_4}$ $\xrightarrow{SOCl_2}$

(c) phenyl–C(=O)–Cl $\xrightarrow[2.\ H_3O^+]{1.\ LiAlH[OC(CH_3)_3]_3}$ $\xrightarrow[H^+]{CH_3OH}$

12.40 Write the structure of the final product of each of the following sequences of reactions.

(a) cyclopentyl–CHO $\xrightarrow[H_2SO_4/\text{acetone}]{CrO_3}$ $\xrightarrow{SOCl_2}$

(b) phenyl–C(=O)–Cl $\xrightarrow{CH_3OH}$ $\xrightarrow[2.\ H_3O^+]{1.\ LiAlH_4}$

(c) phenyl–C(=O)–Cl $\xrightarrow[2.\ H_3O^+]{1.\ LiAlH[OC(CH_3)_3]_3}$ $\xrightarrow[HCl]{Zn(Hg)}$

12.41 Outline the step(s) required to prepare cyclohexanecarboxylic acid from each of the following reactants.
(a) bromocyclohexane (b) cyclohexanol (c) vinylcyclohexane

12.42 Outline the steps required to prepare hexanoic acid from each of the following reactants.
(a) 1-chloropentane (b) hexanal (c) ethyl hexanoate

12.43 Fatty acids from natural sources are long-chain unbranched carboxylic acids that contain an even number of carbon atoms. Outline steps to convert a fatty acid into a homolog containing one additional carbon atom.

12.44 Pivalic acid, $(CH_3)_3CCO_2H$, can be prepared from *tert*-butyl chloride. What method should be used?

Esters

12.45 Write the products of hydrolysis of each of the following esters.

(a) $CH_3-C(=O)-O-CH_2CH_2CH_3$

(b) $CH_3CH_2CH_2CH_2-C(=O)-O-CH_2CH_2CH_2CH_3$

(c) $CH_3CH_2-C(=O)-O-CH_2(CH_2)_6CH_3$

12.46 Write the products of hydrolysis of each of the following esters.

(a) $CO_2CH_2CH_3$ (b) $-CH_2CO_2CH(CH_3)_2$ (c) CO_2CH_3 CO_2CH_3

12.47 What alcohol and acid are required to form each of the following esters?

(a) $CH_3CH_2CH_2-O-C(=O)-CH_3$

(b) $CH_3CH_2-O-C(=O)-CH_2CH_2CH_3$

(c) $CH_3(CH_2)_6CH_2-O-C(=O)-CH_2CH_3$

12.48 What alcohol and acid are required to form each of the following esters?

(a) $CH_2CO_2CH_2CH_3$ (b) $-CO_2CH(CH_3)_2$ CH_3 (c) $CO_2CH_2CH_3$ $CO_2CH_2CH_3$

Claisen Condensation

12.49 What ester is required to form the following compound by a Claisen condensation?

$$CH_3CH_2C(=O)CH(CH_3)CO_2CH_2CH_3$$

12.50 Write the structure of the product of a Claisen condensation reaction of methyl 2-phenylethanoate.

12.51 An intramolecular Claisen condensation reaction of a diester gives the following ketoester. What is the structure of the diester?

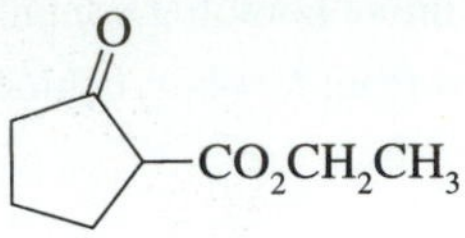

O

$—CO_2CH_2CH_3$

12.52 Under some circumstances, Claisen condensations can take place between two different esters. Explain why the reaction of ethyl acetate and ethyl benzoate occurs readily to give a good yield of a single product. What is the structure of the product?

Spectroscopy

12.53 If you have read Chapter 18 and the material has been assigned by your professor, then Exercises 18.9, 18.19, 18.20, and 18.25, which are based on the structures of carboxylic acids and esters, may be done.

CHAPTER 13

LIPIDS

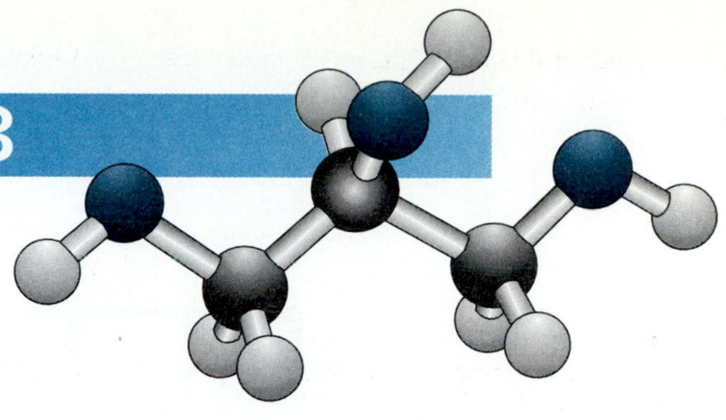

Overview

13.1 Classification of Lipids

Lipids are relatively nonpolar compounds, and they can be separated from more polar cellular substances by their solubility in nonpolar organic solvents. In fact, **lipids** were historically classified as compounds of biological origin that are soluble in organic solvents. The term *lipid* is sometimes used as a synonym for fat (from the Greek, *lipos,* fat). However, fat is only one of the various types of lipids.

The major classes of lipids and their relationship to one another are listed in Figure 13.1. Steroids, which are important hormones in animals, and terpenes, produced in plants, were discussed in Chapters 3 and 4, respectively. The emphasis of this chapter is on the remaining lipids, all of which can be hydrolyzed by aqueous basic solution. The major hydrolysis products of those lipids are the long-chain carboxylic acids with an even number of carbon atoms, called *fatty acids.* The other components resulting from the hydrolysis of these lipids determine the subclasses.

1. **Waxes** are esters of long-chain alcohols and fatty acids.
2. **Triacylglycerols,** also known as *triglycerides,* are esters of glycerol and long-chain fatty acids.
3. **Glycerophospholipids** are composed of glycerophosphate (an ester of glycerol and phosphoric acid), long-chain fatty acids, and certain low molecular weight alcohols.
4. **Sphingolipids** are composed of sphingosine, a long-chain fatty acid present as an amide, and at least one other component. There are two major subclasses of sphingolipids: *Sphingophospholipids* contain a phosphate unit and choline and thus structurally resemble glycerophospholipids. *Glycosphingolipids* contain a carbohydrate unit but not a phosphate unit.

Lipids encompass a range of molecular structures and, as a result, an extraordinary range of biochemical functions. As the molecules making

- Lipids
  - Fatty acids
    - Eicosanoids
    - Triacylglycerols (fats and oils)
    - Glycerophospholipids
      - Plasmalogens
      - Phosphatidates
        - Phosphatidylethanolamines
        - Phosphatidylserines
        - Phosphatidylcholines
        - Other phospholipids
    - Waxes
    - Sphingolipids
      - Ceramides
        - Sphingophospholipids
        - Glycosphingolipids
          - Cerebrosides
          - Gangliosides
  - Steroids
  - Terpenes

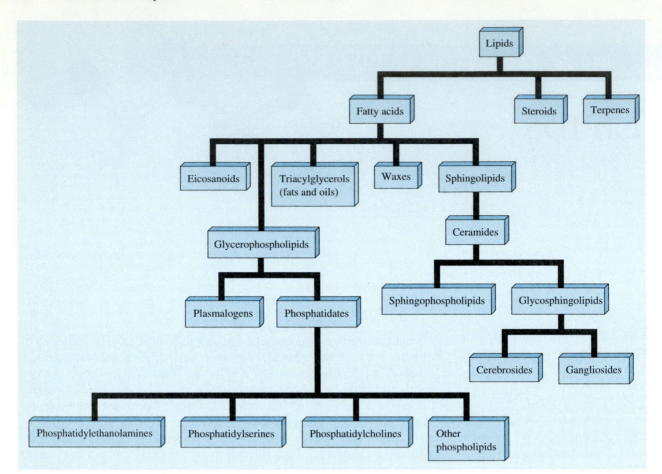

FIGURE 13.1
Classification of Lipids

up dietary fat, certain lipids provide a major source of metabolic energy. As components of biological membranes, lipids provide an insoluble partition between a cell and its watery environment. Finally, as hormones, lipids regulate a wide spectrum of cellular activities.

Some of the functions of lipids are performed by individual molecules such as steroidal hormones. Other functions are the result of aggregates of lipids that operate as units, like glycerophospholipids and sphingolipids in biological membranes. The role of lipids in the composition of biological membranes and how they serve highly specialized functions are discussed in this chapter.

Bees store honey in honeycombs containing wax.

13.2 Waxes

Waxes are esters of fatty acids and long-chain alcohols containing an even number of carbon atoms. They can be represented by a block diagram—a method that we shall use for other classes of lipids as well.

long-chain alcohol	—	long-chain carboxylic acid

Waxes are low-melting solids that coat the surface of plant leaves and fruits and also coat the hair and feathers of some animals. The coating on plants and fruits prevents loss of water and provides some protection against microorganisms. On the surface of feathers, hair, or skin of animals, waxes provide a water-repellent surface. If the wax is removed from the feathers of an aquatic bird as a result of an oil spill, the feathers become wet and the bird loses its buoyancy.

Waxes are usually mixtures of esters. Lanolin, the wax from sheep's wool, is a mixture of esters derived from 32 different alcohols and 36 different carboxylic acids. Waxes also contain small amounts of other compounds, such as alcohols and hydrocarbons. Beeswax, secreted by bees and used as a structural material for the beehive, contains about 20% hydrocarbons. On hydrolysis, beeswax yields several fatty acids and a mixture of alcohols having 24 to 36 carbon atoms. One example of a component ester shown here contains a carboxylic acid with 14 carbon atoms and an alcohol with 26 carbon atoms.

$$CH_3(CH_2)_{12}-C(=O)-O-CH_2(CH_2)_{24}CH_3$$

an ester found in beeswax

EXAMPLE 13-1

Half of the dry weight of a copepod that lives in the waters of British Columbia is the compound $C_{36}H_{62}O_2$. Hydrolysis of this compound yields an unbranched acid, $C_{20}H_{30}O_2$, and a straight-chain alcohol, $C_{16}H_{34}O$. Hydrogenation of the acid yields $C_{20}H_{40}O_2$. Describe the structure of the $C_{36}H_{62}O_2$ compound.

Solution

Since the compound yields an acid and an alcohol when hydrolyzed, it must be an ester. The formula $C_{16}H_{34}O$ corresponds to that of a saturated alcohol ($C_nH_{2n+2}O$). The acid, $C_{20}H_{30}O_2$, must contain five double bonds, since hydrogenation results in the addition of 10 hydrogen atoms. The locations of the double bonds cannot be determined from the information given. A general representation of the wax is as follows:

$$C_{19}H_{29}-C(=O)-O-CH_2(CH_2)_{14}CH_3$$

Problem 13.1

One of the components of carnauba wax, found on the leaves of palm trees, is an ester of an unbranched carboxylic acid containing 26 carbon atoms and an unbranched alcohol containing 30 carbon atoms. Write its structure.

13.3 Fatty Acids

Triacylglycerols, glycerophospholipids, sphingophospholipids, and glycosphingolipids all yield fatty acids on hydrolysis. **Fatty acids** are long-chain carboxylic acids containing an even number of carbon atoms. Over 100 different fatty acids are found in lipids in amounts that vary with the organism. However, those with 14 to 22 carbon atoms are prevalent in mammals, and the 16- and 18-carbon acids are the most abundant.

The most common unsaturated fatty acids also have either 16 or 18 carbon atoms. The double bonds in these molecules have the cis configuration. The po-

Table 13.1 Formulas and Melting Points of Fatty Acids

Name	Formula	Melting Point (°C)
Saturated		
lauric	$CH_3(CH_2)_{10}CO_2H$	44
myristic	$CH_3(CH_2)_{12}CO_2H$	58
palmitic	$CH_3(CH_2)_{14}CO_2H$	63
stearic	$CH_3(CH_2)_{16}CO_2H$	71
arachidic	$CH_3(CH_2)_{18}CO_2H$	77
Unsaturated		
oleic	$CH_3(CH_2)_7CH{=}CH(CH_2)_7CO_2H$	13
linoleic	$CH_3(CH_2)_4CH{=}CHCH_2CH{=}CH(CH_2)_7CO_2H$	−5
linolenic	$CH_3CH_2CH{=}CHCH_2CH{=}CHCH_2CH{=}CH(CH_2)_7CO_2H$	−11
arachidonic	$CH_3(CH_2)_4CH{=}CHCH_2CH{=}CHCH_2CH{=}CHCH_2CH{=}CH(CH_2)_3CO_2H$	−50

sition of the double bond can be indicated by the Greek letter delta (Δ), followed by a superscript to indicate the location of the double bond. Thus, the C-16 fatty acid with a cis double bond between the C-9 and C-10 atoms is called *cis*-Δ^9-hexadecenoic acid. Its common name is palmitoleic acid. The C-18 fatty acid with a cis double bond between the C-9 and C-10 atoms is called *cis*-Δ^9-octadecenoic acid. Its common name is oleic acid.

Fatty acids that contain more than one double bond are said to be **polyunsaturated.** The most common polyunsaturated fatty acids contain 18 carbon atoms and have either two or three double bonds. They are *cis,cis*-$\Delta^{9,12}$-octadecadienoic acid (linoleic acid) and *cis,cis,cis*-$\Delta^{9,12,15}$-octadecatrienoic acid (linolenic acid). These polyunsaturated fatty acids cannot be synthesized by mammals, and they are required in the diet.

The formulas and melting points of some fatty acids are given in Table 13.1. The melting points of the saturated fatty acids increase with increasing chain length because the London forces increase with increasing chain length. The hydrocarbon chains of saturated acids pack together tightly in the solid.

The *cis*-unsaturated acids are "bent" molecules because of the geometry about the double bonds. These "bends" hinder efficient packing, so the London forces between molecules are weaker. As a result, *cis*-unsaturated acids have lower melting points than saturated acids. Increasing the number of cis double bonds lowers the melting point further. Space-filling models of stearic acid and oleic acid are shown in Figure 13.2.

EXAMPLE 13-2

The melting point of palmitoleic acid (*cis*-Δ^9-hexadecenoic acid) is $-1°C$. Compare this melting point with that of palmitic acid and explain the difference.

$CH_3(CH_2)_4CH_2$ and $(CH_2)_7CO_2H$ on the same side of C=C, with H and H on the other side

palmitoleic acid

Stearic acid

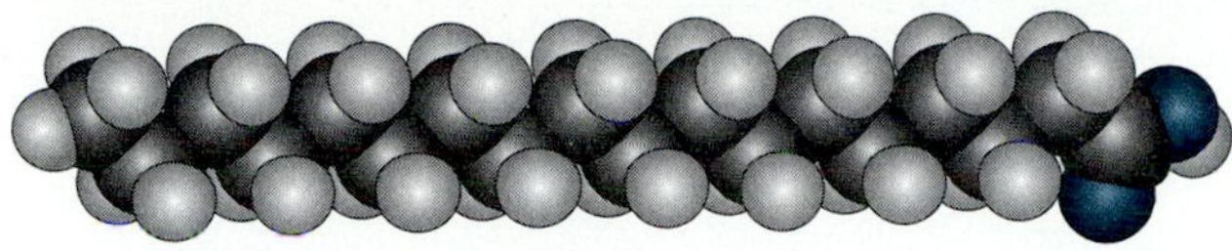

Oleic acid

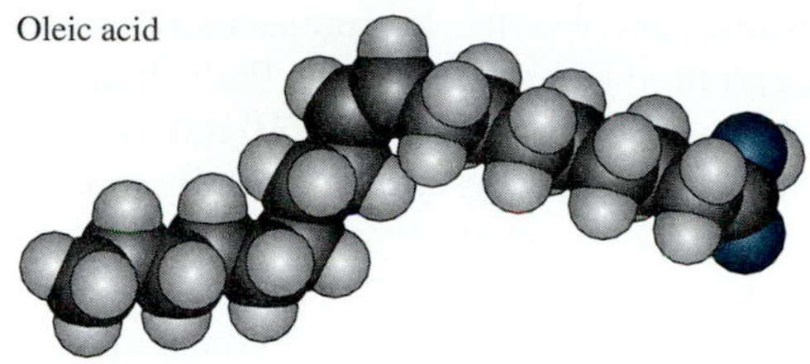

FIGURE 13.2
Space-filling Models of Fatty Acids

Solution

The melting point of palmitic acid is 63°C, which is 64° higher than the melting point of palmitoleic acid. This difference is due to the presence of a cis double bond in palmitoleic acid, which prevents close approach of the hydrocarbon chains and results in weaker London forces.

Problem 13.2

Using data in Table 13.1, estimate the melting point of $CH_3(CH_2)_{22}CO_2H$.

13.4 Triacylglycerols

Triacylglycerols, which are triesters of glycerol and fatty acids, are also known as fats and oils. Triacylglycerols are represented by the block diagram shown in Figure 13.3.

Fats and oils are mixtures of compounds. The acid fraction of these molecules varies in chain length and degree of unsaturation. A single molecule of a fat or oil may contain up to three different acid residues. **Fats** have a high percentage of saturated acids, whereas **oils** have a high percentage of unsaturated acids. Fats are solids or semisolids and are usually obtained from animals. The important acids found in these sources are myristic, palmitic, and stearic acid (Table 13.1). The unsaturated acids found in oils are oleic, linoleic, and linolenic acid; all contain 18 carbon atoms, but they differ in their degree of unsaturation (Table 13.1).

Oils are obtained from vegetable sources such as olives, peanuts, corn, and soybeans. The unsaturated acid residues in the molecules of oils lower their melting points, and they are usually liquids.

Animals accumulate fat (adipose tissue) when their intake of food exceeds their energy output requirements. Vital organs such as kidneys are enclosed in adipose tissue, which provides a protective cushion against a blow. A subcutaneous layer of fat helps insulate the animal against heat loss. Although plants do not generally store fats and oils for energy requirements, some (such as peanuts and olives) produce triacylglycerols in abundance.

Camels store fatty acids in the form of esters to use for metabolic energy.

Prostaglandins

Long-chain fatty acids are widely distributed in natural products. Some have profound effects on biological processes even at extremely low concentrations. Prostaglandins, leukotrienes, and thromboxanes make up a class of hormones called *eicosanoids,* all of which derive from arachidonic acid, a 20-carbon unsaturated acid (see the accompanying figure). However, unlike many hormones, the eicosanoids are not produced by glands and secreted into the blood to affect cells throughout the body. Instead eicosanoids act locally, in the cells that produce them or in neighboring cells. They regulate normal processes, such as smooth muscle contraction, and affect cellular processes that result from diseases. For example, prostaglandins contribute to the symptoms of an illness, such as swelling, nausea, vomiting, and pain.

Prostaglandins are 20-carbon fatty acids that contain a *trans*-substituted, five-membered ring. They are classified as PGA through PGI based on the number and type of their functional groups, such as hydroxyl groups and ketone groups. A subscript on the letter indicates the number of double bonds.

The prostaglandins were originally isolated from the prostate gland. But all nucleated cells produce them, and they can affect any cell type, even at very low concentrations. They can also alter the effects of many hormones. For example, hormone-sensitive enzymes that hydrolyze lipids respond to insulin and to other hormones that regulate the concentration of blood glucose. PGE_1 inhibits these enzymes at a concentration of 10 nM (10^{-8} M).

Prostaglandins affect virtually every aspect of reproduction. They regulate menstruation and control fertility and contraception. PGE_2 stimulates smooth muscle contraction in the uterus. It has been used clinically to induce labor and to abort pregnancies prematurely.

A class of steroid hormones called corticosteroids also inhibits prostaglandin biosynthesis. These steroids act in the following way: arachidonic acid does not occur free in cells, but is esterified to a phospholipid and is incorporated in this form into cell membranes. Arachidonic acid is released by enzymes called phospholipases, and corticosteroids inhibit these enzymes. Corticosteroids such as cortisone are used to reduce inflammation.

Much research aims to develop synthetic prostaglandins for use as therapeutic drugs. Natural prostaglandins cannot be taken orally because they are rapidly degraded and do not survive long enough for effective action. Thus, one research goal is to develop modified prostaglandins that can be administered orally. ■

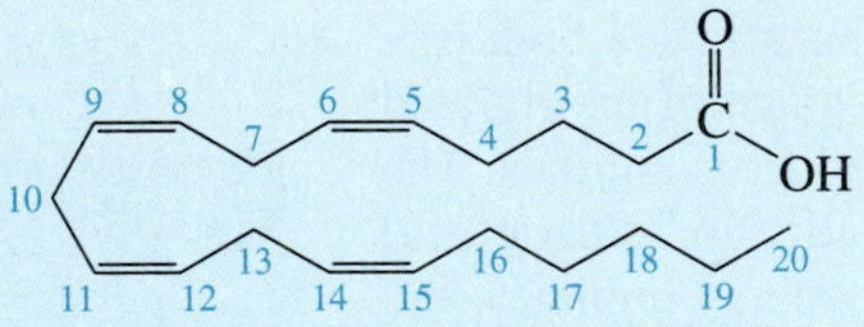

arachidonic acid

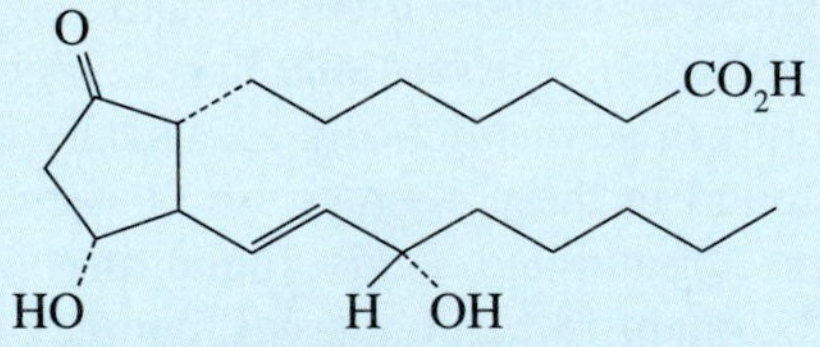

prostaglandin E_1

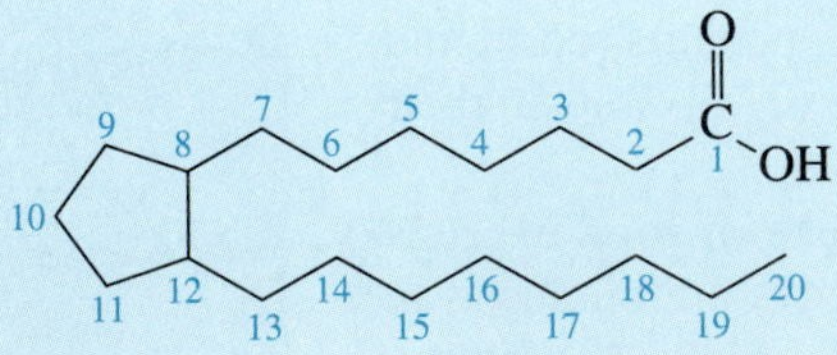

prostanoic acid

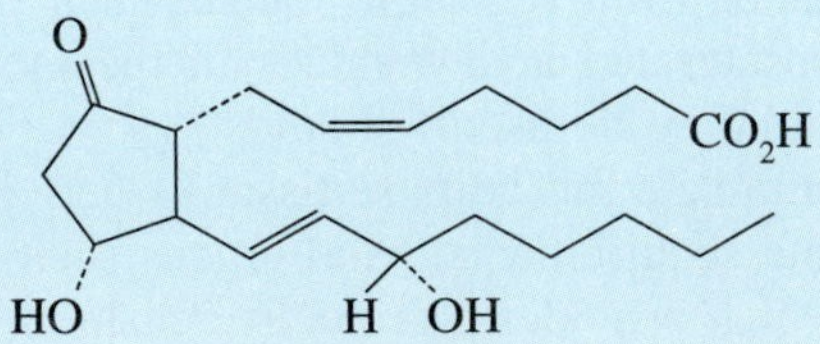

prostaglandin E_2

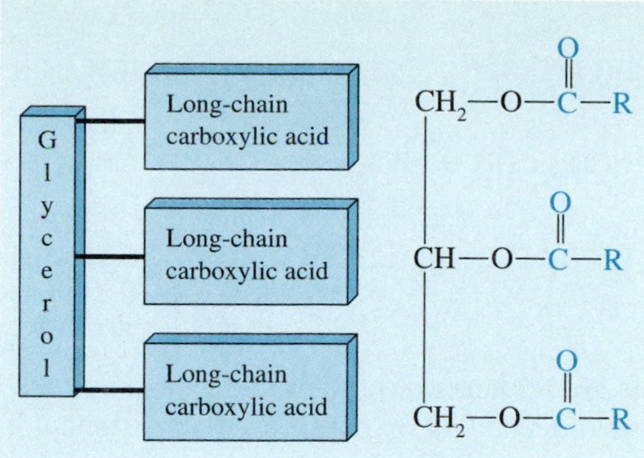

FIGURE 13.3
Representation of a Triacylglycerol

The composition of triacylglycerols varies, as indicated in Table 13.2. The relationship between consumption of saturated fats and arterial disease has been the object of extensive medical research. Unsaturated triacylglycerols are beneficial in preventing arterial deposits.

13.5 Hydrogenation of Vegetable Oils

The catalytic hydrogenation of the carbon-carbon double bonds in vegetable oils occurs in much the same way as the hydrogenation of simple alkenes to give alkanes. Cooking fats such as Crisco® and other commercial products such as margarine are produced by large-scale chemical hydrogenation of vegetable oils. The resulting "fat" product is more saturated than the "oil" reactant.

These non-stick sprays contain oils that are esters of unsaturated fatty acids.

Table 13.2 Composition of Fats and Oils

	Melting Point (°C)	Saturated Fatty Acids (%)				Unsaturated Fatty Acids(%)		
		Myristic	Palmitic	Stearic	Arachidic	Oleic	Linoleic	Linolenic
Animal fats								
butter	32	11	29	9	2	27	4	—
lard	30	1	28	12	—	48	6	—
human fat	15	3	24	8	—	47	10	—
Plant oils								
corn	−20	1	10	3	—	50	34	—
cottonseed	−1	1	23	1	1	23	48	—
linseed	−24	—	6	2	1	19	24	47
olive	−6	—	7	2	—	84	5	—
peanut	3	—	8	3	2	56	26	—
soybean	−16	—	10	2	—	29	51	6

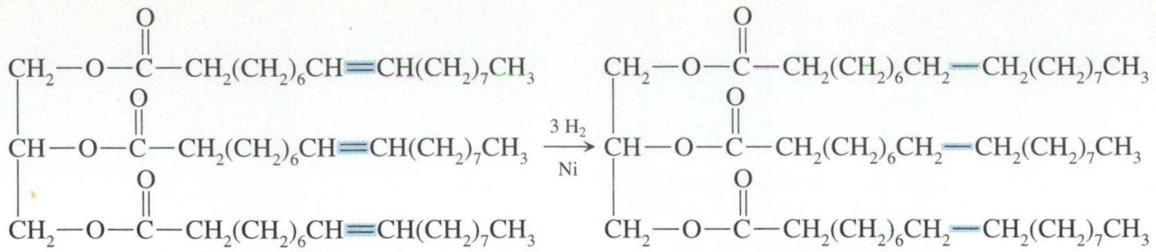

$$\begin{array}{l} CH_2-O-\overset{\overset{\large O}{\|}}{C}-CH_2(CH_2)_6CH{=}CH(CH_2)_7CH_3 \\ | \\ CH-O-\overset{\overset{\large O}{\|}}{C}-CH_2(CH_2)_6CH{=}CH(CH_2)_7CH_3 \\ | \\ CH_2-O-\overset{\overset{\large O}{\|}}{C}-CH_2(CH_2)_6CH{=}CH(CH_2)_7CH_3 \end{array} \xrightarrow[Ni]{3\,H_2} \begin{array}{l} CH_2-O-\overset{\overset{\large O}{\|}}{C}-CH_2(CH_2)_6CH_2-CH_2(CH_2)_7CH_3 \\ | \\ CH-O-\overset{\overset{\large O}{\|}}{C}-CH_2(CH_2)_6CH_2-CH_2(CH_2)_7CH_3 \\ | \\ CH_2-O-\overset{\overset{\large O}{\|}}{C}-CH_2(CH_2)_6CH_2-CH_2(CH_2)_7CH_3 \end{array}$$

glyceryl trioleate (triolein) — glyceryl tristearate (tristearin)

Vegetable oils are mixtures of unsaturated esters, and the extent of the hydrogenation is varied depending on the desired consistency of the product. For example, margarine is obtained by hydrogenating about two-thirds of the double bonds. As a result, the semisolid product has a consistency that keeps it as a solid under refrigeration but allows it to melt on a hot muffin.

During hydrogenation some of the monounsaturated fatty acid residues are isomerized from the normal cis configuration to the trans configuration. The trans isomers have been tentatively associated with an increase in cholesterol level, which in turn is a contributor to heart disease.

EXAMPLE 13-3

Soybean oil is 51% linoleic acid. Draw a structure for one of the possible components of soybean oil.

Solution

Although all oils are mixtures of triacylglycerols, the large percentage of linoleic acid means that there must be a large amount of triacylglycerols containing two or three units of linoleic acid. The structure for the triglyceryl ester of linoleic acid is shown.

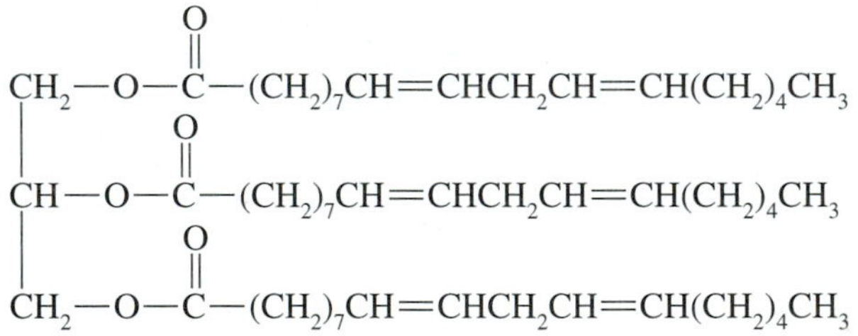

$$\begin{array}{l} CH_2-O-\overset{\overset{\large O}{\|}}{C}-(CH_2)_7CH{=}CHCH_2CH{=}CH(CH_2)_4CH_3 \\ | \\ CH-O-\overset{\overset{\large O}{\|}}{C}-(CH_2)_7CH{=}CHCH_2CH{=}CH(CH_2)_4CH_3 \\ | \\ CH_2-O-\overset{\overset{\large O}{\|}}{C}-(CH_2)_7CH{=}CHCH_2CH{=}CH(CH_2)_4CH_3 \end{array}$$

Problem 13.3

How many moles of hydrogen gas would react with an oil containing one unit each of oleic, linoleic, and linolenic acid? What would be the product?

13.6 Saponification of Triacylglycerols

Because triacylglycerols are esters, they can be hydrolyzed to glycerol and fatty acids. Under basic hydrolysis conditions, the salts of fatty acids result, which are called **soaps.** The best soaps are carboxylate salts made from saturated acids with 14 to 18 carbon atoms. Soaps fabricated as bars are usually sodium salts, whereas the potassium salts, which are softer, are used in shaving creams. The saponification reaction is shown with glyceryl tristearate.

$$\begin{array}{l} CH_2-O-\overset{O}{\overset{\|}{C}}-CH_2(CH_2)_{15}CH_3 \\ | \\ CH-O-\overset{O}{\overset{\|}{C}}-CH_2(CH_2)_{15}CH_3 + 3\,NaOH \longrightarrow \\ | \\ CH_2-O-\overset{O}{\overset{\|}{C}}-CH_2(CH_2)_{15}CH_3 \end{array} \begin{array}{l} CH_2-OH \\ | \\ CH-OH \\ | \\ CH_2-OH \end{array} + 3\,CH_3(CH_2)_{16}-\overset{O}{\overset{\|}{C}}-O^-\,Na^+$$

glyceryl tristearate (tristearin) — glycerol — sodium stearate

Glyceryl tristearate in a mixture of other triacylglycerols is obtained from the beef tallow that is a byproduct of the meat-packing industry. Some soaps are made from triacylglycerols obtained from coconut or palm oil.

The cleansing power of soaps is a direct consequence of the long, nonpolar hydrocarbon chains of the carboxylate ions. They are insoluble in water but cluster to form spherical aggregate structures called micelles. In a **micelle** of carboxylate salts, the nonpolar hydrocarbon chains point toward the interior of the sphere, and the polar carboxylate "heads" lie on the surface of the sphere (Figure 13.4). This arrangement encloses the maximum amount of "hydrocarbon" material for the smallest surface area. As a consequence, the hydrogen-bonded structure of water is disrupted to the smallest extent possible.

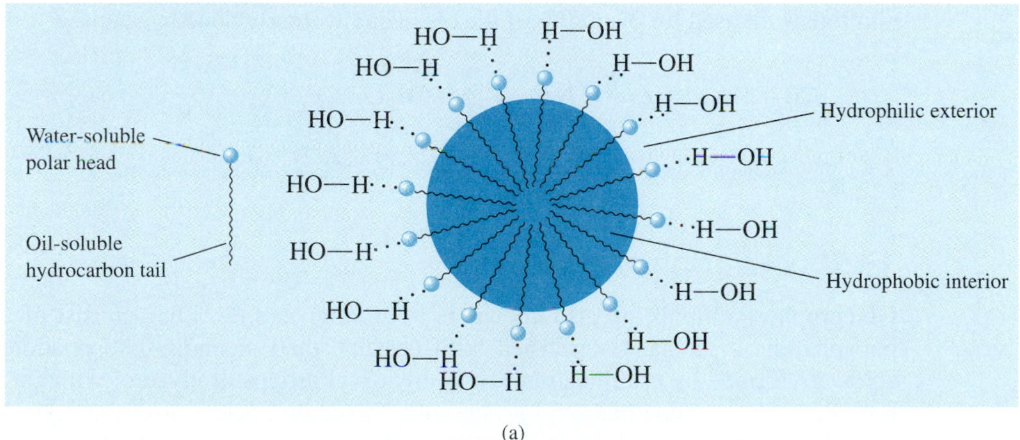

(a)

(b)

FIGURE 13.4
The Micelle of a Soap

In (a) the association of the nonpolar tails of the carboxylate ions is shown as a cross section of the micelle. The polar carboxylate heads are hydrogen-bonded to water molecules. Grease dissolves in the hydrophobic interior. In (b) the three-dimensional shape of the micelle is illustrated with a cut-away portion showing the hydrophobic interior.

The nonpolar hydrocarbon chain of a fatty acid repels water and is called **hydrophobic** ("water fearing"). In contrast, the polar "head" of the carboxylate group forms hydrogen bonds to water and is called **hydrophilic** ("water loving"). London forces between hydrocarbon chains hold the micelle together. The tendency of nonpolar solutes to aggregate in aqueous solution is called the **hydrophobic effect.** The micelle surface, which may contain as many as 100 carboxylate groups, has a large number of negative charges. As a result, individual micelles repel each other and remain suspended in water.

Grease is a hydrophobic substance, which does not dissolve in water because it is nonpolar. However, grease will dissolve in the hydrocarbon region of the micelle and then be washed away in the polar wash water. This process accounts for the cleansing action of a soap.

Detergents contain sulfate esters as one of the cleansing agents.

The micelles of carboxylate salts are ineffective cleansing agents in the presence of the ions in hard water, which contains relatively high concentrations of Ca^{2+} and Mg^{2+} ions. These ions react with the carboxylate ions of soaps and form precipitates (soap scum) that reduce the cleansing power of the soap.

$$2\ CH_3(CH_2)_{16}CO_2^-\ Na^+ + Ca^{2+} \longrightarrow [CH_3(CH_2)_{16}CO_2^-]_2Ca^{2+} + 2\ Na^+$$

sodium stearate (soluble as a micelle) — calcium stearate (insoluble in water)

For this reason, **detergents**—salts of organic sulfate esters—work better than soaps in hard water. Like soaps, detergents have long hydrophobic tails and hydrophilic heads, and form micelles. However, they do not form precipitates with Ca^{2+} and Mg^{2+} ions. These soap substitutes were produced in the late 1950s and today are used for over 80% of the cleansing formerly done by soaps.

$CH_3(CH_2)_{11}—O—SO_3^-\ Na^+$

sodium dodecyl sulfate

$CH_3(CH_2)_9—CH(CH_3)—C_6H_4—SO_3^-\ Na^+$

sodium *p*-(2-dodecyl)benzenesulfonate

13.7 Glycerophospholipids

Glycerophospholipids are also known as *phosphoglycerides.* They consist of a phosphatidic acid that is esterified with certain small alcohols. Phosphatidic acids are formed by esterification of two hydroxyl groups of glycerol with carboxylic acids, and one hydroxyl group with phosphoric acid. (At physiological pH, phosphatidic acids exist in an ionized phosphatidate form.) The C-1 atom of glycerol in a phosphatidic acid is usually esterified with a saturated acid containing 12 to 20 carbon atoms. The acid at C-2 is usually unsaturated. The phosphate group is always bonded to the C-3 atom.

$$\begin{array}{l} CH_2—O—C(=O)—R \\ | \\ R—C(=O)—O—CH \\ | \\ CH_2—O—P(=O)(OH)—OH \end{array}$$

phosphatidic acid

Phosphatidic acids combined with alcohols such as ethanolamine, choline, serine, or inositol are called phosphatidyl esters or phosphatides. Examples of

Table 13.3 Composition of Glycerophospholipids

General Structure	X =	X—OH Name	Type of Glycerophospholipid
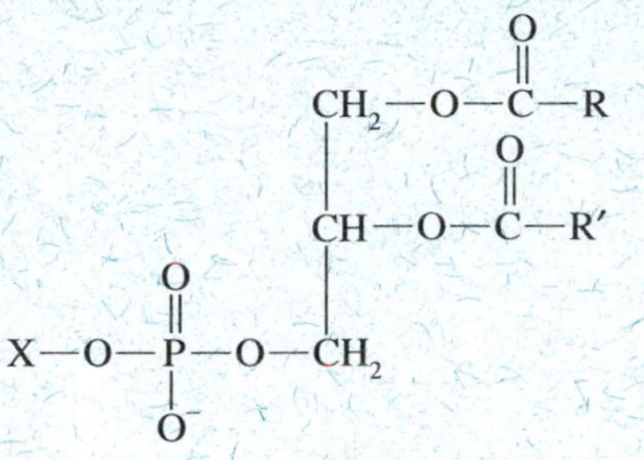 CH_2—O—C(=O)—R; CH—O—C(=O)—R′; X—O—P(=O)(O^-)—O—CH_2	H—	Water	Phosphatidic acid
	$H_3\overset{+}{N}CH_2CH_2$—	Ethanolamine	Phosphatidylethanolamine
	$(CH_3)_3\overset{+}{N}CH_2CH_2$—	Choline	Phosphatidylcholine
	$^-O_2CCHCH_2$— (with NH_3^+ on CH)	Serine	Phosphatidylserine
	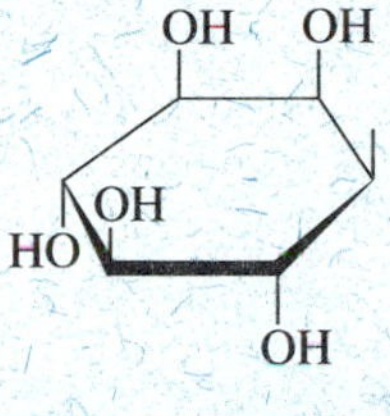 inositol ring (OH, OH, OH, HO, OH)	Inositol	Phosphatidylinositol

the various types of glycerophospholipids are listed in Table 13.3. A block representation of a glycerophospholipid is given in Figure 13.5.

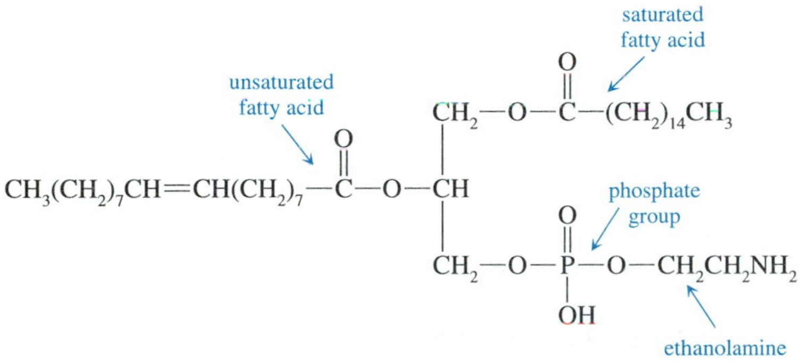

saturated fatty acid

$$CH_2—O—\overset{O}{\overset{\|}{C}}—(CH_2)_{14}CH_3$$

unsaturated fatty acid

$$CH_3(CH_2)_7CH{=}CH(CH_2)_7—\overset{O}{\overset{\|}{C}}—O—CH$$

phosphate group

$$CH_2—O—\overset{O}{\overset{\|}{P}}(OH)—O—CH_2CH_2NH_2$$

ethanolamine

phosphatidylethanolamine

The phosphatides exist as anions at physiological pH (approximately 7). The dissociation of the proton on the phosphate oxygen atom is essentially complete at pH 7.0. Phosphatidylethanolamine is protonated at the nitrogen atom at pH 7.0. Phosphatidylcholine has a positive charge at the nitrogen atom because choline contains a quaternary ammonium ion. As a result, both phosphatidylcholine and phosphatidylethanolamine are dipolar but have no net charge. Phosphatidylserine has three charged sites at pH 7.0. The phosphate oxygen atom bears a negative charge because the proton is dissociated at pH 7.0. In addition, the carboxyl group of serine exists as the carboxylate ion, and the amine group of serine is protonated. Thus, phosphatidylserine bears a net negative charge.

The older name of phosphatidylcholines is **lecithins.** The phosphatidylcholines are found in the protoplasm of body cells, where they serve as emulsifying agents to transport fat molecules from one tissue to another. The nonpolar fat molecules dissolve within the micelle much as grease does in a soap micelle.

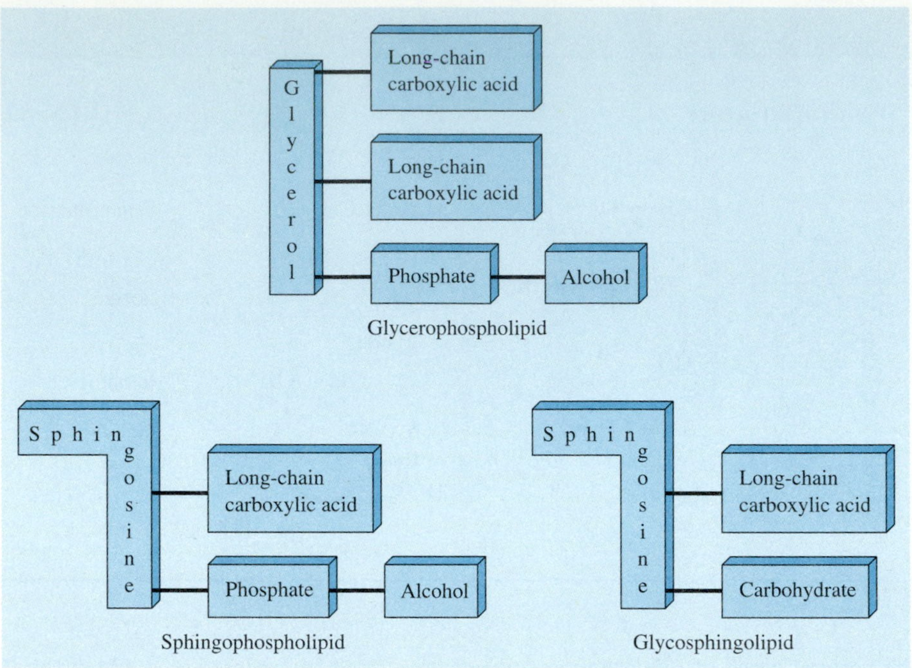

FIGURE 13.5
Representation of Components of Lipids

Phosphatidylethanolamines used to be called **cephalins.** These lipids are found in the heart and liver and in high concentration in brain tissue. Phosphatidylethanolamines are also essential to the blood-clotting mechanism.

All glycerophospholipids have polar sites, which are represented by a circle in the simplified symbol. The two nonpolar hydrocarbon chains of the fatty acids are represented with wavy lines or "tails" attached to the polar "head." These structural features constitute a "head-and-tail" model that is used in representing the structure of cell membranes (Section 13.10).

polar head → ○〰 ← nonpolar tails

representation of a glycerophospholipid

13.8 Sphingophospholipids

Sphingophospholipids contain sphingosine, a long-chain unsaturated amino alcohol, rather than glycerol.

$$CH_3(CH_2)_{12}-CH{=}CH-\underset{|}{CH}-O-H$$

$$H-\underset{\underset{H}{|}}{N}-CH$$

$$CH_2-OH$$

amine (pointing to N–H); primary hydroxyl group (pointing to CH_2–OH)

sphingosine

Conversion of the amine of sphingosine into an amide with a fatty acid yields a ceramide. Esterification of the primary alcohol of a ceramide with phosphoric acid and choline gives a sphingophospholipid. A block diagram of a sphingophospholipid is given in Figure 13.5.

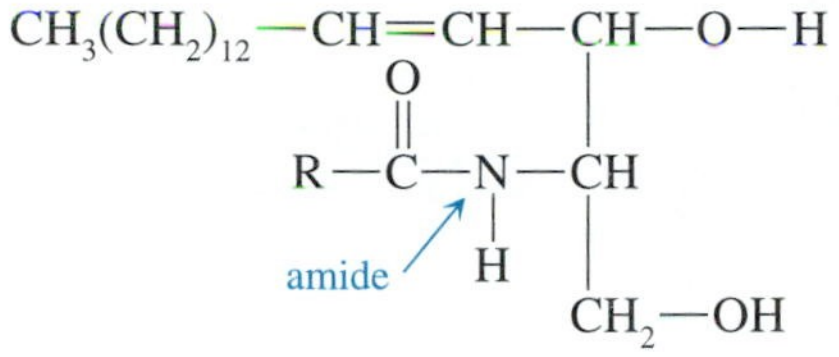

$$CH_3(CH_2)_{12}-CH{=}CH-\underset{\substack{|\\ \displaystyle R-\overset{\displaystyle O}{\overset{\|}{C}}-\underset{\displaystyle H}{\underset{|}{N}}-CH\\ |\\ CH_2-OH}}{CH}-O-H$$

amide

a ceramide

$$CH_3(CH_2)_{12}-CH{=}CH-\underset{\substack{|\\ \displaystyle R-\overset{\displaystyle O}{\overset{\|}{C}}-\underset{\displaystyle H}{\underset{|}{N}}-CH\\ |\\ CH_2-O-\overset{\displaystyle O}{\overset{\|}{P}}-O-CH_2-\overset{\displaystyle NH_3^+}{\overset{|}{\underset{\displaystyle H}{\underset{|}{C}}}}-CO_2^-\\ \displaystyle \underset{\displaystyle O^-}{|}}}{CH}-O-H$$

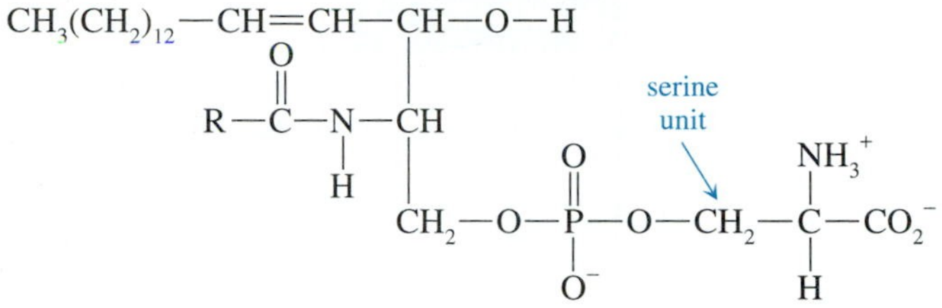

serine unit

a sphingophospholipid

Compare the structures of a sphingophospholipid and a glycerophospholipid (Figure 13.5). Although the components are different, the overall structures are similar. Both compounds have a polar head and two nonpolar tails. However, there are significant chemical differences. Sphingophospholipids have a single amide group and are stable to hydrolysis; glycerophospholipids have two carboxylic esters that are easily hydrolyzed.

Sphingomyelins, compounds found in the myelin sheath surrounding nerve fibers, are sphingophospholipids. Sphingomyelins have acid residues that are 20 to 26 carbon atoms long. These long chains form a very stable coating for nerve fibers.

13.9 Glycosphingolipids

Glycosphingolipids are similar to sphingophospholipids; they contain sphingosine and a fatty acid residue bonded as an amide. However, glycosphingolipids contain no phosphate. A carbohydrate is bonded via a glycosidic linkage to the primary alcohol oxygen atom of the ceramide. A block diagram of a glycosphingolipid is given in Figure 13.5.

$$CH_3(CH_2)_{12}-CH{=}CH-\underset{\substack{|\\ \displaystyle R-\overset{\displaystyle O}{\overset{\|}{C}}-\underset{\displaystyle H}{\underset{|}{N}}-CH\\ |\\ CH_2-O-\text{(sugar ring: H, O)}}}{CH}-O-H$$

monosaccharide or oligosaccharide

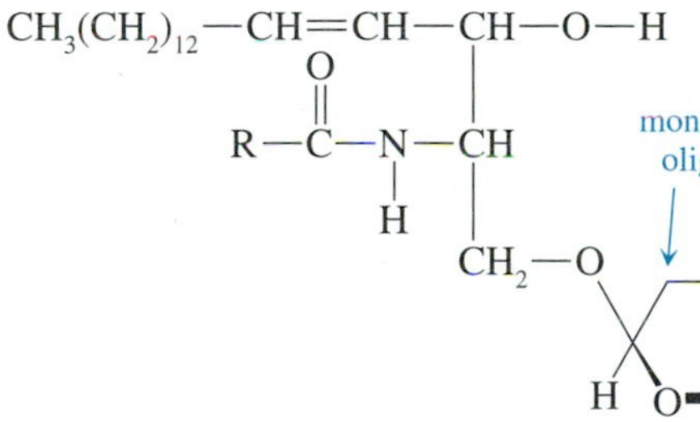

glycosphingolipid

Cerebrosides and gangliosides are glycosphingolipids. Cerebrosides contain only glucose or galactose, whereas gangliosides contain an oligosaccharide. Both are found in the myelin sheath. Cerebrosides are in the white matter of the central nervous system; gangliosides occur in the gray matter of the brain.

Gangliosides are synthesized or degraded by sequential addition or removal of monosaccharide units. The degradation of a ganglioside occurs inside lysosomes, which have the necessary enzymes for the reaction. Tay-Sachs disease is a genetic disease in which gangliosides accumulate in the brain because of a deficiency of the enzymes necessary for their degradation. Retarded development, blindness, and death at an early age are the consequences of this disease, which occurs 100 times more frequently in American Jews than in other Americans.

EXAMPLE 13-4

Plasmalogens are a class of lipids having the following general structure. Identify the components of the lipid and determine which class of lipids it most closely resembles.

$$\begin{array}{l} \qquad\qquad\qquad\quad CH_2{-}O{-}CH{=}CH{-}R \\ \qquad O \\ \qquad \| \\ R{-}C{-}O{-}CH \\ \qquad\qquad\qquad\qquad\qquad\quad O \\ \qquad\qquad\qquad\qquad\qquad\quad \| \\ \qquad\qquad\qquad\quad CH_2{-}O{-}P{-}O{-}CH_2CH_2NH_2 \\ \qquad\qquad\qquad\qquad\qquad\quad | \\ \qquad\qquad\qquad\qquad\qquad\quad OH \end{array}$$

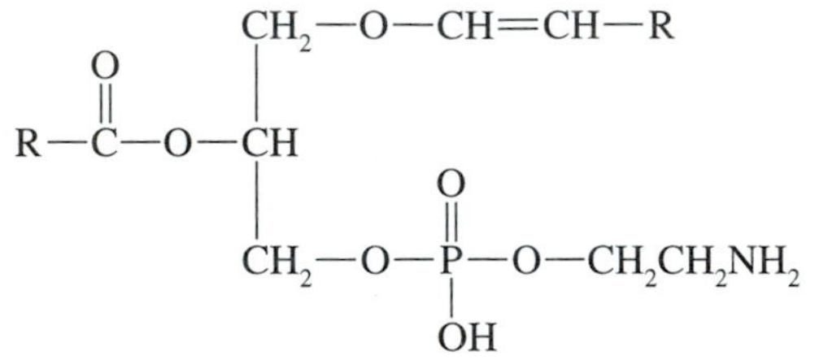

Solution

The center portion of the structure is derived from glycerol. A phosphate ester of ethanolamine appears at the bottom right. An ester of a carboxylic acid occurs at C-2 of glycerol. Only the residue at the top right is different from the other classes of lipids. Instead of an ester the residue is an ether. This structure most closely resembles a phosphatidylethanolamine.

Problem 13.4

Identify the components of the following compound and classify it.

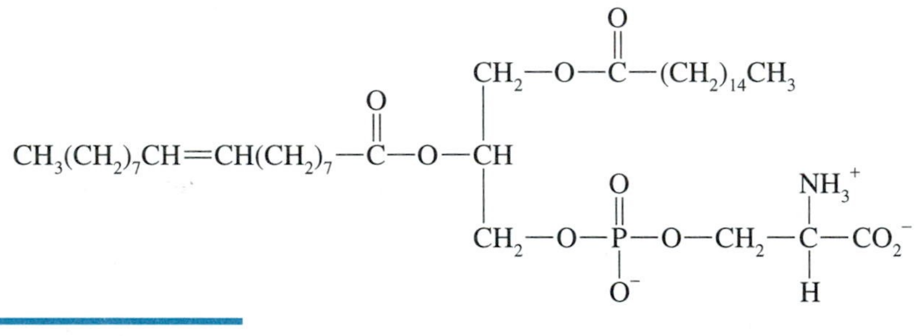

$$\begin{array}{l} \qquad\qquad\qquad\qquad\qquad\qquad\qquad\qquad\qquad\quad O \\ \qquad\qquad\qquad\qquad\qquad\qquad\qquad\qquad\qquad\quad \| \\ \qquad\qquad\qquad\qquad\qquad\qquad\qquad CH_2{-}O{-}C{-}(CH_2)_{14}CH_3 \\ \qquad\qquad\qquad\qquad\qquad\qquad O \\ \qquad\qquad\qquad\qquad\qquad\qquad \| \\ CH_3(CH_2)_7CH{=}CH(CH_2)_7{-}C{-}O{-}CH \\ \qquad\qquad\qquad\qquad\qquad\qquad\qquad\qquad\qquad\quad O \qquad\qquad\qquad NH_3^+ \\ \qquad\qquad\qquad\qquad\qquad\qquad\qquad\qquad\qquad\quad \| \qquad\qquad\qquad\quad | \\ \qquad\qquad\qquad\qquad\qquad\qquad\qquad CH_2{-}O{-}P{-}O{-}CH_2{-}C{-}CO_2^- \\ \qquad\qquad\qquad\qquad\qquad\qquad\qquad\qquad\qquad\quad | \qquad\qquad\qquad\quad | \\ \qquad\qquad\qquad\qquad\qquad\qquad\qquad\qquad\qquad\quad O^- \qquad\qquad\quad\;\; H \end{array}$$

13.10 Biological Membranes

We recall that long-chain carboxylate ions associate in micelles that have the nonpolar hydrocarbon chain in the interior and the polar head on the surface. Phospholipids have two nonpolar hydrocarbon chains and readily form lipid bilayers, as depicted in Figure 13.6, that are about 5 nm thick. The hydrophobic hydrocarbon chains point toward the interior of the bilayer. The hydrophilic heads are in contact with aqueous solutions on each side of the two layers.

Lipid bilayers are one part of biological membranes that separate cells from their environment as well as organelles from one another within the cytoplasm of the cell. The properties of the membrane are determined by the type of fatty acid and polar group in the phospholipids. Cell membranes allow nutrients to enter the cell and substances produced within the cell to get out. Oxygen for metabolic reactions is taken into the cell, while carbon dioxide and aqueous solutions must be discharged from the cell.

Every animal has membranes ideally suited for the environment in which it lives.

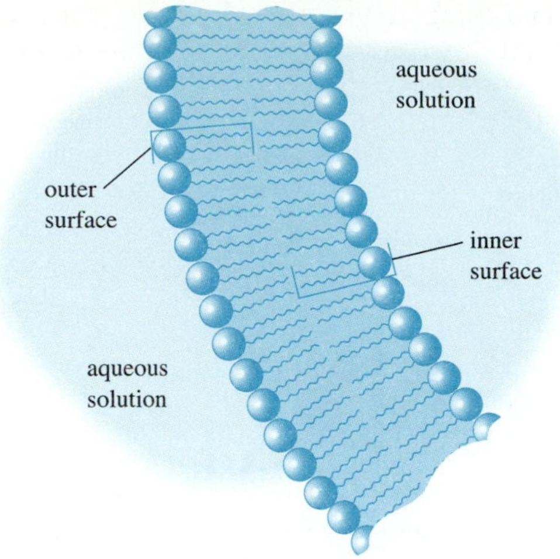

FIGURE 13.6
Cross-Section of a Lipid Bilayer
The circles represent the polar heads of the lipid components of the bilayer that extend toward the aqueous media. The two hydrophobic tails of each lipid extend to the interior.

Composition of Membranes

Cell membranes consist largely of lipids and proteins with smaller amounts of carbohydrates. These additional components can increase the thickness of the membrane to as much as 10 nm. The fraction of each component is related to the function of the membrane, but the average of 40% lipid and 60% protein by mass is representative. However, the myelin covering of nerve fibers is about 80% lipid. This covering is very nonpolar and serves a protective function.

The polar proteins of cell membranes participate in the transport process that occurs across the cell membrane. Mitochondrial membranes are only about 20% lipid, and proteins are the major part of the mitochondrial membrane. The mitochondria play an important role in energy conversions within cells, and many molecules must cross the mitochondrial membrane.

Each membrane has characteristic amounts of the various types of lipids. Membranes in heart and lung tissue have large amounts of phosphatidylglycerols and sphingomyelins. Brain tissue has large amounts of phosphatidylserines. Phosphatidylethanolamines make up 70% of *E.coli* cells but only 50% of red blood cells, whose membranes have mixtures of phosphatidylcholines and phosphatidylserines.

The distributions of phospholipids between the inner and outer surfaces of the lipid bilayer are not the same. Sphingomyelins and phosphatidylcholines make up half of the outer layer of the plasma membrane of human erythrocytes, but only 20% of the inner layer. Phosphatidylethanolamines and phosphatidylserines are the major components of the inner layer.

Proteins in Membranes

Protein molecules are present in the bilayer in three different ways (Figure 13.7). Proteins extending from the edge of the bilayer into the interior of the membrane are **integral proteins.** These proteins may be embedded in only one side of the membrane, but they usually extend through to the other side. Pro-

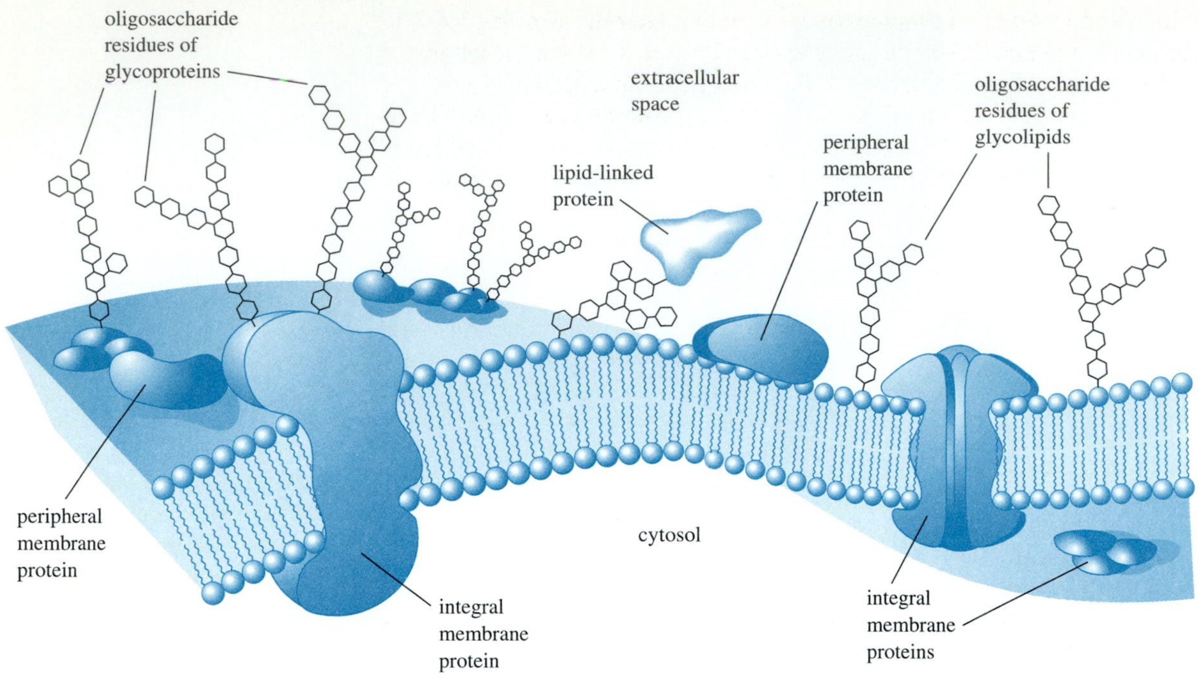

FIGURE 13.7
Proteins, Carbohydrates, and the Structure of a Lipid Bilayer

teins that extend across the membrane are called **transmembrane proteins.** Integral proteins interact with the interior of the membrane by hydrophobic forces between certain nonpolar portions of a protein and the tails of the lipid.

Proteins that are associated with only one surface of the bilayer are called **peripheral proteins.** Peripheral proteins may be bound to the membrane by electrostatic forces or hydrogen bonds. As a result, these proteins may easily be removed from the membrane.

Lipid-linked proteins are covalently bonded to phosphatidylinositols incorporated in the cell membrane. Thus, these proteins are less easily removed than peripheral proteins.

Carbohydrates in Membranes

Membranes contain carbohydrates combined with lipids in glycolipids and combined with proteins in glycoproteins. In mammals these glycolipids and glycoproteins are located on the outer surface of the cell membrane. The carbohydrate portion is hydrophilic and remains directed toward water in the external environment. The protein is anchored in the membrane. The arrangement provides some intercellular recognition. It not only allows the grouping of cells to form tissue but also aids the immune system in recognizing foreign cells.

Membrane Fluidity

The London forces that hold membranes together depend on both the length and geometry of the hydrocarbon chain. The longer the chain, the stronger the London forces and the more rigid the membrane. Unsaturation affects the flexibility of the membrane because unsaturated fatty acids have bends in the chain, and they do not pack together efficiently in the bilayer. As a result, membranes with a high degree of unsaturation are more flexible.

The lipids and proteins in membranes can move laterally (on one side) in the membrane. A phospholipid can move about 10^{-4} cm/s, which means that it can move from one end of a bacterium to the other in about a second.

The lateral mobility of proteins varies. Some proteins are as mobile as lipids, whereas others are essentially immobile. The protein rhodopsin, which is responsible for sight, is very mobile. Other proteins that are anchored to the cytoskeleton remain fixed in place.

13.11 Transport Across Membranes

Molecules and ions pass through a biological membrane to provide food for the cell and to release waste products from the cell. Smaller hydrophobic molecules such as O_2, as well as polar but uncharged molecules such as H_2O, urea, and ethanol, diffuse across the lipid part of the membrane relatively rapidly. However, large polar molecules and most ions, such as Na^+ and K^+, diffuse very slowly. To maintain cellular processes, molecules move across membranes by specialized mechanisms called facilitated diffusion and active transport.

Facilitated Diffusion

Facilitated diffusion occurs without the expenditure of cellular energy and in a direction from high concentration to low concentration. Each substance moves across the membrane faster than in simple diffusion because "carriers" known as channel proteins facilitate the process. These carriers are transmembrane proteins with molecular weights in the range of 9000 to 40,000 amu. There are a variety of carriers, each specific for certain molecules or ions.

A carrier protein meets a specific molecule or ion at one surface of the membrane and forms a complex. Formation of the complex causes a conformational change in the protein that allows the molecule to slip through a "channel" to the other side of the membrane. Once the molecule is released, the protein returns to its original conformation.

The transport of anions in human erythrocytes occurs by facilitated diffusion. A protein exchanges the bicarbonate ion inside the cell for the chloride ion outside the cell. Carbon dioxide produced by cellular metabolism dissolves in aqueous media as the bicarbonate ion. Its concentration within the cell is higher than outside the cell, and it is transported spontaneously to the region of lower concentration. To maintain charge balance, the chloride ion flows into the cell.

Active Transport

Active transport also involves a specific interaction between a component of the cell and the molecule to be transported. However, active transport occurs against the "natural" flow expected from concentration differences. Thus, material moves from a region of low concentration to one of high concentration. Cellular ATP provides the energy needed for active transport.

Active transport proteins have two sites responsible for the transport. One is the site on one face of the membrane, either the inner or outer face depending on the direction of transport required. It recognizes the particular ion or molecule to be transported. The second site is within the membrane and binds ATP to provide energy to the protein, which must change its conformation to move the molecule or ion across the membrane. Because energy is required, such active transport proteins are called pumps. They require energy much like a water pump uses energy to pump water uphill.

Sleep Disorders and Brain Lipids

It is estimated that between 10 and 15% of U.S. citizens suffer from a variety of sleep disorders. As a result, there is a considerable market for sleep-enhancing drugs. However, the ingestion of any medication, even of the over-the-counter variety, is not without risk. Every drug must be metabolized, and the effect of metabolites on the human system, both short term and long term, should always be a concern. Short-term side effects are easily diagnosed, but long-term effects from prolonged use of a drug are not as evident.

The potential for a "natural" sleep-enhancing drug was suggested by the research at Scripps Research Institute, La Jolla, California in 1995. The research group found that lipid-derived amides induce sleep in rats. One such substance is *cis*-9-octadecenamide, the amide of oleic acid.

The compound was isolated from an enemy of the rodent family—the cat. Sleep-deprived cats form *cis*-9-octadecenamide, apparently in an attempt to get some sleep. The compound was then isolated from their cerebrospinal fluid.

Lipids have been implicated in some sleep disorders. For example, individuals on lipid-restricted diets don't sleep well. Furthermore, it is known that metabolism of lipids occurs at a higher rate at night. Thus, researchers sought evidence for a lipid-derived sleep hormone. The apparent answer was found in the Scripps Research Institute studies.

The effects of structural variations of *cis*-9-octadecenamide have been studied. The trans isomer has little effect—not unexpected because configuration often plays an important role in biological reactions. Similarly, changing the position of the double bond to C-8 or C-11 diminished the sleep-inducing properties. Extending the chain to 22 carbon atoms and maintaining the position of the double bond with respect to the terminal methyl group as in *cis*-13-docosenamide also decreased the sleep-inducing properties.

The determination that a substance has an effect on a physiological process and understanding how the interaction occurs require two very different research methods. It was determined that rats could be induced to sleep, even in their normal wake time, by injection of nanomoles of *cis*-9-octadecenamide directly into the brain. The compound is enzymatically hydrolyzed to oleic acid. However, studies have shown that oleic acid itself has no sleep-inducing properties. Ammonia is the other product of the hydrolysis, but it is unknown whether this simple chemical is involved in some as yet undiscovered sleep-inducing process.

The biochemistry of the sleep-inducing properties of *cis*-9-octadecenamide may eventually be understood. However, the development of this chemical as a natural drug may take considerable time. As in the case of any drug, it is necessary to administer the compound in a form that can reach the site where its action can occur. Two of the problems are the survival of the compound in the digestive tract and its transport across cell membranes. ■

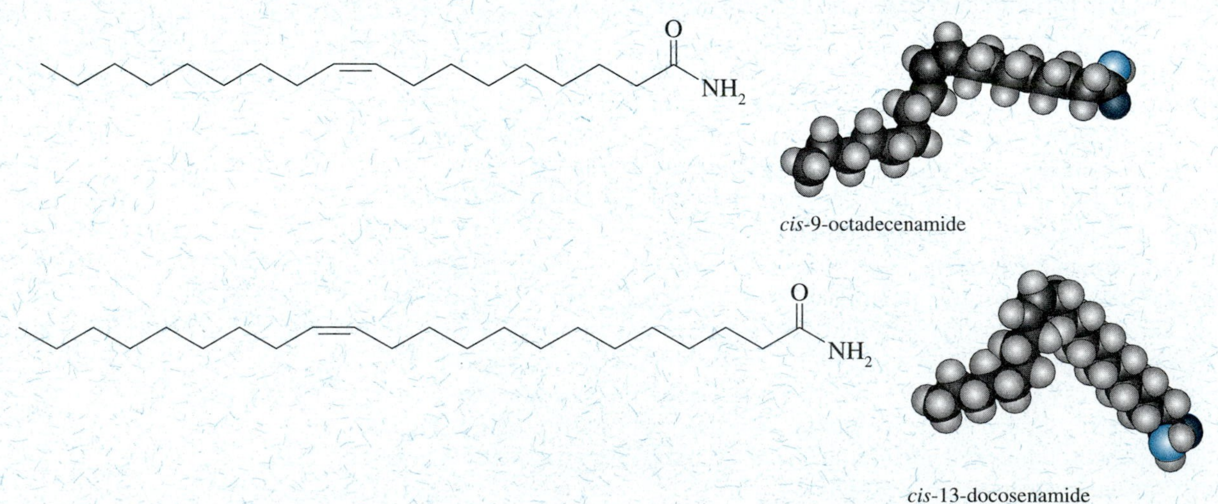

cis-9-octadecenamide

cis-13-docosenamide

One of the most important active transport processes in animal cells is the sodium ion-potassium ion transport system known as the sodium pump. Sodium ions are pumped out of the cell to maintain a concentration of 0.1 M within the cell, while the extracellular sodium ion concentration is 0.14 M. At the same time, potassium ions are pumped into the cell from an extracellular concentration of 0.005 M to provide an intracellular concentration of about 0.15 M. The entire process is controlled because any imbalance in the total concentration of ions would cause a change in the osmotic pressure. The cell would swell if the osmotic pressure increased or shrink if the osmotic pressure decreased.

Summary of Reactions

1. Hydrogenation of Oils (Section 13.5)

$$\begin{array}{l} CH_2-O-\overset{\overset{O}{\|}}{C}-CH_2(CH_2)_6CH=CH(CH_2)_7CH_3 \\ | \\ CH-O-\overset{\overset{O}{\|}}{C}-CH_2(CH_2)_6CH=CH(CH_2)_7CH_3 \\ | \\ CH_2-O-\overset{\overset{O}{\|}}{C}-CH_2(CH_2)_6CH=CH(CH_2)_7CH_3 \end{array} \xrightarrow[\text{Ni}]{3\,H_2} \begin{array}{l} CH_2-O-\overset{\overset{O}{\|}}{C}-CH_2(CH_2)_6CH_2-CH_2(CH_2)_7CH_3 \\ | \\ CH-O-\overset{\overset{O}{\|}}{C}-CH_2(CH_2)_6CH_2-CH_2(CH_2)_7CH_3 \\ | \\ CH_2-O-\overset{\overset{O}{\|}}{C}-CH_2(CH_2)_6CH_2-CH_2(CH_2)_7CH_3 \end{array}$$

2. Saponification of Triacylglycerols (Section 13.6)

$$\begin{array}{l} CH_2-O-\overset{\overset{O}{\|}}{C}-CH_2(CH_2)_{13}CH_3 \\ | \\ CH-O-\overset{\overset{O}{\|}}{C}-CH_2(CH_2)_{13}CH_3 \\ | \\ CH_2-O-\overset{\overset{O}{\|}}{C}-CH_2(CH_2)_{13}CH_3 \end{array} + 3\,NaOH \longrightarrow \begin{array}{l} CH_2-OH \\ | \\ CH-OH \\ | \\ CH_2-OH \end{array} + 3\,CH_3(CH_2)_{14}-\overset{\overset{O}{\|}}{C}-O^-\,Na^+$$

Explorations with Molecular Models

1. Construct a ball-and-stick model corresponding to the following space-filling model of octanoic acid. Make sure that the carbon chain is in a zigzag arrangement. What shape of "container" could be used to store the structure most efficiently? Measure the distance between the "ends" of the structure.

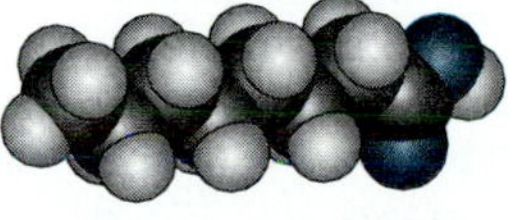

2. Construct a ball-and-stick model corresponding to the following space-filling model of *trans*-4-octenoic acid. Make sure that the carbon chain is in a zigzag arrangement. What shape of "container" could be used to store the structure most efficiently? Measure the distance between the "ends" of the structure. Does this distance differ significantly from the measured distance obtained in Exploration Exercise 1?

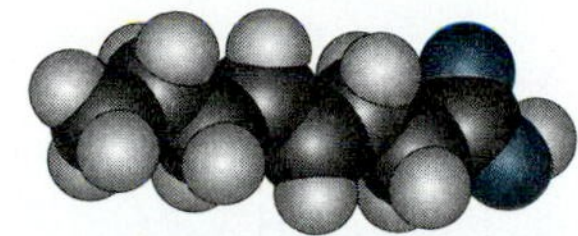

3. Construct a ball-and-stick model corresponding to the following space-filling model of *cis*-4-octenoic acid. Make sure that as many of the carbon atoms as possible are in a zigzag arrangement. Measure the distance between the "ends" of the structure. Does this distance differ significantly from the measured distance obtained in Exploration Exercise 2?

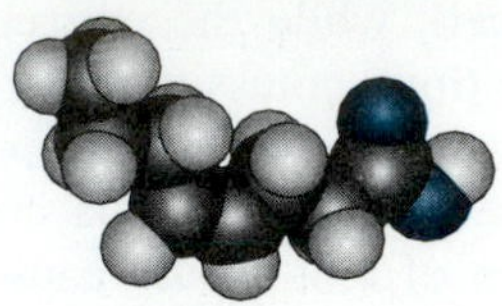

Exercises

Classification of Lipids

13.1 What are the differences and similarities in the structures of glycerophospholipids and sphingophospholipids?

13.2 What are the differences and similarities in the structures of sphingophospholipids and glycosphingolipids?

Waxes

13.3 One of the components of carnauba wax used in floor polish is an ester containing an unbranched carboxylic acid with 20 carbon atoms and an unbranched alcohol with 32 carbon atoms. Write a condensed structural formula for this wax.

13.4 The whale "oil" of the sperm whale is actually a wax with the molecular formula $C_{32}H_{64}O_2$. Hydrolysis of the compound gives palmitic acid. Write the condensed molecular structure of the compound.

Fatty Acids

13.5 Cod liver oil is a triacylglycerol containing palmitoleic acid. Suggest a structure for the acid.

13.6 Estimate the melting point of $CH_3(CH_2)_{20}CO_2H$.

13.7 Steareolic acid is named 9-octadecynoic acid by the IUPAC method. The molecular formula is $C_{18}H_{32}O_2$. Write its structure.

13.8 A compound called hypogeic acid is prepared in the laboratory and is now named 7-hexadecenoic acid. Its melting point is 33°C. What is the geometry at the double bond?

13.9 Why does linoleic acid have a lower melting point than oleic acid?

13.10 The melting point of elaidic acid (*trans*-Δ^9-octadecenoic acid) is 45°C. Compare this value with the melting points of stearic acid and oleic acid and explain the differences.

13.11 10-Undecenoic acid is the antifungal agent in Desenex® and Cruex®. Write its structure. Are stereoisomers possible for this compound?

13.12 Megatomoic acid is the sex attractant of the female black carpet beetle. What is its IUPAC name? How many stereoisomers are possible for this compound?

$$CH_3(CH_2)_7CH{=}CH{-}CH{=}CH{-}CH_2CO_2H$$

megatomoic acid

Soaps and Detergents

13.13 Pentaerythrityl palmitate is a "neutral" detergent. Explain why it has detergent properties.

$$HOCH_2{-}\underset{\displaystyle CH_2OH}{\overset{\displaystyle CH_2OH}{\overset{|}{\underset{|}{C}}H}}{-}CH_2{-}O{-}\overset{\displaystyle O}{\overset{\|}{C}}{-}(CH_2)_{14}CH_3$$

13.14 Benzyldimethyloctylammonium chloride is a "cationic" detergent. Explain why it has detergent properties.

$$C_6H_5{-}CH_2{-}\underset{\displaystyle CH_3}{\overset{\displaystyle CH_3}{\overset{|}{\underset{|}{N^+}}}}{-}CH_2(CH_2)_6CH_3$$

benzyldimethyloctylammonium ion

Triacylglycerols

13.15 Write a balanced equation for the hydrolysis of a fat molecule using a base.

13.16 A sample of one oil is hydrolyzed to produce 50% oleic acid, 35% linoleic acid, and 15% saturated acids. A second oil produces 25% oleic, 50% linoleic acid, and 25% saturated acids. Which oil is more unsaturated?

13.17 Is the following compound more likely a fat or an oil? Identify the component fatty acids.

$$\begin{array}{l} CH_2-O-\overset{O}{\overset{\|}{C}}-(CH_2)_7CH{=}CHCH_2CH{=}CH(CH_2)_4CH_3 \\ | \\ CH-O-\overset{O}{\overset{\|}{C}}-(CH_2)_7CH{=}CH(CH_2)_7CH_3 \\ | \\ CH_2-O-\overset{O}{\overset{\|}{C}}-(CH_2)_7CH{=}CHCH_2CH{=}CHCH_2CH{=}CHCH_2CH_3 \end{array}$$

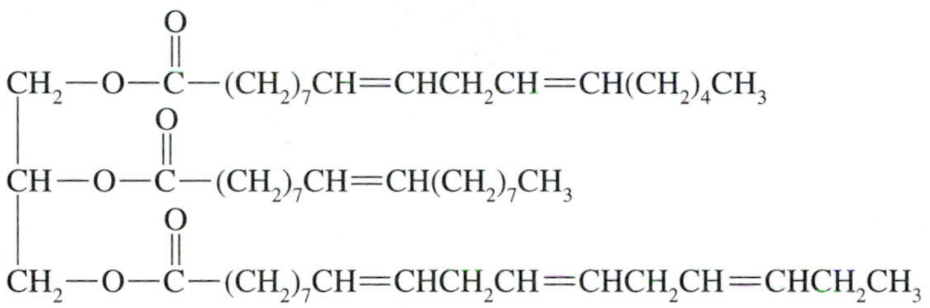

13.18 Is the following compound more likely a fat or an oil? Identify the component fatty acids.

$$\begin{array}{l} CH_2-O-\overset{O}{\overset{\|}{C}}-(CH_2)_{14}CH_3 \\ | \\ CH-O-\overset{O}{\overset{\|}{C}}-(CH_2)_{10}CH_3 \\ | \\ CH_2-O-\overset{O}{\overset{\|}{C}}-(CH_2)_7CH{=}CH(CH_2)_7CH_3 \end{array}$$

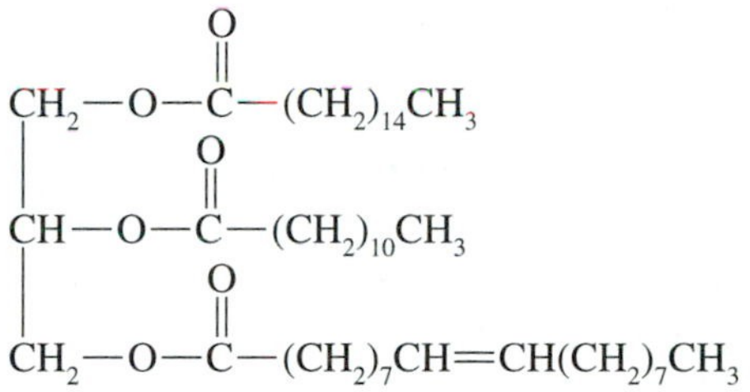

13.19 Draw the structure of a triacylglycerol containing palmitic acid as an ester at the secondary carbon atom and stearic acid as esters at the two primary carbon atoms of glycerol. Can this compound exist in an optically active form?

13.20 Can the following compound be optically active?

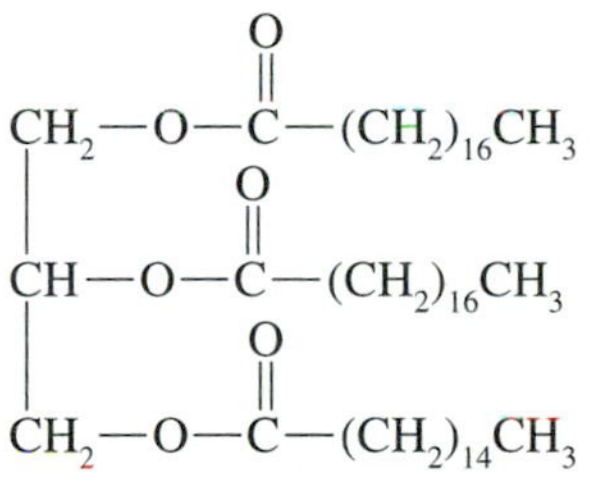

$$\begin{array}{l} CH_2-O-\overset{O}{\overset{\|}{C}}-(CH_2)_{16}CH_3 \\ | \\ CH-O-\overset{O}{\overset{\|}{C}}-(CH_2)_{16}CH_3 \\ | \\ CH_2-O-\overset{O}{\overset{\|}{C}}-(CH_2)_{14}CH_3 \end{array}$$

13.21 How many moles of hydrogen gas will react with the compound shown in Exercise 13.17? What is the name of the product?

13.22 Which oil will react with the larger quantity of hydrogen gas, linseed or olive oil? (See Table 13.2.)

13.23 The iodine value of an oil is equal to the number of grams of iodine that will add to the double bonds in 100 g of the oil. What is the iodine value of glyceryl trioleate (Section 13.5)?

13.24 Which oil has the larger iodine value, corn oil or soybean oil? (See Table 13.2.)

13.25 How many fatty acids result from the hydrolysis of the compound shown in Exercise 13.20 and in what molar ratio?

13.26 Nutmeg contains trimyristin. Based on this name, what fatty acid results from hydrolysis of this compound?

13.27 The saponification number is the number of milligrams of potassium hydroxide (formula weight 56.1 g/mol) that is required to saponify 1.00 g of a fat or oil. The number is usually expressed to three significant figures. What is the saponification number of glyceryl tripalmitate (molecular weight 806 g/mol)?

13.28 The saponification number of a sample of butter fat, a mixture of triacylglycerols, is 236. Calculate the average molecular weight of butter fat. (See Exercise 13.27.)

Glycerophospholipids

13.29 What products result from hydrolysis of a glycerophospholipid?

13.30 Draw the structures of the alcohols found in glycerophospholipids.

13.31 Identify the components of the following glycerophospholipid.

$$\begin{array}{l} \qquad\qquad\qquad\qquad\qquad\qquad\qquad\qquad\quad CH_2-O-\overset{\overset{\displaystyle O}{\|}}{C}-(CH_2)_{16}CH_3 \\ CH_3(CH_2)_7CH_2{=}CH(CH_2)_7-\overset{\overset{\displaystyle O}{\|}}{C}-O-\underset{|}{\overset{|}{C}H} \\ \qquad\qquad\qquad\qquad\qquad\qquad\qquad\qquad\quad CH_2-O-\underset{\underset{\displaystyle O^-}{|}}{\overset{\overset{\displaystyle O}{\|}}{P}}-O-CH_2CH_2\overset{+}{N}(CH_3)_3 \end{array}$$

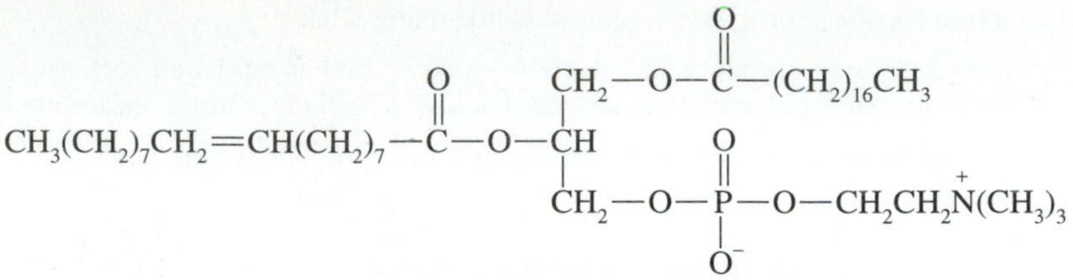

13.32 What are the hydrolysis products of the following glycerophospholipid?

$$\begin{array}{l} \qquad\qquad\qquad\qquad\qquad\qquad\qquad\qquad\quad CH_2-O-\overset{\overset{\displaystyle O}{\|}}{C}-(CH_2)_{14}CH_3 \\ CH_3(CH_2)_7CH_2{=}CH(CH_2)_7-\overset{\overset{\displaystyle O}{\|}}{C}-O-\underset{|}{\overset{|}{C}H} \\ \qquad\qquad\qquad\qquad\qquad\qquad\qquad\qquad\quad CH_2-O-\underset{\underset{\displaystyle O^-}{|}}{\overset{\overset{\displaystyle O}{\|}}{P}}-O-CH_2-\underset{\underset{\displaystyle H}{|}}{\overset{\overset{\displaystyle NH_3^+}{|}}{C}}-CO_2^- \end{array}$$

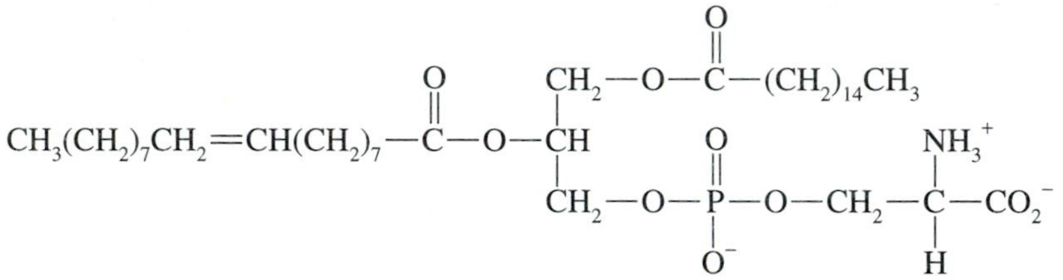

13.33 What charges exist on the polar head of each of the types of glycerophospholipids at physiological pH?

13.34 How many ionizable hydrogen atoms are there in a phosphatidic acid?

Sphingophospholipids

13.35 How do sphingophospholipids differ from glycerophospholipids?

13.36 Why are sphingophospholipids not hydrolyzed as readily as glycerophospholipids?

13.37 Sphingophospholipids are said to have two nonpolar tails. One is a fatty acid residue. What is the structure of the second chain?

13.38 What are sphingomyelins? What structural feature allows them to serve a uniquely important biological function?

Glycosphingolipids

13.39 How are glycosphingolipids similar to sphingophospholipids? In what ways do the two types of compounds differ?

13.40 What is the difference between a cerebroside and a ganglioside? Where are these compounds found?

13.41 What type of bond joins the carbohydrate unit to the sphingosine part of a glycosphingolipid?

13.42 From the structure of glycosphingolipids, predict whether these molecules are more stable in acidic or basic solution.

Biological Membranes

13.43 How does the structure of the fatty acid components affect the rigidity of a cell membrane?

13.44 What kind of forces hold a cell membrane together?

13.45 What relationship exists between the protein content and the permeability of a cell membrane?

13.46 How do a peripheral protein and an integral protein differ?

Diffusion and Transport

13.47 Describe two ways in which materials cross cell membranes.

13.48 Why is an active transport system necessary to maintain the potassium ion content of a cell?

CHAPTER 14

AMINES AND AMIDES

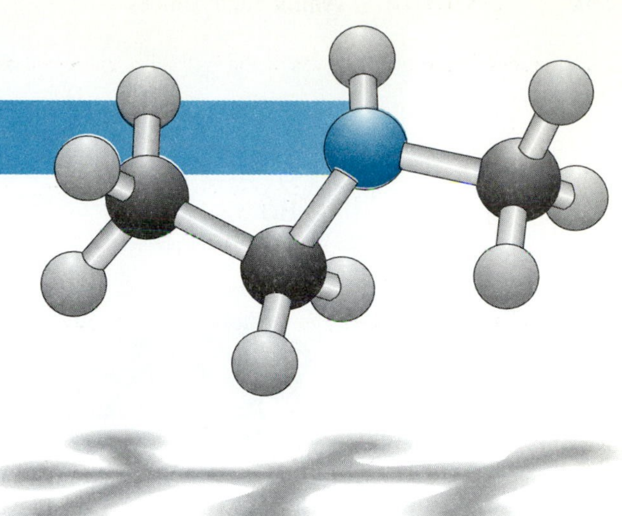

Overview

14.1 Organic Nitrogen Compounds

Nitrogen is the fourth most common atom found in organic compounds of the biological world. It is contained in the amino acids of proteins (Chapter 15), the nucleotides of RNA and DNA (Chapter 16), as well as some vitamins and hormones. Synthetic nitrogen-containing compounds are important industrial products. Among these are polymers such as nylon, many dyes, explosives, and pharmaceutical agents.

The nitrogen atom has five valence electrons and forms a total of three covalent bonds to carbon or hydrogen in neutral organic compounds. A nitrogen atom in functional groups can have single, double, or triple bonds. In this chapter we will consider amines and amides, and, to a lesser extent, imines and nitriles.

$R-\ddot{N}H_2$ an amine; $R-C(=\ddot{O}:)-\ddot{N}H_2$ an amide; $R-C(=\ddot{N}-H)-H$ an imine; $R-C\equiv N:$ a nitrile

Many amines are physiologically active. They affect the brain, spinal cord, and nervous system. These compounds include the neurotransmitters epinephrine, serotonin, and dopamine (Figure 14.1). Epinephrine, commonly called adrenaline, stimulates the conversion of stored glycogen into glucose. Serotonin is a hormone that causes sleep, and serotonin deficiency is responsible for some forms of mental depression. In Parkinson's disease, the dopamine concentration is low.

Proteins, one of the most important and versatile classes of biological compounds, contain an amide functional group. They are polymers consisting of nitrogen-containing molecules called α-amino acids. The amine functional group of one α-amino acid reacts with the carboxyl group of another α-amino acid to form an amide bond. This chemistry will be presented in Chapter 15.

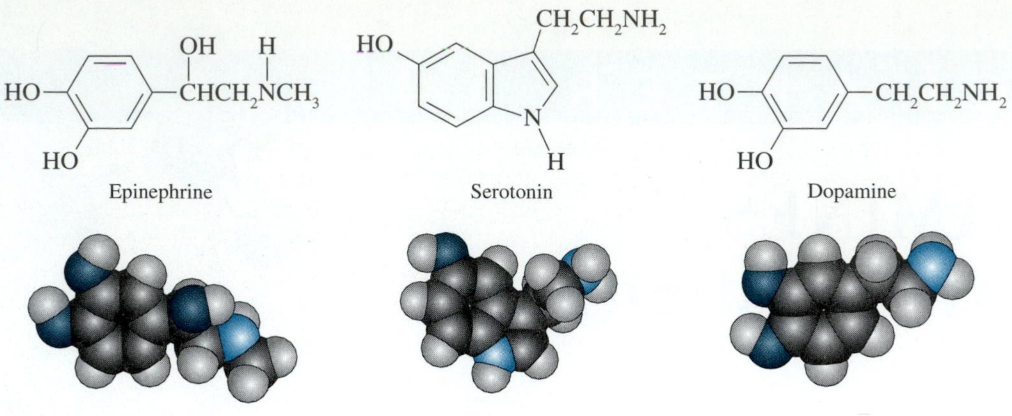

Epinephrine Serotonin Dopamine

FIGURE 14.1
Structures of Neurotransmitters

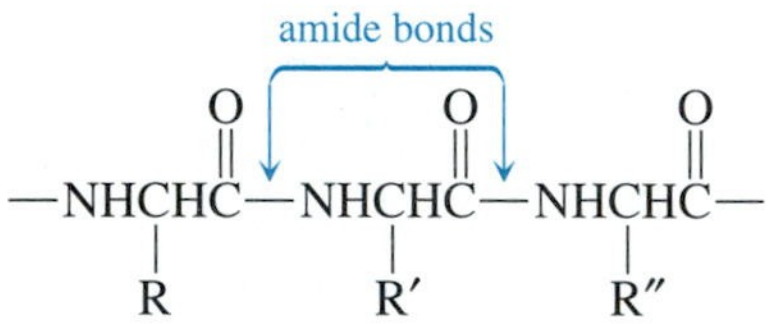

amide bonds

—NHCHC(=O)—NHCHC(=O)—NHCHC(=O)— (R, R′, R″)

EXAMPLE 14-1

Identify the nitrogen-containing functional groups in Valium®.

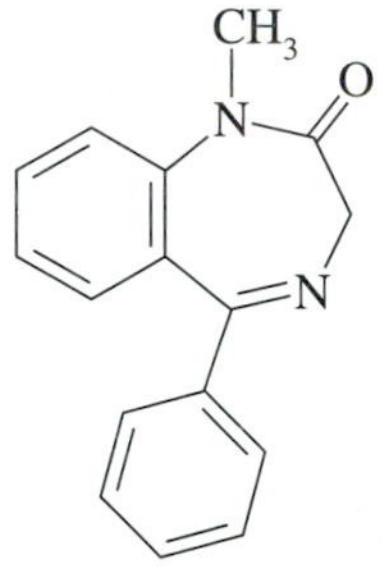

Solution

The nitrogen atom located at the top of this structure is bonded to a carbonyl carbon atom and is part of an amide group. The other nitrogen atom is bonded only to two carbon atoms—one by a double bond. This functional group is an imine.

Problem 14.1

Classify the nitrogen-containing functional groups in Mepivacaine®, a local anesthetic.

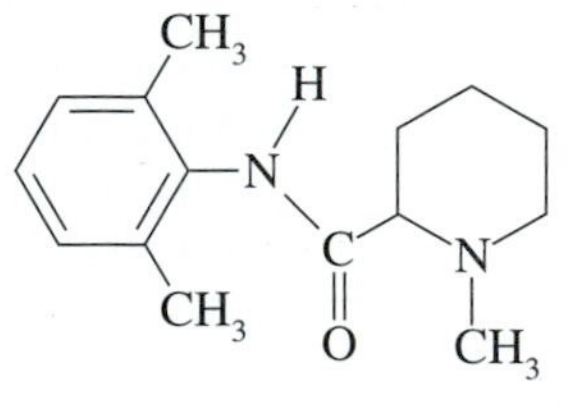

14.2 Structure and Classification of Amines and Amides

In the simplest amine, methylamine (CH_3NH_2), one hydrogen atom of ammonia has been replaced by a methyl group (Figure 14.2). The nitrogen atom of methylamine and other amines has five valence electrons in four sp^3 hybrid orbitals that are directed to the corners of a tetrahedron. Three of these orbitals are half-filled; the fourth contains a nonbonding pair of electrons that plays an important role in the chemical properties of amines.

Amines are classified by the number of alkyl (or aryl) groups attached to the nitrogen atom.

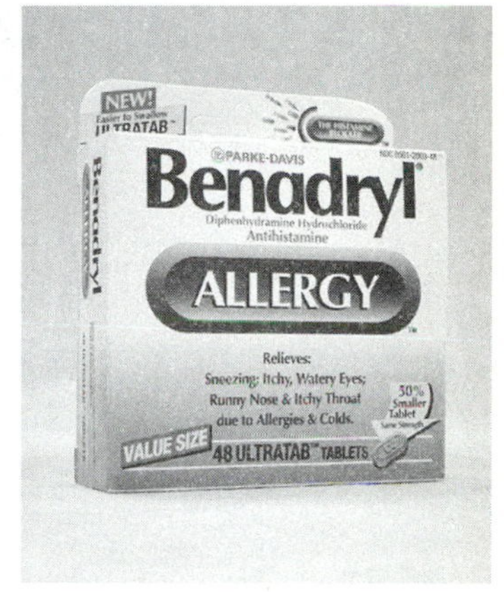

Many common drugs available without prescription contain amines or amides as the active ingredients.

H—N(H)(H):	R—N(H)(H):	R—N(R)(H):	R—N(R)(R):
ammonia	primary (1°) amine	secondary (2°) amine	tertiary (3°) amine

Note that amines are not classified like alcohols. (The classification of alcohols is based on the number of groups attached to the carbon atom bearing the hydroxyl group.) For example, *tert*-butylamine has a *tert*-butyl group attached to an $—NH_2$ group. However, the amine is primary because only one alkyl group is bonded to the nitrogen atom. In contrast, *tert*-butyl alcohol is a tertiary alcohol because the carbon atom bonded to the —OH group is bonded to three alkyl groups. Trimethylamine is a tertiary amine because the nitrogen atom is bonded to three alkyl groups.

$(CH_3)_3C—\ddot{N}H_2$	$(CH_3)_3C—\ddot{O}H$	$(CH_3)_3N:$
tert-butylamine (a primary amine)	*tert*-butyl alcohol (a tertiary alcohol)	trimethylamine (a tertiary amine)

The nitrogen atom of an amine may be contained in a ring, a common feature of nitrogen compounds in nature. The simplest five- and six-membered nitrogen-containing heterocyclic compounds are pyrrolidine and piperidine.

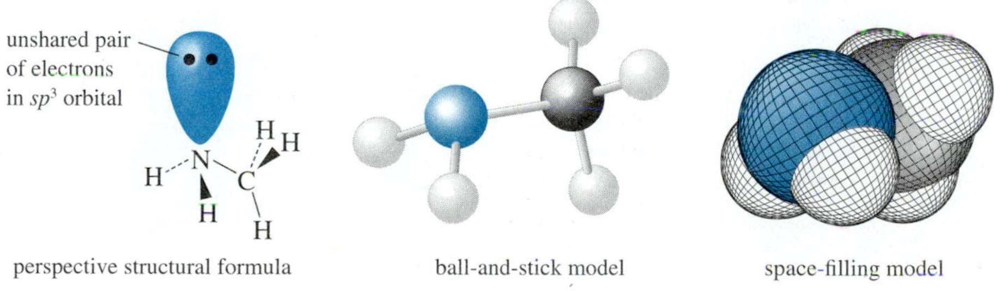

FIGURE 14.2
Structure of Methylamine

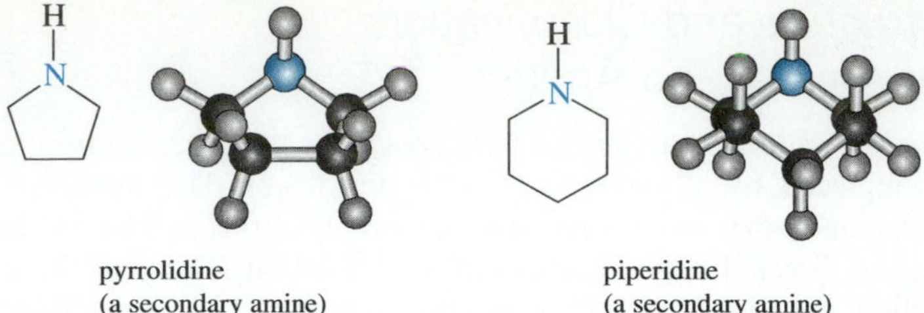

pyrrolidine
(a secondary amine)

piperidine
(a secondary amine)

Amides have an amino group or a substituted amino group bonded to a carbonyl carbon atom. The other two bonds of the nitrogen atom may be to hydrogen atoms, alkyl groups, or aryl groups. Amides are classified based on the number of carbon groups (including the acyl group) bonded to the nitrogen atom.

primary amide — $R{-}C(=O){-}NH_2$ (R—C(=O)—N—H, with H on N)

secondary amide — R—C(=O)—N(H)—R′

tertiary amide — R—C(=O)—N(R″)—R′

The structures of amides resemble those of other carbonyl compounds: the three atoms bonded to carbon are in the same plane (Figure 14.3). The nitrogen atom of an amide has an unshared pair of electrons that is delocalized with the π electrons of the carbonyl group. An amide, such as formamide, can be represented by two resonance structures.

H—C(=Ö:)—N̈H—H $\longleftrightarrow$ H—C(—:Ö:$^-$)=N$^+$H—H

Thus, the bond between carbon and nitrogen has some partial double bond character, which makes rotation about the carbon-nitrogen bond somewhat restricted.

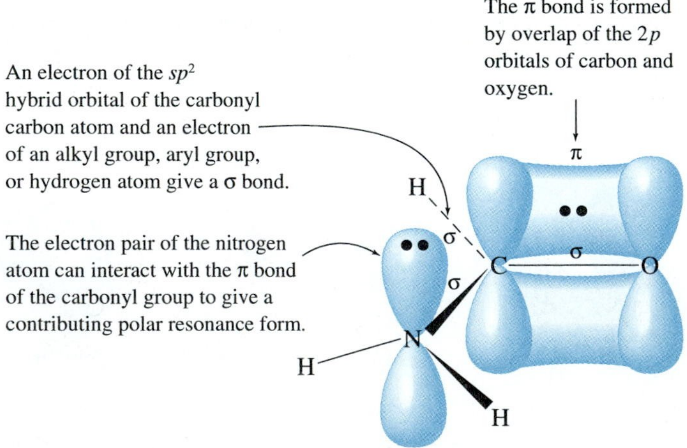

The π bond is formed by overlap of the 2p orbitals of carbon and oxygen.
An electron of the sp^2 hybrid orbital of the carbonyl carbon atom and an electron of an alkyl group, aryl group, or hydrogen atom give a σ bond.
The electron pair of the nitrogen atom can interact with the π bond of the carbonyl group to give a contributing polar resonance form.
π
H
σ
σ
σ
C
O
N
H
H

FIGURE 14.3
Bonding in Formamide

EXAMPLE 14-2

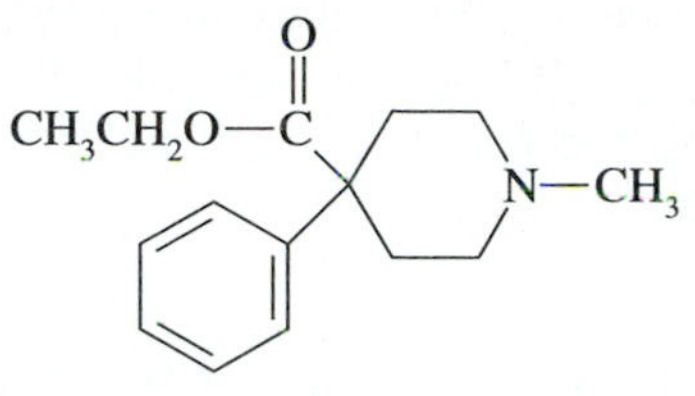

$CH_3CH_2O-C(=O)$ ... $N-CH_3$

Demerol

Classify Demerol®, a synthetic narcotic analgesic, as a primary, secondary, or tertiary amine.

Solution

There are three carbon atoms bonded to the nitrogen atom. Two of the carbon atoms are in the heterocyclic ring. The third carbon atom bonded to the nitrogen atom is a methyl group. Demerol is a tertiary amine.

Problem 14.2

Classify the amine in the illicit drug methamphetamine (speed).

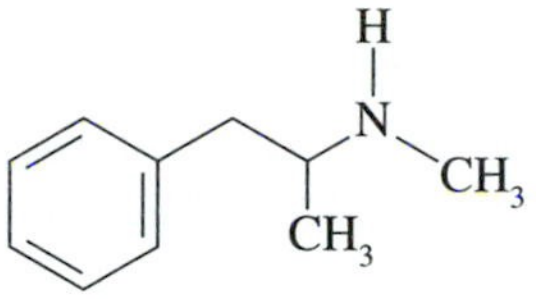

H
N
CH_3
CH_3

14.3 Nomenclature of Amines and Amides

Like all organic compounds, many amines are known by their common names as well as by their IUPAC names. The IUPAC methods of naming amines and amides are related. The amine part of amides is named by the same method used for amines.

Amines

The common name of a primary amine is obtained by naming the alkyl group bonded to the amino group ($-NH_2$) and adding the suffix *-amine*. The entire name is written as one word. The common name for a secondary or tertiary amine is obtained by listing the alkyl groups alphabetically. When two or more identical alkyl groups are present, the prefixes *di-* and *tri-* are used.

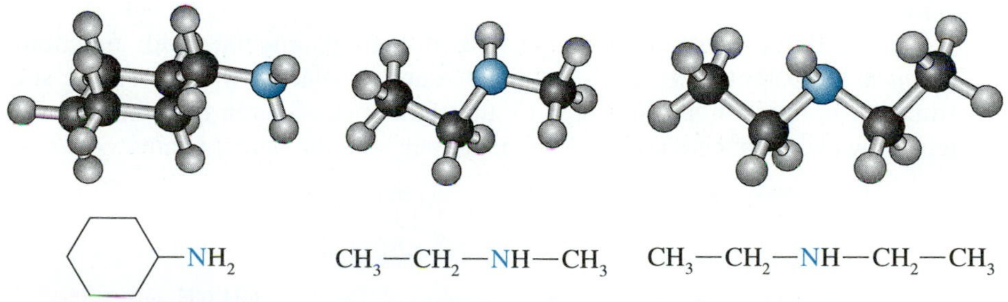

$-NH_2$
cyclohexylamine

$CH_3-CH_2-NH-CH_3$
ethylmethylamine

$CH_3-CH_2-NH-CH_2-CH_3$
diethylamine

For more complex primary amines, the amino group is treated as a substituent. The nitrogen-containing substituent in complex secondary and tertiary amines is named as an *N*-alkylamino (—NHR) or *N,N*-dialkylamino (—NRR′) group. The capital *N* indicates that the alkyl group is bonded to the nitrogen atom and not to the parent chain. The largest or most complicated group is used as the parent molecule.

$NH_2—CH_2—CH_2—CH_2—CO_2H$
γ-aminobutyric acid

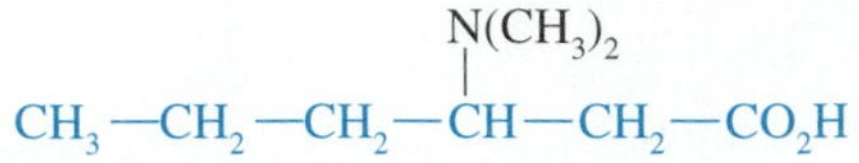

$CH_3—CH_2—CH_2—CH(N(CH_3)_2)—CH_2—CO_2H$
β-(*N,N*-dimethylamino)-caproic acid

Amines are given IUPAC names by rules similar to those used to name alcohols. The longest continuous chain to which the amino group is attached is the parent alkane. The *-e* ending of the alkane is changed to *-amine*. Substituents on the carbon chain are designated by number. The prefix *N-* is used for each substituent on the nitrogen atom.

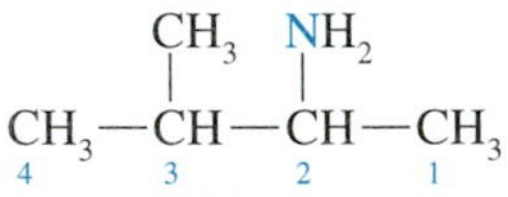

$\overset{4}{CH_3}—\overset{3}{CH}(CH_3)—\overset{2}{CH}(NH_2)—\overset{1}{CH_3}$
3-methyl-2-butanamine

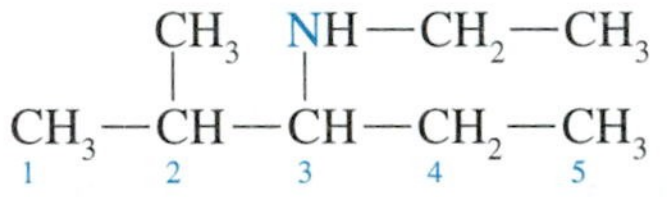

$\overset{1}{CH_3}—\overset{2}{CH}(CH_3)—\overset{3}{CH}(NH—CH_2—CH_3)—\overset{4}{CH_2}—\overset{5}{CH_3}$
N-ethyl-2-methyl-3-pentanamine

Heterocyclic Aromatic Amines

Amines in which the nitrogen atom is part of an aromatic ring are called **heterocyclic aromatic amines.** The position of substituents is indicated by using an individual numbering system for each type of ring. A nitrogen atom is assigned the number 1 and the direction of numbering is selected to provide the lowest possible numbers to additional nitrogen atoms if there is more than one nitrogen atom in the ring.

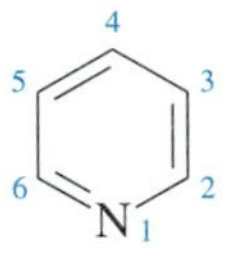

pyridine 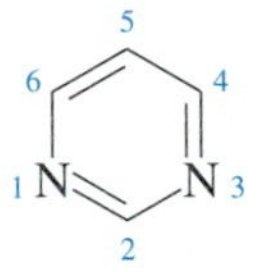   pyrimidine 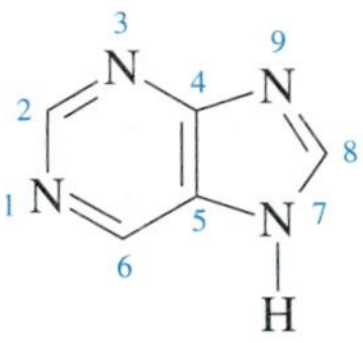   purine 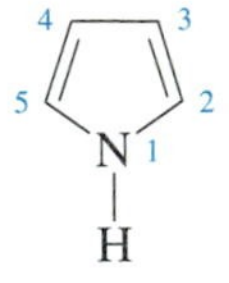   pyrrole 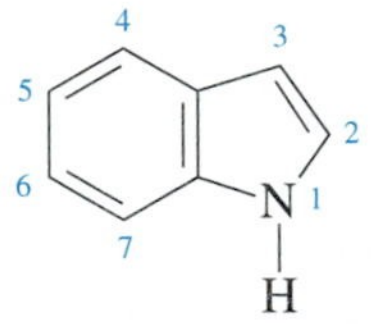   indole

Amides

The common names of amides are formed by dropping the suffix *-ic* of the related acid and adding the suffix *-amide*. When there is a substituent on the nitrogen atom, the prefix *N-* followed by the name of the group bonded to nitrogen is attached to the name. Substituents on the acyl group are designated by Greek letters α, β, γ, and so on, as in the common names of carboxylic acids.

In the IUPAC system, the longest chain that contains the amide functional group is the parent. The final *-e* of the alkane is replaced by *-amide*. The substituents on nitrogen are indicated by the same method as in the common system. However, numbers are used for substituents on the parent chain.

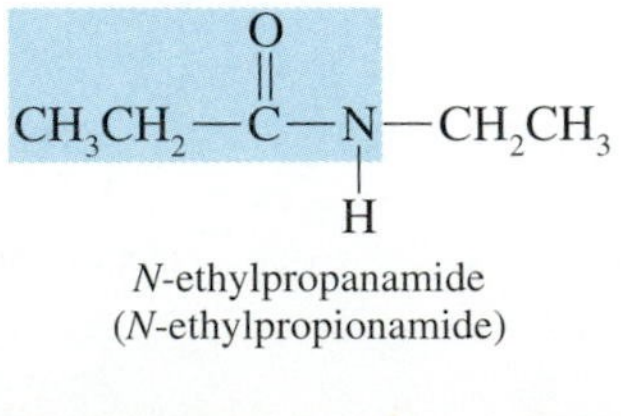

$CH_3CH_2—C(=O)—N(H)—CH_2CH_3$
N-ethylpropanamide
(*N*-ethylpropionamide)

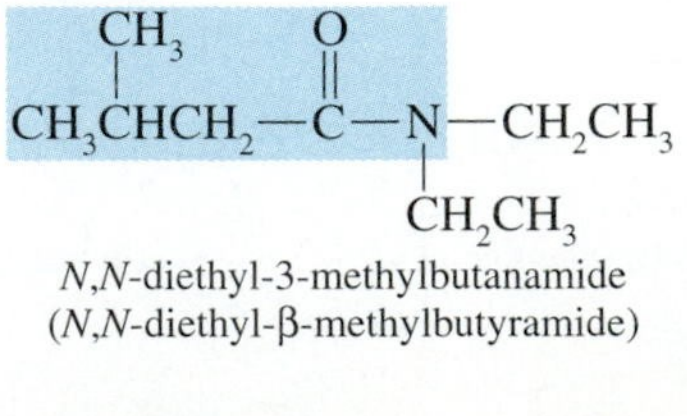

$CH_3CH(CH_3)CH_2—C(=O)—N(CH_2CH_3)—CH_2CH_3$
N,N-diethyl-3-methylbutanamide
(*N,N*-diethyl-β-methylbutyramide)

EXAMPLE 14-3

Write the IUPAC name for baclofen, a muscle relaxant.

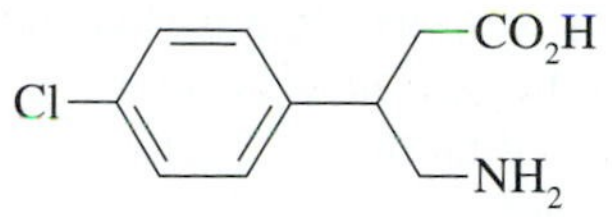

Cl—C_6H_4—CH(CH_2CO_2H)(CH_2NH_2)

Solution

The parent chain of baclofen is butanoic acid. It contains an amino group and an aryl group. The amino group is located at the C-4 atom. The aryl group located at the C-3 atom is named *p*-chlorophenyl. Placing the groups in alphabetical order, the name is 4-amino-3-(*p*-chlorophenyl)butanoic acid. The name of the aryl group is written within parentheses to clearly identify it.

Problem 14.3

2-(3,4,5-Trimethoxyphenyl)ethanamine is the systematic name of mescaline, a hallucinogen. Write its structure.

EXAMPLE 14-4

2-(Diethylamino)-*N*-(2,6-dimethylphenyl)ethanamide is the IUPAC name for lidocaine, a local anesthetic. Write its structure.

Solution

The parent is the two-carbon amide, ethanamide. The group contained within parentheses and preceded by an *N* is 2,6-dimethylphenyl. Place this group on the nitrogen atom of ethanamide.

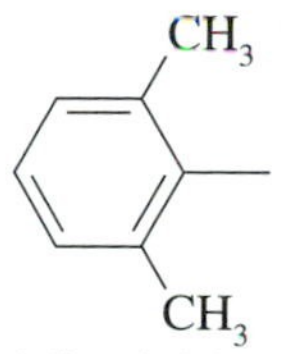

CH_3, CH_3

2,6-dimethylphenyl

CH_3, CH_3, O, C—CH_3, N, H

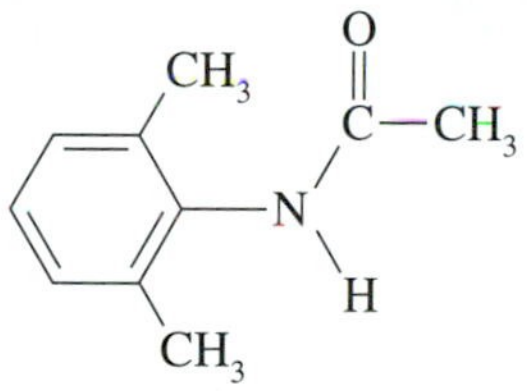

(*N*-aryl substituted amide)

The compound is also an amine because of the diethylamino group bonded to the C-2 atom.

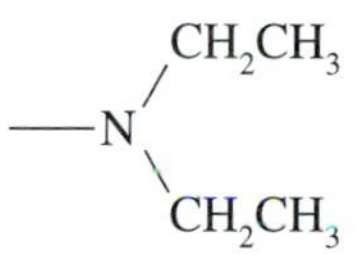

—N(CH_2CH_3)(CH_2CH_3)

diethylamino

CH_3, CH_3, O, C—CH_2—N(CH_2CH_3)(CH_2CH_3), N, H

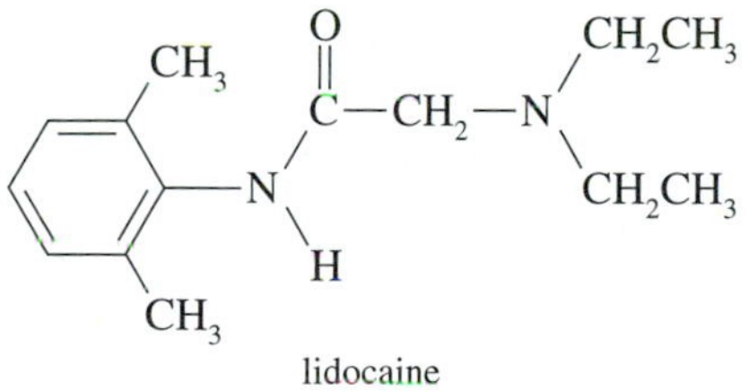

lidocaine

Problem 14.4

Assign the IUPAC name of DEET, an insect repellent.

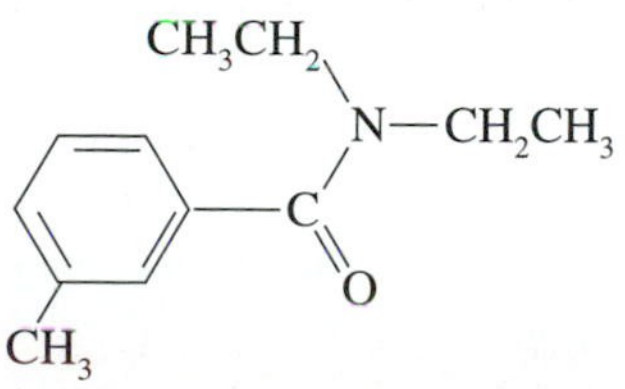

CH_3CH_2, N—CH_2CH_3, C, O, CH_3

14.4 Physical Properties of Amines and Amides

Amines with low molecular weights are gases at room temperature, but amines with higher molecular weights are liquids or solids (Table 14.1). Amines have boiling points that are higher than those of alkanes of similar molecular weight, but lower than those of alcohols.

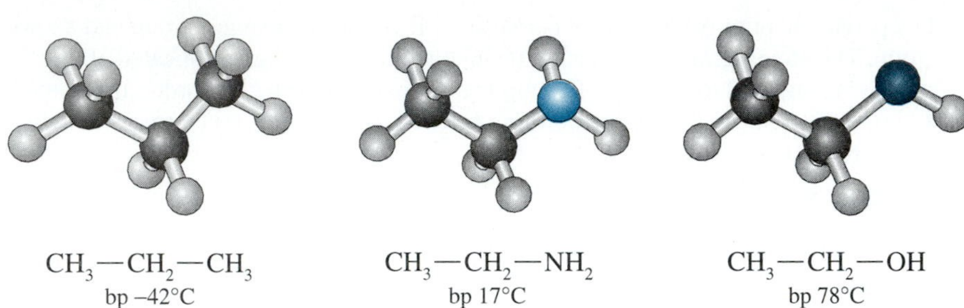

With the exception of formamide, primary amides are solids at room temperature. Substituted amides have lower melting points. All amides have high boiling points.

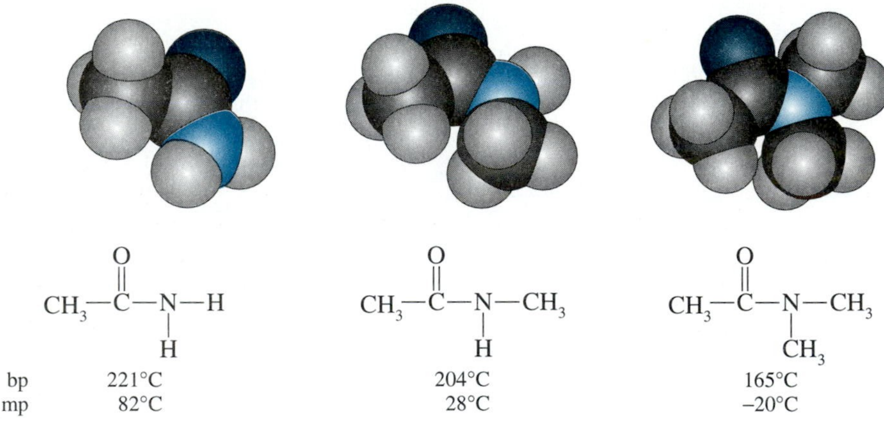

Hydrogen Bonding in Nitrogen Compounds

Primary and secondary amines have higher boiling points than hydrocarbons of comparable molecular weight because they can form intramolecular hydrogen bonds.

a hydrogen bond

$$R—\underset{R}{\overset{H}{N}}\!:\cdots H—\underset{R}{\ddot{N}}—R$$

Tertiary amines have no hydrogen atoms bonded to the nitrogen atom and cannot serve as hydrogen bond donors. As a consequence, they have lower boiling points than primary and secondary amines of comparable molecular weight.

Table 14.1 Boiling Points of Amines

Type of Amine	Name	Boiling Point(°C)
Primary amines	methylamine	−6
	ethylamine	17
	propylamine	49
	isopropylamine	32
	butylamine	77
	isobutylamine	68
	tert-butylamine	45
	cyclohexylamine	135
Secondary amines	dimethylamine	7
	ethylmethylamine	37
	diethylamine	56
	dipropylamine	111
Tertiary amines	trimethylamine	3
	triethylamine	89
	tripropylamine	156
Aromatic amines	aniline	184
	pyridine	116

Amines have lower boiling points than alcohols because nitrogen is less electronegative than oxygen. As a result, the N—H bond is less polar than the O—H bond, and the N—H···N hydrogen bond in amines is weaker than the O—H···O hydrogen bond in alcohols.

Amides form strong intermolecular hydrogen bonds between the amide hydrogen atom of one molecule and the carbonyl oxygen atom of a second molecule (C=O···H—N). This intermolecular interaction is responsible for the high melting and boiling points of primary amides. Substitution of the hydrogen atoms on the nitrogen atom by alkyl or aryl groups reduces the number of possible intermolecular hydrogen bonds and lowers the melting and boiling points. Tertiary amides cannot form intermolecular hydrogen bonds.

$CH_3-C(=\ddot{O}:)-\ddot{N}(H)-H$ ··· $\ddot{O}$: of $CH_3-C(=\ddot{O}:)-\ddot{N}(H)-H$ — hydrogen bond

$CH_3-C(=\ddot{O}:)-\ddot{N}(H)-CH_3$ ··· $\ddot{O}$: of $CH_3-C(=\ddot{O}:)-\ddot{N}(H)-CH_3$ — hydrogen bond

$CH_3-C(=\ddot{O}:)-\ddot{N}(CH_3)-CH_3$ and $CH_3-C(=\ddot{O}:)-\ddot{N}(CH_3)-CH_3$ — no hydrogen bond is possible

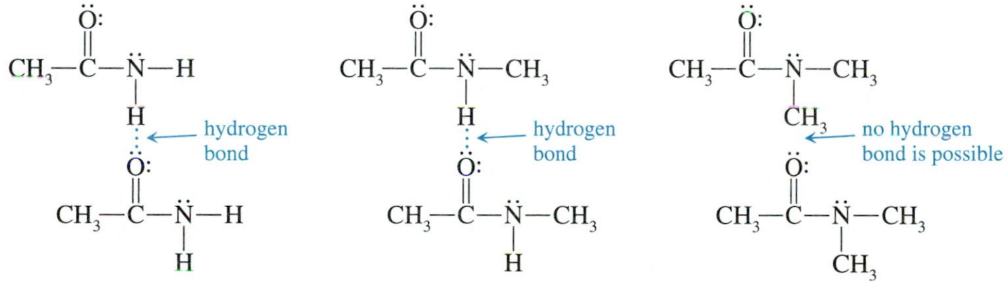

Solubility in Water

Primary and secondary amines function as both hydrogen bond donors and acceptors, and they readily form hydrogen bonds with water. Amines with five or fewer carbon atoms are miscible with water. Even tertiary amines are soluble in water because the nonbonding pair of electrons of the nitrogen atom is a hydrogen bond acceptor of a hydrogen atom of water.

$$R_3N{:}\cdots H{-}\ddot{O}{:}{-}H$$

a hydrogen bond

As we have seen for other types of compounds, the solubility of amines decreases with increasing molecular weight because the functional group is a less significant part of the structure.

Amides having low molecular weights are soluble in water because hydrogen bonds form between the amide group and water. Even low molecular weight tertiary amides are water soluble because the carbonyl oxygen atom can form hydrogen bonds to the hydrogen atoms of water.

Odors of Amines

Amines with low molecular weights have sharp penetrating odors similar to ammonia. Amines with higher molecular weights smell like decaying fish. Two compounds responsible for the odor of decaying animal tissue are appropriately given the common names putrescine and cadaverine.

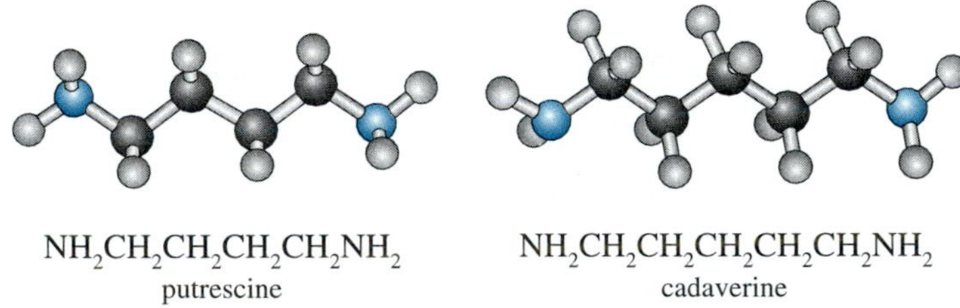

$NH_2CH_2CH_2CH_2CH_2NH_2$
putrescine

$NH_2CH_2CH_2CH_2CH_2CH_2NH_2$
cadaverine

14.5 Basicity of Nitrogen Compounds

The nitrogen atom in both amines and amides has an unshared pair of electrons. However, there is a substantial difference in the chemistry of these two classes of compounds. In this section we consider the ability of these nitrogen compounds to donate an electron pair to a proton—that is, the basicity of the compounds. In the following section we will examine the donation of the electron pair to an electrophile other than a proton.

Amines

Amines are fairly strong bases. They accept a proton from water to form an ammonium ion and a hydroxide ion. The equilibrium constant for the reaction of a base with water is the **base ionization constant,** symbolized by K_b. The K_b values of alkyl-substituted amines are smaller than 10^{-3}. The reaction of methylamine with water is typical.

$$CH_3NH_2 + H_2O \rightleftharpoons CH_3NH_3^+ + OH^-$$

$$K_b = \frac{[OH^-][CH_3NH_3^+]}{[CH_3NH_2]}$$

Amphetamines

There are recurring structures in groups of physiologically active compounds that contain nitrogen. One such structure is 2-phenylethanamine, which is required for binding at certain receptor sites that stimulate the central nervous system. The adrenal medulla produces the hormones epinephrine and norepinephrine, which contain a 2-phenylethanamine unit. Epinephrine, also known as adrenaline, makes glucose available to tissues under conditions of excitement. Norepinephrine maintains the muscle tone of blood vessels and hence controls blood pressure.

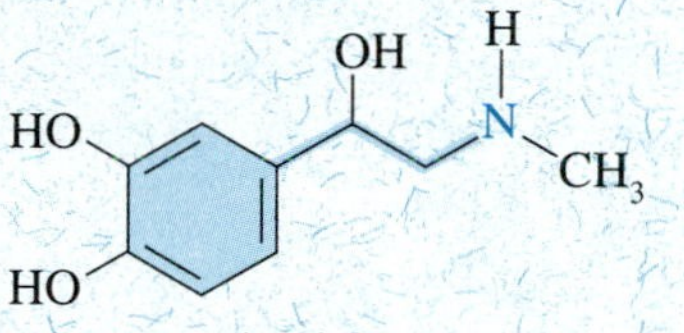

epinephrine

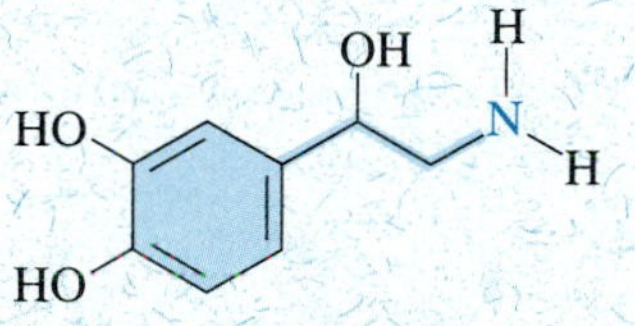

norepinephrine

Amphetamines are drugs known as "uppers" that mimic the action of the naturally occurring phenylethanamines epinephrine and norepinephrine. Amphetamine (also known as benzedrine) has a structure similar to epinephrine. It is a moderate appetite suppressant and stimulates the cortex of the brain, which effectively counters fatigue. For that reason this illegal drug has been used by long-distance truck drivers and students pulling "all-nighters" before exams.

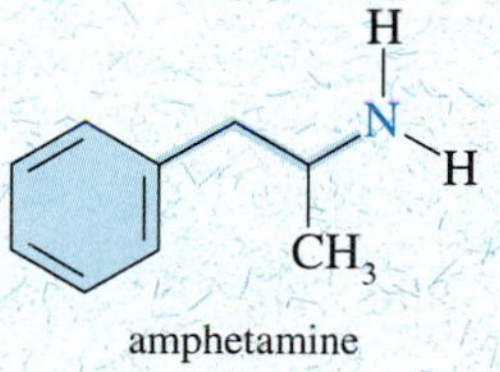

amphetamine

The structurally related illegal drugs methamphetamine and methoxyamphetamine are "speed" and STP, respectively. These drugs can produce severe physiological reactions. In addition, once the drug wears off, the user tends to "crash" into a state of physical and mental exhaustion.

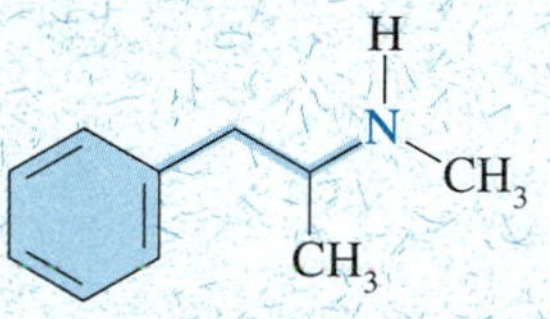

methamphetamine

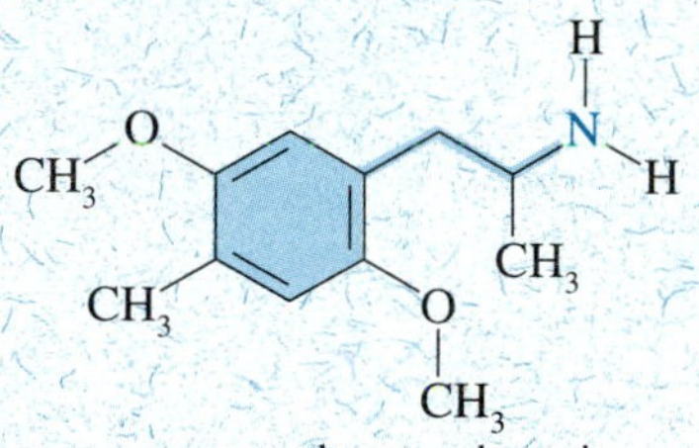

methoxyamphetamine

Phenylpropanolamine is an over-the-counter drug used as an appetite suppressant by those on crash diets. Its use is not recommended by physicians, and the drug can cause health hazards for individuals suffering from hypertension.

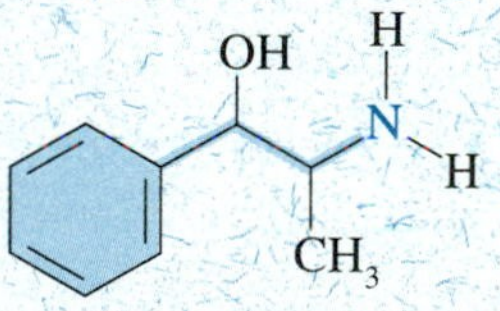

phenylpropanolamine

Although phenylethanamine derivatives generally act as stimulants, methylphenidate (Ritalin®) seems to have a calming effect. This drug has been used in the treatment of hyperactivity in young boys. The use of this drug is controversial. ■

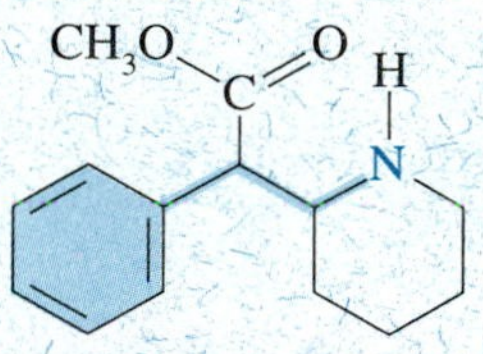

methylphenidate (Ritalin)

The base ionization constants for several amines are given in Table 14.2. We recall that alkyl groups are electron-donating toward carbocations (Section 4.9), and that they are electron-donating in electrophilic aromatic substitution reactions (Section 5.7). Similarly, alkyl-substituted amines are slightly stronger bases than ammonia because the inductive donation of electrons to the nitrogen atom by alkyl groups makes the unshared pair of electrons more available to a proton.

Aryl-substituted amines are much weaker bases than ammonia and alkyl-substituted amines. Their K_b values are less than 10^{-9} (Table 14.2). For example, the K_b value of aniline is 10^{-6} times the K_b value of cyclohexylamine.

$$C_6H_{11}NH_2 + H_2O \rightleftharpoons C_6H_{11}NH_3^+ + OH^- \qquad K_b = 4.6 \times 10^{-4}$$

$$C_6H_5NH_2 + H_2O \rightleftharpoons C_6H_5NH_3^+ + OH^- \qquad K_b = 4.3 \times 10^{-10}$$

Aryl-substituted amines are weaker bases than ammonia because the unshared pair of electrons of the nitrogen atom is resonance delocalized over the π orbital system of the benzene ring. As a result, the unshared electron pair of nitrogen is less available for bonding with a proton.

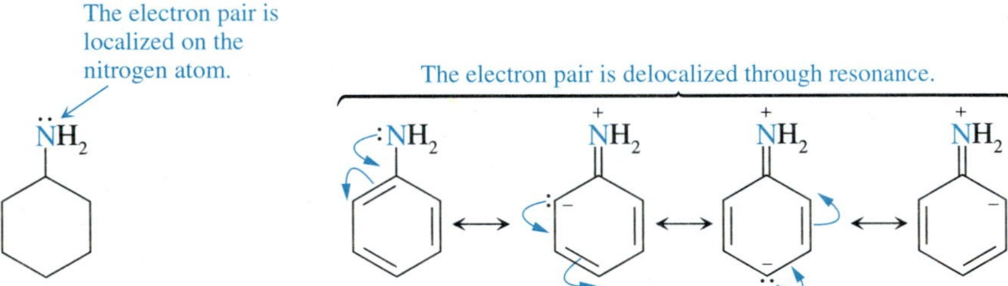

Heterocyclic Amines

The basicity of heterocyclic amines varies over a wide range and reflects both the hybridization of the nitrogen orbital containing the lone pair electrons and the effects of delocalization. Pyridine is a substantially weaker base than alkylamines. The electron pair of pyridine occupies an sp^2-hybridized orbital, and

Table 14.2 Basicity of Amines and Acidity of Ammonium Ions

	K_b	K_a	pK_b	pK_a
ammonia	1.8×10^{-5}	5.5×10^{-10}	4.74	9.26
methylamine	4.6×10^{-4}	2.2×10^{-11}	3.34	10.66
ethylamine	4.8×10^{-4}	2.1×10^{-11}	3.20	10.80
dimethylamine	4.7×10^{-4}	2.1×10^{-11}	3.27	10.73
diethylamine	3.1×10^{-4}	3.2×10^{-11}	3.51	10.49
triethylamine	1.0×10^{-3}	1.0×10^{-11}	3.00	11.00
cyclohexylamine	4.6×10^{-4}	2.2×10^{-11}	3.34	10.66
aniline	4.3×10^{-10}	2.3×10^{-5}	9.37	4.63

lies closer to the nitrogen nucleus than the electron pair in the sp^3-hybridized orbital of alkylamines. As a result, pyridine is a weaker base (larger pK_b) than an alkylamine.

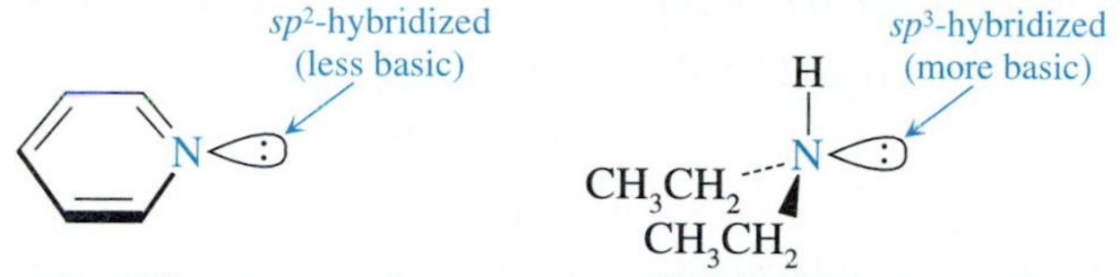

sp^2-hybridized (less basic)

sp^3-hybridized (more basic)

pyridine
$pK_b = 8.75$

diethylamine
$pK_b = 3.51$

Pyrrole is an exceedingly weak base. The pair of electrons of the nitrogen atom interacts with the four electrons of the two carbon-carbon double bonds to give an aromatic six-π-electron system similar to that of benzene. Thus, the electron pair is not readily available for protonation because it is required to maintain the sextet of electrons required for aromaticity in the ring. Pyrrolidine has a pK_b similar to acyclic amines.

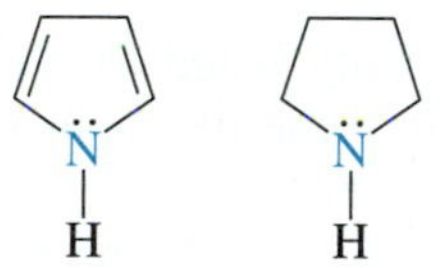

pyrrole
$pK_b = 15$

pyrrolidine
$pK_b = 2.7$

Imidazole is an important aromatic ring found in many biological molecules. One of its nitrogen atoms resembles that of pyrrole and is not basic. The second nitrogen atom, which is structurally similar to the nitrogen atom of pyridine, acts as a base. However, imidazole is about 100 times more basic than pyridine. The increased basicity results from resonance stabilization of the positive charge of the conjugate acid.

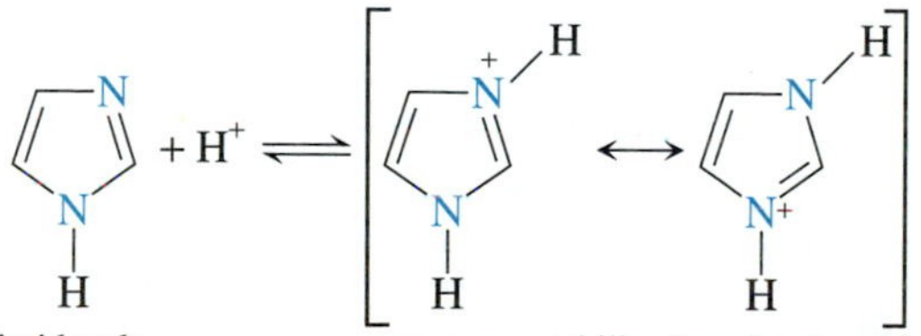

imidazole
$pK_b = 7.05$

resonance-stabilized conjugate acid

Amides

In contrast to amines, amides are extremely weak bases. (The K_b of ethanamide is approximately 10^{-15}!) This difference in basicity is due to the carbonyl group, which draws electron density away from the nitrogen atom. Thus, the unshared electron pair of the amide nitrogen atom is delocalized and not readily available to react with a proton. An amide is polar, planar, and resonance stabilized.

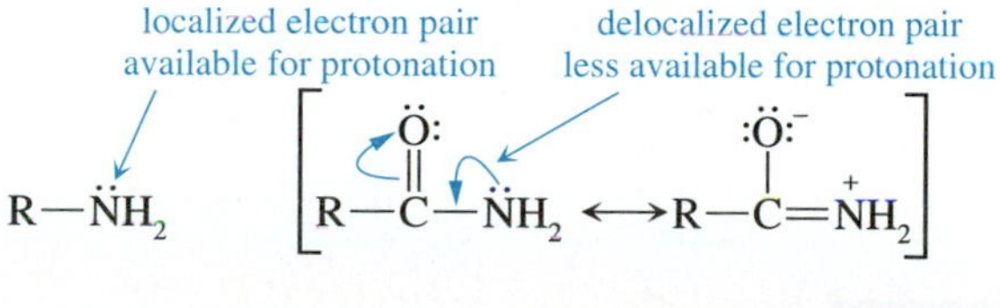

localized electron pair available for protonation

delocalized electron pair less available for protonation

Acidity of Ammonium Ions

When an amine is protonated, the conjugate acid product is a positively charged, substituted ammonium ion. The ionization constant of the conjugate acid of methylamine, the methylammonium ion, is derived from the following reaction:

$$CH_3NH_3^+ + H_2O \rightleftharpoons CH_3NH_2 + H_3O^+$$

$$K_a = \frac{[H_3O^+][CH_3NH_2]}{[CH_3NH_3^+]}$$

There is an inverse relationship between the K_a value of an acid and the K_b value of its related conjugate base. The K_a value of the methylammonium ion and the K_b value of methylamine illustrate this relationship. The K_a value for the methylammonium ion is relatively small; the K_b value for methylamine is relatively large. The values of K_a and K_b for a conjugate acid-base pair are related as follows:

$$(K_a)(K_b) = K_w = 1 \times 10^{-14}$$

pK_b and pK_a

The basicity of an amine is usually listed as a pK_b, the negative logarithm of K_b. For an amine with $K_b = 10^{-4}$, the pK_b is 4. The pK_b values of strong bases are small. Thus, as pK_b increases, base strength decreases.

It is also common practice to indicate the relative base strength of amines in terms of the pK_a of their conjugate acids. (Recall that p$K_a = -\log K_a$.) The sum of the pK_a and pK_b for a conjugate acid-base pair is 14. Thus, for an amine base with p$K_b = 5$, its conjugate ammonium ion has p$K_a = 9$.

EXAMPLE 14-5

Estimate the pK_b of pheneinamine, an antihistamine.

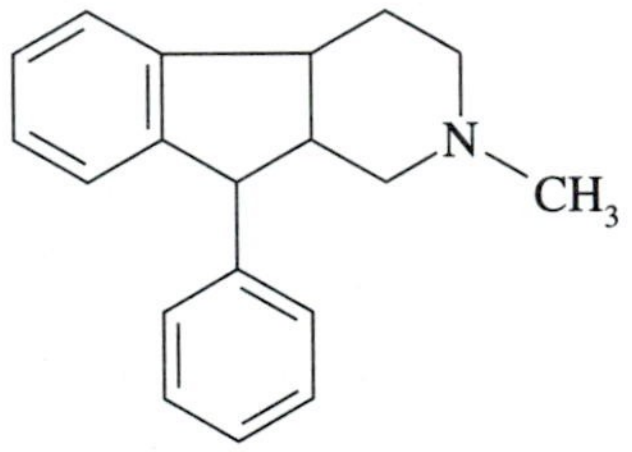

N
CH₃

Solution

The compound is a tertiary amine and the nitrogen atom is bonded only to alkyl groups. Thus, its pK_b should be similar to that of a simpler tertiary amine like triethylamine, which is 3.0 from Table 14.2.

Problem 14.5

Which of the two nitrogen atoms of chlorpromazine (Thorazine®), an antipsychotic drug, is the more basic?

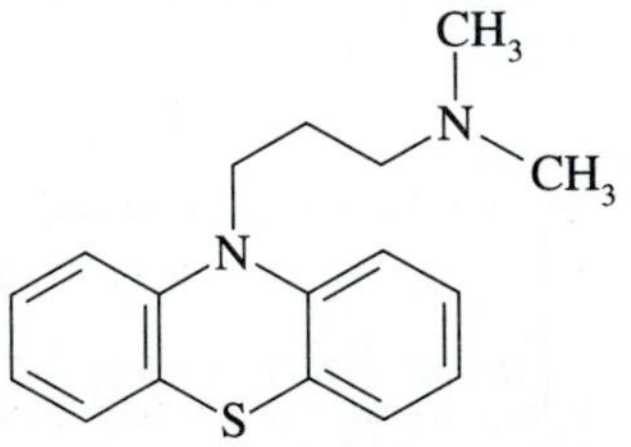

CH_3
N
CH_3
N
S

14.6 Solubility of Ammonium Salts

When an amine is added to a solution of a strong acid such as hydrochloric acid, the amine nitrogen atom is protonated to produce an ammonium salt.

$$RNH_2 + HCl \longrightarrow RNH_3^+ + Cl^-$$

Ammonium salts are more soluble than amines because the nitrogen atom of an ammonium salt has a positive charge. This property is used by drug companies to manufacture compounds that will be soluble in body fluids. Drugs containing an amino group are often prepared as ammonium salts to improve their solubility in body fluids. For example, the solubility of procaine (Novocain®) is only 0.5 g/100 mL, but the solubility of its ammonium salt is 100 g/100 mL.

O, O, N, CH_2CH_3, CH_2CH_3, NH_2

procaine

The ammonium salts of many drugs are more stable and less prone to oxidation than the amine itself. Procaine is most stable at pH 3.6 and becomes less stable as the pH is increased. The ammonium salts have higher melting points than the amines and have virtually no odor. For example, ephedrine melts at 79°C and has a fishy odor. Its hydrochloride salt, used in cold and allergy medications, melts at 217°C and has no odor.

Amines can be separated from other substances by converting them to ammonium salts. Consider the separation of 1-chlorooctane from 1-octanamine, a mixture that can result from a synthesis of the amine by reaction of 1-chlorooctane with ammonia. Both compounds are insoluble in water. Adding HCl to a solution containing both compounds converts the 1-octanamine into its ammonium salt, whereas 1-chlorooctane is not affected.

$$\underset{\text{(insoluble in water)}}{CH_3(CH_2)_6CH_2NH_2} + HCl \longrightarrow \underset{\text{(soluble in water)}}{CH_3(CH_2)_6CH_2NH_3^+} + Cl^-$$

$$\underset{\text{(insoluble in water)}}{CH_3(CH_2)_6CH_2Cl} + HCl \not\longrightarrow \text{no reaction}$$

The 1-chlorooctane is physically separated from the aqueous acid solution. Then the acid solution is neutralized with sodium hydroxide to form the free amine. The amine can then be physically separated from the aqueous solution.

$$\underset{\text{(soluble in water)}}{CH_3(CH_2)_6CH_2NH_3^+} + OH^- \longrightarrow \underset{\text{(insoluble in water)}}{CH_3(CH_2)_6CH_2NH_2} + H_2O$$

Amides are not sufficiently basic to be protonated in aqueous solutions. Thus, their solubility is unaffected by pH. They behave as neutral compounds.

14.7 Nucleophilic Reactions of Amines

We described some reactions of amines in earlier chapters. These reactions occur because the nonbonding electron pair of the nitrogen atom makes amines nucleophilic. We will review each type of reaction in this section.

Reaction with Carbonyl Compounds

In Chapter 10, we described the addition-elimination reaction of amines with carbonyl compounds. An amine adds to the carbonyl carbon atom to give a tetrahedral intermediate. This product is unstable and loses water to form an imine. In general, imines are less stable than carbonyl compounds. Thus, the reaction is favorable only if water is removed from the reaction mixture. Most imines are not stable; they rapidly hydrolyze in aqueous solution to give carbonyl compounds.

$$\underset{\text{amine}}{R{-}NH_2} + \underset{\text{aldehyde}}{R'(H)C{=}O} \rightleftharpoons \underset{\text{reactive tetrahedral intermediate}}{R{-}NH{-}C(R')(H){-}OH} \rightleftharpoons \underset{\text{imine}}{R{-}N{=}C(R')(H)} + H_2O$$

Reaction with Acyl Derivatives

In Chapter 12, we noted that an amide can be made by treating an amine with an acid halide. We recall that acid halides are very reactive acyl derivatives of acids; amides are very stable.

$$\underset{\text{amine}}{R{-}NH_2} + \underset{\text{acid chloride}}{R'(Cl)C{=}O} \longrightarrow \underset{\text{reactive tetrahedral intermediate}}{R{-}NH{-}C(R')(Cl){-}OH} \longrightarrow \underset{\text{an amide}}{R{-}NH{-}C({=}O){-}R'} + HCl$$

$$C_5H_9{-}NH_2 + CH_3{-}\overset{O}{\overset{\|}{C}}{-}Cl \xrightarrow{\text{pyridine}} C_5H_9{-}NH{-}\overset{O}{\overset{\|}{C}}{-}CH_3$$

Only ammonia and primary or secondary amines form amides. For this reason, pyridine, which cannot form an amide, is often used as a base to react with the HCl formed in the reaction.

Reaction of Amines with Alkyl Halides

We described nucleophilic substitution reactions of alkyl halides in Chapter 7. Primary and secondary alkyl halides react with nucleophiles by an S_N2 mechanism. Amines are nucleophiles that can displace a halide ion from a primary or secondary alkyl halide to form an ammonium halide salt that is subsequently neutralized.

$$R{-}\ddot{N}H_2 + R'{-}X \xrightarrow{-X^-} R{-}\overset{+}{N}H_2{-}R' \xrightarrow{OH^-} R{-}\ddot{N}H{-}R'$$

The initial product of the nucleophilic substitution reaction is a secondary ammonium ion. It can lose a proton in an equilibrium reaction with the reactant primary amine.

$$R{-}\ddot{N}H_2 + R{-}\overset{\overset{H}{|}}{\underset{\underset{H}{|}}{N^+}}{-}R' \rightleftharpoons R{-}NH_3^+ + R{-}\underset{\underset{H}{|}}{\ddot{N}}{-}R'$$

The secondary amine then can continue to react with the alkyl halide to give a tertiary amine and eventually a quaternary ammonium ion.

$$\underset{\text{primary amine}}{R{-}\ddot{N}H_2} \xrightarrow{R'X} \underset{\text{secondary amine}}{R{-}\underset{\underset{H}{|}}{\ddot{N}}{-}R'} \xrightarrow{R'X} \underset{\text{tertiary amine}}{R{-}\underset{\underset{R'}{|}}{\ddot{N}}{-}R'} \xrightarrow{R'X} \underset{\text{quaternary ammonium ion}}{R{-}\overset{\overset{R'}{|}}{\underset{\underset{R'}{|}}{N^+}}{-}R'}$$

Quaternary ammonium salts are ammonium salts that have four alkyl or aryl groups bonded to a nitrogen atom. Some quaternary ammonium salts containing a long carbon chain are **invert soaps.**

$$CH_3(CH_2)_n{-}\overset{\overset{R'}{|}}{\underset{\underset{R'}{|}}{N^+}}{-}R'$$

an invert soap

Some mouthwash products contain quaternary ammonium ions.

Invert soaps differ from soaps and detergents because the polar end of the ion in the micelle is positive rather than negative. Like soaps, the long hydrocarbon tail associates with nonpolar substances, and the polar head dissolves in water. Thus, invert soaps act by the same cleansing mechanism described in Chapter 13 for soaps and detergents.

Invert soaps are widely used in hospitals. They are active against bacteria, fungi, and protozoa, but they are not effective against spore-forming microorganisms. Benzalkonium chlorides are one type of invert soap. The alkyl groups of these compounds contain from 8 to 16 carbon atoms.

$$CH_3(CH_2)_{14}CH_2{-}\overset{\overset{CH_3}{|}}{\underset{\underset{CH_3}{|}}{N^+}}{-}CH_2{-}C_6H_5$$

a benzalkonium ion

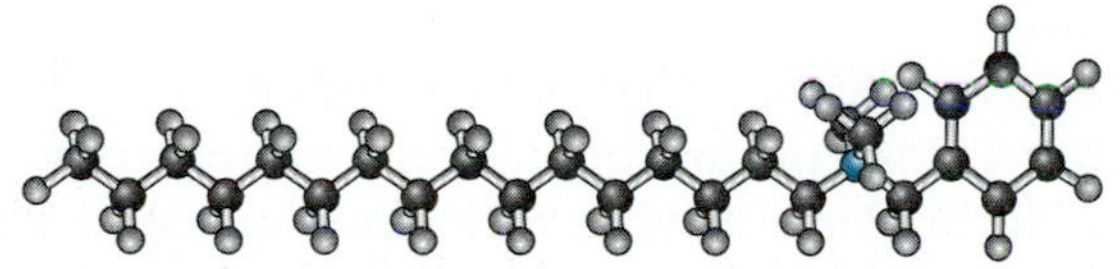

EXAMPLE 14-6

Flecainide, an antiarrhythmic drug, is an amide. Draw the structures of the compounds that could be used to produce the drug. What possible complications might occur with this combination of reactants?

CF_3CH_2O H O N C—NH—CH_2— OCH_2CF_3

Solution

Mentally separate the amide into two components by breaking the bond between the nitrogen atom and the carbonyl carbon atom. Place a hydrogen atom on the nitrogen atom. Place a chlorine atom on the carbonyl carbon atom.

CF_3CH_2O O C—Cl OCH_2CF_3 H N NH_2—CH_2—

Note that the "amine" is actually a diamine: one part is a primary amine, the other part a secondary amine. Thus, the diamine could react at either nitrogen atom and form two isomeric amides. The primary amine is more reactive because the secondary amine is more sterically hindered.

Problem 14.6

Acetaminophen, the analgesic in Tylenol®, is an amide. Draw the structures of the compounds that could be used to produce the drug. What possible complications might occur with this combination of reactants?

H O HO— —N—C—CH_3

14.8 Synthesis of Amines

Many of the general methods to synthesize amines have already been discussed in preceding sections and chapters. Except for the displacement reaction of an alkyl halide by ammonia or an amine, the remaining methods involve compounds that already have a nitrogen atom contained in a functional group, which is then transformed into an amine functional group.

Alkylation of Amines by Alkyl Halides

In Section 14.7 we saw that a nucleophilic substitution reaction of ammonia with an alkyl halide yields a mixture of products resulting from multiple alkylation. The chances for multiple alkylation can be diminished somewhat by selecting the proper reaction conditions. For example, if the reaction of an alkyl halide with ammonia is carried out with excess ammonia, an alkyl halide can be converted to a primary amine. When the concentration of ammonia is greater than the concentration of the primary amine product, the probability is reduced that the primary amine will continue to react with the alkyl halide.

$$\text{C}_6\text{H}_{11}\text{—CH}_2\text{—Br} \xrightarrow{\text{NH}_3} \text{C}_6\text{H}_{11}\text{—CH}_2\text{—NH}_2$$

Reduction of Imines

We recall that the carbonyl group of either aldehydes or ketones is reduced to an alcohol by either catalytic hydrogenation or metal hydrides. Imines are the nitrogen analogs of carbonyl compounds, and they are reduced in the same way.

$$\text{C}_6\text{H}_5\text{—CH=N—C}_6\text{H}_{11} \xrightarrow[\text{Ni}]{\text{H}_2} \text{C}_6\text{H}_5\text{—CH}_2\text{—NH—C}_6\text{H}_{11}$$

Imines do not have to be prepared and isolated for subsequent reduction. A mixture of a carbonyl compound and ammonia or the appropriate amine reacts in the presence of hydrogen gas and a metal catalyst. The imine initially formed is reduced to an amine. The overall process is called **reductive amination.**

$$\text{C}_6\text{H}_5\text{—CHO} + \text{CH}_3\text{NH}_2 \xrightarrow[\text{Ni}]{\text{H}_2} \text{C}_6\text{H}_5\text{—CH}_2\text{—NHCH}_3$$

Reduction of Amides

Reduction of amides is one of the most frequently used methods of preparing amines. The method is very versatile because primary, secondary, and tertiary amines are easily prepared from the corresponding class of amide. Amides are prepared by acylation of amines using activated acyl derivatives such as acid chlorides or acid anhydrides (Section 12.6). Subsequent reduction of the amide with $LiAlH_4$ followed by acidic workup produces the amine.

$$\text{R—C(=O)—NH}_2 \xrightarrow[\text{2. H}_3\text{O}^+]{\text{1. LiAlH}_4} \text{R—CH}_2\text{—NH}_2$$

$$\text{R—C(=O)—NH—R} \xrightarrow[\text{2. H}_3\text{O}^+]{\text{1. LiAlH}_4} \text{R—CH}_2\text{—NH—R}'$$

$$\text{R—C(=O)—N(R}')\text{—R}' \xrightarrow[\text{2. H}_3\text{O}^+]{\text{1. LiAlH}_4} \text{R—CH}_2\text{—N(R}')\text{—R}'$$

Reduction of Nitriles

Nitriles can be prepared from primary alkyl halides by a direct S_N2 displacement reaction using sodium cyanide as the nucleophile (Section 7.6). Then the nitrile is reduced to a primary amine with lithium aluminum hydride.

$$\text{CH}_3\text{—C(CH}_3)_2\text{—CH}_2\text{—CH}_2\text{—Br} \xrightarrow[\text{2. LiAlH}_4 \quad \text{3. H}_3\text{O}^+]{\text{1. CN}^-} \text{CH}_3\text{—C(CH}_3)_2\text{—CH}_2\text{—CH}_2\text{—CH}_2\text{—NH}_2$$

1-bromo-3,3-dimethylbutane → 4,4-dimethyl-1-pentanamine

Reduction of Nitro Compounds

There is no synthetic procedure to introduce an amino group onto an aromatic ring in one step. However, it is possible to substitute an amino group onto an aromatic ring in two steps (Section 5.10). First the ring is nitrated. Then the nitro group is reduced to an amino group.

$$\text{C}_6\text{H}_5\text{CH(CH}_3)_2 \xrightarrow[\text{H}_2\text{SO}_4]{\text{HNO}_3} \text{O}_2\text{N-C}_6\text{H}_4\text{-CH(CH}_3)_2 \xrightarrow[\text{HCl}]{\text{Sn}} \text{H}_2\text{N-C}_6\text{H}_4\text{-CH(CH}_3)_2$$

14.9 Hydrolysis of Amides

Hydrolysis of an amide breaks the carbon-nitrogen bond and produces an acid and either ammonia or an amine. This reaction resembles the hydrolysis of esters, which we discussed in Chapter 12. There are, however, important differences. The hydrolysis of esters occurs relatively easily, whereas amides are very resistant to hydrolysis. Amides are hydrolyzed only by heating for hours with a strong acid or strong base. When amide hydrolysis is carried out in basic solution, the salt of the carboxylic acid forms; one mole of base is required per mole of amide. When amide hydrolysis is carried out under acidic conditions, the ammonium salt of the amine is formed, and one mole of acid is required per mole of amide.

$$\text{C}_6\text{H}_5\text{C(=O)NHCH}_3 + \text{OH}^- \longrightarrow \text{C}_6\text{H}_5\text{C(=O)O}^- + \text{CH}_3\text{NH}_2$$

$$\text{C}_6\text{H}_5\text{C(=O)NHCH}_3 + \text{HCl} + \text{H}_2\text{O} \longrightarrow \text{C}_6\text{H}_5\text{C(=O)OH} + \text{CH}_3\text{NH}_3^+$$

The great stability of amides toward hydrolysis has an important biological consequence, because amino acids in proteins are linked by amide bonds. Because amides are stable, proteins do not readily hydrolyze at physiological pH and at body temperature in the absence of a specific enzyme catalyst. However, in the presence of specific enzymes, the hydrolysis of amides is rapid. These reactions will be discussed in Chapter 15.

EXAMPLE 14-7

What are the products of the hydrolysis of phenacetin by a base? Phenacetin was formerly used in APC analgesic tablets consisting of aspirin, phenacetin, and caffeine.

$$\text{CH}_3\text{-C(=O)-N(H)-C}_6\text{H}_4\text{-OCH}_2\text{CH}_3$$

Solution

The functional group on the right side of the benzene ring is an ether, which does not react with base (Section 9.6). The functional group on the left is an amide.

Hydrolysis of an amide breaks the bond between the nitrogen atom and the carbonyl group. The acid fragment is acetic acid. The amine fragment is a substituted aniline containing an ether substituent.

$$CH_3{-}\overset{O}{\overset{\|}{C}}{-}\overset{H}{\overset{|}{N}}{-}C_6H_4{-}OCH_2CH_3$$

Hydrolysis occurs here.

Because a base is used in the hydrolysis, the acid product is present in the reaction mixture as the acetate ion. The amine is *p*-ethoxyaniline.

$$CH_3{-}\overset{O}{\overset{\|}{C}}{-}O^- \qquad NH_2{-}C_6H_4{-}OCH_2CH_3$$

Problem 14.7

What are the products of the hydrolysis of nubucaine by an acid? Nubucaine is a local anesthetic.

$$O{=}C(NHCH_2CH_2N(CH_2CH_3)_2) \text{ on quinoline ring bearing } OCH_2CH_2CH_2CH_3$$

14.10 Synthesis of Amides

Carboxylic acids react to form an amide and water when heated to a high temperature with ammonia, a primary amine, or a secondary amine. Tertiary amines do not form amides because they have no hydrogen atom bonded to the nitrogen atom.

$$R{-}\overset{O}{\overset{\|}{C}}{-}OH + NH_3 \longrightarrow R{-}\overset{O}{\overset{\|}{C}}{-}NH_2 + H_2O$$

$$R{-}\overset{O}{\overset{\|}{C}}{-}OH + R'NH_2 \longrightarrow R{-}\overset{O}{\overset{\|}{C}}{-}NHR' + H_2O$$

$$R{-}\overset{O}{\overset{\|}{C}}{-}OH + R'R''NH \longrightarrow R{-}\overset{O}{\overset{\|}{C}}{-}\underset{R''}{\underset{|}{N}}{-}R' + H_2O$$

The high temperature of this direct reaction often affects other functional groups in the molecule. An amide can be synthesized at lower temperatures by the reaction of an acyl chloride with ammonia, a primary amine, or a secondary amine (Section 14.7).

$$C_6H_{11}{-}\overset{O}{\overset{\|}{C}}{-}Cl + (CH_3)_3C{-}NH_2 \xrightarrow{\text{pyridine}} C_6H_{11}{-}\overset{O}{\overset{\|}{C}}{-}NH{-}C(CH_3)_3$$

Polyamides

Many commercial products are high molecular weight polyamides. Perhaps the most famous of these is nylon. Synthetic polyamides such as nylon are produced from diamines and dicarboxylic acids by condensation polymerization. One type of nylon is made from adipic acid and hexamethylenediamine (1,6-diaminohexane).

$$HO-\overset{O}{\overset{\|}{C}}-(CH_2)_4-\overset{O}{\overset{\|}{C}}-OH \qquad NH_2-(CH_2)_6-NH_2$$

adipic acid — hexamethylenediamine

The product of the first condensation reaction is an amide that also contains a free amino group and a free carboxylic acid group. The amine end of this molecule can react with another molecule of adipic acid to produce another amide linkage. The carboxylic acid end of the molecule can react with another molecule of hexamethylenediamine.

This end can react with an amine group. — This end can react with a carboxylic acid group.

$$HO-\overset{O}{\overset{\|}{C}}-(CH_2)_4-\overset{O}{\overset{\|}{C}}-NH-(CH_2)_6-NH_2$$

This sequence of reactions occurs again and again to produce a polyamide. The polyamide formed from adipic acid and hexamethylenediamine, shown in the drawing below, is called nylon 6,6. The "6,6" refers to the six-carbon diacid and six-carbon diamine reactants.

Polyamides containing aromatic rings (aramides) have many special properties. The presence of aromatic rings in the polymer produces a stiff and tough fiber. One commercially important aramide is Kevlar®, a polyamide made from terephthaloyl chloride and *p*-phenylenediamine.

$$Cl-\overset{O}{\overset{\|}{C}}-C_6H_4-\overset{O}{\overset{\|}{C}}-Cl \qquad NH_2-C_6H_4-NH_2$$

terephthaloyl chloride — *p*-phenylenediamine

The structure of the polymer Kevlar is shown in the drawing below. It is used in place of steel in bullet-resistant vests. These vests are so light and flexible that they can be worn inconspicuously under normal clothing.

Law enforcement personnel often are protected by bullet-resistant vests.

An aramide called Nomex® has a structure that resembles that of Kevlar. The monomers in Nomex are meta rather than para isomers. Nomex is used in flame-resistant clothing for fire fighters and race car drivers; it is so strong that it can also be used in flame-resistant building materials. ■

$$HO-\overset{O}{\overset{\|}{C}}-(CH_2)_4-\overset{O}{\overset{\|}{C}}\left[NH-(CH_2)_6-NH-\overset{O}{\overset{\|}{C}}-(CH_2)_4-\overset{O}{\overset{\|}{C}}\right]_n NH-(CH_2)_6-NH_2$$

Nylon 6,6

$$HO-\overset{O}{\overset{\|}{C}}-C_6H_4-\overset{O}{\overset{\|}{C}}\left[NH-C_6H_4-NH-\overset{O}{\overset{\|}{C}}-C_6H_4-\overset{O}{\overset{\|}{C}}\right]_n NH-C_6H_4-NH_2$$

Kevlar

Summary of Reactions

1. Reaction of Amines with Carbonyl Compounds (Section 14.7)

$$CH_3CH_2CH(CH_3)CH_2C(=O)-H \underset{H_3O^+}{\overset{CH_3CH_2NH_2}{\rightleftharpoons}} CH_3CH_2CH(CH_3)CH_2C(=NCH_2CH_3)-H \quad + H_2O$$

2. Reaction of Amines with Acyl Derivatives (Section 14.7)

$$CH_3CH_2CH(CH_3)CH_2C(=O)-Cl + CH_3CH_2NH_2 \xrightarrow{\text{pyridine}} CH_3CH_2CH(CH_3)CH_2C(=O)-NHCH_2CH_3$$

3. Reaction of Amines with Alkyl Halides (Section 14.7)

$$CH_3-CH(CH_3)-CH_2-Br \underset{\text{excess}}{\xrightarrow{CH_3NH_2}} CH_3-CH(CH_3)-CH_2-NH-CH_3$$

4. Synthesis of Amines (Section 14.8)

$$C_6H_5C(=O)CH_3 \underset{H_2/Ni}{\xrightarrow{CH_3NH_2}} C_6H_5CH(NH-CH_3)CH_3$$

$$C_6H_5C(=O)NHCH_3 \underset{2.\ H_3O^+}{\xrightarrow{1.\ LiAlH_4}} C_6H_5CH_2NHCH_3$$

$$CH_3CH_2CH(CH_3)CH_2C\equiv N \underset{2.\ H_3O^+}{\xrightarrow{1.\ LiAlH_4}} CH_3CH_2CH(CH_3)CH_2CH_2-NH_2$$

$$o\text{-}(NO_2)C_6H_4CH_2CH_3 \xrightarrow{Sn/HCl} o\text{-}(NH_2)C_6H_4CH_2CH_3$$

5. Hydrolysis of Amides (Section 14.9)

$$CH_3CH_2CH(CH_3)CH_2C(=O)-NHCH_3 \underset{2.\ OH^-}{\xrightarrow{1.\ H_3O^+}} CH_3CH_2CH(CH_3)CH_2C(=O)-OH + CH_3NH_2$$

6. Synthesis of Amides (Section 14.10)

$$C_6H_5N(H)H + CH_3-C(=O)-Cl \xrightarrow{\text{pyridine}} C_6H_5N(H)C(=O)CH_3$$

Explorations with Molecular Models

1. Construct the following ball-and-stick model of an amine. Classify the amine. Determine the value of x in the general molecular formula $C_nH_{2n+x}N$ that represents amines.

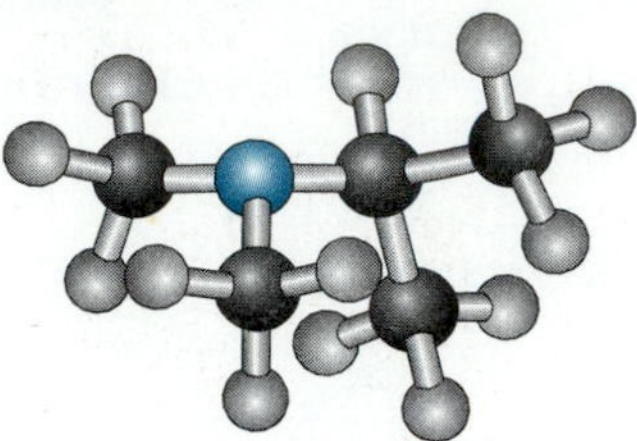

2. Construct the following ball-and-stick model of an amine. Classify the amine. Name the structure. Arrange the structure to view the stereogenic center with the carbon-hydrogen bond directed away from your eye. Determine whether the structure is *R* or *S*. Sighting along the C-2 to C-3 bond with the C-2 atom closest to your eye, draw a Newman projection of the structure.

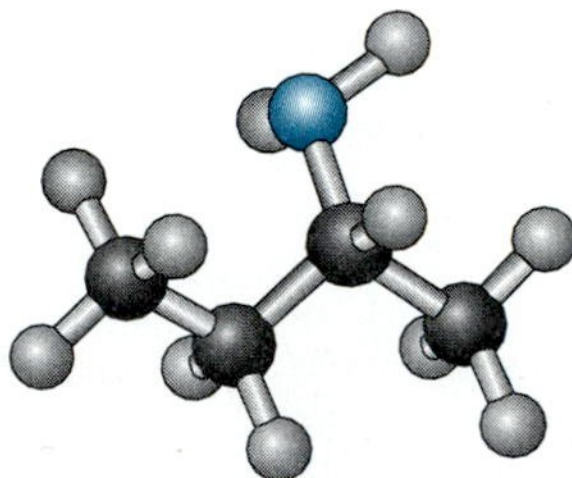

3. Construct the following ball-and-stick model of triethylamine. Make sure that the nitrogen atom is in a staggered arrangement with respect to the hydrogen atoms of each methyl group. Consider the position of the lone pair electrons of the nitrogen atom, which make the nitrogen atom a nucleophilic species. How is the steric environment affected by the methyl groups?

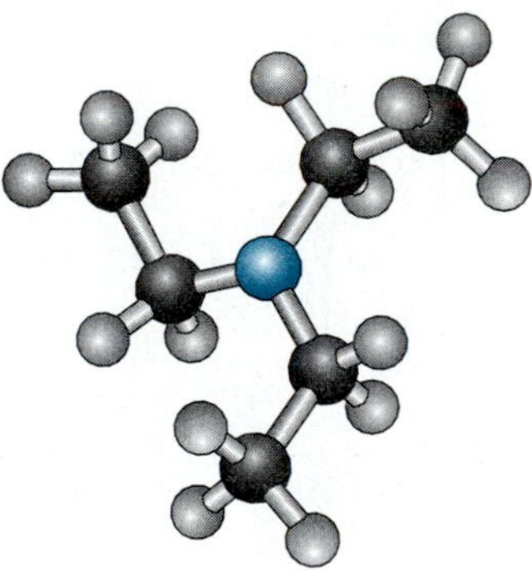

4. Construct the following ball-and-stick model of a bicyclic amine. Consider the position of the lone pair electrons of the nitrogen atom, which make the nitrogen atom a nucleophilic species. How does the steric environment of the nitrogen atom of this structure differ from that of triethylamine?

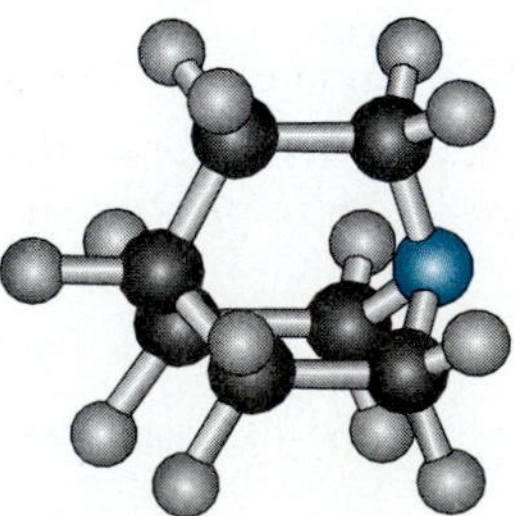

5. Construct the following ball-and-stick model of *N*-methylpiperidine. What term is used to describe the location of the methyl group in this conformation? Push the nitrogen atom downward while pushing the C-4 atom upward to obtain a second chair conformation. What term is used to describe the location of the methyl group in this conformation? Which of these two conformations is the more stable?

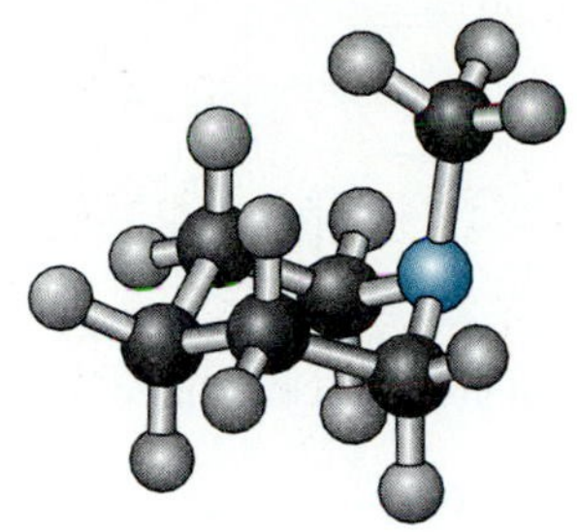

Exercises

Classification of Amines and Amides

14.1 Classify each of the following amines or amides according to their degree of substitution.

(a) $CH_3CH_2—N(H)—CH_3$ (b) $CH_3CH_2—N(CH_3)—CH_2CH_2OH$

(c) cyclohexane bearing CH_3 and NH_2 on the same carbon (d) $CH_3CH_2—N(H)—C(=O)$—cyclopentyl

(e) six-membered lactam ring with C=O and N—CH_3 (f) $C_6H_5—C(=O)NH_2$

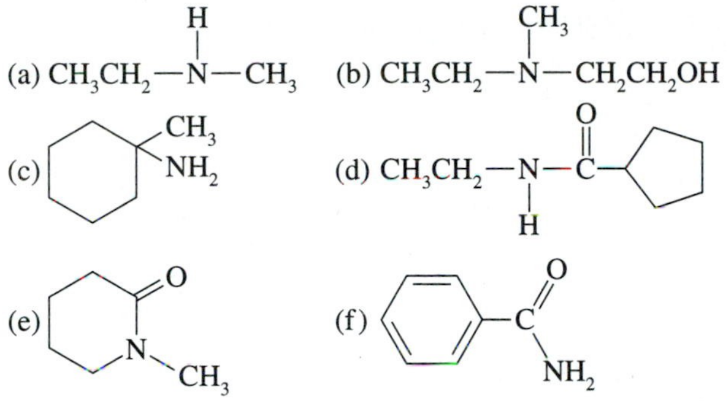

14.2 Classify each of the following amines or amides according to their degree of substitution.

(a) $CH_3CH_2—N(CH_3)—CH_2CH_3$ (b) $CH_3CH_2—N(H)—CH{=}CH_2$

(c) cyclopentyl—CH_2NH_2 (d) $CH_3CH_2—N(CH_3)—C(=O)$—phenyl

(e) six-membered lactam ring with C=O and N—H (f) $C_6H_5—C(=O)NHCH_2CH_3$

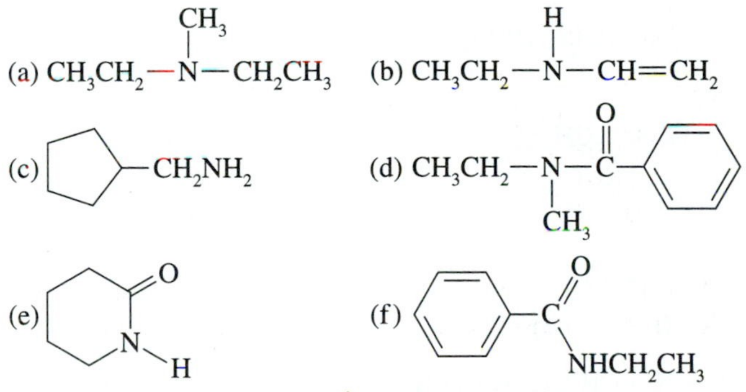

14.3 Classify the nitrogen-containing functional group in each of the following structures.

(a) acetaminophen, the analgesic in Tylenol

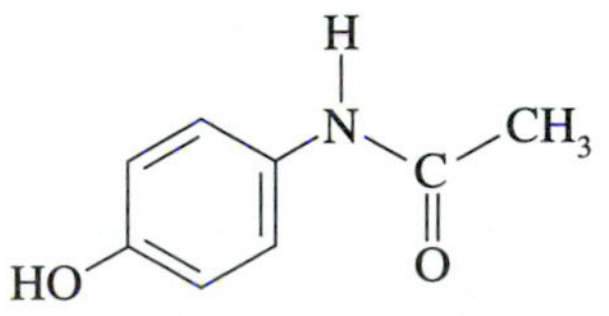

HO—C_6H_4—N(H)—C(=O)—CH_3

(b) coniine, part of the hemlock poison drunk by Socrates

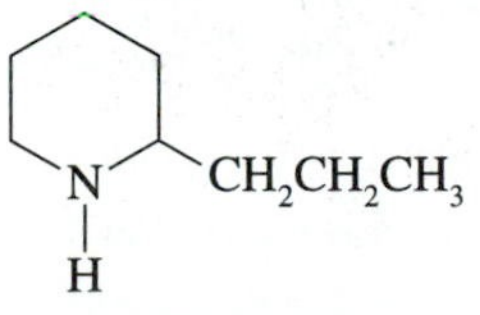

piperidine ring (N—H) with $CH_2CH_2CH_3$ on the carbon adjacent to N

14.4 Classify the nitrogen-containing functional group in each of the following structures.

(a) DEET, an insect repellant (b) phencylidine, a hallucinogen

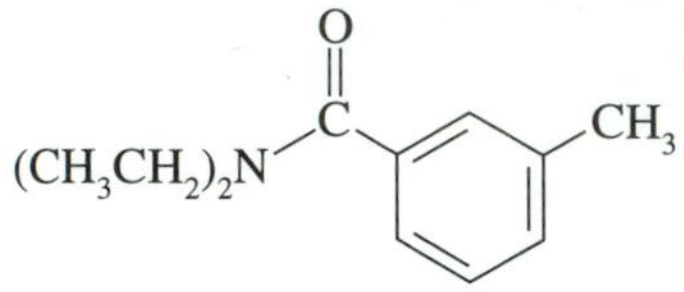

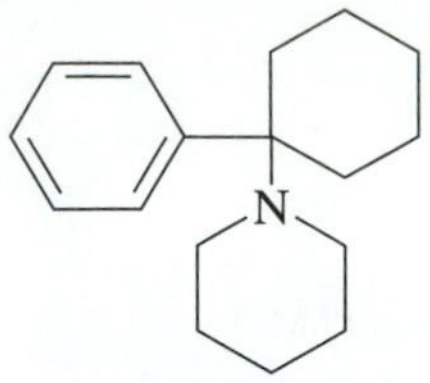

$(CH_3CH_2)_2N$—C(=O)—(C$_6$H$_4$)—CH_3

14.5 Classify the nitrogen-containing functional groups in encainide, an antiarrhythmic drug.

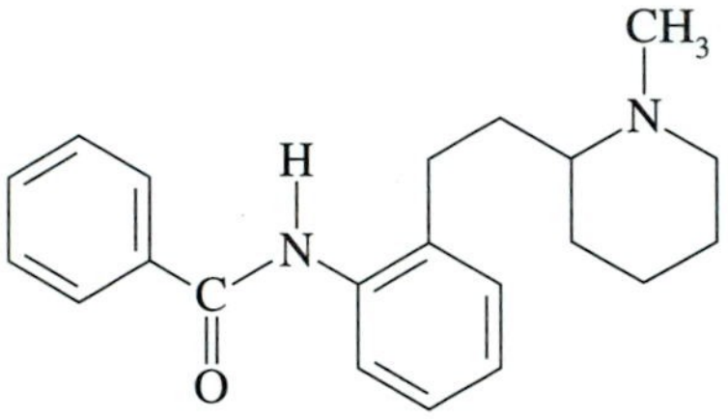

14.6 Classify the nitrogen-containing functional groups in practolol, an antihypertensive drug.

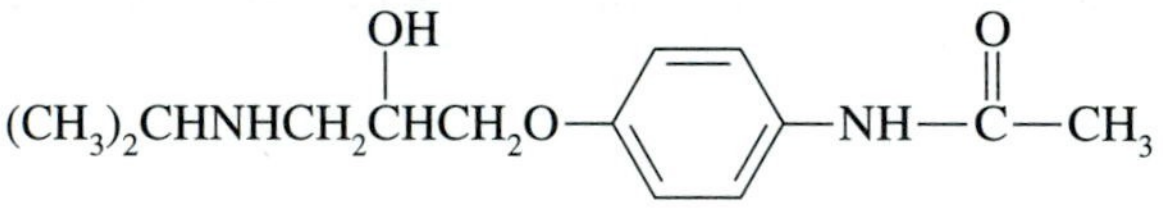

$(CH_3)_2CHNHCH_2CH(OH)CH_2O$—(C$_6H_4$)—NH—C(=O)—$CH_3$

Nomenclature

14.7 Give the IUPAC name for each of the following compounds.

(a) $CH_3—N(CH_3)—CH_2CH_2CH_2CH_3$ (b) $CH_3CH_2—N(H)—CH_2CH(CH_3)CH_2CH_3$

(c) (d)

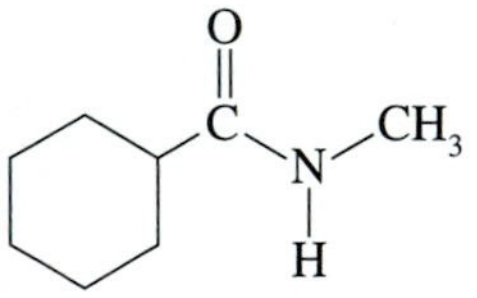

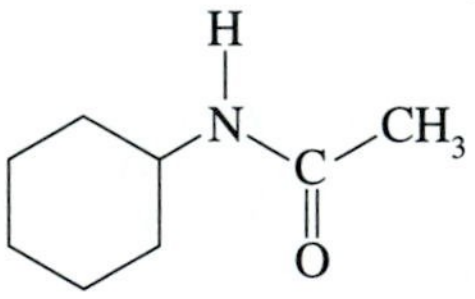

14.8 Give the IUPAC name for each of the following compounds.

(a) $CH_3—N(CH_3)—CH_2CH_2CH(OH)CH_3$ (b) $CH_3CH_2—N(H)—CH_2CH(CH_3)CH_2CO_2H$

(c) (d)

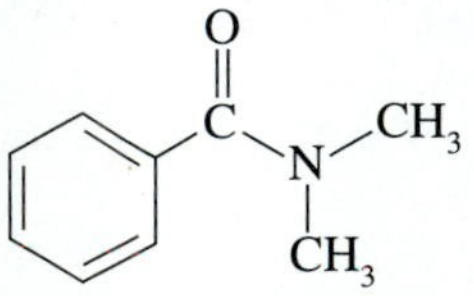

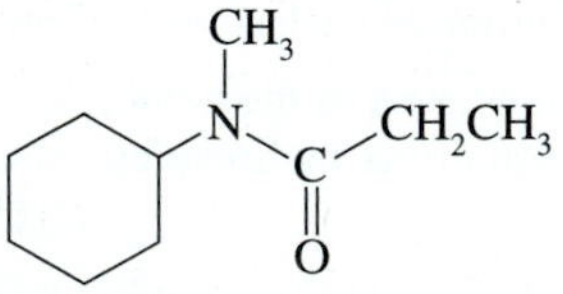

14.9 An antidepressant drug is named *trans*-2-phenylcyclopropyl amine. Draw its structure.

14.10 *trans*-4-(Methylamino)cyclohexanecarboxylic acid is the IUPAC name for tranexamic acid, a drug that aids blood clotting. Draw its structure.

14.11 Name the following compound produced by the marine acorn worm.

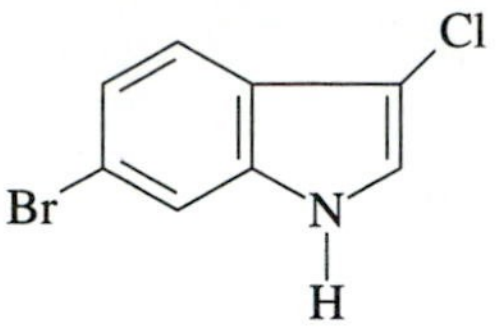

14.12 Draw the structure of each of the following compounds.
(a) 2-ethylpyrrole (b) 3-bromopyridine (c) 2,5-dimethylpyrimidine

Isomers of Amines

14.13 Draw and name all isomers with the molecular formula C_2H_7N.

14.14 Draw and name all isomers with the molecular formula C_3H_9N.

14.15 Draw and name all isomers for primary amines with the molecular formula $C_4H_{11}N$.

14.16 Draw and name all isomers for tertiary amines with the molecular formula $C_5H_{13}N$.

Properties of Amines

14.17 The boiling points of the isomeric compounds propylamine and trimethylamine are 49 and 3.5°C, respectively. Explain this large difference.

14.18 The boiling point of 1,2-diaminoethane is 116°C. Explain why this compound boils at a much higher temperature than propylamine (49°C) although its molecular weight is similar.

Basicity of Amines

14.19 The pK_b values for cyclohexylamine and triethylamine are 3.34 and 2.99, respectively. Which compound is the stronger base?

14.20 The K_b values for dimethylamine and diethylamine are 4.7×10^{-4} and 3.1×10^{-4}, respectively. Which compound is the stronger base?

14.21 Estimate the K_b of each of the following.

(a) 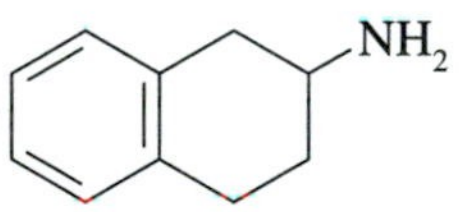 (b) 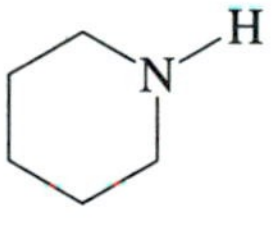 (c)

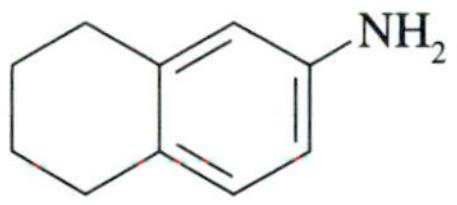

14.22 Estimate the K_b of each of the following.

(a) 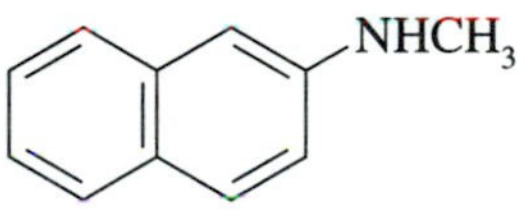 (b) 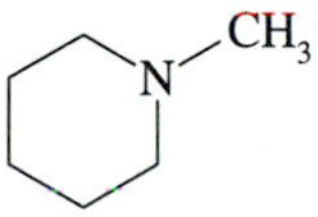 (c)

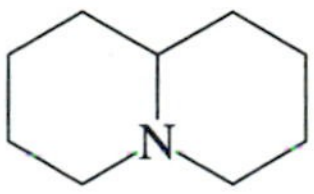

14.23 Explain why the pK_b of aniline (9.4) is different from the pK_b of *p*-nitroaniline (13.0).

14.24 Explain the difference in the pK_b of the following bases.

$N{\equiv}CCH_2CH_2NH_2$ (6.2) $N{\equiv}CCH_2NH_2$ (8.7)

14.25 Physostigmine is used in 0.1 to 1.0% solutions to decrease the intraocular pressure in treatment of glaucoma. Rank the three nitrogen atoms in the molecule in order of increasing basicity.

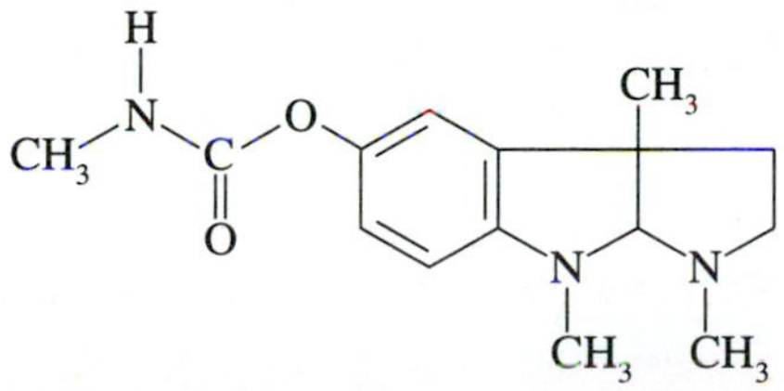

14.26 Nubucaine is a local anesthetic that is administered as the hydrochloride salt. Which nitrogen atom is protonated?

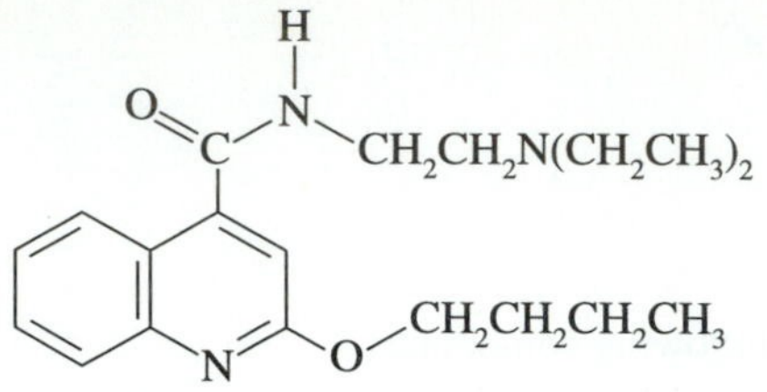

H
O=C–N–$CH_2CH_2N(CH_2CH_3)_2$
N O $CH_2CH_2CH_2CH_3$

Reactions of Amines

14.27 Draw the structure of the compound formed when benzylmethylamine reacts with each of the following reagents.
(a) excess methyl iodide (b) acetyl chloride (c) hydrogen iodide

14.28 Draw the structure of the compound formed when piperidine reacts with each of the following reagents.
(a) allyl bromide (b) benzoyl chloride (c) acetic anhydride

Synthesis of Amines

14.29 Write the structure of the product of each of the following reactions.

(a) $CH_3—\overset{O}{\overset{\|}{C}}—CH_2CH_3 + CH_3CH_2—NH_2 \xrightarrow[Ni]{H_2}$

(b) 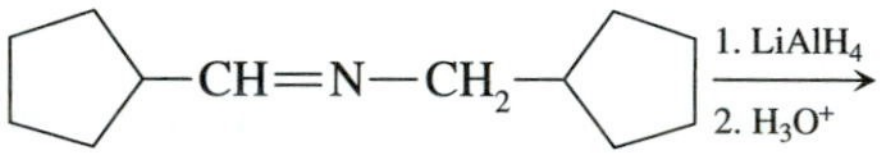 (cyclopentyl)—$CH{=}N—CH_2$—(cyclopentyl) $\xrightarrow[2.\ H_3O^+]{1.\ LiAlH_4}$

(c) 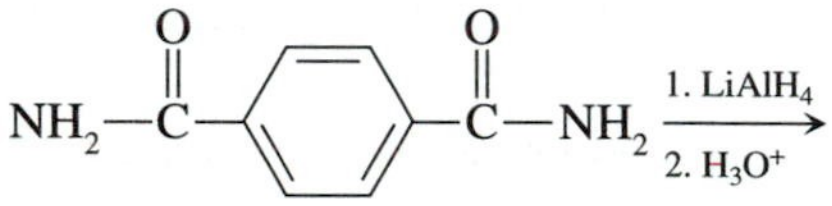 $NH_2—\overset{O}{\overset{\|}{C}}$—(benzene ring)—$\overset{O}{\overset{\|}{C}}—NH_2 \xrightarrow[2.\ H_3O^+]{1.\ LiAlH_4}$

14.30 Write the structure of the product of each of the following reactions.

(a) $CH_3—\overset{O}{\overset{\|}{C}}—CH_2CH_2CH_3 + CH_3—NH_2 \xrightarrow[Ni]{H_2}$

(b) $CH_3—\overset{NCH_3}{\overset{\|}{C}}—CH_2CH_2CH_3 \xrightarrow[2.\ H_3O^+]{1.\ LiAlH_4}$

(b) 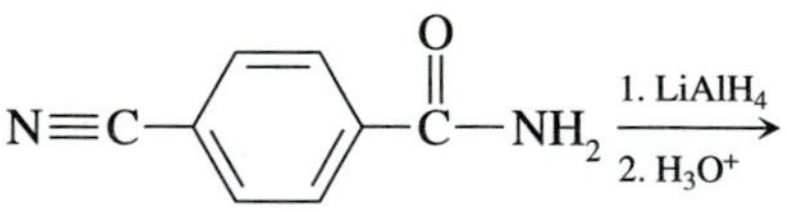 $N{\equiv}C$—(benzene ring)—$\overset{O}{\overset{\|}{C}}—NH_2 \xrightarrow[2.\ H_3O^+]{1.\ LiAlH_4}$

14.31 Write the structure of the final product of each of the following sequences of reactions.

(a) $CH_3(CH_2)_4CO_2H \xrightarrow[pyridine]{SOCl_2} \xrightarrow{NH_3} \xrightarrow[2.\ H_3O^+]{1.\ LiAlH_4}$

(b) 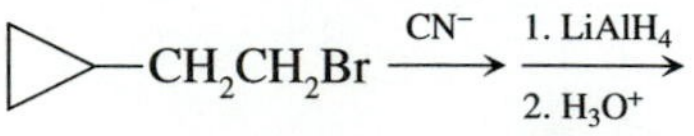 (cyclopropyl)—$CH_2CH_2Br \xrightarrow{CN^-} \xrightarrow[2.\ H_3O^+]{1.\ LiAlH_4}$

(c) $CH_3(CH_2)_3CO_2CH_3 \xrightarrow[2.\ H_3O^+]{1.\ LiAlH_4} \xrightarrow{PBr_3} \xrightarrow[(excess)]{NH_3}$

14.32 Write the structure of the final product of each of the following sequences of reactions.

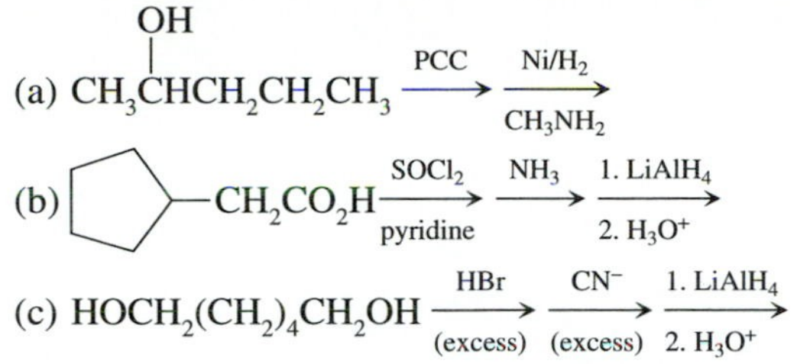

(a) $CH_3CH(OH)CH_2CH_2CH_3 \xrightarrow{PCC} \xrightarrow[CH_3NH_2]{Ni/H_2}$

(b) cyclopentyl$-CH_2CO_2H \xrightarrow[pyridine]{SOCl_2} \xrightarrow{NH_3} \xrightarrow[2.\ H_3O^+]{1.\ LiAlH_4}$

(c) $HOCH_2(CH_2)_4CH_2OH \xrightarrow[(excess)]{HBr} \xrightarrow[(excess)]{CN^-} \xrightarrow[2.\ H_3O^+]{1.\ LiAlH_4}$

14.33 Outline the steps required to convert benzoic acid into *N*-ethylbenzylamine.

14.34 Outline the steps required to convert benzyl chloride into 2-phenylethanamine.

Reactions of Amides

14.35 Write the products of the reaction of each of the following compounds with strong aqueous acid.

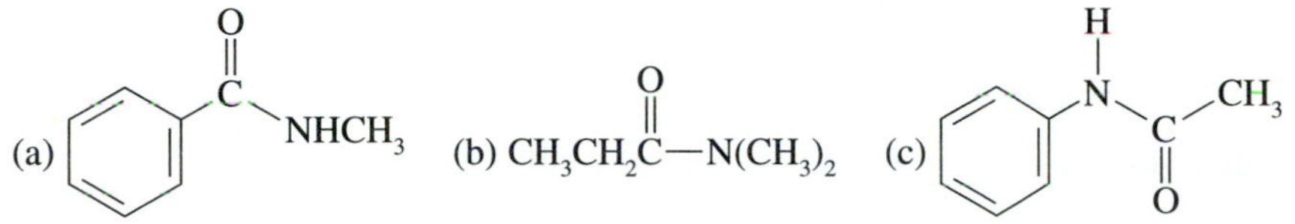

(a) $C_6H_5C(=O)NHCH_3$ (b) $CH_3CH_2C(=O)-N(CH_3)_2$ (c) $C_6H_5-NH-C(=O)CH_3$

14.36 Write the products of the reaction of each of the following compounds with strong aqueous base.

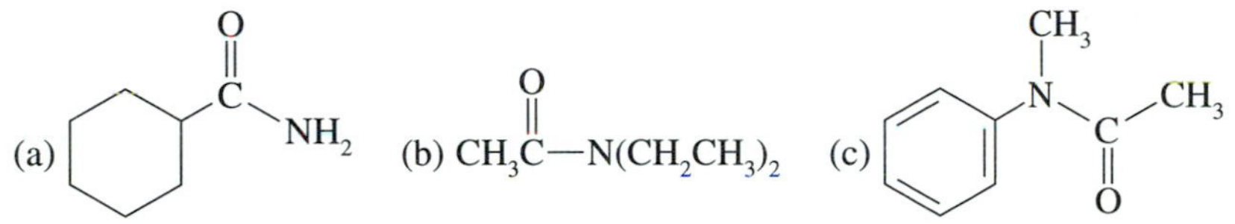

(a) cyclohexyl$-C(=O)NH_2$ (b) $CH_3C(=O)-N(CH_2CH_3)_2$ (c) $C_6H_5-N(CH_3)-C(=O)CH_3$

14.37 Write the product of reduction of each compound in Exercise 14.35 by lithium aluminum hydride followed by hydrolysis with aqueous acid.

14.38 Write the product of reduction of the following cyclic amides (lactams) by lithium aluminum hydride followed by hydrolysis with aqueous acid.

(a) 3,4-dihydro-2(1H)-quinolinone (benzene-fused six-membered lactam, N–H) (b) six-membered lactam with N–CH_3 (c) benzene-fused five-membered lactam with N–CH_3

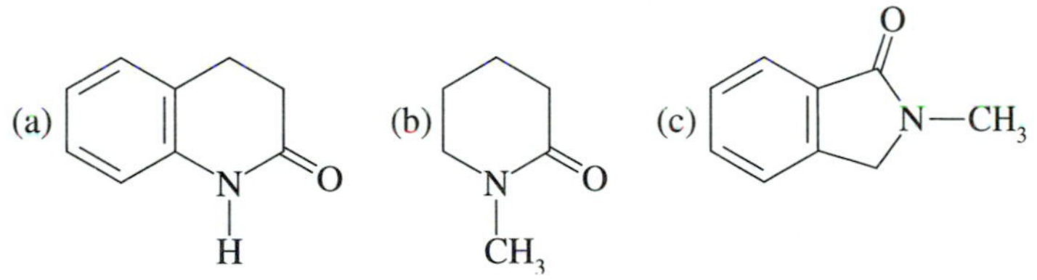

Synthesis of Amides

14.39 Write the structure of the product of each of the following reactions.

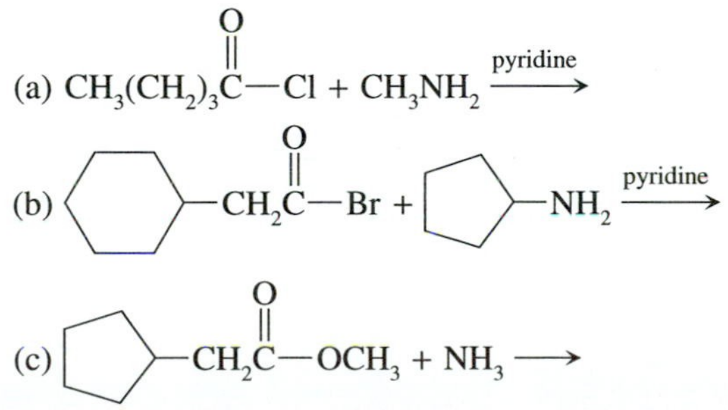

(a) $CH_3(CH_2)_3C(=O)-Cl + CH_3NH_2 \xrightarrow{pyridine}$

(b) cyclohexyl$-CH_2C(=O)-Br +$ cyclopentyl$-NH_2 \xrightarrow{pyridine}$

(c) cyclopentyl$-CH_2C(=O)-OCH_3 + NH_3 \longrightarrow$

14.40 Write the structure of the product of each of the following reactions.

(a) $CH_3(CH_2)_3\overset{\overset{\displaystyle O}{\|}}{C}-OCH_2CH_3 + CH_3CH_2NH_2 \longrightarrow$

(b) 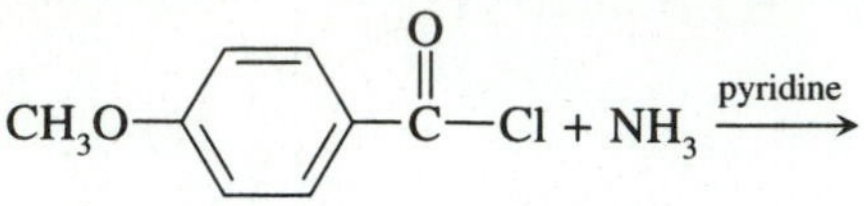 $CH_3O-C_6H_4-\overset{\overset{\displaystyle O}{\|}}{C}-Cl + NH_3 \xrightarrow{\text{pyridine}}$

(c) $C_5H_9-CH_2\overset{\overset{\displaystyle O}{\|}}{C}-SCH_3 + CH_3NH_2 \longrightarrow$

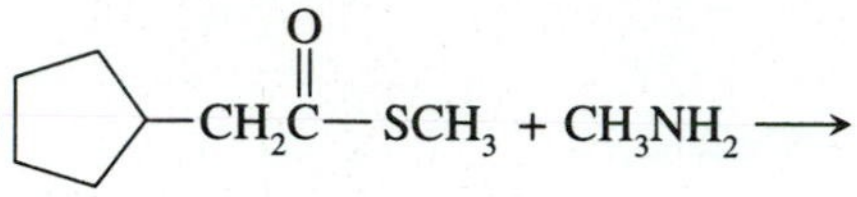

14.41 Select two reactants that could be used to prepare crotamiton, which is used to treat scabies.

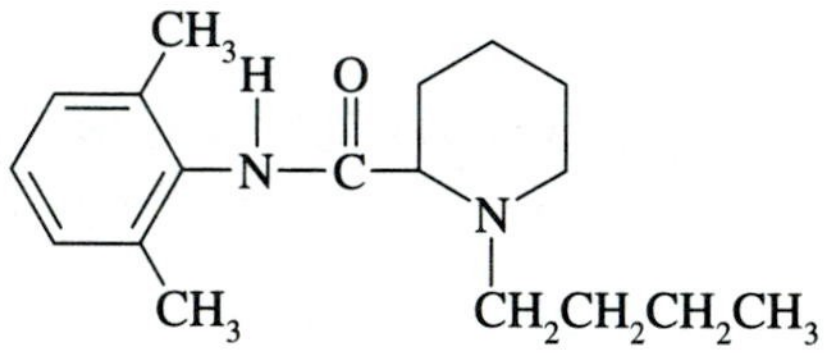

CH_3, CH_3 (on benzene ring) $-\overset{H}{\underset{}{N}}-\overset{\overset{\displaystyle O}{\|}}{C}-$ (piperidine ring) N $-CH_2CH_2CH_2CH_3$

14.42 Select two reactants that could be used to prepare bupivacaine, a local anesthetic.

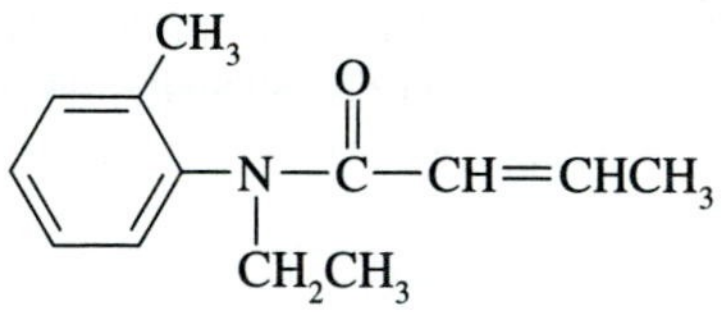

CH_3 (on benzene ring) $-\underset{\underset{\displaystyle CH_2CH_3}{|}}{N}-\overset{\overset{\displaystyle O}{\|}}{C}-CH{=}CHCH_3$

CHAPTER 15

AMINO ACIDS, PEPTIDES, AND PROTEINS

Overview

15.1 Proteins and Polypeptides

From amoebas to zebras, the proteins of all organisms contain α-amino acids bonded together by amide bonds between amino and carboxylic acid functional groups. In proteins, amide bonds are called **peptide bonds.**

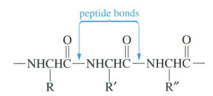

peptide bonds

$$-\mathrm{NH\underset{\substack{|\\ R}}{C}H\overset{\substack{O\\ \|}}{C}-NH\underset{\substack{|\\ R'}}{C}H\overset{\substack{O\\ \|}}{C}-NH\underset{\substack{|\\ R''}}{C}H\overset{\substack{O\\ \|}}{C}}-$$

Each protein has a unique amino acid composition, and all the amino acids are strung together in a specific order that is required for its specialized function.

The name *protein* is derived from the Greek *proteios*, meaning "preeminence" or "holding first place," which reflects the crucial role that proteins play in virtually all cellular processes. The name was suggested in 1839 by the Dutch chemist Gerardus Johannes Mulder, who could not have known how prophetic his suggested name would be. Proteins have an extraordinary range of functions. Proteins called enzymes catalyze nearly all of the chemical reactions in cells. Proteins are required for the transport of most substances across cell membranes. They are the major structural substances of skin, blood, muscle, hair, and other tissues of the body. Proteins in the immune system, called antibodies, resist the effects of foreign substances that enter the body.

Proteins are polymers of 50 or more α-amino acids; some proteins contain more than 8000 amino acid units. **Polypeptides** are smaller molecules that contain fewer than about 50 amino acids. Some important hormonal polypeptides with physiological functions such as pain relief and control of blood pressure contain as few as nine amino acid units.

This chapter begins by describing the structure and properties of the 20 amino acids isolated from proteins. Then we will consider the structure and properties of polypeptides and proteins. We will also describe both the method of synthesizing polypeptides and proteins as well as analytical methods to determine their structure.

15.2 Amino Acids

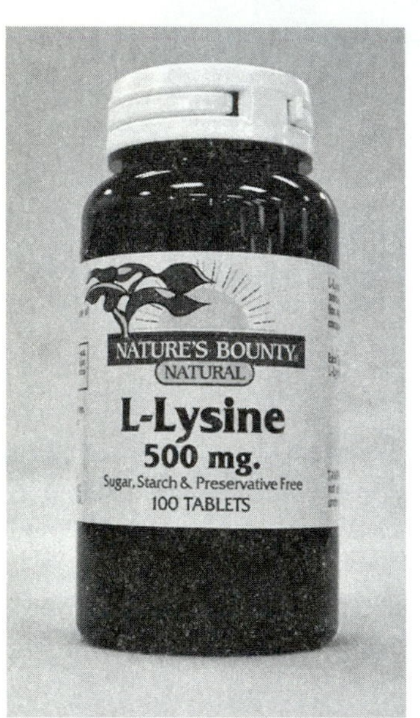

Amino acids are available in pure form and may be used as dietary supplements.

Amino acids contain both an amino group and a carboxylic acid group. About 250 have been found in natural sources; however, only about 20 of them occur in large amounts in proteins (Table 15.1). The amino acids of proteins in all cells are α-amino acids. They have an amino group bonded to the α-carbon atom of a carboxylic acid. The Fischer projection formula for an α-amino acid is as follows:

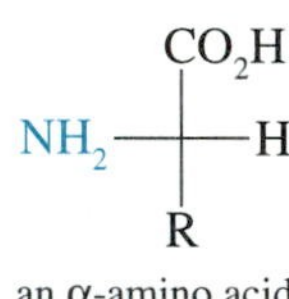

an α-amino acid

In this structure, the R group is called the side chain. There are 20 different R groups in amino acids isolated from proteins. Of these 20 α-amino acids, 19 are chiral; they have the L configuration. The one that is not chiral is glycine: its R group is hydrogen.

Classification of Amino Acids

The amino acids in proteins are primary amines except for proline, which is a secondary amine. Three-letter abbreviations of the amino acids, used as a shorthand to describe protein structure, are given in Table 15.1. (An alternate one-letter shorthand method exists, but will not be used in this text.)

The amino acids are classified by their side-chain R groups as neutral, basic, and acidic. **Neutral amino acids** contain one amino group and one carboxyl group. The neutral amino acids are further divided according to the polarity of the R group. Serine, threonine, and tyrosine are neutral amino acids that are also alcohols. Phenylalanine, tyrosine, and tryptophan contain aromatic rings. Cysteine and methionine contain a sulfur atom. The remaining neutral amino acids have hydrocarbon side chains.

The three **basic amino acids,** lysine, arginine, and histidine, have an additional basic nitrogen-containing functional group in the side chain. The two **acidic amino acids,** aspartic acid and glutamic acid, have carboxylic acid side chains. The acidic amino acids also have close relatives, asparagine and glutamine, that have neutral amides in the side chains.

Amino acids are also classified by the tendency of their side chains to interact favorably or unfavorably with water. Those amino acids with polar side chains are said to be **hydrophilic;** that is, water-loving. Those whose side chains are nonpolar are said to be **hydrophobic.** Hydrophobic amino acids have alkyl or aromatic groups that do not form hydrogen bonds to water.

Table 15.1 Structures of α-Amino Acids in Proteins

Nonpolar R Groups

Glycine (Gly)	$H{-}CH(NH_2){-}COOH$
Alanine (Ala)	$CH_3{-}CH(NH_2){-}COOH$
Valine (Val)	$CH_3{-}CH(CH_3){-}CH(NH_2){-}COOH$
Leucine (Leu)	$CH_3{-}CH(CH_3){-}CH_2{-}CH(NH_2){-}COOH$
Isoleucine (Ile)	$CH_3{-}CH_2{-}CH(CH_3){-}CH(NH_2){-}COOH$
Proline (Pro)	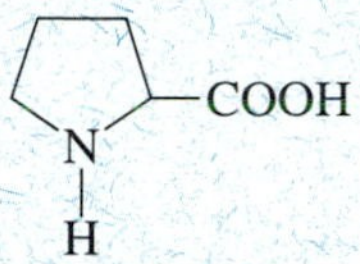 five-membered ring containing N—H, with —COOH on the carbon next to N
Phenylalanine (Phe)	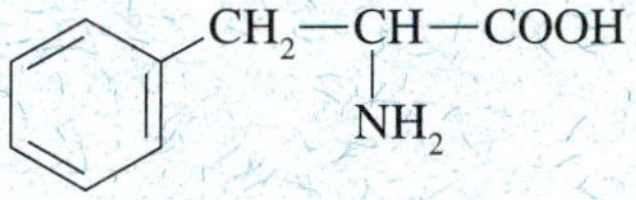 $C_6H_5{-}CH_2{-}CH(NH_2){-}COOH$
Methionine (Met)	$CH_3{-}S{-}CH_2CH_2{-}CH(NH_2){-}COOH$

Polar But Neutral R Groups

Serine (Ser)	$HO{-}CH_2{-}CH(NH_2){-}COOH$
Threonine (Thr)	$CH_3{-}CH(OH){-}CH(NH_2){-}COOH$
Cysteine (Cys)	$HS{-}CH_2{-}CH(NH_2){-}COOH$
Tyrosine (Tyr)	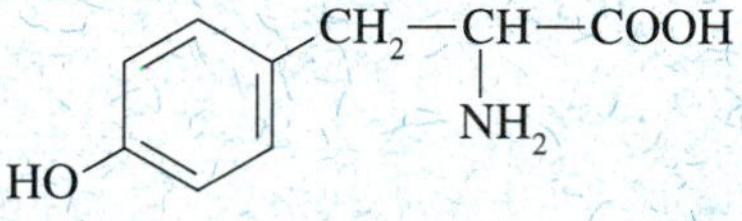 $HO{-}C_6H_4{-}CH_2{-}CH(NH_2){-}COOH$
Asparagine (Asn)	$NH_2{-}C({=}O){-}CH_2{-}CH(NH_2){-}COOH$
Glutamine (Gln)	$NH_2{-}C({=}O){-}CH_2CH_2{-}CH(NH_2){-}COOH$
Tryptophan (Trp)	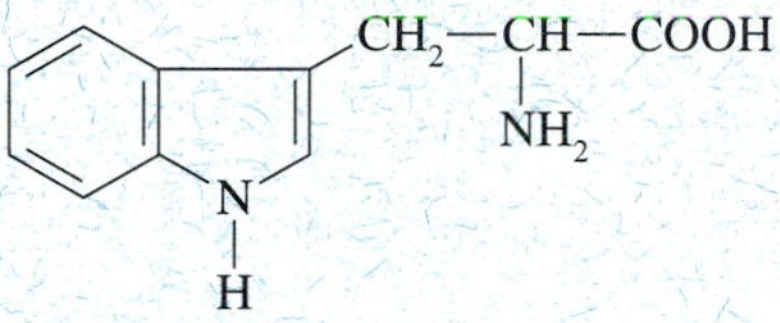 indolyl (ring N—H)$\,{-}CH_2{-}CH(NH_2){-}COOH$

Acidic R Groups

Glutamic acid (Glu)	$HO{-}C({=}O){-}CH_2CH_2{-}CH(NH_2){-}COOH$
Aspartic acid (Asp)	$HO{-}C({=}O){-}CH_2{-}CH(NH_2){-}COOH$

Basic R Groups

Lysine (Lys)	$NH_2CH_2CH_2CH_2CH_2{-}CH(NH_2){-}COOH$
Arginine (Arg)	$NH_2C({=}NH)NH{-}CH_2CH_2CH_2{-}CH(NH_2){-}COOH$
Histidine (His)	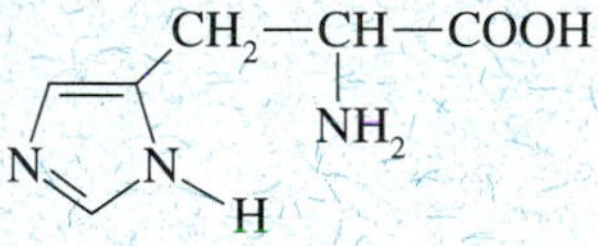 imidazolyl (N, N—H)$\,{-}CH_2{-}CH(NH_2){-}COOH$

15.3 Acid-Base Properties of Amino Acids

Although the structures of α-amino acids shown in Table 15.1 are uncharged molecules, the properties of amino acids resemble those of ionic compounds. For example, ethylamine is a gas and acetic acid is a liquid at room temperature. In contrast, glycine is a solid.

	$CH_3CH_2NH_2$	CH_3CO_2H	$NH_2CH_2CO_2H$
mp	−84°C	16°C	232°C

Additionally, amino acids have low solubilities in organic solvents, but are moderately soluble in water, unlike most organic compounds.

Ionic Forms of Amino Acids

When a neutral amino acid dissolves in water at pH 7, it contains a carboxylate group and an ammonium group. Thus the molecule exists as a dipolar ion sometimes called a **zwitterion.** The dipolar ion is amphoteric; that is, it has the properties of both an acid and a base.

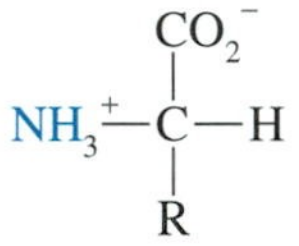

$$\begin{array}{c} CO_2^- \\ | \\ NH_3^+\!-\!C\!-\!H \\ | \\ R \end{array}$$

zwitterion

In basic solution, the conjugate base of a neutral amino acid forms by loss of a proton from the ammonium ion, but the carboxylate group remains. It has a negative charge because the carboxylate ion has a −1 charge and the amino group is uncharged. In acidic solution, a carboxyl group forms by protonation and the ammonium ion remains. This species is the conjugate acid of the original amino acid and has a +1 charge.

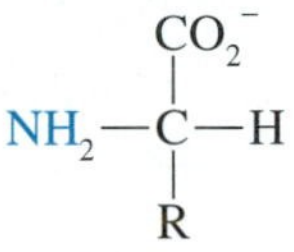

$$\begin{array}{cc} \begin{array}{c} CO_2^- \\ | \\ NH_2\!-\!C\!-\!H \\ | \\ R \end{array} & \begin{array}{c} CO_2H \\ | \\ NH_3^+\!-\!C\!-\!H \\ | \\ R \end{array} \end{array}$$

conjugate base 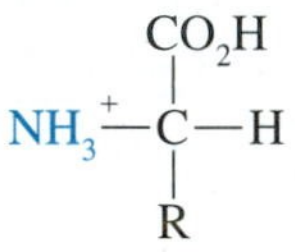 — conjugate acid

EXAMPLE 15-1

Write the structure of the zwitterion and the conjugate base of alanine (see alanine in Table 15.1).

Solution

The zwitterion is written by removing a proton from the carboxyl group and adding a proton to the nitrogen atom. The conjugate base is written by removing a proton from the ammonium ion site of the dipolar ion. The nitrogen atom becomes neutral, whereas the carboxylate ion retains a negative charge.

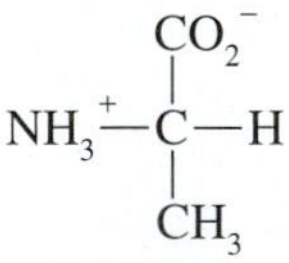

$$\begin{array}{cc} \begin{array}{c} CO_2^- \\ | \\ NH_3^+\!-\!C\!-\!H \\ | \\ CH_3 \end{array} & \begin{array}{c} CO_2^- \\ | \\ NH_2\!-\!C\!-\!H \\ | \\ CH_3 \end{array} \end{array}$$

zwitterion 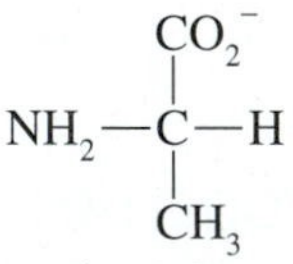 — conjugate base

Problem 15.1

Write the structure of the zwitterion and the conjugate acid of serine (see serine in Table 15.1).

pK_a Values of α-Amino Acids

The pK_a values of the carboxyl and α-ammonium groups of amino acids depend on the structure of the amino acid. The pK_a values of glycine are 2.35 and 9.78.

$$\underset{\text{conjugate acid}}{\overset{+}{NH_3}-CH_2-CO_2H} + H_2O \rightleftharpoons \underset{\text{zwitterion}}{\overset{+}{NH_3}-CH_2-CO_2^-} + H_3O^+ \qquad K_a = 4.5 \times 10^{-3} \quad pK_a = 2.35$$

$$\underset{\text{zwitterion}}{\overset{+}{NH_3}-CH_2-CO_2^-} + H_2O \rightleftharpoons \underset{\text{conjugate base}}{NH_2-CH_2-CO_2^-} + H_3O^+ \qquad K_a = 1.6 \times 10^{-10} \quad pK_a = 9.78$$

The pK_a values of the carboxyl groups of amino acids range from 1.8 for histidine to 2.6 for phenylalanine. The pK_a values of the α-ammonium groups range from 8.8 for asparagine to 10.8 for cysteine (Table 15.2).

When an amino acid dissolves in an aqueous solution, several species exist. When the pH of the solution equals the pK_a of the ionizing group, the concentrations of the acid form and its conjugate base are equal. For example, when the pH of a solution of glycine is 2.35, the concentrations of the zwitterion and the conjugate acid are equal. When the pH of the glycine solution is increased to 9.78, the concentrations of the zwitterion and the conjugate base are equal. At pH values between 2.35 and 9.78, the zwitterion is the major form of the amino acid in solution.

Table 15.2 pK_a Values of α-Amino Acids at 25 °C

Amino Acid	pK_a ($-CO_2H$)	pK_a ($-NH_3^+$)	pK_a (side chain)
glycine	2.35	9.78	
alanine	2.35	9.87	
valine	2.29	9.72	
leucine	2.33	9.74	
isoleucine	2.32	9.76	
methionine	2.17	9.27	
proline	1.95	10.64	
phenylalanine	2.58	9.24	
tryptophan	2.43	9.44	
serine	2.19	9.44	
threonine	2.09	9.10	
cysteine	1.89	10.78	8.38
tyrosine	2.20	9.11	10.07
asparagine	2.02	8.80	
glutamine	2.17	9.13	
aspartic acid	1.99	10.00	3.96
glutamic acid	2.13	9.95	4.32
lysine	2.16	9.20	10.80
arginine	1.82	8.99	12.48
histidine	1.81	9.15	6.00

EXAMPLE 15-2

In what ionic form does serine (see Table 15.2) exist in 0.1 M HCl?

Solution

A 0.1 M solution of HCl has a pH of 1.0. The pK_a values of serine are 2.19 and 9.44. Thus, at pH = 1, serine will exist as the conjugate acid.

$$\overset{+}{N}H_3-\underset{\underset{CH_2OH}{|}}{\overset{\overset{CO_2H}{|}}{C}}-H$$

Problem 15.2

In what ionic form does alanine (see Table 15.2) exist in 0.01 M NaOH?

15.4 Isoionic Point

The pH at which the concentration of the zwitterion is at a maximum is the **isoionic point,** abbreviated pH_i. At this pH the amino acid has no *net* charge. In a more basic solution, when the pH is greater than the pH_i, the conjugate base predominates. In a more acidic solution, when the pH is less than the pH_i, the conjugate acid predominates. The isoionic points of some amino acids are given in Table 15.3. The isoionic points of the neutral amino acids are close to 7. Those of acidic and basic amino acids are significantly less than and greater than 7, respectively.

Table 15.3 Isoionic Points of Amino Acids

Amino Acid	pH_i
glycine	6.06
alanine	6.10
valine	5.96
leucine	5.98
isoleucine	6.02
methionine	5.74
proline	6.30
phenylalanine	5.48
tryptophan	5.89
serine	5.68
threonine	5.60
cysteine	5.07
tyrosine	5.66
asparagine	5.41
glutamine	5.65
aspartic acid	2.77
glutamic acid	3.22
lysine	9.74
arginine	10.76
histidine	7.59

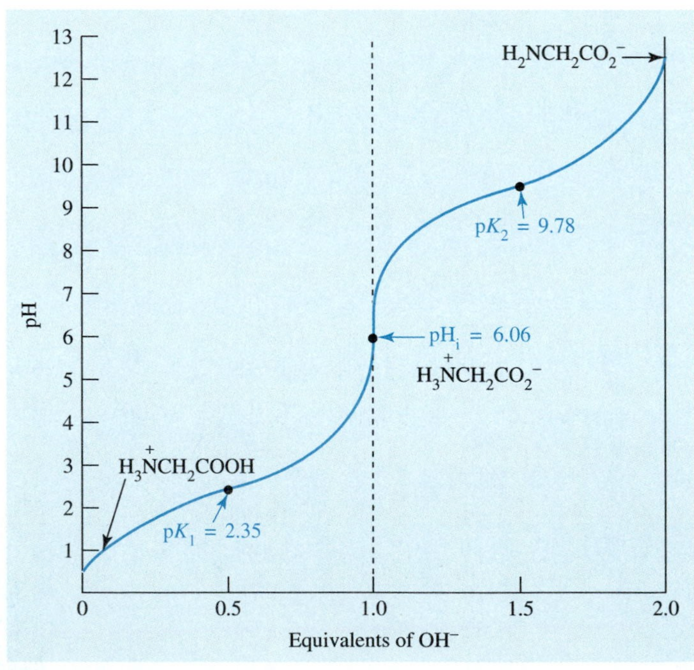

FIGURE 15.1
Titration Curve of Glycine

Titration Curves of Amino Acids

The pK_a values of the carboxyl group and the ammonium group of an amino acid, as well as the pH_i, can be determined by titrating the conjugate acid with base. Figure 15.1 shows a typical titration curve for glycine. As base is added, some of the conjugate acid is converted to the zwitterion, and the pH increases. The pH at which the $-CO_2H$ group is one-half neutralized is equal to pK_1. After one equivalent of base has been added, the zwitterion is the major form in solution, and the pH at this point is pH_i. Addition of a second equivalent of base starts to convert the $-NH_3^+$ group to the conjugate base of the amino acid. The pH at which half-neutralization has occurred is equal to pK_2.

Isoionic Points of Proteins

Because proteins are made of amino acids, a protein has an isoionic point that depends on its amino acid composition. At its isoionic point, a protein has no net charge, and its solubility is at a minimum. As a consequence, a protein tends to precipitate from solution at its isoionic pH. For example, casein, a protein in milk, has a net negative charge at the pH of milk, which is 6.3. The isoionic point of casein is 4.6 because it has many glutamic acid and aspartic acid residues. If milk is made more acidic, casein precipitates because the carboxylate ions of the side chains of glutamic acid and aspartic acid become protonated. Casein, which is used in making cheese, is obtained by adding an acid to milk or by adding bacteria that produce lactic acid.

Separation of Amino Acids and Proteins

Mixtures of amino acids can be separated and identified by a technique called **electrophoresis** (Figure 15.2). In this technique, a paper strip saturated with a

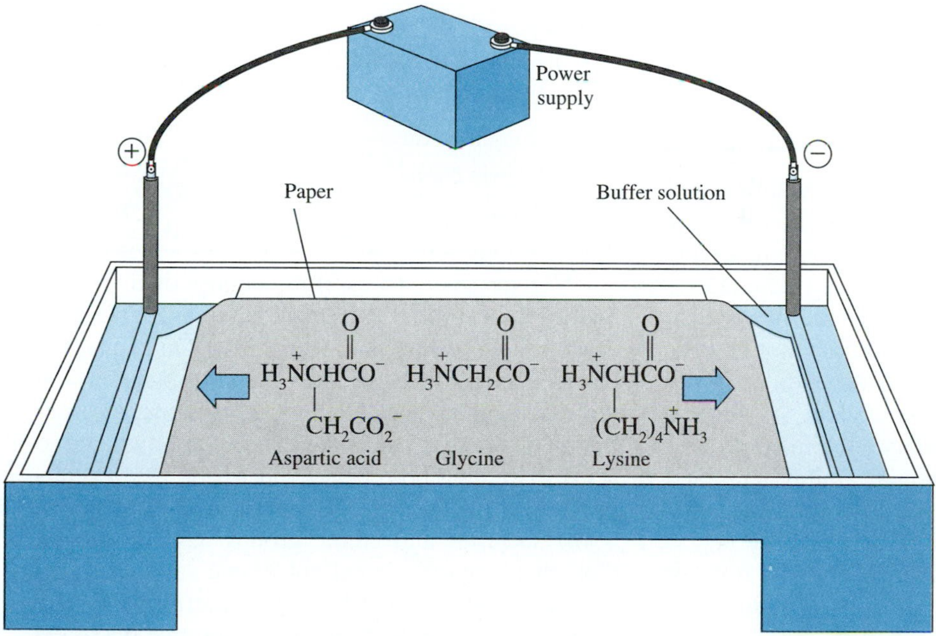

FIGURE 15.2
Electrophoresis of Amino Acids

Essential Amino Acids and Dietary Protein

Adequate amounts of about 10 of the 20 amino acids are synthesized in the body. The remaining amino acids must be obtained from food because either there is no biochemical pathway to produce them or the pathway produces inadequate amounts. Amino acids that must be obtained from food are called **essential amino acids.** Of course, all amino acids are necessary, but the term *essential* is reserved for those amino acids that must be obtained in the diet. The essential amino acids and their estimated minimum daily requirements are listed in Table A.

Table A Essential Amino Acids and Daily Requirements

	Minimum Daily Requirement (g)	
Amino Acid	**Women**	**Men**
isoleucine	0.45	0.7
leucine	0.6	1.1
lysine	0.5	0.8
methionine	0.55	1.0
phenylalanine	1.1	1.4
threonine	0.3	0.5
tryptophan	0.15	0.25
valine	0.65	0.8

Tyrosine is not listed as an essential amino acid because phenylalanine is converted into tyrosine by the body. Histidine is essential for growth in infants and may be essential for adults as well. The rate of synthesis of histidine cannot meet the needs of a growing body, so histidine may or may not be considered an essential amino acid depending on the age and state of health of the individual.

The composition of dietary protein must provide the proper mix of essential amino acids and other amino acids to supply those processes that form essential body protein. If one or more necessary amino acids are not available at the time of synthesis of a vital protein, then the protein is not made.

Dietary proteins are rated in terms of biological value on a percentage scale (see Table B). Complete proteins have a high biological value—they supply all of the amino acids in the amounts required for normal growth. Note that hen's eggs, cow's milk, and fish provide proteins of high biological value. Plant proteins vary more in biological value than animal proteins. However, not all plant proteins are deficient in the same amino acids. Gliadin, a wheat protein, is low in lysine; zein, a corn protein, is low in both lysine and tryptophan. Societies that eat large amounts of corn or wheat products must have other sources of lysine.

Table B Biological Value of Dietary Protein

Food	Biological Value (%)
whole hen's egg	94
whole cow's milk	84
fish	83
beef	73
soybeans	73
white potato	67
whole grain wheat	65
whole grain corn	59
dry beans	58

Vegetarians must carefully choose their food so that all the essential amino acids are available on a daily basis. For example, wheat is low in lysine, but beans are high in lysine as well as tryptophan. On the other hand, wheat is high in cysteine and methionine, whereas beans are low in these two amino acids. By eating both beans and wheat, the vegetarian increases the percentage of usable proteins. Some societies and ethnic groups have developed diets that provide a good nutritional supply of proteins even without the benefit of nutritionists. The American Indian thrived on a diet that included both corn and beans—a mixture that we call succotash. Rice and black-eyed peas of the South as well as corn tortillas and beans in Mexico provide a reasonable balance of amino acids.

The diets in some areas of the world today fall below the minimum daily requirement of protein due to economic conditions and, in some cases, social and religious customs. As income decreases, the more costly animal protein is replaced by cereal grains and other incomplete protein sources. If a variety of plant proteins are not available, a number of diseases in young children result. Kwashiorkor is a protein deficiency disease that develops in young children after weaning, when their diet is changed to starches. The disease is characterized by bloated bellies and patchy skin. After a certain point, death is inevitable. Some forms of mental retardation also result from incomplete nutrition. ■

buffer solution at a selected pH bridges two vessels containing the buffer. A sample of the amino acid mixture is placed at the center of the paper as a "spot," and an electric potential is applied between the two vessels. If the buffer pH equals the isoionic point of an amino acid, the dipolar ion predominates, and it does not migrate. An amino acid with a negative charge at that pH migrates toward the positive electrode, whereas an amino acid with a positive charge at that pH migrates toward the negative electrode. After a while, the original "spot" of the amino acid sample separates into two or more spots, each corresponding to an amino acid that was present in the original mixture.

This bear has a protein diet of high biological value.

Proteins can also be separated by electrophoresis. Electrophoretic separation of proteins is an important tool in clinical laboratories. Because proteins have different charges and molecular weights, they move at different rates in the electrophoresis apparatus. Electrophoresis is commonly used to analyze blood serum. For example, the identification of certain enzymes in the blood is used as a diagnostic tool for heart attacks.

15.5 Peptides

When the α-amino group of one amino acid is linked to the carboxyl group of a second amino acid by an amide bond, the product is called a **peptide.** If the peptide contains two amino acid units, it is a **dipeptide;** a peptide that contains three amino acids is called a **tripeptide.** In general, a prefix di-, tri-, etc., indicates the number of amino acids in a peptide. But a peptide that contains, say, 14 amino acids is more likely called a 14-peptide than a tetradecapeptide. Peptides that contain only a "few" amino acids are called **oligopeptides.**

A peptide has two ends: the end with a free α-amino group is called the **N-terminal amino acid residue;** the end with the free carboxyl group is called the **C-terminal amino acid residue.** Peptides are named from the N-terminal amino acid to the C-terminal amino acid. The two isomeric dipeptides containing alanine and serine (Figure 15.3) illustrate the importance of designating the order of linkage of the amino acids.

$$\underbrace{NH_2-\underset{\displaystyle CH_2OH}{\underset{|}{CH}}-\overset{\displaystyle O}{\overset{\|}{C}}}_{\text{serine residue}}-\underbrace{NH-\underset{\displaystyle CH_3}{\underset{|}{CH}}-\overset{\displaystyle O}{\overset{\|}{C}}-OH}_{\text{alanine residue}}$$

serylalanine (Ser-Ala)

$$\underbrace{NH_2-\underset{\displaystyle CH_3}{\underset{|}{CH}}-\overset{\displaystyle O}{\overset{\|}{C}}}_{\text{alanine residue}}-\underbrace{NH-\underset{\displaystyle CH_2OH}{\underset{|}{CH}}-\overset{\displaystyle O}{\overset{\|}{C}}-OH}_{\text{serine residue}}$$

alanylserine (Ala-Ser)

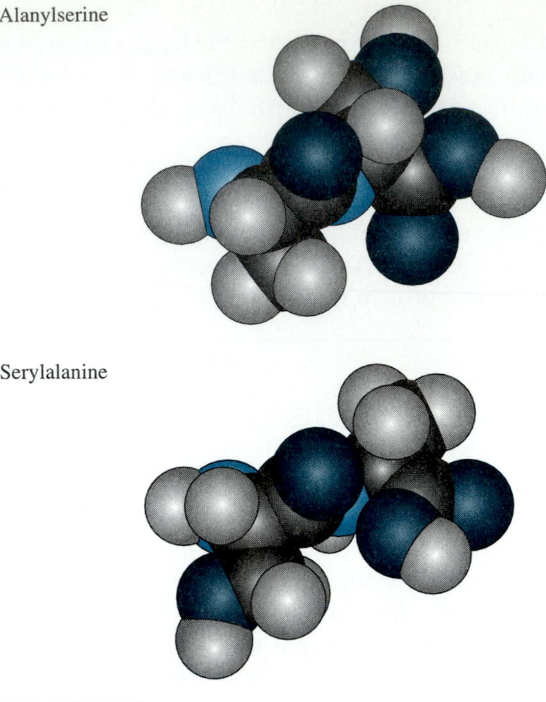

FIGURE 15.3
Structures of Isomeric Dipeptides

The number of isomeric peptides containing one molecule each of n different amino acids is equal to $n!$, where

$$n! = 1 \times 2 \times 3 \times \ldots \times (n - 1) \times n$$

Thus, there are six possible isomers of a tripeptide with three different amino acids. The isomeric tripeptides with the amino acids glycine, alanine, and valine are Gly-Ala-Val, Gly-Val-Ala, Val-Gly-Ala, Val-Ala-Gly, Ala-Gly-Val, and Ala-Val-Gly. For peptides containing one molecule each of 20 different amino acids there are 2,432,902,008,176,640,000 isomers! Polypeptides and proteins may actually contain two or more molecules of same amino acid, and the $n!$ formula does not apply. However, the number of isomers is still astronomically large.

Biological Functions of Peptides

Cells contain many relatively small peptides that have diverse functions. Some are hormones with physiological functions such as pain relief and control of blood pressure (Table 15.4). These oligopeptides are produced and released in small amounts. They are rapidly metabolized, but their physiological action is necessary for only a short time. For example, the 14-peptide somatostatin, which inhibits the release of other hormones such as insulin, glucagon, and secretin, has a biological half-life of less than 4 minutes.

Enkephalins are peptides that bind specific receptor sites in the brain to reduce pain. The enkephalin receptor sites have a high affinity for opiates, including morphine, heroin, and other structurally similar substances. Hence, enkephalin receptors are commonly called *opiate receptors*. Opiates mimic the enkephalins that are normally present in the body to mitigate pain.

Table 15.4 Peptide Hormones

Hormone	Amino Acid Residues	Molecular Weight	Function
tuftsin	4	501	stimulates phagocytosis
met-enkephalin	5	645	analgesic activity
angiotensin II	8	1031	affects blood pressure
oxytocin	8	986	affects uterine contractions
vasopressin	8	1029	an antidiuretic
bradykinin	9	1069	produces pain
somatostatin	14	1876	inhibits release of other hormones
gastrin	17	2110	promotes pepsin secretion
secretin	27	2876	stimulates pancreatic secretions
glucagon	29	3374	stimulates glucose production from glycogen
calcitonin	32	3415	decreases calcium level in blood
relaxin	48	5500	relaxation of uterine muscles
insulin	51	5700	affects blood sugar level

Peptides are produced in many tissues. For example, angiotensin II is made in the kidneys. It causes constriction of the blood vessels and thus increases blood pressure. Angiotensin II is the most potent vasoconstrictor known, and the production of excess angiotensin II is responsible for some forms of hypertension.

Oxytocin and vasopressin are structurally similar nonapeptides formed by the pituitary gland. Oxytocin causes the contraction of smooth muscle, such as that of the uterus. It is used to induce delivery or to increase the effectiveness of uterine contractions. Vasopressin is one of the hormones that regulate the excretion of water by the kidneys, and it affects blood pressure. The structures of oxytocin and vasopressin differ by only two amino acids. They are both cyclic peptides that result from a disulfide bond between what would otherwise be the N-terminal amino acid cysteine and another cysteine five amino acid residues away. The C-terminal amino acid exists as an amide in both compounds.

S———————————S
| |
Cys—Tyr—Ile—Gln—Asn—Cys—Pro—Leu—Gly—NH_2

oxytocin

S———————————S
| |
Cys—Tyr—Phe—Gln—Asn—Cys—Pro—Arg—Gly—NH_2

vasopressin

The structural difference between oxytocin and vasopressin may seem small at first glance. But, in fact, the difference is enormous. Residue 3 in oxytocin is isoleucine and residue 3 in vasopressin is phenylalanine. This is a relatively small difference because both residues are nonpolar and about the same size. However, residue 8 in oxytocin is leucine, a nonpolar amino acid with a *sec*-butyl side chain, whereas residue 8 in vasopressin is arginine, an amino acid with a basic side chain that has a positive charge at pH 7. Because of this difference in charge, the receptor for oxytocin has a weak affinity for vasopressin, and the receptor for vasopressin has a very low affinity for oxytocin.

EXAMPLE 15-3

Identify the terminal amino acids of tuftsin, a tetrapeptide that stimulates phagocytosis and promotes the destruction of tumor cells. Write the amino acid sequence using three-letter abbreviations for the amino acids. Also write the complete name without abbreviations.

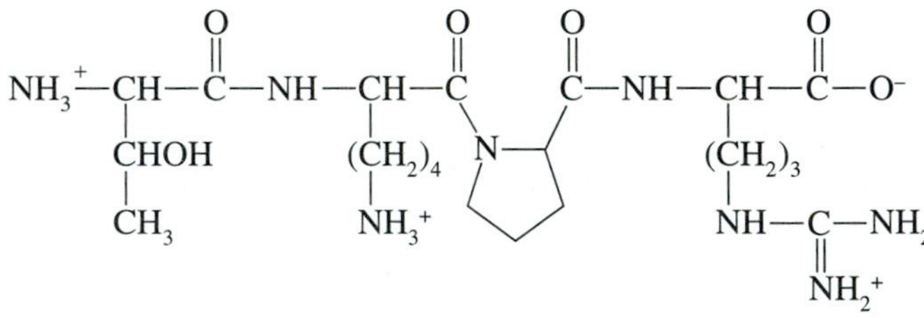

$$\overset{+}{NH_3}-CH[CH(OH)CH_3]-\overset{O}{\overset{\|}{C}}-NH-CH[(CH_2)_4NH_3^+]-\overset{O}{\overset{\|}{C}}-N(\text{pyrrolidine ring})-\overset{O}{\overset{\|}{C}}-NH-CH[(CH_2)_3-NH-C(=NH_2^+)-NH_2]-\overset{O}{\overset{\|}{C}}-O^-$$

Solution

The residue on the left is the N-terminal amino acid threonine. The residue on the right is the C-terminal amino acid arginine. The internal amino acids are the basic amino acid lysine and the secondary amino acid proline. The abbreviated name is Thr-Lys-Pro-Arg. The complete name is threonyllysylprolylarginine.

Problem 15.3

Determine the number of isomeric tripeptides containing one alanine and two glycine residues. Write representations of the isomers using three-letter abbreviations.

15.6 Synthesis of Peptides

The synthesis of peptides and polypeptides is an important aspect of research in biochemistry and in the biotechnology industry. A highly specialized set of reagents has been developed for such syntheses to give high yields. Several steps are required because two amino acids cannot be simply combined to give a desired dipeptide. For example, two amino acids, such as alanine and glycine, yield a mixture of dipeptides. Each amino acid could form bonds with its own kind to form Gly-Gly and Ala-Ala or to a different amino acid to give Gly-Ala and Ala-Gly. Also, the amino acids in the reaction mixture can continue to react with the dipeptide products to yield oligopeptides.

The synthesis of a specific dipeptide requires modification of both amino acids. One amino acid is protected at its carboxyl group, by a reagent we will call P_C, and has the amino group available for peptide bond formation. The second amino acid is protected at the amino group, by a reagent we will call P_N, and has the carboxyl group available for peptide bond formation. Then only one condensation reaction is possible.

$$NH_2-\underset{R}{CH}-\overset{O}{\overset{\|}{C}}-OH \xrightarrow{P_C} NH_2-\underset{R}{CH}-\overset{O}{\overset{\|}{C}}-P_C$$

$$NH_2-\underset{R'}{CH}-\overset{O}{\overset{\|}{C}}-OH \xrightarrow{P_N} P_N-NH-\underset{R'}{CH}-\overset{O}{\overset{\|}{C}}-OH$$

$$P_N-NH-\underset{R'}{CH}-\overset{O}{\overset{\|}{C}}-OH + NH_2-\underset{R}{CH}-\overset{O}{\overset{\|}{C}}-P_C \xrightarrow{-H_2O} P_N-NH-\underset{R'}{CH}-\overset{O}{\overset{\|}{C}}-NH-\underset{R}{CH}-\overset{O}{\overset{\|}{C}}-P_C$$

The carboxyl group is protected by converting it to a methyl or ethyl ester. Because esters are more reactive toward nucleophiles such as hydroxide ion than are amides, the C-terminus is easily "unprotected" at the end of the synthesis.

$$NH_2-\underset{R}{CH}-\overset{O}{\overset{\|}{C}}-OH + CH_3CH_2OH \xrightarrow{HCl} NH_2-\underset{R}{CH}-\overset{O}{\overset{\|}{C}}-OCH_2CH_3 + H_2O$$

Several protecting groups have been developed to protect the amino terminus of an amino acid; the *tert*-butoxycarbonyl (Boc) derivative is typical. Reaction of an amino acid with di-*tert*-butyl dicarbonate gives a Boc-amino acid.

$$\underset{\text{di-}tert\text{-butyl dicarbonate}}{(CH_3)_3CO-\overset{O}{\overset{\|}{C}}-O-\overset{O}{\overset{\|}{C}}-OC(CH_3)_3} + NH_2-\underset{R'}{CH}-\overset{O}{\overset{\|}{C}}-OH \longrightarrow \underset{\text{a Boc-amino acid}}{(CH_3)_3CO-\overset{O}{\overset{\|}{C}}-NH-\underset{R'}{CH}-\overset{O}{\overset{\|}{C}}-OH}$$

Note that the carbonyl group of the Boc group is bonded to both an oxygen atom and a nitrogen atom. This functional group is a *tert*-butyl carbamate that is easily hydrolyzed without affecting the amide or even the ester groups. The Boc group can be removed with trifluoroacetic acid. Both the amide groups and the ester of a protected carboxyl group are unaffected by the reaction conditions. The byproducts of the reaction, CO_2 and 2-methylpropene, are gases.

$$(CH_3)_3CO-\overset{O}{\overset{\|}{C}}-NH-\underset{R'}{CH}-\overset{O}{\overset{\|}{C}}-OCH_2CH_3 \xrightarrow{CF_3CO_2H} NH_2-\underset{R'}{CH}-\overset{O}{\overset{\|}{C}}-OCH_2CH_3 + (CH_3)_2C{=}CH_2 + CO_2$$

The protecting groups of both the amino and the carboxyl groups are sensitive to acid and base. Therefore, the condensation reaction to produce a dipeptide bond must be carried out under neutral conditions. Dicyclohexylcarbodiimide (DCCI) is a special dehydrating agent used to condense two amino acids and form an amide bond by removing water. The reagent does not affect other functional groups of the amino acids.

$$(CH_3)_3CO-\overset{O}{\overset{\|}{C}}-NH-\underset{R'}{CH}-\overset{O}{\overset{\|}{C}}-OH + NH_2-\underset{R}{CH}-\overset{O}{\overset{\|}{C}}-OCH_2CH_3 \xrightarrow{DCCI}$$

$$(CH_3)_3CO-\overset{O}{\overset{\|}{C}}-NH-\underset{R'}{CH}-\overset{O}{\overset{\|}{C}}-NH-\underset{R}{CH}-\overset{O}{\overset{\|}{C}}-OCH_2CH_3$$

The resulting dipeptide that is protected at both the carboxyl and amino groups is then unprotected by hydrolysis of the Boc group at the N-terminal amino acid. The ester linkage of the carboxyl group is unaffected.

$$(CH_3)_3CO-\overset{\overset{\large O}{\|}}{C}-NH-\overset{\overset{\large R'}{|}}{CH}-\overset{\overset{\large O}{\|}}{C}-NH-\overset{\overset{\large R}{|}}{CH}-\overset{\overset{\large O}{\|}}{C}-OCH_2CH_3 \xrightarrow{CF_3CO_2H} NH_2-\overset{\overset{\large R'}{|}}{CH}-\overset{\overset{\large O}{\|}}{C}-NH-\overset{\overset{\large R}{|}}{CH}-\overset{\overset{\large O}{\|}}{C}-OCH_2CH_3$$

This dipeptide is still protected at the carboxyl group and can only react at the free amino group. Reaction with another Boc-amino acid and DCCI yields a tripeptide. Ultimately, after the proper number of reaction sequences, the final polypeptide is liberated by hydrolysis with base.

15.7 Structure Determination of Proteins

The first item of business when analyzing the structure of a protein is determination of its amino acid composition. The components of a polypeptide or protein are called **amino acid residues.** The amino acid composition of a polypeptide is determined in two steps. First, the protein is hydrolyzed. Then the hydrolysis products are separated by chromatography and identified. It is relatively easy to determine the amino acid composition of a peptide or protein because well-established techniques using automated instruments are available for that purpose.

A protein or peptide is hydrolyzed by heating it for 24 hours in 6 N HCl at 100°C. Complete hydrolysis produces the constituent amino acids of the peptide or protein. For example, hydrolysis of the pentapeptide leucine enkephalin gives two molar equivalents of glycine and one each of leucine, tyrosine, and phenylalanine. The composition is represented as Gly_2,Leu,Tyr,Phe.

$$\text{leucine enkephalin} \longrightarrow 2\ \text{Gly} + \text{Leu} + \text{Tyr} + \text{Phe}$$

The linear sequence of amino acid residues in a peptide or protein is called the **primary structure.** The amino acid sequence is determined by a combination of methods that include

1. partial hydrolysis and fragment overlap analysis;
2. selective enzymatic hydrolysis;
3. end group analysis.

Partial Hydrolysis

When a peptide is heated with HCl for short time intervals, hydrolysis reactions yield oligopeptides having random sizes. For example, heating leucine enkephalin in HCl might yield the tripeptides Phe-Leu-Gly and Gly-Phe-Leu and the dipeptide Ala-Gly. The amino acid sequence can be obtained by aligning the common partial sequences.

```
Ala-Gly
    Gly-Phe-Leu
        Phe-Leu-Gly
-------------------
Ala-Gly-Phe-Leu-Gly
```

Enzymatic Hydrolysis

A protein or peptide is specifically cleaved in cells by enzymes called **proteases.** These enzymes are also used in the laboratory to determine the structure of peptides and proteins. Chymotrypsin and trypsin are two common proteases. Chymotrypsin hydrolyzes peptide bonds on the C-terminal side of the aromatic amino acids phenylalanine, tyrosine, and tryptophan. Trypsin hydrolyzes peptide bonds on the C-terminal side of the basic amino acids lysine and arginine.

$$\text{Lys-Glu-Tyr-Leu} \xrightarrow{\text{chymotrypsin}} \text{Lys-Glu-Tyr} + \text{Leu}$$

$$\text{Lys-Glu-Tyr-Leu} \xrightarrow{\text{trypsin}} \text{Lys} + \text{Glu-Tyr-Leu}$$

Based on such information, enzymatic hydrolysis can be used to determine the structure of leucine enkephalin. We know from its total hydrolysis that the compound contains phenylalanine, an aromatic amino acid. The structure of leucine enkephalin can be partially established by the hydrolysis with chymotrypsin. The pentapeptide cleaves to give a tripeptide with a C-terminal phenylalanine residue and a dipeptide with an N-terminal leucine residue.

$$\text{leucine enkephalin} \xrightarrow{\text{chymotrypsin}} \text{Ala-Gly-Phe} + \text{Leu-Gly}$$

Phenylalanine must have been bonded to leucine, the N-terminal amino acid of the other fragment. Based on the known structures of the two fragments, the overall structure must be Ala-Gly-Phe-Leu-Gly.

End Group Analysis

Partial hydrolysis of a polypeptide chain, either by HCl or by specific enzymes, produces peptides of varying lengths. The sequence of these peptides must be determined in order to determine the overall sequence. The sequence of the peptide fragments produced by partial hydrolysis can often be deduced by end group analysis. Consider a tripeptide that may have six isomeric structures. If the identity of the N-terminal amino acid is known, then only two isomeric arrangements are possible for the other two amino acid residues. Subsequent identification of the C-terminal amino acid then establishes the complete structure.

Enzymatic End Group Analysis

Some enzymes hydrolyze peptide bonds at either the N- or the C-terminal amino acid, nibbling their way down the chain until they have digested the entire molecule. For example, **carboxypeptidases** sequentially remove peptides from the C-terminal end of a polypeptide chain. In contrast, **aminopeptidases** sequentially hydrolyze peptides from the N-terminal amino acid. By identifying the amino acids produced by a carboxypeptidase or an aminopeptidase at various time intervals, the sequence of amino acids can be determined. For example, hydrolysis of the pentapeptide leucine enkephalin catalyzed by carboxypeptidase first liberates glycine. The tetrapeptide remaining in solution continues to react yielding leucine, followed by phenylalanine, and so on.

$$\text{leucine enkephalin} \xrightarrow{\text{carboxypeptidase}} \text{tetrapeptide} + \text{Gly}$$

$$\text{tetrapeptide} \xrightarrow{\text{carboxypeptidase}} \text{tripeptide} + \text{Leu}$$

When leucine enkephalin is hydrolyzed by an aminopeptidase, the first amino acid released is alanine. The remaining tetrapeptide then yields glycine, followed by phenylalanine, and so on.

$$\text{leucine enkephalin} \xrightarrow{\text{aminopeptidase}} \text{Ala} + \text{tetrapeptide}$$

$$\text{tetrapeptide} \xrightarrow{\text{aminopeptidase}} \text{Gly} + \text{tripeptide}$$

Carboxypeptidase and aminopeptidase can only be used to determine a few residues in a polypeptide chain. Because these enzymes hydrolyze peptide bonds continuously and at different rates, the reaction mixture rapidly becomes difficult to analyze.

Chemical End Group Analysis

The identity of the N-terminal amino acid of a polypeptide can be determined by a method invented by Pehr Edman that is called the **Edman degradation.** In the Edman degradation, the polypeptide is treated with phenyl isothiocyanate—the Edman reagent—which reacts with the N-terminal amino acid to give an *N*-phenylthiourea derivative. This derivative forms by addition of the terminal N—H bond across the C═N of the phenyl isothiocyanate. After the adduct has formed, anhydrous trifluoroacetic acid is added to the reaction mixture. This reagent cleaves the polypeptide at the N-terminal residue. Under these conditions, the peptide bonds in the protein do not break (Figure 15.4). A com-

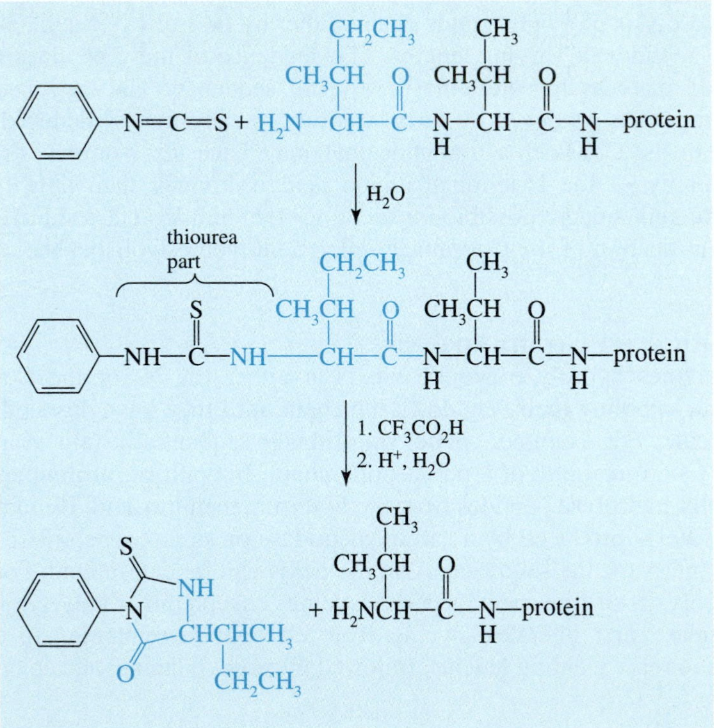

FIGURE 15.4
Use of the Edman Reagent in End Group Analysis

plex cyclization reaction occurs to give a substituted phenylthiohydantoin. This ring contains the carbonyl carbon atom, the α-carbon atom, and the amino-nitrogen atom. The R group of the amino acid is attached to the ring. Comparison of the product with the phenylthiohydantoin of known amino acids establishes the identity of the amino acid. This entire process can be carried out automatically by an instrument called an automatic sequenator.

Because the Edman degradation does not cleave the peptide bond in the protein, it can be repeated to identify the amino acids sequentially from the N-terminal amino acid end of the molecule. The yield of the Edman degradation approaches 100%, and sequences of 30 residues of a polypeptide can be determined from 5 picomole (5×10^{-12} mol) samples. This means that the sequence of a peptide with 30 amino acid residues, with a molecular weight of about 3000, can be determined from a 15 nanogram sample!

EXAMPLE 15-4

There are several enkephalins. Predict the products of the chymotrypsin-catalyzed hydrolysis of the following enkephalin.

Tyr-Gly-Gly-Phe-Leu

Solution

Chymotrypsin catalyzes hydrolysis of peptide bonds on the C-terminal side of aromatic amino acids in peptides and proteins. The enkephalin contains both phenylalanine and tyrosine, so chymotrypsin cleaves the peptide in two places. Consider each step separately. Tyrosine is the N-terminal amino acid, and hydrolysis at its carboxyl end results in free tyrosine.

$$\text{Tyr-Gly-Gly-Phe-Leu} \xrightarrow{\text{chymotrypsin}} \text{Tyr} + \text{Gly-Gly-Phe-Leu}$$

The phenylalanine in the tetrapeptide is bonded to the C-terminal amino acid, leucine. Hydrolysis at the carboxyl group of phenylalanine frees leucine. A tripeptide results.

$$\text{Gly-Gly-Phe-Leu} \xrightarrow{\text{chymotrypsin}} \text{Gly-Gly-Phe} + \text{Leu}$$

The products of the reaction are tyrosine, leucine, and the tripeptide Gly-Gly-Phe.

Problem 15.4

Predict the products of the trypsin-catalyzed hydrolysis of the following pentapeptide.

Ala-Lys-Gly-Arg-Leu

EXAMPLE 15-5

β-Endorphin, a 31-peptide, has analgesic effects and promotes the release of growth hormone and prolactin. Treating β-endorphin with phenyl isothiocyanate followed by hydrolysis with anhydrous trifluoroacetic acid yields the following phenylthiohydantoin. What does this information reveal about the structure of the peptide?

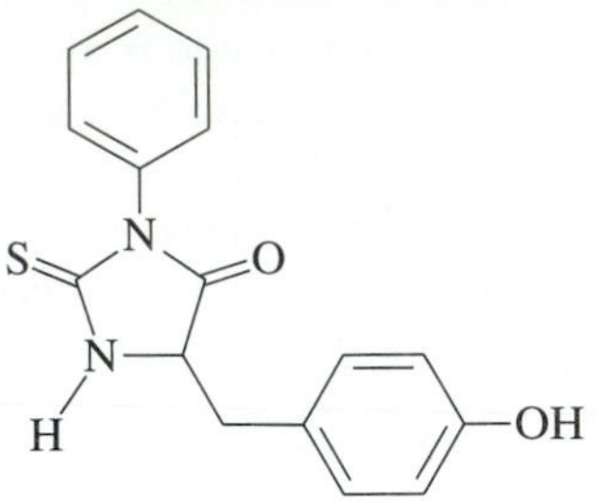

Solution

The procedure described is an Edman degradation, which removes the N-terminal amino acid from the peptide. Be careful not to confuse the two aromatic rings in the phenylthiohydantoin. The aromatic ring bonded to the nitrogen atom between the C=O and the C=S is the phenyl group of the phenyl isothiocyanate. The group bonded to the ring between the nitrogen atom and the C=O is the R group of the amino acid. The N-terminal amino acid of β-endorphin is tyrosine.

Problem 15.5

Draw the structure of the phenylthiohydantoin obtained by reaction of the following tetrapeptide with phenyl isothiocyanate.

$$NH_2-\underset{CH_2OH}{CH}-\overset{O}{\overset{\|}{C}}-NH-\underset{CH_3}{CH}-\overset{O}{\overset{\|}{C}}-NH-\underset{CH(CH_3)_2}{CH}-\overset{O}{\overset{\|}{C}}-NH-\underset{CH_2SH}{CH}-\overset{O}{\overset{\|}{C}}-OH$$

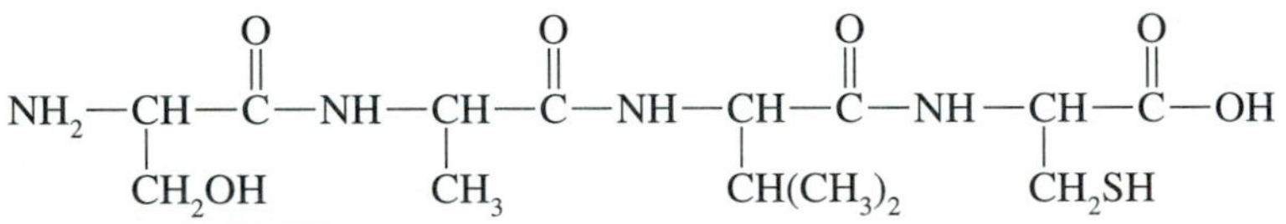

15.8 Structure of Proteins

The biological activity of a protein depends on the three-dimensional shape of the molecule, called its **native state** or **native conformation.** Any alteration of structure caused by breaking any type of bond destroys the physiological function of the protein. Protein structure is described at four levels: primary, secondary, tertiary, and quaternary. Each of these divisions is somewhat arbitrary because it is the total structure of the protein that controls function. Nevertheless, it is useful to consider the levels of structure individually. An overview of the four levels of protein structure is shown in Figure 15.5.

Both the fur and the horns of this mountain goat consist of proteins.

Primary Structure

The linear sequence of amino acids in a protein and the location of disulfide bonds is called its **primary structure.** The peptide bond, which is largely responsible for the primary structure of a protein, is a very strong bond. The peptide bond is pictured as a resonance hybrid of two contributing structures, and the carbon-nitrogen bond has partial double bond character. The peptide bond is planar, and there is restricted rotation about the carbon-nitrogen bond.

$$\left[R-\overset{\ddot{O}:}{\overset{\|}{C}}-\underset{H}{\ddot{N}}-R' \longleftrightarrow R-\overset{:\ddot{O}:^-}{C}=\underset{H}{N^+}-R' \right]$$

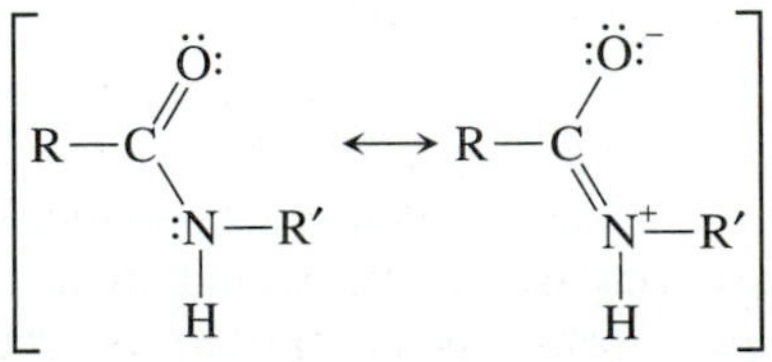

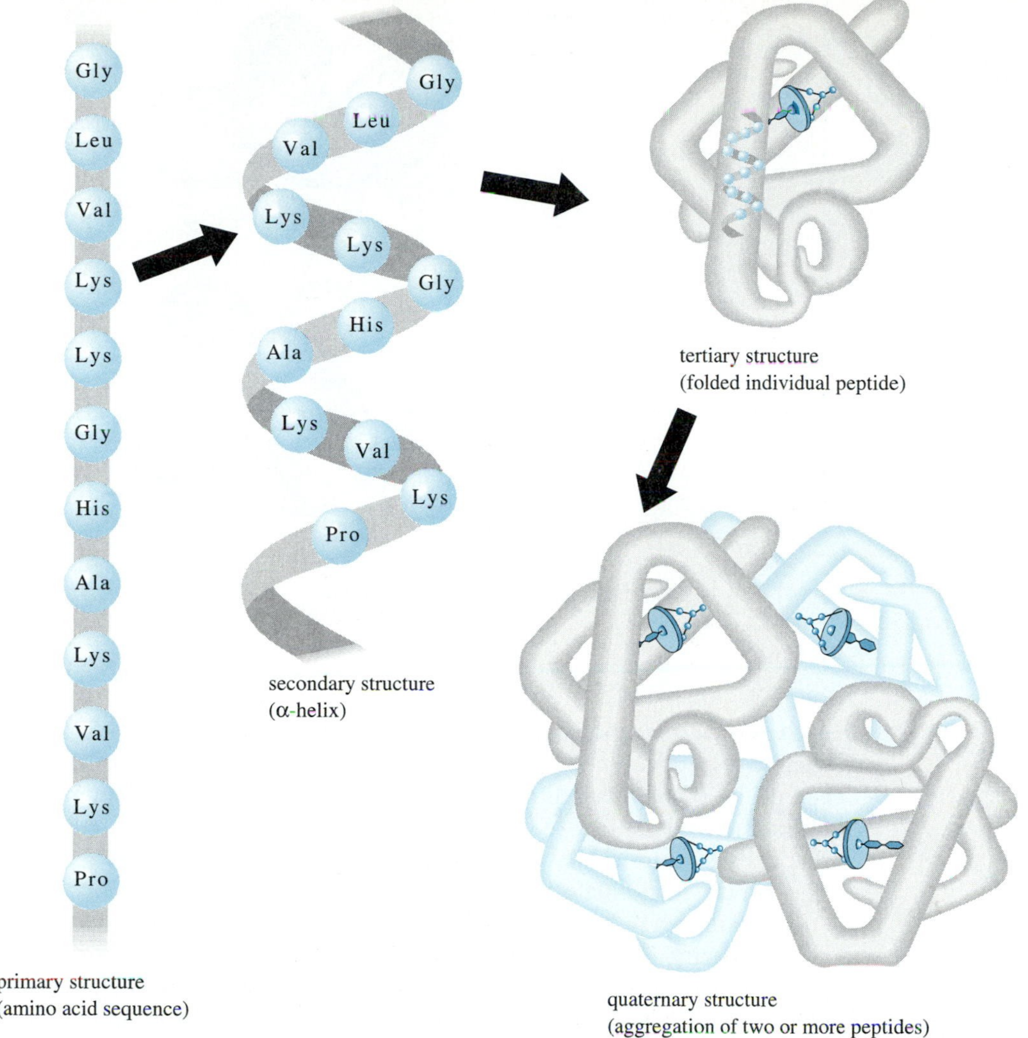

FIGURE 15.5
The Four Levels of Protein Structure

Two conformations are possible about the planar peptide bond. The trans conformation shown for the amide bond of *N*-methylacetamide is more stable than the alternate cis conformation due to steric effects (Figure 15.6). Proteins usually exist in such trans conformations as well for the same reason. However, this conformation is also ideally suited to form hydrogen bonds between amide groups, which are responsible for the secondary structure of proteins.

The bond between the α-carbon atom of peptides and the carbonyl carbon atom is a rotationally free single bond. Similarly, the single bond between the nitrogen atom and the α-carbon atom of the next amino acid is also rotationally free. There is, in addition, free rotation about the bonds between the α-carbon atoms and the R groups. Thus, a protein chain consists of rigid peptide units connected to one another by freely rotating single bonds.

Human hemoglobin consists of heme, which is a complex heterocyclic compound, and globin. Globin is composed of four protein chains: two

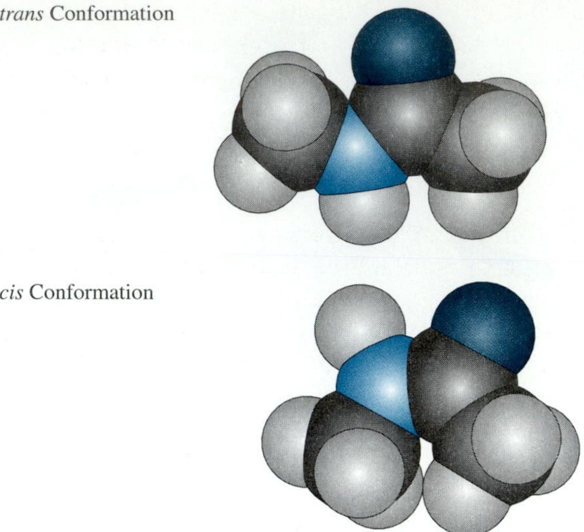

FIGURE 15.6
Conformation of the Peptide Bond

α chains and two β chains. There are 141 and 146 amino acids in the α and β chains, respectively. In some people the sixth amino acid from the N-terminal end of the β chain is valine rather than glutamic acid. This difference of a single amino acid out of 146 in the chain changes the shape of the red blood cells. The cells tend to sickle, and their passage through blood vessels is restricted. The associated circulatory problems are known as *sickle cell anemia.*

Differences in hemoglobin in various animals do not affect its oxygen-carrying capacity. The β chains of gorilla and human hemoglobins differ only at position 104. In gorillas the amino acid is lysine whereas in humans it is arginine, another basic amino acid. The pig β chain differs from human hemoglobin at 17 sites, and that of the horse at 26 sites. In spite of the different composition of hemoglobin in animals, there are nine positions that contain the same amino acids in all hemoglobin molecules. These amino acid residues are important to the oxygen-binding function of hemoglobin.

The **disulfide bond,** which is a covalent bond between two sulfur atoms, is also considered part of the primary structure of a protein. A disulfide bond results from the oxidation of the —SH (sulfhydryl) groups of two cysteine molecules to form cystine.

$$
\begin{array}{cccccc}
 & CO_2H & & CO_2H & & CO_2H \quad\quad CO_2H \\
 & | & & | & & | \quad\quad\quad\quad\quad | \\
NH_2- & C-H & NH_2- & C-H & \xrightarrow{[O]} NH_2- & C-H \quad NH_2-C-H \\
 & | & & | & & | \quad\quad\quad\quad\quad | \\
 & CH_2-S-H & H-S- & CH_2 & & CH_2-S-S-CH_2 \\
 & \text{cysteine} & & & & \text{cystine}
\end{array}
$$

Many proteins containing fewer than 100 amino acid residues have a high cysteine content. Many of these cysteine residues exist in an oxidized form, linking them intramolecularly to other cysteine residues. As a result, the protein conformation is much less flexible. Intrachain disulfide bonds occur in oxytocin and vasopressin (Section 15.5). Disulfide bonds can also link a cysteine residue in one polypeptide chain with a cysteine residue in another polypeptide chain such as occurs in insulin, which consists of two polypeptide chains linked by two disulfide bonds.

Secondary Structure

The specific spatial arrangement of the amino acid residues close to one another in the polypeptide chain is called the **secondary structure** (Figure 15.5). Because the bonds separating the planar and rigid peptide units can rotate freely, these peptide units can be oriented at angles to each other and can exist in a variety of conformations. However, proteins contain many functional groups that can form hydrogen bonds. Although individual hydrogen bonds are weaker than peptide and disulfide bonds, there are many hydrogen bonds and they help stabilize the conformation of proteins. Intramolecular hydrogen bonding between the amide hydrogen atom of one peptide unit and the carbonyl oxygen atom of another peptide unit is very common.

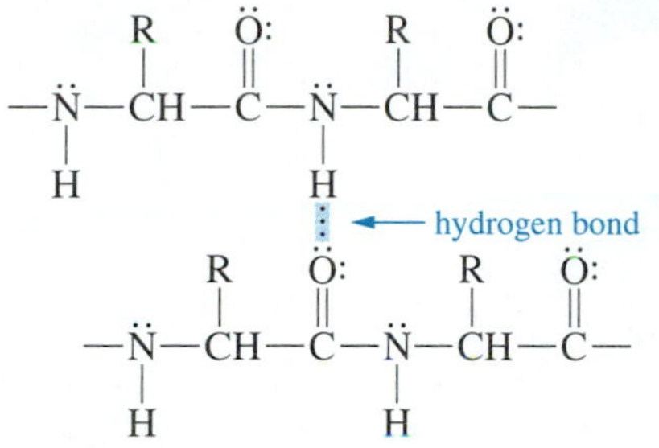

```
        R    :Ö:      R    :Ö:
        |    ||       |    ||
  —N̈—CH—C—N̈—CH—C—
   |          |
   H          H
              :  ← hydrogen bond
           R    :Ö:      R    :Ö:
           |    ||       |    ||
     —N̈—CH—C—N̈—CH—C—
      |          |
      H          H
```

As a result, many proteins exist as chains coiled into a spiral known as a helix. The helix could be either right- or left-handed, but for proteins consisting of L-amino acids the right-handed (or α) helix is more stable than the left-handed helix. The spiral is held together by hydrogen bonds between the proton of the N—H group of one amino acid and the oxygen atom of the C═O group of another amino acid in the next turn of the helix (Figure 15.7).

In addition to the intrachain hydrogen bonding that occurs in the α helix, there are proteins with interchain hydrogen bonding. The β-pleated sheet consists of extended protein chains arranged in an alternating sequence with chains running in the opposite directions (Figure 15.7). Interchain hydrogen bonds occur in fibrin (the blood-clotting protein), myosin (a protein of muscle), keratin (the protein of hair), and fibroin (the protein of silk).

Only a limited number of types of amino acids with small side chains can form a β-pleated sheet structure. For example, silk contains approximately 45% glycine, 30% alanine, and 12% serine. The primary structure consists of glycine separated by either alanine or serine units.

-Gly-Ala-Gly-Ser-Gly-Ala-Gly-

The hydrogen atoms and side chains of the amino acids extend above and below the plane of the sheet. The chains can closely associate without the side chains bumping into each other. Furthermore, the sheets can stack closely together when the groups are small.

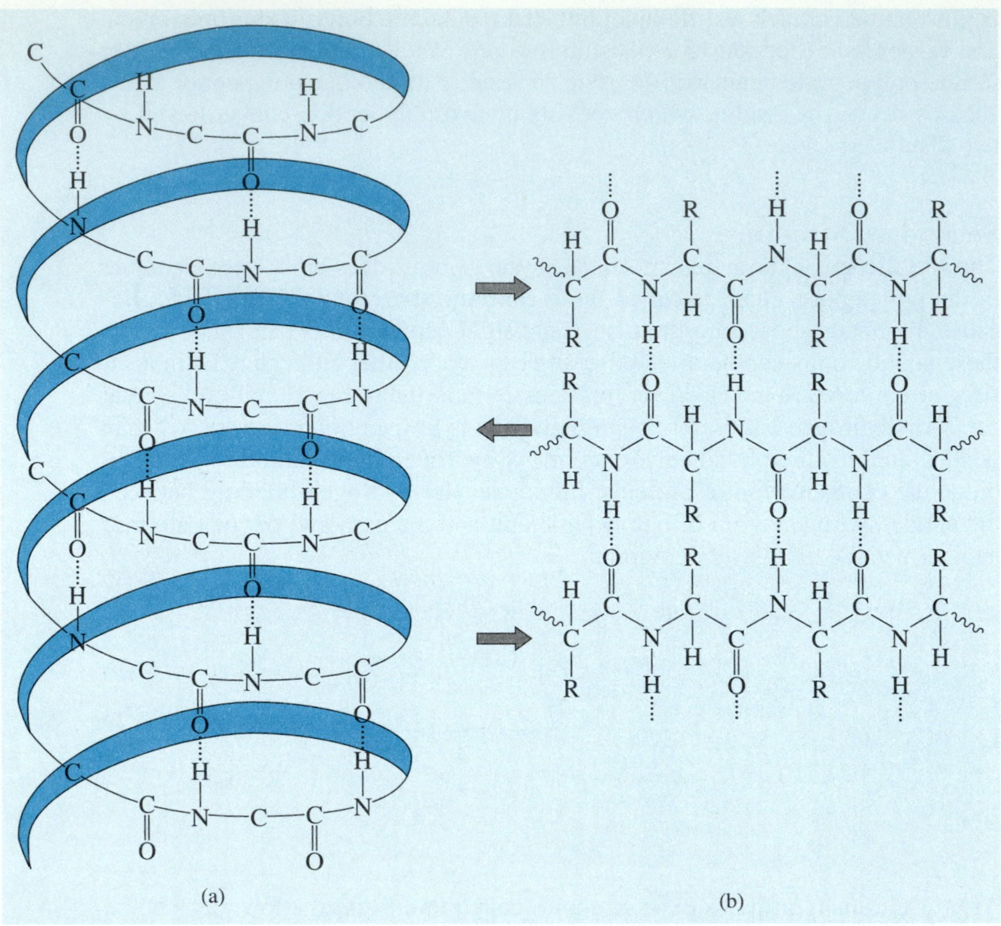

FIGURE 15.7
Hydrogen Bonding in Proteins

(a) The intramolecular hydrogen bonds between coils of an α-helix are shown only on the "front." (b) The intermolecular hydrogen bonds between the chains of proteins in a β-pleated sheet occur between chains oriented in opposing directions, as shown by the arrows.

Tertiary Structure

The three-dimensional shape of the protein is its **tertiary structure.** This spatial arrangement brings together a number of amino acid residues that are far apart in the polypeptide chain. The proximity of amino acids in the tertiary structure is responsible for the activity of many enzymes.

The three-dimensional folded shape of some proteins (Figure 15.5) depends on their primary and secondary structures, which together make long-range interaction possible between amino acids. The forces of attraction between amino acids include ionic bonds, hydrogen bonds, and hydrophobic interactions.

Hydrogen bonds form between various amino acid side chains, such as the —OH groups of serine, threonine, and tyrosine. The side chains of acidic and basic amino acids can also form hydrogen bonds. However, these amino acids may also form an intrachain ionic bond called a **salt bridge.** These ionic attractive forces between the carboxylate groups and the ammonium groups of side chains pull portions of chains together.

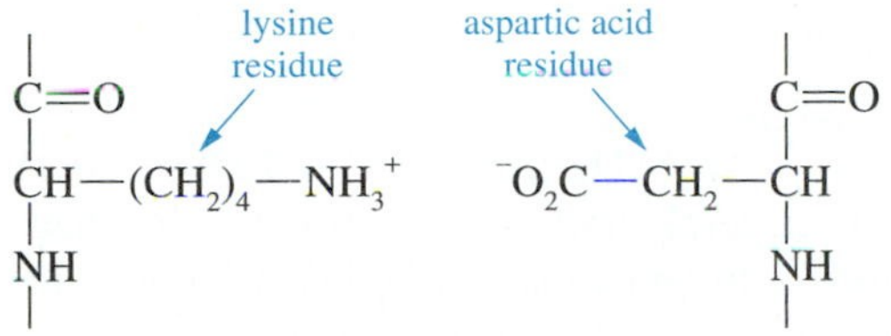

lysine residue — aspartic acid residue

$$-C(=O)-CH(-NH-)-(CH_2)_4-NH_3^+ \quad {}^-O_2C-CH_2-CH(-NH-)-C(=O)-$$

salt bridge

Proteins contain many nonpolar side chains that are repelled by water. As a result, they tend to associate with one another on the "inside" of a folded protein molecule, out of contact with water. The tendency of nonpolar side chains to collect out of contact with the solvent is called the **hydrophobic effect.** The hydrophobic interactions in proteins are similar to those in the micelle of a soap (Section 13.6) or the bilayer of lipids in membranes (Section 13.10). Hydrophobic interactions among nonpolar side chains in proteins are weak, but abundant, and are primarily responsible for maintaining the folded conformation of a protein.

The hydrophobic portions of the protein associate within the interior of the folded structure. Polar or charged (hydrophilic) groups are located at the surface near water molecules. Thus, there is no single reason for a specific shape of a protein. Each shape is unique and is the result of both composition and the sequence of the polypeptide chain.

Quaternary Structure

The **quaternary structure** of a protein is the association of several protein chains or subunits into a closely packed arrangement (Figure 15.5). Each of the subunits has its own primary, secondary, and tertiary structure. The subunits fit together because of their shape and are held together by forces other than covalent bonds. Single-chain proteins have no quaternary structure.

The subunits in a quaternary structure must be specifically arranged for the entire protein to function properly. Any alteration in the structure of the subunits or how they are associated causes marked changes in biological activity.

Hemoglobin consists of two pairs of different proteins, each protein enfolding a molecule of heme. The two identical α chains and two identical β chains are arranged tetrahedrally in a three-dimensional structure (Figure 15.5). These units are held together by hydrophobic interactions, hydrogen bonding, and salt bridges. The four protein subunits of hemoglobin do not behave independently. When one heme molecule binds O_2, the conformation of the surrounding protein chain is slightly altered. Changes in conformation at one site caused by a change at a spatially separated site of a protein molecule are called **allosteric effects.** As a result of allosteric effects, each heme in the other subunits then can bind more easily to additional oxygen molecules. As each oxygen binds, there are further conformational changes in the other protein chains that enhance their binding capability. As a consequence, once oxygenation occurs at one heme, there is cooperation at all other sites in hemoglobin, which then can carry four oxygen molecules.

Cholesterol and Lipoproteins

Lipoproteins are complex particles composed of several types of proteins and lipids, including triacylglycerols, cholesterol, and phospholipids. They transport lipids in human plasma and regulate the cholesterol level in the blood. Plasma lipoproteins share a common structure. They have a hydrophobic core of triacylglycerols and cholesterol esters surrounded by a shell of phospholipids, cholesterol, and proteins. Lipoproteins are divided into three classes according to their composition and density: high density lipoproteins (HDLs), low density lipoproteins (LDLs), and very low density lipoproteins (VLDLs). The density of a lipoprotein depends on the lipid to protein ratio (see the accompanying figure). Because proteins are more dense than lipids, increasing the protein component of a lipoprotein increases its density.

The densities of VLDLs range from 1.006 to 1.018 g/mL. VLDLs consist of about 95% lipid and 5% protein. LDLs have densities of 1.019–1.063 g/mL; they are about 75% lipid and 25% protein. HDLs have densities of 1.063–1.21 g/mL; they are about 50% lipid and 50% protein.

The VLDLs are the principal carriers of triacylglycerols, whereas LDLs carry 80% of the blood serum cholesterol. HDLs carry the remaining cholesterol. The functions of LDLs and HDLs in cholesterol transport are quite different. Low density lipoproteins carry cholesterol to cells, where it is incorporated in cell membranes or used for the synthesis of other molecules. HDLs carry excess cholesterol away from cells to the liver for processing and excretion from the body. Individuals with high HDL levels can efficiently remove unneeded cholesterol from blood serum. This is an important function because the accumulation of cholesterol ester deposits in arteries results in atherosclerosis; that is, hardening of the arteries. If the HDL level is too low, excess cholesterol is deposited on the walls of the arteries. One consequence of this condition is coronary heart disease, which often occurs in men between 35 and 50 years of age. High concentrations of HDLs tend to diminish the risk of heart disease.

The average concentration of HDLs is 45 mg/100 mL for men and 55 mg/100 mL for women. This difference in HDL concentration may partly explain why proportionately fewer young women have heart attacks than men. The concentration of HDLs appears to increase if a person exercises. For example, the HDL concentration in male long-distance runners may be as high as 75 mg HDL/100 mL. ■

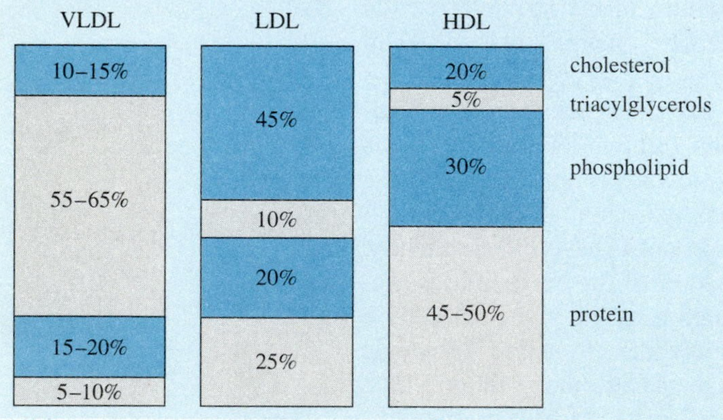

Composition of Lipoproteins

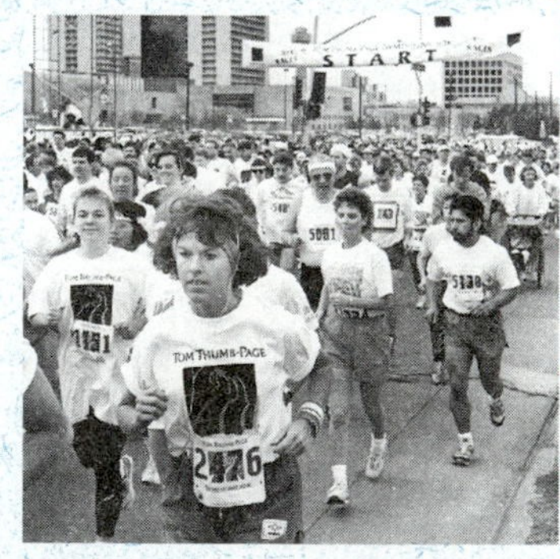

Runners have higher levels of HDLs than individuals that don't exercise.

Explorations with Molecular Models

1. Prepare a ball-and-stick model corresponding to the following space-filling model of an amino acid. Identify the amino acid. Arrange the structure so that the carbon chain is vertical, with the C-1 atom at the top of the structure. Make sure that the hydrogen atom and amino group bonded to the C-2 atom are directed toward you. Is this structure D or L? Arrange the structure to view the stereogenic center, sighting with the carbon-hydrogen bond directed away from your eye. Determine whether the structure is *R* or *S*.

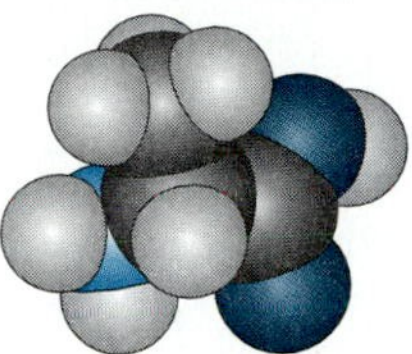

2. Prepare a ball-and-stick model corresponding to the following space-filling model of an amino acid. Identify the amino acid. Arrange the structure so that the carbon chain is vertical, with the C-1 atom at the top of the structure. Make sure that the hydrogen atom and amino group bonded to the C-2 atom are directed toward you. Is this structure D or L? Write the Fischer projection formula representing this structure.

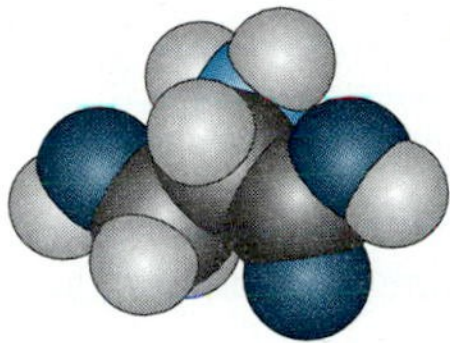

3. Prepare a ball-and-stick model corresponding to the following space-filling model of an amino acid. Rotate the structure about the C-2 to C-3 bond to obtain a conformation that has eclipsed carbon atoms. Arrange this conformation in the orientation required to generate a Fischer projection formula. Does the compound have the D or L configuration? Write the Fischer projection formula.

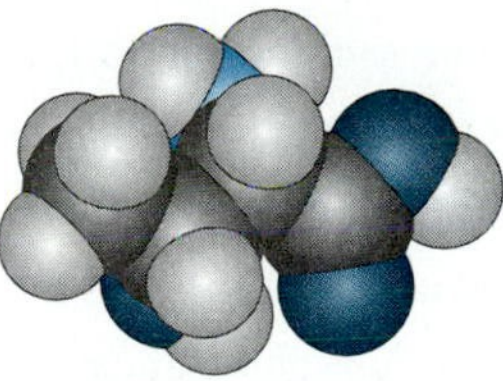

4. Prepare a ball-and-stick model corresponding to the following space-filling model of a dipeptide. Name the compound.

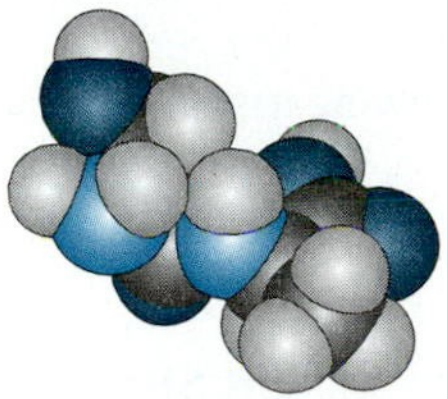

Exercises

Amino Acids

15.1 A number of D-amino acids are found in bacteria. D-Glutamic acid is found in bacterial cell walls. Draw a projection formula for the amino acid.

15.2 Gramicidin S is a cyclic peptide antibiotic that contains D-phenylalanine. Draw a projection formula for the amino acid.

15.3 The following compound is an unusual amino acid found in collagen. From what amino acid could this compound be derived?

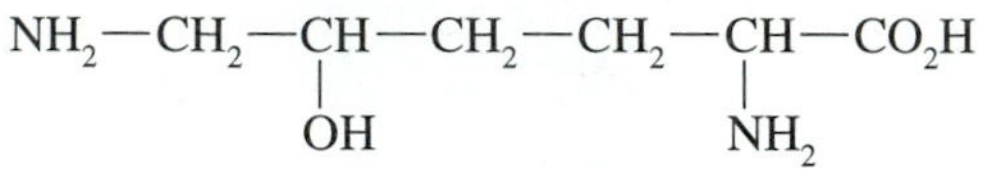

$$NH_2{-}CH_2{-}\underset{\displaystyle OH}{\underset{|}{CH}}{-}CH_2{-}CH_2{-}\underset{\displaystyle NH_2}{\underset{|}{CH}}{-}CO_2H$$

15.4 The following antibacterial agent is contained in garlic. From what amino acid might it be derived?

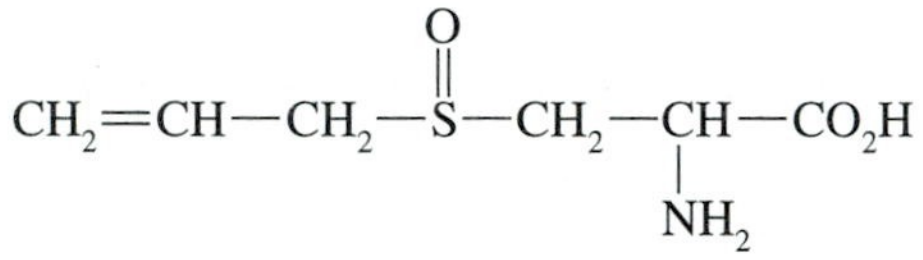

$$CH_2{=}CH{-}CH_2{-}\overset{\displaystyle O}{\overset{\|}{S}}{-}CH_2{-}\underset{\displaystyle NH_2}{\underset{|}{CH}}{-}CO_2H$$

Acid-Base Properties

15.5 Draw the structures of alanine and glutamic acid at pH = 1 and pH = 12.

15.6 Write the structures for the zwitterions of cysteine and valine.

15.7 How could you distinguish between a solution of asparagine and a solution of aspartic acid?

15.8 Would you expect an aqueous solution of lysine to be neutral, acidic, or basic? Explain.

Isoionic Points

15.9 Estimate the isoionic points of the following tripeptides.

(a) Ala-Val-Gly (b) Ser-Val-Asp (c) Lys-Ala-Val

15.10 Estimate the isoionic points of the following tripeptides.

(a) Glu-Val-Ala (b) Arg-Val-Gly (c) His-Ala-Val

15.11 Examine the structures of oxytocin and vasopressin in Section 15.5. Which should have the higher isoionic point?

15.12 Estimate the isoionic point of the following enkephalin.

Tyr-Gly-Gly-Phe-Leu

15.13 The isoionic point of chymotrypsin is 9.5. What does this value indicate about the composition of chymotrypsin?

15.14 The isoionic point of pepsin is 1.1. What does this value indicate about the composition of pepsin?

Peptides

15.15 Write the complete formula and the condensed formula for alanylserine.

15.16 How does glycylserine differ from serylglycine?

15.17 Identify the amino acids contained in the following tripeptide. Name the compound.

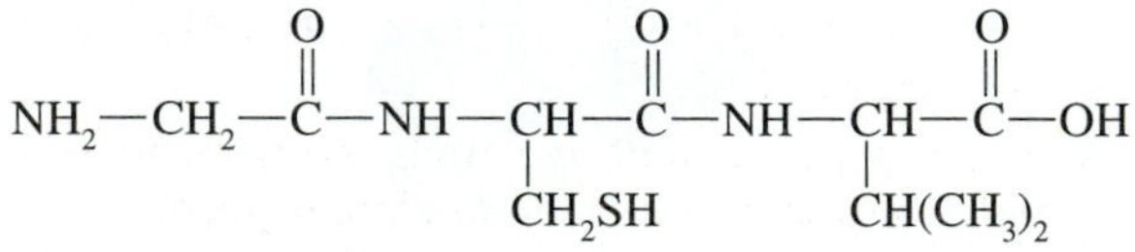

$$NH_2{-}CH_2{-}\overset{\displaystyle O}{\overset{\|}{C}}{-}NH{-}\underset{\displaystyle CH_2SH}{\underset{|}{CH}}{-}\overset{\displaystyle O}{\overset{\|}{C}}{-}NH{-}\underset{\displaystyle CH(CH_3)_2}{\underset{|}{CH}}{-}\overset{\displaystyle O}{\overset{\|}{C}}{-}OH$$

15.18 Identify the amino acids contained in the following tripeptide. Name the compound.

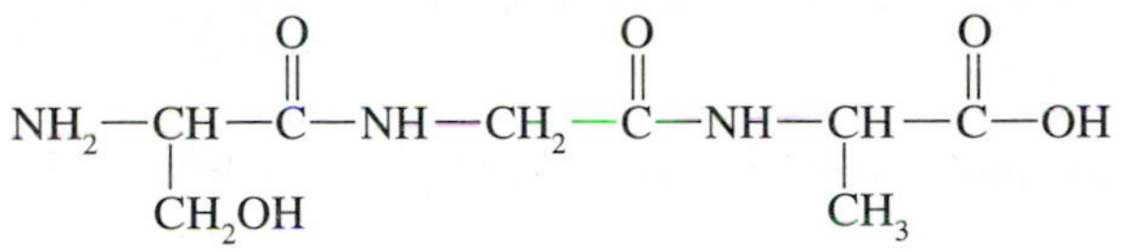
O
O
O
NH2—CH—C—NH—CH2—C—NH—CH—C—OH
CH2OH
CH3

15.19 Thyrotropin-releasing hormone (TRH) causes the release of thyrotropin from the pituitary gland, which then stimulates the thyroid gland. Examine its structure and comment on one unusual structural feature.

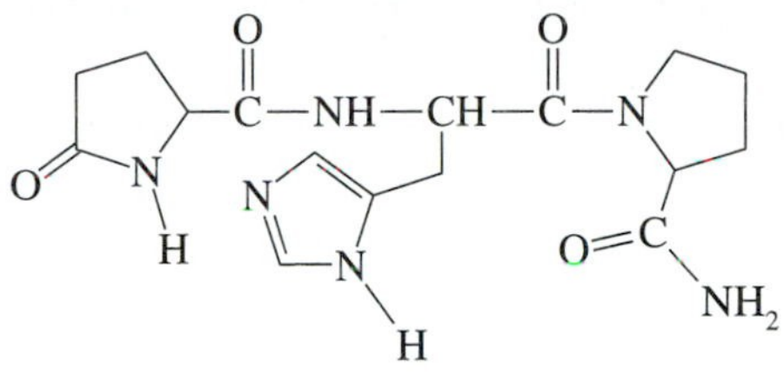
O
O
C—NH—CH—C—N
O
N
H
N
N
H
O=C
NH2

15.20 The tripeptide glutathione, which is important in detoxifying metabolites, has an unusual structural feature. Identify it.

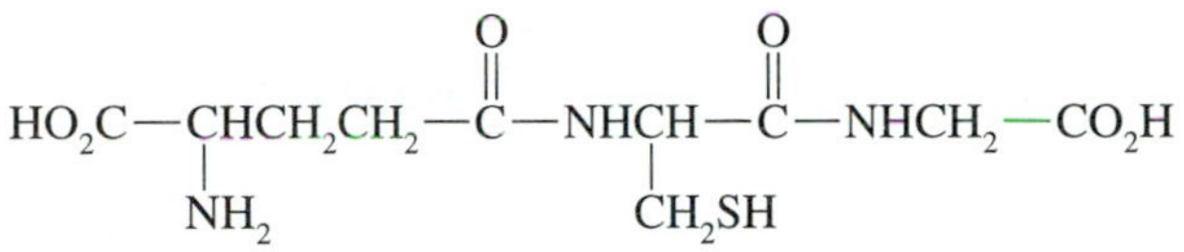
O
O
HO2C—CHCH2CH2—C—NHCH—C—NHCH2—CO2H
NH2
CH2SH

15.21 How many isomeric compounds with the composition Gly_2,Ala_2 are there?

15.22 How many isomeric compounds with the composition Gly_2,Ala,Leu are there?

Hydrolysis and Structure Determination

15.23 Assuming that only dipeptides are formed by partial hydrolysis, what is the minimum number that must be identified to establish the structure of a pentapeptide?

15.24 Assuming that only tripeptides are formed by partial hydrolysis, what is the minimum number that must be identified to establish the structure of an octapeptide?

15.25 The tetrapeptide tuftsin is hydrolyzed to produce Pro-Arg and Thr-Lys. Does this information establish the structure of tuftsin?

15.26 Assume that the octapeptide angiotensin II is hydrolyzed to produce Pro-Phe, Val-Tyr-Ile, Asp-Arg-Val, and Ile-His-Pro. What is its structure?

Enzymatic Hydrolysis

15.27 Which of the following tripeptides will be cleaved by trypsin? If cleavage occurs, name the products.

(a) Arg-Gly-Tyr (b) Glu-Asp-Gly (c) Phe-Trp-Ser (d) Ser-Phe-Asp

15.28 Which of the following tripeptides will be cleaved by trypsin? If cleavage occurs, name the products.

(a) Asp-Lys-Ser (b) Lys-Tyr-Cys (c) Asp-Gly-Lys (d) Arg-Glu-Ser

15.29 Indicate which of the tripeptides in Exercise 15.27 will be cleaved by chymotrypsin and name the products.

15.30 Indicate which of the tripeptides in Exercise 15.28 will be cleaved by chymotrypsin and name the products.

15.31 The tetrapeptide tuftsin is hydrolyzed by trypsin to produce Pro-Arg and Thr-Lys. Does this information establish the structure of tuftsin?

15.32 The pentapeptide met-enkephalin is hydrolyzed by chymotrypsin to give Met, Tyr, and Gly-Gly-Phe. Does this information establish the structure of met-enkephalin?

15.33 The nonapeptide known as the sleep peptide is hydrolyzed by chymotrypsin to produce Ala-Ser-Gly-Glu and Ala-Arg-Gly-Tyr and Trp. What two structures are possible for the sleep peptide?

15.34 The sleep peptide is hydrolyzed by trypsin to produce Gly-Tyr-Ala-Ser-Gly-Glu and Trp-Ala-Arg. What is the structure of the sleep peptide?

End Group Analysis

15.35 Hydrolysis of tuftsin with an aminopeptidase yields Thr. Using the information in Exercise 15.25, what is the structure of tuftsin?

15.36 Hydrolysis of met-enkephalin with a carboxypeptidase yields Met. Using the information in Exercise 15.32, what is the structure of met-enkephalin?

15.37 A structure determination of insulin using the Edman method yields two phenylthiohydantoin products. What does this information indicate about the primary structure of insulin?

15.38 Cholecystokinin, a 33-peptide, plays a role in reducing the desire for food, and its production is stimulated by food intake. Its N-terminal amino acid is lysine. Draw the structure of the phenylthiohydantoin product.

15.39 Reaction of angiotensin II with the Edman reagent yields the following product. What information has been established?

S, H, N, N, CH_2CO_2H, O

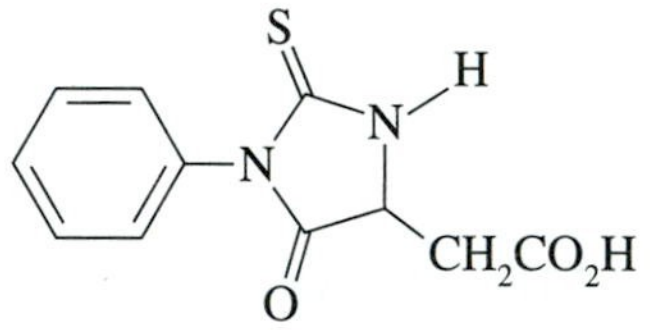

15.40 Corticotropin is released when the blood level of corticosteroids is diminished. Reaction of corticotropin with the Edman reagent yields the following product. What information has been established?

S, H, N, N, CH_2OH, O

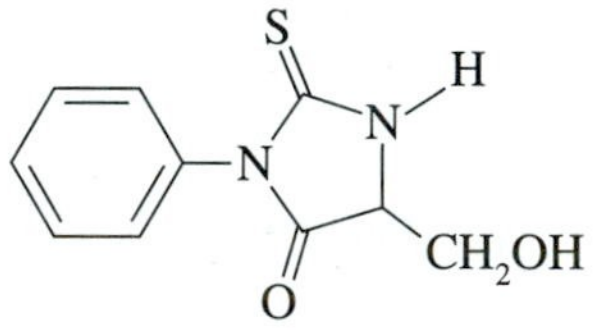

Proteins

15.41 Which amino acids can form salt bridges in proteins?

15.42 Which amino acids have R groups that form hydrogen bonds?

15.43 Which of the following amino acids are likely to exist in the interior of a protein dissolved in an aqueous solution?

(a) glycine (b) phenylalanine (c) glutamic acid

15.44 Which of the following amino acids are likely to exist in the interior of a protein dissolved in an aqueous solution?

(a) lysine (b) cysteine (c) glutamine

15.45 Noting that proline is a secondary amine, explain how proline can disrupt the α helix of a protein.

15.46 Examine the structures of valine and glutamic acid and suggest a reason why human hemoglobin is affected by the substitution of valine for glutamic acid at position 6 in the β chain.

CHAPTER 16

NUCLEIC ACIDS

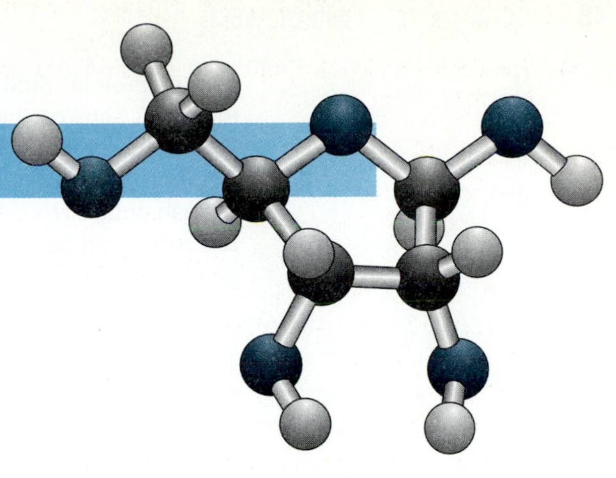

Overview

16.1 The Control of Life Processes

Even in ancient times, the Greeks recognized that organisms pass on traits to their offspring. However, it was not until the 1860s that the experiments of Gregor Mendel firmly established the fundamental principles of heredity. It took almost a hundred years to show that DNA is the molecular basis of heredity.

DNA was discovered in 1869 by Friedrich Meischer shortly after the work of Mendel, but its role in life processes was not recognized until the middle of the next century. Biologists learned that the cell nucleus contains structures called chromosomes that are responsible for hereditary traits and that these chromosomes are passed along in cellular division because the chromosomes are "doubled" in the division process. Chromosomes consist of protein and DNA, and it was thought more likely that the genetic information was encoded in the proteins. DNA is a high molecular weight polymer of nucleotides that consist of deoxyribose, phosphate, and one of four possible nitrogen-containing bases. This number of components was regarded as insufficient to contain the molecular information for life processes. Proteins, on the other hand, contain 20 amino acids that could be combined to give a variety of structures, which could be the carriers of genetic information. However, the four bases of DNA are arranged in a sequence forming three-letter "words" or triplets placed along a backbone of deoxyribose and phosphate. The series of triplets provides the specific "language" that passes on hereditary information.

In 1928, based on work with several species of bacteria and how they infected mice, Frederick Griffith, a British medical officer, showed that it was not the protein of the bacteria that was responsible for assimilation of genetic information. It was not until 1944 that the American bacteriologist Oswald Avery showed that DNA was responsible for Griffith's observations. Although Avery's work was treated with skepticism, over the next decade a number of scientists obtained additional evidence implicating DNA as the genetic material of the cell. Studies of viruses that infect bacteria closed the case. Viruses are essentially only DNA with a protective coat of protein. Studies using radioisotopes to label alternately the protein and the DNA showed that it is the DNA that enters the bacteria and takes over its metabolic machinery. Thus, it is the DNA that contains the genetic information for its replication. Eventually it became clear that not just viruses but all life forms contain DNA that is species specific, and that the molecular diversity of life forms can be explained even though DNA contains only four types of nitrogen-containing bases in the form of nucleotides.

DNA and Replication

Each organism transmits hereditary information using **chromosomes** located in the nucleus of the cell. Chromosomes contain all the primary genetic information that determines the characteristics of the living organism. Thus, carrot seeds produce carrots and not cabbages. Mice give birth to mice and not elephants. The number of chromosomes and their molecular structure is unique for each species. The number of chromosomes is always an even number; they occur as pairs within the nucleus.

Reproduction of all species involves a replication of DNA.

Each chromosome contains a series of hundreds to thousands of genes, which, in turn, contain information for a limited function, such as the synthesis of a particular enzyme. The location of genes on the various chromosomes of a species can be "mapped." The mapping process is not a simple task but one with considerable potential benefits. Genes are responsible for some 5000 known genetic diseases, and the locations of only about 250, such as the gene for cystic fibrosis, have been identified. Technology is being developed to alter genes artificially to change genetic directions, but it remains to be seen how successful medical intervention at the level of the gene will be.

DNA can make exact copies of itself in a process called replication, which is the central process for the continuance of life. DNA also directs the synthesis of three types of ribonucleic acids (RNAs) in a process called transcription. The three types of RNA molecules—ribosomal RNA, messenger RNA, and transfer RNA—control the synthesis of all proteins in a process called translation.

We begin this chapter by examining the structures of nucleotides, which are components of nucleic acids. Then we will discuss the structures of the nucleic acids and their role in the control of all life processes.

16.2 Nucleic Acids Are Polymers

The nucleic acids DNA and RNA are polymers of nucleotides. Nucleotides consist of three units: a sugar, a nitrogen base, and a phosphate group. The sugar in RNA is the β-anomer of D-ribose, which accounts for the name ribonucleic acid. The sugar in DNA is the β-anomer of D-2-deoxyribose, hence the name deoxyribonucleic acid. The difference in these two structures is the absence of a hydroxyl group at C-2 in D-2-deoxyribose (Figure 16.1).

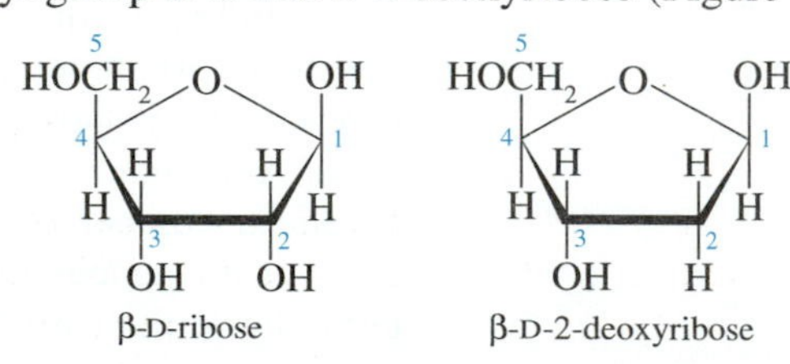

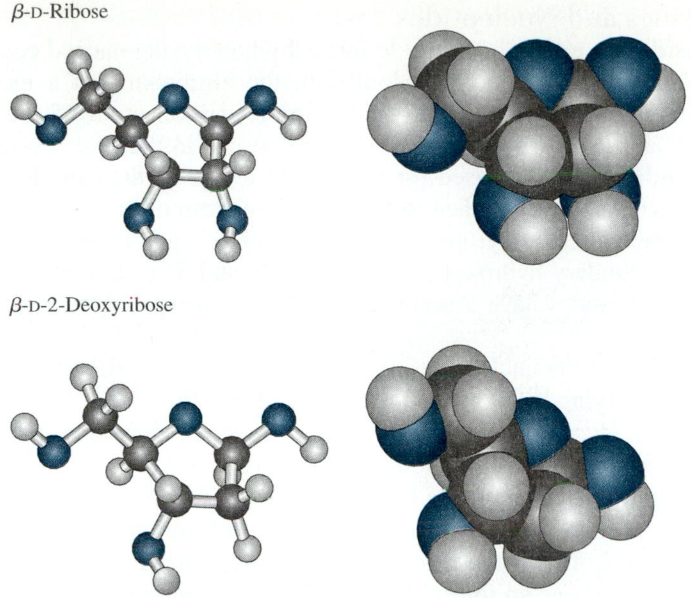

FIGURE 16.1
Structures of D-Ribose and D-2-Deoxyribose

Two purine bases—adenine (A) and guanine (G)—are found in both DNA and RNA. The capital letters are shorthand notations used to represent the bases.

adenine (A) guanine (G)

Three pyrimidine bases—cytosine (C), thymine (T), and uracil (U)—are found in nucleic acids. Cytosine appears in both RNA and DNA. Thymine is found exclusively in DNA; uracil is found only in RNA. Note that thymine and uracil differ only by a methyl group at the 5 position.

cytosine (C) thymine (T) (found in DNA) uracil (U) (found in RNA)

In summary, DNA and RNA contain four bases each: two purines and two pyrimidines. In DNA the bases are A, G, C, and T; in RNA the bases are A, G, C, and U.

Nucleosides and Nucleotides

A **nucleoside** is a nitrogen glycoside formed when the hemiacetal center of the sugar and an —NH of the base join with the elimination of a molecule of water. The configuration is β at the anomeric carbon atom of the sugar in both DNA and RNA (Figure 16.2). The ring atoms in the base are numbered; primed numbers are often used for the carbon atoms of the carbohydrate. The nitrogen base is always attached to the 1′ carbon atom of the carbohydrate, and there is a primary hydroxyl group located at the 5′ carbon atom. Ribonucleosides have secondary hydroxyl groups at the 2′ and 3′ carbon atoms, whereas deoxyribonucleosides have a secondary hydroxyl group only at the 3′ carbon atom.

The names of the nucleosides are listed in Table 16.1. Ribonucleosides are named by modifying the name of the nitrogen base. Purine derivatives have the ending *-osine;* pyrimidine derivatives end in *-idine.* The same convention is used for deoxyribonucleosides along with the prefix *deoxy-*.

A **nucleotide** is an ester of a nucleoside and phosphoric acid. The hydroxyl group at the 5′ carbon atom acts as the alcohol to form the ester (Figure 16.2). The two protons of the monophosphate ester are ionized at physiological pH, and the ester exists as an ion with a −2 charge in solution. The names of the nucleotides are listed in Table 16.1.

EXAMPLE 16-1

Classify the following structure, identify its components, and name it.

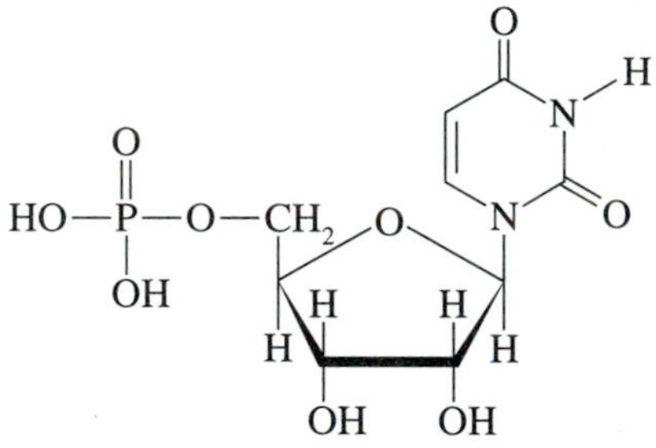

Solution

The compound contains a phosphate group and is a nucleotide. The sugar is ribose, which means that the compound is a ribonucleotide. The base is uracil, which can be found only in ribonucleotides. From Table 16.1 we find that the name of the compound is uridylic acid.

Problem 16.1

Classify the following structure, identify its components, and name it.

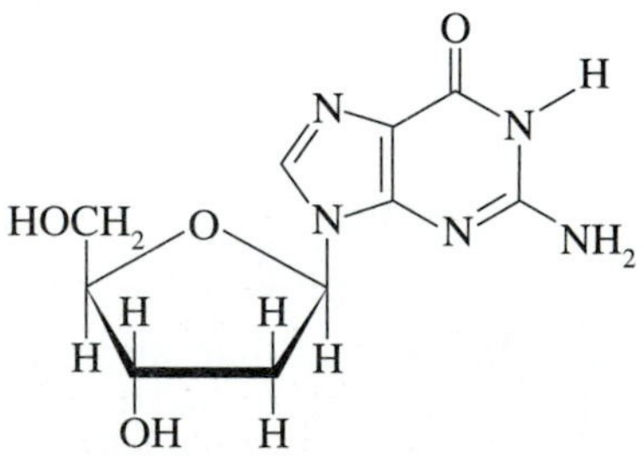

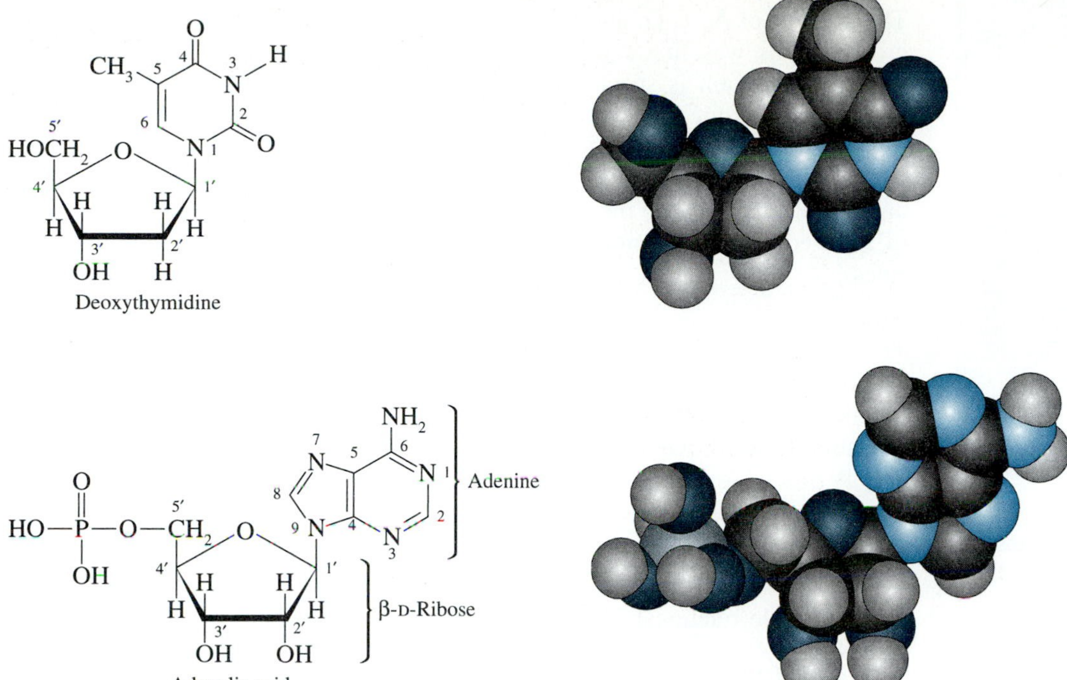

FIGURE 16.2
Structures of Nucleosides and Nucleotides

Table 16.1 Names of Nucleosides and Nucleotides in DNA and RNA

Base	Nucleoside	Nucleotide
DNA		
adenine (A)	deoxyadenosine	deoxyadenylic acid
guanine (G)	deoxyguanosine	deoxyguanylic acid
thymine (T)	deoxythymidine	deoxythymidylic acid
cytosine (C)	deoxycytidine	deoxycytidylic acid
RNA		
adenine (A)	adenosine	adenylic acid
guanine (G)	guanosine	guanylic acid
uracil (U)	uridine	uridylic acid
cytosine (C)	cytidine	cytidylic acid

16.3 Polynucleotides

The composition and structure of molecules have been a focus throughout this text. Although nucleic acids are complex structures, some of the same considerations presented earlier are important to understanding their chemical reactions. As in the case of proteins, it is convenient to discuss the structure of polynucleotides in terms of primary and secondary features.

The determination of DNA structure is a common process used to identify the original tissue and fluids left at crime scenes.

Composition of Polynucleotides

The amount of A, T, G, and C in the DNA of several animals, plants, and microorganisms was first reported by the biochemist Erwin Chargaff in 1947. DNA composition is species specific, and thus the idea that diversity of life forms results from differences in DNA became better established. But a more important result of Chargaff's analysis is the interesting pattern that exists between the quantities of the bases (Table 16.2). Regardless of the species, the amount of C is equal to the amount of G within experimental error; the amount of A is equal to the amount of T. These equalities are known as Chargaff's rules.

$$\%\,C = \%\,G \qquad \%\,A = \%\,T$$

Although the amount of each of the pairs is different from one life form to another, the sum of the percentages of A and G, the two purine bases, equals the sum of the percentages of C and T, the two pyrimidine bases. The relationship is

$$\%\,A + \%\,G = \%\,C + \%\,T$$

We will return to these relationships in DNA later in the next section. There are no simple relationships in the content of the four bases in the three types of RNA.

Primary Structure of Polynucleotides

Nucleotides, bonded through phosphodiester linkages, make up the primary structure of both RNA and DNA molecules (Figure 16.3). Phosphodiesters are formed between the 3′ hydroxyl group of one nucleotide and the 5′ phosphate ester of another nucleotide. Thus, DNA and RNA both have a backbone of alternating sugar and phosphate units. Note that each phosphodiester has one acidic hydrogen atom; hence the name nucleic acids. However, at physiological pH the nucleic acids are ionized. The differences in structure are in the sugar (deoxyribose in DNA and ribose in RNA) and in the order of the bases. Recall that thymine exists only in DNA, and uracil exists only in RNA.

Table 16.2 Composition of DNA of Various Species

	Mole %				Mole Ratios	
Species	**A**	**T**	**G**	**C**	**A/T**	**G/C**
Escherichia coli	26.0	23.9	24.9	25.2	1.09	0.99
yeast	31.3	32.9	18.7	17.1	0.97	1.05
sea urchin	32.8	32.1	17.7	18.4	1.02	0.96
herring	27.8	27.5	22.2	22.6	1.01	0.98
salmon	29.7	29.1	20.8	20.4	1.02	1.02
ox	29.0	29.1	20.7	20.7	1.01	1.00
pig	29.8	29.1	20.7	20.7	1.02	1.00
rat	28.6	28.4	21.4	21.5	1.01	1.00
human	30.9	29.4	19.9	19.8	1.01	1.00

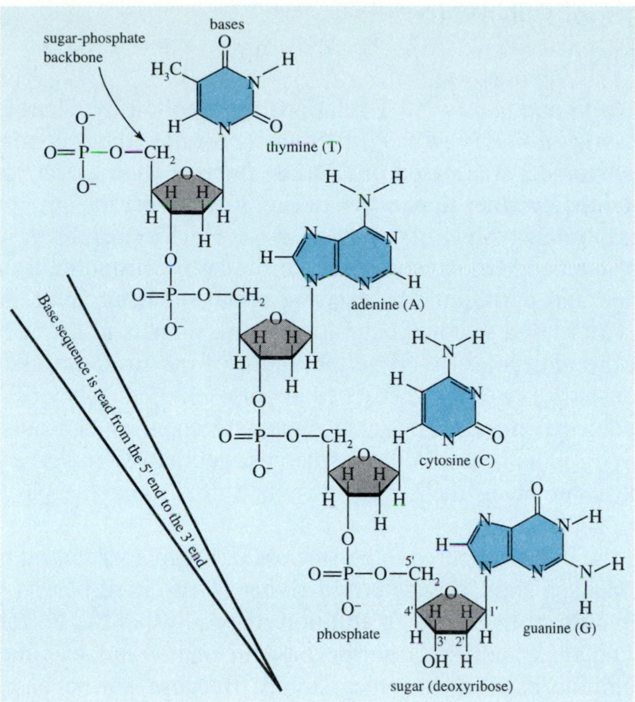

FIGURE 16.3
Structure of a Polynucleotide

As in the case of protein structures, a convention is used to list the sequence of nucleotides. The direction is from the 5′ end, which exists as a phosphate ester, to the 3′ end, which has the free hydroxyl group. For RNA the convention is

$$5' \xrightarrow{-\mathrm{A}-\mathrm{C}-\mathrm{G}-\mathrm{U}-} 3'$$

The same process is used for DNA, but a lowercase *d* is placed at the left of the first base to indicate the deoxyribose in the backbone. The presence of thymine rather than uracil also indicates that the nucleic acid is deoxyribonucleic acid.

$$5' \xrightarrow{-\mathrm{dT}-\mathrm{C}-\mathrm{G}-\mathrm{A}-} 3'$$

When discussing the structure of DNA, its replication, transcription into RNA, and translation into protein, we will use some structural symbols for the bases. The significance of their shapes will become apparent. A sequence of nucleotides is represented by a linear series of these symbols.

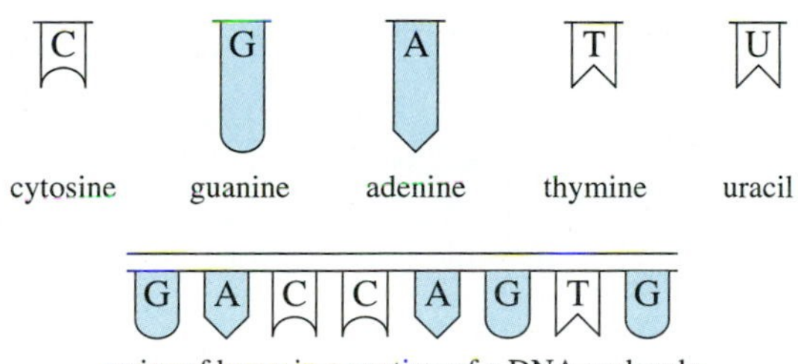

16.4 The Double Helix of DNA

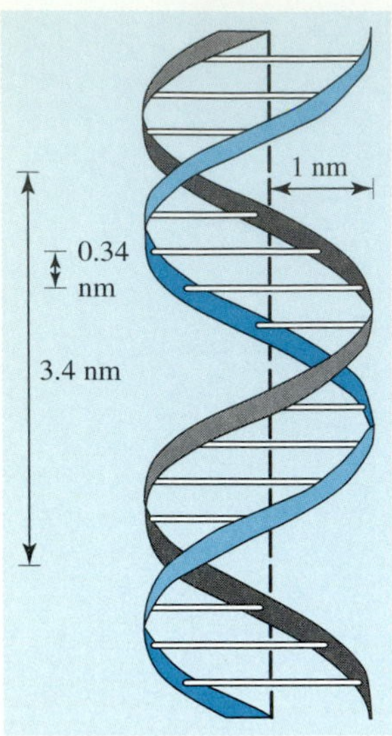

FIGURE 16.4
The Double Helix

The 2-deoxyribose-phosphate backbones are represented by the two strands that run in opposite directions. The bridges between the two strands are the paired nitrogen-containing bases. The bridges are 0.34 nm apart. One complete turn of the helix occurs every 3.4 nm.

The % C = % G and % A = % T relationship reported by Chargaff was used by J. D. Watson and F. H. C. Crick in 1952 to postulate the secondary structure of DNA. Their model was based on a crude X-ray "photograph" of DNA that had been obtained by other scientists working to decipher the structure of DNA. This photograph was typical of a pattern expected for a uniform helix-shaped substance. Watson and Crick estimated that the helix was about 2 nm wide and that the purine and pyrimidine bases were about 0.34 nm apart. At this point Watson and Crick tried to construct various wire models to account for the dimensions of the molecule. Based on the width of the structure, they suggested that DNA consists of two strands twined about each other. They designed a satisfactory model that placed the sugar-phosphate chain on the outside and the purine and pyrimidine bases on the inside. The model resembled a twisted ladder, with the sides being the sugar-phosphate chain and the bases being the rungs (Figure 16.4).

Because the bases appeared to be stacked 0.34 nm apart based on the X-ray photograph, Watson and Crick determined that 10 layers of bases (rungs) occur for each 360° turn of the helix. In addition, they felt that the uniform width of the helix could result only if a purine base in one strand was located across from a pyrimidine base in the other strand. Because purine bases are about twice as wide as pyrimidine bases, two purine bases would cause a bulge in the helix if they were located across from each other.

Finally, Watson and Crick incorporated into the model the stoichiometric relationship between bases in DNA established by Chargaff. They proposed that the bases across from one another are hydrogen bonded. Cytosine forms three hydrogen bonds to guanine, and adenine forms two hydrogen bonds to thymine (Figure 16.5). The matching of bases through hydrogen bonds is called **complementary base pairing.** For each C molecule in DNA there is a molecule of the complementary base G; for each A molecule there is a complementary base T. Note that this requirement also means that for each purine base there is a pyrimidine base. Thus, the Watson-Crick model is consistent with the experimentally observed relationship in the bases of DNA.

The individual strands of polydeoxyribonucleotides run in opposite directions and are said to be antiparallel. As a result, the ribose units in the double helix face in opposite directions and the phosphodiester linkages are "above" the ribose unit in one strand but "below" in the other strand (Figure 16.5). The following diagram uses shorthand representations that summarize the stoichiometric relationships between bases.

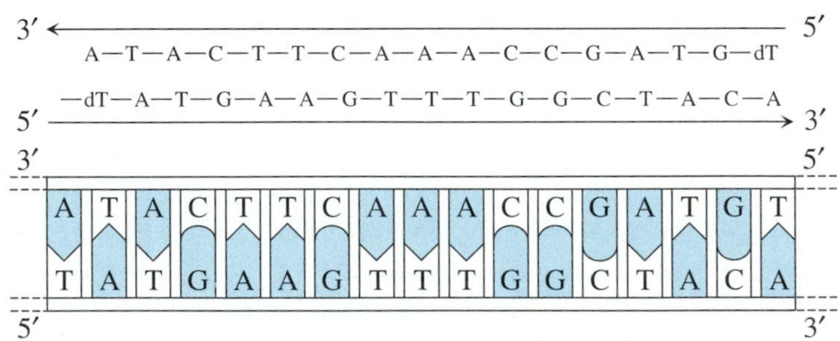

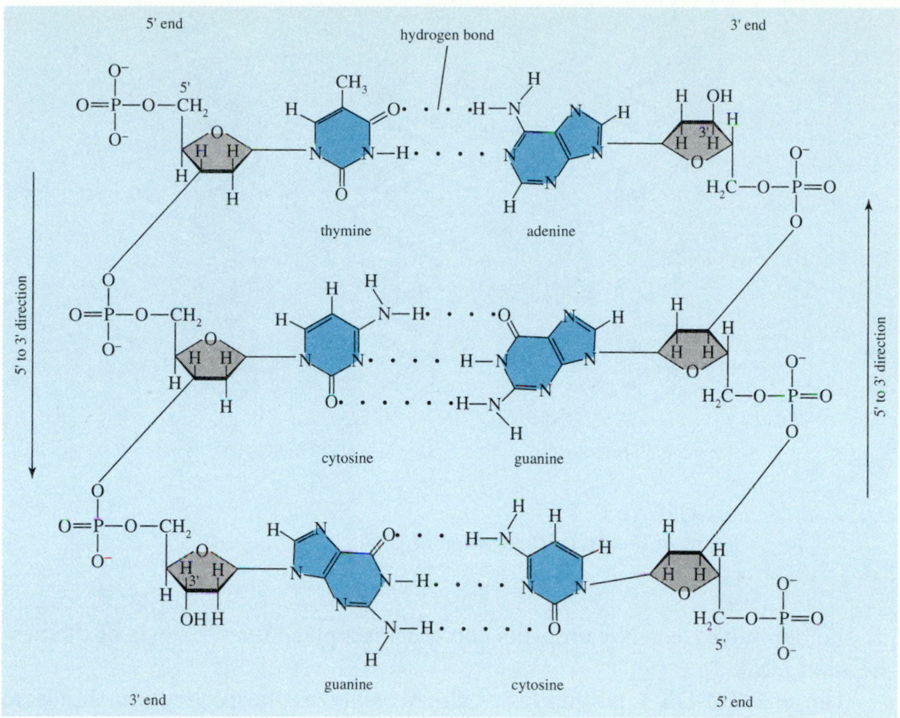

FIGURE 16.5
Hydrogen Bonding in DNA

The pairing of a purine with a pyrimidine base by the proper number of hydrogen bonds not only stabilizes the structure but also results in a helix with a constant diameter. Because the strands run in the opposite directions, the deoxyribose units and their phosphodiesters are oriented differently in the two strands.

16.5 DNA Replicates Itself

Replication is a copying process by which DNA is supplied to the new cells formed by cell division. DNA is replicated in the cell nucleus. As new cells are formed, replication of DNA is controlled by the hydrogen-bonding properties of DNA. Each strand of DNA serves as a pattern or template for producing a new complementary strand. This new complementary strand is identical to the original complementary strand. Since DNA is a double-stranded molecule, each strand replicates its own complementary strand. Studies using radioisotopes have established that replication of DNA occurs by a **semiconservative process;** that is, each new DNA molecule has one strand of the parent DNA molecule of the previous generation. The second strand is formed in the replication process (Figure 16.6).

Although the replication process is complex in detail, some general features can be described. In replication, DNA must unwind so that the new complementary chain can form. The enzyme DNA helicase works its way in between the two strands and then moves along the helix, temporarily separating the strands as it goes. This process exposes the bases at a point in the double helix called the **replication fork** (Figure 16.7). The replication fork is not a sin-

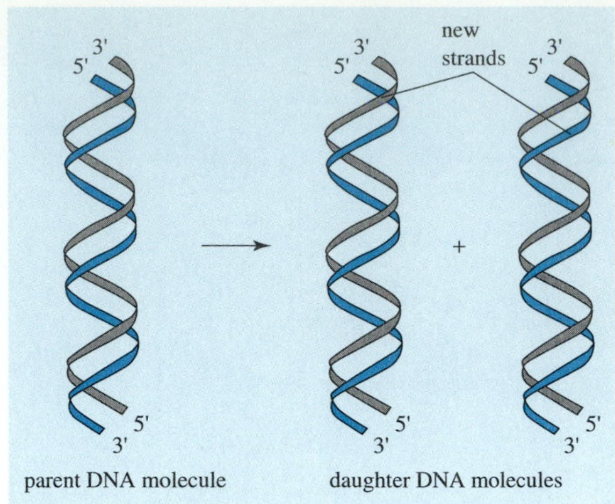

FIGURE 16.6
Semiconservative Replication of DNA

gle site but moves as DNA unravels and creates a place for synthesis of the new strand of DNA.

The enzyme DNA polymerase, which catalyzes the replication, binds to each unwound strand. This enzyme recognizes each base in the DNA chain and matches it with a free nucleotide according to the base pairing rules. Then it binds the nucleotides into a growing strand by bonding the 5′ position of one unit to the 3′ position of another unit to form the backbone. This process is known as 5′→3′ replication. Note that two DNA polymerases must enter the fork and move in opposite directions on each strand to replicate DNA.

EXAMPLE 16-2

Part of the nucleotide sequence in one chain of DNA follows. Write a representation of the complementary DNA chain.

$$5' \xrightarrow{\text{—dA—T—C—G—}} 3'$$

Solution

The complementary base pairs in DNA are A with T and G with C. For each A in one chain there is a T in the other. For each C in one chain there is a G in the other. The complementary chain is

$$3' \xleftarrow{\text{—T—A—G—dC—}} 5'$$

Problem 16.2

Part of the nucleotide sequence in one chain of DNA follows. Write a representation of the complementary DNA chain.

$$5' \xrightarrow{\text{—dC—T—A—G—C—}} 3'$$

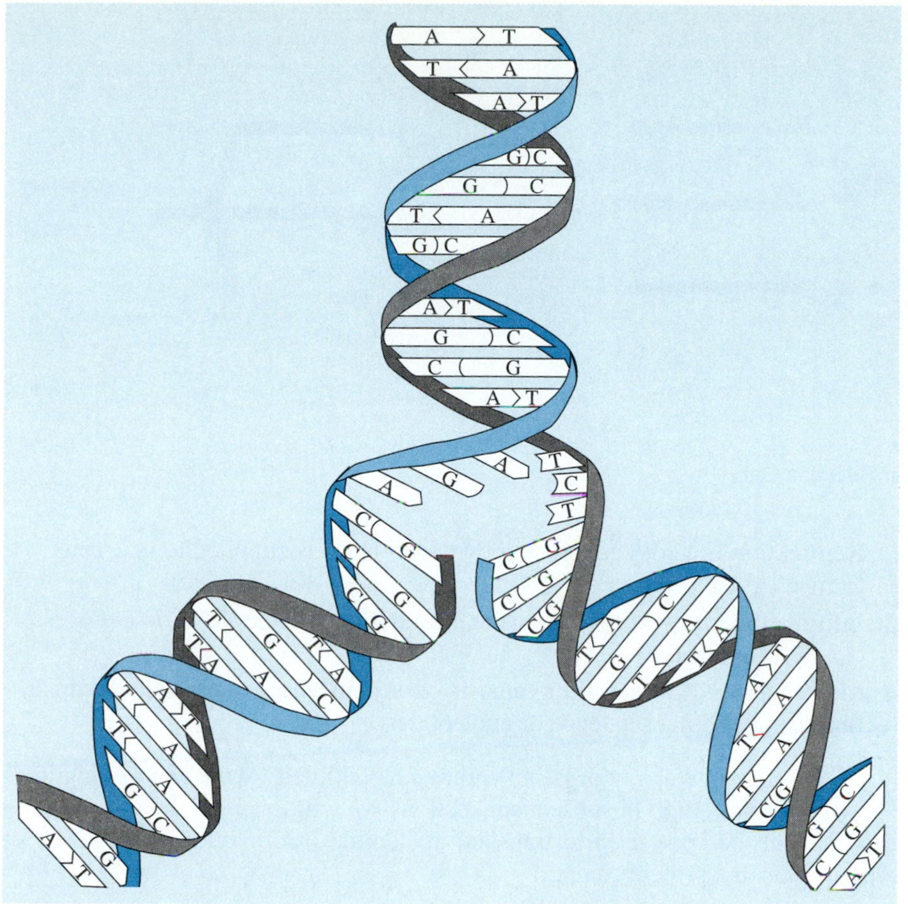

FIGURE 16.7
Model for Replication of DNA

The two strands have been separated at the replication fork by DNA helicase (not shown). The sequence of base pairs is shown using geometric representation of the bases.

16.6 The Flow of Genetic Information

Experiments have established that genes, which are segments of DNA, are responsible for control of the synthesis of proteins. In the following sections we will learn how a nucleotide sequence in DNA codes for formation of an amino acid sequence in a protein. In addition, we will see how DNA in the nucleus controls protein synthesis that occurs in the cytoplasm. The "message" is translated from one molecular language into another by an intermediate molecule, mRNA, that is formed in the nucleus and moves to the cytoplasm.

The flow of genetic information is commonly called the **central dogma.** We summarize the tenets of this dogma before considering details of protein synthesis.

1. DNA stores and transmits all hereditary information, including the instructions for the synthesis of all proteins.
2. DNA is replicated in the cell nucleus when cells are ready to divide.

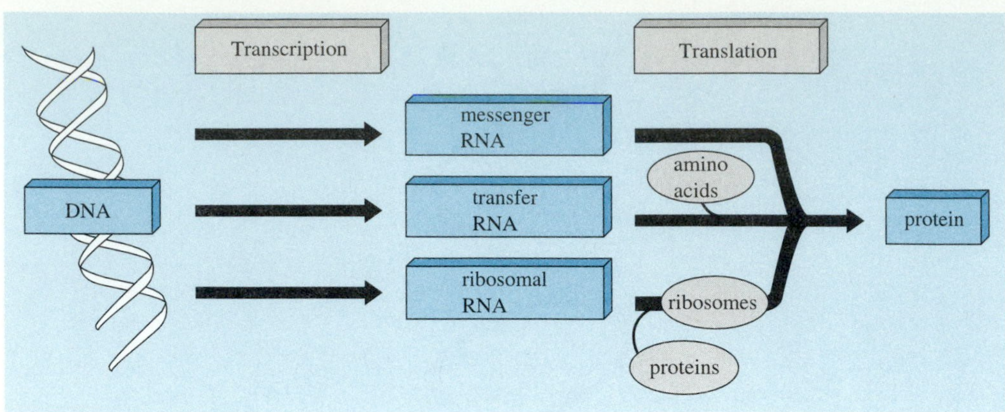

FIGURE 16.8
Information Flow in a Cell

Replication is a copying process by which DNA is supplied to new cells formed by cell division.

3. Information for protein synthesis is passed to messenger RNA in a process called **transcription.**
4. Messenger RNA directs the synthesis of protein by a process called **translation** using a triplet sequence of nucleotides.

The series of steps in protein synthesis as outlined by the central dogma is shown in Figure 16.8. In subsequent sections we will examine the transcription process and the code used to translate nucleotide information into a series of amino acids in a protein chain.

16.7 DNA Produces Several Types of RNA

Protein synthesis requires three types of RNA, all synthesized by DNA: messenger RNA (mRNA), transfer RNA (tRNA), and ribosomal RNA (rRNA). All three types of RNA are formed by a transcription process in which DNA is a template for base pairing. Transcription copies only certain selected genes into RNA. Moreover, the information contained in those genes resides on only one strand of the double helix. The code on one strand leads to the formation of a required protein, and it is very unlikely that the complementary code on the other strand could lead to another useful protein. The strand that contains the gene that is transcribed into mRNA is the **sense strand.**

Transcription is distinguished from replication in several ways. First, RNA polymerase rather than DNA polymerase is the catalyst. Second, the polymerase incorporates uracil rather than thymine in RNA. Finally, because only one strand of DNA is transcribed, the RNA molecules formed are single strands rather than double strands. This process is represented in Figure 16.9. A shorthand representation of transcription is as follows:

$$5' \xrightarrow[\text{—dA—T—C—G—}]{\text{DNA strand}} 3'$$

$$3' \xleftarrow[\text{RNA strand}]{\text{—U—A—G—C—}} 5'$$

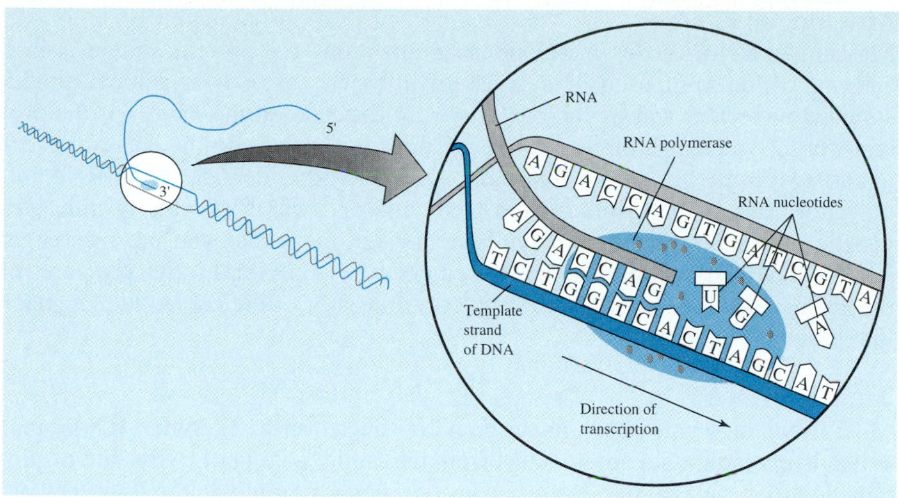

FIGURE 16.9
Transcription and Formation of mRNA

One strand of DNA serves as a template to form mRNA. Note that uracil is placed in mRNA by adenine, whereas the base pair of adenine in DNA is thymine. RNA polymerase as shown is much smaller than the actual molecule, which encompasses about 50 nucleotides at a time.

Steps in Transcription

RNA polymerase locates the proper position in DNA by recognizing a sequence of nucleotides called the **promoter region.** This region precedes that of the nucleotide sequence to be transcribed. RNA polymerase occupies a region of approximately 50 nucleotides, so that it is associated with the promoter region and about a dozen base triplets.

RNA polymerase forces open the DNA double helix and moves along the sense strand in the 3′ to 5′ direction. It then pairs the appropriate nucleotide from the nucleotide pool in the nucleus to its complement in the DNA sense strand. The nucleotides are joined in the growing RNA chain. After about 10 nucleotides are added, that portion is separated and "dangles" while the remaining RNA continues to be synthesized.

Eventually, transcription is completed, and the RNA chain is separated from DNA. This event as well as the release of the RNA polymerase occurs because the enzyme reaches a sequence of nucleotides beyond those required for transcription that triggers the chemical reactions required for release.

Ribosomal RNA

Ribosomal RNA combines with about 50 proteins to form complex structures called **ribosomes,** whose molecular weight is about 3 million. Ribosomes are the sites of protein synthesis. The ribosomes are like protein factories, but they cannot manufacture products without directions and workers. Each ribosome consists of two subunits. In a cell the smaller unit consists of one molecule of rRNA and about 30 proteins. The larger unit consists of three rRNA molecules and about 50 proteins. It is this larger unit that contains the enzymes for the synthesis of peptide chains from amino acids as well as the sites where synthesis occurs.

Messenger RNA

Messenger RNA carries instructions or directions for protein synthesis in a code or **codon** composed of a series of nitrogen bases. It is a long, single-stranded molecule, and its molecular weight depends on the length of the protein whose synthesis it directs. There are three nucleotides in the code for every amino acid in the protein to be formed. However, the identity of the RNA initially formed by transcription from DNA and that which eventually translates genetic information in protein synthesis are not the same. The pre-messenger RNA is modified in cells. The mRNA enters the cytoplasm through the pores of the nuclear envelope prepared to serve as a translator of the nucleotide language into protein sequences.

Transfer RNA

Most tRNA molecules have fewer than 100 nucleotides. **Transfer RNA** binds and delivers individual amino acids from the amino acid pool to the site of protein synthesis. A specific tRNA carries one type of amino acid. However, several different tRNA molecules can transport the same amino acid (Section 16.6).

Each tRNA has an amino acid attachment site and a template recognition site. Each 3′ end has a -C-C-A sequence that is the amino acid attachment site. The amino acid bonds to adenylic acid. The base sequence of the template recognition site is discussed in Section 16.9.

In 1965 R. Holley of Cornell University determined the nucleotide sequence of a tRNA for alanine. He suggested that the single-stranded molecule forms intrachain hydrogen bonds with some complementary bases. He postulated a "cloverleaf" model for the tRNA of alanine (Figure 16.10) having the maximum number of hydrogen bonds. The base pairings have been confirmed

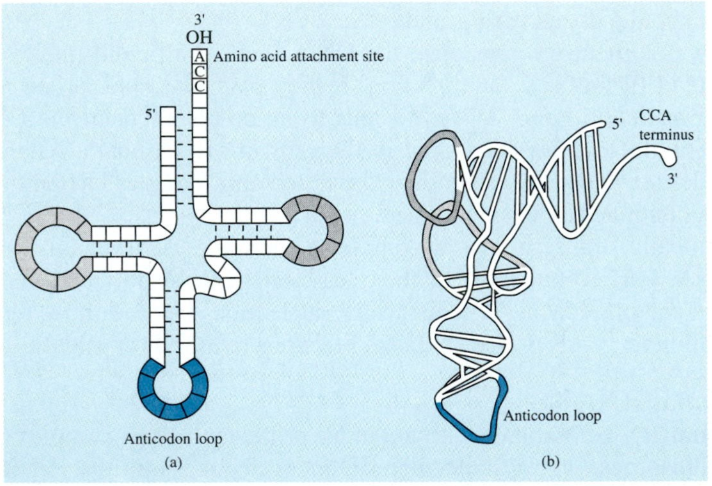

FIGURE 16.10
Structure of tRNA

(a) A two-dimensional simplified "cloverleaf" structure in which the squares represent nucleotides. Base pairing occurs between a number of bases to give the characteristic loops. (b) The three-dimensional shape has the anticodon loop at the bottom of the structure. The 3′ terminus of every tRNA contains the sequence CCA.

experimentally, and the shape of the three-dimensional molecule has been established.

One outside bend of a loop of transfer RNA contains a three-nucleotide sequence called an **anticodon,** which is the complement of a codon located in mRNA. The interaction of the codon of mRNA and the anticodon of tRNA is described in Section 16.8.

EXAMPLE 16-3

What portion of an mRNA chain will be produced from the following portion of a DNA strand?

$$5' \xrightarrow{\text{-dG-C-A-T-}} 3'$$

Solution

Base pairing in RNA will result in C in RNA opposite G in DNA and G in RNA opposite C in DNA. The base T in DNA will pair with A in RNA, but A in DNA will pair with U in RNA.

$$3' \xleftarrow{\text{-C-G-U-A-}} 5'$$

Problem 16.3

What sequence of bases in DNA is required to produce the following section of a messenger RNA chain?

$$5' \xrightarrow{\text{-G-C-A-U-}} 3'$$

16.8 The Genetic Code

We have been referring to DNA as a storehouse of information that contains a code that can be translated into information for protein synthesis. Both replication and transcription have been described only in terms of a series of nucleotides without regard for any relationship between neighboring nucleotides. However, the nucleotide sequence itself is a code that contains information much like the simple series of dots and dashes of the Morse code represent words. All genetic information is coded with only four bases.

George Gamow, of the University of Colorado, suggested that the four bases in mRNA are read as three-letter code words or codons. With three-letter words there are $4 \times 4 \times 4 = 64$ possible codons (Table 16.3). Codons are the three-base sequences present only in mRNA. Transfer RNA, which associates with mRNA in the ribosomes in the translation process, contains a complementary base sequence that is an anticodon. In 1962 the relationship between codons in mRNA and amino acids was determined. *Escherichia coli* ribosomes were bound to synthetic mRNA molecules containing only uridylic acid. The ribosomes associated only with the tRNA of phenylalanine and formed polyphenylalanine. Therefore, the base sequence U-U-U specifies phenylalanine.

The relationship between each codon and the amino acid incorporated into a protein has been established (Table 16.3). There is more than one codon for most amino acids, so the genetic code is redundant or degenerate. A group of codons referred to as **synonyms** can specify the same amino acid. The reason for the existence of synonymous codons is not clear, but there must be advan-

Table 16.3 Codons and Specified Amino Acids

First Base	U		C		A		G		Third Base
	Second Base								
U	UUU	Phe	UCU	Ser	UAU	Tyr	UGU	Cys	U
U	UUC	Phe	UCC	Ser	UAC	Tyr	UGC	Cys	C
U	UUA	Leu	UCA	Ser	UAA		UGA		A
U	UUG	Leu	UCG	Ser	UAG		UGG	Trp	G
C	CUU	Leu	CCU	Pro	CAU	His	CGU	Arg	U
C	CUC	Leu	CCC	Pro	CAC	His	CGC	Arg	C
C	CUA	Leu	CCA	Pro	CAA	Gln	CGA	Arg	A
C	CUG	Leu	CCG	Pro	CAG	Gln	CGG	Arg	G
A	AUU	Ile	ACU	Thr	AAU	Asn	AGU	Ser	U
A	AUC	Ile	ACC	Thr	AAC	Asn	AGC	Ser	C
A	AUA	Ile	ACA	Thr	AAA	Lys	AGA	Arg	A
A	AUG	Met	ACG	Thr	AAG	Lys	AGG	Arg	G
G	GUU	Val	GCU	Ala	GAU	Asp	GGU	Gly	U
G	GUC	Val	GCC	Ala	GAC	Asp	GGC	Gly	C
G	GUA	Val	GCA	Ala	GAA	Glu	GGA	Gly	A
G	GUG	Val	GCG	Ala	GAG	Glu	GGG	Gly	G

tages to the cell in having multiple ways of specifying its need for a particular amino acid.

Because the codons are automatically read three nucleotides at a time, there is no need for spacing between them as there is between words in a sentence or for punctuation marks such as commas. However, it is still necessary to know where the code starts and stops in the instructions to make a particular protein.

The codons UAA, UAG, and UGA are involved in termination of the protein chain and are like the period at the end of a sentence. This feature of the genetic code is discussed in the next section, as is the initiation process.

EXAMPLE 16-4

What are the codons for alanine? How are they related?

Solution

Table16.3 shows the codons GCU, GCC, GCA, and GCG for alanine. These four codons are related by identical letters in the first two positions. Only the third letter differs.

Problem 16.4

What are the codons for cysteine? How are they related?

16.9 Protein Synthesis and Translation

Amino acids attached to transfer RNA (tRNA) are brought to ribosomal RNA (rRNA), the site of protein synthesis. The translation of the codons of mRNA into a sequence of amino acids requires the following substances:

1. Amino acids
2. ATP to activate the amino acids
3. Transfer RNA
4. Messenger RNA
5. Ribosomal RNA
6. Guanosine triphosphate (GTP).

This body builder depends on very complex biological processes that form the proteins of muscles.

Amino Acids Are Activated for Reaction

To prepare an amino acid for protein synthesis, ATP reacts with the carbonyl carbon atom to yield an activated acyl derivative in which adenosine monophosphate (AMP) is bonded to the amino acid. The product is called an aminoacyl adenosine monophosphate or an aminoacyl adenylate (Figure 16.11). The reaction is catalyzed by enzymes known as aminoacyl synthetases, and a specific enzyme is required to activate each amino acid.

Esterification of tRNA

Aminoacyl adenosine monophosphates of amino acids react with a molecule of tRNA. Although each tRNA has a unique structure, they all terminate in the sequence CCA. The carbonyl carbon atom of the amino acid forms an ester with a hydroxyl group of the terminal adenylic acid.

After formation of the proper aminoacyl RNA molecule, the anticodon of tRNA associates with the codon of mRNA. Thus the correct amino acid is brought to the codon site of mRNA, where base pairing occurs.

Site of Protein Synthesis

The two ribosome units that provide the site for protein synthesis are not bound together until they are assembled at mRNA. Messenger RNA and the

Amino acid + Adenosine triphosphate (ATP) → (Anhydride bond) + PP_i

+ tRNA → + Adenosine monophosphate (AMP)

FIGURE 16.11
Formation of Activated Amino Acid and Activated tRNA

Chemistry and Medicine

The immune system consists of two parts—the innate immune system and the adaptive immune system. Each plays a role in combating pathogens, which is a general term for any invader that the body does not recognize as "self." The innate immune system comprises those responses that are neither specific nor inducible. That is, the response is not directed against any specific pathogen, nor is it based on any previous encounter with the pathogen. The adaptive immune system has a "memory" of previous invasions, and it has stored long-lived cells that can be reactivated to fight that pathogen specifically when it enters at a future date. This system provides antibodies that respond to specific foreign molecules called antigens. Antigens may be glycoproteins on the membrane surface of specific bacteria or the protein coat of a virus. Further details of this system are beyond the scope of this text.

Unfortunately, our immune system sometimes needs a little help and, in the cases of some viral infections, a lot of help. The development of drugs to fight both bacterial and viral infections is the province of the chemist. The drugs ideally must be specific in their action and have minimal side effects on the chemistry of the human body. Selectivity is then the issue. A drug must kill a pathogen in the presence of other vital cells.

Bacterial infections are the easiest to deal with using drugs. There are substantial differences between bacterial and mammalian cells. Drugs are sought based on this difference. For example, some bacteria require *p*-aminobenzoic acid to synthesize the vitamin folic acid, which functions as a coenzyme. Sulfa drugs, one of the earliest of antibacterial drugs, are substituted sulfanilamides. These compounds structurally resemble *p*-aminobenzoic acid.

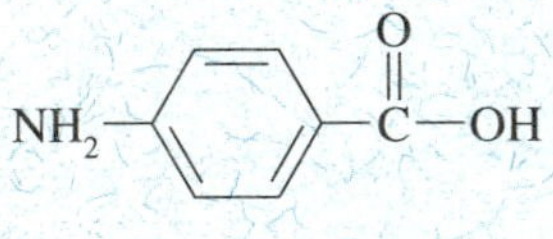

p-aminobenzoic acid

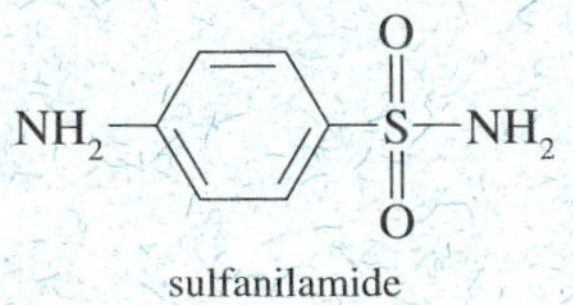

sulfanilamide

The bacteria mistake sulfanilamide for *p*-aminobenzoic acid and incorporate the wrong molecule into a compound that resembles folic acid but has no activity as a coenzyme. As a consequence, the bacteria die. The sulfa drugs are selective because humans do not synthesize folic acid. We rely on food such as green vegetables to obtain our supply. The bacteria cannot do this.

Other drugs, such as streptomycin, erythromycin, and puromycin, are structurally modified nucleosides that inhibit protein synthesis at the ribosome level. Again, there are differences in how bacteria and humans synthesize proteins. These drugs are effective because they inhibit bacterial protein synthesis without adversely affecting our own protein synthesis. Streptomycin interferes with the binding of formylmethionine-tRNA and prevents the initiation of protein synthesis. Erythromycin inhibits translocation. Puromycin causes premature chain termination.

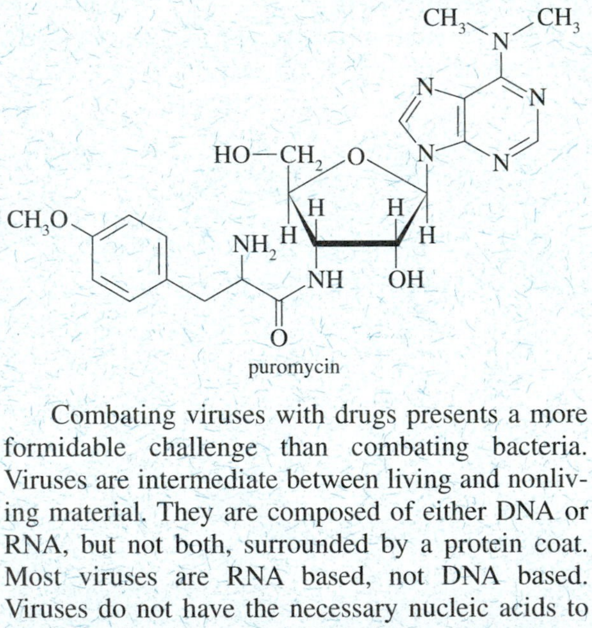

puromycin

Combating viruses with drugs presents a more formidable challenge than combating bacteria. Viruses are intermediate between living and nonliving material. They are composed of either DNA or RNA, but not both, surrounded by a protein coat. Most viruses are RNA based, not DNA based. Viruses do not have the necessary nucleic acids to replicate themselves or the amino acids and enzymes necessary to support life. Because they must have a host cell in order to reproduce, viruses are not considered "living."

Viruses are very small (0.05 to 0.2 μm) and come in a variety of shapes. Both the shape of the virus and its protein coat affect how it functions in infecting cells. Viruses attack cells and take over the chemical machinery of the cell. The protein coat contains an enzyme that breaks down the cell membrane of the host cell. The viral DNA or RNA then "squirts" into the cell. Each virus has a unique membrane-dissolving enzyme, so that it attacks only selected types of cells. Thus there are viruses that will attack only blood cells, only heart muscle, or only the spinal cord.

Once inside the cell, the viral nucleic acids take over the operation of the cell, which then stops making its own DNA, RNA, and protein. The cell starts to replicate the nucleic acids of the virus as well as the viral protein. Many copies of the virus are made by a single cell, which is eventually destroyed because it no longer produces the materials necessary for its own survival. The new viruses are released and attack other cells. If the process is not stopped, the organism gets sick and may die.

A virus can infect only a limited number of types of cells. Apparently, it recognizes certain receptor sites on the surface of the cell. Unfortunately for that cell, the lock and key arrangement between it and a cell leads to the cell's demise. Some viruses are less specific and can infect a range of species. For example, swine flu virus can infect humans as well as hogs. The rabies virus infects a range of mammals, including squirrels, raccoons, dogs, and humans.

Those few viruses that are DNA based take over cells by simply using the nucleotides and enzymes available. However, if the virus is RNA based, the process is a bit more complicated. Remember that RNA is produced in cells under the direction of DNA. A host cell has no directions or enzymes for making the RNA of the virus. Thus, the virus also has to have some conspirator molecules to use the cell machinery. There are two ways in which an RNA virus can replicate itself. One process involves a protein enzyme called RNA replicase, which catalyzes the manufacture of RNA from directions encoded on the viral RNA. Thus, some RNA molecules can reproduce themselves within a cell. Some viruses carry the necessary RNA transcriptase, whereas others use the enzyme in the host cell. Both the polio virus and the rabies virus operate by using RNA transcriptase.

The second mechanism occurs in viruses whose RNA directs the synthesis of a protein enzyme called DNA polymerase or reverse transcriptase. As a consequence, RNA is responsible for the synthesis of DNA, which then forms more RNA. Viruses that operate by this mechanism are called **retroviruses.**

Many viral infections can be prevented by immunization with a vaccine. The vaccine contains an active or inactive viral form that stimulates the body's natural immune system to produce antibodies against the virus. These antibodies deactivate invading active viruses before they can enter host cells. Smallpox has been virtually eradicated by a worldwide program of vaccination. Similarly, the incidence of measles and polio has been greatly reduced in the United States and many other countries by use of vaccines.

There are some virucidal agents, but their effectiveness is limited. They must stop the reproduction of viral nucleic acids inside host cells but not prevent the normal replication of the host's DNA. Any of the steps, including attachment to the membrane of the cell, replication of the DNA, or ejection from the cell, are targets that are considered for each specific virus. For example, in 1996 the protein CD4 on which HIV binds when it infects a cell was identified. Thus, molecules that could bind with this site might prevent infection by preventing viral entry.

The majority of research on virucidal agents has centered on synthetic nucleosides that could trick the virus into incorporating them rather than a structurally similar nucleoside. The goal is to make the virus produce a DNA that is useless and hence serves as a roadblock or deadend. Acyclovir (ACV) is effective in treating herpes simplex type 2, a virus that causes genital infections. It resembles deoxyguanosine but lacks two carbon atoms of the deoxyribose ring.

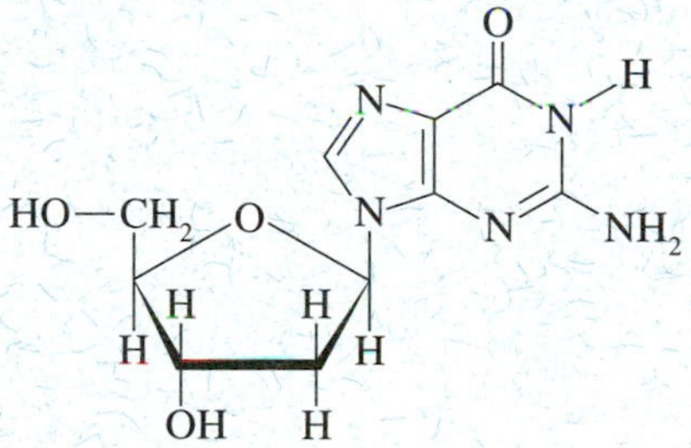

deoxyguanosine

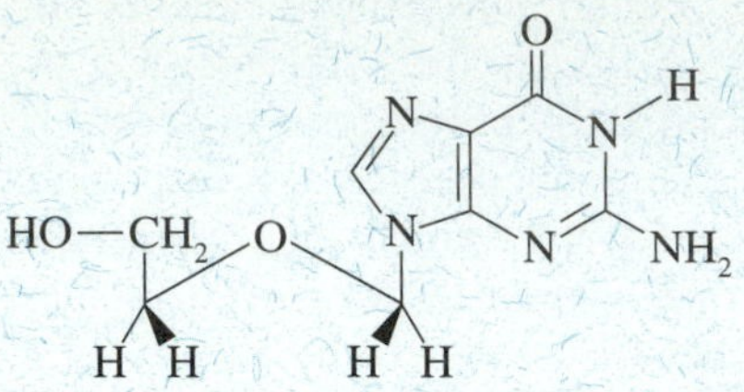

acyclovir

The drug is converted into its triphosphate ester and is taken up by an enzyme in DNA synthesis by the herpes virus. The triphosphate is formed at the position that corresponds to the 5′ position of deoxyguanosine. However, the enzyme-ACV complex does not have a 3′ group and replication cannot occur. The enzyme is tied up, and as a consequence the virus cannot be reproduced. The selectivity of ACV is based on the difference in how uninfected cells behave. The formation of the triphosphate is not catalyzed, and as a result ACV is not taken up by the enzyme in preparation for DNA synthesis.

AIDS, which is the result of HIV infection, is a virally transmitted disease that has reached epidemic proportions in some areas of Africa. The virus attacks helper T cells that are part of the immune system. Several drugs are available to treat HIV-infected individuals. These include AZT, DDI, and DDC, all of which are nucleosides. AZT resembles deoxythymidine, DDC resembles deoxycytidine, and DDI resembles deoxyguanosine.

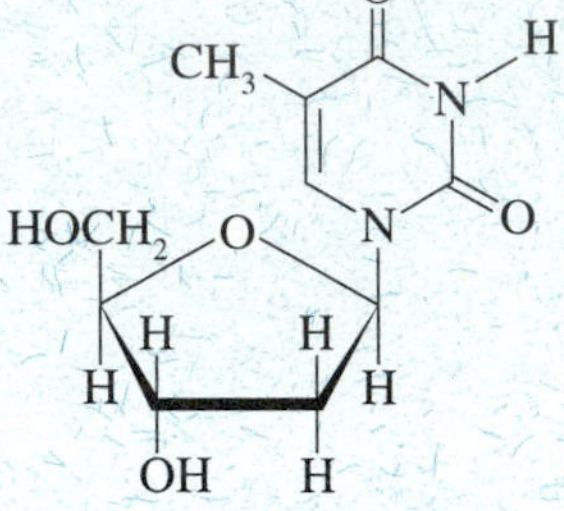

deoxythymidine

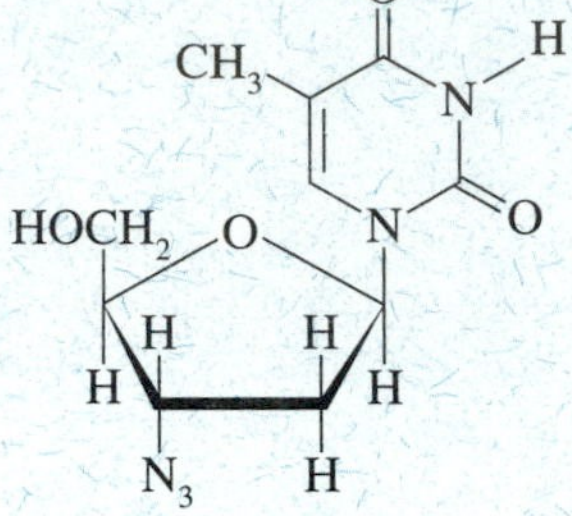

AZT

AZT and the other nucleosides interfere with reverse transcriptase, the enzyme that HIV requires to convert its RNA into DNA. The enzyme incorporates the drug into the growing DNA chain. However, the nucleoside lacks a hydroxyl group at the 3′ position. As a result, the growth process stops. Unfortunately the drug is not selective and it also affects bone marrow cells responsible for forming blood cells. In addition, AZT does not cure HIV, and there is evidence that the virus is becoming drug resistant. Obviously, the development of a "cure" for AIDS is a formidable task and, unlike the cures found for other diseases, will require more than a simple pill. ■

two ribosome units associate starting at the 5′ end, and the assembled ribosome then traverses the mRNA chain to the 3′ end. The larger ribosome unit contains the site of attachment of tRNA with its associated amino acid. However, the complete ribosome is required for the specific coupling of tRNA anticodons with the mRNA codons. After one ribosome starts its journey, other ribosomes start theirs. Thus, several ribosomes simultaneously translate the same mRNA molecule. Each ribosome is at a different position along the mRNA chain, and several proteins are in various stages of growth (Figure 16.12).

Protein synthesis consists of several coordinated enzyme-catalyzed steps. The processes occur at two sites in the larger ribosome units called the **donor (P) site** and the **acceptor (A) site.** A growing protein chain, which is bonded to the tRNA of the C-terminal amino acid, is located at the donor site, where it is associated with its appropriate codon by hydrogen bonds. An amino acid bonded to its tRNA via its carboxyl group enters the acceptor site and associates with the appropriate codon by hydrogen bonds. Then the growing protein

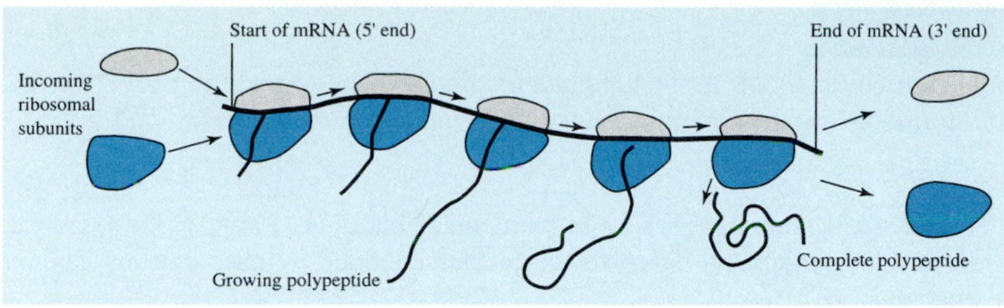

FIGURE 16.12
Ribosomes and mRNA in Protein Formation
The assembled ribosomes move from the 5′ to the 3′ end of mRNA and function independently as they synthesize a series of protein chains.

chain is transferred by an enzyme-catalyzed reaction from the donor site to the amino acid “waiting” at the acceptor site. Thus, the protein chain is synthesized in order from the N-terminal amino acid toward the C-terminal amino acid. Because the growing peptide chain is now attached to the tRNA in the acceptor site, the free tRNA leaves the donor site. The ribosome then moves to the next trinucleotide sequence by a process called **translocation,** which is catalyzed by the enzyme translocase with the expenditure of energy provided by guanosine triphosphate. As a result of the change in the environment of the ribosome, the tRNA with its attached protein chain is now located in a donor site. Another aminoacyl tRNA then enters the acceptor site, and the entire process is repeated until the protein is complete.

Initiation and Termination

Protein synthesis requires both initiation and termination steps. **Initiation** is the start of the growth of the protein chain. **Termination** is the conclusion of the synthesis of the protein chain and its release from the ribosome. The process of initiation is complex. However, for a number of mRNA molecules, the first amino acid in the chain is a modified methionine called formylmethionine (fMet), and it is specified by the codon AUG. Thus initially the growing peptide chain is merely the terminal fMet occupying a donor site. Termination of the protein synthesis is indicated by the termination codons UAA, UGA, or UAG. There are no tRNA molecules with the anticodon to pair with these trinucleotide sequences. The codons instead bind some proteins called release factors. Binding of these factors activates the enzyme needed to hydrolyze the peptide from the tRNA and the donor site. The peptide chain is then released from the ribosome.

Protein Modification

Most proteins formed by translation of mRNA are not the actual proteins required by the cell and are modified after release from the ribosome. One such modification is the hydrolysis of the formyl group of formylmethionine or of methionine itself. In some cases the protein chain is cleaved, as in the conversion of proinsulin into insulin. Other modifications include joining the protein to another group, such as a lipid or a carbohydrate.

EXAMPLE 16-5

What anticodon exists in the template recognition site of an aminoacyl tRNA to bind at the site having the codon sequence AGU in mRNA?

Solution

The anticodon must consist of complementary bases (U for A, C for G, A for U) in order to bind aminoacyl tRNA to the codon sequence by base pairing. The base sequence must be UCA.

Problem 16.5

What portion of an mRNA will associate with an aminoacyl tRNA having the sequence CAU in the template recognition site?

EXAMPLE 16-6

What peptide is synthesized from the following base sequence in mRNA?

-G-C-U-G-A-A-U-G-G-

Solution

Reading the letters as sequences of three-letter words, we have the codons GCU, GAA, and UGG. These codons specify alanine, glutamic acid, and tryptophan, respectively. The peptide will be Ala-Glu-Trp.

Problem 16.6

What possible codon sequences lead to the synthesis of the amino acid sequence Phe-Ile-Gly in a protein?

Explorations with Molecular Models

1. Prepare a ball-and-stick model corresponding to the following space-filling model of a carbohydrate that is a component of some antiviral drugs. Identify the carbohydrate. What is the configuration at the anomeric center?

2. Prepare a ball-and-stick model corresponding to the following space-filling model of a nucleoside. Identify the component carbohydrate and nitrogen base.

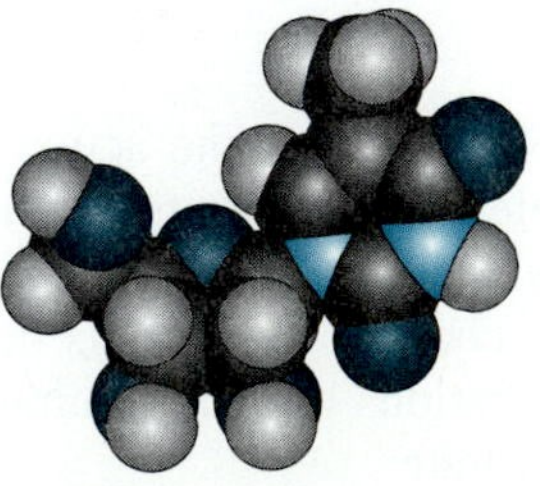

Exercises

DNA and RNA

16.1 How are all DNA molecules structurally alike?

16.2 How do DNA molecules differ?

16.3 What group bridges the sugars in DNA and RNA?

16.4 DNA and RNA are acidic materials. Explain why.

16.5 Indicate whether each of the following is a purine or pyrimidine base.

(a) uracil (b) cytosine (c) adenine (d) thymine

16.6 Which bases are found in DNA? Which bases are found in RNA?

Nucleosides and Nucleotides

16.7 Write the structural formula for adenosine monophosphate.

16.8 Write a structural formula for a nucleoside containing thymine and deoxyribose.

16.9 What is the configuration of the *N*-glycosidic bond of nucleotides?

16.10 What hydroxyl group of ribonucleosides is phosphorylated to give ribonucleotides?

16.11 How many free hydroxyl groups exist in a polynucleotide of ribose? How many stereogenic centers are there in a ribonucleoside?

16.12 How many free hydroxyl groups exist in a polynucleotide of deoxyribose? How many stereogenic centers are there in a deoxyribonucleoside?

16.13 Identify the components of the following substance.

O
N H
O N
HO—P—O—CH₂ O N N NH₂
OH H H
H H
OH H

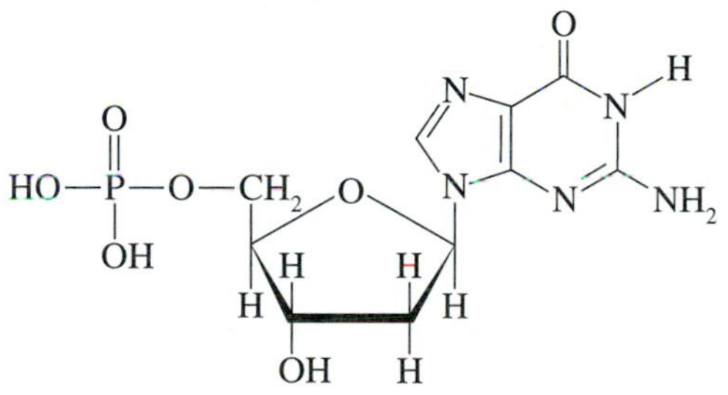

16.14 Identify the components of the following substance.

O
H
N
HOCH₂ O N O
H H
H H
OH OH

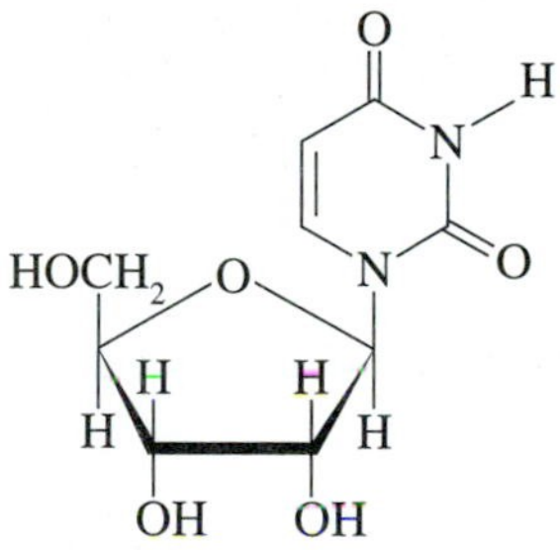

The Double Helix

16.15 Do any of the following pairs occur in the DNA molecule? If not, explain why.

(a) G-A (b) G-T (c) C-U

16.16 Do any of the following pairs occur in the DNA molecule? If not, explain why.

(a) A-A (b) C-T (c) A-C

16.17 What forms the backbone structure of each strand of the DNA double helix?

16.18 Where are the base pairs located in the double helix?

Replication

16.19 What is the complementary strand for a DNA strand with the sequence dG-C-A-T-C-A-G?

16.20 What is the complementary strand for a DNA strand with the sequence dA-T-C-A-G-T-A?

Transcription

16.21 What mRNA would be formed from the base sequence in Exercise 16.19?

16.22 What mRNA would be formed from the base sequence in Exercise 16.20?

16.23 What is the complementary base in mRNA for A in DNA?

16.24 What base in mRNA is the complement of T in DNA?

16.25 A portion of DNA has the sequence C-C-C-T-G-T-A-C-A-C-C-T. What base sequence will form in mRNA? What peptide will form?

16.26 A portion of DNA has the sequence G-G-G-T-G-C-A-G-A-C-C-A. What base sequence will form in mRNA? What peptide will form?

Translation

16.27 How is a codon involved in the translation process?

16.28 How is an anticodon involved in the translation process?

16.29 What amino acid does each of the following codons in mRNA place in a protein chain?

(a) GUU (b) CCC (c) UUU (d) ACG

16.30 What amino acid does each of the following codons in mRNA place in a protein chain?

(a) UGU (b) GCA (c) AGG (d) CUA

16.31 How are acceptor and donor sites in the ribosome used in protein synthesis?

16.32 What amino acid initiates protein synthesis?

16.33 What anticodons in tRNA will base-pair with the codons of Exercise 16.29?

16.34 What anticodons in tRNA will base-pair with the codons of Exercise 16.30?

16.35 What base sequence in mRNA is necessary to form Val-Asp-Ala-Gly?

16.36 What base sequence in mRNA is necessary to form Ser-Glu-Pro-Phe?

CHAPTER 17

SYNTHETIC POLYMERS

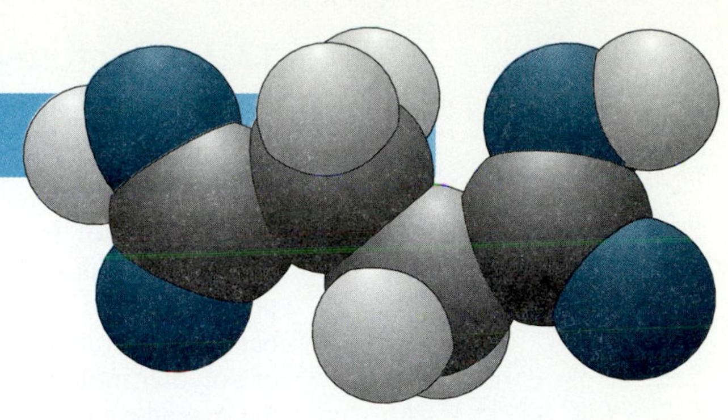

Overview

17.1 Natural and Synthetic Macromolecules

In most of our study of organic chemistry, we have focused on the chemical reactions, structure, and physical properties of "small" molecules. We examined very large molecules, or **macromolecules,** in only three chapters. Repeated condensation reactions of small molecules called monomers form polysaccharides, proteins, and polynucleotides, which are the three classes of macromolecules found in living organisms.

Some naturally occurring macromolecules are important commercial products. For example, wood and cotton are carbohydrates; wool and silk are proteins. But synthetic polymers far outstrip natural polymers in commercial importance. The chemical industry has developed many synthetic macromolecules with diverse properties and a wide variety of uses. These synthetic macromolecules are indispensable in a modern society. They include the rubber of tires, PVC of pipes and floor tile, and synthetic fibers such as nylon. The annual production of raw polymers by the U.S. chemical industry totals more than 30 million tons.

Synthetic polymers have a wide range of properties. For example, certain transparent polymers can be molded into precise shapes in the manufacture of corrective lenses. The polymer rubber used in tires must be flexible enough to be distorted from one shape to another. Synthetic fibers used for clothing have to feel good against the body and be able to hold a dye.

The physical properties of synthetic macromolecules result from the number and kind of monomer units, as well as resulting intermolecular interactions and intramolecular interactions, such as London forces, dipole-dipole forces, and hydrogen bonding. We recall that the properties of proteins and other natural macromolecules result from intermolecular interactions. For example, the strength of structural proteins such as collagen and cellulose is due to the many intermolecular hydrogen bonds. Hydrogen bonding can also be an important feature to consider in designing synthetic macromolecules.

17.2 Structure and Properties of Polymers

By synthesizing appropriate monomers and learning how to polymerize them, organic chemists can prepare macromolecules to meet required specifications. The number of monomer units—a few hundred to several thousand—in a polymer affects its physical properties. However, no synthetic polymerization process can be stopped precisely after a specific number of monomers have been incorporated into the polymer. All polymerization reactions give mixtures of polymer molecules with a range of molecular weights. Therefore, we refer to the average molecular weight of a polymer. The average molecular weight of synthetic polymers is in the 10^5 to 10^6 range.

The types of monomers incorporated in a polymer strongly influence the flexibility and shape of the polymer. For example, polymers whose monomers contain aromatic rings are less flexible than those whose monomers are acyclic. In some polymers, the chains are cross-linked by covalent bonds. Cross-linking creates larger macromolecules with more rigid structures. Cross-links are also important in naturally occurring polymers. For example, the proteins in wool fibers are cross-linked by many disulfide bonds.

London Forces in Polymers

Plastic bags of polyethylene have sufficient strength to store heavy items.

As in the case of proteins, the intermolecular and intramolecular interactions of polymer chains are extremely important in determining physical properties. London forces are largely responsible for the folded or coiled conformations of a polymer. Intermolecular London forces between individual chains help to hold them together. Both intramolecular and intermolecular London forces increase with the polarizability of functional groups in the polymer.

Only London forces affect the properties of polyethylene. This molecule, which can be produced as a linear polymer, resembles a giant alkane (Figure 17.1). The attractive forces between pairs of hydrogen atoms on adjacent chains of a polymer are very small. However, thousands of those interactions exist in a polymer, producing a large total interaction energy.

The linear polymer of ethylene is called high density polyethylene (HDPE). It has a high density and is high melting (135°C) because parts of the molecules "line up" in a tightly packed, orderly array. HDPE is used to make materials as simple as bottle caps for milk containers and as complex as cabinets for televisions and computers.

Ethylene can also be polymerized with branches off the main chain. The resulting polymer is called low density polyethylene (LDPE). LDPE is less dense than HDPE because the branches prevent the main chains from packing closely. The more open structure not only is less dense but also has smaller London forces. Because the intermolecular forces between chains are smaller, LDPE has a lower melting point (120°C) and is a more flexible material. It is used to make plastic bags and flexible bottles for consumer products such as soft drinks and bleach. Containers made of LDPE are also used for windshield wiper fluid, antifreeze, and engine oil.

Structural units such as aromatic rings have polarizable electrons that create strong London forces. Hence, polymers with aromatic rings have higher tensile strength than polymers with acyclic units, because there are strong London forces between aromatic rings in neighboring polymer chains. Furthermore, the

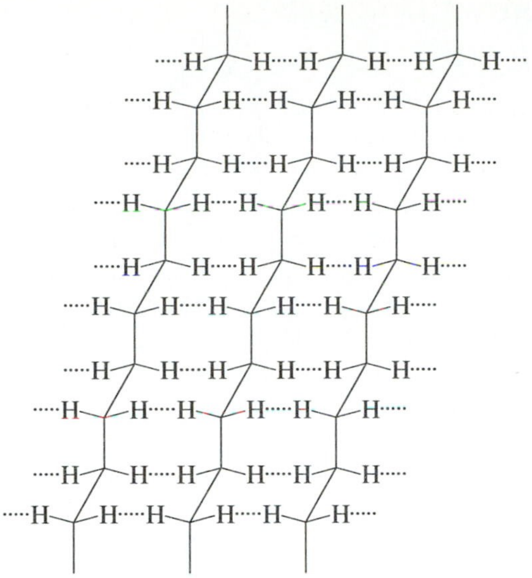

FIGURE 17.1

London Forces in Polyethylene

The dotted lines represent London attractive interactions between the chains of the polymers, which have a surface of surrounding hydrogen atoms. Additional chains (not shown) may be above and below the represented plane.

aromatic rings reduce the flexibility of the polymer chain and the number of possible conformations. Chains of sp^3-hybridized carbon atoms in polymers such as polyethylene are more flexible. They can exist in gauche and anti conformations around each carbon-carbon bond. The allowed motions of the chains affect the properties of the polymer.

Hydrogen Bonding in Polymers

We noted that intermolecular hydrogen bonding dramatically affects the properties of naturally occurring macromolecules. Some synthetic polymers also have extensive hydrogen bonding between polymer chains. The substantial strength of polyamides such as nylon 66 and Kevlar is due to hydrogen bonding. The amount of hydrogen bonding in nylon 66 is affected by the flexibility of the chain and the conformation around the carbon-carbon bonds. The maximum number of hydrogen bonds is formed only if nylon 66 is in the all-anti conformation (Figure 17.2). Kevlar, used in bulletproof vests, is extensively hydrogen-bonded because the aromatic ring and the amide bond restrict the conformation to the one best suited to form the maximum number of hydrogen bonds.

EXAMPLE 17-1

How would the properties of an addition polymer formed from 3-methyl-1-pentene differ from those of a polymer formed from propene?

Solution

The polymer of 3-methyl-1-pentene has large branched *sec*-butyl groups off the main chain, whereas propene has relatively small methyl groups as branches.

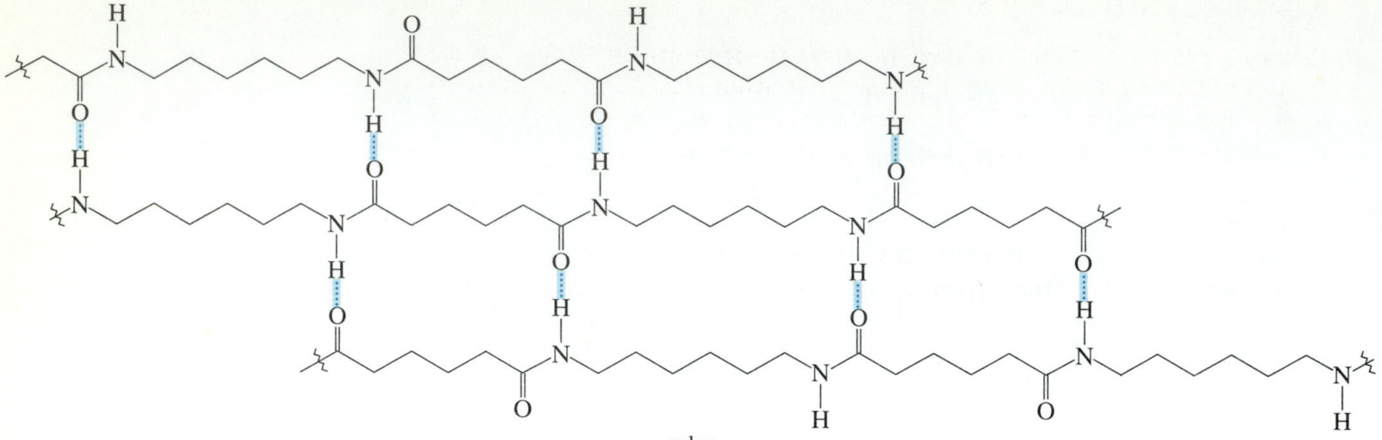

nylon

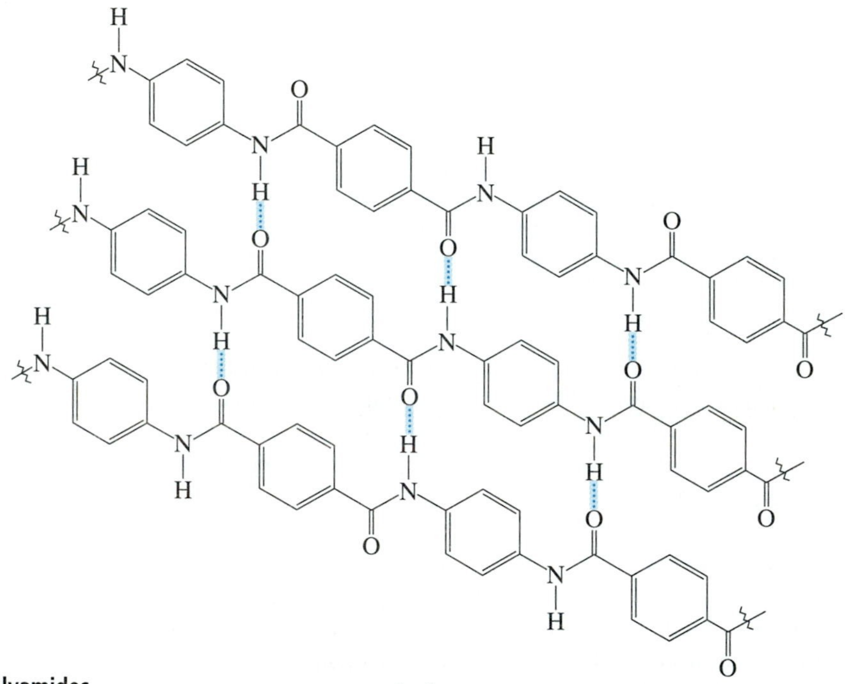

kevlar

FIGURE 17.2
Hydrogen Bonds in Polyamides

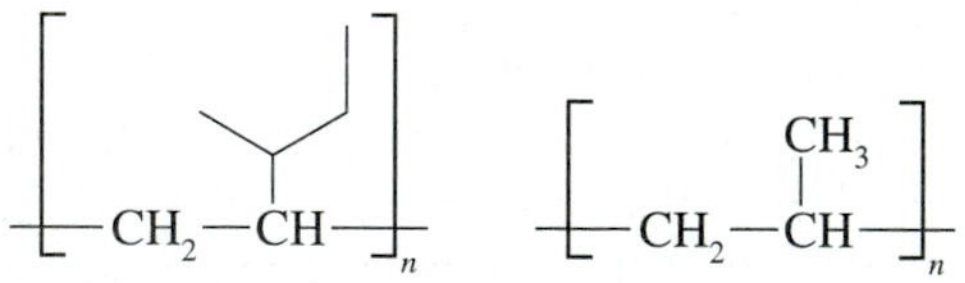

$$\left[-CH_2-CH(CH(CH_3)CH_2CH_3)-\right]_n \qquad \left[-CH_2-CH(CH_3)-\right]_n$$

As a result, the polymer of 3-methyl-1-pentene has a more open structure. The bulkier chains cannot pack closely, and the intermolecular forces between chains are smaller. The polymer of 3-methyl-1-pentene should have a lower melting point and be a more flexible material.

Problem 17.1

There are three isomeric benzenedicarboxylic acids. Each reacts with ethylene glycol to produce a polyester. Which one should produce the polyester that packs most efficiently and hence has the largest intermolecular attractive forces?

17.3 Classification of Polymers

Polymers can be classified by their macroscopic physical properties, by the method of polymerization, or by a number of structural features, such as cross-linking and stereochemistry. The three major classes based on physical properties are elastomers, plastics, and fibers.

Elastomers

Elastomers are elastic materials that regain their original shape if they are distorted. Some common elastomers are rubber, a naturally occurring polymer of isoprene, and neoprene, a synthetic polymer of 2-chloro-1,3-butadiene. These elastomers contain carbon-carbon double bonds separated by intervening units containing two sp^3-hybridized carbon atoms.

repeating unit

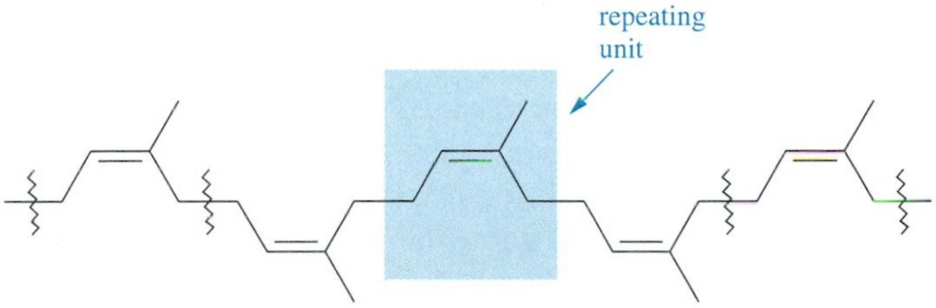

polyisoprene

An elastomer's properties depend on both the groups bonded to the sp^3-hybridized carbon atoms and the geometry of the polymer chain around the double bond. Elastomers are amorphous materials. The individual chains of the polymers are random coils that are tangled in an irregular way. The coils "straighten out" when they are stretched. When the force is released, the elastomer returns to its coiled state because the intermolecular forces are greatest in this arrangement.

The flexibility of an individual chain of an elastomer depends on the structure of the intervening unit of sp^3-hybridized carbon atoms. Some rotation can occur around the σ bonds. The double bonds provide rigidity because they restrict rotation around π bonds. Polymer chains of both *E* and *Z* configurations are known. For example, natural rubber is polyisoprene with a *Z* configuration, whereas gutta-percha, an industrial polymer of isoprene, has an *E* configuration. This stereochemical difference is reflected in the properties of the two polymers. Rubber is an elastomer, but gutta-percha is less flexible. In gutta-percha the trans double bonds give a zigzag arrangement resembling that of saturated fatty acids (Section 13.3). We recall that the regularity of the zigzag chains of fatty acids allows adjacent chains to nestle together, resulting in large London forces. Similar large forces exist in gutta-percha. Natural rubber has a "bent" chain similar to that of an unsaturated fatty acid. This arrangement of atoms gives a more open, less regular relationship among polymer chains. As a result, the polymer chains are not closely packed, and they can slide past one another as the elastomer is distorted.

Plastics

In polymer chemistry, the term *plastic* is used for those polymers that harden upon cooling and can be molded or extruded into shapes that remain after cooling. **Thermoplastics** are polymers that reversibly soften when heated, becoming sufficiently fluid to be molded. **Thermosetting polymers** can be molded when they are first prepared. However, after being heated they "set," hardening irreversibly. If heated to a high temperature, thermosetting polymers decompose rather than melt.

The difference between thermoplastics and thermosetting polymers is related to cross-linking. The polymer chains of thermoplastics are not cross-linked. When a thermoplastic is heated, the kinetic energy of the polymer chains increases, overcoming the intermolecular forces and causing the polymer to melt. Polyethylene is a thermoplastic in which the London forces between hydrocarbon chains are the only intermolecular forces. Thermosetting polymers have extensive cross-links between polymer "chains" that result in much larger polymer molecules.

Fibers

Some thermoplastics are prepared as thin filaments that can be spun into fibers similar to natural fibers. The length of the polymer molecule must be at least 500 nm, which corresponds to a minimum average molecular weight of 10^4. The structure of the polymer chains must also provide sufficiently strong intermolecular forces to give the fiber an adequate tensile strength.

Filaments of thermoplastics are prepared by two methods. If the thermoplastic is stable in the molten state, it may be passed through tiny pores in a die called a spinneret and then cooled. For less stable thermoplastics the polymer is dissolved in a volatile solvent and forced through the spinneret. The solvent evaporates and a filament precipitates. Regardless of the method of formation, the fiber is then drawn out to several times its length after it has cooled. The **cold drawing** orients the molecules along the axis of the fiber. The resultant intermolecular forces between polymer molecules increase the tensile strength of the fiber.

EXAMPLE 17-2

Assign the polymer represented by the following structure to one of the three classes of polymers. Identify a compound whose physical properties it most closely resembles.

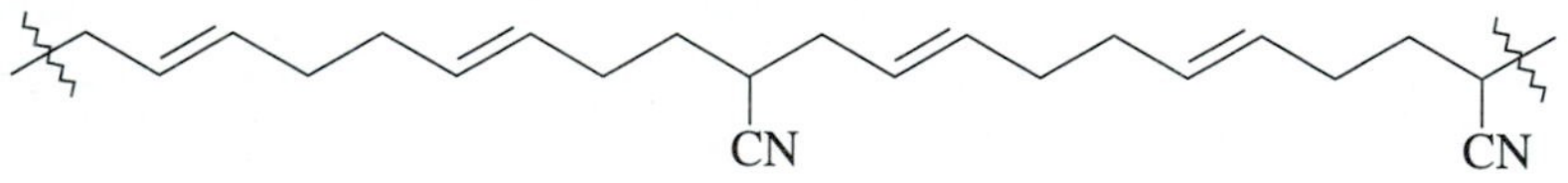

CN CN

Solution

The polymer is an elastomer. The sp^3-hybridized carbon atoms between double bonds provide some flexibility to the elastomer. However, the trans arrangement of the double bonds, which resembles that of gutta-percha, allows chains to pack efficiently and leads to less flexibility in the elastomer.

Problem 17.2

What type of plastic is best suited to make the handles for cooking utensils for the home? What type of plastic is most likely to be used for the frames of eyeglasses?

17.4 Methods of Polymerization

Polymers can be divided into two broad classes called addition polymers and condensation polymers. **Addition polymers** result from the successive addition reactions of one alkene or a mixture of alkenes by radical, cationic, or anionic mechanisms. **Condensation polymers** result from condensation reactions of monomers that contain two or more functional groups, such as an alcohol and a carboxylic acid or an amine and a carboxylic acid. These functional groups react in condensation reactions to eliminate a small molecule, such as water.

Addition polymers are also called **chain-growth polymers.** The polymer chain grows when the reactive intermediate formed in an initiation step adds to another monomer. The initiating species may be a radical, carbocation, or carbanion. For example, in radical polymerization a peroxide such as dibenzoyl peroxide yields a benzoyl radical.

$$C_6H_5C(=O)O{-}OC(=O)C_6H_5 \longrightarrow 2\ C_6H_5C(=O)O\cdot$$

This radical reacts with a monomer to give another radical, which then reacts with another unit of monomer.

$$C_6H_5C(=O)O\cdot + CH_2{=}CH_2 \longrightarrow C_6H_5C(=O)O{-}CH_2CH_2\cdot$$

$$C_6H_5C(=O)O{-}CH_2CH_2\cdot + CH_2{=}CH_2 \longrightarrow C_6H_5C(=O)O{-}CH_2CH_2{-}CH_2CH_2\cdot$$

The successive additions of monomers give a growing chain that always has a reactive end. The number of polymer chains formed therefore depends on the concentration of intermediates initially formed. The monomers cannot react until they encounter one of the growing chains with a reactive site.

Cationic polymerization involves carbocations rather than radicals. A Lewis acid such as BF_3, $Al(CH_2CH_3)_3$, $TiCl_4$, or $SnCl_4$ is used to react with the alkene to form a carbocation, which in turn reacts with another alkene molecule to form another cation. Consider the reaction with 2-methylpropene (isobutylene) as the monomer. The Lewis acid that acts as an electrophile is represented by E^+.

$$E^+ + H_2C{=}C(CH_3)_2 \longrightarrow E{-}CH_2{-}C^+(CH_3)_2$$

$$E{-}CH_2{-}C^+(CH_3)_2 + CH_2{=}C(CH_3)_2 \longrightarrow E{-}CH_2{-}C(CH_3)_2{-}CH_2{-}C^+(CH_3)_2$$

Note that addition occurs in the Markovnikov manner, with the electrophile adding to the less substituted carbon atom. Subsequent reactions continue, with the carbocation adding to the less substituted carbon atom. As a consequence, the more stable tertiary carbocation is formed each time. The structure of the polymer is represented as

$$\left[-CH_2-\overset{\overset{CH_3}{|}}{\underset{\underset{CH_3}{|}}{C}}- \right]_n$$

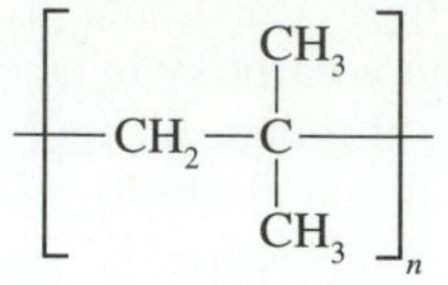

Low molecular weight polyisobutylene is used in lubricating oil and in adhesives for removable paper labels. Higher molecular weight polyisobutylene is used to produce inner tubes for bicycle and truck tires.

Anionic polymerization is initiated by a carbanion that behaves as a nucleophile. One example is the butyl anion, which is provided by butyllithium. The lithium compound has a very polar bond, and the carbon atom has a partial negative charge. In the following reactions, the butyl group is represented as Bu—. The monomer is acrylonitrile.

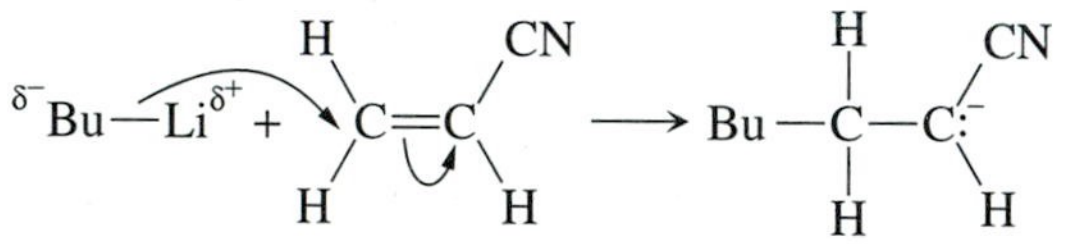

$$^{\delta-}Bu-Li^{\delta+} + (H)_2C=C(H)(CN) \longrightarrow Bu-\overset{\overset{H}{|}}{\underset{\underset{H}{|}}{C}}-\ddot{C}^{-}(H)(CN)$$

Addition occurs at the less substituted carbon atom because the resulting carbanion is resonance stabilized.

$$Bu-CH_2-\ddot{C}H^{-}-C\equiv N: \longleftrightarrow Bu-CH_2-CH=C=\ddot{N}:^{-}$$

Continued reaction of the nucleophilic carbanion "reactant" gives a carbanion "product," and the length of the polymer chain increases.

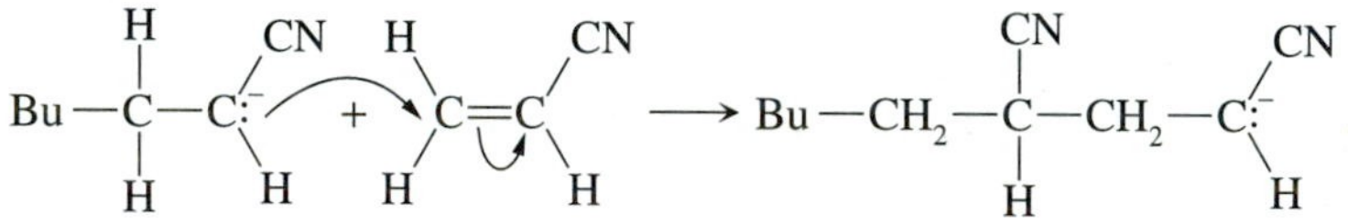

$$Bu-\overset{\overset{H}{|}}{\underset{\underset{H}{|}}{C}}-\ddot{C}^{-}(H)(CN) + (H)_2C=C(H)(CN) \longrightarrow Bu-CH_2-\overset{\overset{CN}{|}}{\underset{\underset{H}{|}}{C}}-CH_2-\ddot{C}^{-}(H)(CN)$$

Subsequent reactions continue to form another resonance-stabilized carbanion. The structure of the polymer is represented as

$$\left[-CH_2-\overset{\overset{CN}{|}}{\underset{\underset{H}{|}}{C}}- \right]_n$$

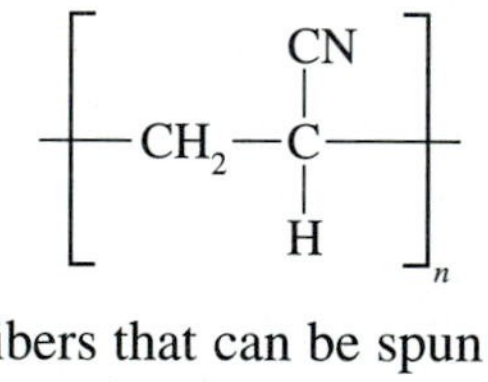

Polyacrylonitrile is used in fibers that can be spun to give the textiles Orlon® or Acrilan®. Some rugs are also produced using this polymer.

Condensation polymers are also called **step-growth polymers.** In reactions between two units, such as a diol and a diacid, an ester forms between one alcohol site and one acid site. This ester also has an alcohol site and an acid site at the ends of the molecule. Monomers can continue to react in condensation reactions with this product.

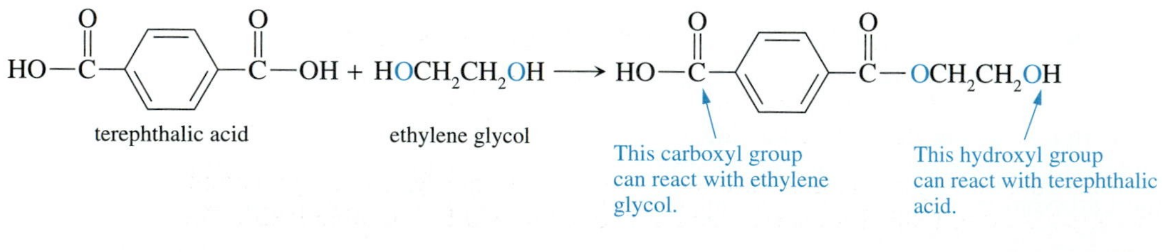

$$\underset{\text{terephthalic acid}}{HO-\overset{\overset{O}{||}}{C}-C_6H_4-\overset{\overset{O}{||}}{C}-OH} + \underset{\text{ethylene glycol}}{HOCH_2CH_2OH} \longrightarrow HO-\overset{\overset{O}{||}}{C}-C_6H_4-\overset{\overset{O}{||}}{C}-OCH_2CH_2OH$$

This carboxyl group can react with ethylene glycol.

This hydroxyl group can react with terephthalic acid.

However, subsequent condensation reactions are not restricted to the reaction of the monomers with the ends of the growing polymer chain. The monomers in the reaction mixture can continue to react with each other to start additional chains. So the monomers in a step-growth polymerization generate many low molecular weight oligomers rather than a smaller number of steadily growing, high molecular weight chains. Formation of true polymers occurs only after the monomer is used up. At this point large increases in the chain length result from the reaction of the ends of the oligomers with each other. Thus, in step-growth polymerization, the polymer is formed in "blocks" that result in a substantially higher molecular weight product.

17.5 Addition Polymerization

Addition polymerization occurs by a chain reaction in which one carbon-carbon double bond adds to another. Monomers continue to react with the end of the growing polymer chain in an addition polymerization reaction until the reactive intermediate is destroyed in a termination reaction.

Disproportionation and dimerization are two possible termination reactions. In disproportionation, a hydrogen atom at a carbon atom α to the radical center is abstracted by a radical in another chain. This produces a double bond in one polymer molecule, and the other polymer molecule becomes saturated. Because no new radical intermediates are formed, the next propagation step cannot occur.

$$2\,R{-}(CH_2CH_2)_n\underset{}{\overset{H}{\overset{|}{C}}}H{-}CH_2\cdot \longrightarrow R{-}(CH_2CH_2)_nCH{=}CH_2 + R{-}(CH_2CH_2)_n\overset{H}{\overset{|}{C}}H{-}CH_2{-}H$$

In the dimerization reaction, two radicals combine to form an even longer polymer chain. Again, the destruction of radicals prevents a continuation of propagation.

$$2\,R{-}(CH_2CH_2)_nCH{-}CH_2\cdot \longrightarrow R{-}(CH_2CH_2)_nCH_2CH_2{-}CH_2CH_2(CH_2CH_2)_n{-}R$$

The probability that the reactive sites of two growing polymer chains will react in either of these bimolecular termination reactions is very small. A bimolecular reaction of one chain with a monomer molecule, which is present in higher concentration and consumed throughout the reaction, is more likely.

Regulation of Chain Length

The average molecular weight of an addition polymer is controlled by the number of times the propagation steps occur before the chain is terminated. However, the length of the chain can also be controlled by using either chain-transfer agents or inhibitors.

Chain-transfer agents control the chain length of a polymer by interrupting the growth of one chain and then initiating the formation of another chain. Thiols are common chain-transfer agents.

$$R{-}CH_2CH_2\cdot + R{-}S{-}H \longrightarrow \underset{\text{one terminated chain}}{R{-}CH_2CH_2{-}H} + R{-}S\cdot$$

$$\underset{\text{initiation of another chain}}{R{-}S\cdot + CH_2{=}CH_2} \longrightarrow R{-}S{-}CH_2CH_2\cdot$$

A chain-transfer agent must be sufficiently reactive to transfer a hydrogen atom, but the resulting radical must be reactive enough to add to a double bond. The rate of polymerization continues and monomer molecules continue to be consumed. However, the average molecular weight of the product is smaller because more chains are formed by the chain-transfer process.

Inhibitors react with the radical site of a growing polymer chain to give a less reactive radical. Benzoquinone is a typical inhibitor used in free radical polymerization reactions.

R· + benzoquinone ⟶ [resonance-stabilized radical]

benzoquinone

The resonance-stabilized radical is less reactive and does not effectively participate in chain-propagation steps. Eventually it is destroyed by disproportionation or dimerization reactions.

Chain Branching

In practice, the linear polymer normally indicated for alkenes is not the major product of the free radical process. (Cationic polymerization is usually used to prepare linear addition polymers of alkenes.) The product chains have many alkyl branches, which most often are the four-carbon-atom butyl groups produced by **short-chain branching.** These products are the result of an intramolecular hydrogen abstraction by way of a six-membered transition state that generates a secondary radical from a primary radical.

primary radical — secondary radical

The polymerization continues at the new radical site, and a butyl group branch is located on the chain.

Reaction occurs here. — $+ CH_2{=}CH_2 \longrightarrow$ — Reaction continues here. — butyl side chain

Long-chain branching occurs by a random process. Intermolecular hydrogen atom abstraction can occur between the terminal radical of one chain and any of the hydrogen atoms located in another chain. In this case, one chain is terminated and the polymerization continues at a site within the other chain.

The length of the resulting branch depends on the site of hydrogen abstraction. Long-chain branches usually result. Short-chain branching is more common than long-chain branching because intramolecular reactions are more probable than intermolecular reactions.

17.6 Copolymerization of Alkenes

The addition polymers that we have discussed are homopolymers, made up of repeating units derived from a single unsaturated monomer. **Copolymers** incorporate two different monomers in the polymer chain. They are formed in reactions of a mixture of two monomers. Copolymerization of various combinations of monomers provides many more possible structures and a greater variety of materials that might have desirable physical properties than homopolymerization.

Two monomers can react randomly to give a random copolymer. The exact composition depends on the reaction conditions and on the concentrations of the two monomers.

$$n\text{A} + n\text{B} \longrightarrow -\text{A}-\text{A}-\text{B}-\text{A}-\text{B}-\text{B}-\text{B}-\text{A}-\text{A}-\text{B}-$$

(a random copolymer)

Few pairs of monomers give totally random copolymers. In fact, monomers are usually selected to avoid random copolymers. Monomers are chosen so that one monomer at the end of the growing polymer chain prefers to react with the other monomer in the mixture, and vice versa. In short, it is desirable to have a monomer at the end of a chain that reacts preferentially with the other monomer in the reaction mixture. Such polymers are called alternating copolymers.

$$n\text{A} + n\text{B} \longrightarrow -\text{B}-\text{A}-\text{B}-\text{A}-\text{B}-\text{A}-\text{B}-\text{A}-\text{B}-\text{A}-$$

(an alternating copolymer)

Perfect alternating copolymers are difficult to form. However, a nearly perfect alternating copolymer does result from the reaction of styrene with maleic anhydride.

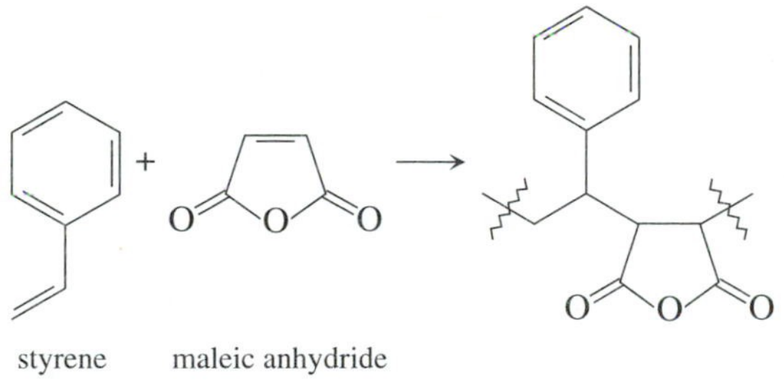

styrene + maleic anhydride ⟶

Maleic anhydride reacts with itself very slowly, and its homopolymer is difficult to form. Styrene readily reacts to form a homopolymer. However, a styrene group at the end of a growing polymer chain reacts faster with maleic anhydride than with styrene. After the addition of styrene to maleic anhydride, a radical is produced that does not react with maleic anhydride. As a result, the next alkene that is added is styrene. Monomers that provide perfect alternating copolymers are highly desirable because the product can be reproduced.

17.7 Cross-linked Polymers

Atoms bonded between polymer chains are called **cross-links.** They form during polymerization of the monomers or in separate reactions after formation of the polymer.

p-Divinylbenzene has two alkene functional groups, each of which can become part of a different polymer chain in a copolymerization reaction with styrene. One alkene group of *p*-divinylbenzene is incorporated in a chain whose major components are styrene units.

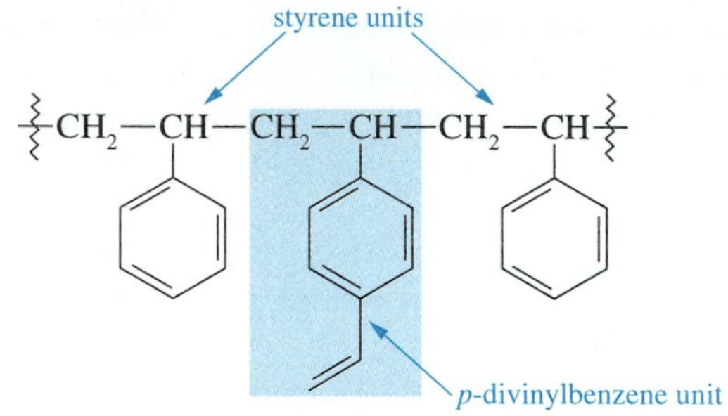

At some point in the reaction, the other alkene group reacts in a chain propagation process that develops a second chain. Thus, divinylbenzene becomes part of each polymer chain and forms a link between the two chains (Figure 17.3). The degree of cross-linking and the space between the divinylbenzene units depend on the amounts of the two monomers used.

The importance of cross-links in determining the properties of a polymer was accidentally discovered by Charles Goodyear in his study of the properties of rubber. Natural and synthetic rubbers can be used to make rubber bands, but are too soft and tacky for many applications, such as tires. The resilience of rubber decreases when it is heated because the polyisoprene chains slide past

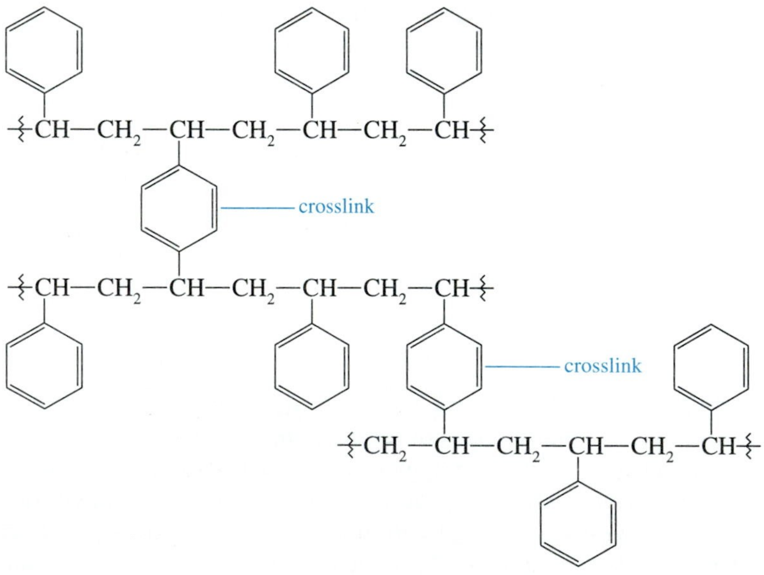

FIGURE 17.3
Cross-links in an Addition Polymer

each other more easily when stretched at higher temperatures. When tension is released, natural rubber does not regain its original structure.

In 1839 Charles Goodyear found that heating natural rubber with a small amount of sulfur produces a material with different properties. He called this process **vulcanization.** The sulfur reacts with the polyisoprene to replace some C—H bonds with disulfide bonds. As a result, the polymer chains become connected by cross-links that may contain one, two, or more sulfur atoms. These cross-links increase the rigidity of the rubber because more of the chains are linked into a larger molecule. The freedom of movement of one chain relative to another is diminished. After distortion, the vulcanized rubber returns to its original molded shape. The amount of sulfur—3 to 10% by weight—controls the flexibility and hardness of the rubber.

17.8 Stereochemistry of Addition Polymerization

Addition polymerization of some alkenes generates stereogenic centers along the entire backbone of the polymer. The relationship of these centers to one another affects the physical properties of the polymer. Consider the polymer formed from propene. If the methyl groups are all on the same side of the backbone of the zigzag chain, the polymer is **isotactic.** If the methyl groups are in a regular alternating sequence on opposite sides of the backbone, the polymer is **syndiotactic.** If the methyl groups are randomly oriented, the polymer is **atactic.** The three forms of polypropylene are shown in Figure 17.4.

The regularity of structure in isotactic and syndiotactic polymers is responsible for their higher melting points, so they can be used to manufacture objects that will be exposed to boiling water. Atactic polymers formed in radical chain polymerization (which also have branches that result from hydrogen abstraction processes) have lower melting points. Both isotactic and syndiotactic forms of polymers are produced with catalysts designed by K. Ziegler of Germany and G. Natta of Italy. These catalysts yield polymers with no chain branching. The

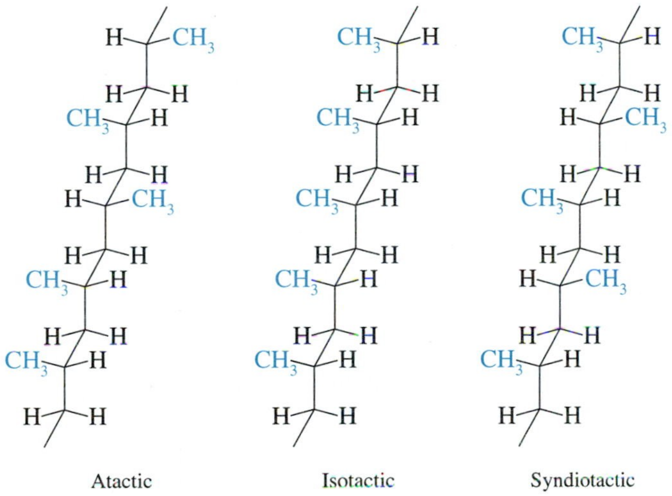

Atactic

Isotactic

Syndiotactic

FIGURE 17.4
Stereochemistry of Addition Polymers

development of methods to form stereochemically regular linear polymers revolutionized polymer science.

The Ziegler-Natta catalysts are organometallic compounds that contain a transition metal. For example, triethylaluminum and titanium(III) chloride combine to give such a catalyst. The structure of the catalyst and its function in the polymerization process are beyond the scope of this text. However, each catalyst coordinates alkene monomers and allows them to react stereoselectively.

Diene Polymers

Conjugated dienes can form addition polymers by a 1,4-addition reaction. The remaining double bond of each monomer unit occurs at every fourth carbon atom along the chain. Natural rubber, for example, is a polymer of 2-methyl-1,3-butadiene (isoprene) with cis stereochemistry at all of the double bonds. The polymer is obtained from the latex synthesized under the bark of some trees that grow in southeast Asia. Gutta-percha is a trans isomer of natural rubber that is produced by trees of a different genus.

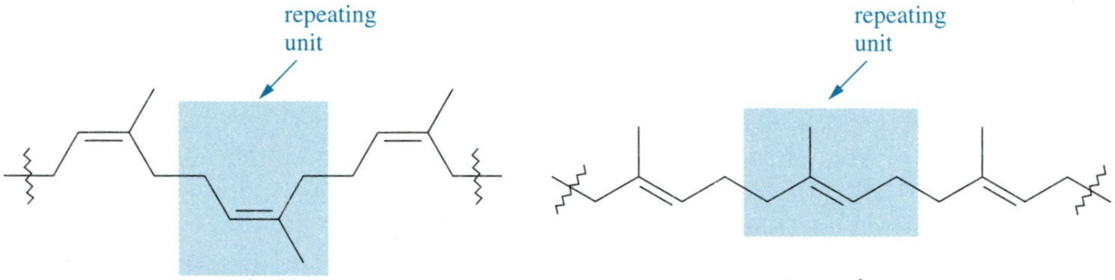

repeating unit

natural rubber
(all *cis*)

repeating unit

gutta-percha
(all *trans*)

The different properties of natural rubber and gutta-percha reflect both the geometries around the double bonds and the molecular weights. As a result of the cis arrangement of the chain in natural rubber, the adjacent molecules cannot fit close to one another. Thus, natural rubber has random coils that can be stretched out when the material is pulled. After the tension is released, the material returns to its original structure. Gutta-percha molecules, on the other hand, can pack more closely because the trans arrangement of the double bonds and the favored anti conformation around the saturated carbon atoms provide a chain with a regular zigzag arrangement. So gutta-percha is a hard, inflexible material. It is used in covers for golf balls and in casings for electrical cables.

Early attempts to polymerize isoprene in industrial processes to prepare synthetic rubber were not successful because the reactions were not stereospecific. However, a variety of Ziegler-Natta catalysts are now available. One catalyst that contains titanium stereospecifically gives polyisoprene with cis double bonds, and another catalyst containing vanadium gives polyisoprene with trans double bonds.

The polymerization of 2-chloro-1,3-butadiene was one of the reactions considered by U.S. industry to replace rubber made from natural sources located in areas of the world that could be cut off in a crisis such as war. This diene structurally resembles isoprene, with a chlorine atom replacing the methyl group of isoprene. Free radical polymerization gives a mixture of cis and trans double bonds as well as a mixture of 1,2- and 1,4-addition products. Polymerization of 2-chloro-1,3-butadiene using a Ziegler-Natta catalyst yields Neoprene®, a compound with trans double bonds.

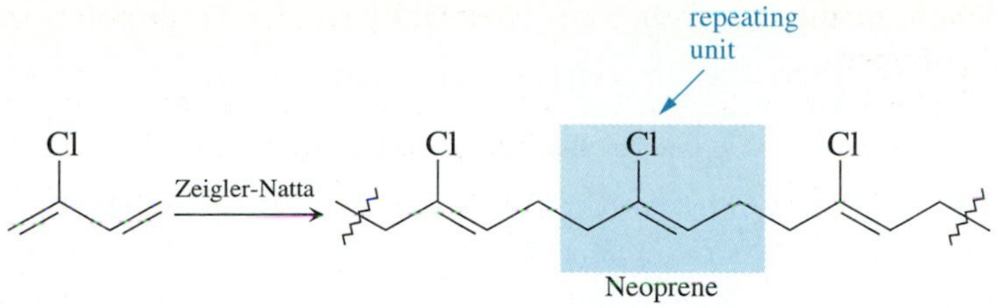

Neoprene

Neoprene resists oxidizing agents better than natural rubber does. Neoprene is therefore used to manufacture materials such as gaskets and industrial hoses.

EXAMPLE 17-3

The free radical polymerization of 1,3-butadiene yields some sections of the polymer that contain a vinyl group. Explain the origin of this group.

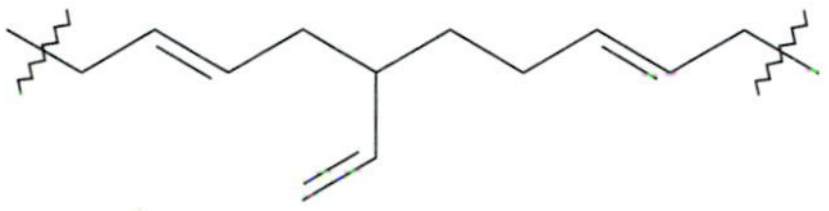

Solution

The presence of a vinyl group bonded to the main chain means than the other vinyl group of 1,3-butadiene is incorporated in the chain. Thus, the polymerization of this unit involves a 1,2-addition similar to that of a simple alkene.

Problem 17.3

Draw the structure of the product of ozonolysis of natural rubber. Would this structure differ from the ozonolysis product of gutta-percha?

17.9 Condensation Polymers

A condensation reaction is a reaction between two reactants that yields one larger product and a second, smaller product such as water. This type of reaction has been illustrated in the reactions of many functional groups containing oxygen or nitrogen. Products of condensation reactions include ethers, acetals, esters, imines, and amides.

We now consider condensation reactions that yield polymers. Two functional groups are required in a monomer so that after one functional group reacts, the other is available to link to another monomer. The functional groups in monomers may be arranged in two ways in a condensation polymerization.

A single compound may contain two different functional groups, such as an amino group and a carboxylic acid group. Reaction of the amino group of one molecule with the carboxylic acid of another molecule gives an amide that still has a free amino group and a free carboxylic acid group.

$$NH_2{-}A{-}\overset{O}{\overset{\|}{C}}{-}OH + NH_2{-}A{-}\overset{O}{\overset{\|}{C}}{-}OH \longrightarrow NH_2{-}A{-}\overset{O}{\overset{\|}{C}}{-}NH{-}A{-}\overset{O}{\overset{\|}{C}}{-}OH$$

Continued reaction of the carboxylic acid end with the amino group of another monomer or of the amino group end with the carboxylic acid group of another monomer yields a homopolymer.

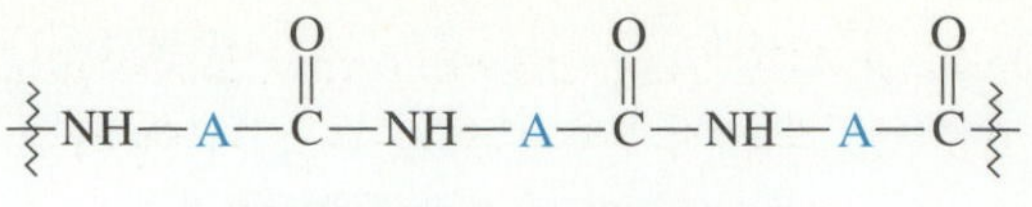

$$\sim NH-A-\overset{O}{\overset{\|}{C}}-NH-A-\overset{O}{\overset{\|}{C}}-NH-A-\overset{O}{\overset{\|}{C}}\sim$$

(a homopolymer of an amino acid)

Condensation reactions also result from the copolymerization of two monomers. Each monomer contains two of the same functional groups. Examples include the reaction of a monomer that is a dicarboxylic acid with a monomer that is a diol. The functional groups on one monomer can only react with the functional groups on the other monomer.

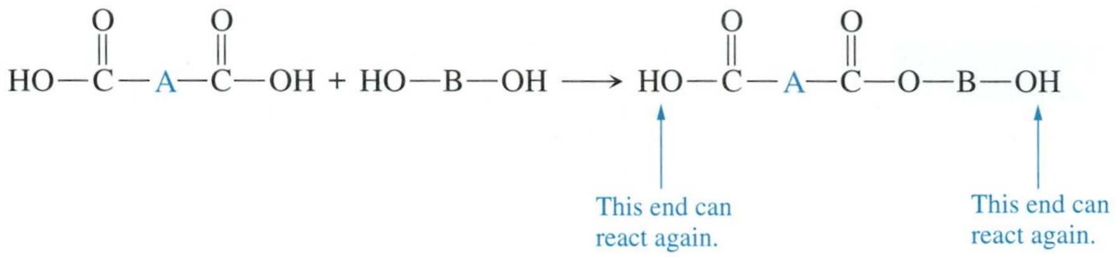

$$HO-\overset{O}{\overset{\|}{C}}-A-\overset{O}{\overset{\|}{C}}-OH + HO-B-OH \longrightarrow HO-\overset{O}{\overset{\|}{C}}-A-\overset{O}{\overset{\|}{C}}-O-B-OH$$

This end can react again. This end can react again.

Continued reaction of the carboxylic acid end with a hydroxyl group of the diol monomer or of the hydroxyl group end with a carboxylic acid group of the dicarboxylic acid monomer yields an alternating copolymer.

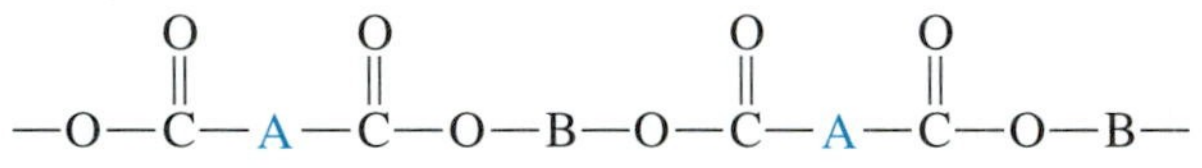

$$-O-\overset{O}{\overset{\|}{C}}-A-\overset{O}{\overset{\|}{C}}-O-B-O-\overset{O}{\overset{\|}{C}}-A-\overset{O}{\overset{\|}{C}}-O-B-$$

(an alternating copolymer)

A monomer can contain two different functional groups, but such monomers are not widely used. First, these monomers are more difficult to prepare without uncontrolled polymerization during their synthesis. Second, the monomer can be used only in one possible polymerization reaction. Condensation polymers formed from two different monomers are more common. The synthesis of each monomer is usually straightforward and less expensive. Each monomer can be used in reactions with other monomers. For example, any of a series of dicarboxylic acids can react with any of another series of diols.

17.10 Polyesters

Polyesters account for approximately 40% of the synthetic fibers produced in the United States. Poly(ethylene terephthalate), also known as PET, is the major polyester. It is a copolymer of ethylene glycol and terephthalic acid.

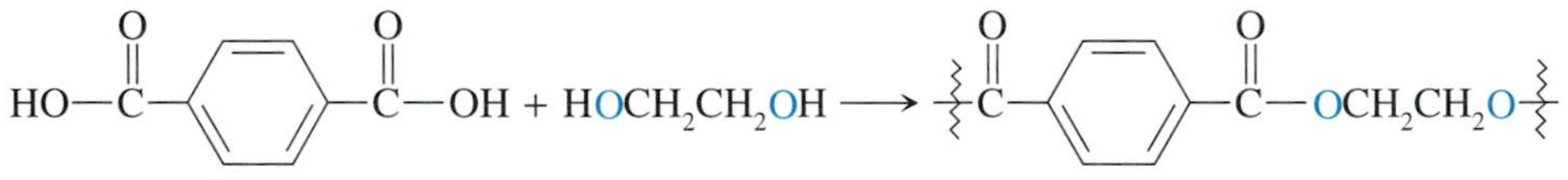

$$HO-\overset{O}{\overset{\|}{C}}-C_6H_4-\overset{O}{\overset{\|}{C}}-OH + HOCH_2CH_2OH \longrightarrow \sim\overset{O}{\overset{\|}{C}}-C_6H_4-\overset{O}{\overset{\|}{C}}-OCH_2CH_2O\sim$$

terephthalic acid ethylene glycol PET

However, PET and many other polyesters are produced industrially by transesterification reactions. A **transesterification** reaction involves exchange of an alkoxy group of an ester for a second alkoxy group provided by an alcohol. The reaction has an equilibrium constant close to 1, and the position of the equilibrium can be controlled by changing the experimental conditions. Ethylene glycol can serve as the nucleophile displacing alcohol units from esters. PET is prepared by the reaction of dimethyl terephthalate with ethylene glycol at 150°C. Neither reactant is volatile at this temperature, but the second product is methanol, which boils at 65°C. As methanol forms, it is continuously vaporized from the reaction mixture, driving the polymerization reaction to completion.

$$CH_3O-\overset{O}{\overset{\|}{C}}-C_6H_4-\overset{O}{\overset{\|}{C}}-OCH_3 + HOCH_2CH_2OH \longrightarrow$$

$$CH_3O-\overset{O}{\overset{\|}{C}}-C_6H_4-\overset{O}{\overset{\|}{C}}-OCH_2CH_2OH + CH_3OH$$

Many containers for products such as juice, milk, and soft drinks are made of PET.

Cyclic anhydrides, such as phthalic anhydride and maleic anhydride, also react with glycols to form polyesters. The anhydride is a difunctional molecule that reacts with the difunctional glycol to give linear alternating copolymers.

phthalic anhydride

However, when a triol reacts with an anhydride, a cross-linked polymer results. For example, the reaction of phthalic anhydride with 1,2,3-propanetriol (glycerol) initially occurs selectively with the primary hydroxyl groups to give a linear polymer.

$$\sim\overset{O}{\overset{\|}{C}}-C_6H_4-\overset{O}{\overset{\|}{C}}-O-CH_2-\underset{OH}{\underset{|}{CH}}-CH_2-O-\overset{O}{\overset{\|}{C}}-C_6H_4-\overset{O}{\overset{\|}{C}}-O-CH_2-\underset{OH}{\underset{|}{CH}}-CH_2-O\sim$$

Reaction of phthalic anhydride with the secondary hydroxyl groups is so slow that continued polymerization can be carried out as a second step. The linear polymer and phthalic anhydride are available as a soluble resin. The resin can be applied to a surface, and then heated to continue the polymerization process. The resulting cross-linked polymer is an insoluble, hard, thermosetting plastic called glyptal (Figure 17.5).

17.11 Polycarbonates

Organic carbonates are esters of carbonic acid. However, because carbonic acid is unstable, carbonates cannot be produced from carbonic acid and an alcohol.

$$HO-\overset{O}{\overset{\|}{C}}-OH \qquad R-O-\overset{O}{\overset{\|}{C}}-O-R$$

carbonic acid (a dialkyl carbonate)

Dialkyl carbonates can be made from the reaction of alcohols with phosgene, a highly toxic gas. The second chlorine atom of phosgene increases the elec-

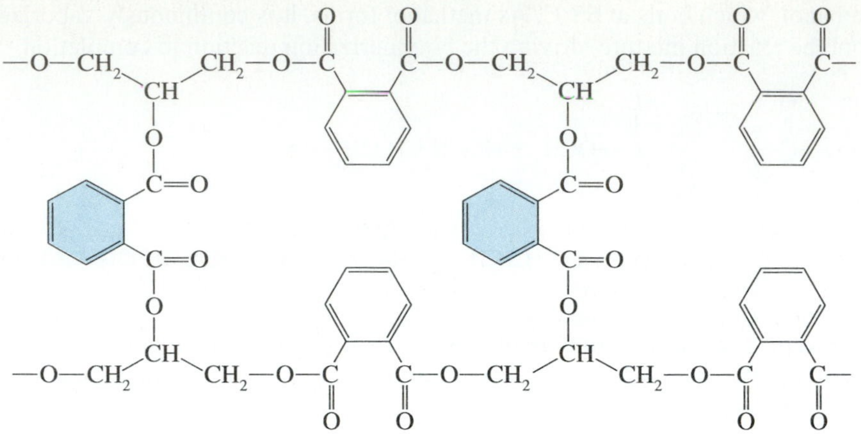

FIGURE 17.5

Cross-linking in a Condensation Polymer

The reaction of 2 moles of 1,2,3-propanetriol for every 3 moles of phthalic anhydride gives a cross-linked polymer called glyptal.

trophilicity of the carbonyl carbon atom. As in the reaction of an alcohol with an acid chloride, a base is required to neutralize the HCl byproduct.

$$2\,ROH + Cl{-}\overset{\overset{\displaystyle O}{\|}}{C}{-}Cl \longrightarrow R{-}O{-}\overset{\overset{\displaystyle O}{\|}}{C}{-}O{-}R + 2\,HCl$$

phosgene; a diakyl carbonate

Although a polymeric carbonate could be produced in the reaction of a diol with phosgene, these products are usually obtained by a transesterification reaction with a dialkyl (or diaryl) carbonate. The reaction of diethyl carbonate with a phenol called bisphenol A gives a polycarbonate known as Lexan®.

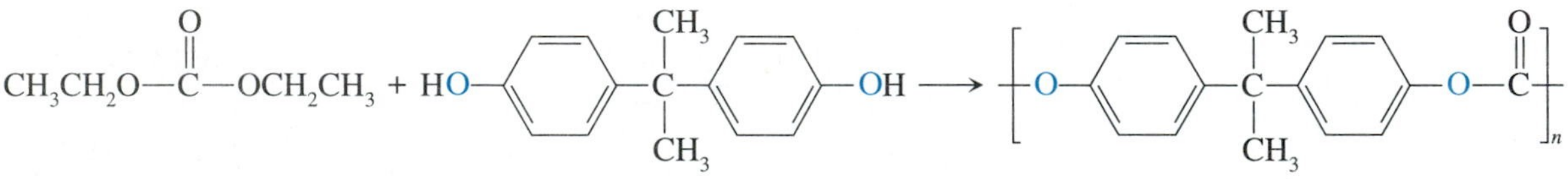

$$CH_3CH_2O{-}\overset{\overset{\displaystyle O}{\|}}{C}{-}OCH_2CH_3 + HO{-}C_6H_4{-}C(CH_3)_2{-}C_6H_4{-}OH \longrightarrow \left[O{-}C_6H_4{-}C(CH_3)_2{-}C_6H_4{-}O{-}\overset{\overset{\displaystyle O}{\|}}{C}\right]_n$$

bisphenol A; Lexan

Lexan has very high impact strength and is strong enough to be used in crash helmets. It is also used to manufacture telephone housings. Because Lexan can be produced as a clear, colorless polymer, it is used in bulletproof windshields and in the visors of astronauts' helmets.

17.12 Polyamides

We recall that amides are best made by the reaction of acid chlorides and amines. Therefore, polyamides can be made by reaction of a monomer with two acid chloride functional groups and a monomer with two amine groups. However, the high reactivity of acid chlorides with nucleophiles such as water requires special precautions to preserve this reagent. Thus, these compounds are not usually used in industrial laboratories.

An alternate method for the synthesis of amides is the direct heating of an amine with a carboxylic acid. The first product is an ammonium salt, which loses water when heated to form the amide.

$$R'{-}CO_2H + R{-}NH_2 \longrightarrow R'{-}CO_2^-\,R{-}NH_3^+ \longrightarrow R'{-}\overset{\overset{\displaystyle O}{\|}}{C}{-}NH{-}R + H_2O$$

Polyamides can be made from the reaction of diacids with diamines. A diammonium salt is formed by proton transfer reactions. When the salt is heated to 250°C, water is driven off and a polyamide forms.

Nylon is a common name for polyamides. The most common polyamide is formed by the reaction of adipic acid, a six-carbon diacid, and 1,6-hexanediamine (hexamethylene diamine), a six-carbon diamine. The ammonium salt formed is called a nylon salt.

$$HO{-}\overset{\overset{\displaystyle O}{\|}}{C}{-}(CH_2)_4{-}\overset{\overset{\displaystyle O}{\|}}{C}{-}OH + NH_2{-}(CH_2)_6{-}NH_2 \longrightarrow$$

$$\underset{\text{nylon salt}}{{}^-O{-}\overset{\overset{\displaystyle O}{\|}}{C}{-}(CH_2)_4{-}\overset{\overset{\displaystyle O}{\|}}{C}{-}O^-\;{}^+NH_3{-}(CH_2)_6{-}NH_3^+} \xrightarrow{250\ ^\circ C}$$

$$\underset{\text{nylon 6,6}}{\left[\overset{\overset{\displaystyle O}{\|}}{C}{-}(CH_2)_4{-}\overset{\overset{\displaystyle O}{\|}}{C}{-}NH{-}(CH_2)_6{-}NH\right]_n}$$

This nylon is called nylon 6,6 (or nylon 66) to indicate that the polyamide is made by the reaction of a six-carbon diamine and a six-carbon diacid.

A polyamide can be produced from a single monomer containing both an amine and a carboxylic acid. However, a related cyclic structure called a lactam can also be converted into a polyamide. When the lactam ring is hydrolyzed, an amino acid is produced that can be polymerized. When ε-caprolactam is heated with a catalytic amount of a nucleophile such as water, the nucleophile attacks the carbonyl carbon atom and opens the ring. The amino group of the resulting amino acid is nucleophilic and reacts with another molecule of the lactam. Subsequent reaction of the amino group of the dimer with the lactam yields a trimer. Continued reaction yields a six-carbon homopolymer called nylon 6.

$$\text{ε-caprolactam (N–H)} \xrightarrow{H_2SO_4} \underset{\text{nylon 6}}{{-}NH{-}(CH_2)_5{-}\overset{\overset{\displaystyle O}{\|}}{C}\left[NH{-}(CH_2)_5{-}\overset{\overset{\displaystyle O}{\|}}{C}\right]_n NH{-}(CH_2)_5{-}\overset{\overset{\displaystyle O}{\|}}{C}{-}}$$

Nylons are used in many products. As a fiber, nylon is used in clothing, rope, tire cord, and parachutes. Because nylon has a high impact strength and resistance to abrasion, it can even be used to make bearings and gears.

Parachutes are made of very light but structurally strong nylon.

Polyurethanes in Treatment of Cancer

Solid tumors require a large blood supply from capillaries to maintain their rapid growth. Thus, any treatment that decreases or eliminates that blood supply can retard the growth of solid tumors in organs for which surgery is difficult. In 1994 a potential use of polymers to block tumor-feeding capillaries was developed. A polyurethane that is soluble in ethanol is injected through a catheter that has been threaded along the patient's arteries to the site of the tumor. Because the polyurethane is insoluble in aqueous tissue fluids such as blood, it precipitates as particles that are sufficiently large to lodge in the capillaries that supply blood to the tumor, blocking the flow of blood. Eventually the polyurethane will hydrolyze and the monomers can be eliminated from the body.

One polyurethane used to "starve" tumors is formed from reaction of 2,2,4-trimethylhexane 1,6-diisocyanate and 1,4-butanediol (tetramethylene glycol).

$$O{=}C{=}N{-}CH_2{-}C(CH_3)_2{-}CH_2{-}CH(CH_3){-}CH_2{-}CH_2{-}N{=}C{=}O$$

2,2,4-trimethylhexane 1,6-diisocyanate

$$HO{-}CH_2{-}CH_2{-}CH_2{-}CH_2{-}OH$$

1,4-butanediol (tetramethylene glycol)

Some quantity of a second diol is used to provide a method of monitoring the delivery of the polyurethane to the site of the cancer. The 1-glyceryl ester of 3,5-diacetamido-2,4,6-triiodobenzoic acid is a diol that when incorporated in the polyurethane can be "seen" because the iodine atoms can be detected by X-rays. ■

$$HO{-}CH_2{-}CH(OH){-}CH_2{-}O{-}C({=}O){-}C_6I_3(NHCCH_3)_2 \quad (\text{2,4,6-triiodo; 3,5-}NH{-}C({=}O){-}CH_3)$$

17.13 Polyurethanes

A urethane is an ester of a carbamic acid. Like carbonic acid, carbamic acids are unstable. They decompose to an amine and carbon dioxide. Therefore, urethanes cannot be made by esterification of a carbamic acid.

$$R{-}NH{-}C({=}O){-}OH \qquad R{-}NH{-}C({=}O){-}O{-}R'$$

(a carbamic acid) (a carbamate or urethane)

However, carbamates can be made by addition of an alcohol to an isocyanate. The reaction formally results in addition of the alcohol across the carbon-nitrogen double bond.

$$R{-}N{=}C{=}O + R'{-}OH \longrightarrow R{-}NH{-}C({=}O){-}O{-}R'$$

(an isocyanate) (a carbamate or urethane)

Polyurethane can be prepared by the reaction of a diisocyanate with a diol. The major diisocyanate used is toluene diisocyanate, which has the isocyanate groups at positions ortho and para to the methyl group. When ethylene glycol is added to the diisocyanate, a typical condensation polymerization occurs to give a polyurethane.

$$O{=}C{=}N{-}C_6H_3(CH_3){-}N{=}C{=}O + HO{-}CH_2CH_2{-}OH \longrightarrow$$

toluene diisocyanate

$$\left[O{-}CH_2CH_2{-}O{-}\overset{O}{\overset{\|}{C}}{-}\overset{H}{\overset{|}{N}}{-}C_6H_3(CH_3){-}\overset{H}{\overset{|}{N}}{-}\overset{O}{\overset{\|}{C}} \right]_n$$

The major use of polyurethanes is in foams. Gases are blown into the liquid polymer to produce bubbles that are trapped as the material cools. When the resulting material is spongy, it is used for cushions. If monomers are selected to give cross-links, the more rigid foams that form are used for thermal insulation in building construction.

Explorations with Molecular Models

1. Prepare a partial structure of the polymer shown here as a ball-and-stick model. What monomer is required to produce the polymer? Describe the stereochemistry of the polymer.

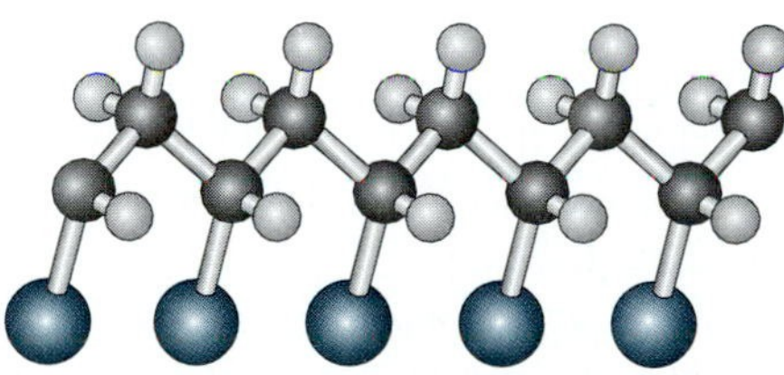

2. Prepare a partial structure of the polymer shown here as a space-filling model. What monomer is required to produce the polymer?

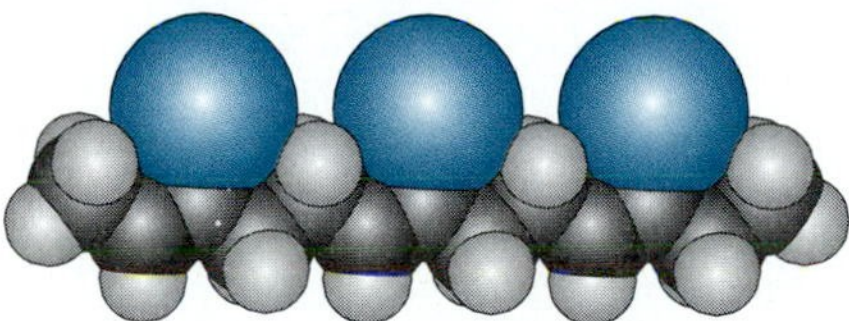

Exercises

Properties of Polymers

17.1 Explain why the polymer of 2-methylpropene is a sticky elastomer with a low melting point.

17.2 How would the properties of the polymer of the following diamine and adipic acid differ from those of nylon 6,6?

$$CH_3{-}NH{-}(CH_2)_6{-}NH{-}CH_3$$

17.3 Explain how 1,2,4,5-benzenetetracarboxylic acid dianhydride could be used to make a thermosetting polyester.

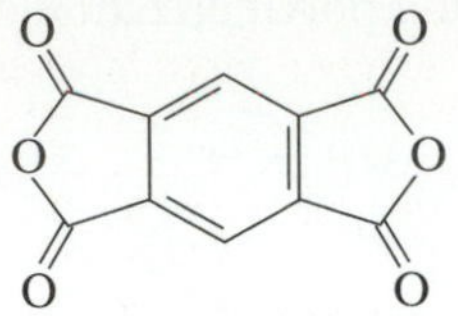

17.4 How would the properties of the copolymer of 1,4-butanediol with terephthalic acid differ from those of PET?

Addition Polymers

17.5 Vinyl acetate is used to make a polymer used in chewing gum. Draw a bond-line representation of the polymer.

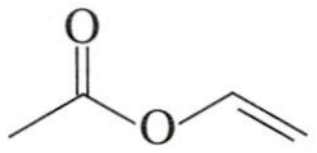

17.6 Draw a bond-line structure of polyvinyl alcohol. Explain why the polymer is prepared by the hydrolysis of polyvinyl acetate.

17.7 What monomer is required to prepare the following polymer?

$$-CFCl-CF_2-CFCl-CF_2-CFCl-CF_2-CFCl-CF_2-$$

17.8 Hexafluoropropene is a monomer used to prepare a polymer called Viton®. Draw a representation of the polymer.

Chain-Transfer Reactions

17.9 Draw the structure of the branch formed by a short-chain transfer reaction in the formation of polystyrene.

17.10 Explain why formation of a polymer of 1-hexene under free radical conditions would produce some molecules with methyl groups bonded to the main chain.

Copolymers

17.11 Draw a representation of an alternating polymer of isoprene and 2-methylpropene.

17.12 Styrene and 1,3-butadiene form a random polymer. What is the probability that a 1,3-butadiene unit will react with a growing polymer chain with styrene at its end?

17.13 Some hair sprays contain a solution of a copolymer made from the following monomers. Draw a representation of the polymer. Why does the copolymer hold hair in place?

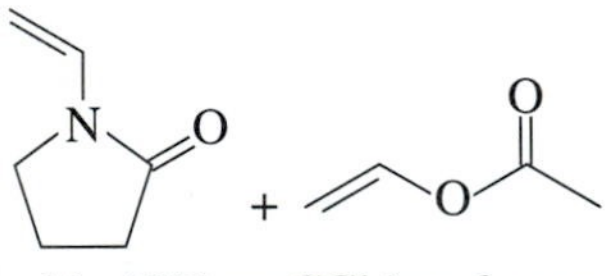

17.14 Saran® is a copolymer of vinylidene chloride ($CH_2{=}CCl_2$) and a smaller amount of vinyl chloride. Draw a representation of the polymer.

Cross-linked Polymers

17.15 What is the difference between the number of cross-links in the rubber used in tires and the rubber used in gloves?

17.16 Draw a representation of the polyester formed from butenedioic anhydride (maleic anhydride) and 1,2-propanediol. Explain how this polymer could be cross-linked by reacting it with styrene.

Stereochemistry of Polymerization

17.17 Which of the following alkenes can be polymerized to give isotactic and syndiotactic structures?

(a) 1-chloroethene (b) 1,1-dichloroethene (c) 2-methylpropene (d) styrene

17.18 Are syndiotactic or isotactic forms of polypropylene optically active?

17.19 3-Methyl-1-pentene reacts with a Ziegler-Natta catalyst to give an isotactic polymer. What relationship exists between the alkyl groups on the polymer chain?

17.20 Ethylene and *cis*-2-butene form a syndiotactic copolymer in a reaction catalyzed by a vanadium catalyst. Draw a representation of the polymer.

Condensation Polymers

17.21 What monomers are required to prepare the following polymers?

(a)

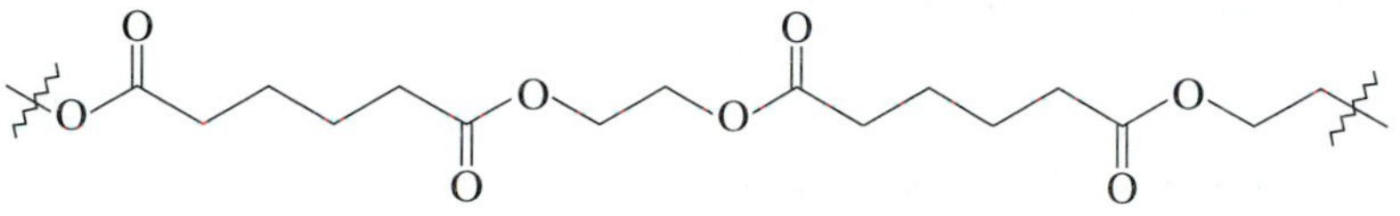

(b) 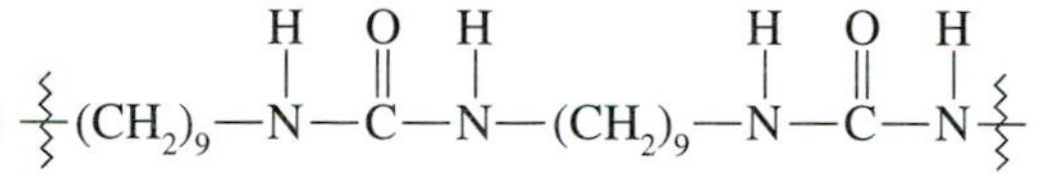 $-(CH_2)_9-N(H)-C(=O)-N(H)-(CH_2)_9-N(H)-C(=O)-N(H)-$

(c)

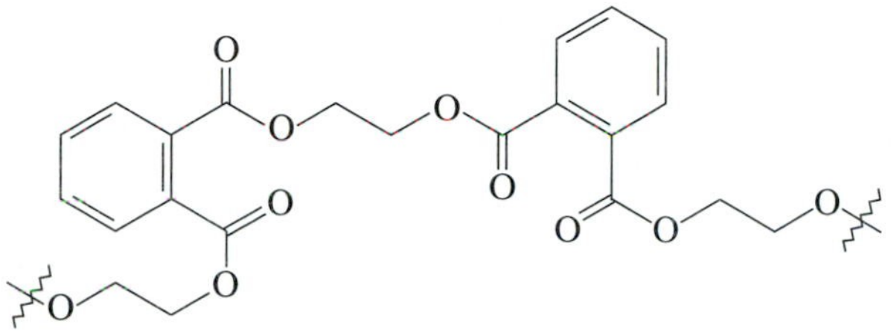

17.22 What monomers are required to prepare the following polymers?

(a)

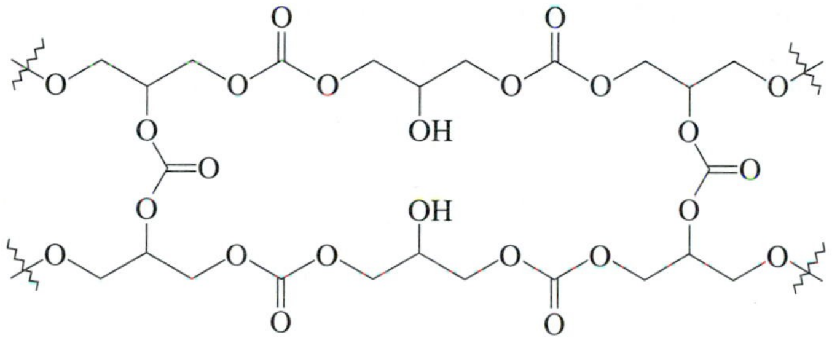

(b)

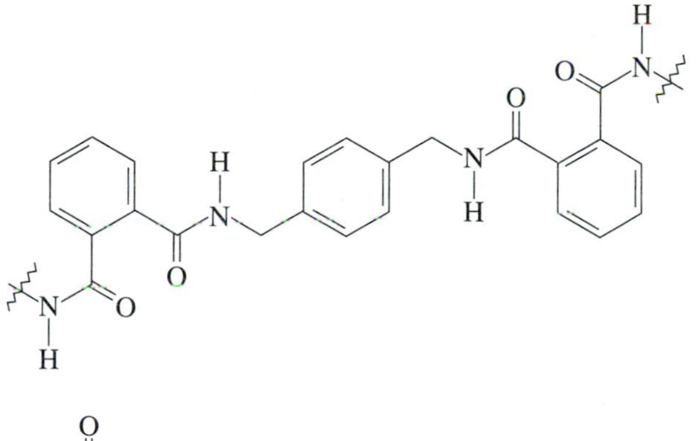

(c)

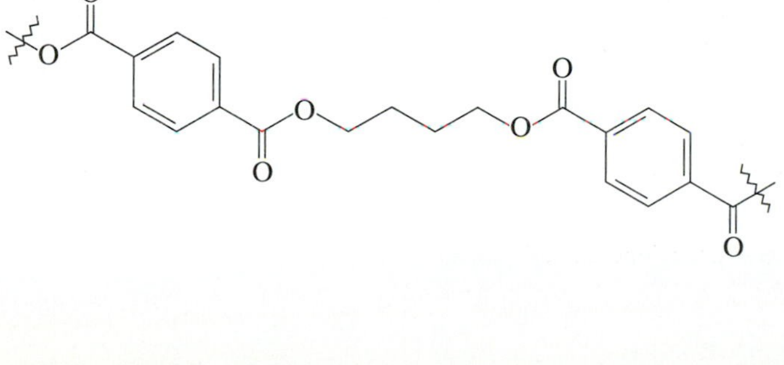

Polyesters

17.23 A homopolymer of lactic acid can be used to make body implants. Write a bond-line representation of the polymer.

$$CH_3CH(OH)CO_2H$$

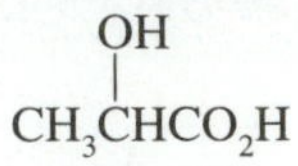

lactic acid

17.24 A polymer of β-propiolactone is obtained by using a catalytic amount of hydroxide ion. Draw the structure of the polymer. Why does the polymerization reaction continue?

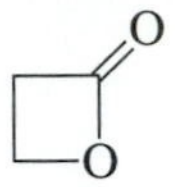

β-propiolactone

17.25 Kodel® is a polymer of terephthalic acid and *trans*-1,4-di(hydroxymethyl)cyclohexane. Draw a representation of the polymer.

17.26 What monomers are used to prepare the following polyester? Identify an unusual feature of this polyester.

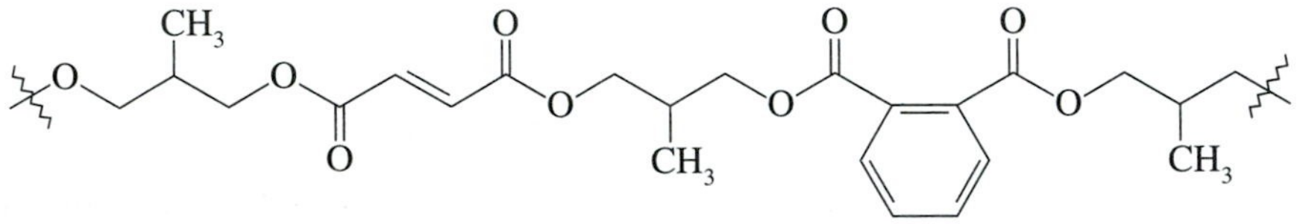

CH_3 O O O O O O O O CH_3 CH_3

Polyamides

17.27 The following structure represents a group of polyamides called Qiana®. The value of x is 8, 10, or 12. What are the component monomers? How does the value of x affect the properties of the polymer?

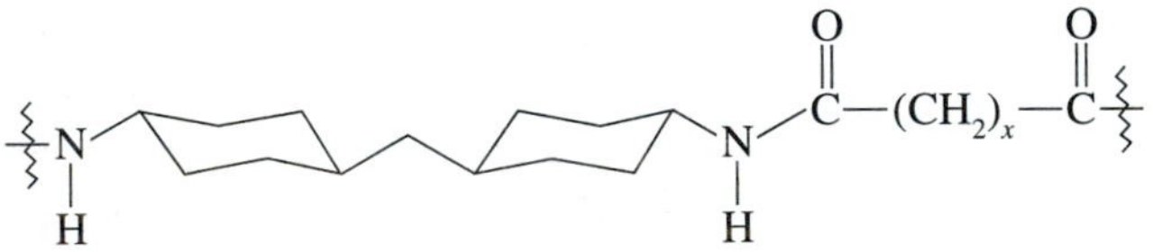

N H H N O C $-(CH_2)_x-$ C O

17.28 A polyamide contains the following structural unit, which is prepared from the reaction of a lactam. Draw the structure of the lactam.

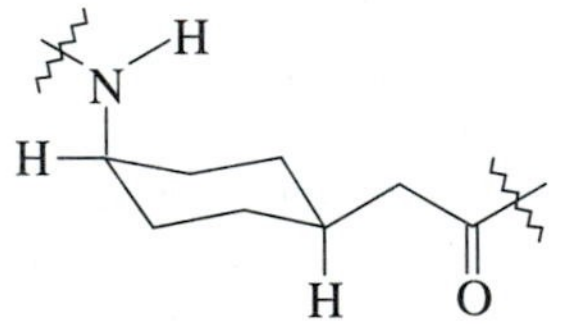

N H H H O

Polyurethanes

17.29 Explain why the addition of glycerol to the polymerization of toluene diisocyanate and ethylene glycol produces a stiffer foam.

17.30 An oligomer of tetramethylene glycol reacts with toluene diisocyanate to form a polyurethane called Lycra®. Draw a representation of the polyurethane.

$$H-O\left[(CH_2)_4-O\right]_nH$$

CHAPTER 18

SPECTROSCOPY

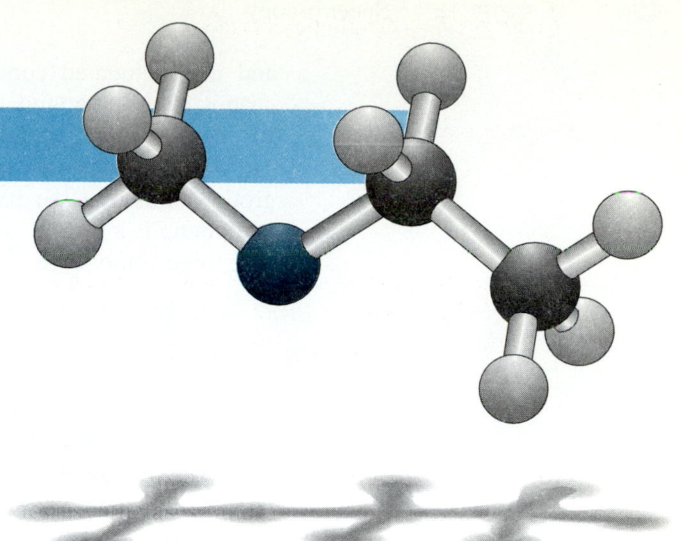

Overview

18.1 Structure Determination

How does a chemist know the molecular structure of a compound obtained either from a natural source or as the product of a chemical reaction? The question is of paramount importance because without a knowledge of structure one cannot explain either the physical or the chemical properties of molecules. At one time, organic chemists determined the structure of an organic compound by chemical reactions that related the unknown compound to other known compounds. To determine the structures of complex molecules, it was often necessary to use reactions that systematically degraded the large molecule into smaller molecules. Then the chemist could "reason backward" to postulate what the structure of the original compound must have been to yield the observed products. Thus, structure determination by chemical reactions is a time-consuming process that requires experimental skills and deductive reasoning. For example, consider the problem of determining the structure of a relatively simple compound with molecular formula $C_5H_{10}O$. Eighty-eight isomers are possible, including ethers, alcohols, aldehydes, and ketones. Many chemical reactions would be required to identify the functional group as well as the hydrocarbon skeleton. Structure determination by chemical reactions also has one severe limitation: each reaction destroys part of the sample of the unknown compound.

Spectroscopic determination requires only minuscule amounts of a compound compared to chemical methods, and spectroscopic methods require much less time. Chemists now determine structure by nondestructive spectroscopic techniques and recover the sample unchanged after they determine its spectrum. Spectroscopic methods provide information about the kinds of atoms present in the compound and how they are connected. In this chapter we examine ultraviolet spectroscopy, infrared spectroscopy, and nuclear magnetic resonance spectroscopy. Each type of spectroscopy provides a different kind of information.

Ultraviolet spectroscopy provides information about the π system in a compound, so this method allows us to distinguish between conjugated and nonconjugated compounds. Infrared spectroscopy reveals the functional groups in a compound. For example, consider again the determination of the structure of a compound with molecular formula $C_5H_{10}O$. Infrared spectroscopy can eliminate some of the 88 possible structures by identifying the functional group containing the oxygen atom. If infrared spectroscopy reveals that the compound is a ketone, the number of possible isomers drops to a more manageable three compounds.

$$CH_3{-}CH_2{-}\overset{\overset{\displaystyle O}{\|}}{C}{-}CH_2{-}CH_3 \qquad CH_3{-}\overset{\overset{\displaystyle O}{\|}}{C}{-}CH_2{-}CH_2{-}CH_3 \qquad CH_3{-}\overset{\overset{\displaystyle O}{\|}}{C}{-}\underset{\underset{\displaystyle CH_3}{|}}{CH}{-}CH_3$$

Nuclear magnetic resonance spectroscopy provides information about the carbon-hydrogen framework of a compound. To further characterize the $C_5H_{10}O$ compound known to be a ketone by infrared spectroscopy, we must determine how the carbon atoms are distributed on either side of the carbonyl carbon atom. To determine the structure of the ketone, we need to "see" either the structurally nonequivalent carbon nuclei or the nuclei of hydrogen atoms bonded to them. Nuclear magnetic resonance spectroscopy (NMR) provides this information. In this chapter we will first consider how to determine the molecular structure using information about structurally different hydrogen nuclei. With information about the hydrogen atoms, we can deduce how the carbon atoms are arranged in the structure. Then we will see how NMR spectroscopy can detect structural differences among the carbon atoms themselves.

18.2 Spectroscopy

Spectroscopy is a study of the interaction of electromagnetic radiation with molecules. Electromagnetic radiation encompasses X-rays, ultraviolet, visible, infrared, microwaves, and radio waves. Electromagnetic radiation is described as a wave that travels at the speed of light (3×10^8 m/s). Waves are characterized by a wavelength (λ, Greek lambda) and a frequency (ν, Greek nu). The wavelength is the length of one wave cycle, such as from trough to trough (Figure 18.1). The wavelength is expressed in the metric unit convenient for each type of electromagnetic radiation. The frequency is the number of waves that move past a given point in a unit of time. Frequency is expressed in hertz (Hz). Wavelength and frequency are inversely proportional, and are related by

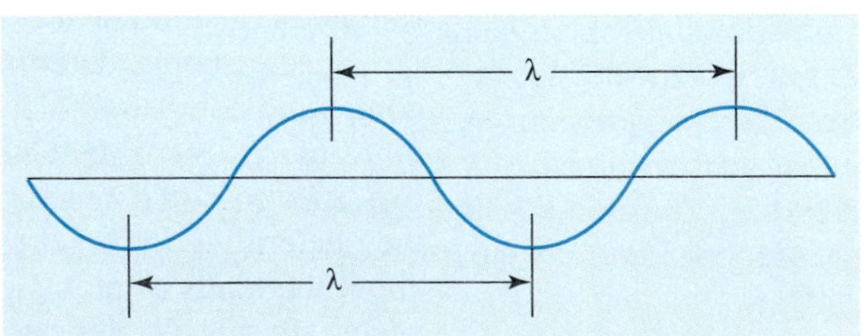

FIGURE 18.1
Electromagnetic Radiation
The wavelength of electromagnetic radiation is the distance between any two peaks or troughs of the wave.

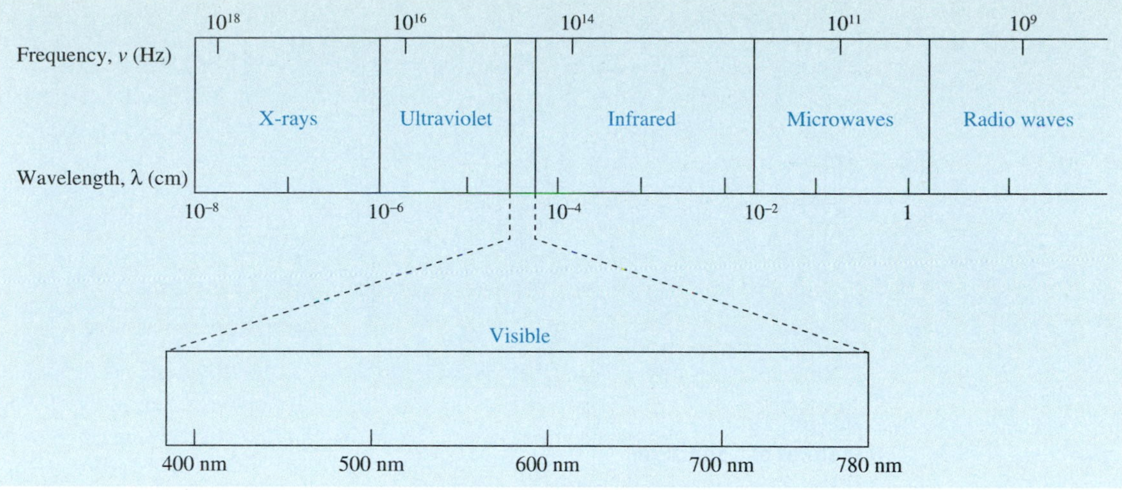

FIGURE 18.2
The Electromagnetic Spectrum

The regions of the spectrum are characterized by a frequency and a wavelength. Usually the wavelength or the reciprocal of the wavelength, the wavenumber, is used to identify absorptions by organic molecules. The relationship of the visible spectrum to other spectral regions is shown within the expansion of that region.

$\lambda = c/\nu$, where c is the speed of light. Thus, as the wavelength of the electromagnetic radiation increases, the corresponding frequency decreases.

The energy (E) associated with electromagnetic radiation is quantized. The relationship is given by

$$E = h\nu = \frac{hc}{\lambda} = hc(1/\lambda)$$

where h is the proportionality constant known as Planck's constant. Thus, the energy of electromagnetic radiation is directly proportional to its frequency but inversely proportional to its wavelength. The energy of electromagnetic radiation is also directly proportional to the quantity $1/\lambda$, known as the **wavenumber.** The frequency of ultraviolet radiation is higher than that of infrared radiation. Alternatively expressed, the wavelength of ultraviolet radiation is shorter than the wavelength of infrared radiation. Because ultraviolet radiation has a high frequency, it has higher energy than infrared radiation (Figure 18.2).

Molecules can absorb only certain discrete amounts of energy. To change the energy of a molecule from E_1 to E_2, the energy difference $E_2 - E_1$ is provided by a characteristic electromagnetic radiation with a specific frequency (or wavelength). The energy absorbed by the molecule can change its electronic or vibrational energy. For example, ultraviolet radiation causes changes in the electron distribution in π bonds; infrared radiation causes bonds to stretch and bond angles to bend.

In the various types of spectroscopy, radiation is passed from a source through a sample that may or may not absorb certain wavelengths of the radiation. The wavelength is systematically changed, and a detector determines which wavelengths of light are absorbed (Figure 18.3). At a wavelength that corresponds to the energy $E_2 - E_1$ necessary for a molecular change, the radia-

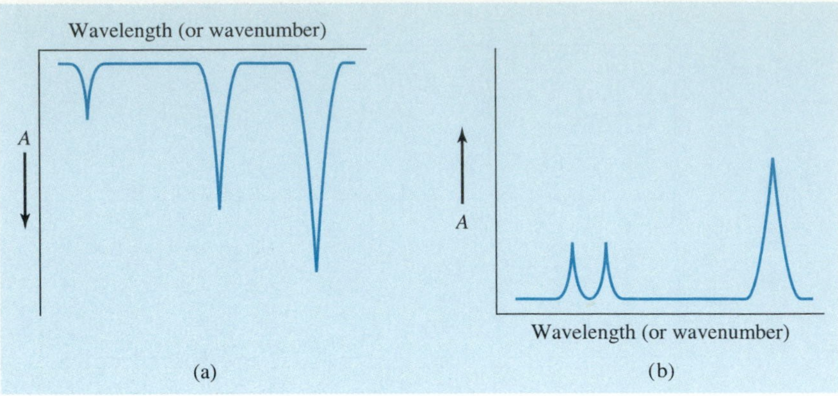

FIGURE 18.3

Features of a Spectrum

The portion of the spectrum where no absorption occurs is the base line, which may be located at the top or bottom of a graph. Absorptions are then recorded as a series of peaks or "dips" from the base line. In an infrared spectrum (a) the base line is at the top. In an ultraviolet spectrum (b) or a nuclear magnetic resonance (NMR) spectrum, the base line is at the bottom.

tion emitted by the source is absorbed by the molecule. The amount of light absorbed by the molecule (absorbance) is plotted as a function of wavelength. At most wavelengths the amount of radiation detected by the detector is equal to that emitted by the source—that is, the molecule does not absorb radiation. At such wavelengths a plot of absorbance on the vertical axis versus wavelength yields a horizontal line (Figure 18.3). When the molecule absorbs radiation of a specific wavelength, the amount of radiation that arrives at the detector is less than that emitted by the source. This difference is recorded as an absorbance.

18.3 Ultraviolet Spectroscopy

The ultraviolet region of the electromagnetic spectrum spans wavelengths from 200 to 400 nm (1 nm = 10^{-9} m). In the ultraviolet region of the electromagnetic spectrum a molecule with conjugated double bonds absorbs energy. Sigma bonds as well as isolated carbon-carbon double bonds require electromagnetic radiation of higher frequency to absorb energy.

Ultraviolet (UV) spectra are simple in appearance. A UV spectrum is a plot of the absorbance of light on the vertical axis and the wavelength of light in nanometers (nm) on the horizontal axis (Figure 18.4). The wavelength corresponding to the UV "peak" is called the λ_{max}. The intensity of the absorption depends on the structure of the compound and the concentration of the sample in the solution. Concentrations in the 10^{-3} to 10^{-5} M range are typically used to obtain a spectrum.

The energy absorbed by conjugated systems moves bonding π electrons into higher energy levels. The specific wavelength of ultraviolet light required to "excite" the electrons in a conjugated molecule depends on the structure of the compound.

Ultraviolet spectroscopy provides information about the extent of the conjugation: as the number of double bonds increases, the wavelength of absorbed light also increases. For 1,3-butadiene, the λ_{max} = 217 nm. With an increased

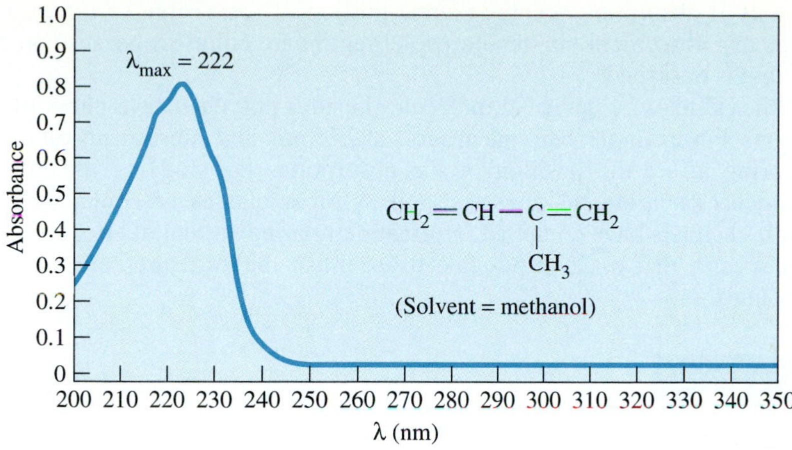

FIGURE 18.4
Ultraviolet Spectrum of Isoprene
The ultraviolet spectrum of isoprene dissolved in methanol is representative of the spectra of conjugated dienes. The maximum absortion occurs at 222 nm.

number of conjugated double bonds, compounds absorb at longer wavelengths; that is, a lower energy is required for electronic excitation. For example 1,3,5-hexatriene and 1,3,5,7-octatetraene have λ_{max} at 268 and 304 nm, respectively.

Although the extent of the conjugation is the primary feature in affecting the λ_{max}, the degree of substitution causes changes that are useful in structure determination. For example, the presence of an alkyl group such as methyl appended to one of the carbon atoms of the conjugated system causes approximately a 5 nm shift to longer wavelengths. Thus, 2-methyl-1,3-butadiene (isoprene) absorbs at 222 nm, compared to 217 for 1,3-butadiene.

Some naturally occurring compounds with extensively conjugated double bonds absorb at such long wavelengths that the λ_{max} occurs in the visible region (400 to 800 nm) of the spectrum. *β*-Carotene, which is contained in carrots, absorbs at 455 nm in the blue-green region of the spectrum. Because blue-green light is absorbed, the light that is transmitted to our eyes is yellow-orange; that is, we see the complement of the absorbed light. Thus, the color of a compound provides qualitative information about its λ_{max} (Table 18.1). A compound is col-

Table 18.1 Absorbed Light and Reflected Color

Absorbed wavelength (nm)	Reflected color
400 (violet)	yellow-green
450 (blue)	orange
510 (green)	purple
590 (orange)	blue
640 (red)	blue-green
730 (purple)	green

ored only if absorption occurs in some portion of the visible spectrum. Compounds that absorb only in the ultraviolet region are colorless because no "visible" light is absorbed.

Other kinds of conjugated molecules besides polyenes have ultraviolet absorptions. For example, benzene absorbs at 254 nm, and substituents on the aromatic ring affect the position of the absorption. By studying the effect of substituents on aromatic rings, as well as other classes of conjugated compounds, chemists have compiled information relating structural effects on ultraviolet spectra that make it possible to establish the structures of "unknown" compounds.

EXAMPLE 18-1

Predict the λ_{max} of 2,4-dimethyl-1,3-pentadiene.

Solution

First draw the structure of the compound and determine the structural features that can affect the position of the ultraviolet absorption.

The compound contains two conjugated double bonds. Thus, the compound should absorb near 217 nm, as for butadiene. However, there are two branching methyl groups as well as the C-5 methyl group bonded to the unsaturated carbon atoms of the butadiene-type system. Thus, the compound should absorb at 217 + 3(5) = 232 nm.

Problem 18.1

Naphthalene and azulene are isomeric compounds that have extensively conjugated π systems. Naphthalene is a colorless compound, but azulene is blue. Deduce information about the absorption of electromagnetic radiation by these two compounds.

naphthalene azulene

18.4 Infrared Spectroscopy

Bonded atoms in a molecule do not remain at fixed positions with respect to each other. Molecules vibrate at various frequencies that depend on molecular structure. Similarly, the angle between two atoms bonded to a common central atom expands and contracts by a small amount at a frequency that depends on molecular structure. These vibrational and bending frequencies correspond to the frequencies of light in the infrared region in the electromagnetic spectrum.

For every type of bond or bond angle there is a specific wavelength at which the molecule absorbs infrared radiation. However, the number of different absorptions is large for even the simplest of organic molecules. The infrared spectrum of 1-methylcyclopentene is shown in Figure 18.5. The wavelength, given on the bottom of the graph, is plotted against percent transmittance of light by the sample. An absorption corresponds to a "peak" pointed toward the bottom of the graph. Because the wavelength of absorbed light is inversely proportional to its energy, absorptions that occur at high wavelength (toward the

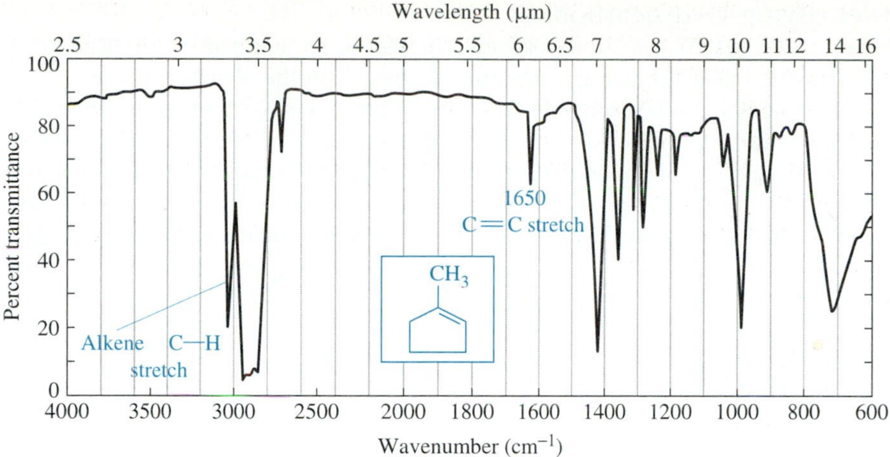

FIGURE 18.5

Infrared Spectrum of 1-Methylcyclopentene

right of the graph) represent molecular vibrations that require low energy. The plot also gives the corresponding value of the wavenumber, $1/\lambda$, for the absorption. The energy of absorbed light is directly proportional to the wavenumber. Thus, absorptions that occur at higher wavenumber (toward the left of the graph) represent molecular vibrations that require high energy.

The infrared spectrum of an organic molecule is complex, and a peak-by-peak analysis is very difficult. However, the total spectrum is characteristic of the compound and can be used to clearly establish the identity of an "unknown" compound. If the spectrum of the "unknown" has all of the same absorption peaks—both wavelength and intensity—as a compound of known structure, then the two samples are identical. If the "unknown" has one or more peaks that differ from the spectrum of a known, then the two compounds are not identical or some impurity in the "unknown" sample causes the extra absorptions. On the other hand, if the "unknown" lacks even one absorption peak that is present in the known structure, the unknown definitely has a different structure than the known.

Characteristic Group Vibrations

Although the infrared spectrum of an organic compound is complex, distinctive bands appear in spectra of compounds with common functional groups. Distinctive absorptions corresponding to the vibration of specific bonds or functional groups are called **group vibrations.** Thus, we can use the presence or absence of these absorptions to characterize a compound. For example, the absorption at 1650 cm^{-1} in 1-methylcyclopentene is due to the stretching of a carbon-carbon double bond (Figure 18.5). Although the exact position of the carbon-carbon double bond absorption varies slightly for various alkenes, they are all in the 1630–1670 cm^{-1} region. Isomers such as 3-methylcyclopentene or cyclohexene have carbon-carbon double bonds that have absorptions in the 1650 cm^{-1} region. However, the spectra will differ in some other areas, meaning that the compounds are not identical to 1-methylcyclopentene.

In the following subsections, we will consider a few characteristic group vibrations. The small differences in group vibrations that result from minor but important differences in structure will not be discussed in this text.

Identifying Hydrocarbons

When we first started to study organic chemistry, we learned that hydrocarbons are classified as saturated and unsaturated based on the absence or presence of multiple bonds. Multiple bonds decrease the number of hydrogen atoms in a molecular formula below the number given in C_nH_{2n+2}, the molecular formula for a saturated acyclic hydrocarbon. However, the molecular formulas of hydrocarbons do not unambiguously indicate the presence or absence of a multiple bond. Both 1-octene and cyclooctane have the same molecular formula, C_8H_{16}.

$CH_3(CH_2)_5CH{=}CH_2$

1-octene   cyclooctane

The structural features present in 1-octene that are absent in cyclooctane are a carbon-carbon double bond and sp^2-hybridized C—H bonds. Thus, the characteristic group absorptions of these features will be present in the spectrum of 1-octene and absent in the spectrum of cyclooctane.

Now also consider the difference between 1-octyne and an isomeric bicyclic hydrocarbon.

$CH_3(CH_2)_4CH_2C{\equiv}CH$

1-octyne   (a bicyclic hydrocarbon)

The structural features present in 1-octyne and absent in the bicyclic hydrocarbon are a carbon-carbon triple bond and an sp-hybridized C—H bond. Thus, the characteristic group absorptions of these features will be present in the spectrum of 1-octyne and absent in the spectrum of the bicyclic hydrocarbon.

The energy of the infrared radiation absorbed by a C—H bond depends on the hybridization of the hybrid orbital (Table 18.2). The bond strengths of carbon-hydrogen bonds are in the order of $sp^3 < sp^2 < sp$, because the increased s character of the hybrid orbital keeps the bonding electrons closer to the carbon atom. Thus, the energy required to stretch the bond also increases.

The sp^3-hybridized C—H bonds in saturated hydrocarbons like octane (Figure 18.6) absorb infrared radiation in the 2850–3000 cm^{-1} region. The sp^2-

Table 18.2 Characteristic Infrared Group Frequencies

Class	Group	Wavenumber (cm^{-1})
alkane	C—H	2850–3000
alkene	C—H	3080–3140
	C=C	1630–1670
alkyne	C—H	3300–3320
	C≡C	2100–2140
alcohol	O—H	3400–3600
	C—O	1050–1200
ether	C—O	1070–1150
aldehyde	C=O	1725
ketone	C=O	1700–1780

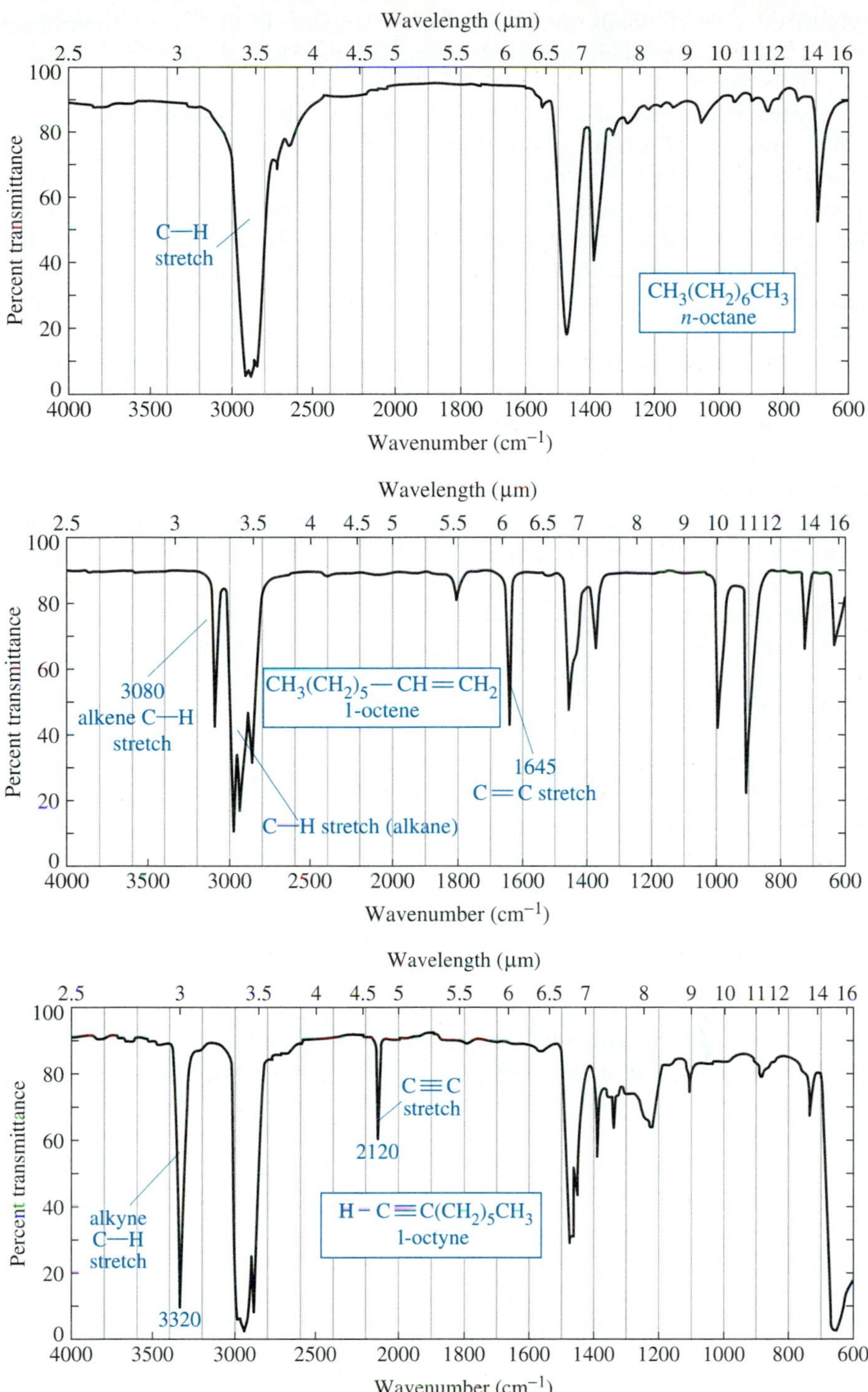

FIGURE 18.6
Infrared Spectra of Hydrocarbons

hybridized C—H bonds in alkenes such as 1-octene absorb energy at 3080 cm^{-1}. This peak is well separated from the absorptions associated with the sp^3-hybridized C—H bonds in this molecule (Figure 18.6). An *sp*-hybridized C—H bond in a molecule such as 1-octyne (Figure 18.6) absorbs infrared radiation at 3320 cm^{-1}.

Hydrocarbons can also be classified based on absorptions due to the carbon-carbon bond. Carbon-carbon bond strength increases in the order single < double < triple. Thus, the wavenumber position (cm^{-1}) of the absorption corresponding to stretching these bonds increases in the same order. Saturated hydrocarbons all contain many carbon-carbon single bonds that absorb in the 800–1000 cm^{-1} region, but the intensity is very low. Carbon-carbon single bonds present in unsaturated compounds also absorb in the same region. Many other molecular bond stretching vibrations and bond angle bending modes occur in the same region and are much more intense. Thus, this region has limited diagnostic value. Moreover, we already know that most organic compounds have carbon-carbon single bonds.

Unsaturated hydrocarbons are identified by the absorption for the carbon-carbon double bond, which occurs in the 1630–1670 cm^{-1} region. The intensity of the absorption decreases with increased substitution. Terminal alkenes have the most intense absorptions. The double bond in 1-octene absorbs at 1645 cm^{-1} (Figure 18.6).

The absorption for carbon-carbon triple bonds occurs in the 2100–2140 cm^{-1} region. Terminal alkynes have the most intense absorptions; internal (disubstituted) alkynes have lower intensity absorptions. The triple bond in 1-octyne absorbs at 2120 cm^{-1} (Figure 18.6).

Identifying Oxygen-Containing Compounds

Many functional groups contain oxygen. These functional groups have the characteristic infrared absorptions given in Table 18.2. The characteristic group frequencies of aldehydes and ketones are from 1700 to 1780 cm^{-1}. The carbon-oxygen double bond of carbonyl compounds requires more energy to stretch than does the carbon-oxygen single bond of ethers and alcohols. Thus, aldehydes and ketones absorb infrared radiation at higher wavenumber positions (1700–1780 cm^{-1}) than alcohols and ethers (1050–1200 cm^{-1})

The absorption for a carbonyl group is extremely intense and is easily detected because it is in a region of the infrared spectrum that is devoid of conflicting absorptions. Note that carbon-carbon double bond stretching vibrations are at a lower wavenumber position than those of carbonyl compounds. A typical spectrum of a ketone is shown for 2-heptanone in Figure 18.7. The carbonyl stretching vibration occurs at 1712 cm^{-1}.

The position of the carbonyl group absorption of acyl derivatives depends on the inductive and resonance effects of atoms bonded to the carbonyl carbon atom. We recall that a carbonyl group is represented by two contributing resonance forms.

$$-\overset{:\ddot{O}}{\overset{\|}{C}}- \longleftrightarrow -\overset{:\ddot{O}:^-}{\overset{|}{\underset{+}{C}}}-$$

Because less energy is required to stretch a single bond than a double bond, any structural feature that stabilizes the contributing polar resonance form with a

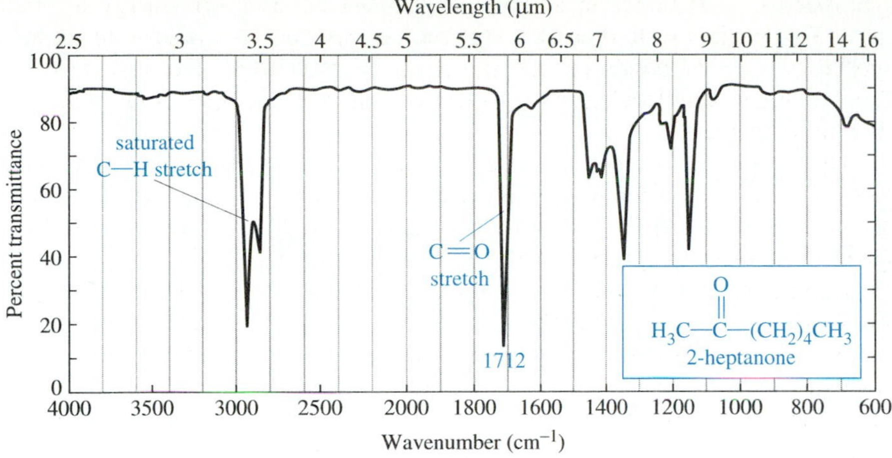

FIGURE 18.7
Infrared Spectrum of 2-Heptanone

carbon-oxygen single bond will cause the infrared absorption to occur at lower wavenumber position. Thus, any group that donates electrons by resonance causes a shift in the absorption to lower wavenumbers. We recall that the nitrogen atom of amides is very effective in donation of electrons to the carbonyl carbon atom (Section 14.5).

Thus, the double bond character of the carbonyl group decreases. As a result, an amide carbonyl group absorbs in the 1650–1690 cm^{-1} region, which is at a lower wavenumber than for ketones.

The carbon-oxygen single bond stretching vibration of alcohols and ethers appears in a region complicated by many other absorptions. However, the absorption of a carbon-oxygen single bond is more intense than the absorption of carbon-carbon single bonds. The presence of a hydroxyl group is better established by the oxygen-hydrogen stretching vibration that occurs as an intense broad peak in the 3360 cm^{-1} region. The absorption is illustrated in the spectrum of 1-butanol (Figure 18.8).

Ethers can be identified by a process of elimination. If a compound contains an oxygen atom and the infrared spectrum lacks absorptions characteristic of a carbonyl group or a hydroxyl group, we can conclude that the compound is an ether.

EXAMPLE 18-2

Explain how you could distinguish between the following two compounds by infrared spectroscopy.

$—CH{=}CH_2$ $\qquad$ $—C{\equiv}CH$

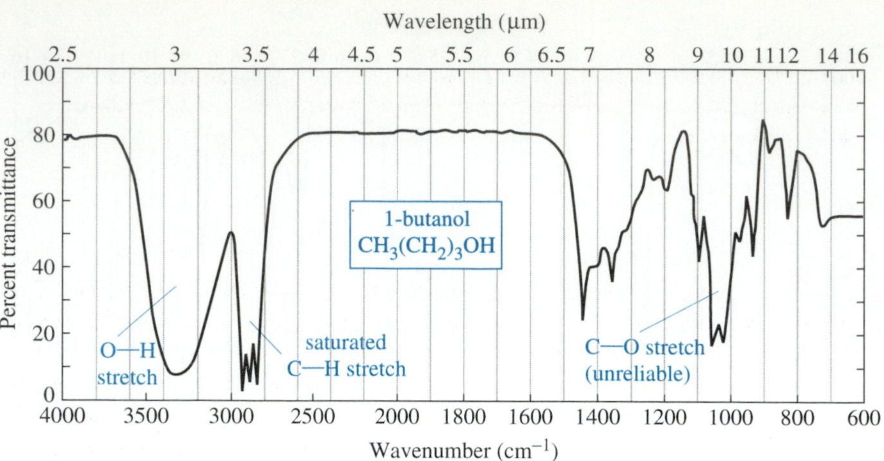

FIGURE 18.8
Infrared Spectrum of 1-Butanol

Solution

The compounds are isomeric. The compound on the left is a diene, which has two differently substituted double bonds. Thus, this compound may have two absorptions in the 1630–1670 cm^{-1} region corresponding to the two carbon-carbon double bond stretching vibrations. There should also be a C—H stretching absorption in the 3080–3140 cm^{-1} region for the hydrogen atoms bonded to sp^2-hybridized carbon atoms.

The compound on the right is an alkyne. Thus, this compound has an absorption in the 2100–2140 cm^{-1} region that corresponds to the carbon-carbon triple bond stretching vibration. There is also a C—H stretching absorption in the 3300–3320 cm^{-1} region for the hydrogen atom bonded to the *sp*-hybridized carbon atom.

Problem 18.2

How can the following two compounds be distinguished by infrared spectroscopy?

O—H　　　　CH_3 / O

I　　　　II

EXAMPLE 18-3

The carbonyl group of an acid chloride absorbs at 1800 cm^{-1}. Explain why this value is at a higher wavenumber than for an aldehyde or ketone.

$$R-\overset{\overset{\ddot{O}:}{\|}}{C}-\ddot{\underset{..}{Cl}}:$$

Solution

We recall that a chlorine atom is not effective in donating electrons by resonance but does inductively withdraw electrons. The chlorine atom of acid chlorides destabilizes the contributing polar resonance form. As a consequence, the carbonyl group of an acid chloride has more double bond character and absorbs at a higher wavenumber than the carbonyl group of an aldehyde or ketone.

Problem 18.3

The carbonyl stretching vibration of ketones is at a longer wavelength than the carbonyl stretching vibration of aldehydes. Suggest a reason for this difference.

18.5 Nuclear Magnetic Resonance Spectroscopy

Many atomic nuclei behave as if they are spinning—that is, they have a nuclear spin. For example, the ^1H nucleus spins about its axis in either of two directions, described as clockwise and counterclockwise or alternatively as α and β. Because the nucleus is charged, a magnetic moment results from the spinning nucleus. Thus, the hydrogen nucleus is a tiny magnet, with two possible orientations in the presence of an external magnetic field. The magnetic moment of the nucleus may be aligned with the external magnetic field or against it, with the former being of lower energy. If a spinning hydrogen nucleus with its magnetic moment aligned with the external field is irradiated with electromagnetic radiation in the radio-frequency range, the absorbed energy causes the nucleus to "flip" and spin in the opposite direction (Figure 18.9). Thus, absorption of energy results in a higher energy state for the hydrogen nucleus. The process is called **nuclear magnetic resonance** (NMR). Some nuclei, such as ^{12}C, do not have a nuclear spin and cannot be detected by NMR. However, the ^{13}C isotope, which does have a nuclear spin, is detectable by NMR (Section 18.7).

NMR spectroscopy uses electromagnetic radiation in the radio-frequency range. The energy associated with this radiation is very small. The energy required depends on the strength of the external magnetic field. Increasing the strength of the external magnetic field increases the energy difference between the two spin states. An NMR experiment can be done by selecting a magnetic field strength and then slowly varying the radio frequency to find the proper frequency to make the hydrogen nucleus "flip." Alternatively, a constant radio frequency may be used while the magnetic field strength is varied to cause a difference in the energy of the nuclei corresponding to the energy of the electromagnetic radiation. In practice, this latter technique is the easiest to do experimentally.

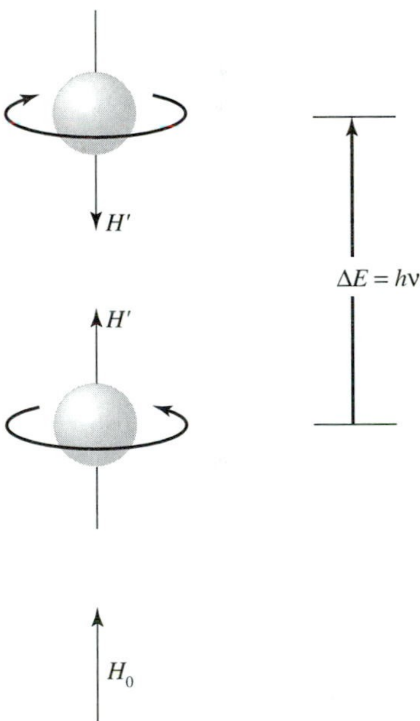

FIGURE 18.9
Absorption of Electromagnetic Radiation by a Nucleus
When the magnetic moment of a spinning nucleus (H′) is aligned with the magnetic field of an NMR spectrometer (H_0) a low energy state results. Absorption of a specific frequency causes a change in the spin of the nucleus and results in a magnetic moment opposed to the magnetic field of the instrument.

Chemical Shift

The magnetic field strength required to "flip" the spin of various hydrogen nuclei within a molecule differs. If all hydrogen nuclei absorbed the same electromagnetic radiation in an NMR experiment at the same magnetic field strength, then only a single absorption would be observed. As a consequence we would only know that the molecule contained hydrogen atoms.

The hydrogen nuclei in organic molecules are surrounded by electrons that also have spins. The electrons thus set up small local magnetic fields that are opposed to the applied external magnetic field. The local fields affect the magnetic environment of the hydrogen nuclei. When a local field opposes the external magnetic field, we say that the nucleus is **shielded.** The effective field felt by the nucleus is the applied magnetic field minus the local magnetic field generated by the electrons.

$$H_{\text{effective}} = H_{\text{applied}} - H_{\text{local}}$$

The local magnetic fields due to the spins of the electrons differ throughout the molecule because the bonding characteristics differ. Thus, the degree of shielding of each hydrogen nucleus is unique, and distinct resonances are obtained for each structurally nonequivalent hydrogen atom in a molecule. At a constant radio frequency, the external magnetic field required is larger for the more shielded nucleus.

The strengths of the local magnetic fields for various hydrogen atoms are about 10^{-6} times that of the applied magnetic field. Thus, the magnetic fields required to flip various structurally different hydrogen nuclei in a molecule differ on the order of parts per million (ppm). Rather than using absolute values of the field strength, a relative scale is used. An NMR chart is labeled on the horizontal axis with a **delta scale,** in which one delta unit (δ) is 1 ppm of the magnetic field used. The resonance for the hydrogen atoms of tetramethylsilane, $(CH_3)_4Si$, is used as a reference for the magnetic resonance spectra of hydrogen compounds. This resonance is defined as 0 δ. By convention, an absorption that occurs at lower field (higher energy) than tetramethylsilane (TMS) appears to the left of the TMS absorption and is assigned a positive δ value.

By examining the NMR spectrum, you can tell at a glance how many sets of structurally nonequivalent hydrogen atoms are contained in a molecule. Consider the NMR spectrum of 1,2,2-trichloropropane shown in Figure 18.10. The spectrum consists of two peaks pointed to the top of the graph. The resonances occur at 2.2 and 4.0 δ. There are two different sets of hydrogen atoms in 1,2,2-trichloropropane. Each set of hydrogen atoms gives rise to one peak.

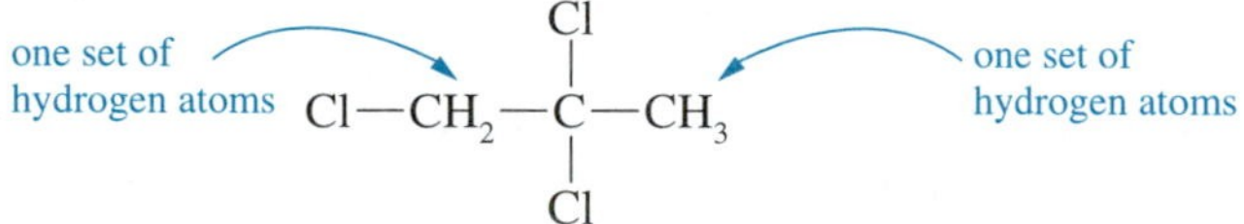

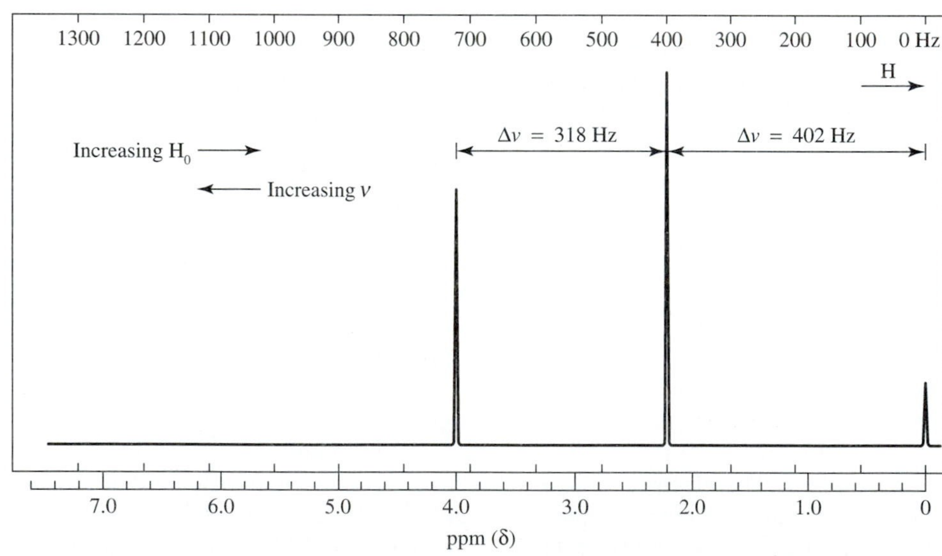

FIGURE 18.10
NMR Spectrum of 1,2,2-Trichloropropane
The three equivalent hydrogen atoms bonded to the C-3 atom absorb at 2.2 δ. The two equivalent hydrogen atoms bonded to the C-1 atom absorb at 4.0 δ.

For most organic molecules, the various hydrogen resonances appear between 0 and 10 δ. This range is conveniently divided into regions that reflect certain structural characteristics. Hydrogen atoms bonded to sp^2 carbon atoms absorb at lower fields than hydrogen atoms bonded to saturated sp^3 carbon atoms. For example, hydrogen atoms bonded to sp^2 carbon atoms absorb in the 5.0–6.5 δ region. Hydrogen atoms bonded to saturated carbon atoms without any directly bonded substituents absorb in the 0.7–1.7 δ range. The hydrogen atoms bonded to an aromatic ring occur in the general region of 6.5–8.0 δ.

The exact position of absorption for hydrogen atoms bonded to either sp^3 or sp^2 carbon atoms also depends on the number and type of substituents also bonded to the atom. Hydrogen atoms bonded to carbon atoms that are also bonded to electronegative atoms such as oxygen, nitrogen, or halogens absorb at lower fields. Examples of typical chemical shifts are listed in Table 18.3.

Relative Peak Areas

The set of hydrogen atoms bonded to the C-1 atom of 1,2,2-trichloropropane has an absorption at 4.0 δ, and the set of hydrogen atoms bonded to the C-3 atom has its absorption at 2.2 δ. This assignment is made based on the generalization that hydrogen atoms bonded to a carbon atom that is also bonded to an electronegative atom absorb at a lower field (higher δ). However, there is another method to confirm this assignment—"proton counting." The area under each resonance peak is proportional to the relative number of hydrogen atoms of each kind.

The relative area of a resonance peak is obtained from an electronic integrator used after the spectrum has been recorded. The area is proportional to the vertical displacement of a "stairstep" superimposed on the resonance peak. These vertical distances have a ratio equal to the ratio of the number of hydrogen atoms. Thus, the ratio of the integrated intensities of the two resonances of 1,2,2-trichloropropane is 3:2 (Figure 18.11).

Table 18.3 Chemical Shifts of Hydrogen Atoms

Partial Structural Formula	Chemical Shift (ppm)	Partial Structural Formula	Chemical Shift (ppm)
$-CH_3$	0.7–1.3	$Br-\overset{|}{\underset{|}{C}}-H$	2.5–4.0
$-CH_2-$	1.2–1.4		
$-\overset{|}{\underset{|}{C}}-H$	1.4–1.7	$I-\overset{|}{\underset{|}{C}}-H$	2.0–4.0
$>C=\underset{|}{C}-CH_3$	1.6–1.9	$-O-\overset{|}{\underset{|}{C}}-H$	3.3–4.0
$-\overset{O}{\overset{\|}{C}}-CH_3$	2.1–2.4	$>C=\underset{|}{C}-H$	5.0–6.5
$-C\equiv C-H$	2.5–2.7	$Ar-H$	6.5–8.0
$Cl-\overset{|}{\underset{|}{C}}-H$	3.0–4.0	$-\overset{O}{\overset{\|}{C}}-H$	9.7–10.0

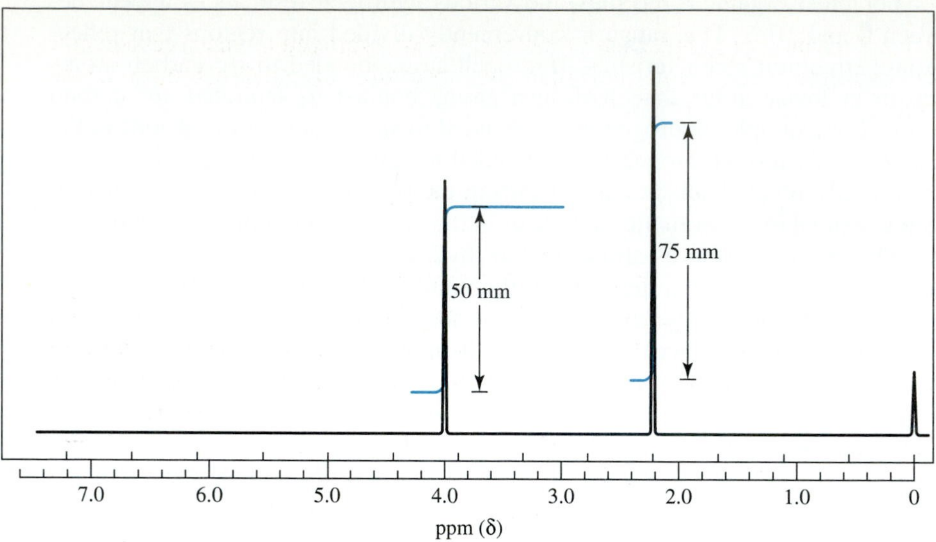

FIGURE 18.11
Integrated Intensities of an NMR Spectrum
The area of each resonance peak is proportional to the number of hydrogen atoms. The vertical distance of the "stair steps" shown is proportional to those areas. The ratio of 75 to 50 mm is 3:2, which corresponds to the number of hydrogen atoms bonded to the C-3 and C-1 atoms, respectively, in 1,2,2-trichloropropane.

EXAMPLE 18-4

Predict the chemical shift of the resonances of 1,2-dichloro-2-methylpropane. What are the relative intensities of the absorptions?

Solution

Write the structure of the compound to determine the number of sets of nonequivalent hydrogen atoms.

$$CH_3-\underset{\displaystyle Cl}{\overset{\displaystyle CH_3}{\overset{|}{\underset{|}{C}}}}-CH_2-Cl$$

One set of hydrogen atoms is located at the C-1 atom, which has one chlorine atom bonded to it. These hydrogen atoms should have a resonance near 4 δ. The six hydrogen atoms of the two equivalent methyl groups should have a resonance near 1 δ. The ratio of the two sets of hydrogen atoms is 6 : 2. The ratio of the relative areas at 1 and 4 δ should be 3 : 1.

Problem 18.4

How many sets of nonequivalent hydrogen atoms are contained in each of the following isomeric ketones?

(a) $CH_3-CH_2-\overset{\displaystyle O}{\overset{\|}{C}}-CH_2-CH_3$ (b) $CH_3-\overset{\displaystyle O}{\overset{\|}{C}}-CH_2-CH_2-CH_3$

(c) $CH_3-\overset{\displaystyle O}{\overset{\|}{C}}-\underset{\displaystyle CH_3}{\underset{|}{CH}}-CH_3$

18.6 Spin-Spin Splitting

In 1,2,2-trichloropropane, each of the two sets of hydrogen atoms is responsible for a single peak. Now consider the spectrum of 1,1,2-tribromo-3,3-dimethylbutane (Figure 18.12).

$$\begin{array}{c} \quad CH_3 \;\; Br \\ \quad | \qquad | \\ CH_3{-}C{-}CH{-}CHBr_2 \\ | \\ CH_3 \end{array}$$

1,1,2-tribromo-3,3-dimethylbutane

The nine equivalent hydrogen atoms of the three equivalent methyl groups give rise to the intense peak at 1.2 δ. The single hydrogen atoms bonded to the C-1 and C-2 atoms are nonequivalent. The resonance of the hydrogen atom at the C-1 atom is located at 6.4 δ as a consequence of the two electronegative bromine atoms bonded to that carbon atom. The resonance of the hydrogen atom at the C-2 atom is located at 4.4 δ because only one bromine atom is bonded to the carbon atom. Thus, the intensities of the absorptions and the chemical shifts of the hydrogen atoms are as expected based on molecular structure.

Both the 4.4 δ and 6.4 δ absorptions of 1,1,2-tribromo-3,3-dimethylbutane are "split" as shown in the inserts containing expanded representations of the resonances. Each area contains two peaks called **doublets.** The phenomenon of multiple peaks is common in NMR spectroscopy. Other common multiplets include **triplets** and **quartets,** meaning that resonances are split into three and four peaks, respectively. This characteristic of a set of peaks associated with a given type of nucleus is called its **multiplicity.**

Multiple absorptions for a set of equivalent hydrogen atoms is known as **spin-spin splitting.** It results from the interaction of the nuclear spin(s) of one or more nearby "neighboring" hydrogen atom(s) with that set of equivalent hy-

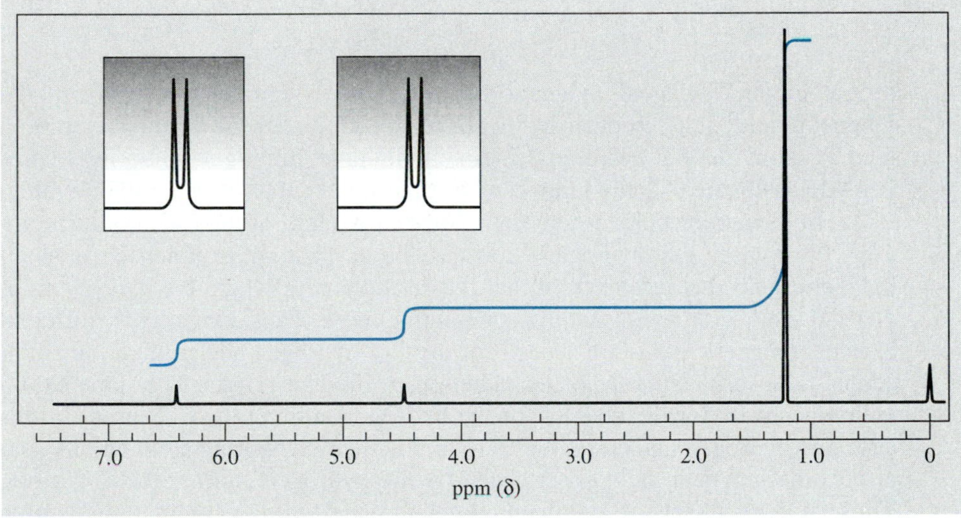

FIGURE 18.12

NMR Spectrum of 1,1,2-Tribromo-3,3-dimethylbutane

The inserts show the doublets for the resonances at 4.4 and 6.4 δ. The total integrated area of each doublet is proportional to one hydrogen atom.

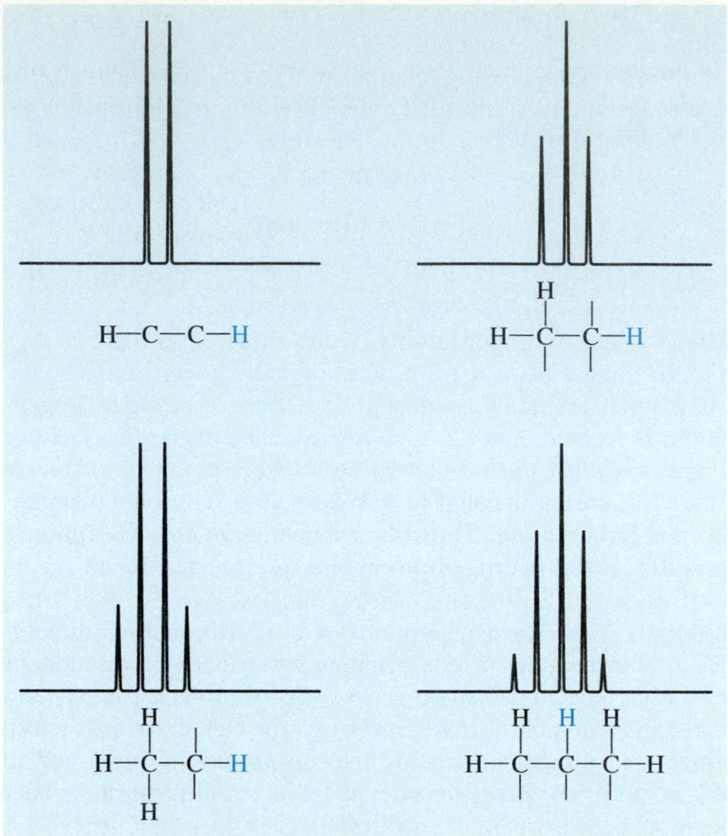

FIGURE 18.13
Common NMR Multiplets
The resonance of a common single hydrogen atom in different environments is shown. The number of equivalent neighboring hydrogen atoms is responsible for the multiplicity of the resonance.

drogen atoms. The small magnetic field of nearby hydrogen nuclei affects the magnetic field "felt" by other hydrogen nuclei. Consider the hydrogen atom on the C-1 atom of 1,1,2-tribromo-3,3-dimethylbutane. It has a neighboring hydrogen nucleus at the C-2 atom that can be spinning in either of two possible directions. In those molecules where the hydrogen nucleus at the C-2 atom is spinning clockwise, the hydrogen atom at the C-1 atom experiences a small magnetic field that differs from that felt in molecules where the hydrogen nucleus at the C-2 atom is spinning counterclockwise. Thus, two slightly different external magnetic fields are needed for the C-1 hydrogen nuclei in various molecules to absorb electromagnetic radiation. A doublet results. The same explanation accounts for the doublet for the hydrogen atom at the C-2 atom. In this case, the hydrogen nucleus at the C-1 atom is the neighboring atom and its spin affects the magnetic field experienced by the hydrogen atom at the C-2 atom. Thus, in general, sets of hydrogen atoms on neighboring carbon atoms **couple** with each other. If hydrogen atom A couples and causes splitting of the resonance for hydrogen atom B, the resonance for hydrogen atom B is also split by hydrogen atom A. The resonance for the nine methyl hydrogen atoms of 1,1,2-

tribromo-3,3-dimethylbutane is not "split" because the neighboring quaternary carbon atom has no hydrogen atoms.

A set of one or more hydrogen atoms that has n equivalent neighboring hydrogen atoms has $n + 1$ peaks in the NMR spectrum. The appearance of several sets of multiplets resulting from $n = 1$ through $n = 4$ is shown in Figure 18.13 for some common structures. The areas of the component peaks of a doublet are equal; the areas of the component peaks of other multiplets are not.

To understand the relative peak areas of multiplets resulting from more than one neighboring hydrogen atom, let's consider the spectrum of 1,1,2-trichloroethane (Figure 18.14).

```
      Cl  H
      |   |
  Cl—C—C—Cl
      |   |
      H   H
```

1,1,2-trichloroethane

The doublet near 4 δ corresponds to the two hydrogen atoms bonded to the C-2 atom. These hydrogen atoms have one neighboring hydrogen nucleus that spins in either of two directions we can designate as α or β. As a consequence, the C-2 hydrogen atoms in various molecules experience two different magnetic fields, resulting in a doublet absorption. Now let's consider the triplet resonance for the C-1 hydrogen atom. The spins of the neighboring two hydrogen nuclei at the C-2 atom can be $\alpha\alpha$, $\alpha\beta$, $\beta\alpha$, and $\beta\beta$. Because the magnetic fields generated by either $\alpha\beta$ or $\beta\alpha$ sets of spins are equivalent, the hydrogen atom at the C-1 atom experiences three different magnetic fields in the ratio of 1:2:1. This ratio is the same as the ratio of the component peaks of the observed triplet.

Extension of the possible combinations of spins of n neighboring equivalent hydrogen atom accounts for the area ratios listed in Table 18.4.

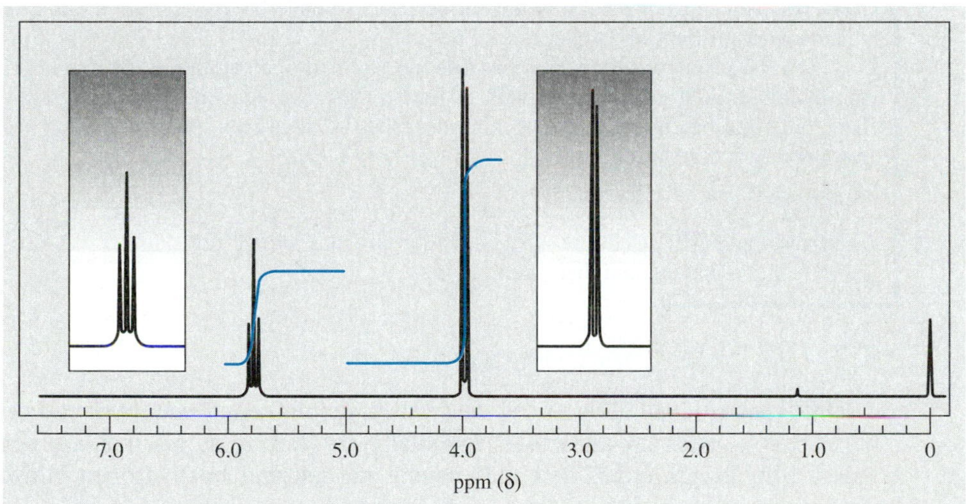

FIGURE 18.14

NMR Spectrum of 1,1,2-Trichloroethane

The insets show the doublet for the 4.0 δ resonance of the hydrogen atoms bonded to the C-2 atom and the triplet for the 5.8 δ resonance of the hydrogen atom bonded to the C-1 atom. Note that the integrated intensities of the 4.0 and 5.8 δ resonances are in the ratio of 2:1.

Table 18.4 Number of Peaks of Multiplets and Area Ratios

Number of equivalent adjacent hydrogens	Total number of peaks	Area ratios
0	1	1
1	2	1:1
2	3	1:2:1
3	4	1:3:3:1
4	5	1:4:6:4:1
5	6	1:5:10:10:5:1
6	7	1:6:15:20:15:6:1

EXAMPLE 18-5

Describe the NMR spectrum of 2-chloropropane.

Solution

First draw the structure of 2-chloropropane to determine the number of sets of non-equivalent hydrogen atoms and the number of neighboring hydrogen atoms that can couple with each set.

$$CH_3-\underset{\displaystyle H}{\overset{\displaystyle Cl}{C}}-CH_3$$

The six hydrogen atoms located on the equivalent C-1 and C-3 atoms are equivalent. They have a resonance in the 1 δ region. The single hydrogen atom at the C-2 atom has a resonance near 4 δ because a chlorine atom is bonded to that carbon atom. The relative areas of the 1 and 4 δ peaks are 6 : 1.

The 4 δ resonance is a septet because the C-2 hydrogen atom has six neighboring hydrogen atoms that couple with it. Each of the two equivalent methyl groups have only one neighboring hydrogen atom—the C-2 hydrogen atom. Thus, the 1 δ resonance is a doublet.

Problem 18.5

Describe the NMR spectrum of 1,3-dichloropropane citing the number of resonances, their δ values, and their multiplicities.

18.7 ^{13}C NMR Spectroscopy

^{13}C NMR spectroscopy allows us to detect the structural environment of carbon atoms. This is often an advantage, especially for carbon atoms that are not bonded to hydrogen atoms and thus cannot be detected by hydrogen NMR spectroscopy.

NMR spectra can be easily obtained for many isotopes with half-integer spins, such as ^{19}F and ^{31}P, because their natural abundance is 100%. The detection of the isotope ^{13}C is more difficult because it has an abundance of only 1%. However, by using larger samples than are used for ^{1}H NMR spectroscopy and specialized instrumentation, it is possible to obtain ^{13}C NMR spectra.

Magnetic Resonance Imaging

In the 1940s, equipment for NMR spectroscopy was developed by physicists. Their studies were limited by the difficulty of obtaining homogeneous magnetic fields of sufficient strength to distinguish between environmentally different sites for a nucleus such as hydrogen. The method was not widely used because the instruments had to be "built" in each laboratory. By the 1960s, commercially available instruments were widely used in universities and in industry to determine routinely the structures of compounds. The field strengths increased and the effective cost of the instrumentation decreased to such a point that instruments were placed in teaching laboratories. Solutions of samples contained within a tube the size of a pencil could be easily placed within a chamber located between the poles of a magnet. By the 1980s, instruments were developed in which a patient's body could be placed between the poles of large electromagnets. These instruments are now routinely available for medical diagnosis in most cities in the United States.

NMR has developed as a valuable method in diagnostic medicine and is used to "see" tissues such as the brain and spinal cord. X-ray images, which are still used for evaluating the condition of bones, cannot provide the detailed information about tissues that are available by NMR studies. Although NMR is the method, the public knows of the technique as MRI, which is an acronym for magnetic resonance imaging. This name was selected because it was thought that the term *nuclear* carried with it the concerns of the general public about radiation of radioactive isotopes.

MRI techniques are not used to obtain information about the identity of compounds within the body. Rather, MRI is used to see the general shape and condition of tissues based on the fact that they all contain hydrogen in water located within tissues. Variations from the norm in the water-density patterns may be detected, and medical personnel use this information in diagnosis. For example, the white and gray matter of the brain can be seen in pictures developed by MRI methods. Thus, conditions such as multiple sclerosis can be evaluated.

In addition to the water density of tissues, MRI determines information about the relaxation times of the hydrogen nuclei at each site. A nucleus in an "excited" high energy state must "relax" by giving off its energy and returning to the lower energy state. This phenomenon, which is recorded as relaxation time, occurs by two mechanisms. The transfer

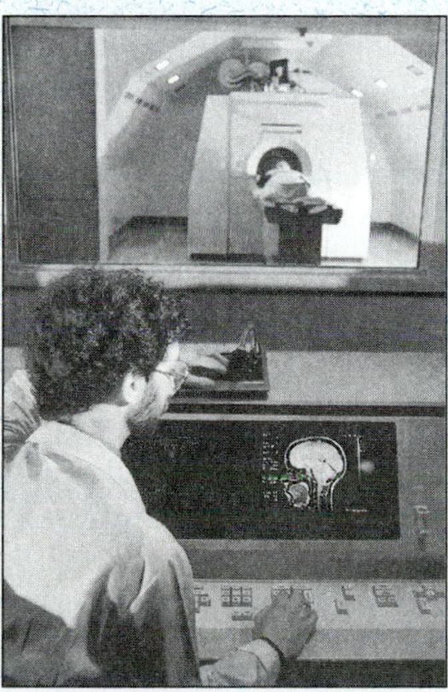

of energy from one nucleus to another is designated as T_2. This process results in no net loss of energy among the collection of nuclei. The transfer of energy from a nucleus to the surroundings as heat is designated as T_1.

Diagnosis using MRI is based on empirical correlations between the images and the knowledge about tissues developed by standard anatomical procedures. However, there are several advantages to MRI methods. The method is noninvasive and requires relatively little time as compared to an operation. In addition, the method provides early warnings of developing abnormalities such as small tumors. The T_1 relaxation times of cancer cells are longer than for normal tissue. Although the methods of determining the relaxation time and how the data are formulated into an image are beyond the scope of this text, their use is standard practice in medical laboratories.

Pictures of sections of the human body such as the head are shown as cross sections. These pictures result from computer analysis of data obtained using magnetic field gradients. There is a linear relationship between the radio frequency used and the strength of the magnetic field required to "see" a hydrogen atom. The behavior of nuclei in different environments then differs with changes in the field strength. By varying the magnetic field along an axis, the computer image generated is a planar cross section. It is these cross sections that are evaluated. A series of cross sections provides a three-dimensional image. ■

Let's consider the location of the ^{13}C isotope in a compound such as 2-butanol. Most of the carbon atoms are ^{12}C, which has no nuclear spin. The probability is equal for the location of ^{13}C at any of the positions in a molecule. The probability of finding a ^{13}C at C-1 of a molecule is 1%. The probability of finding a ^{13}C at C-2 is also 1%, and so on. The probability of finding two or more ^{13}C in the same molecule and simultaneously bonded to each other is very low. For example, the probability of finding ^{13}C in the same molecule at both C-1 and C-2 is only 0.01%. As a result, a ^{13}C NMR spectrum shows a sum of the signals generated by individual atoms at all of the possible sites in a collection of isotopically substituted molecules. The observed spectrum therefore resembles that expected for a molecule with ^{13}C located at every position but without coupling between the ^{13}C isotopes.

Characteristics of ^{13}C Spectra

The ^{13}C spectra of organic compounds are shown using a δ scale relative to the resonance of ^{13}C in TMS. The chemical shift of ^{13}C shows many of the same trends as hydrogen chemical shifts. However, the range of chemical shifts for ^{13}C is very much larger, on the order of 200 ppm (Table 18.5). Thus, the chemical shifts are very sensitive to changes in structural environment. As a result, it is usually possible to "see" distinct signals for every nonequivalent ^{13}C in a molecule.

The resonances for ^{13}C are split by hydrogen atoms. The rules for the multiplicity of a ^{13}C resonance split by hydrogen are the same as for hydrogen coupled to hydrogen. The multiplicity is $n + 1$ for *n* equivalent neighboring hydrogen atoms. The largest coupling is observed for ^{13}C directly bonded to hydrogen—that is, a one-bond coupling. Coupling of ^{13}C with hydrogen atoms farther away is small compared with the ^{13}C chemical shift scale. By using specialized experimental methods, all splitting by hydrogen atoms other than by one-bond coupling can be eliminated.

The ^{13}C spectrum of 2-butanol is shown in Figure 18.15. The four carbon atoms are clearly seen as four regions each consisting of multiplets. The signal

Table 18.5 Chemical Shifts of ^{13}C

Carbon atom	Chemical shift (ppm)
RCH_2CH_3 (CH₃ carbon)	12–15
RCH_2CH_3 (CH₂ carbon)	16–25
R_3CH	12–35
C_6H_5–CH_2R (CH₂ carbon)	20–30
$CH_3C(=O)R$ (CH₃ carbon)	30
RCH_2Cl	40–45
RCH_2Br	27–35
RCH_2OH	50–65
RCH_2OR	50–65
$RCH{=}CH_2$ (CH₂ carbon)	115–120
$RCH{=}CH_2$ (CH carbon)	125–140
C_6H_5 (aromatic C–H)	120–170
$RC(=O)OR$ (carbonyl carbon)	170–175
$RC(=O)H$ (carbonyl carbon)	190–200
$RC(=O)R$ (carbonyl carbon)	205–220

The chemical shift value is for the carbon atom shown in color in the generalized structures.

at lowest field is assigned to C-2 because that atom is deshielded by an oxygen atom. This assignment is confirmed because the signal is a doublet that results from the coupling of the directly bonded hydrogen atom. The triplet at 32.3 δ is assigned to the C-3 atom because it has two hydrogen atoms bonded to it.

The two quartets in the spectrum of 2-butanol correspond to the C-1 and C-4 methyl groups. Each has three hydrogen atoms, which are responsible for the observed quartets. Assignment of the lower field quartet to C-1 is based on its proximity to the oxygen atom, which deshields that carbon atom by an inductive effect. The higher field quartet is due to the C-4 methyl group.

Proton-Decoupled Spectra

There is a convenient method that eliminates the multiplicity of all resonances and reduces the spectrum to singlets for each nonequivalent carbon atom, making it easier to determine structures using ^{13}C NMR. The splitting of a resonance for a ^{13}C atom by hydrogen can be eliminated to generate a singlet by a

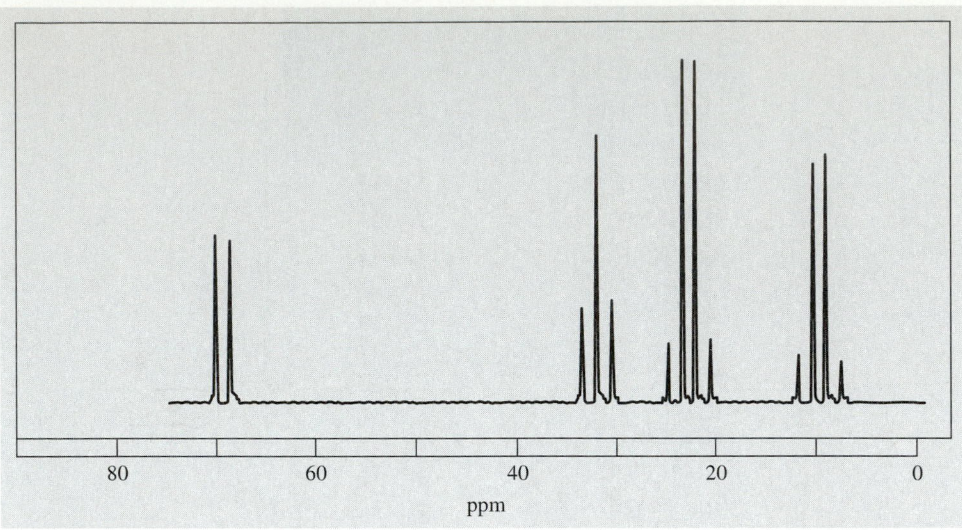

FIGURE 18.15
C^{13} NMR Spectrum of 2-Butanol

technique called **proton decoupling.** The resulting spectrum is called a **proton-decoupled NMR spectrum.**

The proton-decoupled spectrum of 2-butanol is greatly simplified (Figure 18.16). It consists of four signals corresponding to the four nonequivalent carbon atoms. Consequently, we can eliminate the isomers 2-methyl-1-propanol and 2-methyl-2-propanol as possible structures, because the spectra of these compounds would show three and two signals, respectively.

$$CH_3-\overset{\overset{\large CH_3}{|}}{\underset{\underset{\large H}{|}}{C}}-CH_2OH \qquad CH_3-\overset{\overset{\large CH_3}{|}}{\underset{\underset{\large OH}{|}}{C}}-CH_3$$

2-methyl-1-propanol 2-methyl-2-propanol

Counting Carbon Atoms

We note that the intensities of the four nonequivalent carbon atoms of 2-butanol are not equal. Unlike hydrogen NMR spectroscopy, the method used to obtain ^{13}C spectra gives peak intensities that are not proportional to the number of carbon atoms. The signals for carbon atoms bearing more hydrogen atoms tend to be larger than for carbon atoms bearing fewer hydrogen atoms. Carbon atoms without hydrogen atoms, such as quaternary carbon atoms and ketone-carbonyl carbon atoms, have the lowest intensity signals. However, substituents also affect the intensity of the signal of a carbon atom. As a result, we cannot accurately "count" the number of equivalent carbon atoms responsible for a resonance.

The number of equivalent carbon atoms can be determined by comparing the number of signals in a ^{13}C spectrum with the number of carbon atoms in the molecular formula. If some of the carbon atoms in a molecule are equivalent, the number of signals is reduced. As a result, ^{13}C spectroscopy is quite useful in determining the symmetry of a molecule.

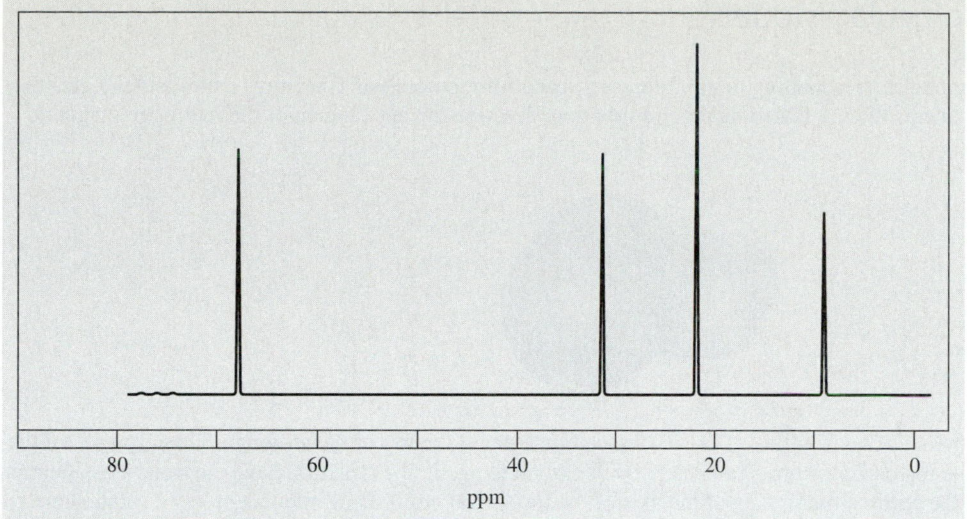

FIGURE 18.16
Proton Decoupled ^{13}C Spectrum of 2-Butanol

The determination of structure using ^{13}C depends on a one-to-one correspondence between the number of sets of equivalent carbon atoms and the number of signals in the spectrum. Although of rare occurrence, two nonequivalent carbon atoms can have the same chemical shift. For example, the ^{13}C spectrum of 1-octene contains only seven signals, even though there are eight nonequivalent carbon atoms.

EXAMPLE 18-6

The isomeric alcohols 3-heptanol and 4-heptanol cannot be easily distinguished by hydrogen NMR spectroscopy. Describe how ^{13}C NMR spectroscopy can be used to distinguish between these isomers.

Solution

First draw the structures of the isomers to determine the number of equivalent and nonequivalent carbon atoms.

$$CH_3CH_2CH_2\underset{\displaystyle OH}{\underset{|}{C}}HCH_2CH_2CH_3 \qquad CH_3CH_2\underset{\displaystyle OH}{\underset{|}{C}}HCH_2CH_2CH_2CH_3$$

4-heptanol 3-heptanol

4-Heptanol has only four sets of nonequivalent carbon atoms as a result of the symmetry of the molecule. The C-1 and C-7 atoms are equivalent; the C-2 and C-6 atoms are equivalent; the C-3 and C-5 atoms are equivalent. The C-4 atom is unique. For 3-heptanol, all carbon atoms are nonequivalent. Thus, the two compounds can be distinguished based on the number of resonances in proton-decoupled spectra of the two compounds, four for 4-heptanol and seven for 3-heptanol.

Problem 18.6

How can a compound of molecular formula $C_4H_{10}O$ be established as an ether or an alcohol using ^{13}C NMR spectroscopy?

Explorations with Molecular Models

1. Construct a ball-and-stick model corresponding to the following space-filling model of 1-bromo-1-chloroethene. Are the hydrogen atoms equivalent or nonequivalent? Based on this information, describe the appearance of the hydrogen magnetic resonance spectrum.

2. Construct a ball-and-stick model of adamantane. Classify the carbon atom at the top of the structure. Classify each of the three carbon atoms bonded to the top carbon atom. How many other carbon atoms in the structure have the same classification as the top carbon atom? Rotate the entire structure about the axis of the top C—H bond. How many degrees of rotation are required until an equivalent depiction of the structure results? Based on the symmetry of adamantane, how many "peaks" should be seen in the hydrogen-decoupled ^{13}C NMR spectrum?

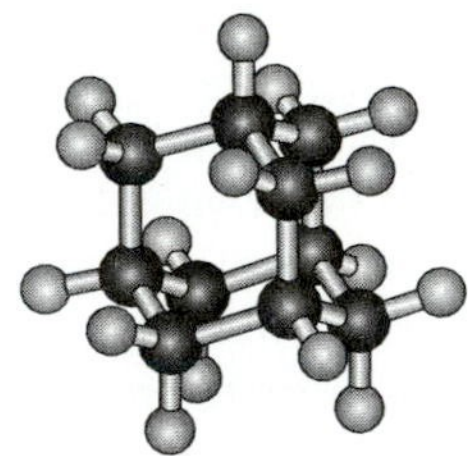

3. Construct a ball-and-stick model of *trans*-1-chloro-4-methylcyclohexane. Based on the symmetry of the structure, how many "peaks" should be seen in the hydrogen-decoupled ^{13}C NMR spectrum?

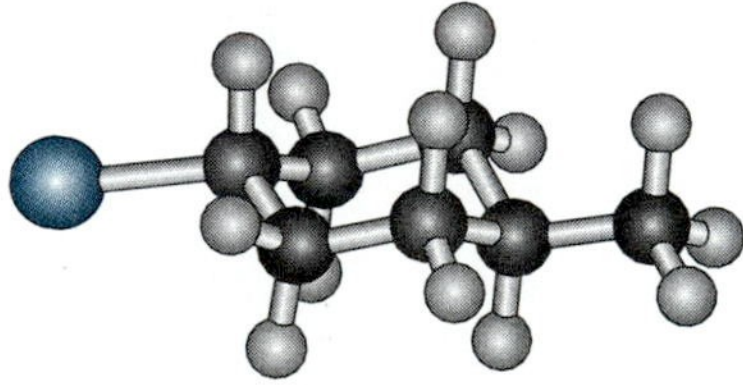

4. Construct a ball-and-stick model corresponding to the following space-filling model of 1,2,3-trimethylbenzene. Based on the symmetry of the structure, how many resonances that correspond to the methyl groups should be observed in the hydrogen NMR spectrum? What should be their relative intensities? How many resonances that correspond to the hydrogen atoms bonded to the benzene ring should be observed in the hydrogen NMR spectrum? What should be their relative intensities? What should be their multiplicities?

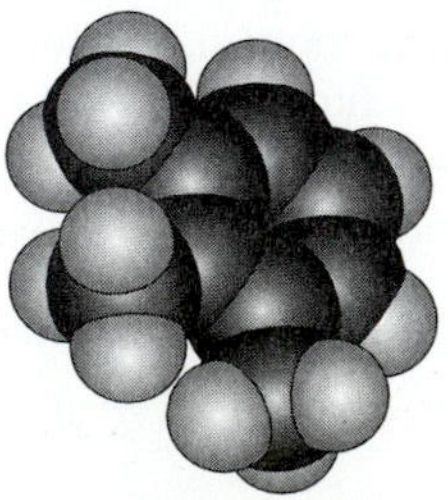

Exercises

Ultraviolet Spectroscopy

18.1 The λ_{max} values of naphthalene, anthracene, and tetracene are 314, 380, and 480 nm, respectively. Suggest a reason for this order of wavelength of absorption. Are any of the compounds colored?

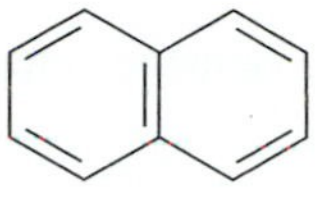

naphthalene 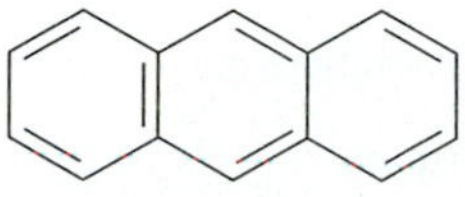 anthracene 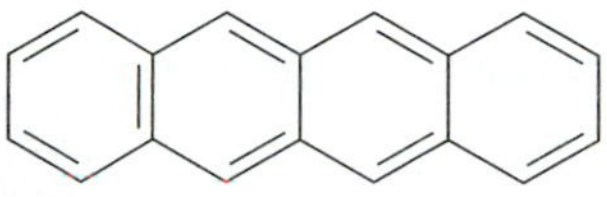 tetracene

18.2 How many conjugated double bonds are contained in lycopene? Compare the conjugation in this compound to that of β-carotene. Based on this information, predict the color of lycopene.

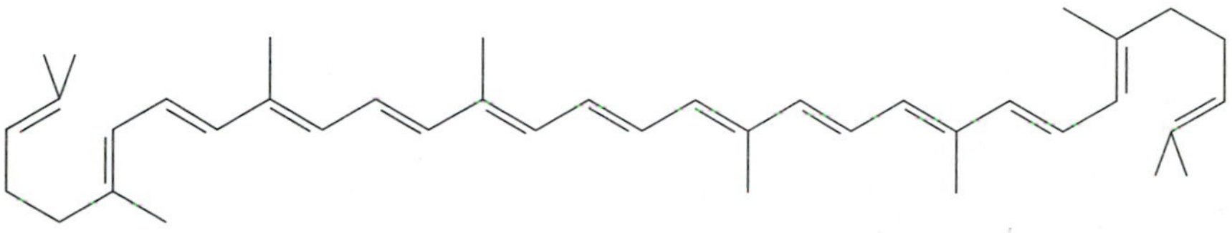

lycopene

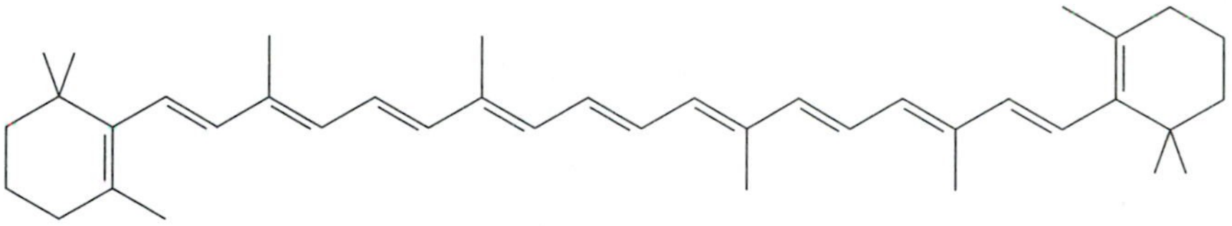

β-carotene

18.3 How might 2,4-hexadiyne be distinguished from 1,4-hexadiyne by ultraviolet spectroscopy?

$CH_3-C\equiv C-C\equiv C-CH_3$ (2,4-hexadiyne) $CH_3-C\equiv C-CH_2-C\equiv CH$ (1,4-hexadiyne)

18.4 One of the following unsaturated ketones has λ_{max} = 225 nm and the other has λ_{max} = 252 nm. Assign each value to the proper structure.

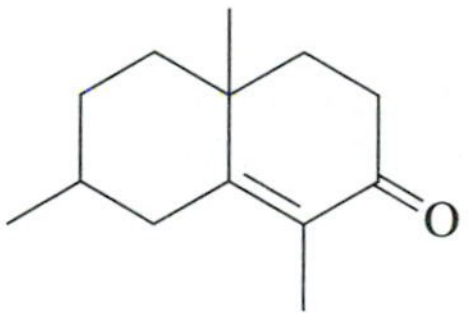

(I) 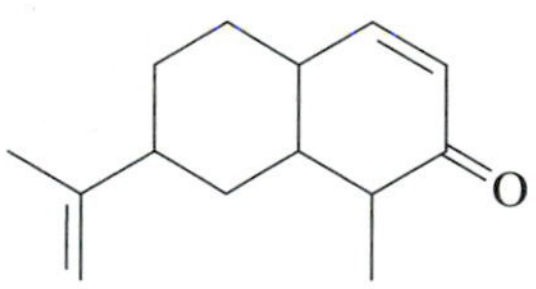 (II)

Infrared Spectroscopy

18.5 How could infrared spectroscopy be used to distinguish between propanone and 3-propen-1-ol?

$CH_3-C(=O)-CH_3$ (propanone) $CH_2{=}CH-CH_2OH$ (3-propen-1-ol)

18.6 How could infrared spectroscopy be used to distinguish between 1-pentyne and 2-pentyne?

$CH_3-CH_2-CH_2-C\equiv C-H$ (1-pentyne) $CH_3-C\equiv C-CH_2-CH_3$ (2-pentyne)

18.7 How could infrared spectroscopy be used to distinguish between the following isomeric compounds?

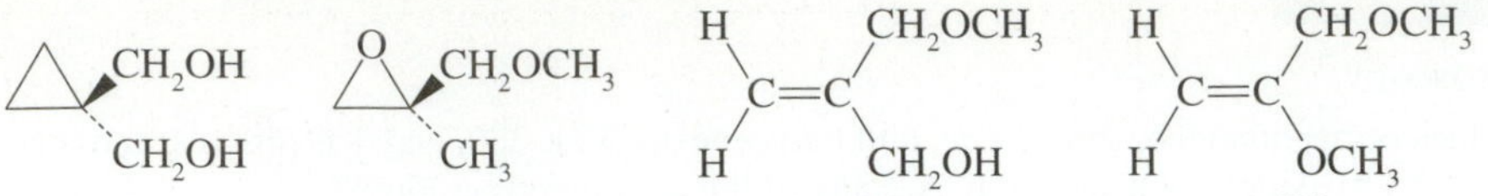

(I) cyclopropane with CH_2OH and CH_2OH; (II) epoxide with CH_2OCH_3 and CH_3; (III) H, H, $C{=}C$, CH_2OCH_3, CH_2OH; (IV) H, H, $C{=}C$, CH_2OCH_3, OCH_3

18.8 How could infrared spectroscopy be used to distinguish between the following isomeric compounds?

(I) 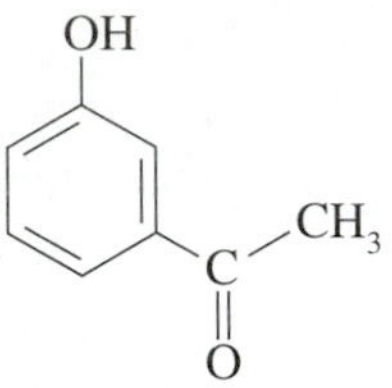 benzene ring with OH and $C(=O)CH_3$; (II) 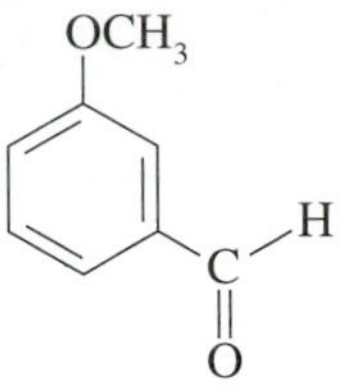 benzene ring with OCH_3 and $C(=O)H$; (III) 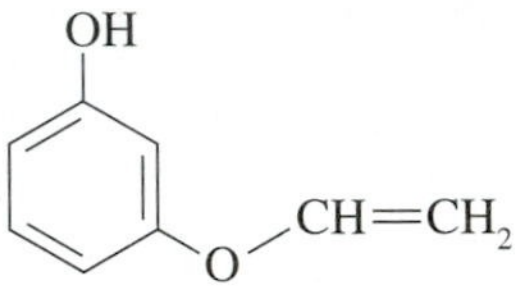 benzene ring with OH and $O{-}CH{=}CH_2$

18.9 An infrared spectrum of a compound with molecular formula $C_4H_6O_2$ has a intense broad band between 3500 and 3000 cm^{-1} and an intense peak at 1710 cm^{-1}. Which of the following compounds best fits these data?

(I) $CH_3CH_2CO_2CH_3$ (II) $CH_3CO_2CH_2CH_3$ (III) $CH_3CH_2CH_2CO_2H$

18.10 Dehydration of 1-methylcyclohexanol gives two isomeric alkenes. The minor product has a more intense C=C stretching vibration than the major product. Assign the structures of these compounds.

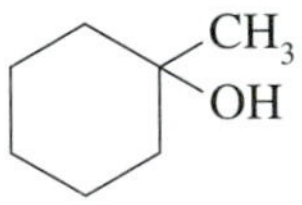

CH_3, OH

1-methylcyclohexanol

Proton Magnetic Resonance Spectroscopy

18.11 The NMR spectrum of a compound with molecular formula $C_3H_6Cl_2$ has only a singlet at 2.8 δ. What is the structure of the compound?

18.12 The NMR spectrum of a compound with molecular formula C_4H_9Br has only a singlet at 1.8 δ. What is the structure of the compound?

18.13 The NMR spectrum of a compound with the formula $C_5H_{11}Br$ consists of a singlet at 1.1 δ (9 H atoms) and a singlet at 3.2 δ (2 H atoms). What is the structure of the compound?

18.14 The NMR spectrum of a compound with the formula $C_7H_{15}Cl$ consists of a singlet at 1.1 δ (9 H atoms) and a singlet at 1.6 δ (6 H atoms). What is the structure of the compound?

18.15 The NMR spectrum of iodoethane has resonances at 1.9 and 3.1 δ. What are the intensities of the two resonances? What are the multiplicities?

18.16 The NMR spectrum of 1-chloropropane has resonances at 1.0 δ, 1.8 δ, and 3.5 δ. What are the intensities of the resonances? What are the multiplicities?

18.17 The NMR spectrum of an ether with molecular formula $C_5H_{12}O$ has singlets at 1.1 and 3.1 δ whose intensities are 3:1. What is the structure of the compound?

18.18 The NMR spectrum of an ether with molecular formula $C_6H_{14}O$ consists of an intense doublet at 1.0 δ and a septet at 3.6 δ. What is the structure of the ether?

18.19 Describe the differences in the NMR spectra of the following two esters.

(I) $CH_3CH_2CO_2CH_3$ (II) $CH_3CO_2CH_2CH_3$

18.20 Two isomeric esters having the molecular formula $C_5H_{10}O_2$ have different NMR spectra. One has the following characteristics: 1.0 δ (doublet, 6 H atoms), 2.0 δ (multiplet, 1 H atom), 4.0 δ (doublet, 2 H atoms), and 8.1 δ (singlet, 1 H atom). The other has the following characteristics: 1.2 δ (triplet, 3 H atoms), 1.4 δ (triplet, 3 H atoms), 2.5 δ (quartet, 2 H atoms), and 4.1 δ (quartet, 2 H atoms). What is the structure of each compound?

18.21 Determine the structure of a compound having the molecular formula C_4H_9Cl based on the following spectrum.

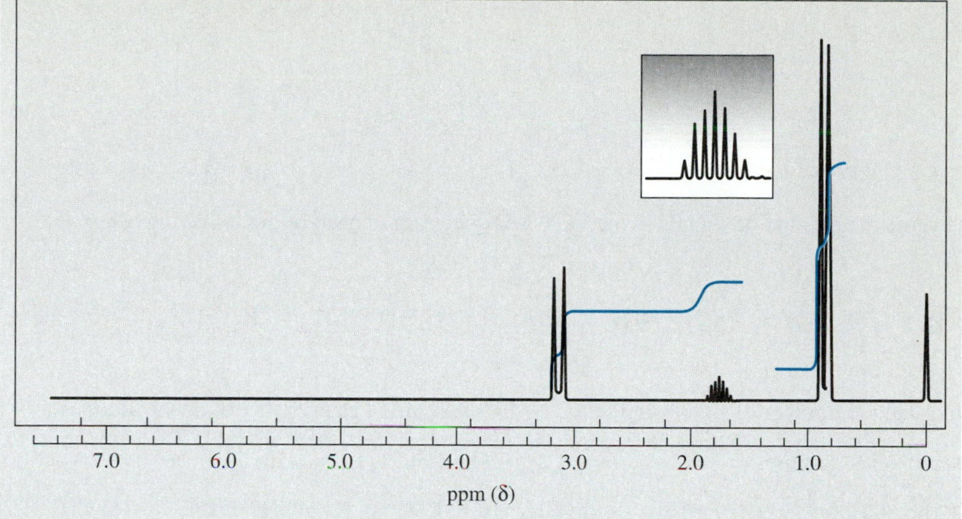

18.22 Determine the structure of a compound having the molecular formula $C_3H_6Cl_2$ based on the following spectrum.

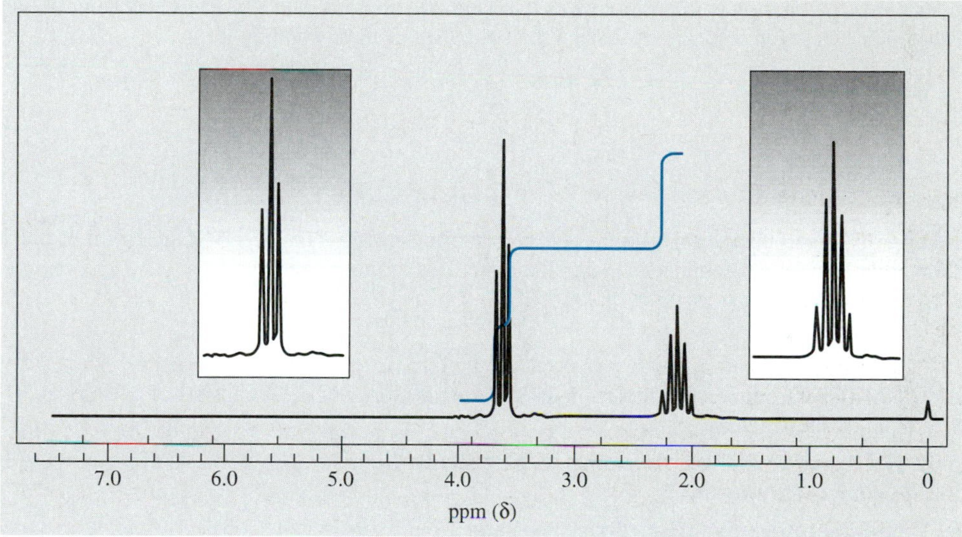

Carbon-13 NMR

18.23 Determine the number of signals in the ^{13}C NMR spectrum of each of the following aromatic compounds.

(a) CH_3, CH_3, CH_3 (b) Br, Br, Br (c) CH_3, Br

18.24 Consider the following isomeric hydrocarbons. One compound has a ^{13}C NMR spectrum with absorptions at 20.8, 30.7, and 31.4 δ. A second compound has a ^{13}C NMR spectrum with absorptions at 16.0, 23.6, 31.3, and 34.2 δ. A third compound has a ^{13}C NMR spectrum with absorptions at 23.0, 26.5, 32.9, 35.1, and 44.6 δ. Assign these features to the appropriate hydrocarbon.

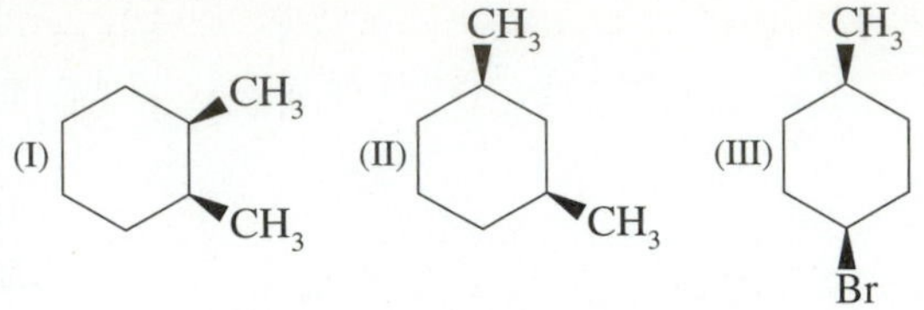

(I) (II) (III)

18.25 Determine the number of signals in the ^{13}C NMR spectrum of each of the following esters.

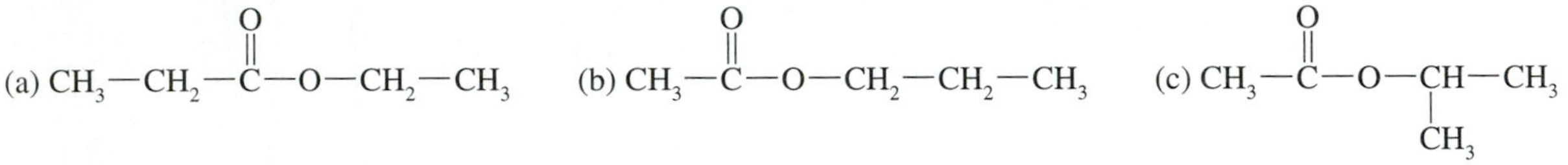

(a) $CH_3-CH_2-C(=O)-O-CH_2-CH_3$ (b) $CH_3-C(=O)-O-CH_2-CH_2-CH_3$ (c) $CH_3-C(=O)-O-CH(CH_3)-CH_3$

18.26 Determine the number of signals in the ^{13}C NMR spectrum of each of the following alcohols.

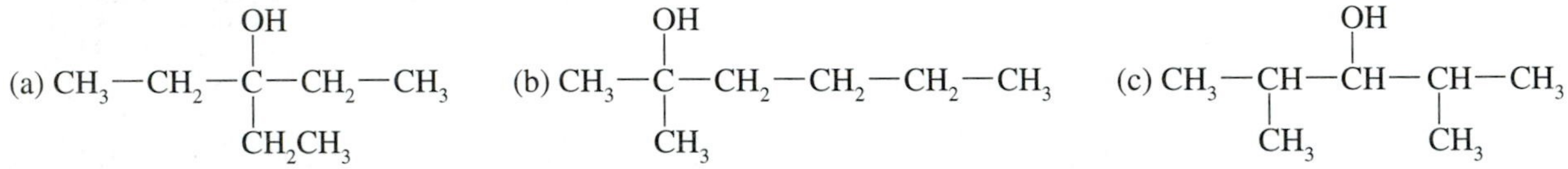

(a) $CH_3-CH_2-C(OH)(CH_2CH_3)-CH_2-CH_3$ (b) $CH_3-C(OH)(CH_3)-CH_2-CH_2-CH_2-CH_3$ (c) $CH_3-CH(CH_3)-CH(OH)-CH(CH_3)-CH_3$

18.27 Dehydration of 1-methylcyclohexanol gives two isomeric alkenes. The ^{13}C NMR spectrum of one has five resonances and the other has seven resonances. Assign the structures of these compounds.

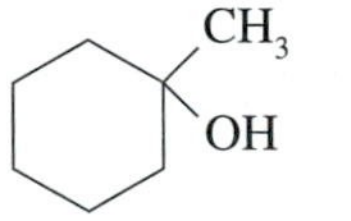

1-methylcyclohexanol

18.28 Hydration of 2-pentyne gives a mixture of two isomeric ketones. The ^{13}C NMR spectrum of one has five resonances and the other has three resonances. Assign the structures of these compounds.

$CH_3-C\equiv C-CH_2-CH_3$

2-pentyne

18.29 The ^{13}C NMR spectrum of a hydrocarbon with molecular formula C_4H_6 consists of a triplet at 30.2 δ and a doublet at 136 δ. What is the structure?

18.30 The ^{13}C NMR spectrum of a hydrocarbon with molecular formula C_6H_{14} consists of a quartet at 19.1 δ and a doublet at 33.9 δ. What is the structure?

18.31 The ^{13}C NMR spectra of both 3-methylpentane and 2,2-dimethylbutane consist of four resonances. Explain how the two compounds can be distinguished based on the multiplicity of the signals.

$CH_3-CH_2-CH(CH_3)-CH_2-CH_3$ $CH_3-C(CH_3)_2-CH_2-CH_3$

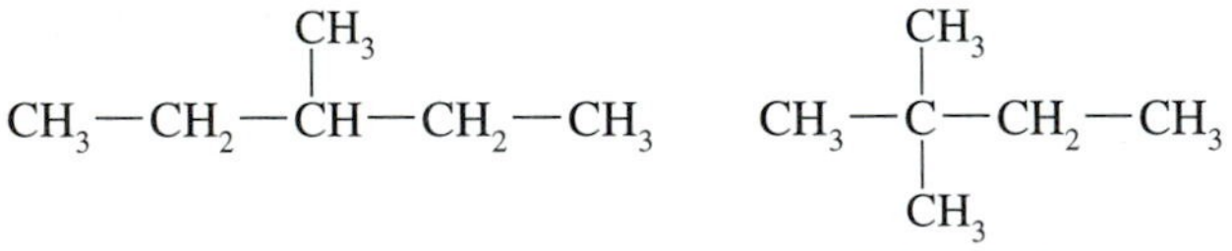

3-methylpentane 2,2-dimethylbutane

18.32 The ^{13}C NMR spectra of both 1-butanol and 2-butanol consist of four resonances. Explain how the two compounds can be distinguished based on the multiplicity of the signals.

$CH_3-CH_2-CH_2-CH_2-OH$ $CH_3-CH_2-CH(OH)-CH_3$

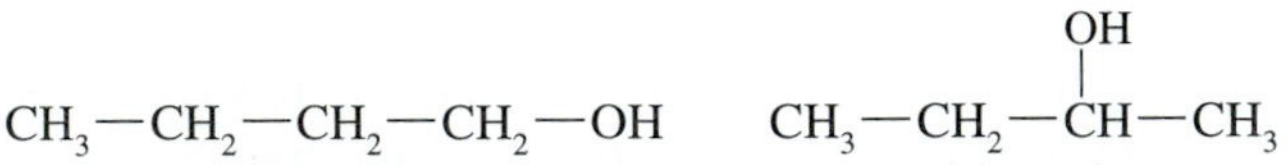

1-butanol 2-butanol

18.33 The proton-decoupled ^{13}C NMR spectrum of a compound with molecular formula C_8H_{16} is shown. What structures are consistent with this spectrum? Could a proton-coupled ^{13}C spectrum distinguish between these possibilities?

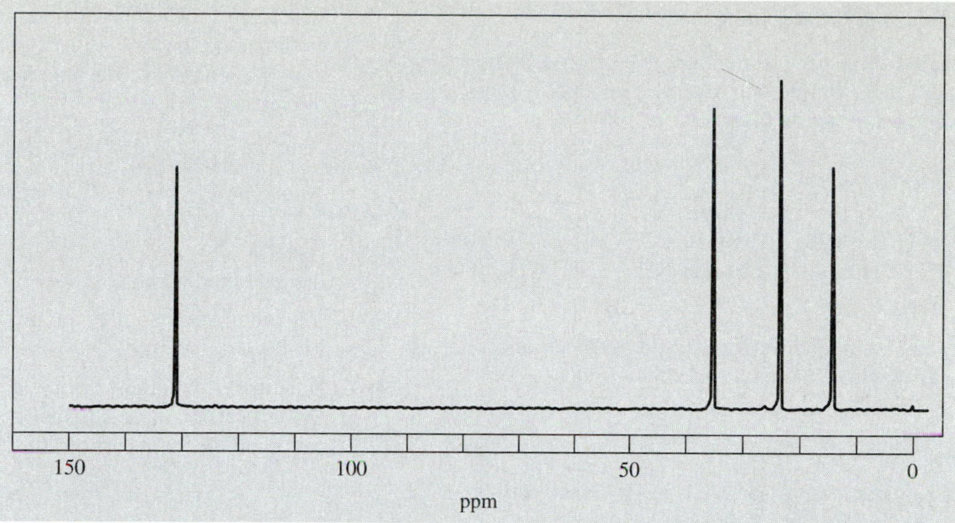

18.34 The proton-decoupled ^{13}C NMR spectrum of a compound with molecular formula C_4H_9Cl is shown. What structures are consistent with this spectrum? Explain how the proton-coupled ^{13}C spectrum could distinguish between these possibilities.

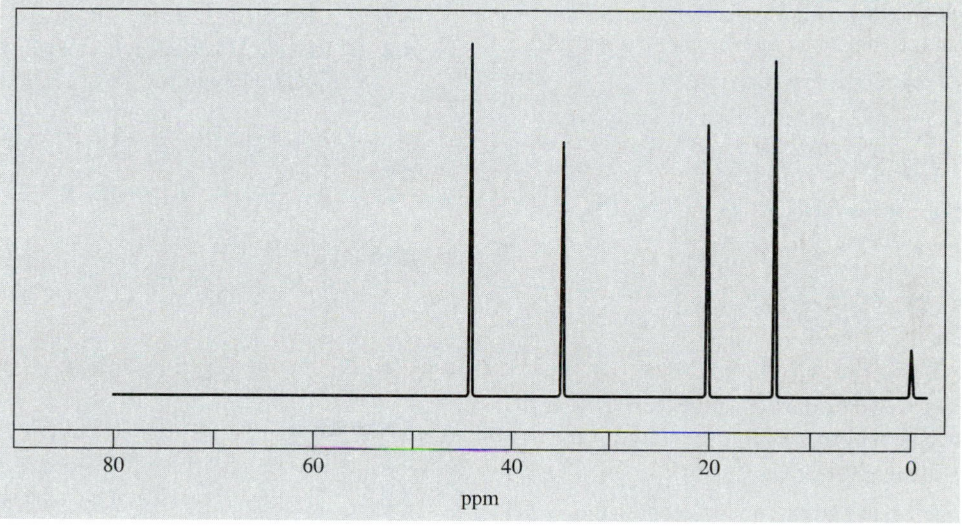

Answers to Problems

Chapter 1

1.1 $^{-}\!:\!\ddot{\underset{..}{O}}\!-\!\ddot{N}\!=\!\ddot{O}\!: \longleftrightarrow :\!\ddot{O}\!=\!\ddot{N}\!-\!\ddot{\underset{..}{O}}\!:^{-}$

1.2 There are three regions containing electon density: a single bond, a double bond, and a nonbonding pair of electrons. The ion is angular.

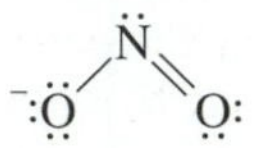

$^{-}\!:\!\ddot{\underset{..}{O}}\!-\!\ddot{N}\!=\!\underset{..}{O}\!:$

1.3 $NH_2{-}CH_2{-}CH_2{-}CH_2{-}CH_2{-}CH_2{-}CH_2{-}NH_2$
$NH_2CH_2CH_2CH_2CH_2CH_2CH_2NH_2$ $NH_2(CH_2)_6NH_2$

1.4 $C_{10}H_9NO_2$

1.5 They are isomers differing only in the location of the carbonyl group and a hydroxyl group.

Chapter 2

2.1 $CHCl_3$ is the more polar. CCl_4 has a higher boiling point because it has a higher molecular weight and therefore stronger London forces.

2.2 The isomers have similar shapes and, as a result, similar London forces. No hydrogen bonding is possible.

2.3 $CH_2{=}CH_2$ is a Lewis base; Br^+ is a Lewis acid.

2.4 Neither oxidation nor reduction occurs. Two hydrogen atoms and an oxygen atom are removed.

2.5 It is a rearrangement reaction.

2.6 The conjugate base is resonance stabilized.

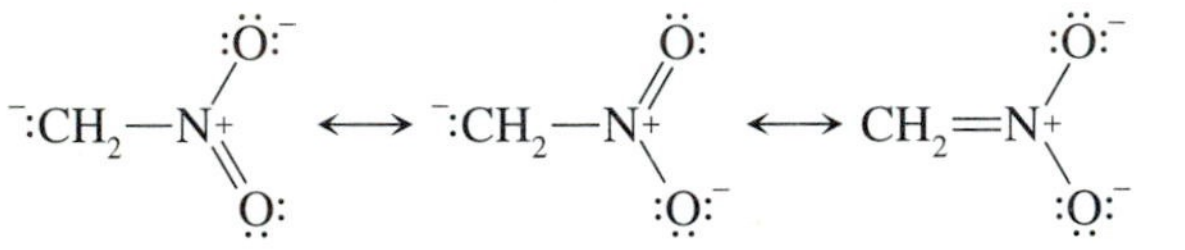

$^{-}\!:CH_2{-}N^{+}(=O)(O^{-}) \longleftrightarrow {}^{-}\!:CH_2{-}N^{+}(=O)(O^{-}) \longleftrightarrow CH_2{=}N^{+}(O^{-})(O^{-})$

Chapter 3

3.1 $C_{100}H_{202}$

3.2 Central carbon atom is quaternary; the other four carbon atoms are primary.

3.3 3,3,5-trimethylheptane

3.4 Two *tert*-butyl groups are adjacent to the —OH group; a methyl group is at the bottom of the structure.

3.5 $C_{10}H_{20}O$

3.6 [structure: bicyclic ether with O, CH_3, H, CH_3CH_2]

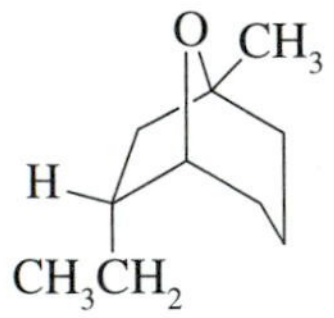

3.7 (a) isobutylcyclopentane (b) propylcyclobutane (c) *cis*-1,5-dibromocyclodecane

3.8 One chlorocyclobutane; five dichlorinated isomers, which are 1,1-dichlorocyclobutane, *cis*-1,2-dichlorocyclobutane, *trans*-1,2-dichlorocyclobutane, *cis*-1,3-dichlorocyclobutane, *trans*-1,3-dichlorocyclobutane.

Chapter 4

4.1 $C_{40}H_{56}$

4.2 disubstituted alkyne

4.3 Yes, geometric isomers are possible about both double bonds.

4.4 Z isomer; the highest priority groups are C_6H_5 on the left carbon atom and the ring with an oxygen group on the right carbon atom.

4.5 (a) 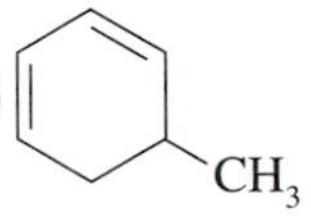 [methylcyclohexadiene, CH_3] (b) 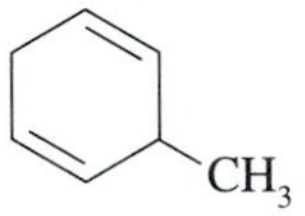 [methylcyclohexadiene, CH_3]

4.6 $CH_2{=}CHCH{=}CHC{\equiv}CC{\equiv}CC{\equiv}CCH{=}CHCH_3$

4.7 $CH_3CH(CH_3)CH_2CH(OH)CH_2CH(CH_3)CH_2CH_3$

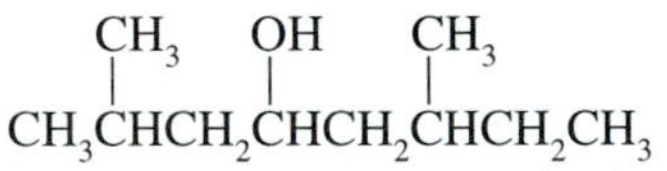

4.8 Reduce 11-tetradecyn-1-ol using lithium metal in liquid ammonia.

4.9 [cyclohexane ring bearing CH_3 and Cl on the same carbon]

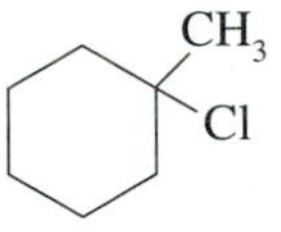

4.10 [cyclopentane ring bearing CH_3 and OH on the same carbon]

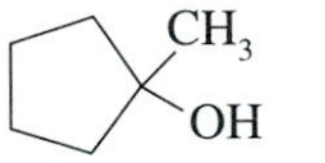

Chapter 5

5.1 $C_{18}H_{12}$; it is aromatic based on the Hückel rule.

5.2 The two nitrogen atoms in the six-membered ring and the nitrogen atom at the top of the structure in the five-membered ring have nonbonded electrons in sp^2 hybrid orbitals. They each contribute one electron to the π system. The other nitrogen atom in the five-membered ring contributes two nonbonded electrons to the π system.

5.3 2,6-dimethylaniline

5.4 4-benzylpyridine

5.5 (a) faster (b) slightly faster (c) faster

5.6 A mixture of ortho and para isomers from the first structure; meta isomer from the second structure.

5.7 (structures: 1,2-diethylbenzene — CH_2CH_3, CH_2CH_3; 1-isopropyl-2-methylbenzene — CH_3, $CH(CH_3)_2$; 1-methyl-2-propylbenzene — CH_3, $CH_2CH_2CH_3$)

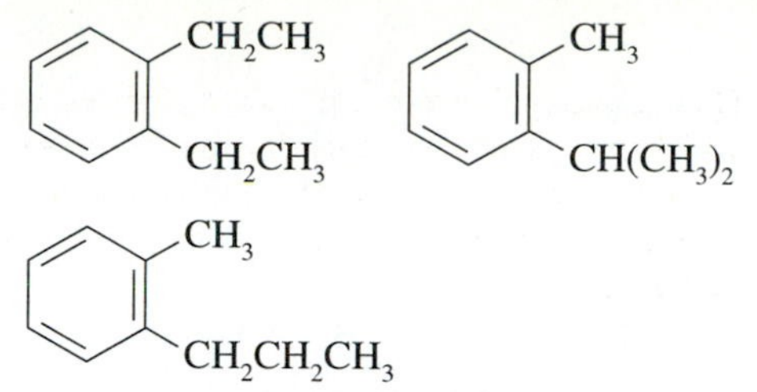

5.8 Do a Friedel-Crafts alkylation with CH_3Cl and $AlCl_3$. Then nitrate with HNO_3 and H_2SO_4 to obtain the para isomer. Finally, oxidize with $KMnO_4$.

5.9 Nitrate bromobenzene with HNO_3 and H_2SO_4 to obtain the para isomer, then reduce it with Sn/HCl. Diazotize *p*-bromoaniline and react the diazonium ion with $Cu_2(CN)_2$.

Chapter 6

6.1 Yes, there is a stereogenic center at the carbon atom of the five-membered ring where it bonds to the pyridine ring.

6.2 H < *p*-chlorophenyl < $—CH_2CO_2H$ < $—CH_2NH_2$

6.3 R

6.4 The first two of the following three isomers have specific rotations of opposite sign. The third compound is meso and is optically inactive.

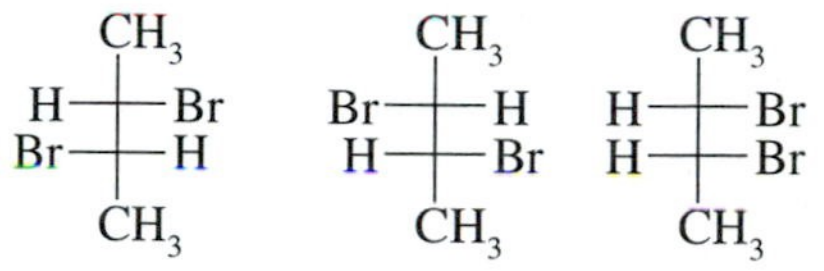

(Fischer projections: CH_3; H—Br; Br—H; CH_3 | CH_3; Br—H; H—Br; CH_3 | CH_3; H—Br; H—Br; CH_3)

6.5 Three stereogenic centers; eight stereoisomers.

Chapter 7

7.1 (a) *trans*-1-fluoro-2-methylcyclohexane

(b) (*Z*)-2,3-dichloro-3-hexene

(c) 2-chlorobenzyl bromide

7.2 Add $NaNH_2$ to 1-butyne and then react with CH_3Br, or make the conjugate base of propyne and react with CH_3CH_2Br.

7.3 Methanol and formamide are protic solvents and form hydrogen bonds with the nucleophile. Dimethyl formamide is aprotic and cannot form hydrogen bonds with the nucleophile.

7.4 Elimination is enhanced because *tert*-butoxide is a stronger base than methoxide. Furthermore, *tert*-butoxide is a more hindered nucleophile, so substitution is less likely.

Chapter 8

8.1 Three secondary alcohol sites and one primary alcohol site.

8.2 3,7-dimethyl-6-octen-1-ol

8.3 (structures: $(CH_3)_2CH$(H)C=C(CH_3)(H) — major; $(CH_3)_2CH$(H)C=C(H)(CH_3) — minor)

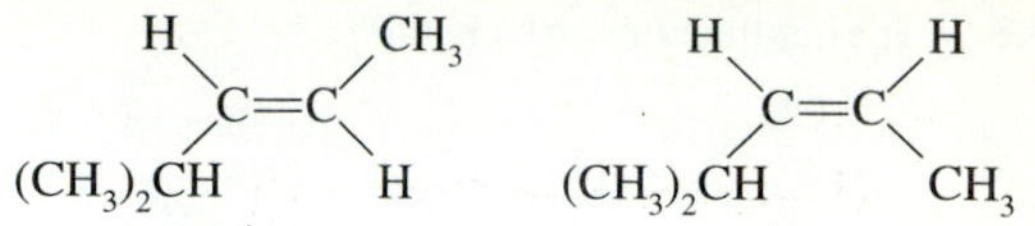

$(CH_3)_2CHCH_2CH{=}CH_2$
minor

8.4 The primary alcohols: 1-pentanol, 2-methyl-1-butanol, 3-methyl-1-butanol, and 2,2-dimethyl-1-propanol.

8.5 $(CH_3)_3C—CH(OH)—CH_3$ $\quad (CH_3)_3C—CH(H)—CH_2—OH$

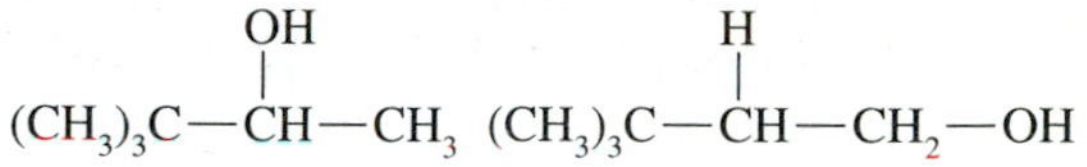

Chapter 9

9.1 (a) ethoxycyclopentane

(b) *trans*-1,3-dimethoxycyclohexane

(c) *cis*-1,3-diethoxycyclopentane

9.2 Alkylate with 2-chloropropane and $AlCl_3$; brominate with Br_2 and $FeBr_3$; make the Grignard reagent with Mg and ether; add D_2O.

9.3 React *tert*-butoxide with benzyl bromide.

9.4 5-bromo-2-pentanol

9.5 (structure: phenyl—$CH(OCH_3)CH_2OH$, printed as OCH_3 over $CHCH_2OH$)

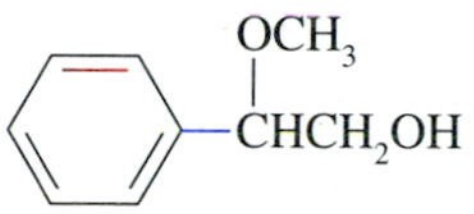

Chapter 10

10.1 (*E*)-2-hexenal

10.2 Compound (2) reacts with Fehling's solution to give a red precipitate; compound (1) does not.

10.3 (a) (b) 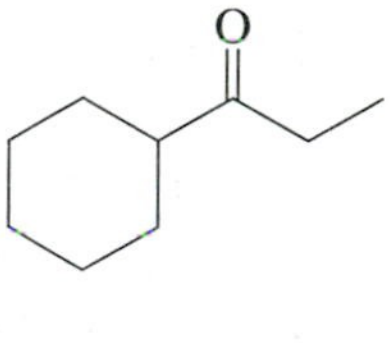 (c) OH (skeletal structures)

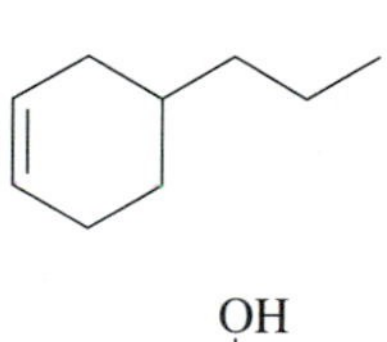

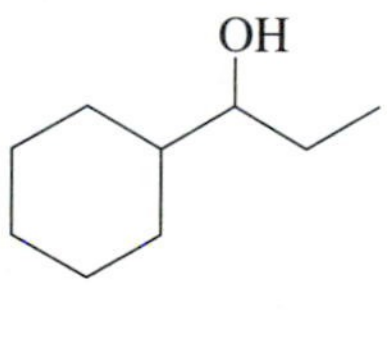

10.4 (chair structures: $(CH_3)_3C$ … OH, CH_3; $(CH_3)_3C$ … CH_3, OH)

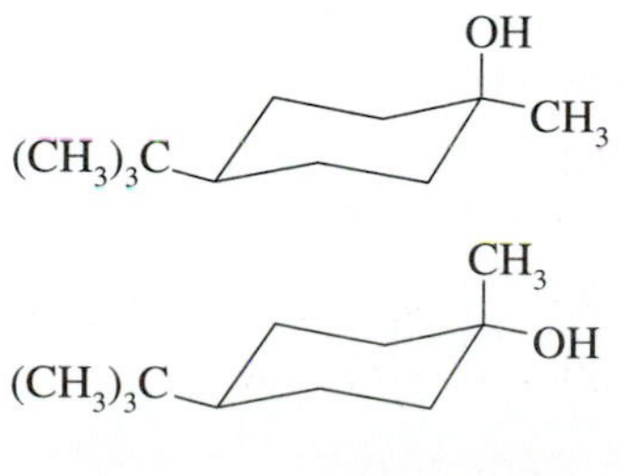

10.5 (a) hemiacetal (b) ketal

10.6 (cyclohexyl methyl ketone: CH_3–C=O on cyclohexane ring) H(H₂N)N–C(=O)NH_2

10.7 acetone

Chapter 11

11.1 They are diastereomers.

11.2 Fischer projections: CHO; H–OH; HO–H; CH_2OH — L-threose. CHO; HO–H; HO–H; CH_2OH — L-erythrose

11.3 CH_2OH; =O; HO–H; H–OH; CH_2OH

11.4 $HOCH_2$; H; O; OH; H; OH; HO; HO; H; H; H

11.5 β-D-mannopyranose

11.6 A stereogenic center is created at C-2, and the alditols produced are diastereomers.

11.7 β-D-galactose and 2-propanol

11.8 Mannose is the aglycone bonded to ribose; the compound is 4-*O*-(α-D-ribopyranosyl)-β-D-mannose

Chapter 12

12.1 (*R*)-3,5-dihydroxy-3-methylpentanoic acid

12.2 $H-\overset{O}{\overset{\|}{C}}-O-CH_2\overset{CH_3}{\overset{|}{C}}HCH_3$

2-methylpropyl methanoate

12.3 Oxaloacetic acid; the carboxyl group nearer to the carbonyl group or the hydroxyl group.

12.4 Add HBr to give 2-bromo-2-methylheptane; make the Grignard reagent and add CO_2.

12.5 furanoic acid (furan ring, O; C(=O)–OH)

Chapter 13

13.1 $CH_3(CH_2)_{24}CO_2CH_2(CH_2)_{28}CH_3$

13.2 $77° + 2(6°) = 89°$

13.3 six moles, tristearate

13.4 Palmitic acid, oleic acid, phosphate, and serine; glycerophospholipid.

Chapter 14

14.1 secondary amide and tertiary amine

14.2 secondary amine

14.3 CH_3O; CH_3O; CH_3O (on benzene ring) –$CH_2CH_2NH_2$

14.4 *N*, *N*-diethyl-3-methylbenzamide

14.5 The one bonded to methyl groups; the other is bonded to two aromatic rings, which decrease the basicity of amines.

14.6 Reaction could occur to give an ester as well.

HO–(benzene ring)–N(H)–H $HO-\overset{O}{\overset{\|}{C}}-CH_3$

14.7 (quinoline ring) O=C–OH; N; $OCH_2CH_2CH_2CH_3$

$^{+}NH_3CH_2CH_2N(C_2H_5)_2$

Chapter 15

15.1 CO_2^-; NH_3^+–C–H; CH_2OH — zwitterion. CO_2H; NH_3^+–C–H; CH_2OH — conjugate acid

15.2 At pH 12 it exists as the conjugate base.

15.3 Ala-Ala-Gly Ala-Gly-Ala Gly-Ala-Ala

15.4 Ala-Lys, Gly-Arg, and Leu

15.5 [structure: 3-phenyl-2-thioxo-imidazolidin-4-one with N–H and CH_2OH substituent; labels S, N, O, N, H, OH]

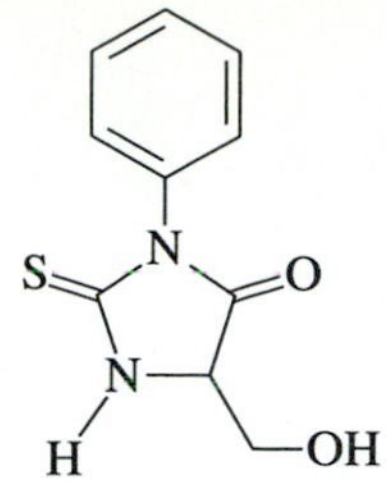

Chapter 16

16.1 Nucleoside of guanine and deoxyribose; deoxyguanosine.

16.2 3′ ← —G—A—T—C—dG— 5′

16.3 3′ ← —C—G—T—A— 5′

16.4 UGU and UGC; only the third letter differs.

16.5 GUA

16.6 There are two possible codons for phenylalanine, three for isoleucine, and four for glycine. Thus, there are $2 \times 3 \times 4 = 24$ possible sequences. One possible sequence using the first codon listed for each amino acid in the table is UUUAUUGGU.

Chapter 17

17.1 The para isomer forms a polymer with the fewest number of bends.

17.2 Thermosetting plastics are used for handles. Thermoplastics are used for frames of eyeglasses.

17.3 [structure: H, O, O, CH_3]

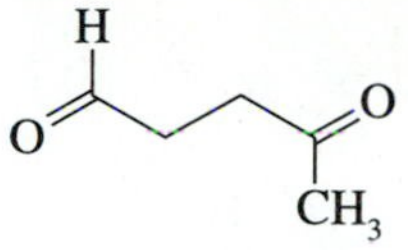

No, both structures would be the same.

Chapter 18

18.1 Azulene has a λ_{max} in the visible region, whereas naphthalene does not.

18.2 Compound I has an intense broad O—H absorption in the 3400–3600 cm^{-1} region. Compound II does not.

18.3 Stabilization of the polar resonance contributor by two alkyl groups increases the single bond character of the carbonyl group in a ketone. As a result, the bond is easier to stretch.

18.4 (a) 2 (b) 4 (c) 3

18.5 Ratio of 2:4 (1:2) for high field resonance ($\delta \simeq 1.2–1.4$) to low field resonance ($\delta \simeq 3.0–4.0$). High field resonance is a quintet; low field resonance is a triplet.

18.6 An ether has two low field resonances due to the carbon atoms bonded to oxygen, whereas an alcohol has only one.

Answers to Odd-Numbered Exercises

Chapter 1

1.1 (a) 5 (b) 7 (c) 4 (d) 6 (e) 7 (f) 7 (g) 6 (h) 5

1.3 (a) H—N̈(H)—Ö(H): (b) H—C(H)(H)—C(H)(H)—H
(c) H—C(H)(H)—Ö(H): (d) H—C(H)(H)—N̈(H)—H
(e) H—C(H)(H)—B̈r: (f) H—C(H)(H)—S̈(H):

1.5 (a) CH_3—C(=Ö:)—Ö—H (b) CH_3—C(=Ö:)—Ö—CH_3
(c) CH_3—C(=Ö:)—N̈(H)—CH_3 (d) CH_3—S̈—CH=CH_2

1.7 (a) :C̈l—C(=Ö:)—C̈l: (b) H—N̈(H)—C(=Ö:)—N̈(H)—H
(c) H—C(H)(H)—C(=Ö:)—S̈—H

1.9 (a) (:C̈l:)(:C̈l:)C=C(:C̈l:)(:C̈l:) (b) (H)(:C̈l:)C=C(:C̈l:)(:C̈l:)

1.11 (a) O, 0; C, 0; N, 0
(b) O, 0; N, +1; C, −1
(c) C, 0; N (left to right) 0, +1, −1

1.13 +1

1.15 $^-$:S̈—C≡N: sulfur has the negative charge
:S̈=C=N̈:$^-$ nitrogen has the negative charge

1.17 CH_3—C(—:Ö:$^-$)=$\overset{+}{N}H_2$

1.19 (a) 180° (b) 109.5° (c) 109.5° (d) 180°

1.21 120°

1.23 (a) left to right, sp^3, sp^2
(b) left to right, sp^3, sp^2, sp^2
(c) left to right, sp^3, sp^2

1.25 sp^2, sp^2

1.27 (a) C_5H_{12} (b) C_4H_{10}
(c) C_4H_8 (d) C_4H_6

1.29 (a) $C_3H_6Cl_2$ (b) $C_3H_6Cl_2$
(c) $C_2H_4Br_2$

1.31 (a) Br—CH_2—CH_2—Br
(b) CH_3—CH_2—CH_2—CH_2—CH_3
(c) CH_3—CH_2—CH_2—SH

1.33 (a) $BrCH_2CH_2Br$
(b) $CH_3CH_2CH_2CH_2CH_3$
(c) $CH_3CH_2CH_2SH$

1.35 (a) H—C(H)(H)—C(H)(H)—C(H)(H)—C(H)(H)—H
(b) H—C(H)(H)—C(H)(H)—C(H)(H)—Cl
(c) H—C(H)(H)—C(Cl)(H)—C(H)(H)—C(H)(H)—H
(d) H—C(H)(H)—C(H)(H)—C(Br)(H)—C(H)(H)—H
(e) H—C(H)(H)—C(H)(H)—C(Br)(H)—Br
(f) H—C(H)(H)—C(Br)(Br)—C(H)(H)—C(H)(H)—C(H)(H)—H

1.37 (a) $C_{10}H_{18}O$ (b) $C_8H_{17}Br$ (c) $C_7H_{16}O$

1.39 (a) $C_6H_{12}S$ (b) $C_{13}H_{20}O$

1.41 (a) amide (b) ketone, double bond

1.43 (a) same (b) same (c) isomers

1.45

(a)
```
   Br Br            H  Br
   |  |             |  |
H—C—C—H         H—C—C—Br
   |  |             |  |
   H  H             H  H
```

(b)
```
   H  H                H     H
   |  |                |     |
H—C—C—O—H         H—C—O—C—H
   |  |                |     |
   H  H                H     H
```

(c)
```
   Br Cl            H  Br
   |  |             |  |
H—C—C—H         H—C—C—Cl
   |  |             |  |
   H  H             H  H
```

(d)
```
   H  H  H              H  Cl H
   |  |  |              |  |  |
H—C—C—C—Cl         H—C—C—C—H
   |  |  |              |  |  |
   H  H  H              H  H  H
```

(e)
```
   H  H                 H  H  H
   |  |                 |  |  |
H—C—C—N—H          H—C—N—C—H
   |  |  |              |     |
   H  H  H              H     H
```

Chapter 2

2.1 (a) The more branched compound has smaller London forces and a lower boiling point.
(b), (c) Hydrogen bonding occurs in first structure.

2.3 Last compound is not polar; bond moments of C—Cl bonds are opposed to each other.

2.5 More hydrogen bonds can be formed with two OH groups in the same molecule.

2.7

```
          H                    H
          |                    |
(a) CH3—S+—CH3      (b) CH3—O+—CH3
          ··                   ··

          H                   H
          |                   |
(c) CH3—N+—H        (d) CH3—O+—H
          |                   ··
          H
```

2.9

```
   H                 H
   |                 |
H—N+—Ö—H        H—N̈—O+—H
   |                 |  ··
   H                 H
              more acidic
```

2.11 (a) CH_3CH_2Cl is Lewis base; $AlCl_3$ is Lewis acid
(b) CH_3CH_2SH is Lewis acid; CH_3O^- is Lewis base
(c) CH_3CH_2OH is Lewis acid; NH_2^- is Lewis base

2.13 CH_3OH; left

2.15 The carboxylic acid group is bonded to a carbon atom that is bonded to an electronegative nitrogen atom, which inductively withdraws electron density from the carboxyl group.

2.17 (a) reduction (b) oxidation (c) oxidation

2.19 (a) oxidation (b) reduction

2.21 (a) addition; CH_3OH required
(b) elimination; water formed
(c) rearrangement

2.23 (a) oxidation (b) addition
(c) oxidation

2.25 oxidation, elimination

2.27 Products are less stable.

2.29 A → X is faster; it has lower activation energy.

2.31 (a) homolytic C—H bond cleavage; homogenic H—Br bond formation
(b) homolytic Br—Br bond cleavage; homogenic C—Br bond formation

2.33 homolytic cleavage reactions

```
          Ö:
          ||
(benzene ring)—C—Ö·
```

2.35 HO^+

Chapter 3

3.1 $C_{31}H_{64}$; $CH_3(CH_2)_{29}CH_3$

3.3

```
(a) CH3—CH2—CH2—CH3

(b) CH3—CH2—CH2—CH—CH2—CH3
                 |
                 CH2—CH3

(c) CH3—CH2—CH—CH2—CH3
             |
             CH3
```

3.5 (a) propyl (b) *sec*-butyl (c) isobutyl

3.7 (a) 2-methylbutyl (b) 3-methylbutyl

3.9 1,1,3,3-tetramethylbutyl

3.11 (a) 2-methylbutane (b) 3-methylhexane
(c) 2-methylpentane (d) 3-methylheptane

3.13 5-(1-ethylpropyl)decane

3.15

```
(a) CH3—CH2—CH—CH—CH2—CH3
             |   |
             CH3 CH3

         CH3
         |
(b) CH3—C—CH—CH2—CH3
         |  |
         CH3CH3

(c) CH3—CH—CH—CH—CH—CH3
         |   |   |   |
         CH3 CH3 CH3 CH3
```

3.17

(a) 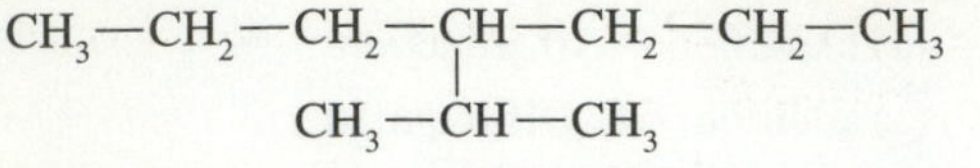 $CH_3—CH_2—CH_2—CH—CH_2—CH_2—CH_3$
$\qquad\qquad\qquad\qquad |$
$\qquad\qquad CH_3—CH—CH_3$

(b) 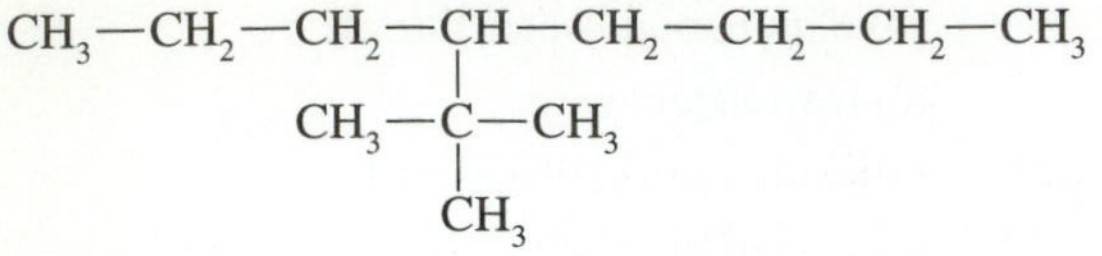 $CH_3—CH_2—CH_2—CH—CH_2—CH_2—CH_2—CH_3$
$\qquad\qquad\qquad\qquad |$
$\qquad\qquad CH_3—C—CH_3$
$\qquad\qquad\qquad\quad |$
$\qquad\qquad\qquad CH_3$

3.19 2-methylhexane, 3-methylhexane

3.21
$\qquad CH_3$
$\qquad |$
$CH_3—C—CH_3$
$\qquad |$
$\qquad CH_3$

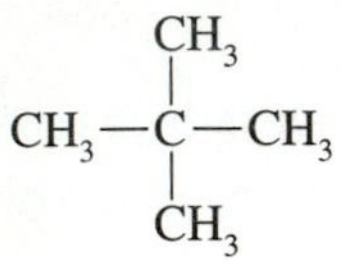

3.23 (a) four primary, one quaternary
(b) three primary, one secondary, one tertiary
(c) three primary, two secondary, one tertiary
(d) four primary, two tertiary

3.25
H
CH_3 CH_3
H H
CH_3

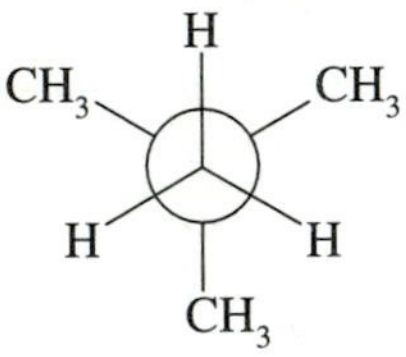

3.27 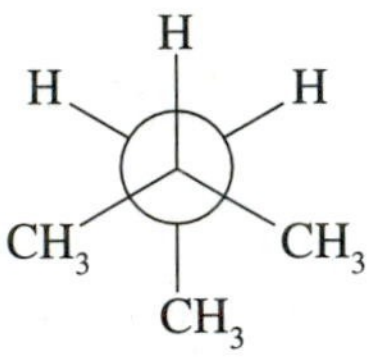 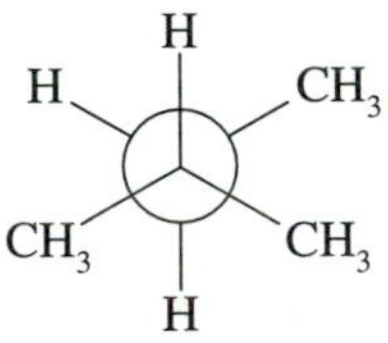
H
H H
CH_3 CH_3
CH_3

H
H CH_3
CH_3 CH_3
H
more stable

3.29 (a) 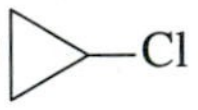 cyclopropyl—Cl (b) 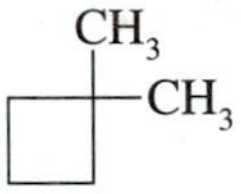 1,1-dimethylcyclobutane (CH_3, CH_3) (c) 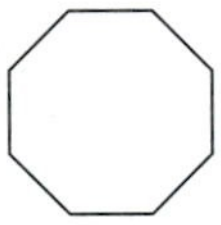 cyclooctane

3.31 (a) cyclodecane
(b) *trans*-1,2-dichlorocyclohexane
(c) 1,1-dichlorocyclohexane

3.33 (a) $C_{10}H_{18}$ (b) $C_8H_{12}O$ (c) $C_7H_{13}N$ (d) C_6H_6

3.35 Smaller for fluorocyclohexane, because fluorine atom is smaller than CH_3 group.

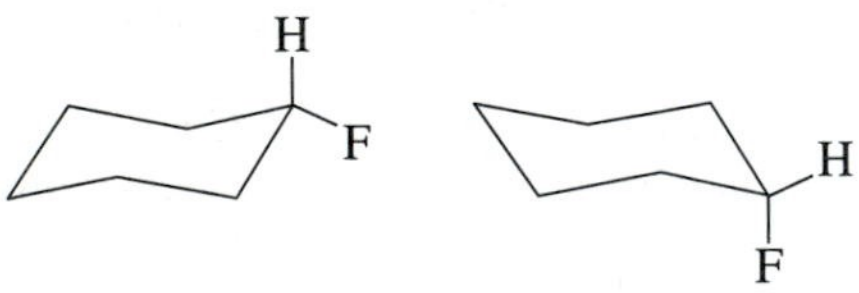

H F — H F

3.37 (a) 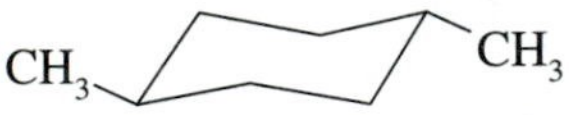 CH_3 CH_3
(b)  CH_3 CH_3

3.39 It could explode as a result of a spark.

3.41 Octane because it is the least branched isomer; 2,2,3,3-tetramethylbutane because it is the most branched isomer.

3.43 (a) two (b) two (c) two (d) one

3.45 $CH_3CH_2CH_2CH_2CH_2\cdot$
$CH_3CH_2CH_2\dot{C}HCH_3$
$CH_3CH_2\dot{C}HCH_2CH_3$

3.47

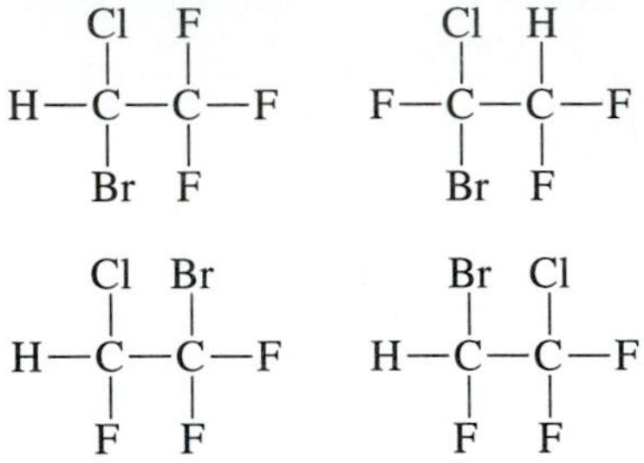

$H—C(Cl)(Br)—C(F)(F)—F$ $\qquad F—C(Cl)(Br)—C(H)(F)—F$

$H—C(Cl)(F)—C(Br)(F)—F$ $\qquad H—C(Br)(F)—C(Cl)(F)—F$

3.49 Methylene chloride because the bond moments of the C—Cl bonds do not cancel.

3.51 The molecular weights are close and the density should be about the same.

3.53 See answers for Chapter 18.

CHAPTER 4

4.1 (a) C_6H_{12} (b) C_5H_8 (c) C_7H_{12}

4.3 (a) $C_{10}H_{16}$ (b) C_9H_{16} (c) C_7H_{14}

4.5 (a) disubstituted
(b) disubstituted
(c) trisubstituted

4.7 (a) trisubstituted
(b) tetrasubstituted

4.9 (a) and (c)

4.11 (b) and (c)

4.13 (a) $—CHClCH_3$
(b) $—CH_2C{\equiv}CH$
(c) $—OCH_3$

4.15 (a) *Z* (b) *E*

4.17 (a) 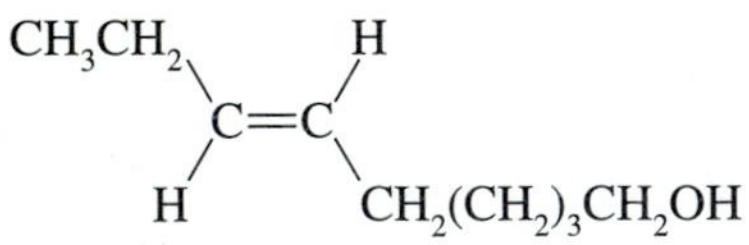 $CH_3CH_2(H)C{=}C(H)CH_2(CH_2)_3CH_2OH$

(b) 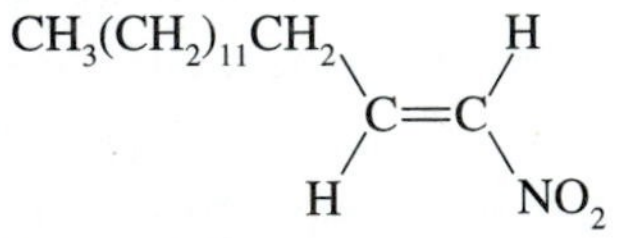 $CH_3(CH_2)_{11}CH_2(H)C{=}C(H)NO_2$

4.19 (a) 2-methyl-2-pentene
(b) (*E*)-3,4-dimethyl-3-hexene
(c) (*E*)-3-methyl-2-pentene

4.21 (a) 4-methylcyclohexene
(b) 3-isopropylcyclopentene
(c) vinylcyclobutane

4.23 (a) $(CH_3)_2C{=}CHCH_2CH_3$

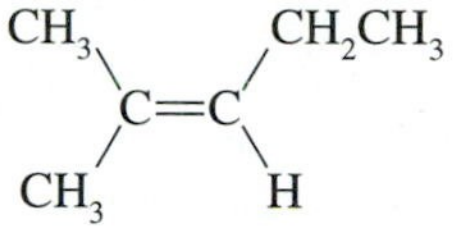

(b) $CH_3CH_2CH{=}CHCH(CH_3)CH_3$ (both H on the same side)

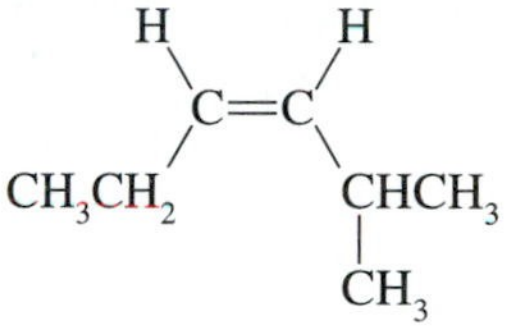

(c) $CH_3CH{=}CHCH_2CH(CH_3)CH_3$

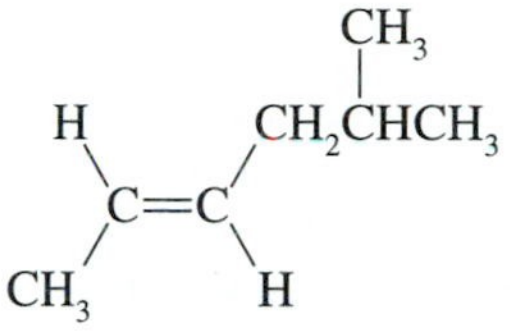

4.25 (a) 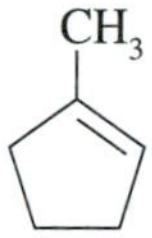 (b) 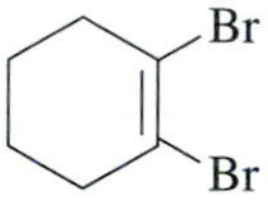 (c)

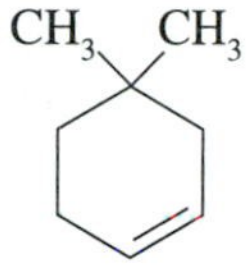

4.27 (a) 1-pentyne
(b) 2,2-dimethyl-3-hexyne
(c) 4-methyl-2-hexyne

4.29 (a) $CH_3C{\equiv}CCH_2CH_2CH_3$
(b) $HC{\equiv}CCH(CH_3)CH_2CH_3$
(c) $CH_3CH_2C{\equiv}CCH(CH_2CH_3)CH_2CH_2CH_3$

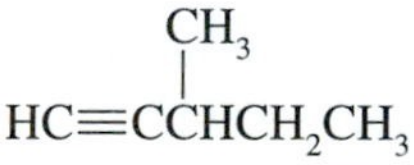

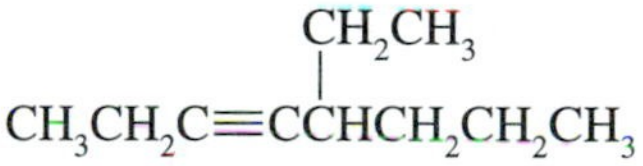

4.31 (a) three (b) four (c) four

4.33 Hydrogenate the related alkyne using a Lindlar catalyst.

4.35 Purple color fades; cyclopentane would not react.

4.37 (a) $O{=}C(H)CH_3$ $O{=}C(CH_2CH_3)CH_2CH_3$
(b) $O{=}C(CH_3)CH_2CH_3$ $O{=}C(CH_2CH_3)CH_3$
(c) $O{=}C(H)CH_3$ $O{=}C(CH_2CH_3)H$

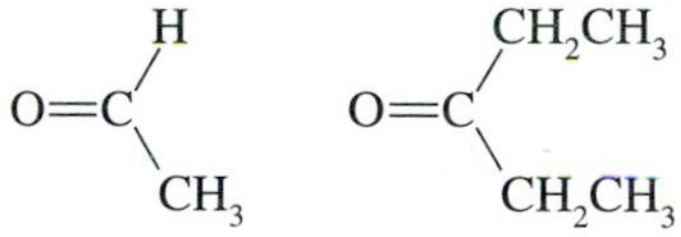

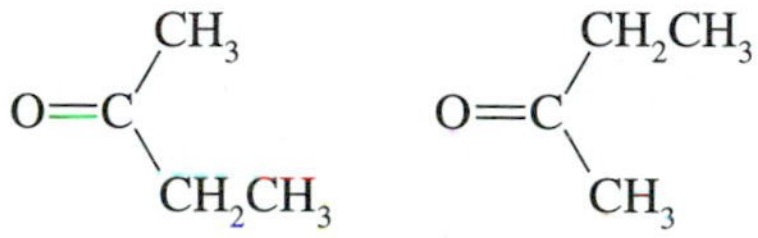

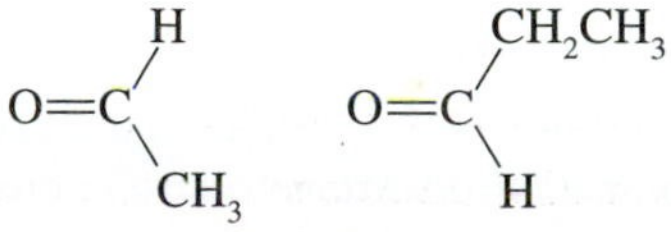

4.39 (a) The Markovnikov product results. ent numbers
(b) Both carbon atoms of the double bond are equivalent.

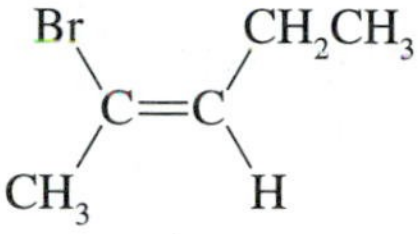

4.41 (a) $H_2C{=}C(Br)CH_2CH_3$
(b) $CH_3(Br)C{=}CHCH_2CH_3$ and $CH_3CH{=}C(Br)CH_2CH_3$
(c) $CH_3CH_2(Br)C{=}CHCH_2CH_3$

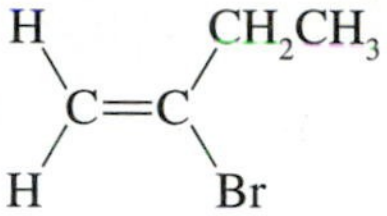

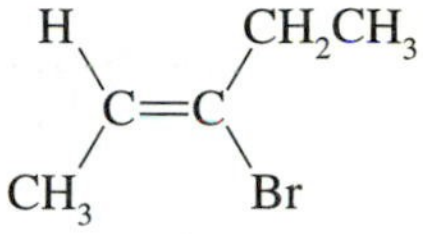

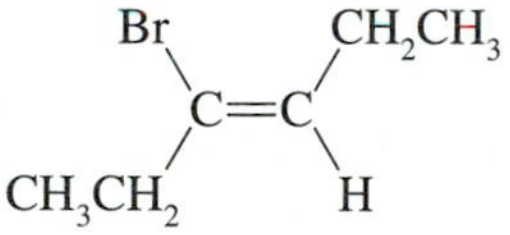

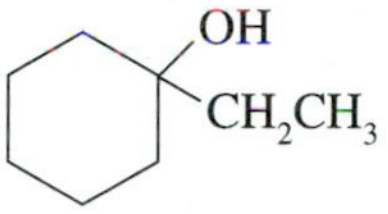

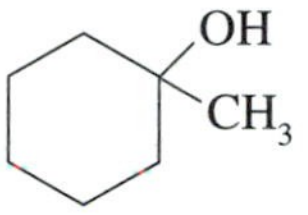

4.43 (a) OH, CH_2CH_3 (b) OH, CH_3 (c) OH

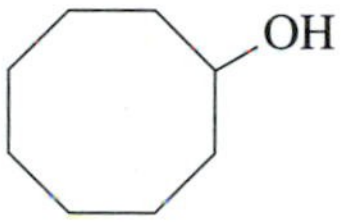

4.45 (a) *trans*-2-Pentene is the major product; *cis*-2-pentene is the minor product.
(b) 3,3-Dimethyl-1-butene is the only product.
(c) *trans*-2-Hexene is the major product; *cis*-2-hexene and 1-hexene are minor products.

4.47 $C_6H_{11}{-}CH_2{-}CHBr_2$

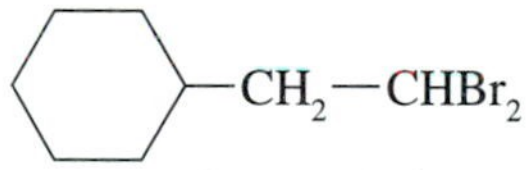

4.49 Only compound (a) has conjugated double bonds.

4.51 (a) O, O, CH_3, CH_3 monoterpene (b) CH_3, CH_2 sesquiterpene

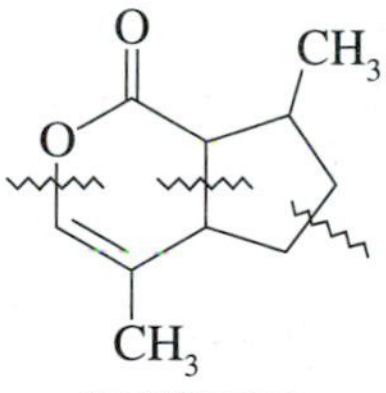

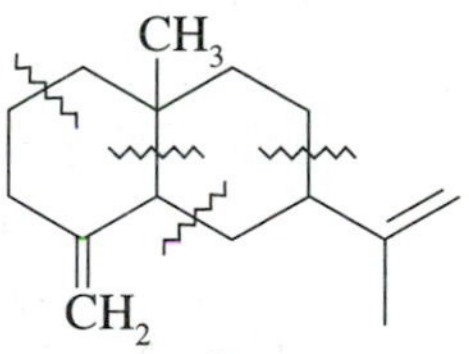

4.53 $\left[CH_2{-}C(CO_2CH_3)(H)\right]_n$

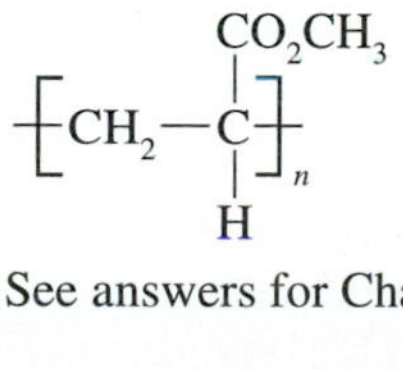

4.55 See answers for Chapter 18.

Chapter 5

5.1 (a) no (b) no (c) no

5.3 [structures: 1-bromonaphthalene; 2-bromonaphthalene]

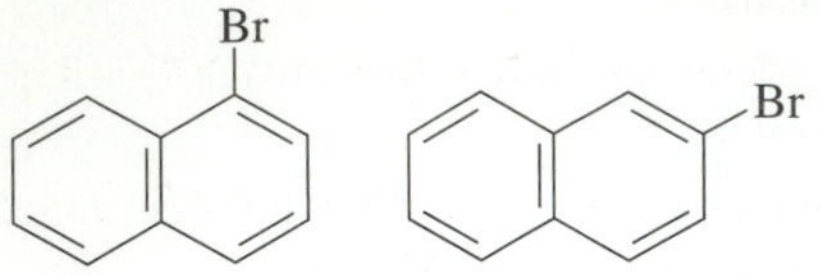

5.5 [structures: pyridazine, pyrimidine, pyrazine, numbered 1–6]

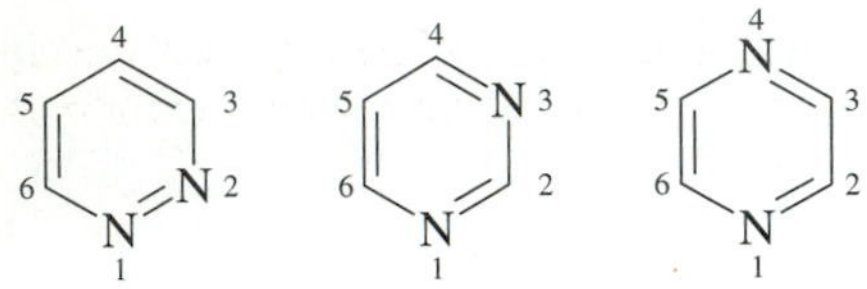

The 1,4-isomer has no dipole moment.

5.7 (a) oxygen—two electrons; nitrogen—one electron

(b) each nitrogen—one electron

(c) sulfur—two electrons; nitrogen—one electron

5.9 (a) furan (b) thiophene

5.11 para isomer

5.13 ethylbenzene, *o*-dimethylbenzene, *m*-dimethylbenzene, *p*-dimethylbenzene

5.15 (a) para (b) meta

5.17 (a) 1,2-diethylbenzene (b) isopropylbenzene

(c) 1,3,5-trimethylbenzene

5.19 (a) 4-chloro-3,5-dimethylphenol

(b) 4-chloro-2-phenylphenol

5.21 [structure: phenol with OH, Cl, NO_2, Cl, Cl substituents]

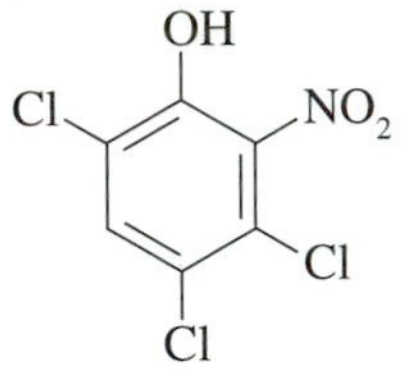

5.23 It will be a deactivating group similar to chlorine. It can supply electrons by resonance to favor ortho,para products.

5.25 *tert*-Butylbenzene will be formed because the isobutyl carbocation rearranges to the more stable *tert*-butyl carbocation.

5.27 Bromination will occur in the ortho,para positions on the left ring having the activating nitrogen atom. The right ring has a deactivating carbonyl group.

5.29 (a) 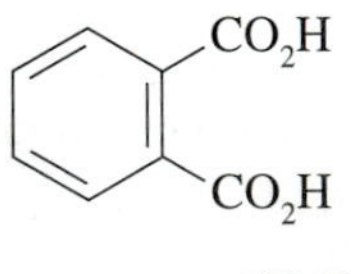 [structure: benzene with two adjacent CO_2H groups] (b) 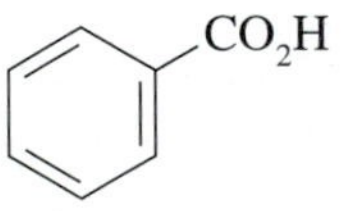 [structure: benzene with CO_2H]

(c) 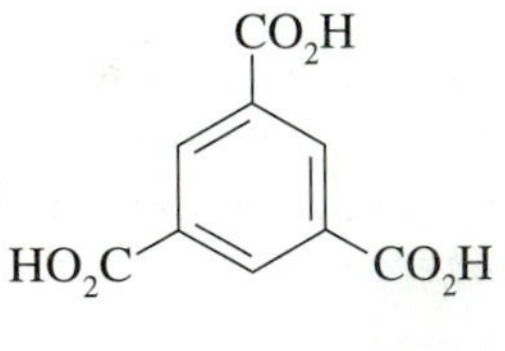 CO_2H HO_2C CO_2H

5.31 [structure: benzene with CH(Br)CH_3] The intermediate secondary radical is resonance-stabilized.

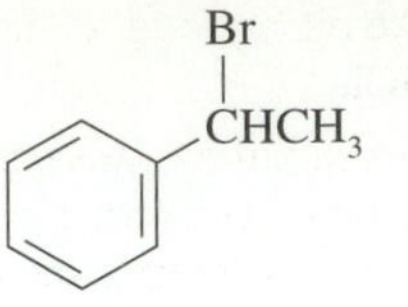

5.33 (a) Nitric acid with sulfuric acid; a mixture of ortho and para isomers of bromonitrobenzene.

(b) Fuming sulfuric acid; *m*-nitrobenzenesulfonic acid.

(c) Bromine and iron(III) bromide; a mixture of ortho and para isomers of bromoethylbenzene.

(d) Chloromethane and aluminum trichloride; a mixture of ortho and para isomers of methylanisole.

5.35 (a) Brominate benzene using bromine and iron(III) bromide. Nitrate the bromobenzene using nitric acid with sulfuric acid to give a mixture of isomers, and separate them by a physical method.

(b) Nitrate benzene using nitric acid with sulfuric acid. Brominate the nitrobenzene using bromine and iron(III) bromide.

(c) Alkylate benzene using bromoethane and aluminum tribromide. Brominate ethylbenzene using bromine and iron(III) bromide to give a mixture of isomers, and separate them by a physical method.

(d) Acetylate benzene using acetyl chloride and aluminum trichloride. Then brominate acetophenone using bromine and iron(III) bromide. Finally, reduce the *m*-bromoacetophenone using zinc/mercury amalgam and HCl.

5.37 (a) Nitrate benzene twice using nitric acid and sulfuric acid to give *m*-dinitrobenzene. Then chlorinate the *m*-dinitrobenzene using chlorine and iron(III) chloride.

(b) Nitrate toluene three times using nitric acid with sulfuric acid.

(c) Nitrate toluene using nitric acid with sulfuric acid to obtain a mixture of the ortho and para isomers of nitrotoluene. Separate the para isomer by a physical method. Brominate twice using bromine and iron(III) bromide.

5.39 (a) Nitrate benzene to give nitrobenzene. Then brominate nitrobenzene to give *m*-bromonitrobenzene. Reduce the nitro group to an amino group using tin and HCl. Convert the amino group into a diazonium group using nitrous acid, and react with hot aqueous acid to give *m*-bromophenol.

(b) Nitrate benzene to give nitrobenzene. Then brominate nitrobenzene to give *m*-bromonitrobenzene. Reduce the nitro group to an amino group using Sn and HCl.

(c) Nitrate toluene to give a mixture of the ortho and para isomers of nitrotoluene. Separate the para isomer by a physical method. Reduce the nitro group to an amino group using tin and HCl. Convert the amino group into a diazonium group using nitrous acid, and react with hot aqueous acid to replace the diazonium group with a hydroxyl group.

5.41 The other ring has a deactivating group (Cl) bonded to it. The oxidation occurs para to the electron donating nitrogen atom.

5.43 See answers for Chapter 18.

Chapter 6

6.1 Only compound (b) has a stereogenic center.

6.3 2-chloropentane, 1-chloro-2-methylbutane, 2-chloro-3-methylbutane

6.5 (a) zero (b) one (c) zero

6.7 (a) zero (b) one (c) one

6.9 (a) —OH —OCH_3 —SH —SCH_3

(b) —CH_2Cl —CH_2Br —Cl —Br

(c) —$CH_2CH{=}CH_2$ —$CH_2C{\equiv}CH$

—$C{\equiv}C{-}CH_3$ —CH_2OCH_3

(d) —CH_2CH_3 —CH_2CH_2Cl

—CH_2OH —OCH_3

6.11 (a) —H —$CH_2OC_6H_4Cl$

—CH_2OCONH_2 —OH

(b) —H —CH_3

—$CH_2OC_6H_3(CH_3)_2$ —NH_2

6.13 (a) Cl H (b) Cl H

(c) H Cl

6.15 (a) *S* (b) *S* (c) *R*

6.17 *R*

6.19 − 53

6.21 *R* and *S* refer to configuration; + and − refer to direction of rotation of plane-polarized light.

6.23 − 19.6, +14.8

6.25 Four; there are two stereogenic centers.

6.27 A plane of symmetry in ribitol bisects the molecule through the center H—C—OH group.

6.29 Addition can occur from top or bottom of the carbon-carbon double bond, so both enantiomers are produced.

6.31 2(*R*),3(*R*) is optically active; 2(*S*),3(*R*) is a meso compound.

6.33 The enzymes would not metabolize the enantiomer, so it would pass through the body.

6.35 The enantiomer cannot interact as effectively with chiral molecules that are responsible for biological activity.

Chapter 7

7.1 (a) fluoroethene

(b) 3-chloro-1-propene

(c) 3-bromo-1-propyne

7.3 (a) Br, CH_3 (b) Br

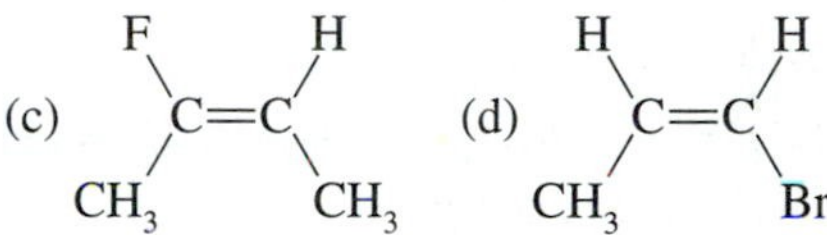

7.5 (a) $CH_3CH_2CH_2CH_2CH_2I$

(b) $NCCH_2CH_2CH_2CN$

(c) CH_3–C_6H_4–$CH_2C{\equiv}CH$

(d) $CH_3CH_2CH(SH)CH_3$

7.7 The cyclic product is formed by an internal displacement of the bromide by the alkoxide formed by reaction of the alcohol with hydride ion.

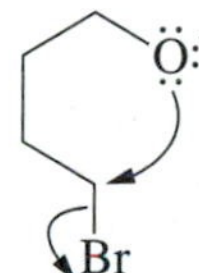

7.9 (a) 1-Bromobutane, because primary alkyl halides react faster than secondary alkyl halides.

(b) 1-Chloropentane, because chlorocyclopentane is a secondary alkyl halide.

(c) 2-Bromo-4-methylpentane, because 2-bromo-2-methylpentane is a tertiary alkyl halide.

7.11 (a) 1-Bromo-1-methylcyclohexane, because it is a tertiary compound.

(b) Isopropyl iodide, because it is a secondary compound.

(c) 3-Bromo-1-pentene, because it is an allylic compound.

7.13 The electron donating methyl group increases the rate of S_N1 reaction, whereas the electron withdrawing nitro group decreases the rate.

7.15 The product has the *S* configuration, because the reaction occurs with inversion of configuration.

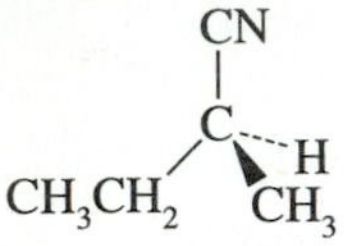

CN
C, H
CH_3CH_2 CH_3

7.17 The alcohol product is a racemic mixture, because the reaction of a secondary benzylic compound occurs by an S_N1 mechanism, giving a resonance-stabilized carbocation.

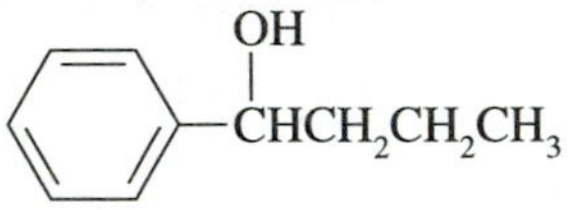

OH
C_6H_5–$CHCH_2CH_2CH_3$

7.19 Multiple S_N2 substitutions of iodide by iodide ion eventually give an equimolar mixture of enantiomers.

7.21 3-Methylcyclohexene forms because the hydrogen atom at C-2 of 1-bromo-2-methylcyclohexane is not located trans to the leaving group.

7.23 In the cis isomer, bromine is axial and there are trans hydrogen atoms at the C-2 and C-6 positions. In the trans isomer, bromine is equatorial and there are no trans hydrogen atoms.

7.25 (*E*)-2-bromo-2-butene

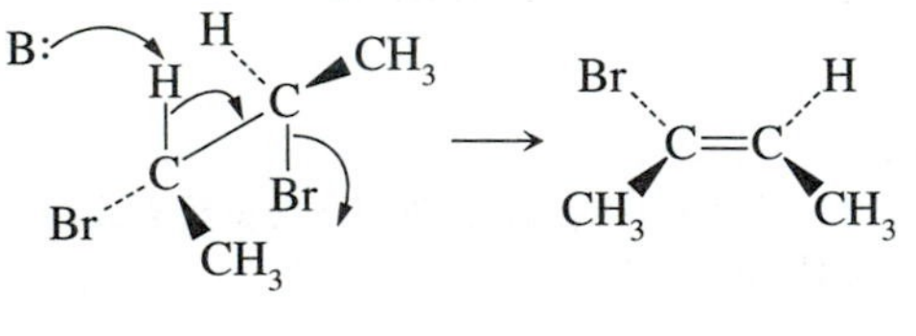

B: H H C CH_3 C Br Br CH_3 ⟶ Br C=C H CH_3 CH_3

CHAPTER 8

8.1 (a) 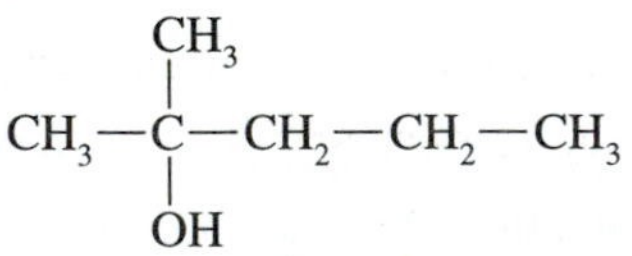 $CH_3—C(CH_3)(OH)—CH_2—CH_2—CH_3$

(b) $HO—CH_2—CH(CH_3)—CH_2—CH_3$

(c) $CH_3—CH(CH_3)—CH(CH_3)—CH_2—OH$

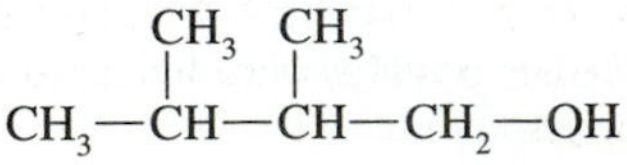

8.3 (a) 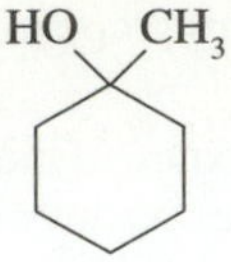 1-methylcyclohexanol (HO, CH_3) (b) 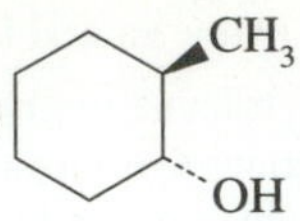 (CH_3, OH) (c) 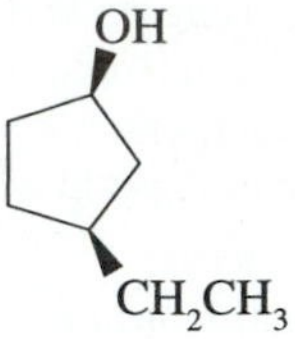 (OH, CH_2CH_3)

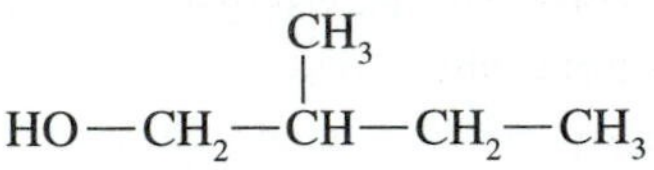

8.5 (a) 3-ethyl-2-hexanol

(b) 2,4,7-trimethyl-5-nonanol

(c) 4-methyl-3-hexanol

8.7 (*E*)-6-nonene-1-ol

8.9 (a) secondary (b) tertiary (c) tertiary

8.11 (a) primary alcohol and phenol (b) primary

8.13 1,2-Hexanediol has a ratio of one —OH group per three carbon atoms, which makes it more soluble than 1-heptanol with one —OH group per seven carbon atoms, because it can form more hydrogen bonds.

8.15 It has more carbon atoms; thus it interferes more with the hydrogen bonding between water molecules.

8.17 It is more acidic because the chlorine atoms are electronegative and pull electron density away from the oxygen-hydrogen bond.

8.19 similar to isopropyl alcohol (1×10^{-18})

8.21 II > I > III

8.23 (a) 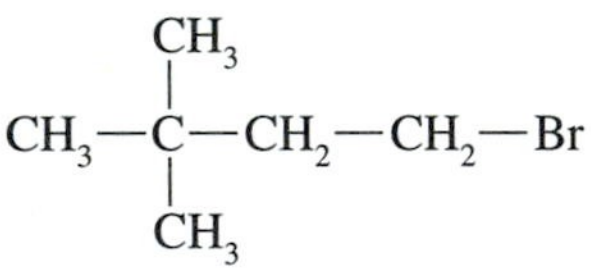 $CH_3—C(CH_3)_2—CH_2—CH_2—Br$

(b) $CH_3—C(CH_3)(Br)—CH_2—CH_3$

(c) $CH_3—CH(CH_3)—CH_2—CH(Br)—CH_3$

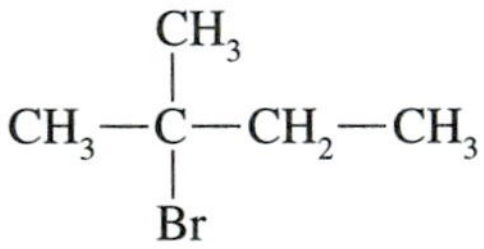

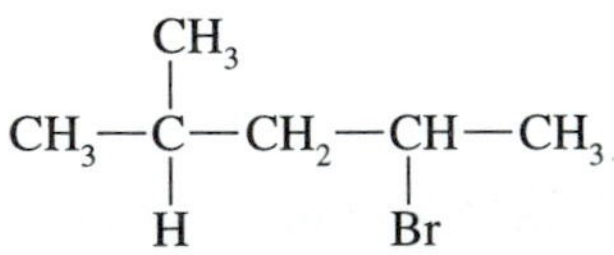

8.25 Reaction occurs at either of two carbocation centers shown in resonance forms:

$CH_3CH{=}CH—CH_2^+ \longleftrightarrow CH_3\overset{+}{C}H—CH{=}CH_2$

8.27 (a) 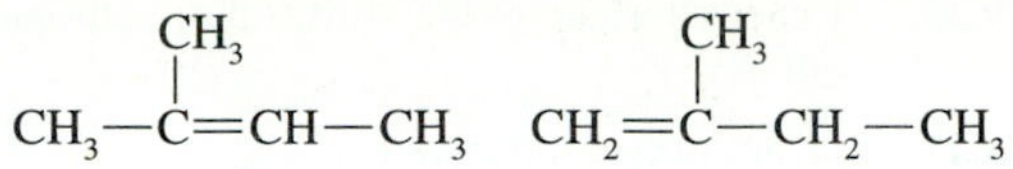 $CH_3-C(CH_3){=}CH-CH_3$ (major) $\quad CH_2{=}C(CH_3)-CH_2-CH_3$

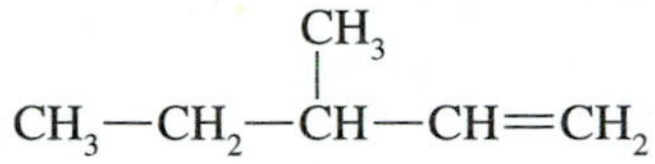

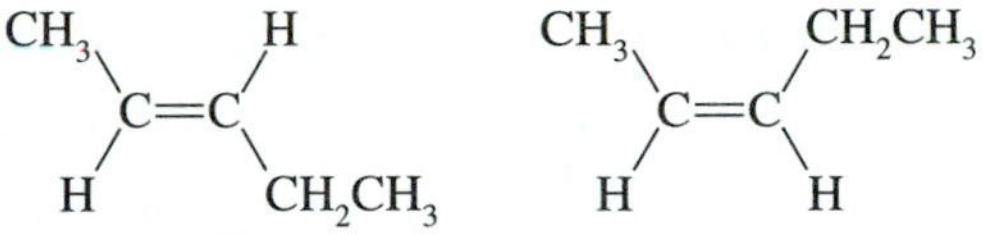

(b) $CH_3-CH_2-CH(CH_3)-CH{=}CH_2$

(c) $CH_3(H)C{=}C(H)CH_2CH_3$ (major; CH_3 and CH_2CH_3 trans) $\quad CH_3(H)C{=}C(H)CH_2CH_3$ (CH_3 and CH_2CH_3 cis)

8.29 *trans*-1-Phenyl-1-propene; product has a conjugated double bond.

8.31 (I) 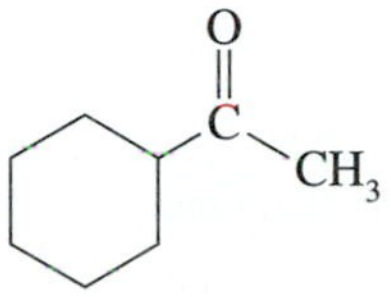 cyclohexyl$-C(=O)-CH_3$ (II) no oxidation product

(III) 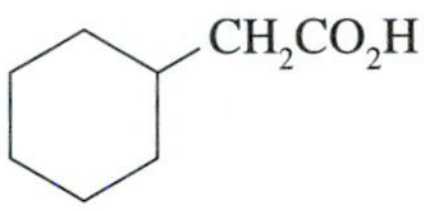 cyclohexyl$-CH_2CO_2H$

8.33 $CH_3CH_2(H)C{=}C(H)CH_2CH_2CH_2CH_2CHO$ (trans)

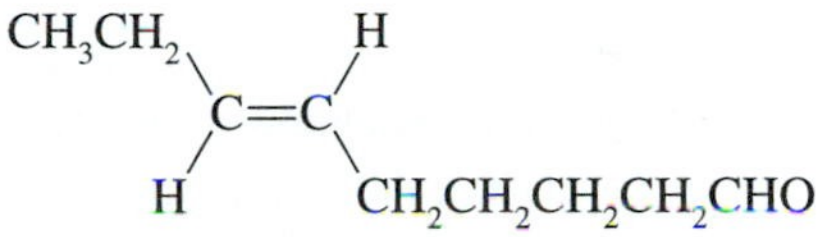

8.35 (c) of 8.23; (a) and (b) of 8.24

8.37 (a) 2-methyl-2-propanol (b) 2-methyl-2-butanol (c) 2-butanol

8.39 (a) 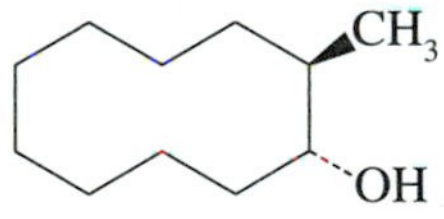 cyclodecane ring with CH_3 (wedge) and OH (dash) on adjacent carbons

(b) 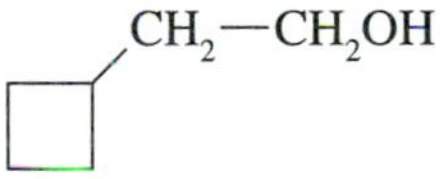 cyclobutyl$-CH_2-CH_2OH$

(c) 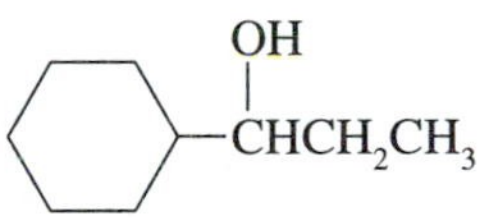 cyclohexyl$-CH(OH)CH_2CH_3$

8.41 (a) 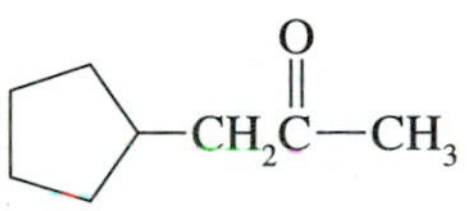 cyclopentyl$-CH_2C(=O)-CH_3$

(b) 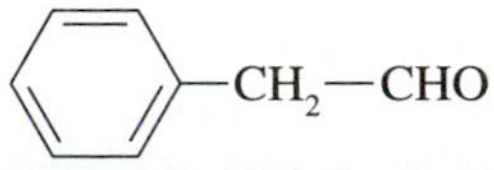 phenyl$-CH_2-CHO$

(c) $CH_3CH_2CH_2CH_2-CO_2H$

8.43 The nitro group withdraws electron density by resonance and stabilizes the phenoxide ion.

8.45 2-chloro-1,4-naphthoquinone (naphthalene-1,4-dione ring with Cl)

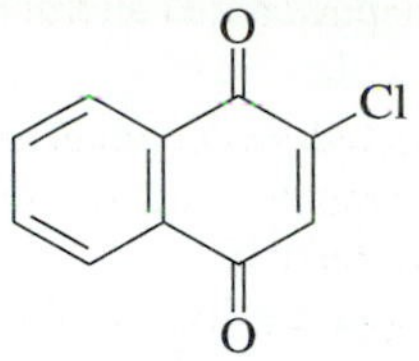

8.47 $CH_3CH_2CH_2CH_2-SH \quad CH_3CH_2CH(SH)CH_3$

$CH_3CH(CH_3)CH_2-SH \quad CH_3C(CH_3)(SH)CH_3$

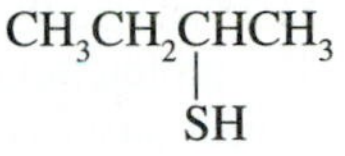

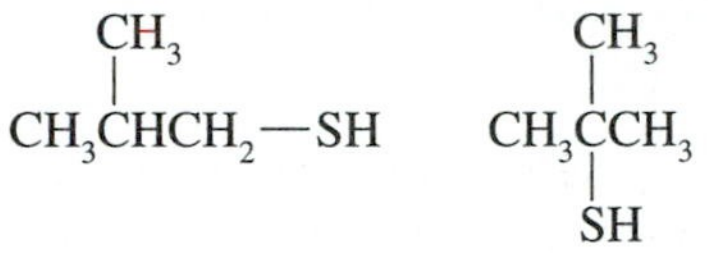

8.49 (a) $CH_3CH_2CH_2-SH$ (b) 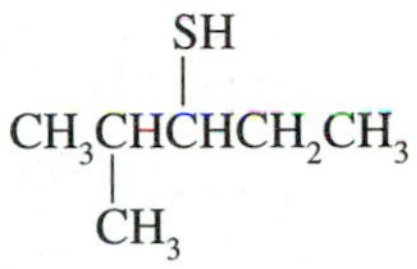 $CH_3CH(CH_3)CH(SH)CH_2CH_3$

(c) 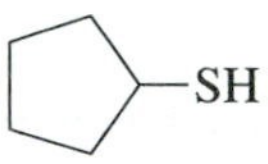 cyclopentyl$-SH$

8.51 The conjugate base $CH_3CH_2CH_2S^-$ forms, and it is not volatile.

8.53 $CH_2{=}C(CH_3)-CH_2CH_2S^- + CH_3Br$ or

$CH_2{=}C(CH_3)-CH_2CH_2Br + CH_3S^-$

8.55 oxidation of CH_3SH

8.57 See answers for Chapter 18.

Chapter 9

9.1 $CH_3CH_2OCH_2CH_3$; $CH_3OCH_2CH_2CH_3$; $CH_3OCH(CH_3)_2$

9.3 methyloxirane (epoxide ring with CH_3) $\quad$ oxetane

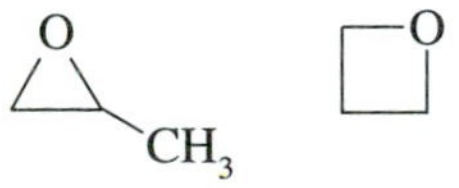

9.5 (a) dicyclopentyl ether (b) phenyl propyl ether (c) cyclopentyl propyl ether

9.7 (a) 2-methoxypentane (b) 2-methoxy-4-methylpentane (c) 3-ethoxyhexane

9.9 divinyl ether

9.11 The carbon/oxygen ratio is 2 : 1 and the number of hydrogen bonds possible is sufficient to allow dissolution to occur.

9.13 Diisopropyl ether is a more spherically shaped molecule and has weaker London forces.

9.15 The ether oxygen atom is protonated and the resulting ion is soluble in the acid.

$CH_3CH_2CH_2—\overset{H}{\overset{|}{O^+}}—CH_2CH_2CH_3$

9.17 Add HBr to form 1-bromo-1-methylcyclohexane. Prepare the Grignard reagent and add D_2O.

9.19 The proton of ethanol will react with the Grignard reagent just as the proton of water does.

9.21 React the alkoxide of 1-ethylcyclopentanol with bromoethane.

9.23 React the following alkoxide with 1-bromobutane.

H
O=C—N—$CH_2CH_2N(CH_2CH_3)_2$
N O$^-$

9.25 Addition of $HgOAc^+$ and capture of the intermediate ion by methanol gives an addition product with $—OCH_3$ and —HgOAc on adjacent carbon atoms. Subsequent reaction with $NaBH_4$ replaces the mercury with hydrogen.

$CH_3CH_2CH_2CH_2\overset{OCH_3}{\overset{|}{C}}H—CH_2—HgOAc \xrightarrow{NaBH_4}$

$CH_3CH_2CH_2CH_2\overset{OCH_3}{\overset{|}{C}}H—CH_2—H$

9.27 (a) $CH_2\overset{OCH_3}{\overset{|}{C}}HCH_3$

(b) $—CH_2CH_2OCH_2CH_3$

(c) $CH_2CH_2CH_2OCH_2CH_2OH$

9.29 $CH_3OCH_2CH_2OCH_2CH_3$

9.31 tetrahydropyran

9.33 CH_3 O CH_3 O

9.35 React ethylene oxide with sodium ethoxide in ethanol.

9.37 2-methoxy-1-propanol

9.39 $CH_3CH_2SCH_2C(CH_3)_2OH$

9.41 See answers for Chapter 18.

Chapter 10

10.1 (a) $CH_3CH_2\overset{CH_3}{\overset{|}{C}}HCHO$ (b) $CH_3CH_2\overset{CH_2CH_3}{\overset{|}{C}}HCH_2CHO$

(c) $CH_3CH_2CH_2\overset{Br}{\overset{|}{C}}HCHO$

10.3 (a) butanal (b) 3,3-dimethylbutanal

(c) 2-ethyl-3-methylpentanal

10.5 (a) 4-chloro-2,3-dimethylheptanal

(b) 6-ethyl-3-methyl-2-nonanone

(c) 5-bromo-2,4,6-trimethylheptanal

10.7 $CH_3—\overset{O}{\overset{\|}{C}}—CH_2CH_2CH_2CH_2—\overset{O}{\overset{\|}{C}}—H$

6-oxoheptanal

10.9 The carbonyl group of butanal is polar because there is a difference in electronegativity between carbon and oxygen. The carbon-carbon double bond of 1-butene is not polar.

10.11 2-Methylpropanal is a more spherical molecule and has weaker London forces.

10.13 Butanol is both a hydrogen bond donor and acceptor, so it is slightly more soluble.

10.15 Red precipitate forms; silver mirror forms.

10.17 (a) $\overset{OH}{\overset{|}{C}}HCH_3$ (b) OH

(c) CH_2OH

10.19 The —OH group produced may be cis or trans to the unsaturated group bonded to the ring.

10.21 There is no steric hindrance to prevent nucleophilic attack on the carbonyl carbon atom.

10.23 The *tert*-butyl group bonded to the carbonyl carbon atom of 2,2-dimethylpropanal sterically hinders the reaction with water.

10.25 OH | C_6H_5—CH—CN

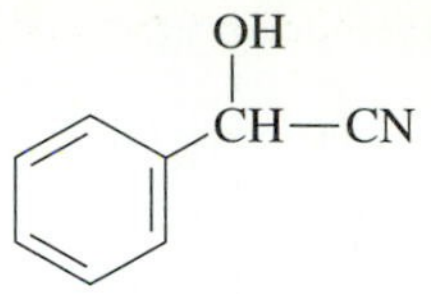

10.27 (a) ethyl Grignard reagent and propanal
(b) isopropyl Grignard reagent and ethanal, or methyl Grignard reagent and 2-methylpropanal
(c) isobutyl Grignard reagent and propanal, or ethyl Grignard reagent and 3-methylbutanal

10.29 Add methyl Grignard reagent and dehydrate the tertiary alcohol to give the cycloalkene.

10.31 (a) acetal (b) ketal (c) hemiacetal

10.33 (a) acetal (b) hemiacetal (c) ketal

10.35 (a) $HOCH_2CH_2OH$ and cyclohexanone
(b) $HOCH_2CH_2CH_2OH$ and cyclopentanone

10.37 (a) $CH_3CH{=}NCH_3$ (b) $(CH_3)_2C{=}NCH_2CH_3$
(c) 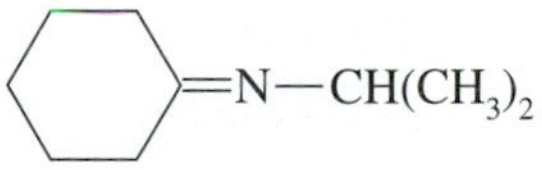 cyclohexylidene $=N-CH(CH_3)_2$

10.39 Geometric isomers about the carbon-nitrogen double bond of the product are possible with cyclopentanecarbaldehyde but not with cyclohexanone, which has two equivalent groups bonded to the original carbonyl carbon atom.

10.41 It does not have an alpha carbon atom.

10.43 (a) 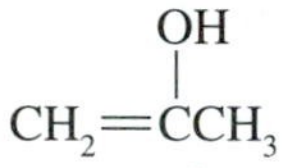 $CH_2{=}C(OH)CH_3$ (b) $CH_3CH_2CH{=}CHOH$
(c) 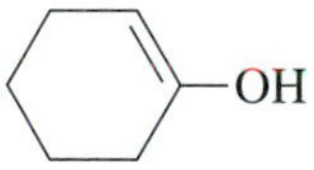 cyclohexene ring —OH

10.45 (a) 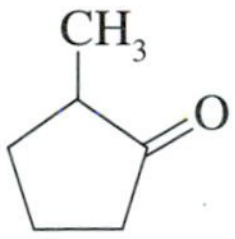 2-methylcyclopentanone (CH_3, =O) (b) 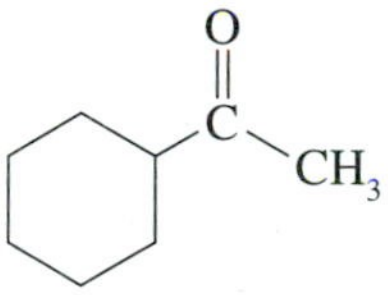 cyclohexyl—C(=O)—CH_3

10.47 $CH_3CH_2CH_2-CH(OH)-CH(CH_2CH_3)-C(=O)-H$

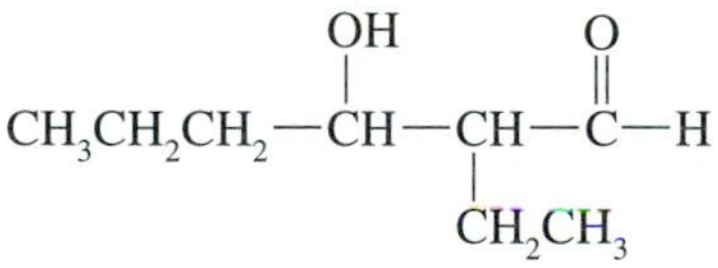

10.49 4-hydroxy-4-phenyl-2-butanone
$C_6H_5-CH(OH)-CH_2-C(=O)-CH_3$

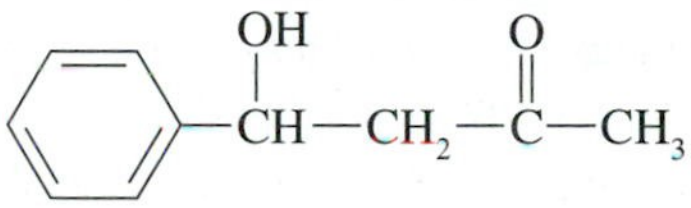

10.51 The C-6 methylene group reacts with the carbonyl carbon atom at C-2. Dehydration of the aldol product gives jasmone.

10.53 See answers for Chapter 18.

Chapter 11

11.1 Aldoses are carbohydrates that contain an aldehyde group, whereas ketoses are carbohydrates that contain a ketone group.

11.3 (a) aldopentose (b) ketohexose (c) ketopentose
(d) aldohexose

11.5 (a) D (b) L (c) L (d) D

11.7

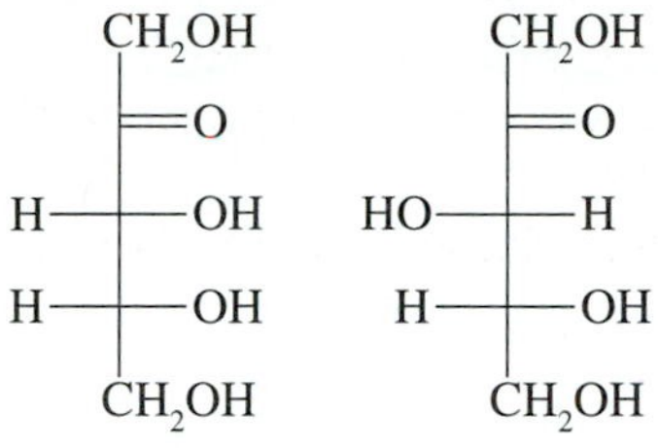

	CH_2OH	
	=O	
H		OH
H		OH
	CH_2OH	

	CH_2OH	
	=O	
HO		H
H		OH
	CH_2OH	

11.9 (a)

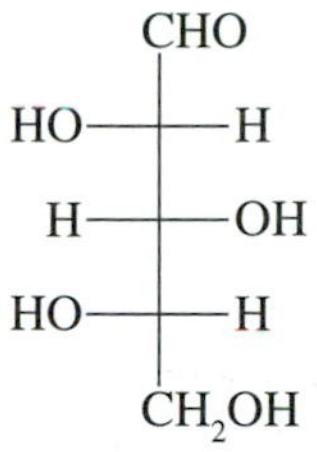

	CHO	
HO		H
H		OH
HO		H
	CH_2OH	

(b)

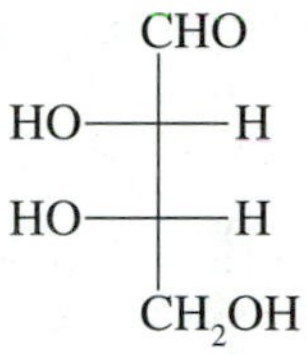

	CHO	
HO		H
HO		H
	CH_2OH	

(c)

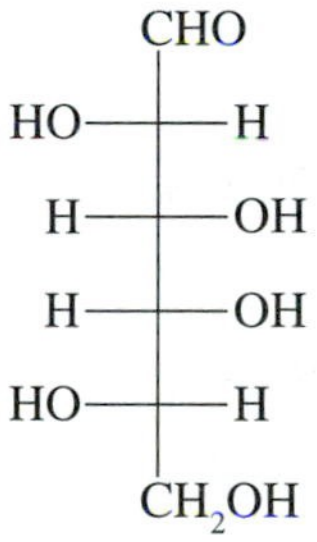

	CHO	
HO		H
H		OH
H		OH
HO		H
	CH_2OH	

11.11 (a)

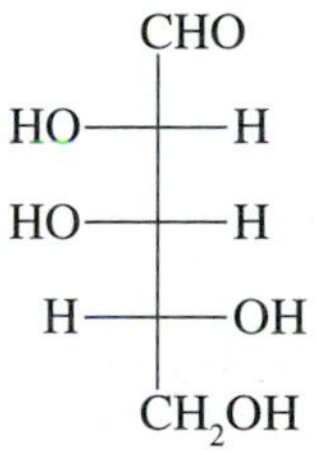

	CHO	
HO		H
HO		H
H		OH
	CH_2OH	

(b)

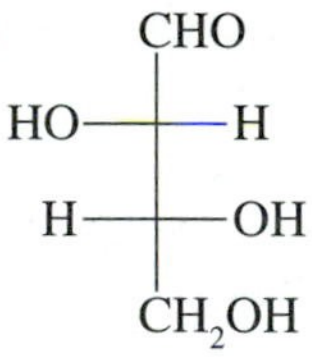

	CHO	
HO		H
H		OH
	CH_2OH	

(c)

	CHO	
H		OH
HO		H
HO		H
H		OH
	CH_2OH	

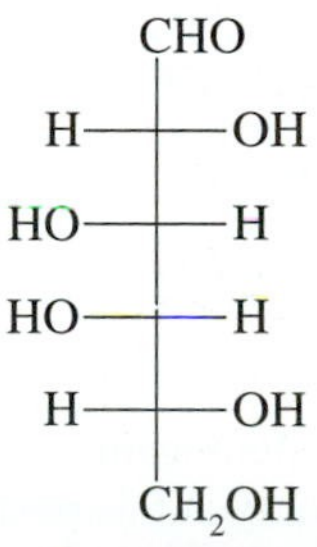

11.13 (a) (b) (c) (d)

11.15 (a) galactose (b) mannose (c) ribose

11.17 (a) α-D-galactopyranose (b) β-D-mannopyranose (c) α-D-ribopyranose

11.19

β-galactopyranose
one axial OH at C-4

β-mannopyranose
one axial OH at C-2

11.21 Yes, because ring opening and reclosure can result in either of two anomers.

11.23 The compounds in (a) and (c) can both mutarotate. The compound in (b) is a glycoside of a ketopentose and cannot mutarotate.

11.25 The observed rotation is closer to that of the β anomer, which must therefore predominate at equilibrium.

11.27 The alditol of erythrose is optically inactive because it has a plane of symmetry bisecting the bond between the C-2 and C-3 atoms.

CH_2OH / H–OH / H–OH / CH_2OH CH_2OH / HO–H / H–OH / CH_2OH

11.29 The C-2 atom becomes a stereogenic center, and stereoisomers that differ only at this position result. The alditols are glucitol and mannitol.

11.31 (a) CO_2H / HO–H / HO–H / H–OH / H–OH / CH_2OH (b) CO_2H / H–OH / HO–H / HO–H / H–OH / CH_2OH

(c) CO_2H / H–OH / H–OH / H–OH / CH_2OH

11.33 The aldaric acid of erythrose is optically inactive because it has a plane of symmetry bisecting the bond between the C-2 and C-3 atoms.

11.35 CHO / HO–H / H–OH / H–OH / H–OH / CH_2OH CH_2OH / =O / H–OH / H–OH / H–OH / CH_2OH

11.37

11.39

OCH_3

CHO

11.41 (a) both glucose; β-1,3

(b) fructose as a pyranose bonded to ribose via a β-2,4 linkage

Chapter 12

12.1 (a) propionic acid (b) caproic acid (c) stearic acid

12.3 (a) ethyl formate (b) methyl butyrate (c) octyl acetate

12.5 (a) cyclodecanecarboxylic acid
(b) 3-cyclopentylpropanoic acid
(c) *trans*-4-methoxycyclohexanecarboxylic acid

12.7 $CH_3CH(CH_3)CH_2-C_6H_4-CH(CH_3)CO_2H$

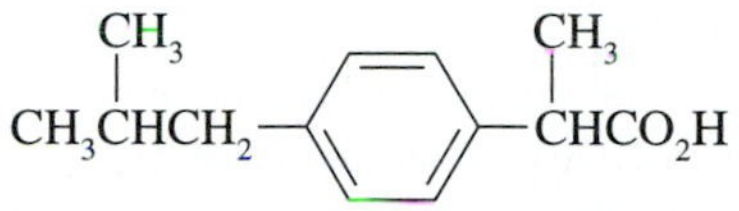

12.9 (a) 5-hydroxypentanoic acid lactone
(b) 4-hydroxypentanoic acid lactone
(c) 9-hydroxynonanoic acid lactone

12.11 **b** and **c** are lactones; **a** is an anhydride

12.12 **a** and **c** are lactams; **b** is a keto amine

12.13 $C_nH_{2n}O_2$

12.15 two; $CH_3CH_2CH_2CO_2H$ and $(CH_3)_2CHCO_2H$

12.17 There are fewer sites in butanol to form hydrogen bonds than in butanoic acid.

12.19 Decanoic acid has stronger London forces due to the extra CH_2 unit.

12.21 They are isomeric esters and have similar London forces.

12.23 formic acid

12.25 The methoxy group is inductively electron withdrawing and decreases the electron density in the O—H bond.

12.27 The substituent group at the para position is electron withdrawing because the sulfur atom is bonded to two oxygen atoms.

12.29 (a) yes (b) no (c) yes

12.31 (a) 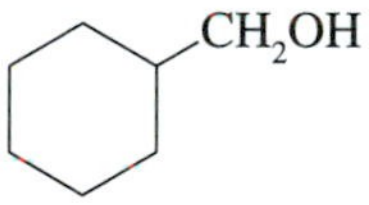 $C_6H_{11}-CH_2OH$ (cyclohexyl)

(b) $CH_3CH_2CH(CH_3)(CH_2)_6CH_2OH$

(c) C_6H_5-CHO

12.33 (a) $CH_3(CH_2)_3CH_2C(=O)-Cl$

(b) $CH_3(CH_2)_3CH_2C(=O)-OCH_3$

(c) $CH_3(CH_2)_3CH_2CH_2OH$

12.35 $HOCH_2CH_2CH_2C(=O)-NHCH_3$

12.37 Loss of water gives a cyclic anhydride.

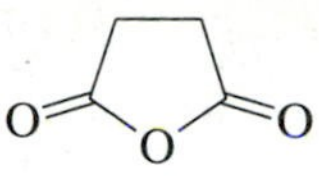

(five-membered ring: O=C–O–C=O with two CH_2 groups)

12.39 (a) 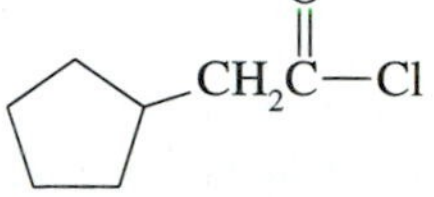 $C_5H_9-CH_2C(=O)-Cl$ (cyclopentyl) (b) $C_6H_5-CH_2Cl$

(c) $C_6H_5-CH(OCH_3)_2$

12.41 (a) Treat with magnesium to form Grignard reagent; add to CO_2 and acidify.
(b) Treat with PBr_3 to form bromocyclohexane; add magnesium to form Grignard reagent; add to CO_2 and acidify.
(c) Use ozonolysis to form cyclohexanecarbaldehyde; oxidize with the Jones reagent.

12.43 Reduce with $LiAlH_4$ to give an alcohol. React with PBr_3 to form a bromoalkane. Add magnesium to form the Grignard reagent. Then add to CO_2 and acidify.

12.45 (a) CH_3CO_2H and $CH_3CH_2CH_2OH$
(b) $CH_3(CH_2)_3CO_2H$ and $CH_3CH_2CH_2CH_2OH$
(c) $CH_3CH_2CO_2H$ and $CH_3(CH_2)_7OH$

12.47 (a) CH_3CO_2H and $CH_3CH_2CH_2OH$
(b) $CH_3(CH_2)_2CO_2H$ and CH_3CH_2OH
(c) $CH_3CH_2CO_2H$ and $CH_3(CH_2)_7OH$

12.49 $CH_3CH_2-C(=O)-OCH_2CH_3$

12.51 diethyl ester of adipic acid

12.53 See answers for Chapter 18.

Chapter 13

13.1 Glycerophospholipids contain glycerol and ester linkages. Sphingophospholipids contain sphingosine and an amide linkage. Both types of lipids contain long-chain fatty acids and a phosphate ester.

13.3 $CH_3(CH_2)_{17}CH_2-C(=O)-O-CH_2(CH_2)_{30}CH_3$

13.5 $CH_3(CH_2)_5CH{=}CH(CH_2)_7CO_2H$

13.7 $CH_3(CH_2)_7C{\equiv}C(CH_2)_7CO_2H$

13.9 Additional cis double bonds prevent close packing of molecules.

13.11 No, because there are two hydrogen atoms bonded to the C-11 atom.

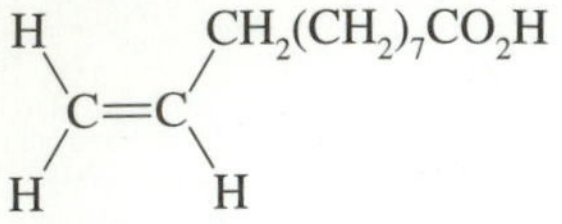

$$\begin{array}{ccc} H & & CH_2(CH_2)_7CO_2H \\ & C{=}C & \\ H & & H \end{array}$$

13.13 The long hydrophobic hydrocarbon chains of these molecules can form the interior of a micelle. The exterior of the micelle contains the neutral polar heads.

13.15

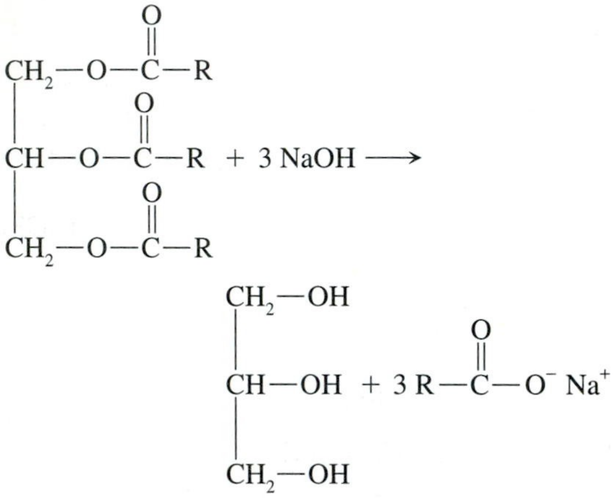

$$\begin{array}{l} CH_2{-}O{-}\overset{O}{\overset{\|}{C}}{-}R \\ | \\ CH{-}O{-}\overset{O}{\overset{\|}{C}}{-}R \; + \; 3\,NaOH \longrightarrow \\ | \\ CH_2{-}O{-}\overset{O}{\overset{\|}{C}}{-}R \end{array}$$

$$\begin{array}{l} CH_2{-}OH \\ | \\ CH{-}OH \; + \; 3\,R{-}\overset{O}{\overset{\|}{C}}{-}O^-\,Na^+ \\ | \\ CH_2{-}OH \end{array}$$

13.17 It is an oil, because all fatty acid components are unsaturated. From top to bottom in the structure, the fatty acids are linoleic, oleic, and linolenic acid.

13.19 No, this compound cannot exist in an optically active form, because it contains no stereogenic centers. The C-2 atom is bonded to two equivalent carbon atoms.

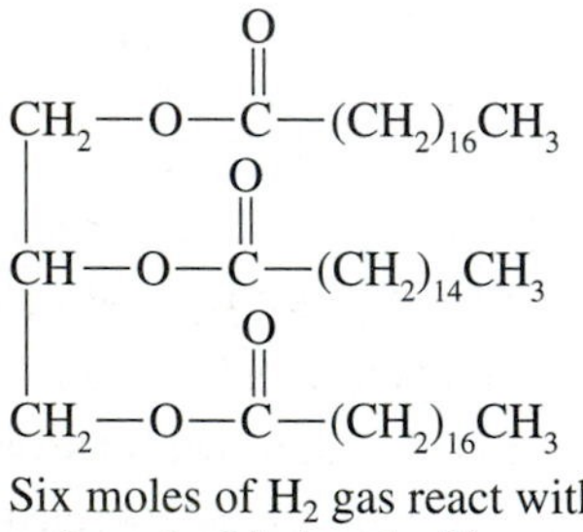

$$\begin{array}{l} CH_2{-}O{-}\overset{O}{\overset{\|}{C}}{-}(CH_2)_{16}CH_3 \\ | \\ CH{-}O{-}\overset{O}{\overset{\|}{C}}{-}(CH_2)_{14}CH_3 \\ | \\ CH_2{-}O{-}\overset{O}{\overset{\|}{C}}{-}(CH_2)_{16}CH_3 \end{array}$$

13.21 Six moles of H_2 gas react with the six carbon-carbon double bonds. The product is tristearin or glyceryl tristearate.

13.23 The molecular weight of glyceryl trioleate ($C_{57}H_{104}O_6$) is 884, and the compound contains three double bonds. The iodine value is 86.

13.25 Stearic acid and palmitic acid are formed in a 2 : 1 ratio.

13.27 The saponification number of tripalmitate is 209.

13.29 glycerol, 2 fatty acids, an alcohol, and phosphoric acid

13.31 stearic acid, oleic acid, glycerol, phosphoric acid, and choline

13.33 Phosphatidyl choline and phosphatidyl ethanolamine both have a positive and a negative charge; phosphatidyl serine has two negative charges and a positive charge; phosphatidyl inositol has a negative charge.

13.35 Sphingophospholipids have sphingosine as the backbone and have an amide linkage to one fatty acid. Glycerophospholipids have glycerol as the backbone and have ester linkages to two fatty acids.

13.37 The unsaturated carbon chain present in sphingosine is the "second chain."

13.39 Both contain sphingosine; glycosphingolipids have a carbohydrate unit instead of a phosphate unit.

13.41 A glycosidic bond results from reaction of the C-1 hydroxyl group of sphingosine with the hemiacetal center of a carbohydrate.

13.43 The saturated acids make the membrane more rigid.

13.45 Permeability is increased by proteins.

13.47 Active transport requires energy and occurs against concentration differences. Facilitated diffusion requires no energy and is concentration dependent.

Chapter 14

14.1 (a) secondary amine (b) tertiary amine (c) primary amine (d) secondary amide (e) tertiary amide (f) primary amide

14.3 (a) secondary amide (b) secondary amine

14.5 secondary amide, tertiary amine

14.7 (a) *N*,*N*-dimethyl-1-butanamine
(b) *N*-ethyl-2-methyl-1-butanamine
(c) *N*-methylcyclohexanecarboxamide
(d) *N*-cyclohexylethanamide

14.9 NH_2

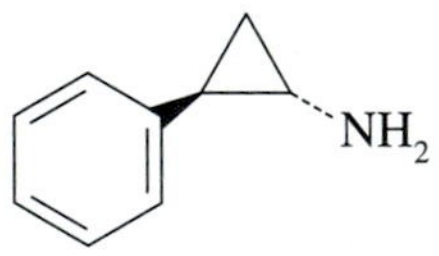

14.11 6-bromo-3-chloroindole

14.13 2 isomers; dimethylamine and ethylamine

14.15 4 isomers; butylamine, *sec*-butylamine, isobutylamine, *tert*-butylamine

14.17 Trimethylamine cannot form intermolecular hydrogen bonds, whereas propylamine can.

14.19 triethylamine

14.21 (a) 10^{-4} (b) 10^{-4} (c) 10^{-10}

14.23 The electron withdrawing nitro group decreases the electron density at the nitrogen atoms and decreases the K_b (increases pK_b).

14.25 amide < tertiary aryl amine < tertiary amine

14.27 (a) $C_6H_5CH_2\overset{+}{N}(CH_3)_3\ I^-$

(b) $C_6H_5CH_2N(CH_3)C(=O)CH_3$

(c) $C_6H_5CH_2\overset{+}{N}H_2CH_3\ I^-$

14.29 (a) $CH_3-CH(NH-CH_2-CH_3)-CH_2-CH_3$

(b) $C_5H_9-CH_2-NH-CH_2-C_5H_9$

(c) $NH_2CH_2-C_6H_4-CH_2NH_2$

14.31 (a) $CH_3(CH_2)_4CH_2NH_2$

(b) $C_3H_5-CH_2CH_2CH_2NH_2$

(c) $CH_3(CH_2)_3CH_2NH_2$

14.33 React with thionyl chloride; react acid chloride with aminoethane; reduce amide with lithium aluminum hydride.

14.35 (a) $C_6H_5-CO_2H + CH_3NH_3^+$

(b) $CH_3CH_2CO_2H + (CH_3)_2NH_2^+$

(c) $C_6H_5-\overset{+}{N}H_3 + CH_3CO_2H$

14.37 (a) $C_6H_5CH_2NHCH_3$

(b) $CH_3CH_2CH_2-N(CH_3)_2$

(a) $C_6H_5-NH-CH_2CH_3$

14.39 (a) $CH_3(CH_2)_3C(=O)-NHCH_3$

(b) $C_6H_{11}-CH_2C(=O)NH-C_5H_9$

(c) $C_5H_9-CH_2C(=O)-NH_2$

14.41 2,6-dimethylaniline (CH_3, NH_2, CH_3 on benzene ring) and piperidine-2-carboxylic acid with $CH_2CH_2CH_2CH_3$ on N (HO_2C, N, $CH_2CH_2CH_2CH_3$)

CHAPTER 15

15.1 Fischer projection: top CO_2H; H—C—NH_2; bottom $CH_2CH_2CO_2H$

15.3 It is formed from lysine by oxidation at the C-5 atom to give a hydroxyl group.

15.5 alanine at pH 1: CO_2H (top), $\overset{+}{N}H_3$—C—H, CH_3 (bottom)

alanine at pH 12: CO_2^- (top), NH_2—C—H, CH_3 (bottom)

glutamic acid at pH 1: CO_2H (top), $\overset{+}{N}H_3$—C—H, $CH_2CH_2CO_2H$ (bottom)

glutamic acid at pH 12: CO_2^- (top), NH_2—C—H, $CH_2CH_2CO_2^-$ (bottom)

15.7 The aspartic acid solution has a lower pH.

15.9 (a) pH near 7 because all amino acids are neutral

(b) pH < 7 because it contains aspartic acid

(c) pH > 7 because it contains lysine

15.11 Vasopressin, because arginine is a basic amino acid.

15.13 It must have a large number of basic amino acids.

15.15 $\overset{+}{N}H_3-CH(CH_3)-C(=O)-NH-CH(CH_2OH)-C(=O)-O^-$

alanine residue; serine residue

alanylserine (Ala-Ser)

15.17 glycine, cysteine, and valine; glycylcysteylvaline

15.19 The N-terminal amino acid, glutamic acid, is cyclized in the form of a lactam. The C-terminal amino acid, proline, is in the form of an amide.

15.21 six

15.23 four

15.25 No; it could be Pro-Arg-Thr-Lys or Thr-Lys-Pro-Arg

15.27 only (a) to give Arg and Gly-Tyr

15.29 (a) no reaction

(b) no reaction

(c) gives Phe, Trp, and Ser

(d) gives Ser-Phe and Asp

15.31 No; it could be Pro-Arg-Thr-Lys or Thr-Lys-Pro-Arg

15.33 The fragments could come from Trp-Ala-Arg-Gly-Tyr-Ala-Ser-Gly-Glu or Ala-Arg-Gly-Tyr-Trp-Ala-Ser-Gly-Glu.

15.35 Thr-Lys-Pro-Arg

15.37 There must be two protein chains, because there are two N-terminal amino acids.

15.39 N-terminal amino acid is aspartic acid.

15.41 acidic and basic amino acids

15.43 Glycine and phenylalanine, which have nonpolar R groups.

15.45 After peptide formation there is no N—H to form hydrogen bonds.

Chapter 16

16.1 All are polymers of nucleotides consisting of deoxyribose units bonded to a nitrogen base and linked by 5′→3′ phosphodiester bonds.

16.3 Phosphate groups forming phosphodiester bonds.

16.5 Adenine is a purine; thymine, uracil, and cytosine are pyrimidines.

16.7 HO—P(=O)(OH)—O—CH_2 (ribose with H, H, H, H, OH, OH; adenine with NH_2)

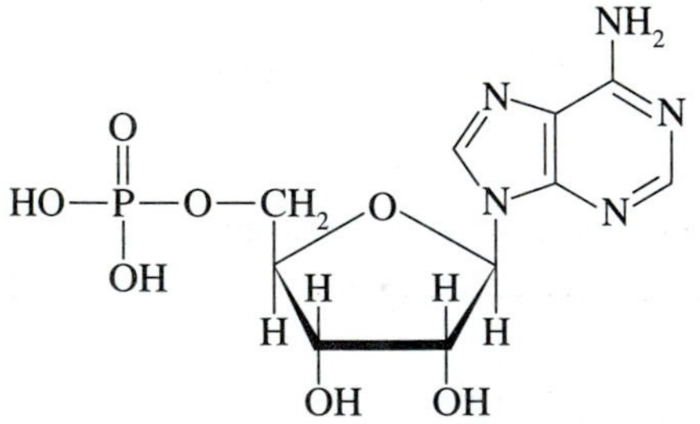

16.9 β

16.11 one, four

16.13 deoxyribose, guanine, and phosphoric acid

16.15 None do.

16.17 deoxyribose and phosphodiester linkages

16.19 C—G—T—A—G—T—dC

16.21 C—G—U—A—G—U—C

16.23 U

16.25 G—G—G—A—C—A—U—G—U—G—G—A
Gly—Thr—Cys—Gly

16.27 It forms a complex with the anticodon of a tRNA to specify an amino acid or an instruction, such as termination of protein synthesis.

16.29 (a) Val (b) Pro (c) Phe (d) Thr

16.31 The growing protein chain is transferred from the donor site to the next amino acid at the acceptor site.

16.33 (a) CAA (b) GGG (c) AAA (d) UGC

16.35 GUUGAUGCUGGU is one possible sequence.

Chapter 17

17.1 The elastomer has alternating quaternary carbon atoms that prevent the polymer chain from packing closely together in a regular array.

17.3 Reaction of the dianhydride at one of the anhydride sites with a diol gives a polyester that retains an anhydride unit, which could react with added monomers to give bridged links between chains.

17.5 OAc OAc OAc OAc

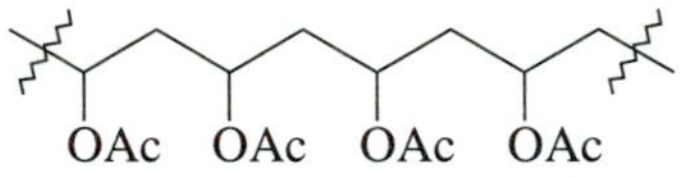

17.7 F F C=C Cl F

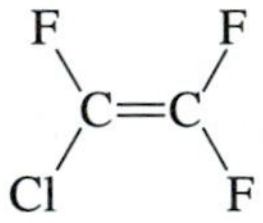

17.9 The short-chain transfer reaction involves intramolecular abstraction of a hydrogen atom, resulting in a new radical site where polymerization continues. The branch formed is 2,4-diphenylbutyl.

17.11

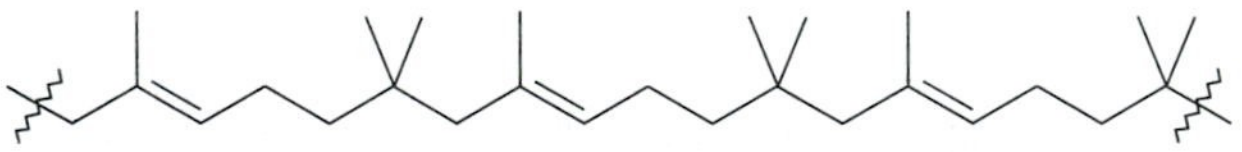

17.13 The polymer has polar functional groups that resemble those in proteins and can form hydrogen bonds to the protein molecules in hair.

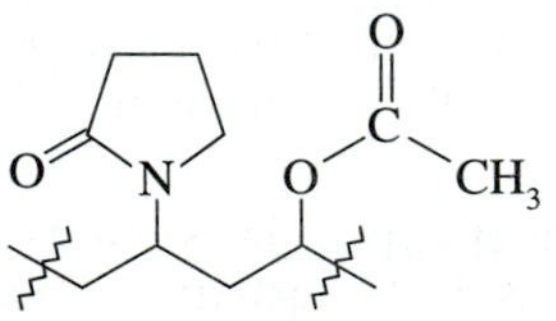

O N O C O CH_3

17.15 The number of cross-links in rubber gloves is fewer, because much greater flexibility is required in gloves than in the rubber of tires.

17.17 Only (a) and (d) can give isotactic and syndiotactic polymers.

17.19 Each tertiary carbon atom of the chain contains a *sec*-butyl group. Each alkyl group is on the same side of the backbone of the zigzag chain.

17.21 (a) HO⁄\⁄OH HO–C(=O)–(CH₂)₄–C(=O)–OH

(b) Cl–C(=O)–Cl $NH_2(CH_2)_9NH_2$

(c) HO⁄\⁄OH (benzene-1,2-dicarboxylic acid: CO_2H, CO_2H)

17.23 Lactic acid is $CH_3CH(OH)CO_2H$.

17.25

17.27 The diamine portion of the polyamide is relatively rigid. The flexibility of the carbon chain of the dicarboxylic acid increases as *x* increases and affects the properties of the polymer.

17.29 The glycerol provides a third hydroxyl group that can react with additional toluene diisocyanate to give a cross-linked polymer.

Chapter 18

18.1 Conjugation is extended by the larger number of rings. Tetracene is colored because it absorbs visible light.

18.3 2,4-Hexadiyne has conjugated triple bonds that should absorb ultraviolet light.

18.5 Propanone will absorb near 1710 cm^{-1}; 3-propen-1-ol will absorb near 1640 cm^{-1} and in the 3400–3600 cm^{-1} region.

18.7 Compound I has an absorption in the 3400–3600 cm^{-1} region but none in the 1640 cm^{-1}. Compound II has no absorption in the 3400–3600 cm^{-1} or in the 1640 cm^{-1} region. Compound III has absorptions in the 3400–3600 cm^{-1} and 1640 cm^{-1} regions. Compound IV has no absorption in the 3400–3600 cm^{-1} region but has an absorption in the 1640 cm^{-1} region.

18.9 Compound III, because it has both a carbonyl group and an OH group.

18.11 $CH_3–CCl_2–CH_3$

18.13 $CH_3–C(CH_3)_2–CH_2–Br$ (1.1 δ; 3.2 δ)

18.15 The peak at 1.9 δ is intensity 3 and is a triplet; the peak at 3.1 δ is intensity 2 and is a quartet.

18.17 $CH_3–C(CH_3)_2–O–CH_3$ (1.1 δ; 3.1 δ)

18.19 Compound I has a quartet of intensity 2 at 2 δ for the hydrogen atoms on the α carbon atoms, whereas compound II has a singlet of intensity 3 in this region.

18.21 $CH_3–CH(CH_3)–CH_2Cl$ (0.90 δ; 3.2 δ; H 1.8 δ)

18.23 (a) four at low field and two at high field

(b) two at low field

(c) four at low field and one at high field

18.25 (a) 5 (b) 5 (c) 4

18.27 CH_3 / CH_2

seven resonances   five resonances

18.29 30.2 δ → ← 136 δ

cyclobutene

18.31 All resonances in 3-methylpentane are multiplets (2 quartets, 1 triplet, 1 doublet). One resonance of 2,2-dimethylbutane is a singlet because there is a quaternary carbon atom; the other resonances are 2 quartets and a triplet.

18.33 The resonance at 130 δ corresponds to an sp^2-hybridized carbon atom. Based on the four resonances, the alkene must be symmetrical. The only two possibilities are 4-octene and 3,4-dimethyl-3-hexene, which can be distinguished based on the multiplicities of the resonances in a proton-coupled ^{13}C spectrum. The low field resonance of 4-octene would be a doublet, whereas the low field resonance of 3,4-dimethyl-3-hexene would be a singlet. *E* and *Z* isomers are possible for both isomers, and they cannot be distinguished based on ^{13}C NMR.

$$\begin{array}{c} H \quad\quad CH_2CH_2CH_3 \\ \diagdown \; \diagup \\ C{=}C \\ \diagup \; \diagdown \\ CH_3CH_2CH_2 \quad\quad H \end{array}$$

$$\begin{array}{c} CH_3 \quad\quad CH_2CH_3 \\ \diagdown \; \diagup \\ C{=}C \\ \diagup \; \diagdown \\ CH_3CH_2 \quad\quad CH_3 \end{array}$$

INDEX

Periodic Table of the Elements

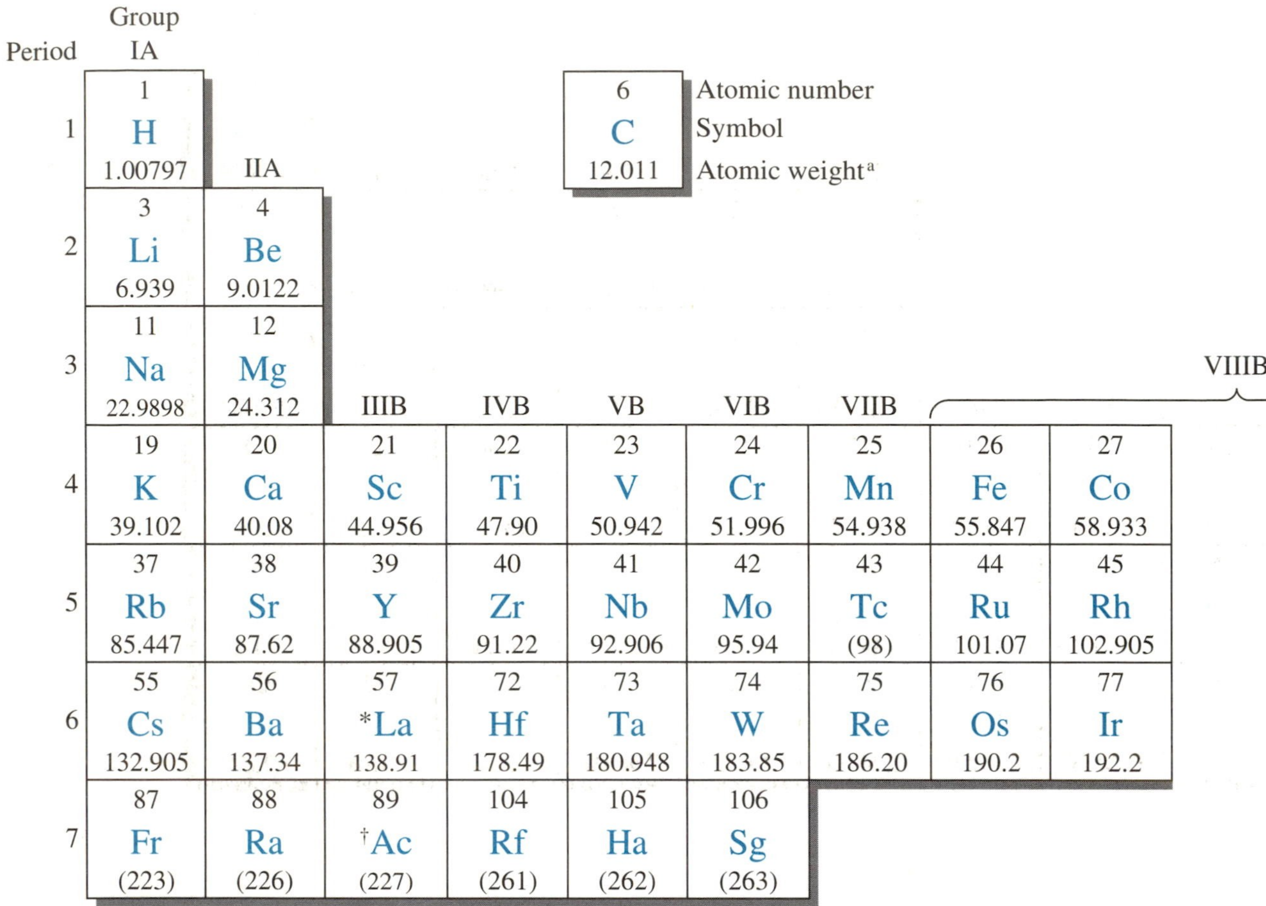

6 C 12.011	Atomic number Symbol Atomic weight[a]

Period	Group IA	IIA	IIIB	IVB	VB	VIB	VIIB	VIIIB	
1	1 H 1.00797								
2	3 Li 6.939	4 Be 9.0122							
3	11 Na 22.9898	12 Mg 24.312							
4	19 K 39.102	20 Ca 40.08	21 Sc 44.956	22 Ti 47.90	23 V 50.942	24 Cr 51.996	25 Mn 54.938	26 Fe 55.847	27 Co 58.933
5	37 Rb 85.447	38 Sr 87.62	39 Y 88.905	40 Zr 91.22	41 Nb 92.906	42 Mo 95.94	43 Tc (98)	44 Ru 101.07	45 Rh 102.905
6	55 Cs 132.905	56 Ba 137.34	57 *La 138.91	72 Hf 178.49	73 Ta 180.948	74 W 183.85	75 Re 186.20	76 Os 190.2	77 Ir 192.2
7	87 Fr (223)	88 Ra (226)	89 †Ac (227)	104 Rf (261)	105 Ha (262)	106 Sg (263)			

*Lanthanide series	58 Ce 140.12	59 Pr 140.907	60 Nd 144.24	61 Pm (147)	62 Sm 150.35
†Actinide series	90 Th 232.038	91 Pa (231)	92 U 238.03	93 Np (237)	94 Pu (242)

[a]Numbers in parentheses are mass numbers of the most stable or best-known isotope of radioactive elements. Names for elements 104 to 106 are still being debated.

	IB	IIB	IIIA	IVA	VA	VIA	VIIA	Noble gases
								2 He 4.0026
			5 B 10.811	6 C 12.011	7 N 14.0067	8 O 15.9994	9 F 18.9984	10 Ne 20.183
			13 Al 26.9815	14 Si 28.086	15 P 30.9738	16 S 32.064	17 Cl 35.453	18 Ar 39.944
28 Ni 58.71	29 Cu 63.54	30 Zn 65.37	31 Ga 69.72	32 Ge 72.60	33 As 74.922	34 Se 78.96	35 Br 79.916	36 Kr 83.80
46 Pd 106.4	47 Ag 107.870	48 Cd 112.40	49 In 114.82	50 Sn 118.69	51 Sb 121.75	52 Te 127.60	53 I 126.904	54 Xe 131.30
78 Pt 195.09	79 Au 196.967	80 Hg 200.59	81 Tl 204.37	82 Pb 207.19	83 Bi 208.980	84 Po (210)	85 At (210)	86 Rn (222)

63 Eu 151.96	64 Gd 157.25	65 Tb 158.924	66 Dy 162.50	67 Ho 164.930	68 Er 167.26	69 Tm 168.934	70 Yb 173.04	71 Lu 174.97
95 Am (243)	96 Cm (247)	97 Bk (249)	98 Cf (251)	99 Es (254)	100 Fm (253)	101 Md (256)	102 No (253)	103 Lr (257)